Elastic Analysis
of Structures

Elastic Analysis of Structures
Classical and Matrix Methods

John B. Kennedy
Murty K. S. Madugula

University of Windsor

1817

HARPER & ROW, PUBLISHERS, New York

Grand Rapids, Philadelphia, St. Louis, San Francisco

London, Singapore, Sydney, Tokyo

Sponsoring Editor: James Cook
Project Editor: Ellen MacElree
Cover Design: Heather A. Ziegler
Text Art: RDL Artset Ltd.
Production Manager: Jeanie Berke
Production Assistant: Beth Maglione
Compositor: TAPSCO, Inc.
Printer and Binder: R. R. Donnelley & Sons Company

Elastic Analysis of Structures: Classical and Matrix Methods

Library of Congress Cataloging-in-Publication Data

Kennedy, John B.
    Elastic analysis of structures: classical and matrix methods/by John B. Kennedy and Murty K. S. Madugula.
        p.   cm.
    Includes index.
    ISBN 0-06-043634-4
    1. Structural analysis (Engineering)   2. Elasticity.   I. Murty, Madugula K. S., 1938–   . II. Title.
TA645.K45   1990
624.1′71—dc20   89-7453
                                                                                CIP

89 90 91 92 9 8 7 6 5 4 3 2 1

to

Nancy, Lynette, Deborah, and Lorraine

JBK

Indira, Suryakala, and Rama

MKSM

Contents

Preface

Many years of teaching, working in industry, and consulting have convinced us that in this age of computers there is a genuine need for a basic text in structural analysis. By basic text we mean one that covers the subject matter that gives the student a real understanding of elementary and more advanced structural theory, leading to a firm grasp and feel for the subject; we also mean one that introduces early the concepts of structural stiffness and flexibility and their relationships to the classical methods of analysis, phasing in matrix methods of structural analysis and an introduction to the finite-element method; this provides a smooth and logical transition from the classical methods to the matrix methods of analyses. The background required of the student is elementary notions of statics and calculus.

Chapters 1 to 3 introduce the historical background of structural analysis, the manner in which structures carry and transmit loads, and the conditions to be met for structural stability and determinacy. The methods of analysis of statically determinate trusses, beams, frames, arches, and cables are presented in Chapters 4, 5, and 6. The deflections of structures are treated in Chapter 7. Results from computer programs for statically indeterminate structures can be checked by approximate methods of analysis given in Chapter 8; these approximate methods also serve in preliminary design of indeterminate structures prior to more accurate analysis.

Chapters 9 and 10 deal with the classical methods of analysis of indeterminate structures; these methods provide a physical feel for the deformation of a structure as well as improve the student's judgment in design. Furthermore, these methods give a rapid solution without help from the computer if a structure is not too complicated. To ensure that the student is introduced to the modern notions of stiffness and flexibility in the analysis of structures, this subject matter is presented wherever appropriate and applicable in Chapters 9 and 10; this layout provides the student with the necessary background for Chapters 11, 12, and 13, which deal with the matrix formulation for the analysis of structures using stiffness and flexibility approaches. An introduction to the finite-element method is included in Chapter 13; this should give the student the required basic material to appreciate this powerful method of analysis and its current use in the design office. Chapter 13 also includes techniques of structural analysis related to energy methods applicable to indeterminate structures. Thus, the contents of Chapter 13 may or may not be required by the curricula of all engineering schools. The influence of moving loads on determinate and indeterminate structures is dealt with in Chapter 14. Chapter 15 presents a useful method to analyze an indeterminate structure based on an analogy with stresses in columns.

This book is primarily intended for the undergraduate student; Chapters 1 through 8 and portions of Chapters 9 through 11 and 14 can be included in a first course on structural analysis; while the remaining portions of Chapters 9 through 11 and 14, and Chapters 12, 13, and 15 are suitable for a follow-up advanced undergraduate course, with the option of omitting the more advanced topics. There are many solved examples of varying degrees of difficulty throughout the book to explain the principles presented. Numerous problems with answers are also included at the end of chapters for students to solve; by so doing, they will gain a thorough understanding of the methods of structural analysis and develop the necessary skills in applying them to analyze the structures conceived in the design office. We believe that with the approach adopted in this book, the student will be able to develop a fundamental understanding of structures and structural behavior with basic expertise in the more modern techniques of matrix analysis; in this way the graduating student is able to perform and contribute efficiently and more quickly as a structural designer. In conformity with the present-day practice, both U.S. customary (English) units and the Système International d'Unités (SI units) have been used throughout the book.

We are indebted to the many different sources that have been used to write this book. The presentation format of the subject matter is based on many years of teaching; it reflects feedback from our students, industry, and practicing structural engineers and reviewers. The authors wish to thank their graduate students for assistance in the preparation of the many drawings and the solved problems; the authors are grateful to their wives for their tolerance and encouragement. Further, the authors are indebted to Professors Knostman, Leger, and Thomas for their critical review of the manuscript. Last, but certainly not least, the authors are indebted to Judy Assef and Anne-Marie Bartlett for their expertise in typing the entire manuscript and their great patience and cooperation in coping with the several drafts of the manuscript.

John B. Kennedy
Murty K. S. Madugula

Chapter 1

Brief History of Structural Engineering

When we build, let it be such a work as our descendants will thank us for; and let us think, as we lay stone on stone, that a time is to come when those stones will be held sacred because our hands have touched them, and that men will say as they look upon them.

See this our fathers did for us.

Ruskin

If the student is to understand structural engineering of today and to use it wisely to illuminate tomorrow, a brief historical review of the subject matter becomes essential. Structural analysis is concerned with predicting the behavior of a structure subjected to forces; such a structure may be defined as a family of bodies arranged and supported so that it can resist and transmit external and internal forces; the transmission of a force is influenced by the material and shape, including the dimensions, of the structural member. No doubt the Egyptians had some empirical rules for determining the safe dimensions of structural members used to erect their great pyramids!

The following is a brief history of structural engineering (analysis and design):

3000 B.C.	Use of beams and columns as structural elements in the Pyramid of Sakkara by Imhotep, who may be classified as the first structural engineer
582 B.C.	Theorem of right triangle in geometry by Pythagoras
287–212 B.C.	Center of gravity of bodies and foundation of statics by Archimedes
100 B.C.–A.D. 300	Introduction of the masonry arch as a structural element by the Romans
circa A.D. 600	Invention of the present system of arabic numerals by Indian mathematicians
1452–1519	Moment of a force, experimental study of strength of structural materials by Leonardo da Vinci
1518–1580	Use of long span trusses by Andrea Palladio
1564–1642	Introduction of science of strength of materials, and of dynamics by Domino Galileo Galilei

1620–1684	Laws of impact, law of conservation of momentum by E. Mariotti
1635–1703	Relation between force and deformation (Hooke's law) by Robert Hooke
1640–1718	Analysis of arches by Lahire
1642–1727	Laws of motion, law of universal gravitation, development of infinitesimal calculus by Isaac Newton
1654–1705	Deflection of an elastically bent bar by James Bernoulli
1667–1748	Principle of virtual displacements (used in determining deflections of structures) by Johann Bernoulli
1700–1782	Vibrations of prismatic bars by Daniel Bernoulli
1707–1783	First book on mechanics using calculus, buckling of columns by Leonhard Euler
1736–1806	Theory of friction, stability of retaining walls and arches, correct location of neutral axis in bent beam by C. A. Coulomb
1736–1813	Generalized coordinates and generalized forces for the solution of complex structural problems by J. L. Lagrange
1773–1829	Elastic modulus by Thomas Young
1785–1836	Theory of bent plates, fundamental equations of the theory of elasticity, analysis of suspension bridges and thin shells by L. M. Navier
1788–1867	Effect of shearing force on deflection, fatigue phenomenon by Poncelet
1781–1886	Further contributions to torsion, flexure and shear in beams, impact and vibration, theory of plasticity by A. Cauchy, S. D. Poisson, G. Lamé, and B. de Saint Venant
1790–1888	Analysis of articulated structures by A. F. Möbius and S. Whipple
1799–1864	Theorem of equality of external and internal work in a strained structure and theorem of three moments by B. P. Clapeyron
1816	Development of photoelasticity by David Brewster
1820–1912	Stability of retained soil systems by W. J. Rankine
1821–1881	Graphical methods of analysis of structures by Karl Culmann
1824–1887	Deflection of plates by G. Kirchhoff
1830–1879	Theory of reciprocal deflections by James Clerk Maxwell and E. Betti
1862	Introduction of singularity functions by A. Clebsch
1863	Method of sections by A. Ritter
1864	Maximum shear stress yield criterion by H. Tresca
1867	Concept of influence lines by E. Winkler
1877	Semigraphical method of deflection of trusses by Williot (extended later by Otto Mohr)
1835–1918	Conjugate-beam method, circles of stress and strain by Otto Mohr
1847–1884	Theorem of least work by Alberto Castigliano
1873	Moment-area method by Charles E. Greene

1848–1931	Theorem of compatibility, tangent modulus theory of buckling, buckling of built-up columns by F. Engesser
1851–1925	Influence lines, tension coefficient method of analysis of trusses by H. Müller-Breslau
1899	Concept of plastic behavior of structural members by J. A. Ewing
1909	Approximate solution of field problems in continuum mechanics by Ritz
1909–1913	Advances in the field of plastic theory by T. Von Kármán, A. Haar, and R. von Mises
1914	Development of plastic hinges in structural analysis by G. V. Kazinczy
1914–1915	Slope-deflection method by A. Bendixen and G. A. Maney
1919	Use of singularity functions for deflections of beams by W. H. Macaulay
1924	Moment-distribution method by Hardy Cross
1930	Column-analogy method by Hardy Cross
1940	Relaxation method by R. V. Southwell
1943	Extension of the Ritz method for solution of torsion problems by R. Courant
1945–1960	Plastic behavior of structures by J. F. Baker, L. S. Beedle, et al.
1950	Matrix-force method of analysis of elastic structures by H. Falkenheiner
1953	Semimatrix scheme of analysis of aeronautical structures by S. Levy
1954	Concept of stiffness in the analysis of stiff-jointed frames by R. K. Livesley
1955–1956	Concepts of framework analysis and continuum analysis by J. H. Argyris, S. Kesley, M. T. Turner, R. W. Clough, H. C. Martin, and L. J. Topp
1960	Introduction of finite-element terminology by R. W. Clough
1960–1988	Significant advances in the finite-element method of analysis by Wilson, Bathe, Zienkiewicz, and others

It is hoped that this brief background in the development of the theory of structural analysis will arouse some interest in the student and will facilitate the understanding of the subject matter.

He who considers things in their first growth and origin will obtain the clearest view of them

Aristotle

Chapter 2

Structural Form and Idealization of Structures

Of all the human artifacts, none can beat the great bridge; for power, for grace, for the reflection of period and the taste of progress, for the illumination of a setting or the honouring of old vitality. Nothing is more masterly than a high Roman aqueduct, striding across Languedoc or the Campagna, nothing more inspiring than the Golden Gate, that supreme expression of the American genius. Sydney Harbour, the Rialto, Tower Bridge, Avignon—across the world the most famous of the bridges have acquired a status so commanding and so symbolic that their several landscapes would now seem mutilated without them.

James Morris

2.1 INTRODUCTION

The structural engineer must be able to accurately predict the overall response of a structure when subjected to forces, loads, displacements, etc., to assure its adequate and reliable performance. Such prediction must be based upon the *structural analysis* of a *model* whose behavior approximates as closely as possible the actual structure being modeled. The loads acting on a structure produce five basic types of force action: tension, compression, bending, shear, and torsion as illustrated in Fig. 2.1, with force defined by magnitude and direction. The effect of these forces is to produce rotational and translational movements. As a consequence, the structural engineer must ensure that the actual structure satisfies the following criteria:

1. *Strength.* The structure, and its component parts, must be sufficiently strong to resist the various loads imposed on it by proportioning members of the structure such that the computed stresses (axial, shearing, and bending) do not exceed the specified allowable values for the selected material(s) of the structural member.
2. *Stiffness.* The structure, and its components, must be sufficiently stiff so as not to deflect or deform excessively when loaded; the stiffness of a member is a function of the physical dimensions of the member and the mechanical properties of the material of the member; the stiffness of the structure as a whole is a function of the stiffnesses of its individual members. The allowable magnitudes of deflection or deformation of a structure are usually specified in the design codes of practice.
3. *Stability.* Here the structure or one of its component members must be adequately proportioned so that it will not buckle (become unstable) when sub-

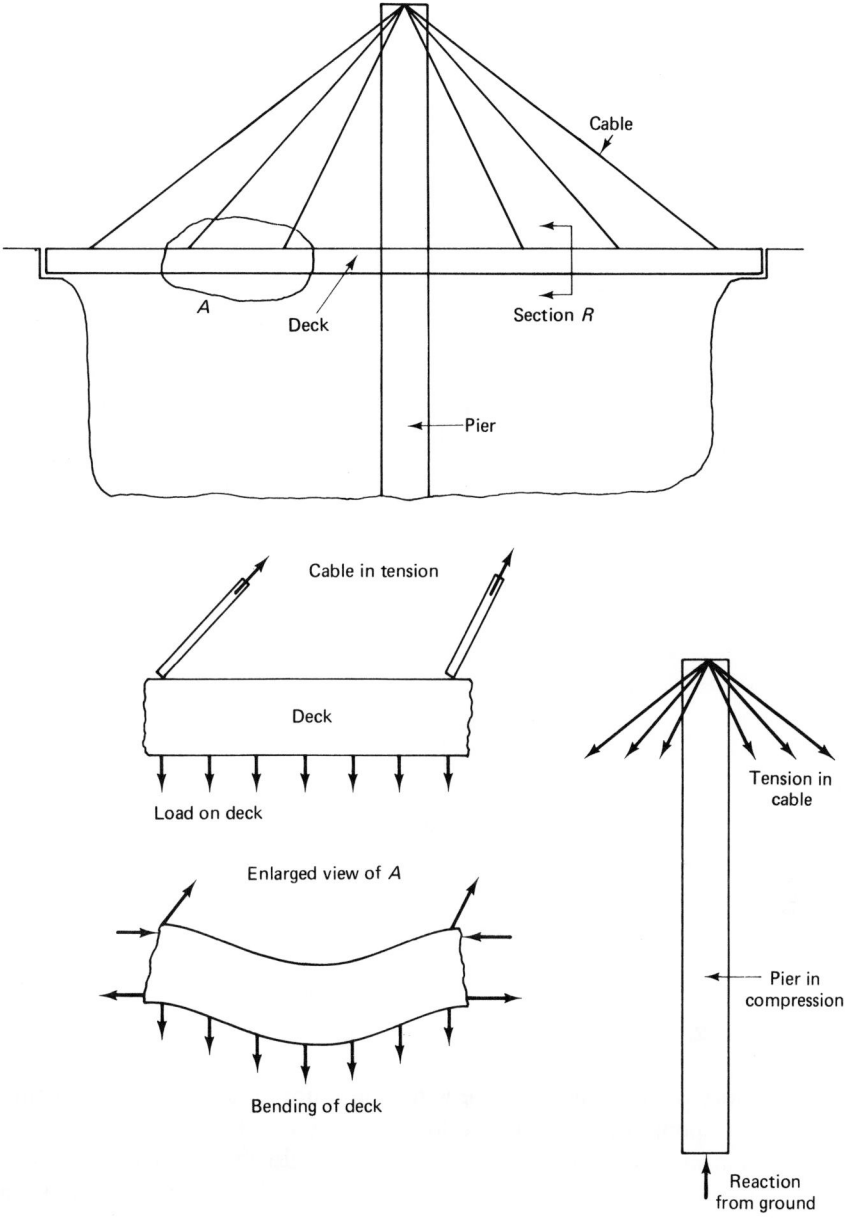

Figure 2.1 Cable stayed bridge.

jected to compressive load; whether a structural member is stable or not depends on its slenderness as well as on the mechanical properties of the material of the member.

Besides meeting the requirements pertinent to each of the above three categories, the structural engineer must exercise good *engineering judgment* to produce a satis-

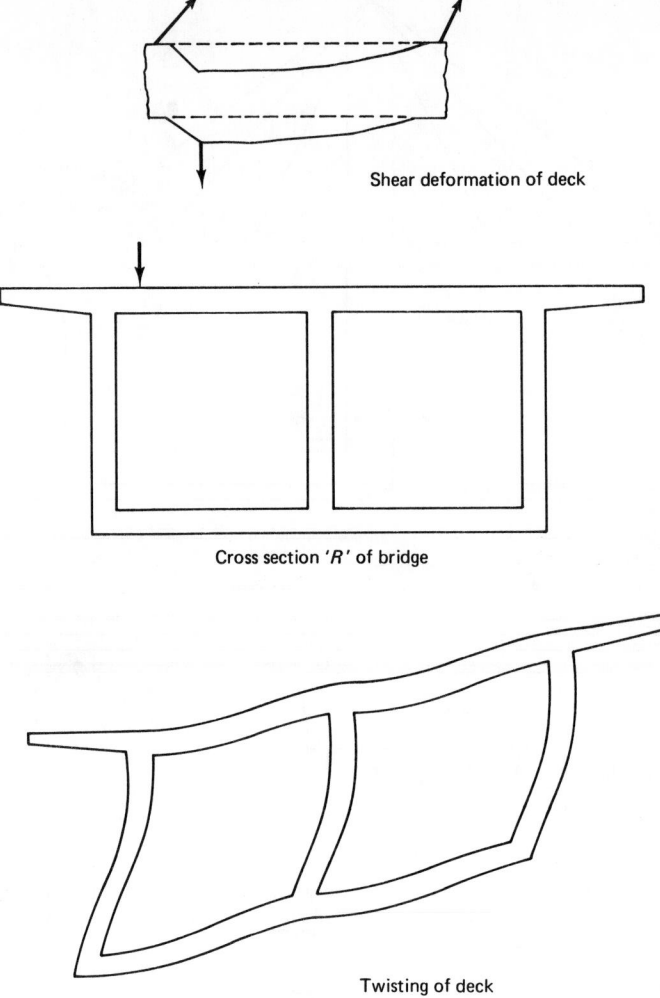

Figure 2.1 (*Continued*)

factory performing structure with minimum cost. It should be mentioned also that the response of a structure to load is influenced by its form or shape; for example, it is the structural form that determines the distribution and magnitude of the design stresses mentioned earlier. It is important for the student to understand at the outset the role that the form or shape of a structure plays in its analysis.

Selection of the most suitable structural form for a particular project is one of the important decisions made in the design of a structure. Quite often, several alternatives are considered before the final structural shape is determined. Such determination is influenced mainly by economics and by a number of other factors such as site and foundation conditions, aesthetic considerations, assessment of loading and methods of design, sociological and functional requirements, quality of workmanship, and materials of construction and their availability.

Figure 2.2a United States Pavilion, Expo '74, Spokane, Washington. Showing a stressed cable network. One-dimensional structure. (Photo courtesy of American Institute of Steel Construction.)

Since the behavior of a structure is influenced by its shape, it is instructive to classify structures as follows:

1. One-dimensional or skeletal structures, for example, beams, trusses, frames, grids, arch ribs, and cable structures (having one dimension, for example, length large in comparison with cross-sectional dimensions).
2. Two-dimensional or surface structures, for example, floor slabs, thin plates, deep beams, or walls and shells (having two dimensions large in comparison with the thickness).
3. Three-dimensional or solid structures, for example, retaining walls, foundations, soil masses, and gravity dams (having significant thickness in comparison with length and width).

Figure 2.2 shows examples of the three classifications of structures.

Figure 2.2b IBM Office building, Minneapolis, Minnesota—concrete waffle slab floor. Two-dimensional structure. (Photo courtesy of Portland Cement Association.)

Figure 2.2c Green Lake hydro project—Sitka Dam, Alaska. Three-dimensional structure. (Photo courtesy of American Society of Civil Engineers.)

2.2 SKELETAL OR ONE-DIMENSIONAL STRUCTURES

The three basic types of skeletal structures are the beam, arch, and cable. They may be varied or combined to assist each other in the same structure. To illustrate the difference between these types, reference is made to Fig. 2.3. The beam (Fig. 2.3*a*) transmits its self-weight and the external vertical load directly to the supports, producing no horizontal reaction on them; whereas the arch (Fig. 2.3*b*) thrusts outward on its supports, which explains the saying that "An arch never sleeps." Conversely, the suspended cable (Fig. 2.3*c*) pulls on its anchorages. Thus an arch is always in a state of compression and thrusting outward; while a suspended cable is in tension and tends to pull its anchorages inward. We will now discuss each of the three types in detail.

2.2.1 The Beam

The simplest structure is the beam, dating back to prehistoric times. The post-and-beam construction (e.g., a block of stone across two others) was dominant in the architecture of ancient Egypt and Greece. The external load normal to the longitudinal

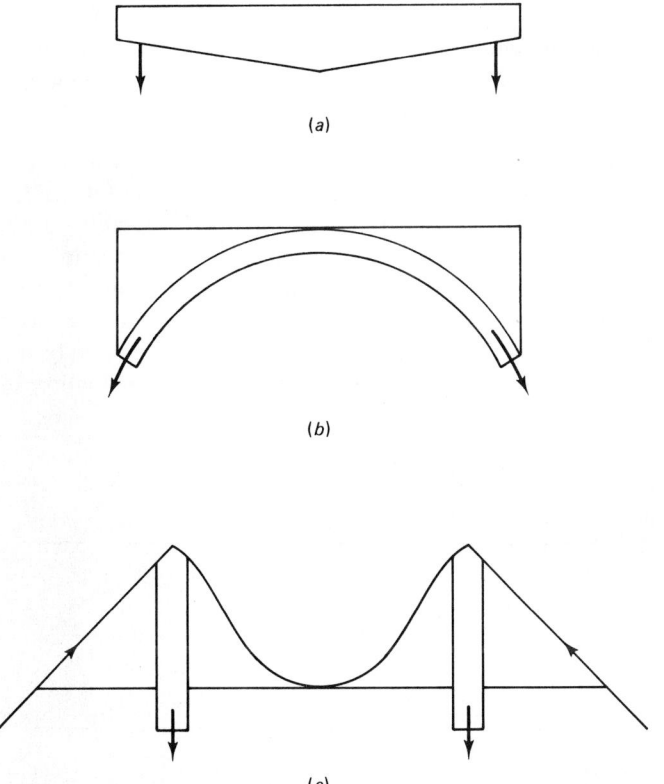

(a)

(b)

(c)

Figure 2.3 Three types of skeletal structures. [Arrows refer to directions of forces on the supports (ground).] (*a*) Beam (or girder bridge). (*b*) Arch bridge. (*c*) Suspended cable (suspension bridge).

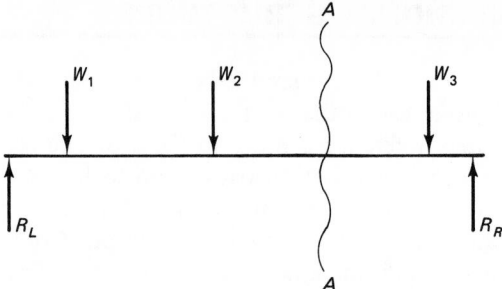

Figure 2.4 Bridging a horizontal gap. (The problem is how to carry loads W_1, W_2, etc., between R_L and R_R.)

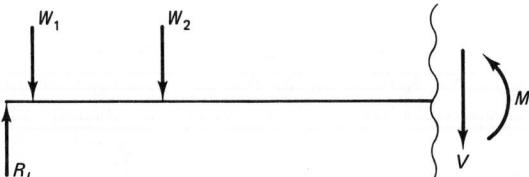

Figure 2.5 Loads W_1, W_2 and left-hand reaction R_L produce a shearing force V and a bending moment M that depend only on these loads and not on the type of structure.

axis of the beam is transmitted along the beam to the supports by *bending* action; see Figs. 2.4, 2.5, and 2.6. This bending or flexural action is created by a couple formed from two equal resultant forces, one in compression (C) and the other in tension (T) (Fig. 2.6). The beam structure, although simple and versatile, is not as efficient as other skeletal structures such as the cable or arch structures; in the beam structure only a few critical sections are stressed to the maximum, and hence the strength of the material is not utilized to the fullest extent. To eliminate some of these disadvantages, the fixed-end beam is used instead of the simply supported beam; Fig. 2.7 shows both types of beam, and it can be observed that there is a reduction in the maximum

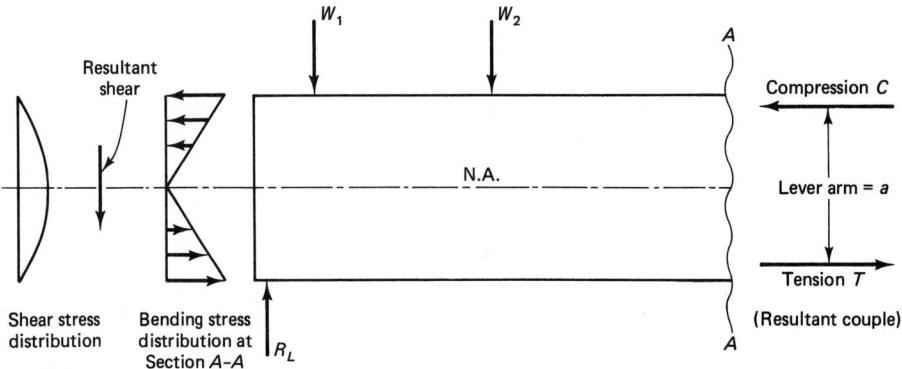

Figure 2.6 Load transfer by bending action (development of internal resisting couple).

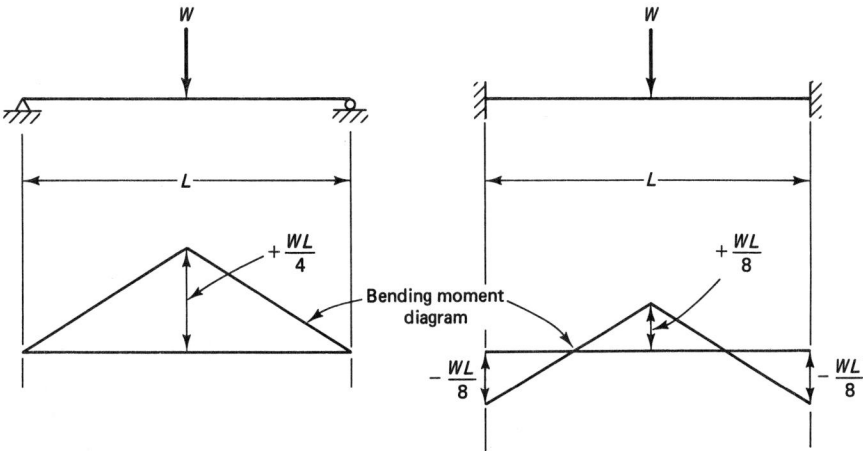

Figure 2.7 Comparison of bending actions between simply supported and fixed-end beams.

bending action due to the fixity of the supports. However, such a high degree of fixity requires massive, rigid, and usually expensive foundations.

To circumvent the need for massive foundations, and to achieve almost the same reduction in the bending action, a continuous beam is quite often used (Fig. 2.8). However, any differential support settlement induces large bending actions in the case of continuous beams as well as the fixed-end beam. An alternative form of beam construction is the balanced cantilever as shown in Fig. 2.9. Here, the bending action of the structure is not influenced by any differential settlement of the supports, and because of the cantilever spans the bending action is considerably reduced when compared with a series of simply supported spans.

A variation of the basic beam form is the frame; this is made up of a number of beam elements (Fig. 2.10a). Multistory frames are but a combination of a series of rectangular frames (Fig. 2.10b). The resistance of multistory frames to load is dependent on the types of connections. For example, one with rigid joints is capable of resisting horizontal loads, such as those due to wind, by the bending action of its

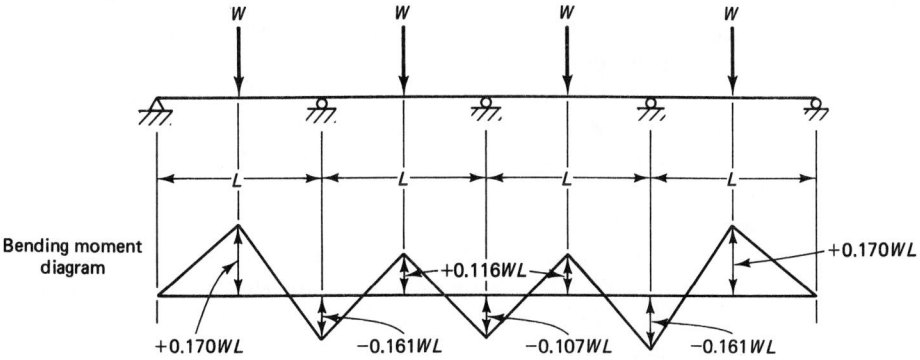

Figure 2.8 Continuous beam and its bending-moment diagram.

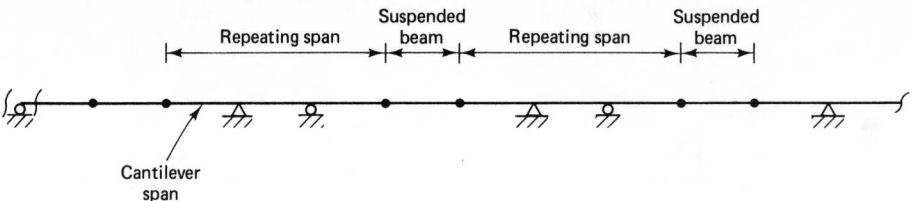

Figure 2.9 Balanced cantilever beam.

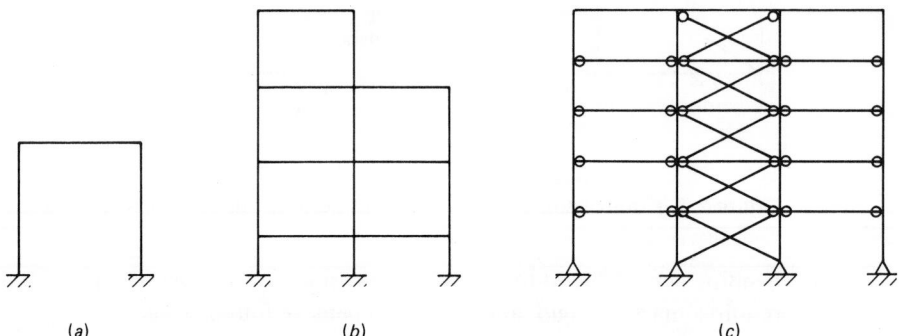

Figure 2.10 Frame—a variation of the basic beam form. (*a*) Rigid rectangular frame. (*b*) Rigid multistory frame. (*c*) Braced multistory frame.

members (Fig. 2.10*a* and *b*); if the beams are fastened to the columns with hinged connections (braced frames), then additional structural members in the form of cross bracing must be provided to resist any horizontal wind loads (Fig. 2.10*c*). It should be mentioned that in general each member of a frame is subjected to axial force, shear force, and a bending moment. Even though frame construction is often less efficient structurally than other forms, it provides clear rectangular openings needed for important functional requirements.

2.2.2 The Cable

While the post-and-beam construction is economical for small and medium-sized spans, it is not well suited for long spans. Consider Fig. 2.6; the beam shown must have adequate depth to provide the required lever arm a between the sections carrying the tensile force T and those that are carrying the compressive force C. As the span increases, these sections become heavier and a stage is reached when the beam cannot support its own weight, however deep it is made—since the greater the depth, the greater the weight it has to carry. However, in practice, the limit of usefulness is reached long before that.

Now a large lever arm can be produced if we curve the member as shown in Fig. 2.11*a*; such a member, with essentially no bending stiffness and no resistance to compression, is called a cable or tension structure constructed of materials that have a high tensile strength; it carries the load by tension along its profile. Figure 2.11*b* shows the free-body diagram of the right half of the cable under its own weight. Since

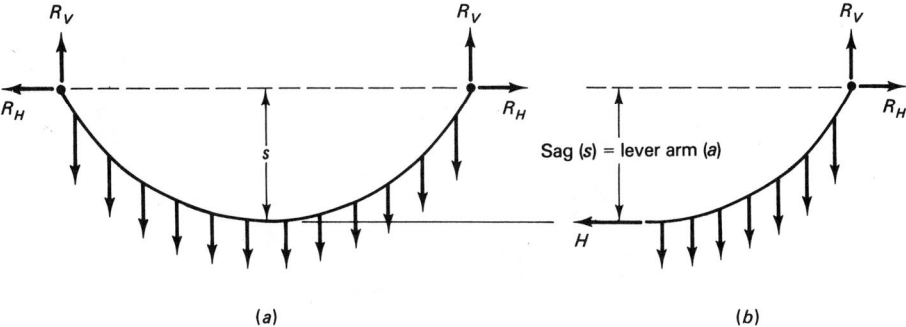

Figure 2.11 (*a*) The cable with axial stiffness but with no bending stiffness. (*b*) Free-body diagram of part of *a*.

the force H is horizontal at midspan of the cable, from static equilibrium, $H = R_H$; these two equal and opposite forces separated by the lever arm *a* form the resisting couple or moment at midspan of the cable. The sag *s* is several times larger than the lever arm *a* generally available in beams; this makes it possible for cable structures to span much larger spans than otherwise economically possible with beams. It is interesting to note that the web of a spider is probably one of the most efficient tension structures ever built, combining maximum strength with minimum weight. The shape of a tension structure is a unique function of the magnitude and position of the applied loads. Figure 2.12 shows a long-span suspension bridge employing the above principles related to cables. Suspension roofs are also commonly used to cover sporting arenas; Fig. 2.13 shows a suspension roof with parallel cables, anchored to buttresses that serve also as banked seats; this type of structure is more functional than a flat roof or a dome (Fig. 2.14) because less space need be enclosed, resulting in economy in heating and air-conditioning costs. The solution to the anchorage problem is simple in a circular arena (Fig. 2.15), where the cables are supported between an inner tension

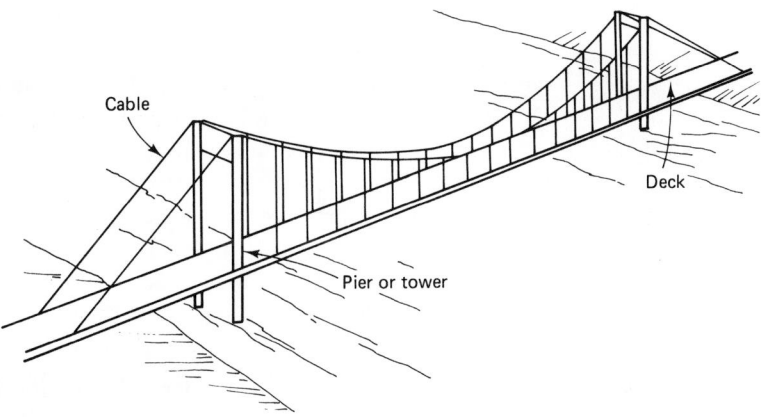

Figure 2.12 Long-span suspension bridge.

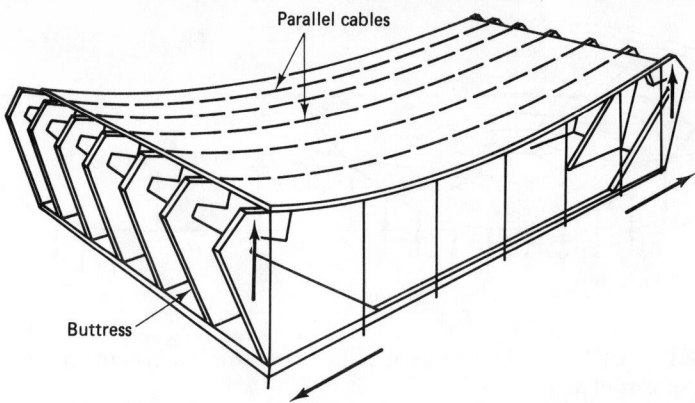

Figure 2.13 A suspension roof with parallel cables, anchored to banked seats.

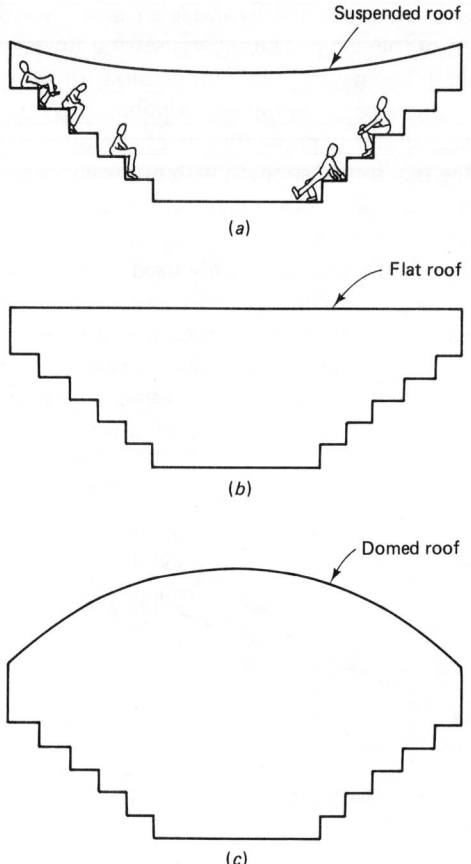

Figure 2.14 (*a*) A suspension roof. (*b*) A flat roof. (*c*) A dome, for covering banked seats in an arena.

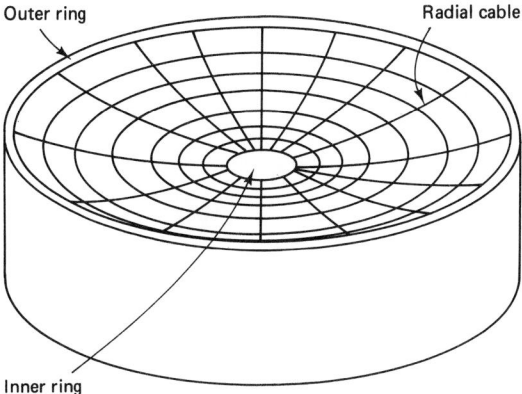

Figure 2.15 A suspension roof with radial cables. The cables are supported between an outer compression and an inner tension ring, and the resulting shape is a dished dome.

ring and an outer compression ring and the reactions are completely absorbed because each circular ring is self-balancing. Another self-balancing support is provided by two crossed arches (Fig. 2.16) that absorb the reaction of the cables; such arches would be subjected to pure compression when the cable roof is designed to acquire a specific shape.

2.2.3 The Arch

An arch can be derived from a tension cable if such a cable is made rigid and then turned upside down as shown in Fig. 2.17. The arch resists the external loads in pure compression if it acquires the exact reverse of the shape assumed by the generating cable under the same loads (compare Fig. 2.17a and b; also Fig. 2.17c and d). If this is not possible, bending moments will be introduced in the arch; and the more deviation there is between the shapes of the arch and the inverted cable, the greater the bending

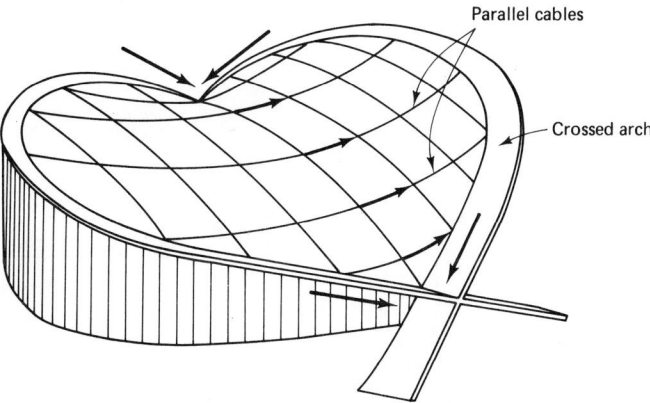

Figure 2.16 A suspension roof with parallel suspension cables, supported by crossed arches.

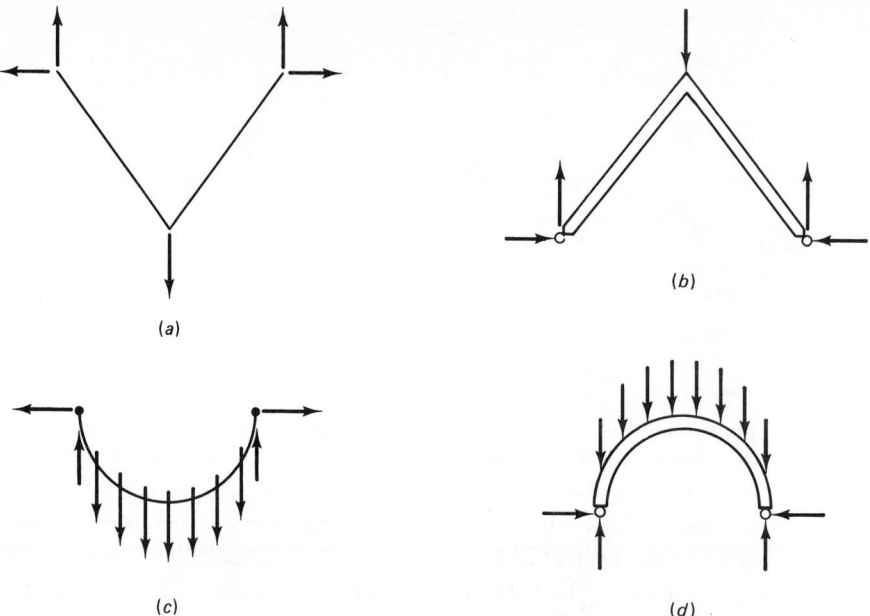

Figure 2.17 The derivation of an arch from a cable. (*a*) Cable supporting a concentrated load. (*b*) Arch corresponding to *a*. (*c*) Cable supporting a uniformly distributed load. (*d*) Arch corresponding to *c*.

action will be in the arch. As was mentioned earlier, the arch is always thrusting outward at its ends; it is the external compressive resistance to this thrust supplied in the form of massive foundations, rock, etc., that gives the arch the ability to carry loads over longer spans.

Another *compression structure,* besides the arch, is the *column;* it is less efficient than a tensile member because it has a tendency to buckle when compressed, unless

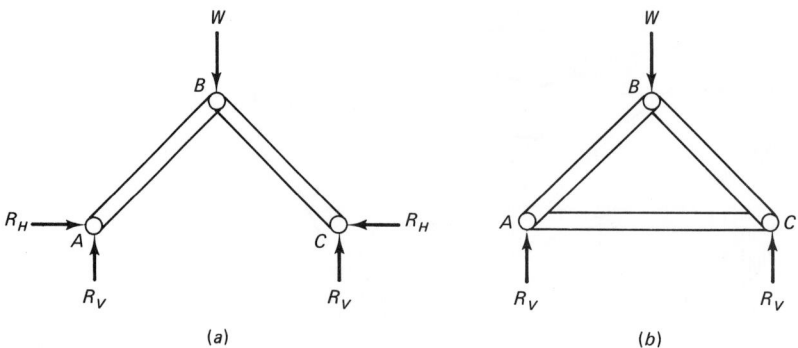

Figure 2.18 The derivation of a truss from a three-hinged arch. (*a*) Three-hinged arch. (*b*) Truss corresponding to *a*.

it is extremely short. Compression structures are built from materials that possess high compressive strength, and they must have sufficient stiffness to prevent buckling.

2.2.4 The Truss

The truss, an efficient structure, composed of a pinned assemblage of members, can carry variable loads through the mechanism of change in the magnitude of the member axial forces. If the arch in Fig. 2.18a is not provided with a horizontal reaction such as R_H, it will collapse. This can be prevented by attaching the member AC as shown in Fig. 2.18b, and the result is a simple truss. When the joints in a truss are pinned with very little resistance to rotation, the forces in the members of the truss are axial (either tension or compression) with no bending in the members.

The resistance to bending of a simply supported truss is provided by compression in the top chord and tension in the bottom chord, the lever arm being the depth of the truss (Fig. 2.19). The empty space between the compression and tension members makes the truss a much lighter structure than a solid beam carrying the same load over the same span. Resistance to shear is provided by the vertical and diagonal members. For trusses with large spans, for example, those used in bridge construction, the compression members may become too long and likely to buckle at a relatively small load. To guard against such premature failure, the compression chord panels are subdivided to reduce the chord's unsupported length, as shown in Fig. 2.20.

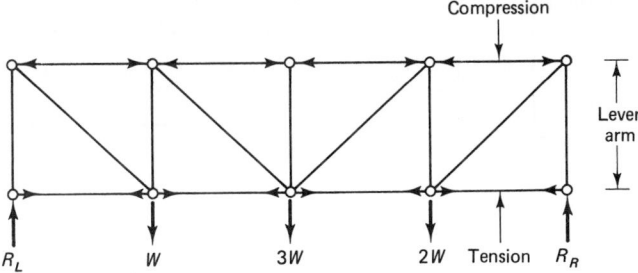

Figure 2.19 Pratt truss.

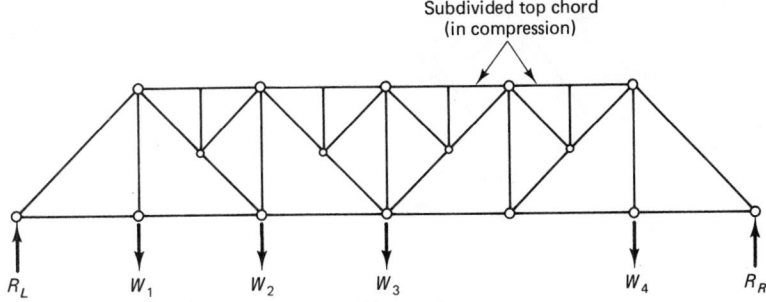

Figure 2.20 Pratt truss with subdivided top chord.

2.2.5 The Horizontal Grid

A closely spaced system of orthogonal or nonorthogonal beams forms a grid that transmits the external load to the supporting columns by means of primary beams or girders (Fig. 2.21a and b). The two-way action of a grid can be demonstrated as follows: Consider the two simply supported identical beams, carrying a load P over a square plan as shown in Fig. 2.21c. Since the two beams interact with each other, they will deflect by the same amount under the load P, and hence each supports half the load, thereby saving material. A similar argument applies when the plan is rectangular or when the beams are of different size; the stiffer beam carries a larger share of the load, the exact proportion depending on the relative stiffness (EI/L^3) of the two beams. Two-way grids are therefore shallower than one-way beams over the same

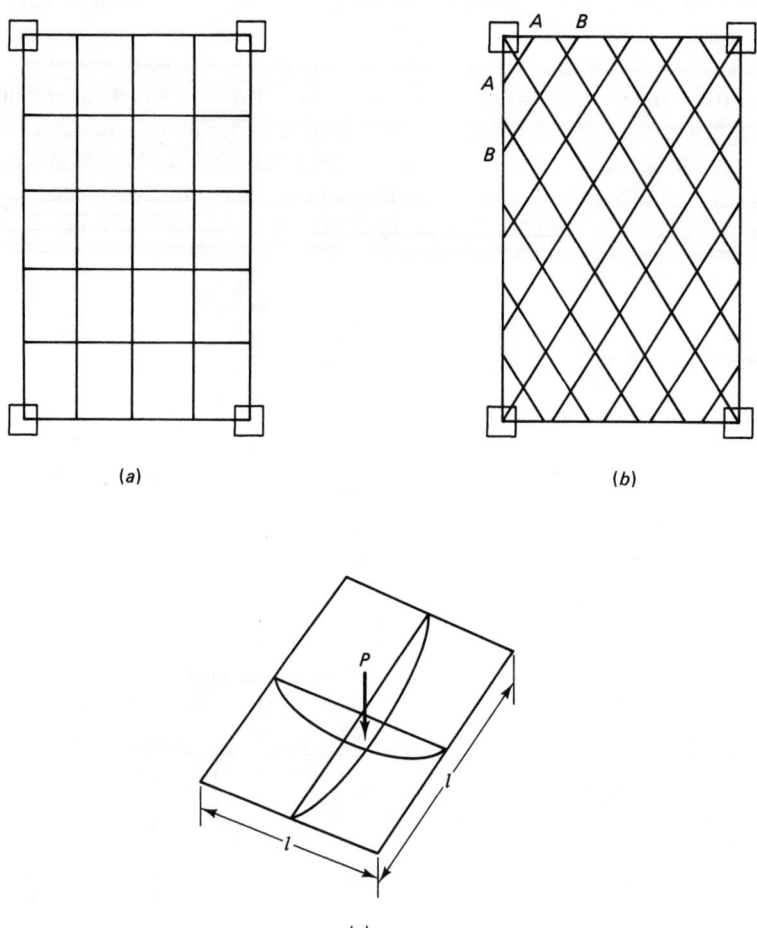

(a)

(b)

(c)

Figure 2.21 A horizontal grid. (a) Orthogonal arrangement. (b) nonorthogonal arrangement. (c) Distribution of load between two beams at right angles.

span. The diagonal (nonorthogonal) system, shown in Fig. 2.21*b* is more expensive to fabricate than the orthogonal system (Fig. 2.21*a*); however, for the supporting system shown in Fig. 2.21*b*, the diagonal system is stiffer for the same overall depth. The short members such as *A-A* and *B-B* add considerably to the relative stiffness of the entire grid, resulting in a shallower structural depth. The beams in the two systems must be designed for both bending and twisting.

The lamella roof shown in Fig. 2.22 is a three-dimensional variation of the diagonal grid. When the members are connected rigidly, they become subjected to a combination of axial compression, bending, and twisting. However, if additional members are added to yield a triangulated frame (Fig. 2.23) with flexible joints, the members become subjected to simple axial tension or compression. Other such space frames are used in the construction of transmission towers.

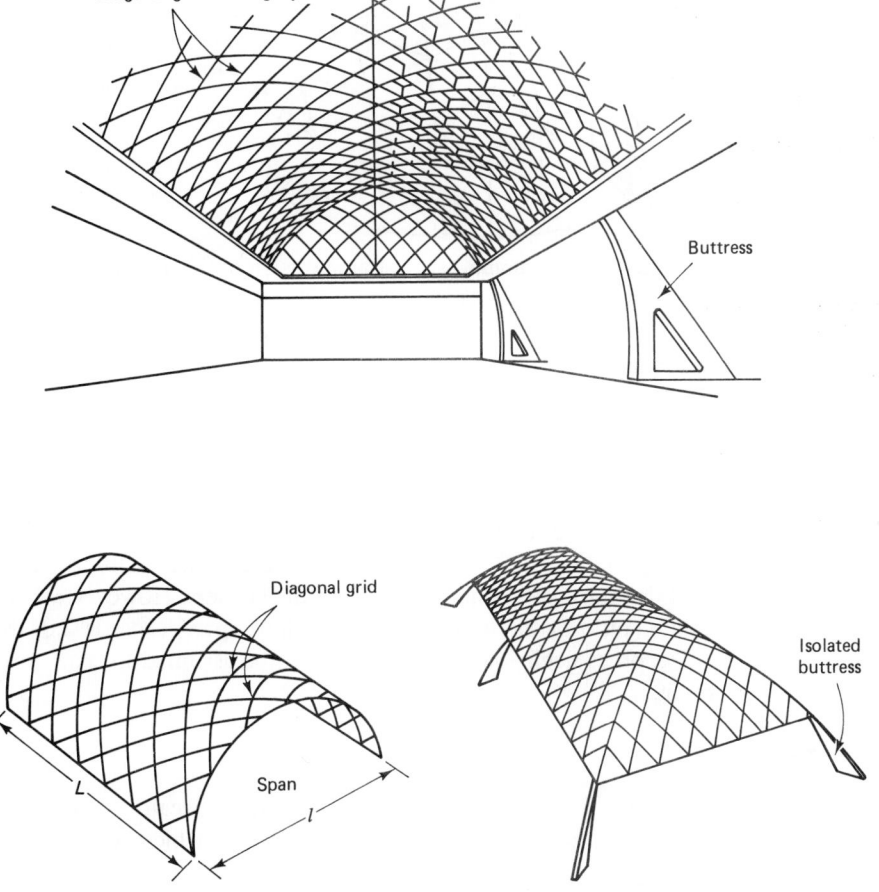

Figure 2.22 Three different views of lamella roof—space-frame version of the diagonal grid.

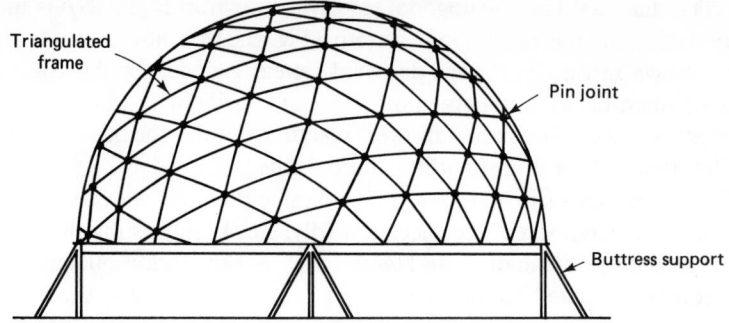

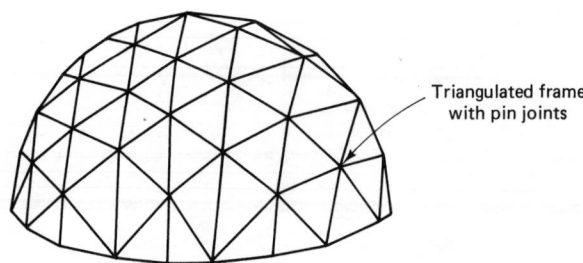

Figure 2.23 Geodesic dome—space-frame version of the triangulated truss.

2.3 SURFACE STRUCTURES

2.3.1 The Thin Plate

If the mesh of the beam grid is progressively reduced, in the limit a plate is formed. The plate carries a load P by two-way action and is subjected to bending as well as twisting; the latter causes the corners of a rectangular plate to curl up,[1] unless they are held down by additional reactions at the corners. However, when the aspect ratio of a plate (ratio of the long span to the short span) exceeds 2.0, the two-way plate action becomes negligible, and the load can be assumed to span entirely in the short direction.

2.3.2 The Shell

This structure has an attractive structural form because of its high efficiency and pleasing lines; the most common shell is the egg which, when subjected to a force,

---

[1] The obtuse corners of a skew (or a parallelogramic) plate press down and the corresponding acute corners curl up.

transmits such a force by in-plane or membrane stresses. Many large structures such as auditoriums, arenas, airline terminals, and churches are roofed by shell structures of many configurations. Water and fuel-storage tanks are shells stressed primarily in tension; while some dams are doubly curved shells, transferring the water pressure into compressive thrusts against rock foundations; airplane fuselages, wings, and space rockets are structural shells with internal beams acting as integral stiffeners. Shell structures are built of materials such as reinforced and prestressed concrete, metals such as steel, and sometimes wood. While shells are highly efficient and aesthetically attractive structures, their use is inhibited by two factors: relatively high cost of construction and difficulty in analysis.

Although an eggshell is considered thin and fragile in proportion to its span, reinforced-concrete shells can be even thinner, as demonstrated in the accompanying table. Such high ratios of span to thickness can be realized only if the designer provides correctly designed supports so that most of the bending and transverse shear are eliminated, thus leaving only in-plane tension, compression, and shear forces within the surface of the shell to be designed for.

Shell structure	Ratio of span to thickness
St. Peter's dome in Rome, built in 1590	13
Average hen's egg	100
Planetarium at Jena, East Germany, built in 1923, the first reinforced-concrete shell	670
CNIT Exhibition Hall, built in Paris in 1958 with a span of 720 ft	1600

The simplest shell form is a cylindrical vault (Fig. 2.24a). Take a piece of paper and form it into a half cylinder; then tie the ends with a string to prevent them from spreading. The paper is much stiffer now than when it was flat, which demonstrates the structural superiority of a shell structure over the plate structure. The tying string at the ends is *essential* to absorb the horizontal reaction of the curved shape; without it the paper would flatten out. The cylindrical vault was used in Roman and Romanesque structures, and is now built in reinforced concrete.

The cylindrical shell is curved in only one direction. If curvature improves the structural performance so much, curvature in both directions may be expected to be even more favorable. We thus obtain the dome, which has a long and honorable history of architectural design (Fig. 2.24b). However, the dome has one important disadvantage in terms of modern technology; it cannot be formed by means of a series of straight pieces (usually in timber and sometimes in steel); thus, since we cannot draw straight lines on a dome, the formwork is quite expensive.

The most important of the shells that can be formed, say, entirely from straight pieces of timber, is the *hyperbolic paraboloid shell,* or *hypar,* for short (Fig. 2.25). This is a remarkably versatile form which can be used in a variety of ways; while the dome is like the top of a hill, the hypar shell is like the saddle of a mountain pass.

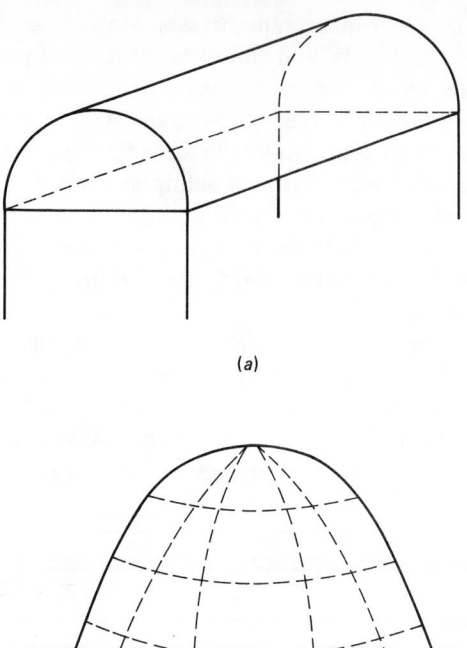

(a)

(b)

Figure 2.24 (*a*) A cylindrical vault. (*b*) A dome.

Hypar shells can be used as saddles (Fig. 2.26), or they can be used with straight boundaries, either alone or in combination.

2.3.3 The Folded Plate

The folded plate consists of straight pieces joined with sharp edges (Fig. 2.27). Because it is subjected to bending, it cannot be made as thin as a shell. Folded plates are rarely constructed in timber because of jointing problems or in metal because of their susceptibility to buckle. However, the flat surfaces in folded-plate structures are easily formed for casting in reinforced-concrete construction.

2.3.4 The Shear Wall

When wind forces act laterally on the side of a multistory building, they are transmitted by the cladding of the building to the vertical framing elements, which in turn transmit them to the horizontal slab elements. These slab elements are usually supported against lateral movements by means of wall elements. The slab elements must be capable of transmitting the applied forces to the wall elements by virtue of their in-plane strengths, or alternatively, bracing must be provided in the horizontal planes. Thus, the forces

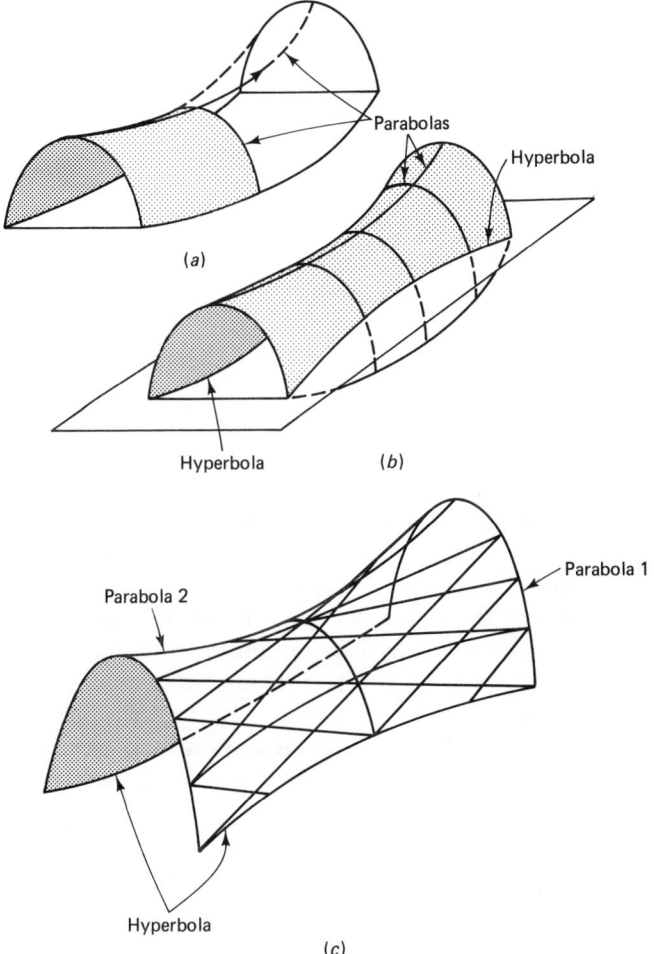

Figure 2.25 A hyperbolic paraboloid (hypar) shell.

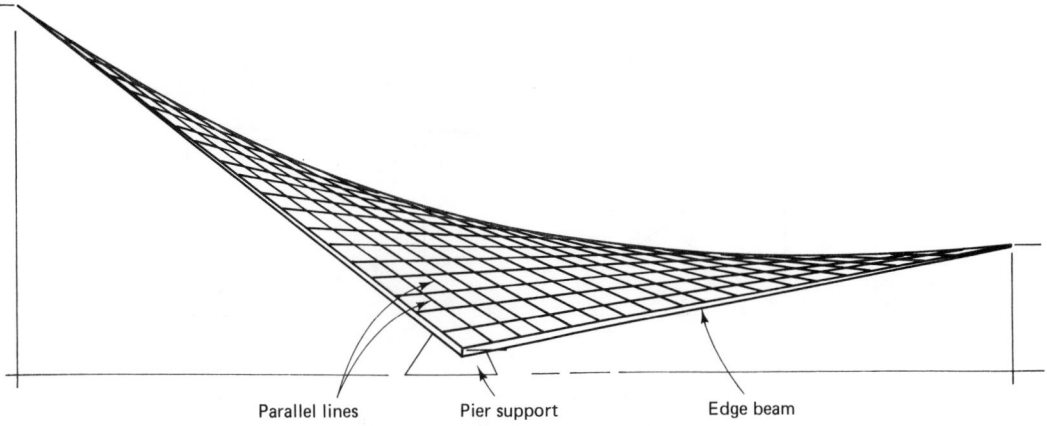

Figure 2.26 A hypar shell (saddle), the simplest hyperbolic paraboloid.

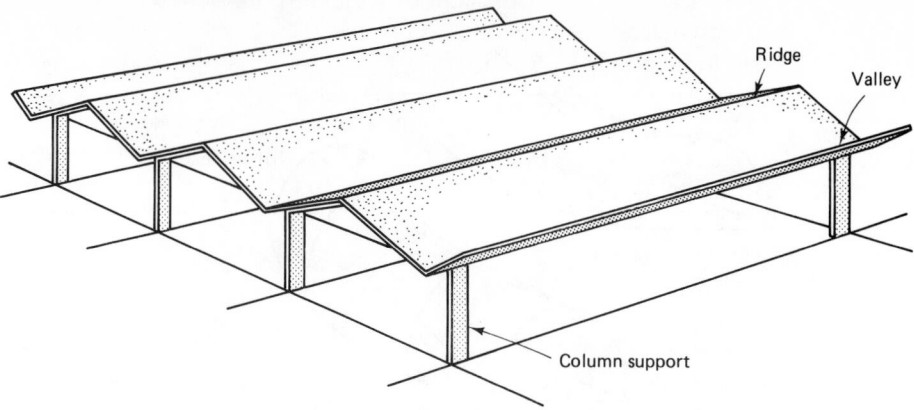

Figure 2.27 Folded-plate structure.

are transferred to the walls which transmit them in turn to the ground; such wall elements are called "shear walls." They can be solid reinforced-concrete walls or suitably braced vertical frames (Fig. 2.28). The main function of a shear wall is to resist changes in geometry due to in-plane shearing action, so that any horizontal force applied to the top of the shear wall will be transmitted by it into the ground. In order for a roof slab to be stable under all directions of in-plane loading, *at least* three shear-wall reactions must be provided (if more shear walls are provided, there is a further increase in structural stiffness). The three shear walls must not all be parallel, nor should their lines of action be concurrent. Elevator shafts and other service shafts

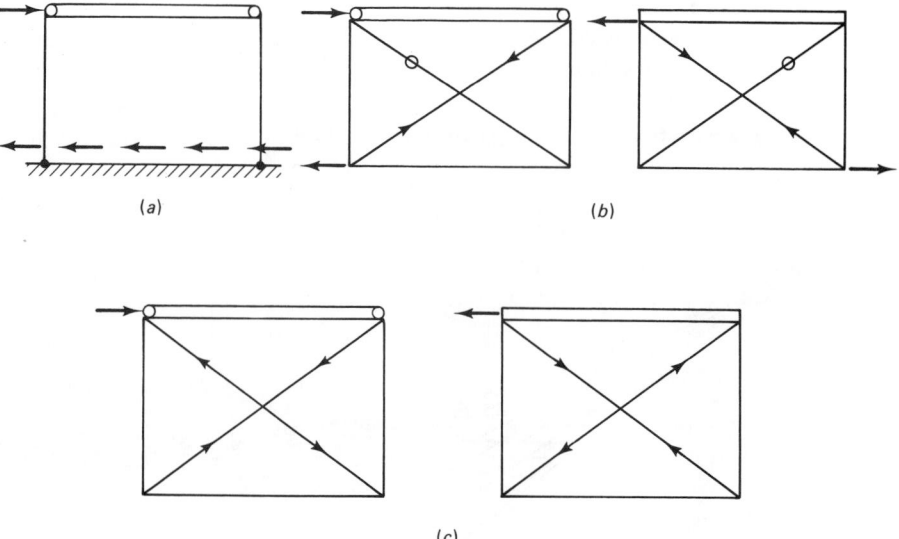

Figure 2.28 Different shear walls. (*a*) Solid slab element. (*b*) Tension system of counterbraced shear walls. (*c*) Tension-compression system of counterbraced shear walls.

provide a very compact system of shear walls. If, however, such shafts are positioned toward one end of a building, another shear wall should be provided toward the other end to help relieve the heavy torsional moments that would otherwise act upon the shaft. An ideal solution is to position the shaft centrally.

2.4 IDEALIZATION OF STRUCTURES

Rarely if ever does an actual structure correspond to the idealized structure that is considered in its analysis. The materials of which the structure is built do not have the exact properties assumed, nor do the dimensions correspond exactly with their theoretical values. Because of the width of members, considerable difference may exist between clear spans and center-to-center spans, which are ordinarily used in analyses. Support details may vary considerably from the idealized type assumed for purposes of analysis. A member may not actually be prismatic, yet it may be assumed to be, to simplify computations. Structural details such as lacing bars and gusset plates introduce effects that might make an analysis very complicated indeed; but because they have little actual effect, they are usually neglected in the analysis of stresses in the main members. Thus, we realize it is necessary to idealize a structure in order to carry out a practical analysis, since real structures are built in three dimensions and are too complex for effective analysis. Experience and judgment are necessary in determining the idealized structure, that is, the *analytical model* that should be used in a given case. In important structures, where doubt exists as to the most logical assumptions to be made in idealizing a structure, it is sometimes desirable to compute stresses on the basis of more than one possible analytical model and to design the structure to resist the stresses corresponding to all the analyses, using proper judgment and experience.

The technique in idealization and modeling of actual structures involves the following sequence: First the real structure and imposed external loads are idealized to create the conceptual or mathematical model; then the response of the idealized structure to the idealized loads is determined by one of the various methods of structural analysis; this response is then related to the real structure. Obviously, the above idealization entails simplifying assumptions that require judicious interpretation of the results as well as an accurate forecast of the behavior of the real structure under the various loading conditions. We should stress that idealization of a real structure must necessarily deal with structural form, connection and support conditions, applied loads, and the material behavior of the structure.

2.4.1 Structural Form

A three-dimensional framed structure is idealized by line diagrams where the lines generally coincide with the actual centroidal axes of the members of the structure. Thus, these members become linear ones (e.g., beams, columns) with negligible lateral dimensions and with their material concentrated along their centroidal axes. Figure 2.29a shows an actual three-dimensional framed structure and Fig. 2.29b the corresponding skeletal structure formed from straight lines. Since the techniques of analyzing

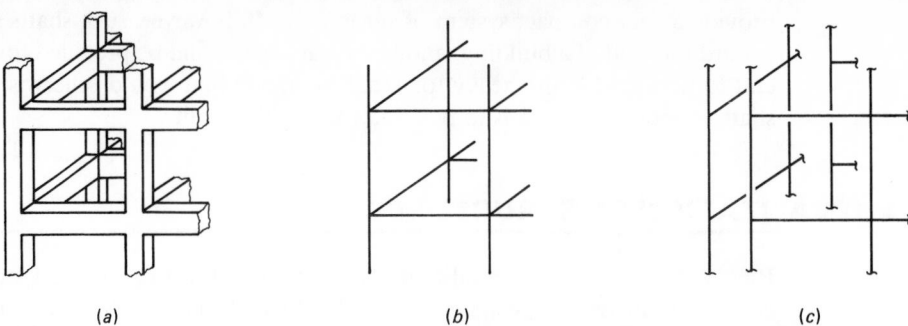

(a)   (b)   (c)

Figure 2.29 (*a*) A three-dimensional framed structure. (*b*) Skeletal structure corresponding to *a*. (*c*) Decomposition of *b* into plane frames.

three-dimensional structures are quite involved, the skeletal structure in Fig. 2.29*b* is decomposed into individual plane frames as shown in Fig. 2.29*c*; each frame is assumed to carry a portion of the total load on the structure and to act independently, even though it is actually joined to other plane frames and therefore interacts with them. As a consequence of this idealization, actual joints in the structure reduce to points (Fig. 2.30). Furthermore, plate structures, for example, a floor slab in a building, are represented by segments of planes; whereas shell structures are shown as surfaces; in both cases no thickness is shown since the thickness of a plate or a shell is generally small in comparison with the other two dimensions.

Consider the reinforced-concrete arch bridge in Fig. 2.31, consisting of a deck, arch, and columns, all cast monolithically; for a preliminary design, the deck is analyzed as a continuous plate element supported by the vertical columns, which are treated as straight-line elements; whereas the concrete arch, supporting the column loads, is analyzed as a curved-line element.

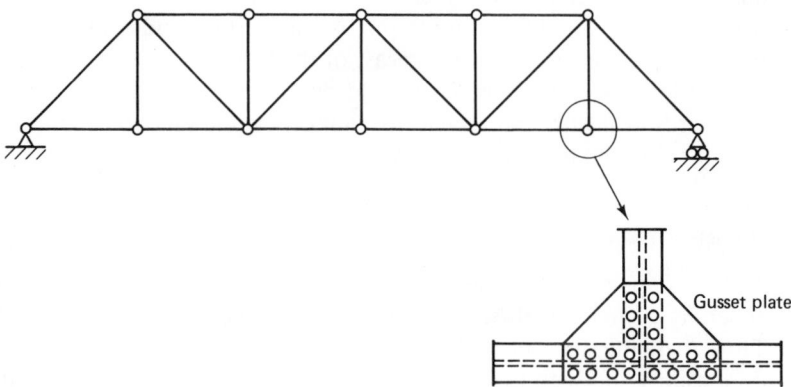

Figure 2.30 Idealization of a joint in a truss.

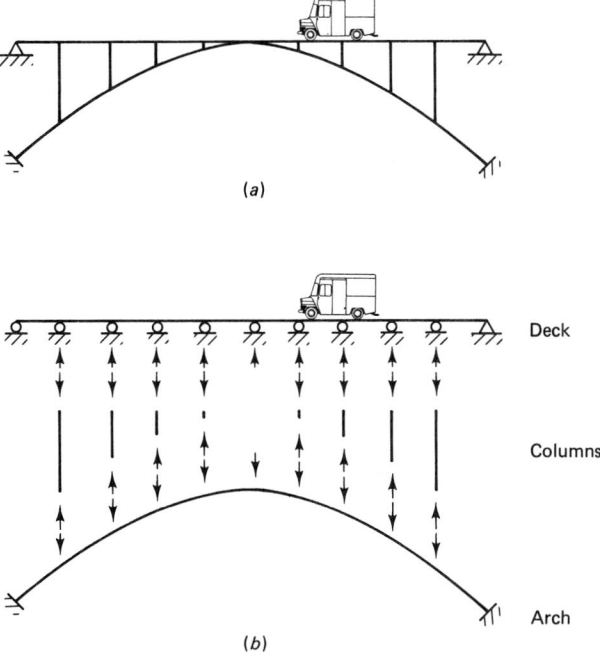

Figure 2.31 (*a*) Arch bridge. (*b*) Decomposition into elements.

2.4.2 Connections and Support Conditions

Connections or joints are usually assumed to be pinned (Fig. 2.30) or rigid (Fig. 2.32). A welded joint is assumed perfectly rigid, although in reality it is not; such an assumed rigid connection will carry any design moment and will maintain the angles between the centerlines of members framing into the joint before and after loading; while a bolted joint is assumed to be a frictionless pin allowing full rotation, although in reality it will not; such an assumed pinned connection will carry shear force and torsional moment, but no bending moment.

The same idealizations apply to supports; these are assumed to be fixed, pinned, or on rollers. A fixed support permits no rotation or translation, while a pinned support allows no translation but permits unlimited rotation in the plane of bending; it also provides torsional restraint. A roller support allows movement only in the direction parallel to the support surface and permits unlimited rotation. The actual supports of the steel frame in Fig. 2.33*a* are idealized in Fig. 2.33*b*. It should be noted that the massive footing support at *A* (assumed fixed) will allow some rotation, while the thin footing at *D* (assumed pinned) will have some restraint against rotation. Such idealizations simplify the analysis considerably, and the errors introduced are generally within tolerable limits. Tables 3.1 and 3.2 give a full description of the various types of supports. The various hinges used in actual concrete structures, for example, are shown in Fig. 2.34.

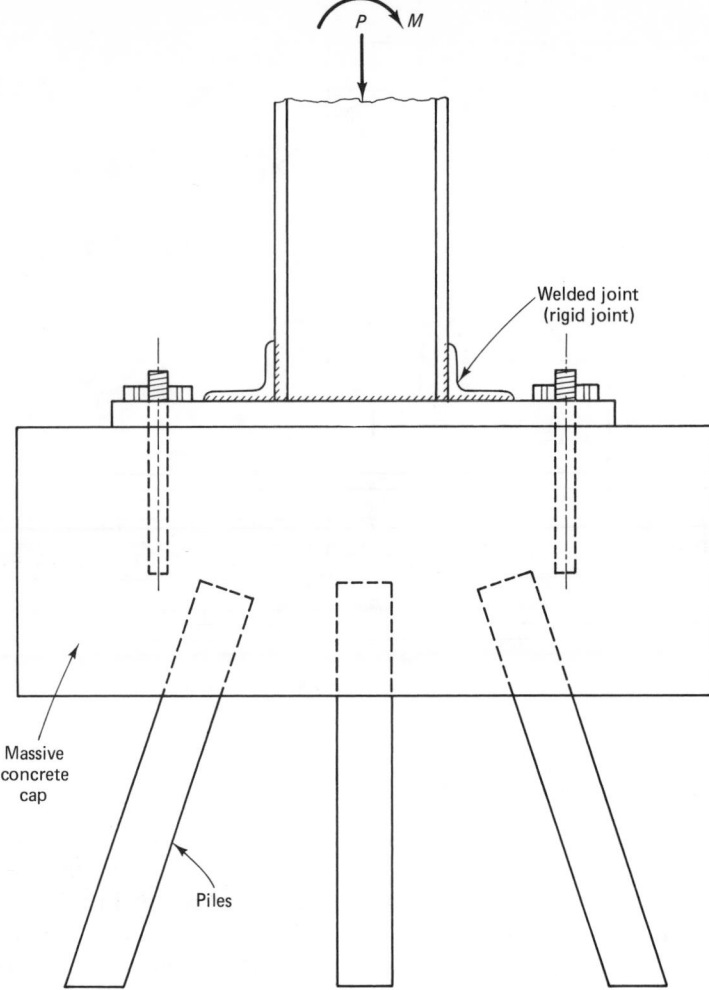

Figure 2.32 Rigid connection.

2.4.3 Loads and Reactions

Structural loads can be caused by gravity, wind pressure, hydrostatic pressure, accelerations, temperature changes, shrinkage, friction, earthquakes, impacts, and vibrations. These various loads and the resulting reactions are generally idealized into equivalent concentrated forces or couples, and distributed forces or couples as shown in Fig. 2.35. A concentrated load or a couple is applied at a point while a distributed load is applied over a length or an area, and a distributed couple is applied over a length. In reality, concentrated forces or couples cannot exist; however, when the area of application of the loads is small in comparison with the overall dimensions of the member, the assumed point application (at the centroid or center of gravity) results in only small errors. Figure 2.36a shows a beam carrying a load and supported by brick walls.

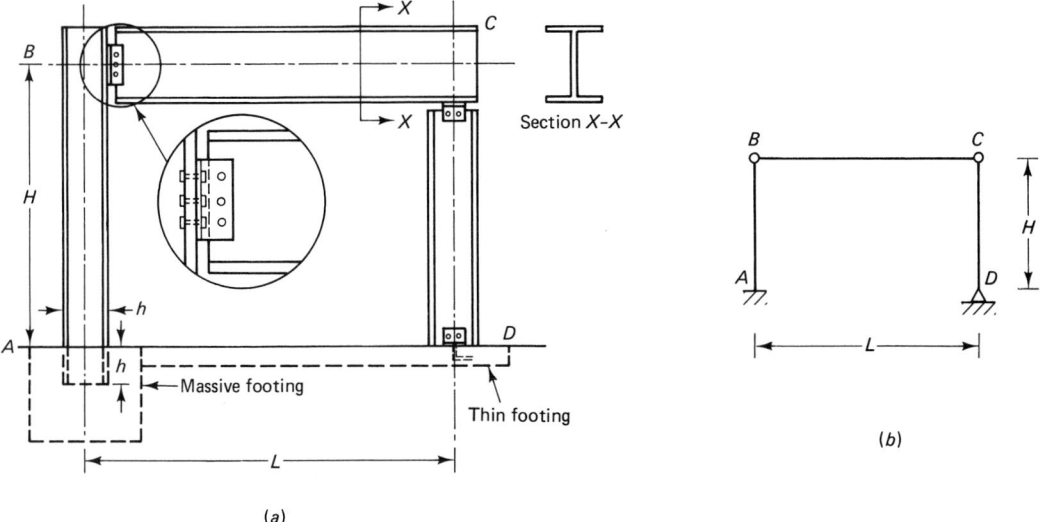

Figure 2.33 Idealization of connections in a steel frame. (a) Actual structure. (b) Idealized structure.

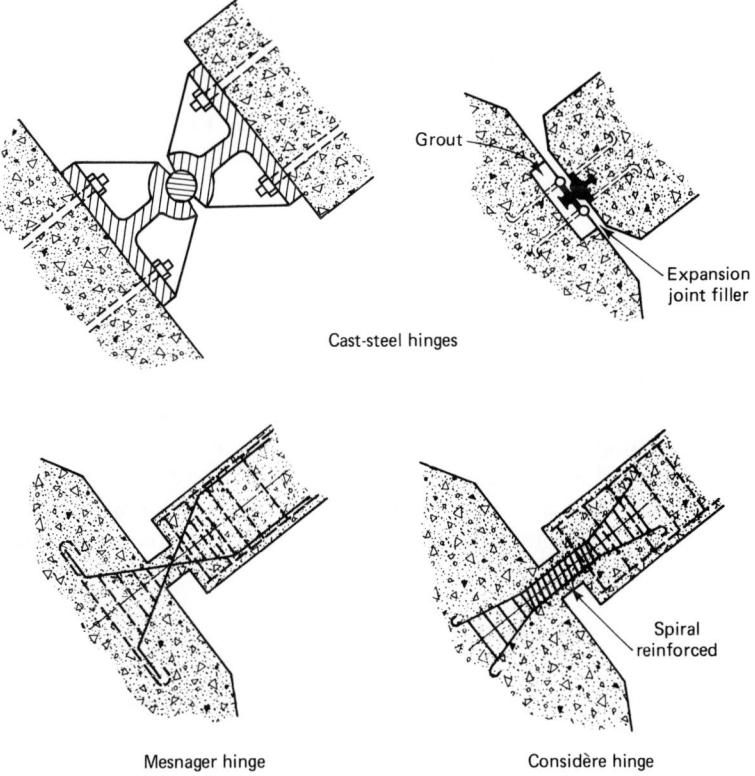

Figure 2.34 Various hinges for concrete structures.

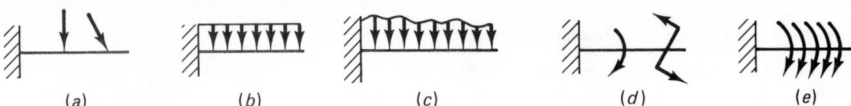

Figure 2.35 Idealized loads. (*a*) Forces. (*b*) Uniformly distributed forces. (*c*) Varying distributed forces. (*d*) Couples. (*e*) Uniformly distributed couples.

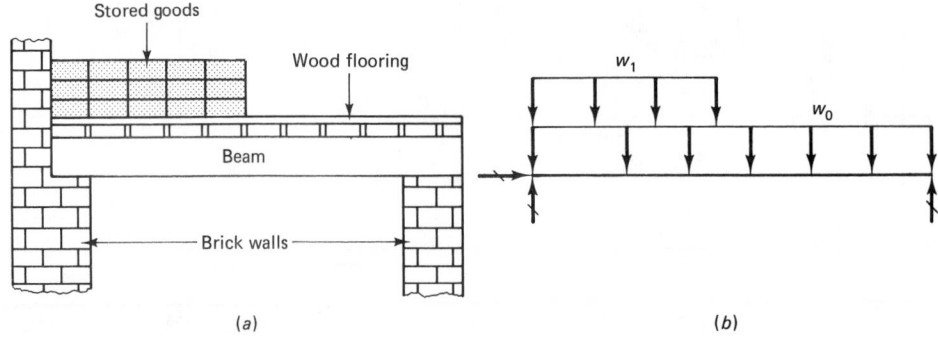

Figure 2.36 Idealization of loads.

Note how the support from the wall is idealized by equivalent concentrated loads (Fig. 2.36*b*); the actual beam is represented by its centerline. The wood flooring and weight of the beam are represented as uniformly distributed load over the entire length of the beam; whereas the superimposed load due to the stored goods is represented by a uniform load acting on that length of beam directly beneath the stored goods.

2.4.4 Material Behavior

It is known that the behavior of all structural materials is nonisotropic at the microscopic level; however, the behavior of steel, for example, becomes isotropic at the macroscopic level and is linearly elastic up to its proportional limit. On the other hand, the behavior of reinforced concrete is nonlinear; it is nevertheless idealized to be linearly elastic if the concrete compressive stress does not exceed $0.5 f'_c$ (where f'_c is the ultimate compressive strength of the concrete) and the steel stress does not exceed $0.6 f_y$ (where f_y is the yield strength of the reinforcing steel). From the foregoing it is seen that idealization of an actual structure is a planning process. Experience and judgment are requisites for decisions leading to the correct ideal structure.

2.5 TRANSMISSION OF FORCES

It is important for the student to understand and appreciate how a structure transmits its load to the supporting foundation. Let us consider a through-truss railroad bridge shown in Fig. 2.37*a*(i); the railroad track is supported by the longitudinal stringers, which, in turn, are supported by the floor beams [Fig. 2.37*a*(ii)]. The ends of these

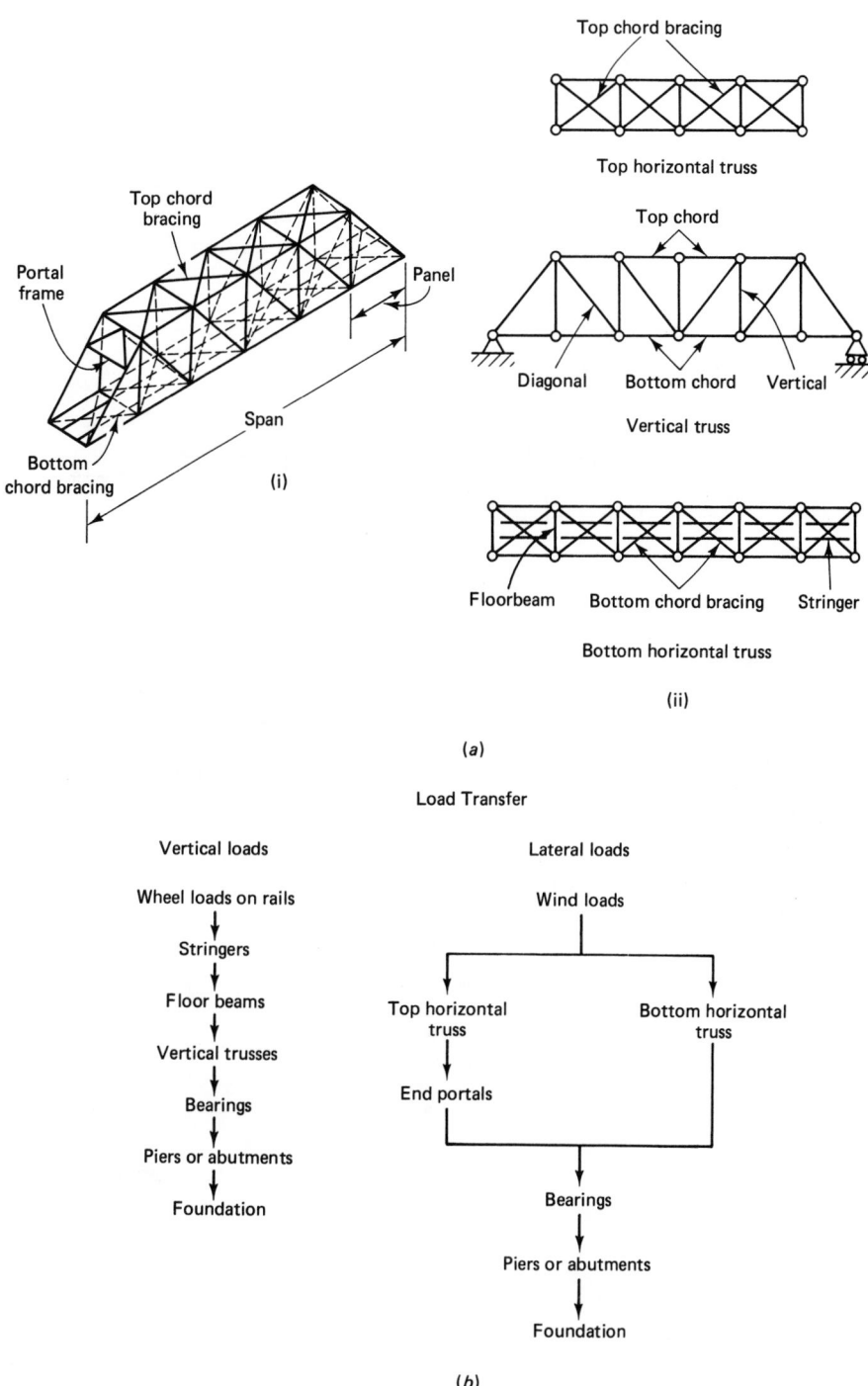

Figure 2.37 Load transfer in a through-truss rail bridge.

floor beams are attached to the bottom panel joints of the vertical longitudinal trusses; the abutments supply the reactions to the vertical trusses. Furthermore, the passing of a train over a bridge coupled with wind loading introduces horizontal (lateral) forces on the vertical trusses at their top and bottom panel joints, and these forces are resisted by the top and bottom horizontal trusses; the loads on the top horizontal truss are resisted by the two end portals that transfer the loads to the bearings on the abutments and hence to the supporting soil. Figure 2.37b shows the path for the transfer of load on a railway truss bridge to the foundation. It should be noted from Fig.

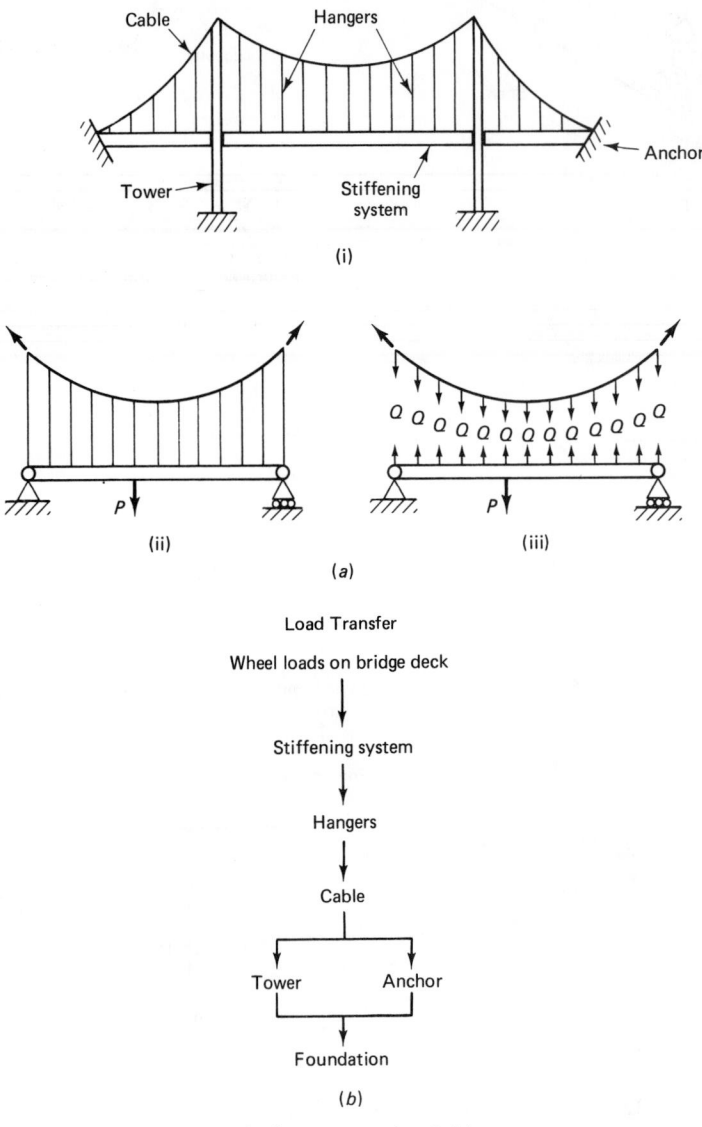

Figure 2.38 Load transfer in a suspension bridge.

2.37a(ii) that the floor beams also act as part of the bottom chord bracing. As another example, let us examine the suspension bridge in Fig. 2.38a. Here, a stiffening system, for example, a girder or a truss, is tied to the road deck so that the flexible cable does not appreciably change in shape while a load moves on the bridge. Figure 2.38b shows the load transfer in such a suspension bridge. A typical framing plan for a steel-framed office building is shown in Fig. 2.39a(i). Note how the girder *CD* is idealized [Fig. 2.39a(ii)] when its ends are connected to the column through the web only [Fig. 2.39a(iii)]. The load transfer for such an office building is shown in Fig. 2.39b. Finally, the mill building structure in Fig. 2.40a(i) transfers its vertical loads and transverse and longitudinal wind loads as shown diagrammatically in Fig. 2.40b. The conceptual model of the gable frame for this building is shown in Fig. 2.40a(ii) with rigid joints to avoid large deformation of the frame; Fig. 2.40a(iii) illustrates a welded joint providing the required degree of rigidity.

When certain conditions regarding the material and geometry of the deformed structure are present, by virtue of the *principle of superposition* the effects of the different loads on a structure can be combined to yield a resultant effect. For example, the bending moments corresponding to various loads on a structure can be superimposed

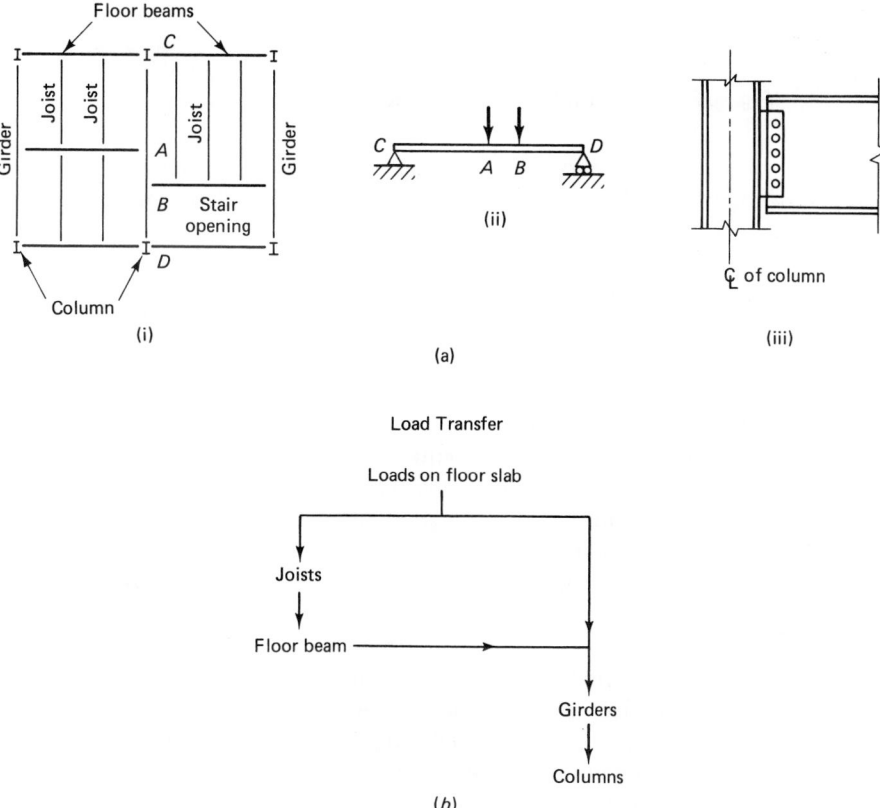

Figure 2.39 Load transfer in a steel-framed office building.

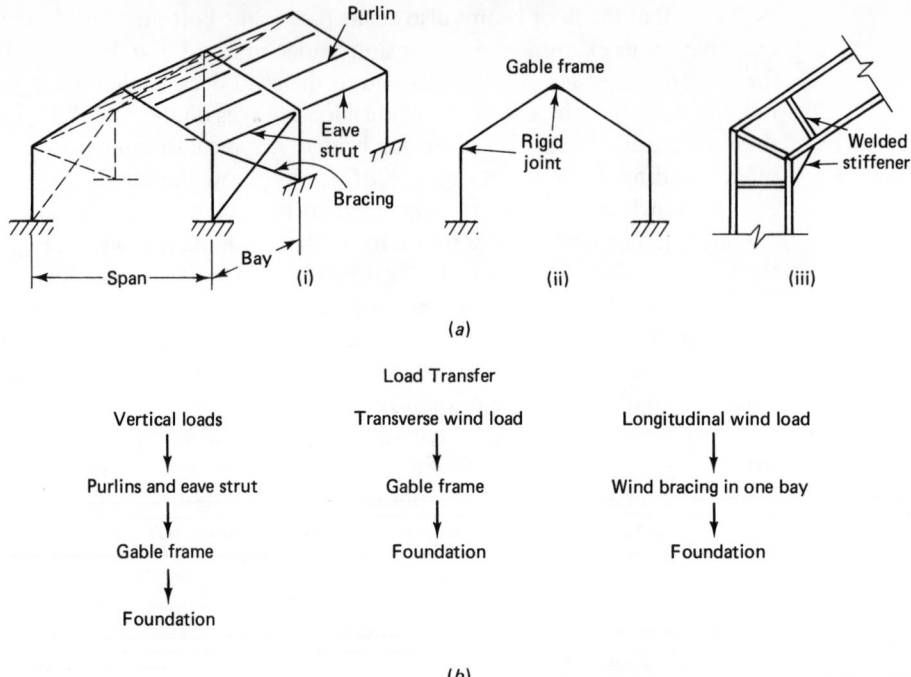

Figure 2.40 Load transfer in a mill building.

to obtain the net or final bending moment. The principle of superposition plays an important role in structural analysis; it is expounded in the next section.

2.6 THE PRINCIPLE OF SUPERPOSITION

2.6.1 Introduction

Superposition is a purely abstract concept, based on the fundamental axiom of logic and science that "the whole equals the sum of its parts," that is, the *separate effects* are superimposed to obtain the *total* effect. The principle of superposition is used widely and continually in the analysis of determinate and indeterminate structures. The principle states that *if several causes (applied loading, temperature change, settlement of supports, prescribed deformation, fabrication error, etc.) act simultaneously on a structural system and if each effect (reactive and internal forces, displacements, strains, stresses, etc.) is directly proportional to its cause, then the total effect is the sum of the individual* effects. The principle is limited to problems where each pair of cause and effect is independent of every other pair, such that there is no coupling between a particular cause and the effects of other causes.

For *statically determinate stable structures,* the *superposition of force effects* is valid, provided deformations are so small that the overall initial geometry of the structure is not changed significantly and the equations of equilibrium may be written with reference to the undeformed geometry. Furthermore, since forces and reactions in statically determinate structures are dependent on the geometry of the structure (but not on the geometry of its cross sections) and independent of deformations, the material properties (factors affecting deformation) need not be considered. In the case of *statically indeterminate structures,* superposition of force effects due to separate causes yields the correct total force effect only if the material of the structure is linear (obeys Hooke's law) and gross distortions are avoided.

When considering superposition of deformation effects, it is important to identify the causes bringing about the deformations. If the *causes are strains,* then deformations due to various strains are clearly superimposable with the proviso that the strain-displacement relations are linear. Furthermore, the absence of deformations causing gross distortions of the structure is sufficient for superposition of deformations due to various strains in both statically determinate and indeterminate structures. However, if the *causes are loads,* then forces and stresses must be related to strains and deformations, and hence the principle of superposition, in this case, can be applied only if, in addition to the above requirements, the material of the structure is linear. Table 2.1 summarizes the conditions necessary and sufficient for superimposing deformation and forces in determinate and indeterminate structures.

2.6.2 Elastic, Inelastic, Linear, and Nonlinear Behavior

When a structure is loaded, the length of each stressed component member changes. If, upon unloading, such component members return to their original length, the structural behavior is said to be *elastic.* On the other hand, if the stressed component members do not regain their original unstressed length upon unloading, the structural behavior is said to be *inelastic.* Furthermore, when the stress-strain response of each component member of a structure is linear, the behavior of the structure is defined as *linear;* otherwise it is said to be *nonlinear.*

Elastic behavior of a structure can be either linear (linear elastic structure) or nonlinear (nonlinear elastic structure). Furthermore, inelastic behavior can be nonlinear (nonlinear inelastic behavior). Figure 2.41 demonstrates the various behaviors; Fig. 2.41*a* shows a linear elastic material up to the proportional limit σ_{pl} (point *A*); if the component is stressed beyond that limit to point *B,* its behavior becomes nonlinear and inelastic with a residual plastic strain ϵ_r upon unloading to zero stress. The nonlinear elasticity of rubber and other rubberlike materials is shown in Fig. 2.41*b*; while there is no residual strain upon unloading, superposition of deformations is not possible in this case; for example, the strain ϵ_c, corresponding to a stress σ_c equal to the sum of σ_a and σ_b, is not equal to the sum of the corresponding strains ϵ_a and ϵ_b. Superposition is applicable to only one case, and that is when full unloading is superimposed on the loading, leading to complete recovery of strain, and the structure assumes its state of no deformation. Figure 2.41*c* shows the behavior of a nonlinear inelastic material; structural concrete behaves in such a manner. It should be pointed out that inelastic behavior can be produced by short-term loading when the proportional limit of the material σ_{pl} has been exceeded, or by long-term loading that produces nonrecoverable creep strains. In general, most materials deform elastically under low stress levels and become inelastic under higher levels of stress (Fig. 2.41*a*).

TABLE 2.1 CONDITIONS OF APPLICABILITY FOR THE PRINCIPLE OF SUPERPOSITION

Superposition of deformation for				Superposition of force for	
Determinate stable structure		Indeterminate stable structure		Determinate stable structure	Indeterminate stable structure
When deformation caused by strains	When deformation caused by loads (force or stress)	When deformation caused by strains	When deformation caused by loads (force or stress)		
1. No gross distortion* of structure 2. Linear strain-displacement relationship	1. No gross distortion of structure 2. Linear strain-displacement relationship 3. Material is linear†	1. No gross distortion of structure 2. Linear strain-displacement relationship	1. No gross distortion of structure 2. Linear strain-displacement relationship 3. Material is linear	1. No gross distortion of structure (Material can be linear or nonlinear)	1. No gross distortion of structure 2. Material is linear

* It is assumed that deformation of structure is very small.

† Material obeys Hooke's law (i.e., Hookean material).

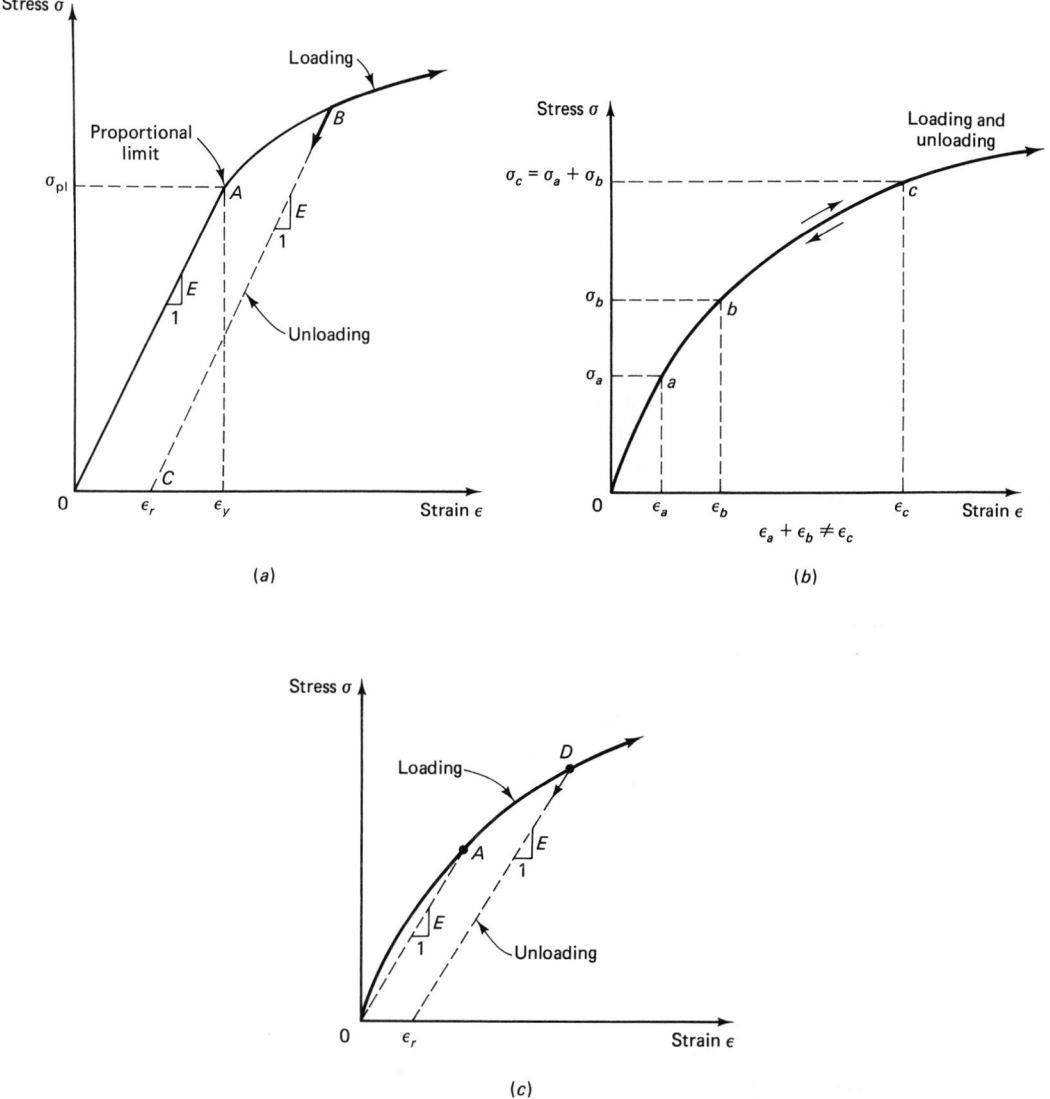

Figure 2.41 Material behavior. (a) Linear elastic behavior (from O to A to O) and nonlinear inelastic behavior (from O to B to C). (b) Nonlinear elastic behavior (from O to C to O). (c) Nonlinear inelastic behavior.

To gain further insight into the principle of superposition, let us discuss a few cases in which superposition is not applicable. Figure 2.42 shows a degenerate three-hinged arch having a *linear elastic material*. Now the vertical external reactions V_a and V_b can be found by the equation of statical equilibrium using the initial geometry of the structure acb and are therefore linear functions of the load P; that is, $V_a = 2P/3$ and $V_b = P/3$. Thus, if the load P is doubled, V_a and V_b will double. However, the external horizontal reaction H can be found only in terms of the deflected geometry

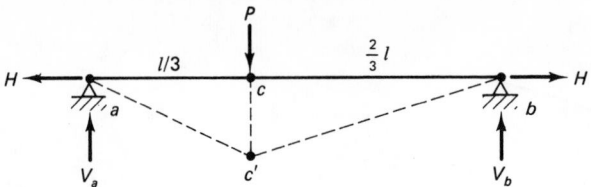

Figure 2.42 A degenerate three-hinged arch (all hinges a, c, b at the same level) (geometrically nonlinear behavior).

of the structure $ac'b$. It can be shown that the deflection cc' is a nonlinear function of the load P and that $H = AP^{2/3}$ where $A = $ constant; thus H is a nonlinear function of the external load P and if P is doubled, say, the horizontal reaction H will be less than doubled. This is an example of a mixed system in which some effects are linear (can be superimposed) and others are nonlinear (cannot be superimposed) even though the material is linearly elastic (Hookean material); thus, a structure of Hookean material behaves nonlinearly when changes in the geometry are caused by the applied loads. Another example for which the principle of superposition is not valid is the beam-column of Hookean material shown in Fig. 2.43. If this structure is subjected to an axial force F of magnitude below its buckling load, it will maintain its undeflected shape; that is, the deflection at a point such as O will be $y_0 = O$ (Fig. 2.43a). If the structure is subjected only to a lateral load P (Fig. 2.43b), the deflection and bending moment at O will be y_0 and M_0, respectively. Now if both F and P act simultaneously on the structure (Fig. 2.43c), the deflection at O will increase to y_0' and the bending moment will increase to $M_0 + Fy_0$; the increased deflection at O ($y_0' > y_0$) is caused by the additional moment Fy_0. Such an additional moment is

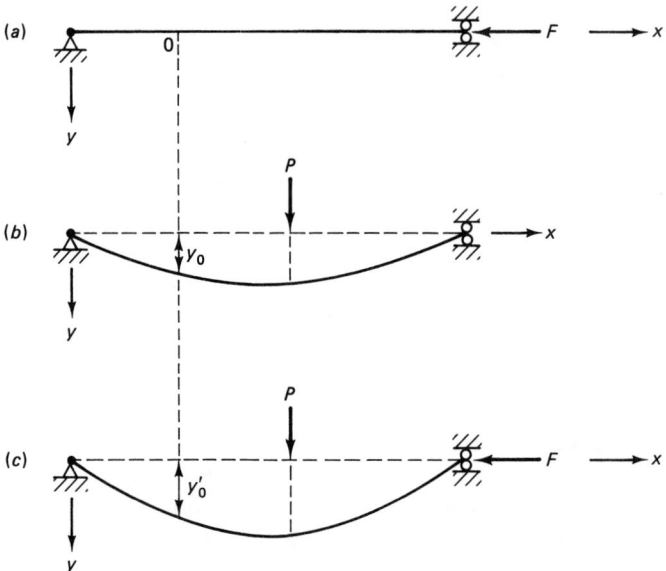

Figure 2.43 Beam-column (geometrically nonlinear behavior).

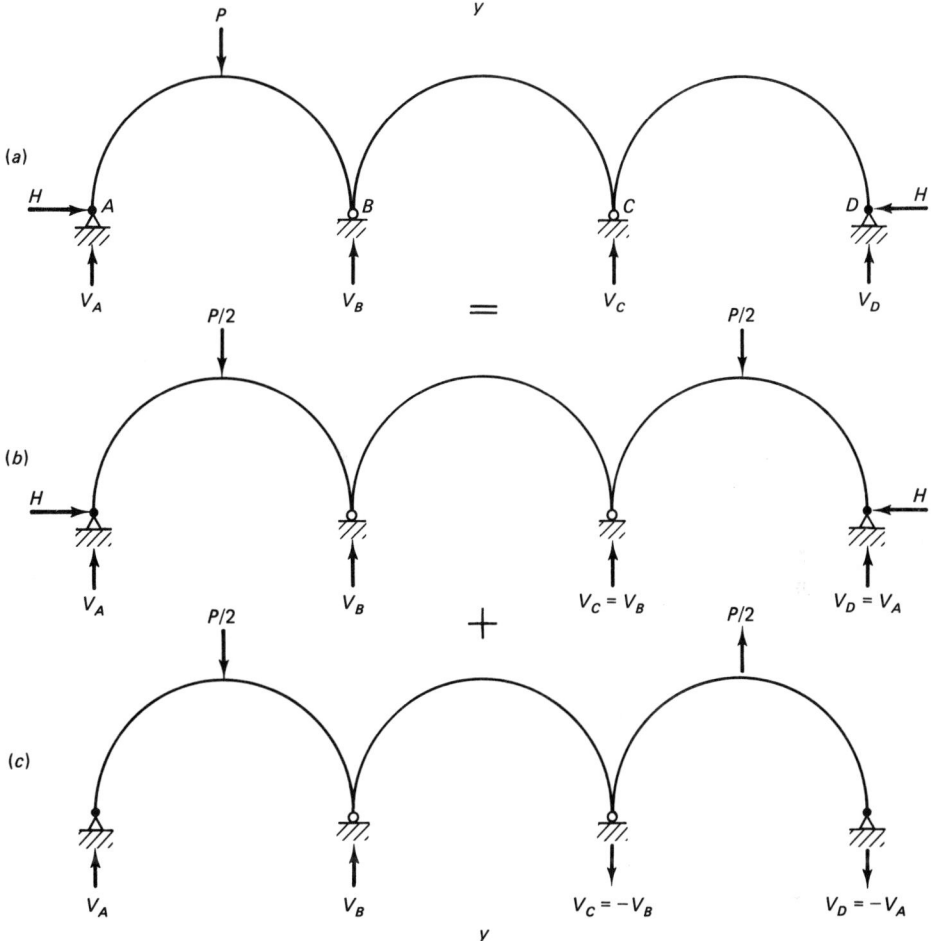

Figure 2.44 Decomposition of (a) an unsymmetrically loaded structure to (b) a symmetrically loaded structure plus (c) an antisymmetrically loaded structure.

not present when the loads F and P act separately (Fig. 2.43a and b), leading to the conclusion that the combined effect of F and P is not equal to the sum of their separate effects. Thus the principle of superposition cannot be applied in this case, since changes in the geometry are caused by the applied load.

2.6.3 Symmetry and Superposition

The analytical approach to many structural problems can be simplified if a linear elastic structure is symmetrical and, in addition, the loading and/or deformation is either symmetrical or antisymmetrical. Thus, for such a structure loaded arbitrarily the deformations and internal forces and stresses can be readily determined by superimposing the solution for the symmetrically loaded structure to that for the antisymmetrically loaded structure. Figure 2.44 illustrates this point; Fig. 2.44a shows a symmetrical structure about the vertical axis y-y and loaded by a concentrated load P; Fig. 2.44b and c show, respectively, the same structure being subjected to symmetrical and antisymmetrical loadings. The analysis of the structure in Fig. 2.44a involves the determination of three indeterminate reactions (redundants) V_B, V_C, and H, while the analysis of the structure in Fig. 2.44b requires the determination of only two redundants V_B and H and that in Fig. 2.44c only one redundant V_B. Thus, it is simpler to analyze the structures in Fig. 2.44b and c and superpose the results than to analyze directly the structure in Fig. 2.44a. Such load decomposition leads to simplification in the analysis of highly indeterminate structures.

Chapter **3**

Stability and Determinacy

Knowledge is proud that he has learned so much
Wisdom is humble that he knows no more.

William Cowper

3.1 INTRODUCTION

In this chapter, the conditions of stability and determinacy of structures are defined. Various types of supports through which a structure transmits its load to the supporting foundation are listed. The number and arrangement of reaction components for statical stability and determinacy of planar and space structures are discussed. However, to understand and appreciate the subject of stability and determinacy, it is prudent first to review the conditions of equilibrium of rigid bodies and the concept of free-body diagrams.

3.2 EQUILIBRIUM OF RIGID BODIES

A rigid body acted upon by a general system of forces will be in equilibrium if there is no resultant force or resultant moment acting on the rigid body, that is,

$$R = \sum F = 0 \quad \text{and} \quad \sum M_0 = 0$$

Resolving each force and each moment in the x, y, z directions:

$$\sum F_x = 0 \qquad \sum F_y = 0 \qquad \sum F_z = 0$$
$$\sum M_x = 0 \qquad \sum M_y = 0 \qquad \sum M_z = 0$$

Classification of force systems acting on rigid bodies is shown in Fig. 3.1. Equilibrium of rigid bodies under each of these force systems will now be discussed.

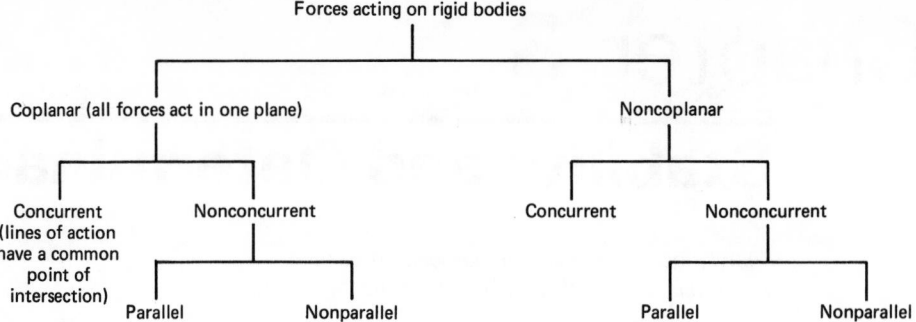

Figure 3.1 Classification of force systems acting on a rigid body.

3.2.1 Coplanar Force Systems

Concurrent Force System

In a concurrent force system the forces lie in one plane and intersect at one common point. If the rigid body is in equilibrium, *two* independent equations are available. If the forces are all acting, say, in the xy plane, and x, y axes are any pair of orthogonal axes passing through the common point of intersection, then *any one* of the following four pairs of equilibrium equations can be used:

$$\Sigma F_x = 0 \quad \text{and} \quad \Sigma F_y = 0$$

or

$$\Sigma F_x = 0 \quad \text{and} \quad \Sigma M_A = 0$$

where point A is *not* located on the y axis, or

$$\Sigma F_y = 0 \quad \text{and} \quad \Sigma M_B = 0$$

where point B is *not* located on the x axis, or

$$\Sigma M_A = 0 \quad \text{and} \quad \Sigma M_B = 0$$

where points A and B are *not* located on a line passing through the point of concurrency.

Nonconcurrent Force Systems

Parallel Forces   A system of parallel forces is in equilibrium if there is neither a resultant force nor a resultant moment, that is, if

$$\Sigma F = 0 \quad \text{and} \quad \Sigma M_0 = 0$$

The equilibrium equations can also be written as

$$\Sigma M_A = 0 \quad \text{and} \quad \Sigma M_B = 0$$

where points A and B are located on a line that is *not* parallel to the lines of action of the forces.

Nonparallel Forces Three independent equations of equilibrium are available for a coplanar nonconcurrent nonparallel force system:

$$\Sigma F_x = 0 \quad \Sigma F_y = 0 \quad \text{and} \quad \Sigma M_O = 0$$

where x and y are any two axes (generally taken perpendicular to each other) and O is any point in the plane.

Two other sets of equilibrium equations are:

$$\Sigma F_x = 0 \quad \Sigma M_A = 0 \quad \text{and} \quad \Sigma M_B = 0$$

where the x axis is any axis lying in the plane of the forces and A and B are any two points on a line that is *not* perpendicular to the x axis, or

$$\Sigma M_A = 0 \quad \Sigma M_B = 0 \quad \text{and} \quad \Sigma M_C = 0$$

where points A, B, C do not lie on the same line.

3.2.2 Noncoplanar Force Systems

Concurrent Force Systems

For equilibrium, the algebraic sum of the moments about any line, such as BL, must be equal to zero and the algebraic sum of the components of the forces in any direction must be zero. Expressed algebraically, *any three* of the following four equations constitute independent equations of equilibrium:

$$\Sigma M_{BL} = 0$$
$$\Sigma F_x = 0$$
$$\Sigma F_y = 0$$

and

$$\Sigma F_z = 0$$

Nonconcurrent Force Systems

Parallel Forces If the forces are parallel to the y axis, the following *three* equations must be satisfied for equilibrium:

$$\Sigma F_y = 0$$
$$\Sigma M_x = 0$$

and

$$\sum M_z = 0$$

where x and z are the other two coordinate axes.

Nonparallel Forces   The noncoplanar nonconcurrent nonparallel force system is the most general force system; the following *six* equations must be satisfied for equilibrium:

$$\sum F_x = 0 \quad \sum F_y = 0 \quad \sum F_z = 0$$
$$\sum M_x = 0 \quad \sum M_y = 0 \quad \text{and} \quad \sum M_z = 0$$

3.3 EQUILIBRIUM OF TWO-FORCE BODIES

Any rigid body acted upon by a system of forces at only *two* points is called a two-force body. Figure 3.2a shows such a two-force body. All the forces acting at A can be replaced by a single resultant force R_1 and those acting at B by the resultant R_2 as shown in Fig. 3.2b. If the rigid body is in equilibrium, resultants R_1 and R_2 must be equal and opposite and collinear as shown in Fig. 3.2b. The solution of problems involving two-force bodies is considerably simplified by making use of the above property. The most common example of a two-force member is the member of a pin-connected truss with forces acting at the two pin ends only (neglecting the self-weight of the member).

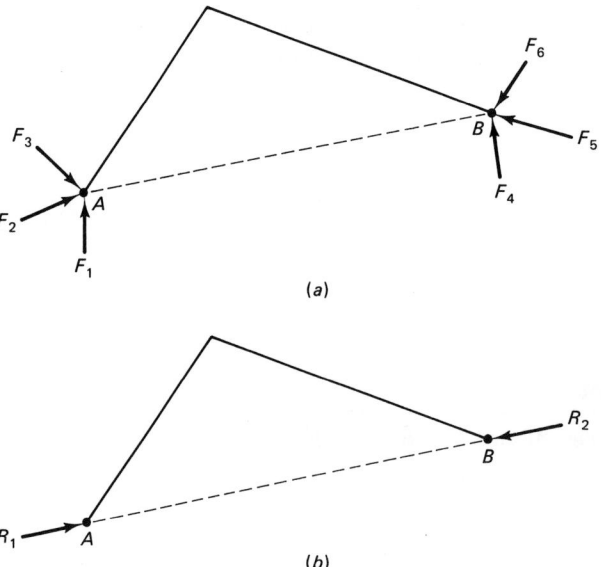

Figure 3.2 A two-force body.

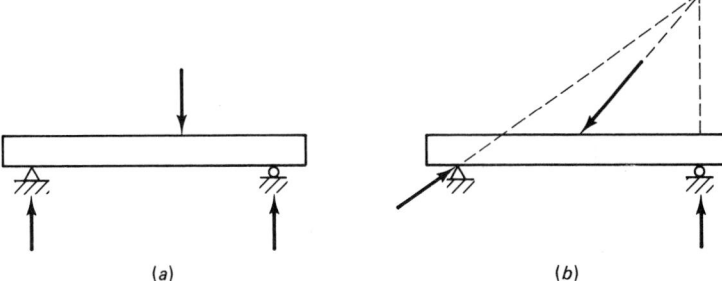

Figure 3.3 A three-force body.

3.4 EQUILIBRIUM OF THREE-FORCE BODIES

Another case of equilibrium that is of special interest is a three-force body, that is, a rigid body acted upon by forces at three points only. For equilibrium, the three forces must be either parallel or concurrent as shown in Fig. 3.3*a* and *b*.

3.5 FREE-BODY DIAGRAMS

A free-body diagram (abbreviated FBD) is defined as a diagram of a rigid body or any portion of a rigid body showing all the forces acting on it. It is very important for the student to realize the importance of drawing free-body diagrams correctly before attempting to solve problems concerning the equilibrium of rigid bodies. The problem will be greatly simplified once a free-body diagram is drawn showing correctly all the forces acting on it. In the following chapters of this textbook, free-body diagrams are extensively used in the analysis of engineering structures.

3.6 STABILITY AND DETERMINACY

A structure is considered geometrically *stable* if for any incipient movement, due to any conceivable system of applied load, an elastic resistance to this movement is *immediately* developed. A structure is considered statically determinate if the external reactions and internal member forces can be determined only by considering the equations of statical equilibrium, that is, without any reference to the elastic properties of the members of the structure.

Before a study of stability and determinacy of structures can be made, it is necessary to list the various types of supports through which a structure transmits its load to the supporting foundation. Tables 3.1 and 3.2 present the various supports for planar (two-dimensional) and space (three-dimensional) structures, respectively; the two tables show the relationships between the restraints imposed by the various types of supports and the corresponding reactive forces and moments generated. Notice also how an actual support in practice is idealized in order to analyze a particular structure.

TABLE 3.1 SUPPORTS AND REACTIONS FOR PLANAR STRUCTURES

Designation of support	Symbolic representation	Restraint(s) provided against	Number of unknown reactions provided
Roller or rocker	Rollers, Rocker, Smooth surface	Linear movement in either direction of the normal to supporting surface	One (V) as shown*
Link or cable	Short cable, Short link	Linear movement in direction of the link's axis	One (R) in direction of link or cable
Hinge or pin	Smooth pin or hinge, Rough surface	Two linear movements (e.g., both horizontal and vertical)	Two (R, θ) (or) Two (H and V)
Fixed or built-in or encastré		Two linear and one rotational movements (i.e., translation and rotation)	Three (V, H, M) (or) Three (R, θ, M)

* Direction of arrow can be reversed. V = vertical reaction, H = horizontal reaction, and M = moment.

TABLE 3.2 SUPPORTS AND REACTIONS FOR SPACE STRUCTURES

Designation of support	Symbolic representation	Restraint(s) provided against	Number of unknown reactions provided
Ball	Ball Smooth surface	Movement in either direction of the normal to supporting surface	One (F_y) as shown*
Ball in a groove	Roller on rough surface Wheel on rail	Movement in either direction of the two normals to the supporting surface	Two (F_y, F_z)
Spherical pin	Rough surface Ball and socket	Three movements (i.e., complete translational restraint)	Three (F_x, F_y, F_z)
Universal joint	Universal joint	Three linear movements and one rotational movement	Four (F_x, F_y, F_z, M_1)

TABLE 3.2 (Continued)

Cylindrical roller pin		Two linear and two rotational movements	Four (F_y, F_z, M_2, M_3)
Cylindrical pin	 Hinge and bearing Pin and bracket	Three linear and two rotational movements	Five (F_x, F_y, F_z, M_2, M_3)
Fixed		Three linear and three rotational movements	Six (F_x, F_y, F_z, M_1, M_2, M_3)

* Direction of arrow can be reversed.

Therefore, the student should study and become completely familiar with the information in Tables 3.1 and 3.2 in order to tackle correctly the analysis of any structure.

3.7 NUMBER AND ARRANGEMENT OF REACTION COMPONENTS FOR STATICAL STABILITY AND DETERMINACY

A rigid planar structure, for example, beam, frame, or truss, can be acted upon by a general system of forces that can be combined into a single force and a single moment. For statical stability and determinacy the number and arrangement of reactions must be such that they keep the external forces in equilibrium. Since three equations of statics are available ($\Sigma F_x = 0$, $\Sigma F_y = 0$, $\Sigma M = 0$) for planar structures, *there must be three components of reaction, which are neither parallel nor concurrent.* If the components of reaction are less than three, the structure is *unstable*. If the number of reaction components is three, and if those components are neither parallel nor concurrent, the structure is externally stable and determinate. If the number of reaction components is greater than three, and if there are no special equations of construction in the structure (see Section 3.9), and if the components are arranged such that they are neither parallel nor concurrent, the structure is *externally stable* and *statically indeterminate externally.* The degree of external indeterminacy is defined as the number of unknown reactions in excess of the number of available equations of statics plus the number of equations of construction (or equations of condition).

For statical stability and determinacy of a rigid *space structure,* the number and arrangement of reactions must be such that they maintain the general system of external forces in equilibrium. Since six equations of statics are available for a space structure ($\Sigma F_x = 0$, $\Sigma F_y = 0$, $\Sigma F_z = 0$, $\Sigma M_x = 0$, $\Sigma M_y = 0$, and $\Sigma M_z = 0$), there must be six components of reaction, the lines of action of which do not intersect a single straight line. (Parallel links are assumed to intersect at infinity and are therefore unstable.) If the components of reaction are less than six, the structure is *unstable*. If the number of reaction components is six and if they do not intersect a single straight line, the structure is *stable and determinate externally.* With the number of reaction components greater than six, such components being arranged so that their lines of action do not intersect a single straight line, and with no special equations of construction, the structure is *stable and indeterminate externally.* For a space structure the degree of external indeterminacy is defined as the number of unknown reactions in excess of the available equations of statics plus the number of equations of construction.

3.8 EXAMPLES OF STABILITY AND DETERMINACY OF PLANAR STRUCTURES

Figure 3.4*a* shows a simple beam with three reaction components that are neither parallel nor concurrent. Therefore, it is stable and statically determinate. Figure 3.4*b* shows a beam on two roller supports. Since the number of reaction components is less than three ($r = 2$), the structure is statically unstable. Under vertical loads, equilibrium will, of course, be maintained, and the structure is then said to be in *unstable*

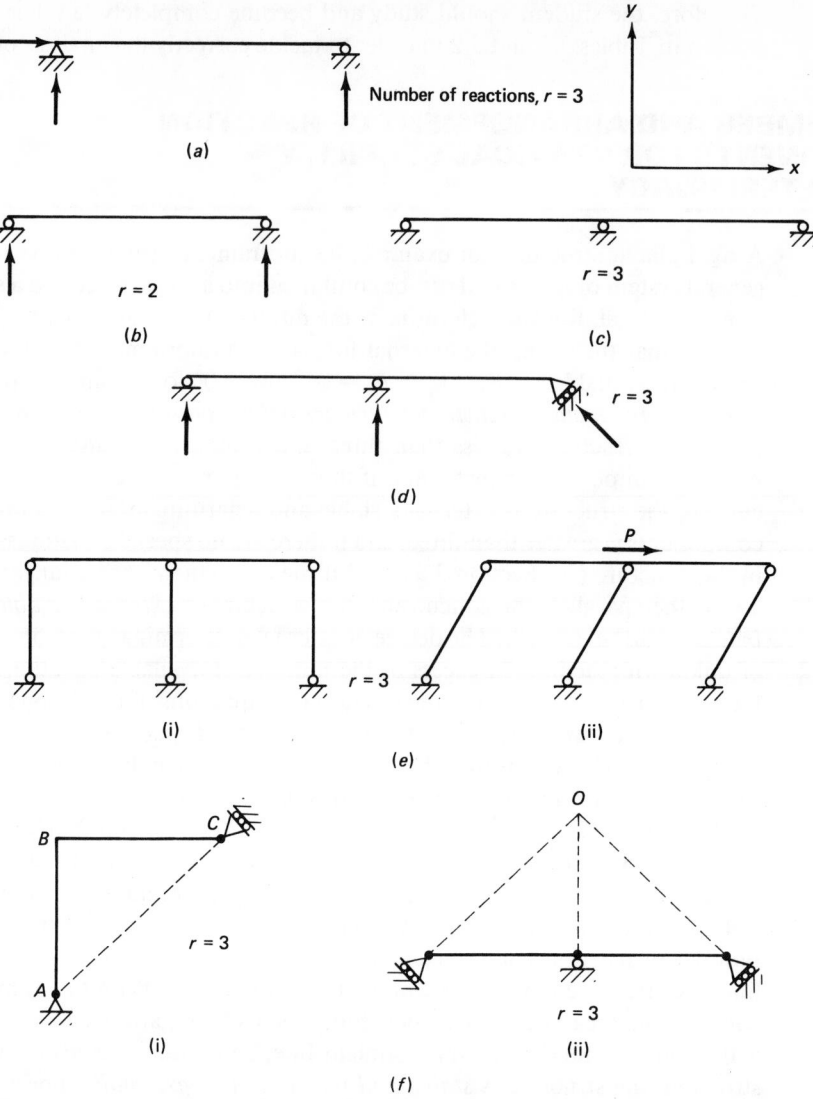

Figure 3.4 Examples of stability and determinacy of planar structures. (*a*) Simple beam. (*b*) Beam on two roller supports. (*c*) Beam on three roller supports. (*d*) Beam on three roller supports. (*e*) Beam on three parallel links. (*f*) Structure on supports with concurrent reactions. (*g*) Cantilever beam. (*h*) Propped cantilever. (*i*) Fixed beam. (*j*) Fixed arch. (*k*) Rigid frame. (*l*) Hinged portal frame. (*m*) Simply supported bent. (*n*) Cantilever bent.

equilibrium. For a structure to be classified as a stable structure, it must be able to resist immediately *any* conceivable system of external loads. Figure 3.4*c* shows a beam on three roller supports. Although the number of reaction components is three, they are all parallel, and therefore the structure is unstable. For the special case of vertical loading, the structure will be in unstable equilibrium and is indeterminate to the first degree, since there are three unknown reactions and only two equations of statics $(\Sigma F_y = 0, \Sigma M = 0)$.

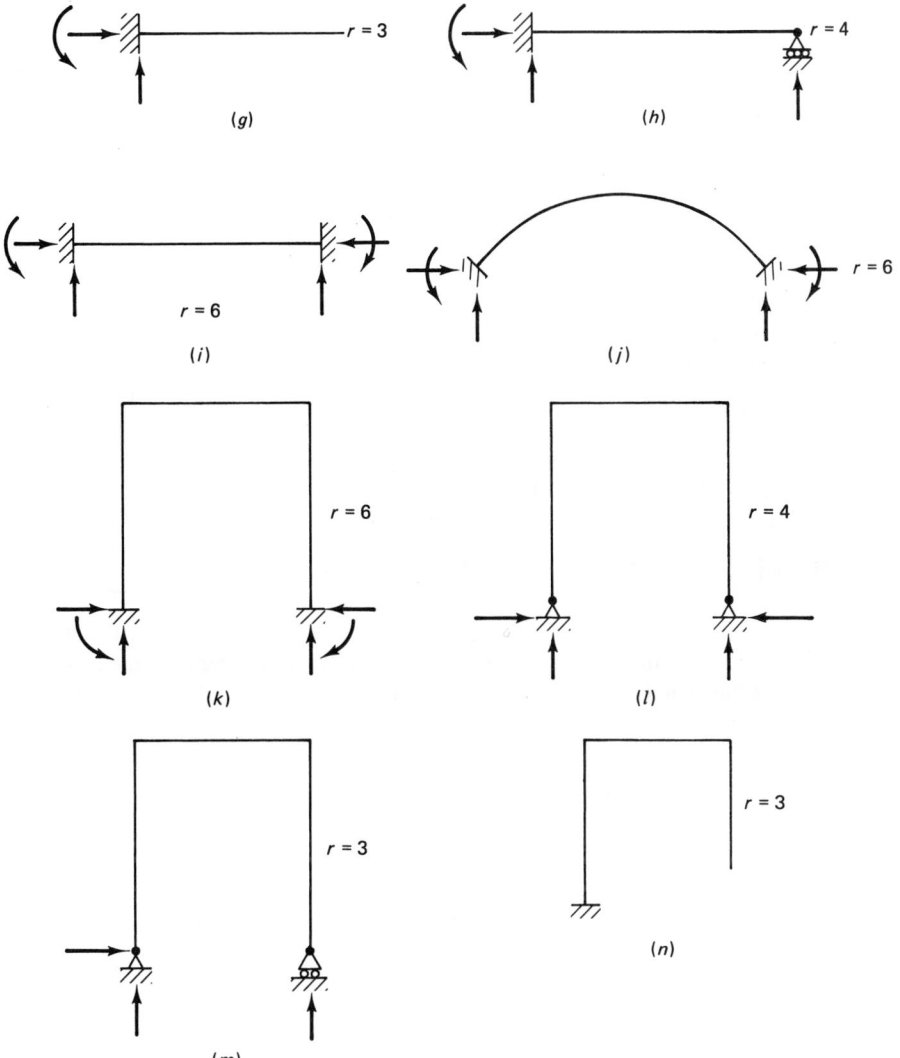

Figure 3.4 (*Continued*)

In Fig. 3.4*d,* the beam is again supported on three rollers, but the lines of action of the reactions are neither parallel nor concurrent. Therefore, the structure is stable and statically determinate.

Figure 3.4*e* shows a beam supported on three parallel links, and like the beam in Fig. 3.4*c,* the structure is unstable. Under vertical loads the structure is in unstable equilibrium and is statically indeterminate to the first degree. Under horizontal loads, the beam shown in Fig. 3.4*c* cannot develop the necessary horizontal reactions for equilibrium, whereas the beam supported on three links will take the shape shown in Fig. 3.4*e*(ii) and will not collapse completely. The inclination of the links is such that the sum of their horizontal components balances the external load. However, this involves considerable distortion of the structure, and since restraint to horizontal forces is not generated *immediately* upon load application, the structure is considered to be unstable.

Figure 3.4 *f*(i) shows a frame on a hinged and roller supports, with the roller support arranged in such a manner that the normal to the surface of the roller (point *C*) passes through the hinge support at *A*. In this case, the reactions are all concurrent at *A*, and therefore the structure is unstable. In Fig. 3.4 *f*(ii) the beam reactions intersect at *O*, and the structure can rotate about *O*, the instantaneous center of rotation; hence the structure is also unstable.

Figure 3.4*g* shows a cantilever beam (with *r* = 3) that is stable and statically determinate. Figure 3.4*h* is a propped cantilever (with *r* = 4) that is stable and statically indeterminate to the first degree. Figure 3.4*i, j,* and *k* are, respectively, a fixed beam, fixed arch, and rigid frame (with *r* = 6). These structures are stable and statically indeterminate to the third degree. A hinged portal frame with *r* = 4 is shown in Fig. 3.4*l*. The structure is stable and statically indeterminate to the first degree. Figures 3.4*m* and *n* show simply supported and cantilever bents with *r* = 3. Both structures are stable and statically determinate.

3.9 EQUATIONS OF CONSTRUCTION (CONDITION)

Quite often construction facilities, such as internal hinges, rollers, and links, are introduced into a redundant structure. Such a facility reduces the degree of indeterminacy of the structure. Consider, for example, a beam *AB* fixed at *A* and supported at *B* on

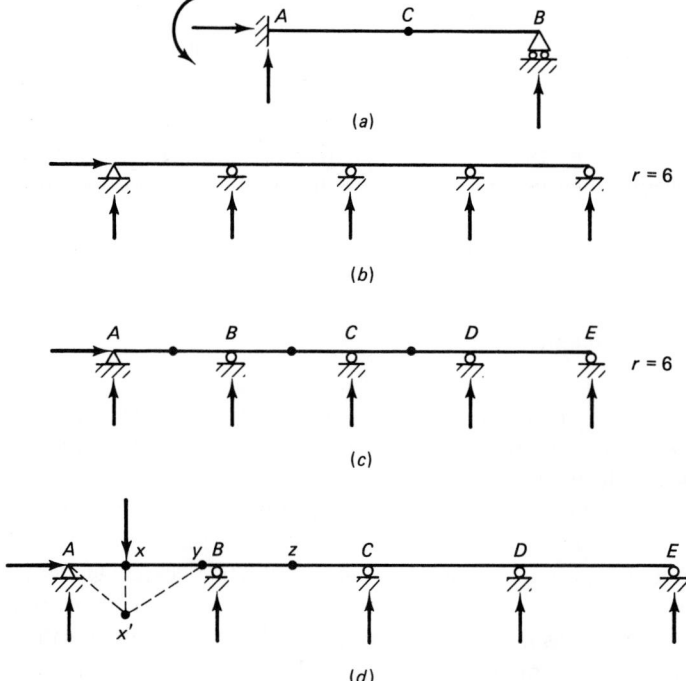

Figure 3.5 Examples of structures with specified condition. (*a*) Propped cantilever with internal hinge at *c*. (*b*) Four-span continuous beam. (*c*) Four-span continuous beam with three internal hinges. (*d*) Four-span continuous beam with improper arrangement of internal hinges.

a roller. The number of reaction components at A is three and that at B is one. Therefore, the total number of unknown reactions is four, while only three equations of statics are available; the structure is therefore statically indeterminate to the first degree (degree of statical indeterminacy = number of unknowns − number of equations of statics available). However, if an internal hinge is introduced at C as shown in Fig. 3.5*a*, it provides an additional equation; that is, the bending moment at C (i.e., moment of all forces acting *either* to the left *or* to the right of C) is zero. The structure thus becomes statically determinate. Such equations are referred to as *equations of construction* or *condition*. In introducing internal hinges, etc., in the structure, one must ensure that the stability of the structure is not violated, hence endangering the structure. Consider, for example, the four-span continuous beam shown in Fig. 3.5*b*. The number of reaction components is six and the structure is indeterminate to the third degree. To make the structure statically determinate, three internal hinges are required; one possible arrangement of these hinges is shown in Fig. 3.5*c*, where the structure is stable. If, on the other hand, the hinges are introduced as shown in Fig. 3.5*d* with hinges x, y in span AB, and a vertical load is applied at x, say, portion AxB, will deform to $Ax'yB$ before any resistance to the applied load can develop; such a state of affairs violates the definition of structural stability, and therefore the structure in Fig. 3.5*d* is said to be unstable.

3.10 STABILITY AND DETERMINACY OF TRUSSES

Trusses consist of straight members joined together at their ends by joints assumed incapable of resisting any moment, that is, frictionless pins. However, the joints of actual trusses used in construction do possess a certain amount of rigidity leading to a certain amount of restraint to rotation. The assumption of joints behaving as frictionless pins greatly simplifies the analysis of a truss; furthermore, extensive theoretical and experimental investigations indicate that the error introduced by this assumption is insignificant when dealing with a truss having the usual proportions and loaded at the joints. We shall deal with the stability and determinacy of planar trusses first, followed by space trusses.

Planar Trusses   The simplest stable plane truss is a triangle with three joints and three bars (Fig. 3.6). Each additional joint can be obtained by adding two bars to the joints already formed. A truss so developed is called a simple plane truss. If the total number of joints is j, the number of bars b in a simple plane truss can be computed as follows: For the first three joints, three bars are required; for each of the $(j - 3)$ additional joints, $2(j - 3)$ bars are required. Thus, the total number of bars $b = 3 + 2(j - 3) = 2j - 3$. If a simple plane truss has less than $(2j - 3)$ bars, the

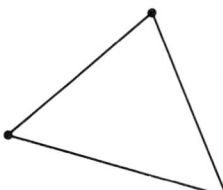

Figure 3.6 Basic planar truss.

truss is *unstable.* If the number of bars is equal to or greater than $(2j - 3)$, the arrangement of bars should be examined for stability and determinacy. As an example, consider the trusses shown in Fig. 3.7a and b. Both trusses have 10 joints, 17 bars, and thus the relationship $b = 2j - 3$ is satisfied; although the total count is the same, the arrangement of bars in the two trusses is different; the diagonal in panel 3 of Fig. 3.7a is removed and added to panel 2 in Fig. 3.7b. If one attempts to determine by statics the bar forces of the truss in Fig. 3.7b, it will be found that the analysis leads to inconsistent and incorrect results. For the loading shown, there is a shear of $P/4$ in panel 3. If a section xx is taken as shown and if either the left or the right portion of the truss is isolated as a free body (Fig. 3.7c), there is an unbalanced vertical force of $P/4$ with no internal bar force to resist this force, and equilibrium cannot be maintained. (We assume, of course, that all truss members carry load *only* along their longitudinal axes.) Panel 3 is, in fact, a *mechanism* and therefore *unstable,* since any small lateral joint movement in that panel does not induce an immediate resistance to the movement. Hence the structure in Fig. 3.7b is said to be unstable. Furthermore, because of the arrangement of the bars, panel 2 is now overstiff, that is, indeterminate, since all member forces cannot be found from statics alone. This leads to the conclusion that for a simple plane truss to be stable and statically determinate, the condition that

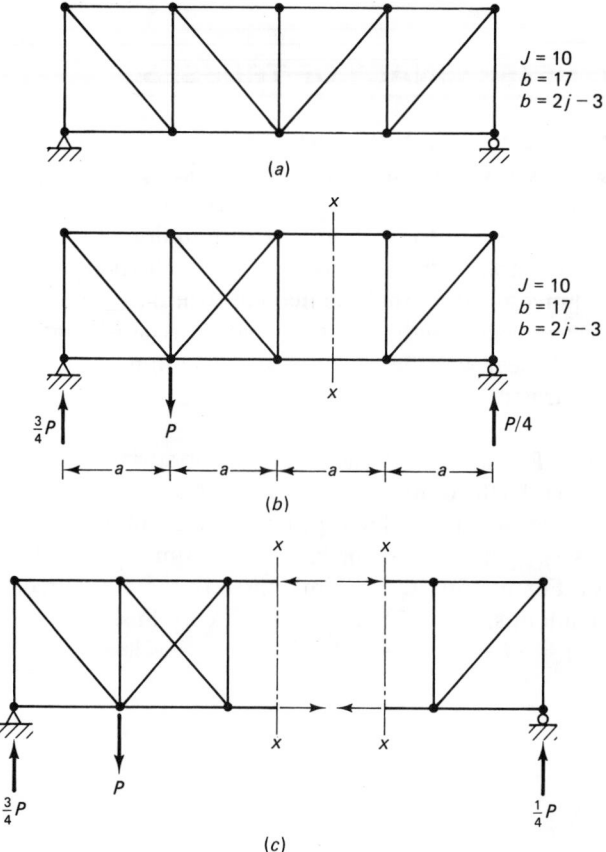

Figure 3.7 Illustration of stability and determinacy of planar trusses.

the number of bars $b = 2j - 3$ is necessary but not sufficient and that the bars must be properly arranged to preclude instability and indeterminacy. If the number of bars b is greater than $(2j - 3)$, the simple plane truss is statically indeterminate (this is also the case when $b = 2j - 3$, but with improper bar arrangement); whether the truss is stable or not depends entirely upon the arrangement of the bars and the manner in which the truss is supported.

Two simple plane trusses can be joined together by three nonparallel nonconcurrent bars to form a compound truss as shown in Fig. 3.8a. Since two bars can be replaced by a common hinge, a compound truss can also be formed by joining together two plane trusses with a common hinge and a bar, as shown in Fig. 3.8b. More simple trusses can be added in the same manner. It is interesting to note that the relationship $b = 2j - 3$ is valid also for compound plane trusses. Trusses that cannot be classified

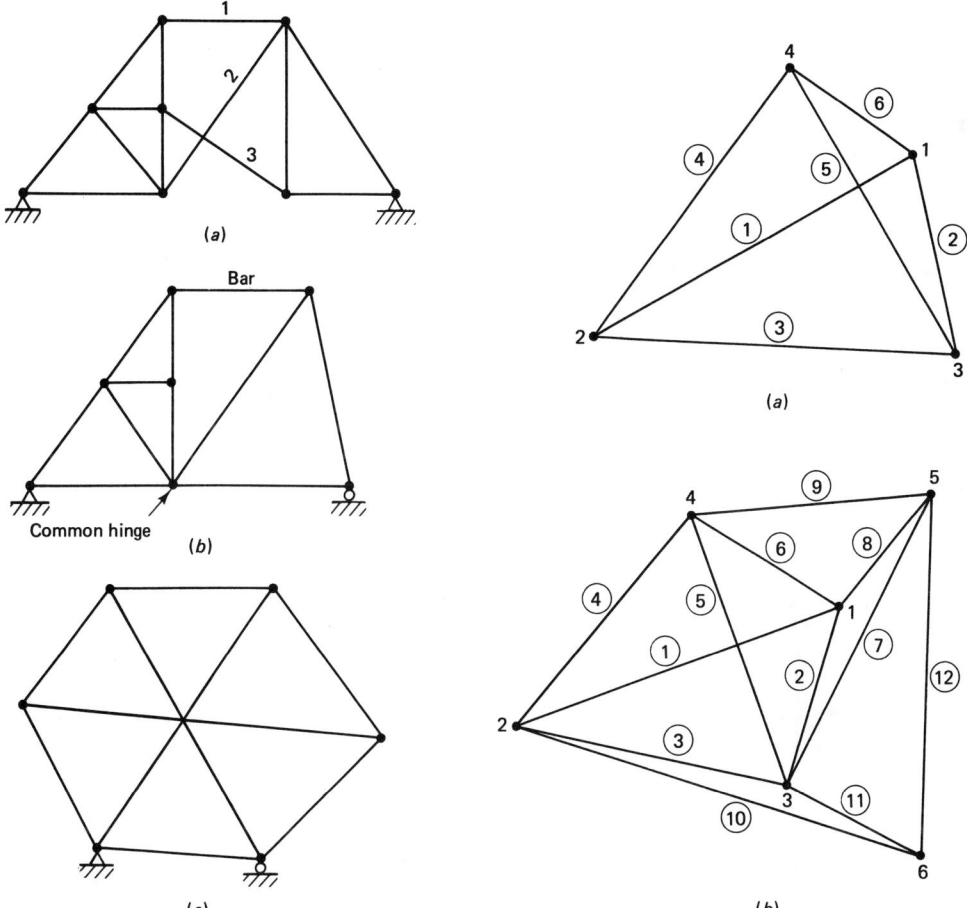

Figure 3.8 Examples of planar compound and complex trusses. (*a*) Compound truss formed by using three bars. (*b*) Compound truss formed by using a common hinge and a bar. (*c*) Complex truss.

Figure 3.9 Development of space trusses. (*a*) Development of simple space truss. (*b*) Development of more complex space truss.

as either simple or compound are termed "complex trusses" (Fig. 3.8c). Complex trusses, although not frequently used, are aesthetically more pleasing than other types and therefore are chosen for sites where appearance is important.

Space Trusses The simplest space truss has four joints and six bars, for example, a tetrahedron in Fig. 3.9a. Each additional joint can be formed by adding three bars (not all lying in one plane) to the joints already formed (Fig. 3.9b). The relationship between the number of bars and the number of joints for a space truss can be derived as follows: Let b = total number of bars and j = total number of joints. Six bars are required for the first four joints. Three bars are required for each of the additional $(j - 4)$ joints. Therefore, $b = 6 + 3(j - 4) = 3j - 6$. Two simple space trusses can be joined together to form a compound space truss by six bars whose lines of action

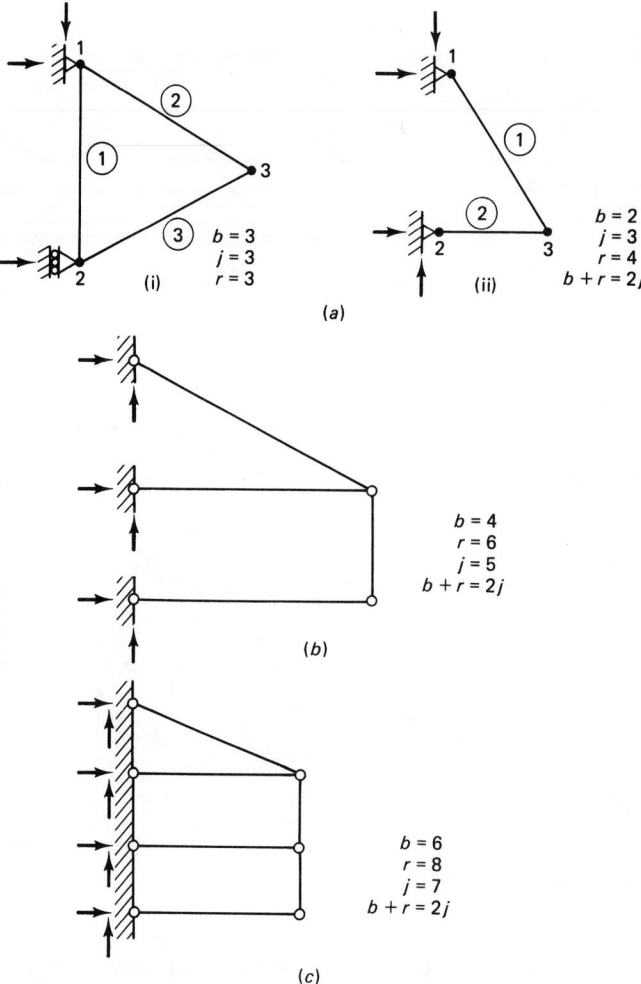

Figure 3.10 Influence of arrangement of bars and supports on the stability and determinacy of trusses. (*a*i) Simple plane truss. (*a*ii) Cantilever truss.

do not intersect a common straight line. For the same arguments used for plane trusses, the count ($b = 3j - 6$) is a necessary but not sufficient condition for stability and determinacy. For example, if $b < 3j - 6$, the space truss is internally unstable; and with $b \geqslant 3j - 6$, the space truss is stable if the arrangement of bars precludes instability.

Cantilever Trusses   Quite often in discussing the problem of stability and determinacy it is necessary to consider the truss and the supports together. Let us examine the simple plane truss shown in Fig. 3.10a(i) where $b = 3, j = 3, r = 3$. The structure is stable and statically determinate both internally and externally. Member ① can be removed and the roller support at 2 can be replaced by a hinged support, giving rise to a cantilever truss shown in Fig. 3.10a(ii). For such a truss, it is necessary to consider the structure and the supports together. If b is the number of bars and r is the total number of reactions, the total number of unknowns $= b + r$. The number of statical equations $= 2j$. (At each joint, there are two available equations of statics, $\Sigma F_x = 0$, $\Sigma F_y = 0$.) The student at first glance might think that there are $2j$ equations for j individual joints plus three equations of statics for the structure as a whole. But the three equations of equilibrium for the structure as a whole are already included in the $2j$ equations for individual joints and thus they are not independent. Examples of statically determinate, stable cantilever trusses are shown in Fig. 3.10b and c. For a cantilever space truss, since three equations of statics are available at each joint, if $b + r = 3j$ and if the arrangement of bars is proper, the structure is stable and statically determinate. But if $b + r < 3j$, the structure is unstable, and when $b + r > 3j$, the structure is statically indeterminate.

3.11 CRITICAL FORMS OF TRUSSES

Some structural geometries and support arrangements must never be used if a designer is to avoid a critical form for a structure. For example, the structures shown in Fig. 3.11a and b are both stable and statically determinate; whereas the one in Fig. 3.11c is unstable in spite of the fact that it has the same number of bars, reactions, and joints. Any vertical load P at point C cannot be resisted by the horizontal members AC and CB, and therefore equilibrium will not be maintained. It is only when point C deflects to C' that the external load P can be balanced by the vertical components of the forces in members $C'A$ and $C'B$. Since the structure $AC'B$ is substantially different from the structure ACB, the latter is said to be unstable. It can be observed from Fig. 3.11a and b that when C is above or below the line joining A to B the resulting structure is stable. Thus, when C is on the line AB, instability occurs and the structure in Fig. 3.11c is said to be the *critical form* of the structures in Fig. 3.11a and b. This instability can also be explained by geometrical consideration. If the bars AC and BC in Fig. 3.11d are disconnected at C, then CA can rotate about A and CB about B with the two arcs rr and ss having only one common tangent at point C. Thus, if the end C of bar CA moves a short distance perpendicular to line AB, the bar CB will not offer any resistance; hence the structural system in Fig. 3.11d is *geometrically unstable* since the shape of the structure can be altered without any change in length of any member. This situation does not exist in the stable structure of Fig. 3.11e since the

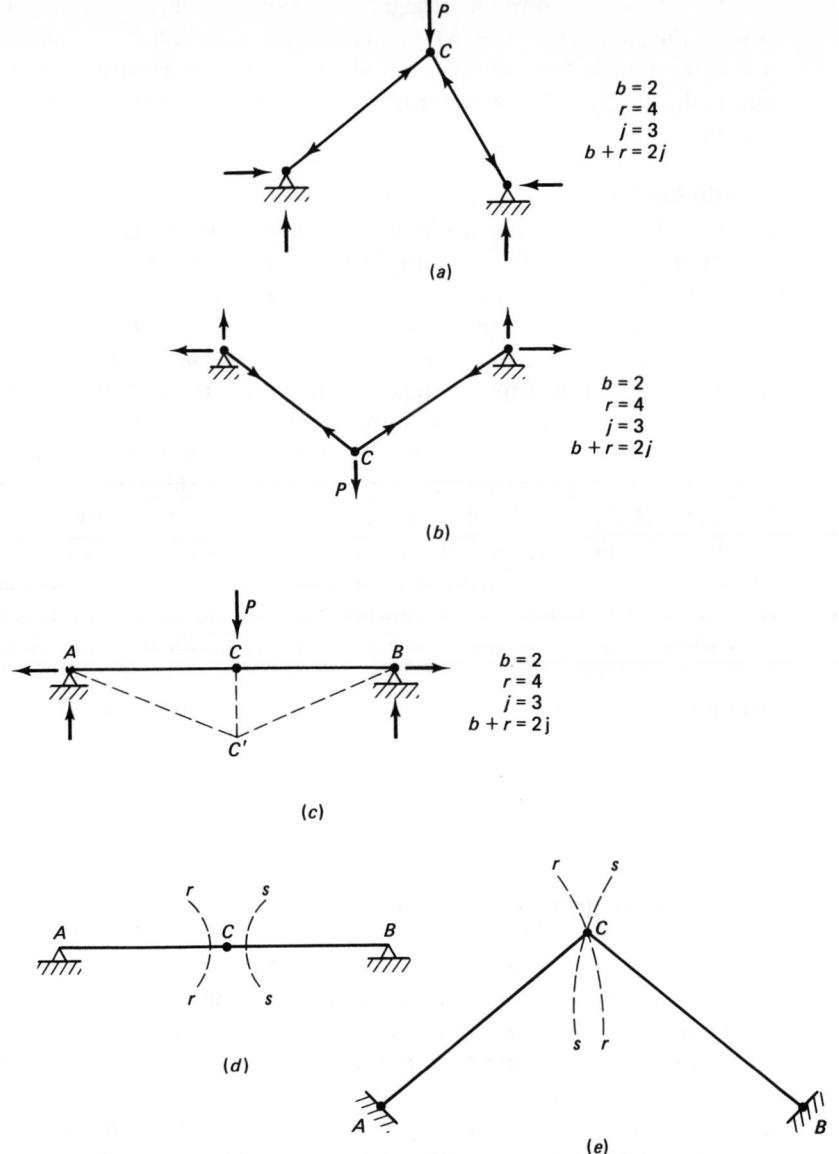

Figure 3.11 Critical forms of trusses. (*a*) Stable statically determinate structure. (*b*) Statically determinate stable structure. (*c*) Unstable structure.

two arcs *rr* and *ss* have no common tangent and any incipient movement of *C* will generate an immediate restraint from the various parts of the structure. Two tests are available for investigating the critical forms of statically determinate trusses: the zero-load test and the coefficient-matrix determinant test.

The Zero-Load Test   This test, first suggested by A. F. Möbius (1790–1868), is applied as follows:

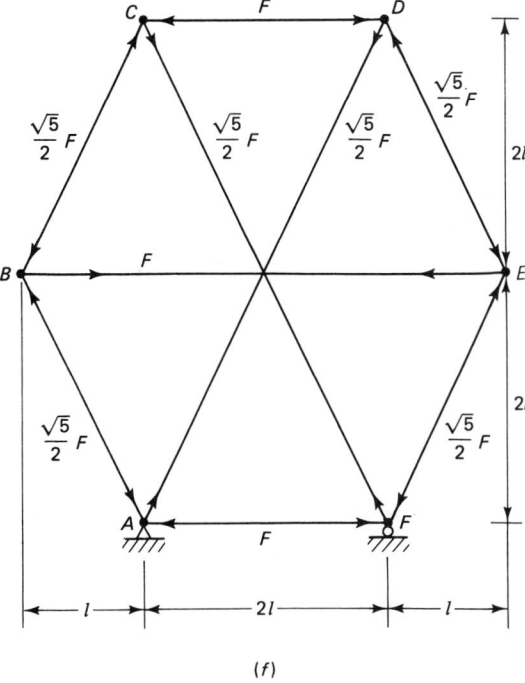

(f)

Figure 3.11 (*Continued*)

1. Assume zero external load and hence zero external reactions.
2. Assume a certain nonzero force in any one member.
3. Determine the forces in all other members of the truss considering the equilibrium of all the joints in the truss.
4. If equilibrium *can be established* at all the joints, it indicates that a nonzero solution exists with zero external load, which leads to the conclusion that the structure is geometrically unstable.
5. If equilibrium *cannot be established* at all the joints, it indicates that a nonzero solution does not exist and the structure is stable.

As an example of the zero-load test, consider the statically determinate complex truss shown in Fig. 3.11*f*. If the force in member *BE* is assumed as *F*, the forces in the other members are found by considering the equilibrium of all the joints. Since the forces in the members shown in Fig. 3.11*f* indicate that a nonzero solution exists, the truss is geometrically unstable.

Coefficient-Matrix Determinant Test   This test examines the geometric instability of statically determinate trusses by calculating the determinant of the coefficient matrix of the equilibrium equations. The test entails the following steps:

1. Write down the equilibrium equations for all the joints [i.e., (2*j*) and (3*j*) equations for planar and space trusses, respectively].

2. Determine the coefficient matrix of the equilibrium equations and compute its determinant.
3. If the determinant is positive and nonzero, the truss is stable; otherwise the truss is unstable.

It should be remarked that this test may entail considerable labor and therefore the services of a computer may be required.

The test for stability and the determination of the number of redundants in planar trusses are illustrated for the structures in Fig. 3.12.

Fig. 3.12*a*

$$b + r = 3 + 3 = 6$$
$$2j = 6$$
$$b + r = 2j$$

Arrangement of bars is proper, and therefore structure is stable and statically determinate.

Fig. 3.12*b*

$$b + r = 4 + 3 = 7$$
$$2j = 8$$
$$b + r < 2j \qquad \therefore \text{ unstable}$$

The truss can easily take the form $AB'C'D$ without requiring any change in length of bars.

Fig. 3.12*c*

$$b + r = 2 + 4 = 6$$
$$2j = 6$$
$$b + r = 2j$$

However, AC and CB are collinear. Small movements of joint C are possible without any deformation of members. Therefore, the structure has a critical form; that is, it is instantaneously unstable.

Fig. 3.12*d*

$$b + r = 2 + 4 = 6$$
$$2j = 6$$
$$b + r = 2j$$

Structure is stable and statically determinate.

Fig. 3.12e

$$b + r = 9 + 3 = 12$$
$$2j = 12$$
$$b + r = 2j$$

The structure is a compound truss. It is stable as long as links ①, ②, ③ are not parallel or concurrent.

Fig. 3.12f

$$b + r = 9 + 3 = 12$$
$$2j = 12$$
$$b + r = 2j$$

Structure is a compound truss. It is stable and statically determinate.

Fig. 3.12g

$$b + r = 9 + 3 = 12$$
$$2j = 12$$
$$b + r = 2j$$

Structure is a stable and statically determinate compound truss.

Fig. 3.12h

$$b + r = 19 + 3 = 22$$
$$2j = 22$$
$$b + r = 2j$$

Structure is a stable and statically determinate compound truss. The two shaded simple trusses are joined together by the links ①, ②, ③. Joint O is then formed by adding bars ④ and ⑤.

Fig. 3.12i

$$b + r = 11 + 3 = 14$$
$$2j = 14$$
$$b + r = 2j$$

Structure is a stable and statically determinate compound truss, consisting of three simple trusses connected together by two hinges and two bars. Trusses Ⓐ and Ⓑ can be assumed connected together at hinge D and by a bar ① shown dashed in Fig. 3.12*j*. Trusses Ⓑ and Ⓒ can be assumed connected together by bars ② and ③ and a fictitious bar ④, shown dashed in Fig. 3.12*j*. Bars ① and ④ can be replaced

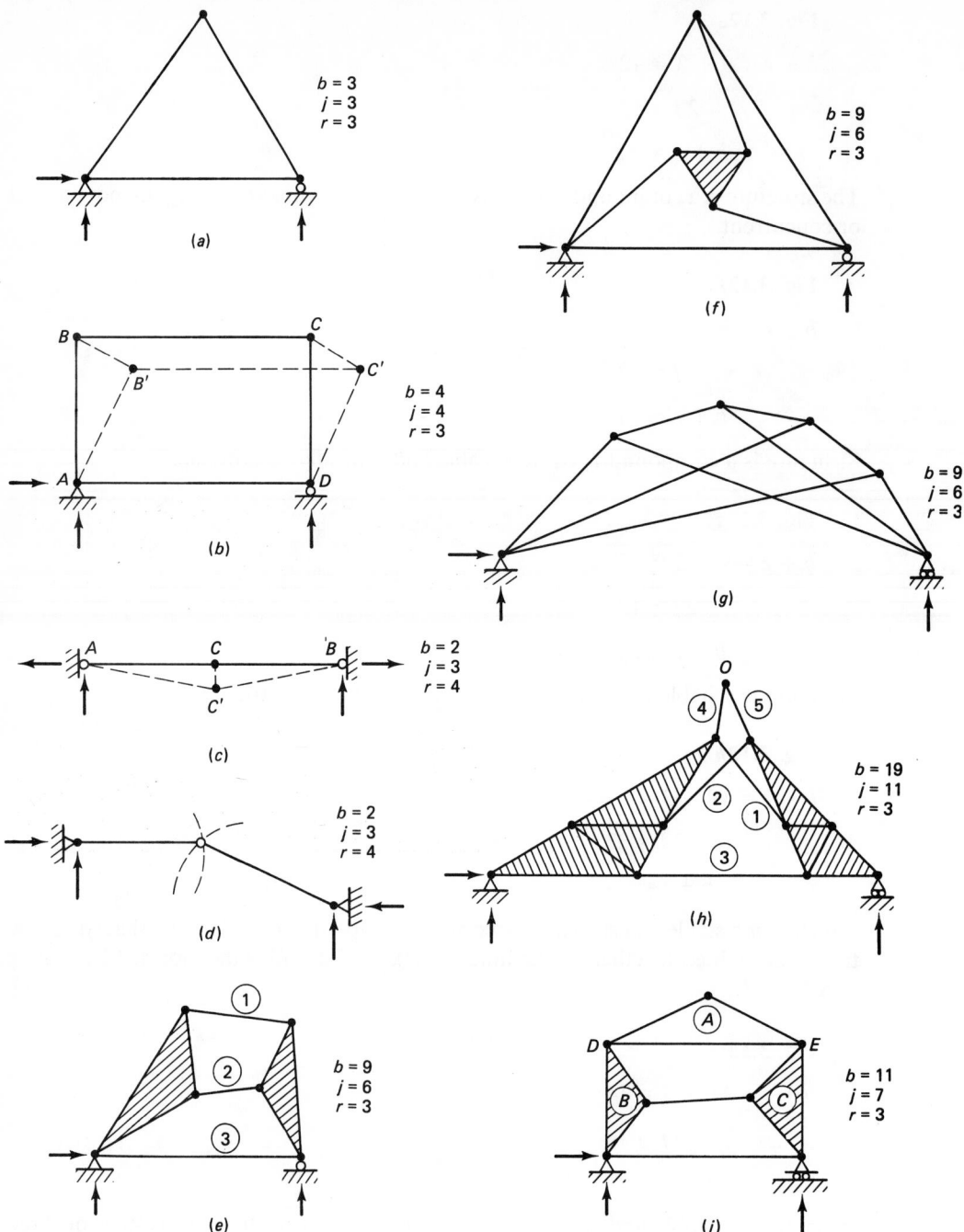

Figure 3.12 Examples for determining stability and determinacy of planar truss.

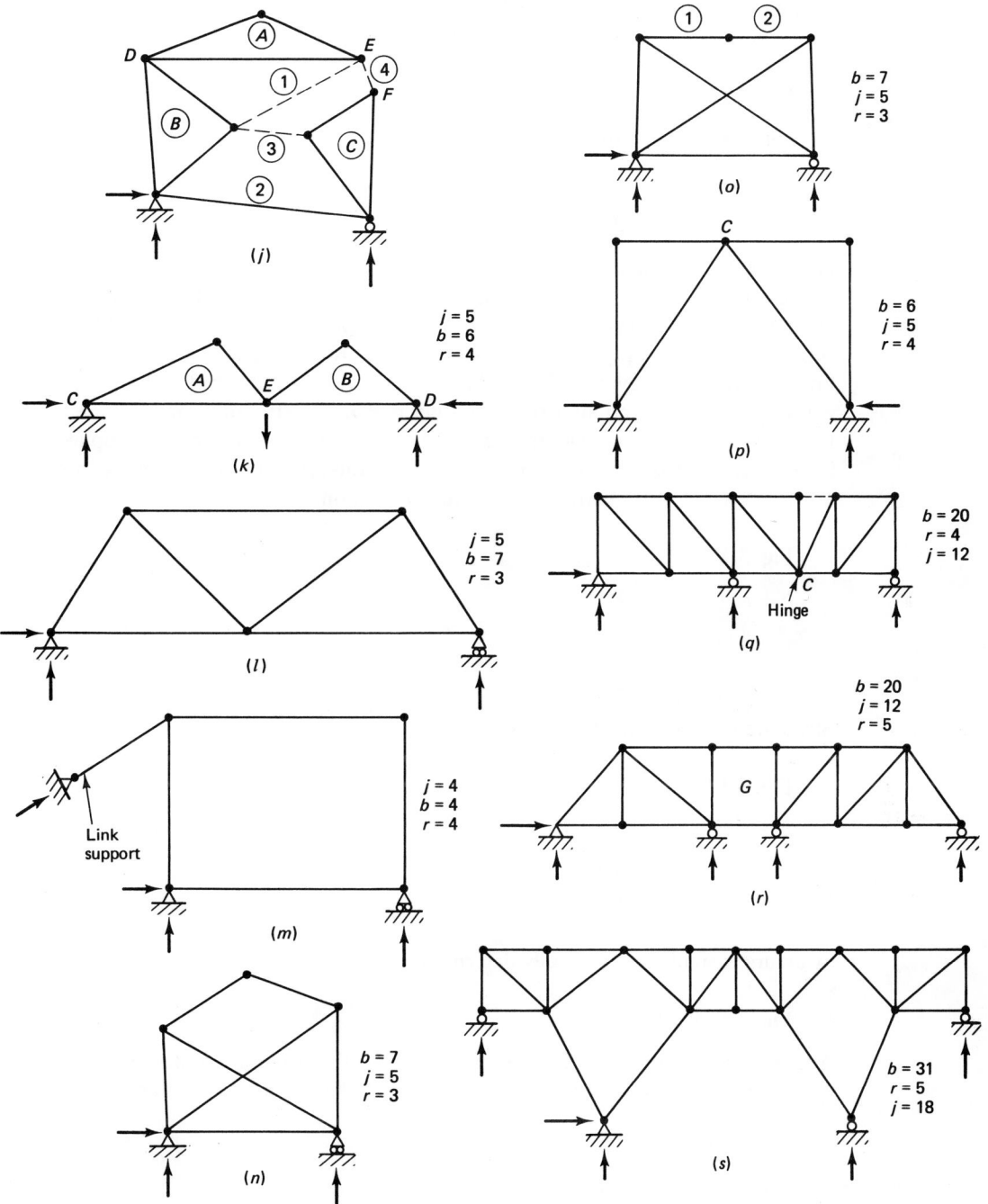

Figure 3.12 (*Continued*)

by a hinge located at the point of intersection of the bars, that is, at E, thus moving joint F to joint E, resulting in the truss shown in Fig. 3.12i.

Fig. 3.12k

$$b + r = 6 + 4 = 10$$
$$2j = 10$$
$$b + r = 2j$$

However, the arrangement of reactions and bars is incorrect. Trusses Ⓐ and Ⓑ are connected together by one hinge only. They should be connected together by a hinge and a bar, as shown in Fig. 3.12l. The hinged support at D can be replaced by a roller support as shown in Fig. 3.12l to make the structure stable and statically determinate. For the structure shown in Fig. 3.12k, if a vertical load is applied at E, the structure will rotate about E and after some deflection at E, the structure may become stable. However, although the structure may not collapse completely under the applied load, it is still classified as a geometrically unstable structure, since the load does not encounter an elastic restraint immediately upon its application.

Fig. 3.12m

$$b + r = 4 + 4 = 8$$
$$2j = 8$$
$$b + r = 2j$$

Structure is stable and statically determinate.

Fig. 3.12n

$$b + r = 7 + 3 = 10$$
$$2j = 10$$
$$b + r = 2j$$

Structure is stable and statically determinate.

Fig. 3.12o

Although b, r, and j have the same values as for the truss in Fig. 3.12n, bars ① and ② are collinear, resulting in a critical form, and the structure is unstable.

Fig. 3.12p

$$b + r = 6 + 4 = 10$$
$$2j = 10$$

Structure is stable and statically determinate.

Fig. 3.12*q*

$$b + r = 20 + 4 = 24$$

$$2j = 24$$

Structure is stable and statically determinate. The four reaction components can be determined from the three equations of statics and one equation of condition (i.e., bending moment about hinge at $C = 0$).

Fig. 3.12*r*

$$b + r = 20 + 5 = 25$$

$$2j = 24$$

The structure is stable and statically indeterminate to the first degree. There are five unknown reaction components and only three equations of statics plus one equation of condition (shear in panel G is zero). It should be noted that the supports on either side of panel G prevent the otherwise unstable rectangular-shaped panel from collapsing.

Fig. 3.12*s*

This is a special type of truss called a "Wichert truss." It is a stable statically determinate structure. However, one cannot determine the reactions for this type of truss without first determining some of the bar forces.

3.12 STATICAL DETERMINACY OF PLANE RIGID JOINTED FRAMES

Plane rigid frames comprise generally linear members (i.e., members whose length is large compared with the other two dimensions) lying in one plane and joined together by bolts, welds, rivets, etc. Such joints are capable of resisting moments and shearing forces; the members themselves are subjected to axial and shearing forces as well as moments. (This definition can be extended to space rigid frames where the members are arranged in a three-dimensional framework.) Thus for each member in a plane rigid frame there are three unknown stress resultants. For example, the beam AB in Fig. 3.13a is subjected to six stress resultants $V_1, H_1, \ldots, M_2$; however, there are also three equations of equilibrium for the beam AB, that is, $\Sigma F_x = 0$, $\Sigma F_y = 0$, and $\Sigma M = 0$. Therefore, each member in a plane rigid frame gives rise to $6 - 3 = 3$ unknowns. Furthermore, at each rigid joint there are three equations of equilibrium ($\Sigma F_x = 0$, $\Sigma F_y = 0$, $\Sigma M = 0$). The rigid frame in Fig. 3.13b has 6 joints, 6 members, and 6 reactions; the number of unknowns = 3×6 (members) + 6 reactions = 24; the number of equations available = 3×6 (joints) = 18. Therefore, the degree of statical indeterminacy = $24 - 18 = 6$. As another example let us examine the degree of indeterminacy for the rigid multistory multibay frame in Fig. 3.13c. The number

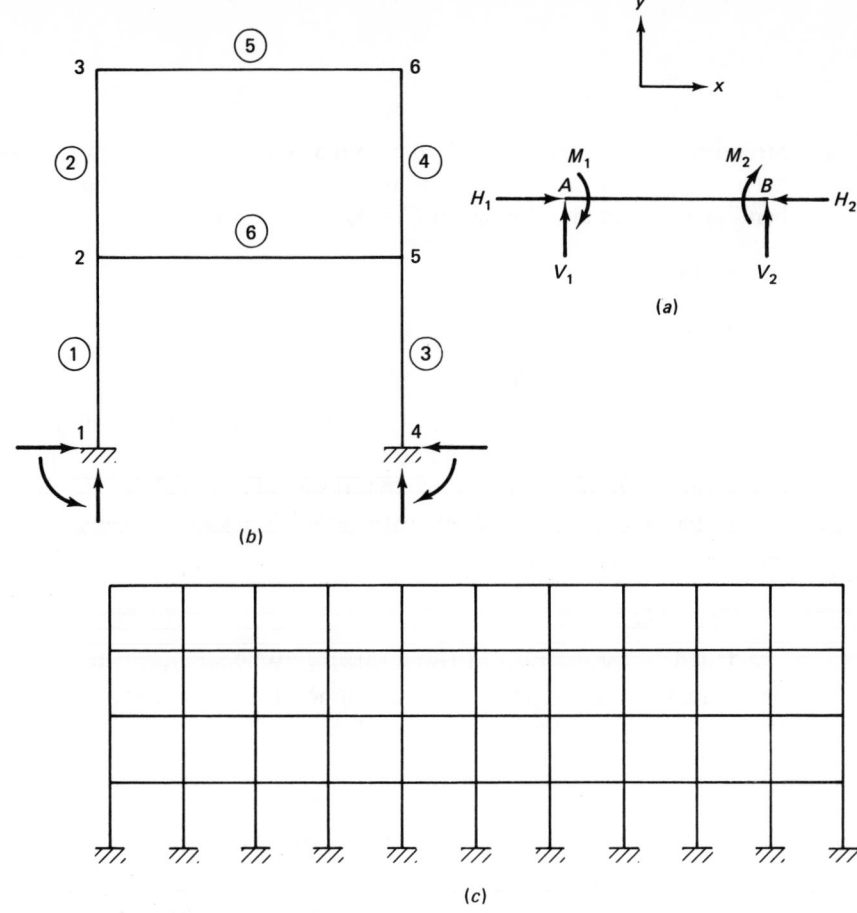

Figure 3.13 Degree of indeterminacy of rigid frames.

of joints = 55, number of members = 84, and number of reactions = $11 \times 3 = 33$; the total number of unknowns = $3 \times 84 + 33 = 285$; the total number of equations of statics = 3×55 (joints) = 165. Therefore, the degree of statical indeterminacy = $285 - 165 = 120$.

3.13 EQUATIONS OF CONSTRUCTION (CONDITION) IN STRUCTURAL FRAMES

Often, internal hinges or sliding joints are introduced in frames for construction reasons, or otherwise, and this has the effect of reducing the degree of indeterminacy. The rigid frame in Fig. 3.14a has three hinges at A, C, and E, where the bending moment is zero; hence three equations of condition are present. If AB, BD, and DE are taken as three members, the number of unknown member forces = $3 \times 3 = 9$; the number of unknown reactions = $2 \times 3 = 6$; hence the total number of unknowns = $9 + 6 = 15$.

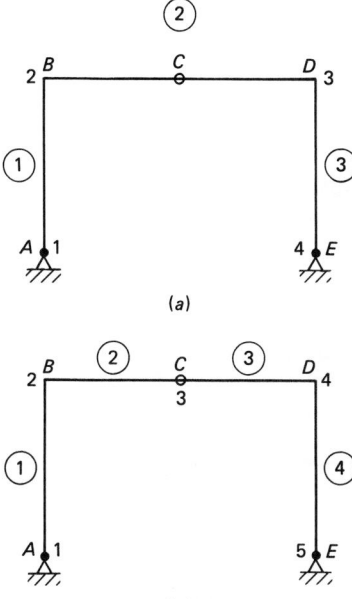

(a)

(b)

Figure 3.14 Structural frames with special conditions.

The number of joints = 4; and the number of equations of statics = $4 \times 3 = 12$; the number of equations of condition = 3; thus, the total number of equations available = $12 + 3 = 15$, from which the degree of indeterminacy = $15 - 15 = 0$; that is, the structure is statically determinate.

Instead of taking BD as one member, BC and CD can be taken as two members and C as an additional joint as shown in Fig. 3.14b. In this case, the number of joints = 5; number of equations of condition = 3; total number of equations = $5 \times 3 + 3 = 18$; number of members = 4; number of reactions = 6; total number of unknowns = $4 \times 3 + 6 = 18$. The degree of indeterminacy = $18 - 18 = 0$, as before.

An alternative scheme to the above is to treat a hinge in a manner to reduce the number of unknown member forces and unknown components of reaction by one. As an example, consider again Fig. 3.14b:

Member AB − one end (A) hinged, the other end (B) fixed

∴ number of unknown member forces = 2

Member BC − one end (B) fixed, the other end (C) hinged

∴ number of unknown member forces = 2

Similarly, unknown member forces for CD and DE are two, respectively.

Number of unknown components of reaction at hinge A = 2

Number of unknown components of reaction at hinge E = 2

Total number of unknowns = $2 + 2 + 2 + 2 + 2 + 2 = 12$

The number of equations of equilibrium available at each of the hinged joints A, C, and E is 2 and the number of equations available at each of the rigid joints B and D is equal to 3. Thus, the total number of available equations $= 3 \times 2 + 2 \times 3 = 12$. The degree of indeterminacy $= 12 - 12 = 0$, as before.

3.14 GENERAL METHOD OF DETERMINING THE DEGREE OF INDETERMINACY OF PLANE RIGID FRAMES

The above procedure can be extended to derive a general expression for the total degree of indeterminacy for structures having combinations of sliding, hinged, and rigid joints; we should remember that the number of unknown force components depends on the type of end condition of the member. Thus the classification in the accompanying table may be used.

Type of end condition of member		Number of unknown member force components	Notation
One end	Other end		
Fixed	Fixed	3	m_3
Fixed	Hinged	2	m_2
Fixed	Sliding (guide)	1	m_1
Hinged	Hinged	1	m_1
Hinged	Sliding (guide)	0	m_0

Total number of unknown member force components

$$= 1 \times m_1 + 2 \times m_2 + 3 \times m_3$$

$$= m_1 + 2m_2 + 3m_3 \tag{3.1}$$

The number of components of reaction depends on the type of support as shown in the following table.

Type of support	No. of reaction components	Notation
Fixed	3	r_3
Hinged	2	r_2
Roller	1	r_1

Total number of reaction components $= 1 \times r_1 + 2 \times r_2 + 3 \times r_3$

$$= r_1 + 2r_2 + 3r_3 \tag{3.2}$$

Furthermore, three equations of statical equilibrium are available at a rigid joint, two equations at a hinged joint, and one equation at a sliding joint, as given in tabular form.

Type of joint	No. of available equations of statical equilibrium	Notation
Fixed	3	j_3
Hinged	2	j_2
Sliding	1	j_1

Total number of available equations of statics $= 1 \times j_1 + 2 \times j_2 + 3 \times j_3$

$$= j_1 + 2j_2 + 3j_3 \tag{3.3}$$

Hence the total degree of indeterminacy (number of redundants) is the excess of unknowns over the available equations. Thus

$$n = (m_1 + 2m_2 + 3m_3) + (r_1 + 2r_2 + 3r_3) - (j_1 + 2j_2 + 3j_3) \tag{3.4}$$

If we wish, we can determine the degree of *external* indeterminacy from

$$n_{\text{ext}} = (r_1 + 2r_2 + 3r_3) - R \tag{3.5}$$

where

$R =$ equations of statics for structure as a whole

$$+ \text{ equations of condition} = e + s \tag{3.6}$$

while the degree of internal indeterminacy is

$$n_{\text{int}} = (m_1 + 2m_2 + 3m_3) + R - (j_1 + 2j_2 + 3j_3) \tag{3.7}$$

The sum of the right-hand sides of Eqs. (3.5) and (3.7) yields the right-hand side of Eq. (3.4).

Equations (3.4), (3.5), and (3.7) will now be applied to determine the degree of indeterminacy of the rigid frame structures in Fig. 3.15.

Fig. 3.15a

$$r_2 = 2 \quad j_2 = 2 \quad j_3 = 2 \quad m_2 = 2 \quad m_3 = 1 \quad e = 3 \quad s = 0$$

$$n_{\text{ext}} = 2r_2 - (e + s)$$

$$= (2)(2) - (3 + 0) = 1$$

$$n_{\text{int}} = 2m_2 + 3m_3 + (e + s) - (2j_2 + 3j_3)$$

$$= (2)(2) + (3)(1) + (3 + 0) - [(2)(2) + (3)(2)]$$

$$= 0$$

$$n = 1 + 0 = 1$$

Fig. 3.15b

$$r_2 = 2 \quad j_2 = 3 \quad j_3 = 2 \quad m_2 = 4 \quad e = 3 \quad s = 1$$

$$n_{\text{ext}} = 2r_2 - (e + s) = (2)(2) - (3 + 1) = 0$$

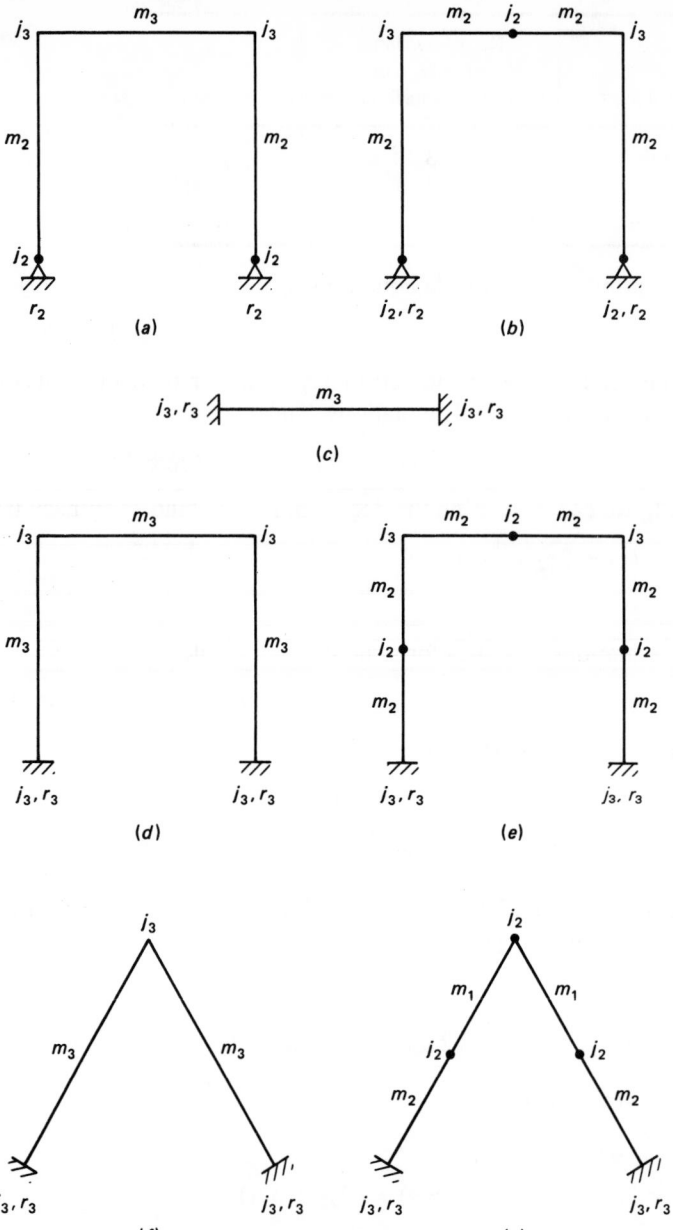

Figure 3.15 Examples for determining degree of indeterminacy of rigid frames. (*a*) Two-hinged portal frame. (*b*) Three-hinged portal frame. (*c*) Fixed beam. (*d*) Fixed portal frame. (*e*) Fixed portal frame with three hinges. (*f*) Fixed bent. (*g*) Fixed bent with internal hinges. (*h*) Two-story bent with hinged and rollered supports. (*i*) Three-span continuous beam with two internal hinges. (*j*) Three-span continuous beam with internal hinge and sliding joint. (*k*) Two-span beam with an internal link.

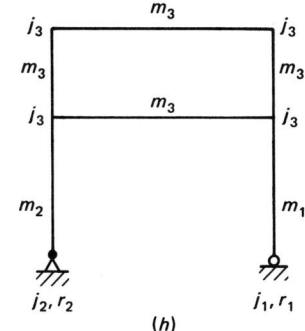

(h)

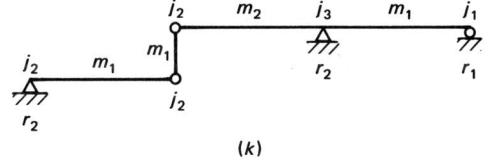

(i)

(j)

(k)

Figure 3.15 (*Continued*)

$$n_{\text{int}} = 2m_2 + (e + s) - (2j_2 + 3j_3)$$

$$= (2)(4) + (3 + 1) - [(2)(3) + (3)(2)]$$

$$= 0$$

$$n = 0 + 0 = 0$$

Fig. 3.15c

$$r_3 = 2 \quad j_3 = 2 \quad m_3 = 1 \quad e = 3 \quad s = 0$$

$$n_{\text{ext}} = 3r_3 - (e + s) = (3)(2) - (3 + 0) = 3$$

$$n_{\text{int}} = 3m_3 + (e + s) - 3j_3$$

$$= (3)(1) + (3 + 0) - (3)(2) = 0$$

$$n = 3 + 0 = 3$$

Fig. 3.15*d*

$r_3 = 2 \quad j_3 = 4 \quad m_3 = 3 \quad e = 3 \quad s = 0$

$n_{ext} = 3r_3 - (e + s) = (3)(2) - (3 + 0) = 3$

$n_{int} = 3m_3 + (e + s) - 3j_3$

$\quad = (3)(3) + (3 + 0) - (3)(4) = 0$

$n = 3 + 0 = 3$

Fig. 3.15*e*

$r_3 = 2 \quad j_2 = 3 \quad j_3 = 4 \quad m_2 = 6 \quad e = 3 \quad s = 3$

$n_{ext} = 3r_3 - (e + s) = (3)(2) - (3 + 3) = 0$

$n_{int} = 2m_2 + (e + s) - (2j_2 + 3j_3)$

$\quad = (2)(6) + (3 + 3) - [(2)(3) + (3)(4)] = 0$

$n = 0 + 0 = 0$

Fig. 3.15*f*

$r_3 = 2 \quad j_3 = 3 \quad m_3 = 2 \quad e = 3 \quad s = 0$

$n_{ext} = 3r_3 - (e + s) = (3)(2) - (3 + 0) = 3$

$n_{int} = 3m_3 + (e + s) - (3j_3)$

$\quad = (3)(2) + (3 + 0) - (3)(3) = 0$

$n = 3 + 0 = 3$

Fig. 3.15*g*

$r_3 = 2 \quad j_2 = 3 \quad j_3 = 2 \quad m_1 = 2 \quad m_2 = 2 \quad e = 3 \quad s = 3$

$n_{ext} = 3r_3 - (e + s) = (3)(2) - (3 + 3) = 0$

$n_{int} = (m_1 + 2m_2) + (e + s) - (2j_2 + 3j_3)$

$\quad = [(2) + (2)(2)] + (3 + 3) - [(2)(3) + (3)(2)] = 0$

$n = 0 + 0 = 0$

Fig. 3.15*h*

$r_1 = 1 \quad r_2 = 1 \quad j_1 = 1 \quad j_2 = 1 \quad j_3 = 4 \quad m_1 = 1 \quad m_2 = 1 \quad m_3 = 4,$

$e = 3, \quad s = 0$

$n_{ext} = (r_1 + 2r_2) - (e + s)$

$\quad = (1 + 2) - (3 + 0) = 0$

$$n_{int} = (m_1 + 2m_2 + 3m_3) + (e + s) - (j_1 + 2j_2 + 3j_3)$$
$$= [(1) + (2)(1) + (3)(4)] + (3 + 0) - [(1) + (2)(1) + (3)(4)]$$
$$= 3$$
$$n = 0 + 3 = 3$$

Fig. 3.15i

$$r_2 = 3 \quad r_3 = 1 \quad j_2 = 2 \quad j_3 = 4 \quad m_2 = 4 \quad m_3 = 1 \quad e = 3 \quad s = 2$$
$$n_{ext} = (2r_2 + 3r_3) - (e + s)$$
$$= [(2)(3) + (3)(1)] - (3 + 2)$$
$$= 4$$
$$n_{int} = (2m_2 + 3m_3) + (e + s) - (2j_2 + 3j_3)$$
$$= [(2)(4) + (3)(1)] + (3 + 2) - [(2)(2) + (3)(4)] = 0$$
$$n = 4 + 0 = 4$$

Fig. 3.15j

$$r_1 = 2 \quad r_2 = 2 \quad j_1 = 1 \quad j_2 = 3 \quad j_3 = 2 \quad m_0 = 1 \quad m_1 = 1 \quad m_2 = 3,$$
$$e = 3, \quad s = 1 + 2 = 3$$
$$n_{ext} = (r_1 + 2r_2) - (e + s)$$
$$= [(2) + (2)(2)] - (3 + 3) = 0$$
$$n_{int} = (m_1 + 2m_2) + (e + s) - (j_1 + 2j_2 + 3j_3)$$
$$= [1 + (2)(3)] + (3 + 3) - [1 + (2)(3) + (3)(2)]$$
$$= 0$$
$$n = 0 + 0 = 0$$

Fig. 3.15k

$$r_1 = 1 \quad r_2 = 2 \quad j_1 = 1 \quad j_2 = 3 \quad j_3 = 1 \quad m_1 = 3 \quad m_2 = 1,$$
$$e = 3, \quad s = 2$$
$$n_{ext} = (r_1 + 2r_2) - (e + s)$$
$$= [1 + (2)(2)] - (3 + 2) = 0$$
$$n_{int} = (m_1 + 2m_2) + (e + s) - (j_1 + 2j_2 + 3j_3)$$
$$= [3 + (2)(1)] + (3 + 2) - [1 + (2)(3) + (3)(1)]$$
$$= 0$$
$$n = 0 + 0 = 0$$

3.15 SOLVED PROBLEM

■ SOLVED PROBLEM 3.1

Using Eqs. (3.5) and (3.7), determine the external and internal conditions of stability and determinacy for the structures shown in Fig. 3.16.

Solution

   Fig. 3.16a

   $r_1 = 0,$ $r_2 = 0,$ $r_3 = 3$

   $m_1 = 0,$ $m_2 = 0,$ $m_3 = 4$

   $j_1 = 0,$ $j_2 = 0,$ $j_3 = 5$

   $R = 3$

From Eq. (3.5):

   $n_{ext} = 3r_3 - R = 6$ stable and indeterminate to sixth degree

From Eq. (3.7):

   $n_{int} = 3m_3 + R - 3j_3 = 0$ determinate

   Fig. 3.16b

   $r_1 = 0,$ $r_2 = 3,$ $r_3 = 0$

   $m_1 = 0,$ $m_2 = 3,$ $m_3 = 9$

   $j_1 = 0,$ $j_2 = 3,$ $j_3 = 8$

   $R = 3$

From Eq. (3.5):

   $n_{ext} = 2r_2 - R = 3$ stable and indeterminate to third degree

From Eq. (3.7):

   $n_{int} = (2m_2 + 3m_3) + R - (2j_2 + 3j_3) = 6$

                    stable and indeterminate to sixth degree

   Fig. 3.16c

   $r_1 = 2,$ $r_2 = 1,$ $r_3 = 0$

   $m_1 = 16,$ $m_2 = 0,$ $m_3 = 0$

   $j_1 = 0,$ $j_2 = 9,$ $j_3 = 0$

   $R = 3$

From Eq. (3.5):

   $n_{ext} = (r_1 + 2r_2) - R = 1$ stable and indeterminate to first degree

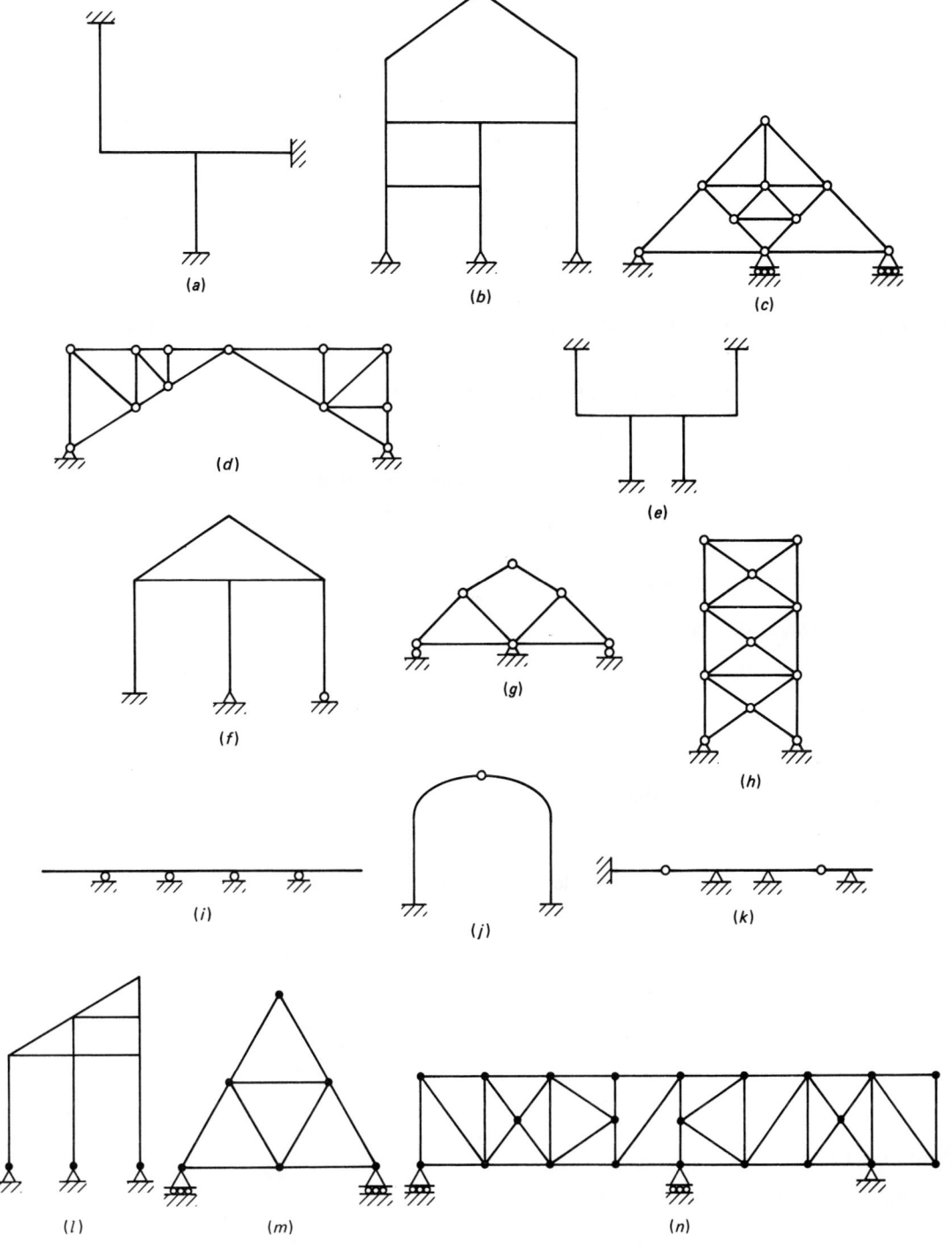

Figure 3.16 Structures for Solved Problem 3.1.

From Eq. (3.7):

$$n_{int} = m_1 + R - 2j_2 = 1 \qquad \text{stable and indeterminate to first degree}$$

Fig. 3.16d

$$r_1 = 0, \qquad r_2 = 2, \qquad r_3 = 0$$
$$m_1 = 20, \quad m_2 = 0, \qquad m_3 = 0$$
$$j_1 = 0, \qquad j_2 = 12, \quad j_3 = 0$$
$$R = 3 + 1 = 4$$

(The bending moment at the central hinge = 0; this is the equation of condition.)
From Eq. (3.5):

$$n_{ext} = 2r_2 - R = 0 \qquad \text{stable and statically determinate}$$

From Eq. (3.7):

$$n_{int} = m_1 + R - 2j_2 = 0 \qquad \text{stable and statically determinate}$$

Fig. 3.16e

$$r_1 = 0, \quad r_2 = 0, \quad r_3 = 4$$
$$m_1 = 0, \quad m_2 = 0, \quad m_3 = 7$$
$$j_1 = 0, \quad j_2 = 0, \quad j_3 = 8$$
$$R = 3$$

From Eq. (3.5):

$$n_{ext} = 3r_3 - R = 9 \qquad \text{stable and statically indeterminate to ninth degree}$$

From Eq. (3.7):

$$n_{int} = 3m_3 + R - 3j_3 = 0 \qquad \text{stable and statically determinate}$$

Fig. 3.16f

$$r_1 = 1, \quad r_2 = 1, \quad r_3 = 1$$
$$m_1 = 1, \quad m_2 = 1, \quad m_3 = 5$$
$$j_1 = 1, \quad j_2 = 1, \quad j_3 = 5$$
$$R = 3$$

From Eq. (3.5):

$$n_{ext} = (r_1 + 2r_2 + 3r_3) - R = 3$$

$$\text{stable and statically indeterminate to third degree}$$

From Eq. (3.7):

$$n_{int} = (m_1 + 2m_2 + 3m_3) + R - (j_1 + 2j_2 + 3j_3) = 3$$

stable and statically indeterminate to third degree

Fig. 3.16g

$$r_1 = 2, \quad r_2 = 1, \quad r_3 = 0$$
$$m_1 = 8, \quad m_2 = 0, \quad m_3 = 0$$
$$j_1 = 0, \quad j_2 = 6, \quad j_3 = 0$$
$$R = 3$$

From Eq. (3.5):

$$n_{ext} = (r_1 + 2r_2) - R = 1 \qquad \text{stable and indeterminate to first degree}$$

From Eq. (3.7):

$$n_{int} = m_1 + R - 2j_2 = -1 \qquad \textit{unstable}$$

Fig. 3.16h

$$r_1 = 0, \quad r_2 = 2, \quad r_3 = 0$$
$$m_1 = 21, \quad m_2 = 0, \quad m_3 = 0$$
$$j_1 = 0, \quad j_2 = 11, \quad j_3 = 0$$
$$R = 3$$

From Eq. (3.5):

$$n_{ext} = 2r_2 - R = 1 \qquad \text{stable and indeterminate to first degree}$$

From Eq. (3.7):

$$n_{int} = m_1 + R - 2j_2 = 2 \qquad \text{stable and indeterminate to second degree}$$

Fig. 3.16i

$$r_1 = 4, \quad r_2 = 0, \quad r_3 = 0$$
$$m_0 = 2, \quad m_1 = 0, \quad m_2 = 0, \quad m_3 = 3$$
$$j_1 = 0, \quad j_2 = 0, \quad j_3 = 4$$
$$R = 3$$

From Eq. (3.5):

$$n_{ext} = r_1 - R = 1 \qquad \textit{unstable} \, (\text{parallel reactions})$$

and statically indeterminate to first degree

From Eq. (3.7):

$$n_{\text{int}} = 3m_3 + R - 3j_3 = 0 \qquad \text{stable and determinate}$$

Fig. 3.16j

$$r_1 = 0, \quad r_2 = 0, \qquad r_3 = 2$$
$$m_1 = 0, \quad m_2 = 2, \quad m_3 = 0$$
$$j_1 = 0, \quad j_2 = 1, \quad j_3 = 2$$
$$R = 3 + 1 = 4$$

(The condition equation is that the bending moment at the central hinge is zero.) From Eq. (3.5):

$$n_{\text{ext}} = 3r_3 - R = 2 \qquad \text{stable and indeterminate to second degree}$$

From Eq. (3.7):

$$n_{\text{int}} = 2m_2 + R - (2j_2 + 3j_3) = 0 \qquad \text{stable and determinate}$$

Fig. 3.16k

$$r_1 = 0, \qquad r_2 = 3, \qquad r_3 = 1$$
$$m_0 = 1, \quad m_1 = 0, \quad m_2 = 4, \quad m_3 = 1$$
$$j_1 = 0, \qquad j_2 = 2, \qquad j_3 = 4$$
$$R = 3 + 2 = 5$$

(The two condition equations are that the bending moments at hinges are zero.) From Eq. (3.5):

$$n_{\text{ext}} = 2r_2 + 3r_3 - R = 4 \qquad \text{stable and indeterminate to fourth degree}$$

From Eq. (3.7):

$$n_{\text{int}} = (2m_2 + 3m_3) + R - (2j_2 + 3j_3) = 0 \qquad \text{stable and determinate}$$

Fig. 3.16l

$$r_1 = 0, \qquad r_2 = 3, \qquad r_3 = 0$$
$$m_1 = 0, \quad m_2 = 3, \quad m_3 = 8$$
$$j_1 = 0, \qquad j_2 = 3, \qquad j_3 = 6$$
$$R = 3$$

From Eq. (3.5):

$$n_{\text{ext}} = 2r_2 - R = 3 \qquad \text{stable and indeterminate to third degree}$$

From Eq. (3.7):

$$n_{\text{int}} = (2m_2 + 3m_3) + R - (2j_2 + 3j_3) = 9$$

stable and indeterminate to ninth degree

Fig. 3.16m

$$r_1 = 2, \quad r_2 = 0, \quad r_3 = 0$$
$$m_1 = 9, \quad m_2 = 0, \quad m_3 = 0$$
$$j_1 = 0, \quad j_2 = 6, \quad j_3 = 0$$
$$R = 3$$

From Eq. (3.5):

$$n_{\text{ext}} = r_1 - R = -1 \qquad \textit{unstable}$$

From Eq. (3.7):

$$n_{\text{int}} = m_1 + R - 2j_2 = 0 \qquad \text{stable and determinate}$$

Fig. 3.16n

$$r_1 = 2, \quad r_2 = 1, \quad r_3 = 0$$
$$m_1 = 43, \quad m_2 = 0, \quad m_3 = 0$$
$$j_1 = 0, \quad j_2 = 22, \quad j_3 = 0$$
$$R = 3$$

From Eq. (3.5):

$$n_{\text{ext}} = (r_1 + 2r_2) - R = 1 \qquad \text{stable and indeterminate to first degree}$$

From Eq. (3.7):

$$n_{\text{int}} = m_1 + R - 2j_2 = 2 \qquad \text{stable and indeterminate to second degree} \qquad \blacksquare$$

PROBLEMS

3.1 to 3.25 Determine the external and internal conditions of stability and determinacy of the plane trusses shown in Figs. P3.1 to P3.25.

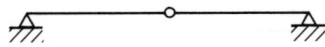

Figure P3.1

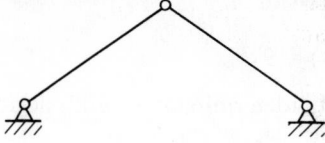

Figure P3.2

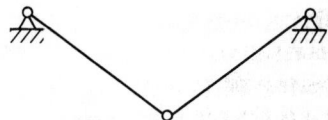

Figure P3.3

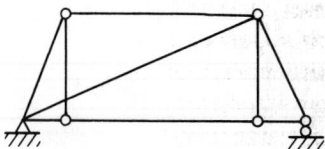

Figure P3.4

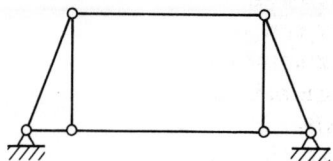

Figure P3.5

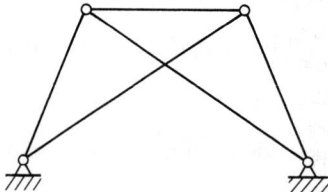

Figure P3.6

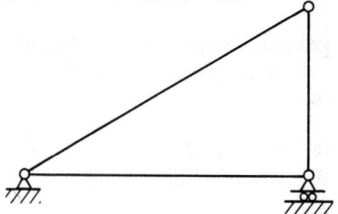

Figure P3.7

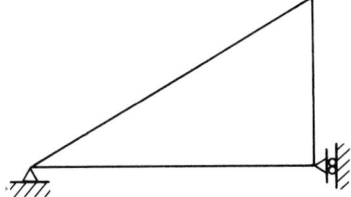

Figure P3.8

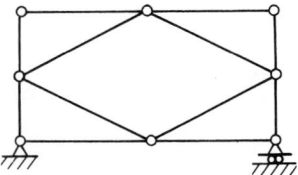

Figure P3.9

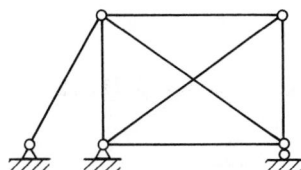

Figure P3.10

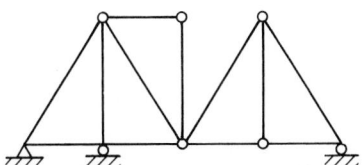

Figure P3.11

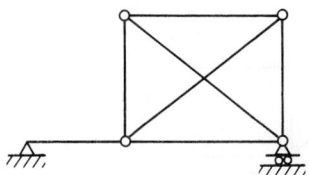

Figure P3.12

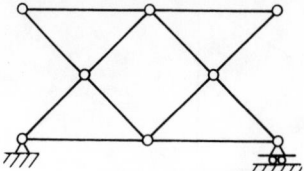

Figure P3.13

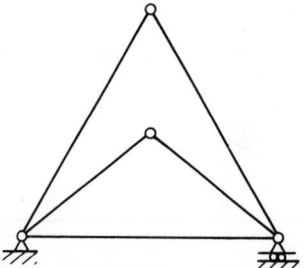

Figure P3.14

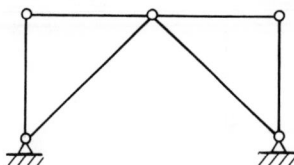

Figure P3.15

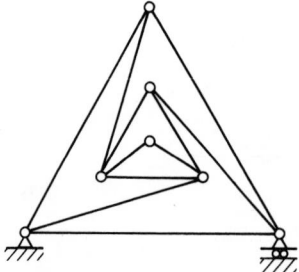

Figure P3.16

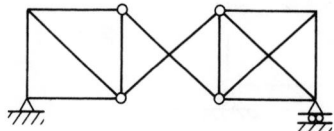

Figure P3.17

Figure P3.18

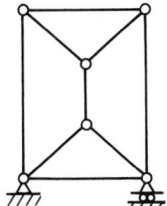

Figure P3.19

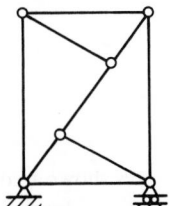

Figure P3.20

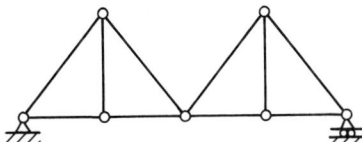

Figure P3.21

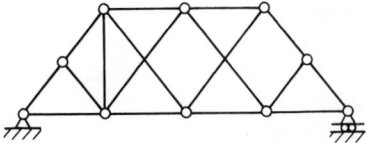

Figure P3.22

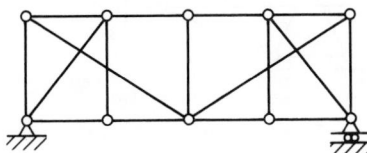

Figure P3.23

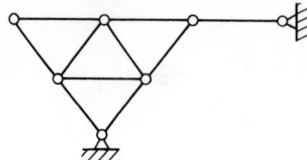

Figure P3.24

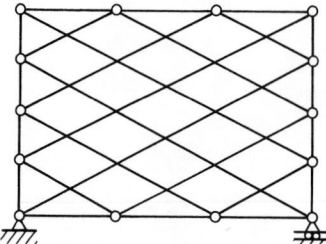

Figure P3.25

3.26 to 3.31 Use the zero-load test for the structures shown to determine whether the trusses shown in Figs. P3.26 to P3.31 have critical form.

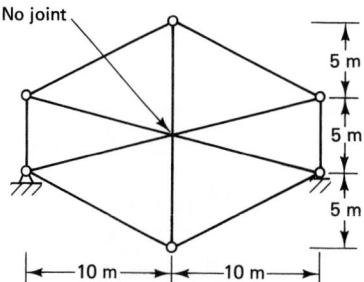

Figure P3.26

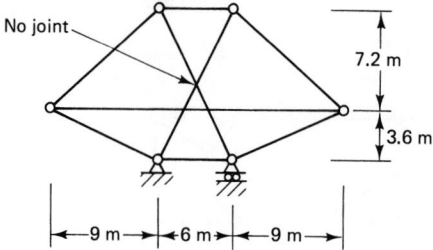

Figure P3.27

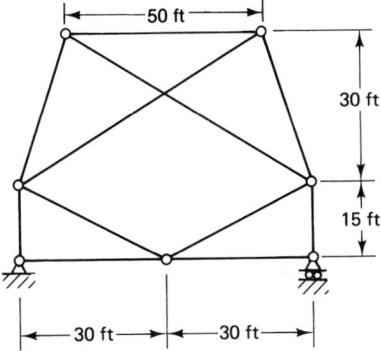

Figure P3.28

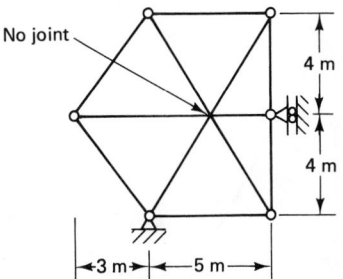

Figure P3.29

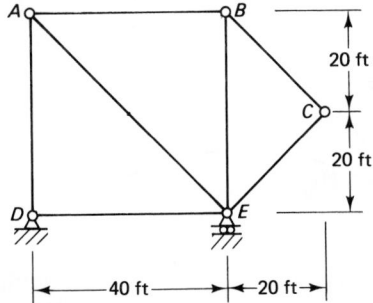

Figure P3.30

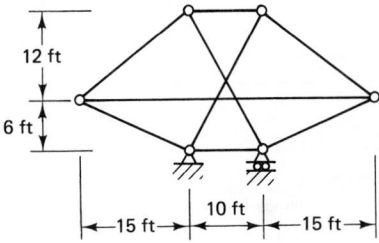

Figure P3.31

3.32 to 3.51 Determine the external and internal conditions of stability and determinacy for beams and plane frames with rigid joints and pin joints shown in Figs. P3.32 to P3.51.

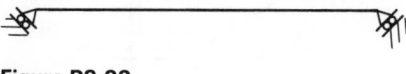

Figure P3.32

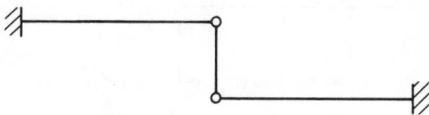

Figure P3.33

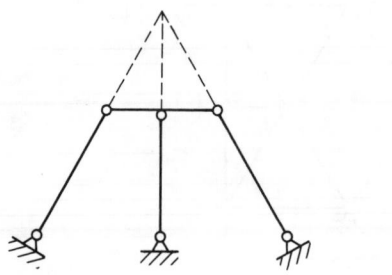

Figure P3.34

Figure P3.35

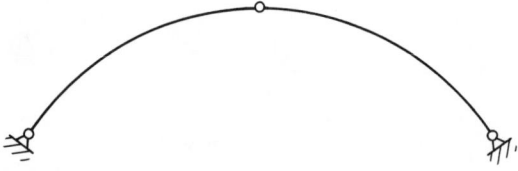

Figure P3.36

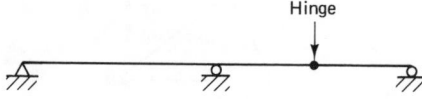

Figure P3.37

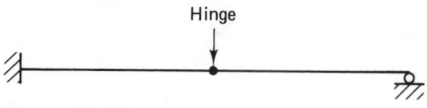

Figure P3.38

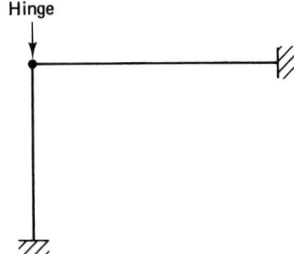

Figure P3.39

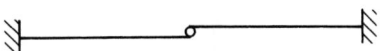

Figure P3.40

Figure P3.41

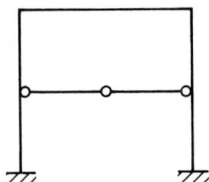

Figure P3.42

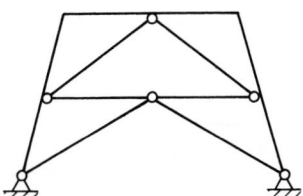

Figure P3.43

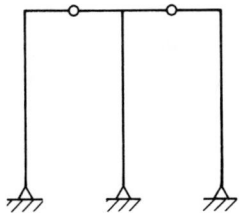

Figure P3.44

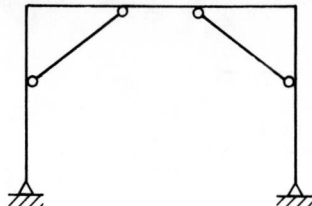

Figure P3.45

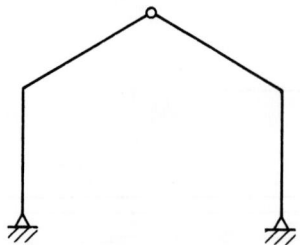

Figure P3.46

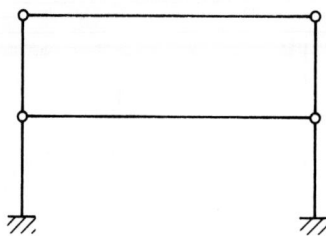

Figure P3.47

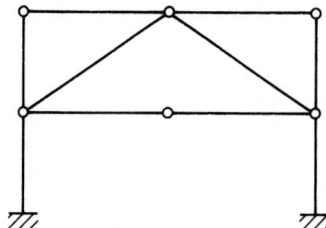

Figure P3.48

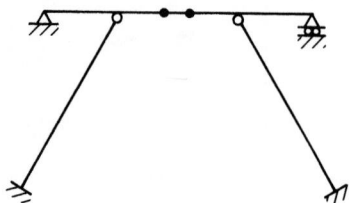

Figure P3.49

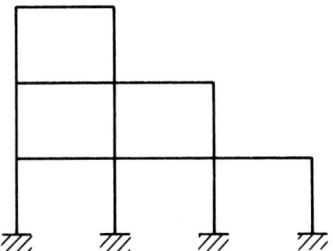

Figure P3.50

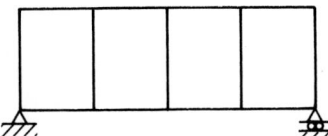

Figure P3.51

Chapter **4**

Trusses

Our world is not measured by the distance from horizon to horizon but by the extent of our understanding.

Jessie L. Beattie

4.1 INTRODUCTION

A truss is a framed structure consisting of a number of members connected at their ends either by rivets or bolts, or by welding. In the analysis of a truss, the following assumptions are made:

1. The joints are frictionless pins and therefore the members are free to rotate at their ends.
2. The loads and reactions are applied at the joints only.
3. The members are straight and there is no eccentricity of connection at the joints.
4. The weight of the members is negligible in comparison with the applied loads.

An analysis based on such a mathematical abstraction of an idealized truss results in purely axial forces in the members of the truss that are either tensile or compressive. These forces are called "primary forces" causing "primary stresses"; the analysis based on the above assumptions is sufficiently accurate for a majority of cases. In cases where greater accuracy is required, stresses due to the following effects should also be included:

1. *Rigidity of connections.* When members are not free to rotate at their ends, bending stresses are induced in the members when they try to rotate; such stresses, called "secondary stresses," are about 5 to 10 percent of "primary stresses."
2. *Off-joint loading.* These also induce bending stresses. One common example is the bending moment due to the self-weight of members and due to wind forces.
3. *Eccentricity of connections.* Although great care is taken to minimize eccentricities at connections, some eccentricities do exist, inducing bending stresses in the truss members.

4.2 HISTORY OF TRUSSES

The origin of trusses dates back several centuries to when trusses of timber were used for simple roof and bridge structures. The Italian architect Andrea Palladio (A.D. 1518–1580) is credited with being the first to use long-span trusses. However, a rational analysis of a truss was not available until the publication of Squire Whipple's book entitled "An Essay on Bridge Building" in 1847. It is interesting to note that this is the first contribution to the theory of structures from the United States. The determination of the bar forces in a truss by considering the equilibrium of a portion of the truss ("method of sections") was first presented by J. W. Schwedler in 1851 and was improved upon by A. Ritter in 1863.

4.3 COMMON PATTERNS OF TRUSSES

Plane trusses can be broadly divided into two categories: (1) bridge trusses and (2) roof trusses. Bridge trusses are characterized by either horizontal or relatively flat top members whereas roof trusses have top members with relatively steep slope. These are shown in Figs. 4.1 and 4.2, respectively. The top member (either horizontal or

American Institute of Steel Construction
Wrigley Building, 400 North Michigan Avenue
Chicago, Illinois 60611

Francis Scott Key Bridge, Baltimore, Maryland—1200-ft main span. (Photo courtesy of American Institute of Steel Construction.)

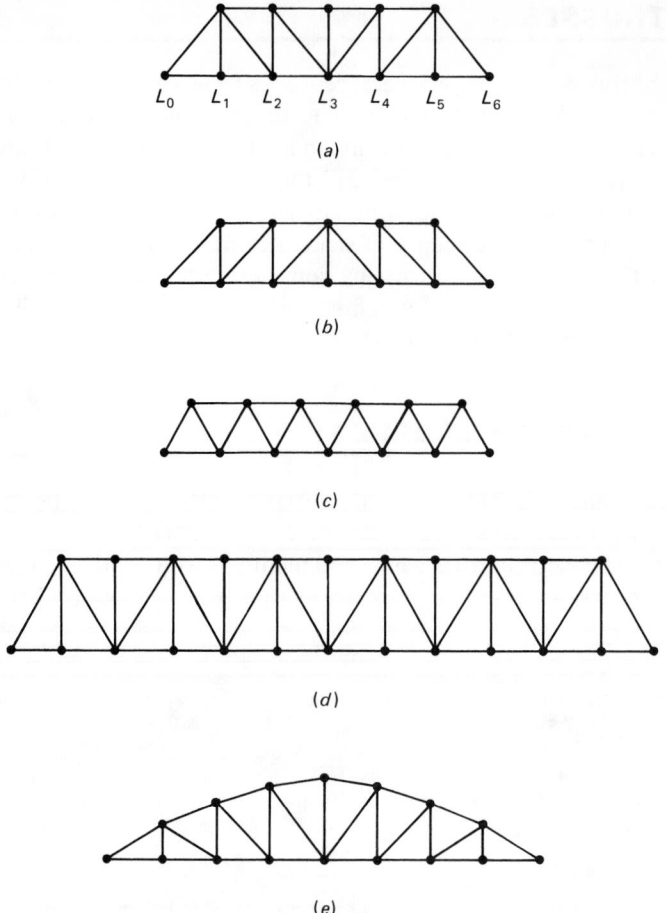

$L_0 \quad L_1 \quad L_2 \quad L_3 \quad L_4 \quad L_5 \quad L_6$

(a)

(b)

(c)

(d)

(e)

Figure 4.1 Examples of bridge trusses. (a) Pratt truss. (b) Howe truss. (c) Warren truss. (d) Warren truss with verticals. (e) Parker truss. (f) Whipple truss. (g) Baltimore truss. (h) Petit or Pennsylvania truss. (i) K truss.

inclined, as the case may be) is called the "top chord." The bottom member is generally, though not always, horizontal and is called the "bottom chord." The members connecting the top and bottom chords are called the "web members." In the case of bridge trusses, the distance between the bottom panel points (e.g., L_0L_1, L_1L_2, etc., in Fig. 4.1a) depends upon the floor beam arrangement and the depth of the truss. The ratio of depth to span in the case of bridge trusses generally varies from 1:12 for light loads to 1:8 for heavy loads. (For roof trusses, the ratio of rise to span depends upon the type of roofing material and snow- and wind-load considerations.) In bridge trusses, it leads to economy when the inclination of web members to the vertical does not exceed an angle of 45°. In the case of roof trusses, the distance between the upper panel points (e.g., U_0U_1, U_1U_2, etc., in Fig. 4.2a) depends on the type of roofing material. (The roofing material is supported by longitudinal purlins which are, in turn, connected to the roof trusses at top panel points U_0, U_1, U_2, etc.)

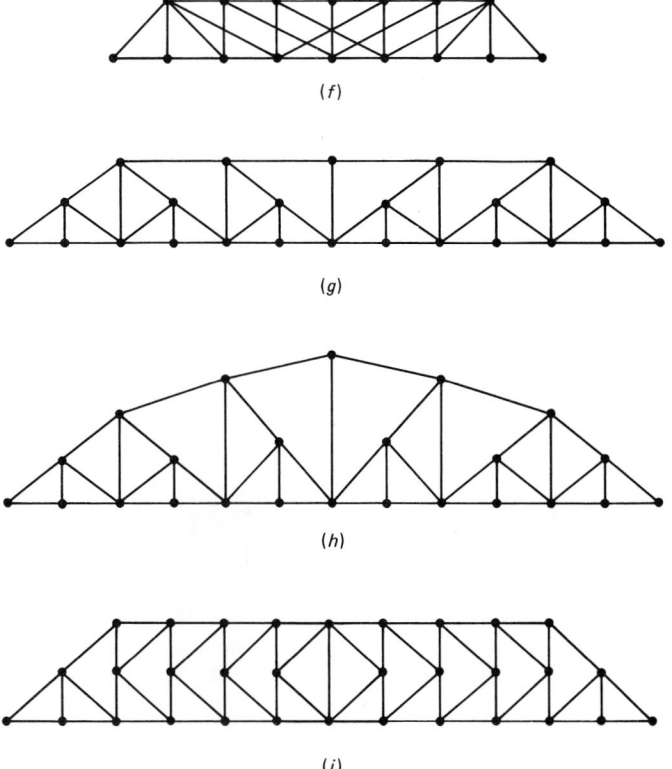

Figure 4.1 (*Continued*)

It is found more economical sometimes to vary the depth of the truss as shown in Fig. 4.1*e*. In case the distance between the bottom panel points is such as to make the bridge floor beams heavy, the bottom panels are subdivided as shown in Fig. 4.1*g* and *h*. To reduce the unsupported length of compression web members, the arrangement shown in Fig. 4.1*i* is sometimes adopted in cases of deep trusses.

4.4 METHODS OF ANALYSIS OF SIMPLE AND COMPOUND PLANE TRUSSES

In this chapter, only statically determinate trusses will be considered. Two methods of analysis that are most common are the method of joints and the method of sections. The method of joints is useful when the forces in all or the majority of members are required, whereas the method of sections is more expedient if the force in a few members only is required. (The method of sections can, of course, be used even if the forces in *all* the members are required.) Sometimes a combination of the two methods is useful, especially in the analysis of "compound trusses." These two methods will now be described in detail. To illustrate better the principles involved in the use of the two methods, the discussion will be confined to plane trusses.

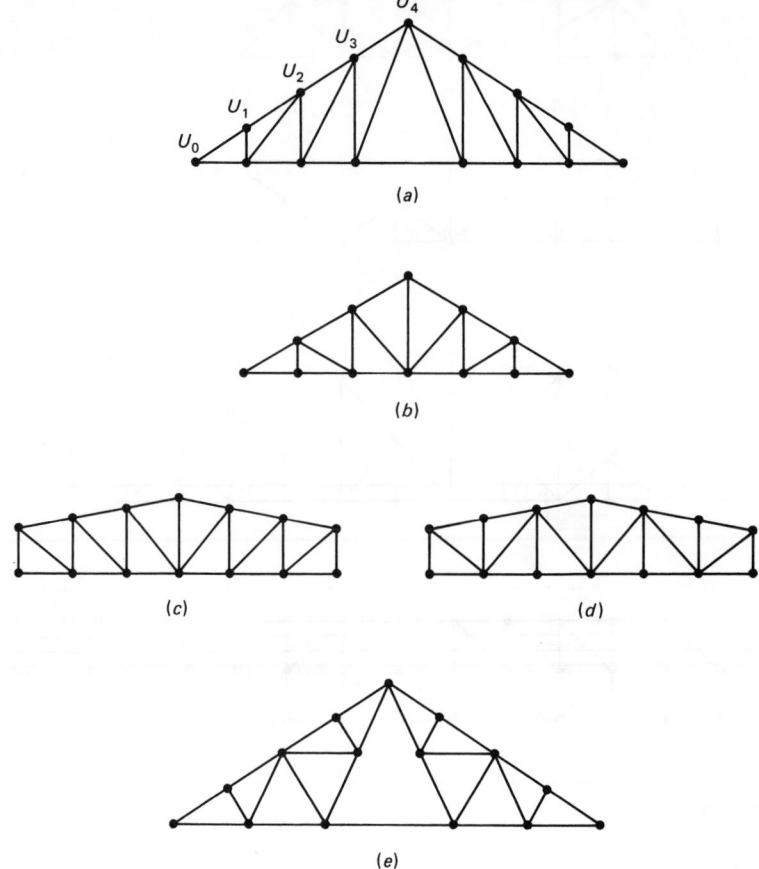

Figure 4.2 Examples of roof trusses. (a) Pratt truss. (b) Howe truss. (c) Flat Pratt truss. (d) Flat Warren truss. (e) Fink truss.

4.4.1 Method of Joints

For a plane truss, two equations of equilibrium, $\Sigma F_x = 0$ and $\Sigma F_y = 0$, are available at each joint. Therefore, two unknown bar forces (which do not have the same line of action) can be determined at each joint. The fundamental principle of the method is to apply the two equations of equilibrium for each joint in succession, until all the unknown bar forces are computed. For a statically determinate truss, since the total number of unknown bar forces and the number of reactions is equal to twice the number of joints (see Chapter 3), all the unknowns can be determined by this method. If all the components of reactions r are determined beforehand by applying the equations of statics for the truss as a whole ($\Sigma F_x = 0$, $\Sigma F_y = 0$, $\Sigma M_z = 0$) and j is the number of joints in the truss, only $(2j - r)$ joint equations are required for the analysis of bar forces and the remaining r equations can be used as a check on the analysis. It is not important that the joints be analyzed in any particular order. The only principle is that a joint at which not more than two unknown bar forces (not having the same line of action) are present can be analyzed by writing the two equations of equilibrium,

$\Sigma F_x = 0$, $\Sigma F_y = 0$. The number of bars in which the forces are already known is immaterial. It is the number of bars whose forces are yet to be computed that matters. If at any joint there are more than two unknown bar forces, all the unknown bar forces cannot be computed from the only two available equations; in this case equations of equilibrium for other joints also will have to be considered and all the equations solved simultaneously, and the method of joints loses much of its usefulness. A combination of the method of joints and the method of sections will be found useful in such cases. As an example of the application of the method of joints, consider the Warren truss, shown in Fig. 4.3a, with a span of 40 m and a depth (or height) of 5 m. The span is divided into four equal panels, each 10 m long. Two concentrated loads act on the truss: a 40-kN load at joint L_1 and a 100-kN load at joint L_2. First, the reactions are determined in the usual manner—the left reaction is found to be 80 kN and the right reaction is 60 kN. (The horizontal reaction at the hinged support is zero since the truss is simply supported and loaded by vertical loads only.) There are 9 joints and 15 bars. Since at each joint two equations of statics are available, there are a total of 18 equations. Since there are only 15 unknown bar forces (the three reactions are already known), the three extra equations can be used as a check on the arithmetic.

Each of joints L_0 and L_4 has two unknowns (not in the same line of action). Therefore, either one of them can be chosen as the starting joint for the analysis. Let us assume that joint L_0 is chosen to be the first joint. The joint is isolated and the free-body diagram (FBD) of the joint is shown in Fig. 4.3b; since the forces in the bars L_0U_1 and L_0L_1 are not known, both are *assumed to be tensile* (positive); a negative answer indicates that the force in that bar is compressive. Since a tensile force in the bar pulls the joint, the arrowheads are shown away from the joint in Fig. 4.3b. It is to be noted that the free-body diagram of the joint, and not of the members, is required.

In this method, instead of computing the angles of inclination of members and dealing with their sines and cosines, work is carried out using the horizontal and vertical projected lengths of members. For the example truss, the slope of the top and bottom chords is zero, and the slope of all the web members is 1 horizontal to 1 vertical as indicated in Fig. 4.3a. Two equations are available at joint L_0 and there are two unknown bar forces $F_{L_0U_1}$ and $F_{L_0L_1}$. Referring to Fig. 4.3b, if the equilibrium equation $\Sigma F_y = 0$ is written first, it is noticed that only $F_{L_0U_1}$ has a component in the y direction, and it can be directly evaluated.

$$\Sigma F_y = 0 \uparrow^+: \qquad F_{L_0U_1} \times \frac{1}{\sqrt{2}} + 80 = 0$$

$$F_{L_0U_1} = -80\sqrt{2} \text{ kN}$$

that is, the force in member L_0U_1 is compressive.

$$\Sigma F_x = 0 \rightarrow^+: \qquad F_{L_0U_1} \times \frac{1}{\sqrt{2}} + F_{L_0L_1} = 0$$

$$(-80\sqrt{2}) \times \frac{1}{\sqrt{2}} + F_{L_0L_1} = 0$$

$$F_{L_0L_1} = +80 \text{ kN} \qquad \text{(tension)}$$

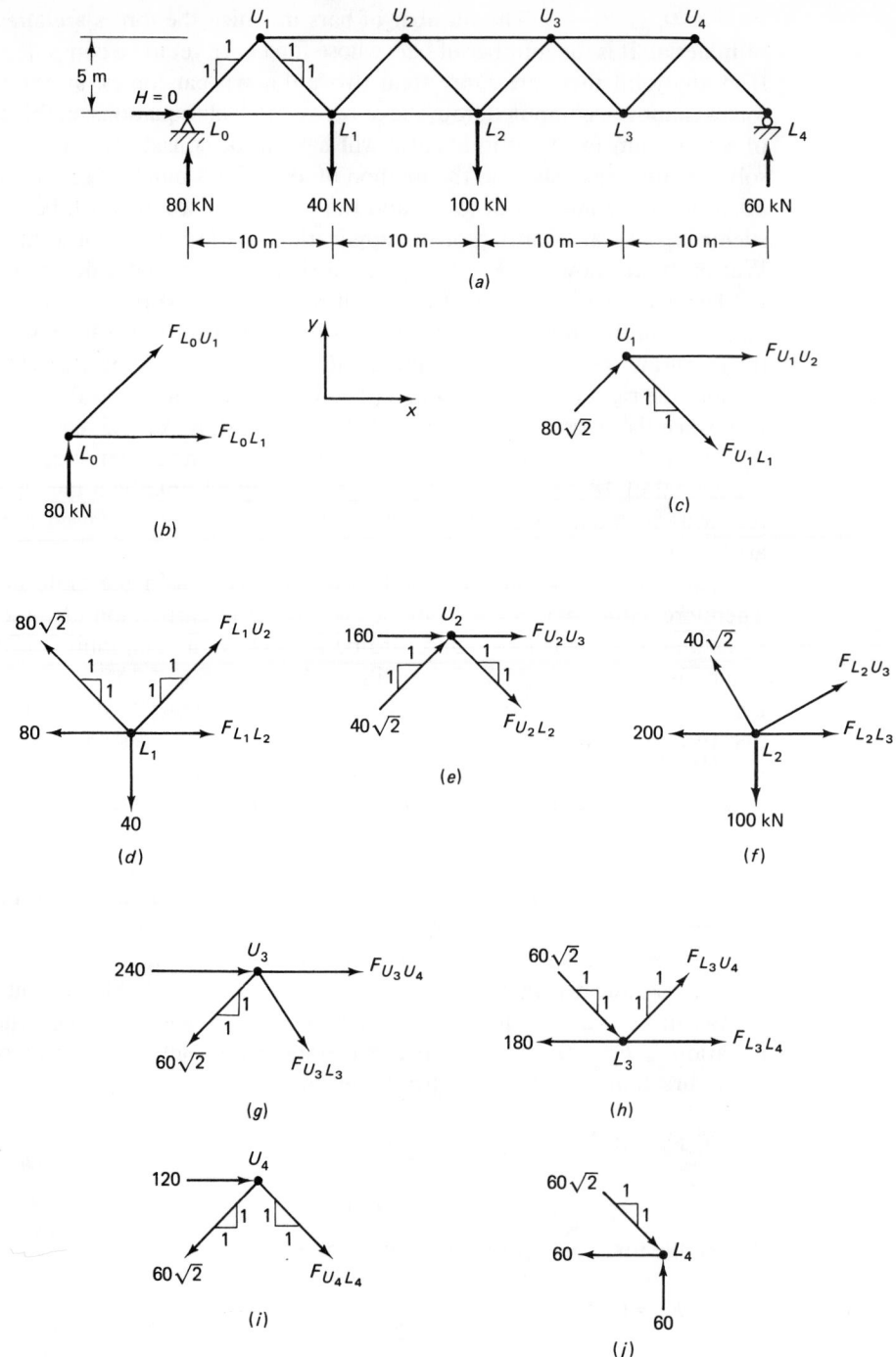

Figure 4.3 Example of the application of the method of joint. (*b*) FBD of joint L_0. (c) FBD of joint U_1. (*d*) FBD of joint L_1. (*e*) FBD of joint U_2. (*f*) FBD of joint L_2. (*g*) FBJ of joint U_3. (*h*) FBD of joint L_3. (*i*) FBD of joint U_4. (*j*) FBD of joint L_4.

It is to be noted that it is not necessary here to solve the two equilibrium equations simultaneously. This is found to be the case for a majority of the joints. However, if both members (whose bar forces are to be evaluated) are inclined to the x and y axes, it becomes necessary to solve the equilibrium equations simultaneously. In the special case when the two inclined members are mutually perpendicular, by resolving the forces in the direction of the line of action of the members (instead of along x and y axes), the two equilibrium equations will each contain one unknown which can then be readily determined.

After joint L_0 is analyzed, there are two joints L_4 and U_1 at which there are two unknown bar forces (at joint U_1 the force in bar U_1L_0 is already known) and one can proceed to either joint U_1 or joint L_4. The sequence of analysis of the joints, as mentioned earlier, is unimportant. The complete analysis of the truss is shown below.

Determination of $F_{U_1L_1}$ and $F_{U_1U_2}$

Refer to Fig. 4.3c:

$$\sum F_y = 0\!\uparrow^+\!: \qquad 80\sqrt{2} \times \frac{1}{\sqrt{2}} - F_{U_1L_1} \times \frac{1}{\sqrt{2}} = 0 \quad F_{U_1L_1} = 80\sqrt{2} \text{ kN} \quad (\text{tension})$$

$$\sum F_x = 0\!\rightarrow^+\!: \qquad 80\sqrt{2} \times \frac{1}{\sqrt{2}} + F_{U_1L_1} \times \frac{1}{\sqrt{2}} + F_{U_1U_2} = 0$$

Substituting $F_{U_1L_1} = 80\sqrt{2}$, $F_{U_1U_2} = -160$ kN (compression).

Determination of $F_{L_1U_2}$ and $F_{L_1L_2}$

Refer to Fig. 4.3d:

$$\sum F_y = 0\!\uparrow^+\!: \qquad 80\sqrt{2} \times \frac{1}{\sqrt{2}} + F_{L_1U_2} \times \frac{1}{\sqrt{2}} - 40 = 0$$

$$F_{L_1U_2} = -40\sqrt{2} \text{ kN} \qquad (\text{compression})$$

$$\sum F_x = 0\!\rightarrow^+\!: \qquad -80 - 80\sqrt{2} \times \frac{1}{\sqrt{2}} + F_{L_1U_2} \times \frac{1}{\sqrt{2}} + F_{L_1L_2} = 0$$

Substituting $F_{L_1U_2} = -40\sqrt{2}$, $F_{L_1L_2} = +200$ kN (tension).

Determination of $F_{U_2L_2}$ and $F_{U_2U_3}$

Refer to Fig. 4.3e:

$$\sum F_y = 0\!\uparrow^+\!: \qquad 40\sqrt{2} \times \frac{1}{\sqrt{2}} - F_{U_2L_2} \times \frac{1}{\sqrt{2}} = 0$$

$$F_{U_2L_2} = +40\sqrt{2} \text{ kN} \qquad (\text{tension})$$

$$\sum F_x = 0\!\rightarrow^+\!: \qquad 40\sqrt{2} \times \frac{1}{\sqrt{2}} + 160 + F_{U_2U_3} + F_{U_2L_2} \times \frac{1}{\sqrt{2}} = 0$$

Substituting $F_{U_2L_2} = 40\sqrt{2}$, $F_{U_2U_3} = -240$ kN (compression).

Determination of $F_{L_2U_3}$ and $F_{L_2L_3}$

Refer to Fig. 4.3f:

$$\sum F_y = 0\!\uparrow\!+: \qquad 40\sqrt{2} \times \frac{1}{\sqrt{2}} + F_{L_2U_3} \times \frac{1}{\sqrt{2}} - 100 = 0$$

$$F_{L_2U_3} = +60\sqrt{2} \text{ kN} \qquad (\text{tension})$$

$$\sum F_x = 0\!\rightarrow\!+: \qquad -200 - 40\sqrt{2} \times \frac{1}{\sqrt{2}} + F_{L_2U_3} \times \frac{1}{\sqrt{2}} + F_{L_2L_3} = 0$$

Substituting $F_{L_2U_3} = 60\sqrt{2}$, $F_{L_2L_3} = +180$ kN (tension).

Determination of $F_{U_3L_3}$ and $F_{U_3U_4}$

Refer to Fig. 4.3g:

$$\sum F_y = 0\!\uparrow\!+: \qquad -60\sqrt{2} \times \frac{1}{\sqrt{2}} - F_{U_3L_3} \times \frac{1}{\sqrt{2}} = 0$$

$$F_{U_3L_3} = -60\sqrt{2} \text{ kN} \qquad (\text{compression})$$

$$\sum F_x = 0\!\rightarrow\!+: \qquad -60\sqrt{2} \times \frac{1}{\sqrt{2}} + 240 + F_{U_3U_4} + F_{U_3L_3} \times \frac{1}{\sqrt{2}} = 0$$

Substituting $F_{U_3L_3} = -60\sqrt{2}$, $F_{U_3U_4} = -120$ kN (compression).

Determination of $F_{L_3U_4}$ and $F_{L_3L_4}$

Refer to Fig. 4.3h:

$$\sum F_y = 0\!\uparrow\!+: \qquad -60\sqrt{2} \times \frac{1}{\sqrt{2}} + F_{L_3U_4} \times \frac{1}{\sqrt{2}} = 0$$

$$F_{L_3U_4} = +60\sqrt{2} \text{ kN} \qquad (\text{tension})$$

$$\sum F_x = 0\!\rightarrow\!+: \qquad -180 + 60\sqrt{2} \times \frac{1}{\sqrt{2}} + F_{L_3U_4} \times \frac{1}{\sqrt{2}} + F_{L_3L_4} = 0$$

Substituting $F_{L_3U_4} = 60\sqrt{2}$, $F_{L_3L_4} = +60$ kN (tension).

Determination of $F_{U_4L_4}$

Refer to the FBD of joint U_4 shown in Fig. 4.3i. There is only one unknown at this joint.

$$\sum F_y = 0\!\uparrow\!+: \qquad -60\sqrt{2} \times \frac{1}{\sqrt{2}} - F_{U_4L_4} \times \frac{1}{\sqrt{2}} = 0$$

$$F_{U_4L_4} = -60\sqrt{2} \text{ kN} \qquad (\text{compression})$$

The other equilibrium equation $\sum F_x = 0$ can be used as a check:

$$\sum F_x = 0 \rightarrow +: \qquad -60\sqrt{2} \times \frac{1}{\sqrt{2}} + 120 + F_{U_4L_4} \times \frac{1}{\sqrt{2}}$$

Substituting $F_{U_4L_4} = -60\sqrt{2}$, $\sum F_x = 0$ (check).

Refer to the FBD of joint L_4 shown in Fig. 4.3 j. There are no unknown bar forces at this joint. Both equilibrium equations can be used as checks:

$$\sum F_x(\rightarrow +) = -60 + 60\sqrt{2} \times \frac{1}{\sqrt{2}} = 0 \qquad (\text{check})$$

$$\sum F_y(\uparrow +) = -60\sqrt{2} \times \frac{1}{\sqrt{2}} + 60 = 0 \qquad (\text{check})$$

4.4.2 Method of Sections

As mentioned earlier, this method is particularly useful if the forces in a few members only are required. The method consists in passing a section through the entire truss, cutting the members in which the forces are required. The truss is thus separated into two segments and unknown forces (assumed tensile) are introduced in the cut members. Either segment can be considered in subsequent analysis. For a coplanar non-concurrent force system, since there are three equations of statics available ($\sum F_x = 0$, $\sum F_y = 0$, $\sum M_z = 0$), three unknown bar forces (all of which do not meet at one point) can be determined by passing one section. If the equilibrium equations are used in the form $\sum F_x = 0$, $\sum F_y = 0$, and $\sum M_z = 0$, it may become necessary to solve them simultaneously. Therefore, it will sometimes be found convenient to use the equilibrium equations in the following form:

$\sum M$ about a point where second and third members meet $= 0$

($\therefore$ solve for the first unknown bar force directly)

$\sum M$ about a point where first and third members meet $= 0$

($\therefore$ solve for the second unknown bar force directly)

$\sum M$ about a point where first and second members meet $= 0$

($\therefore$ solve for the third unknown bar force directly)

The moments are taken about a point where two of the three unknown members meet so that the moments of these unknown forces vanish and the third unknown can be readily determined.

As an example, suppose it is required to determine the forces in bars 1, 2, and 3 of the truss shown in Fig. 4.4a. There is a single concentrated load of 100 kN acting at L_2. The reactions are first determined by considering the equilibrium of the structure as a whole. An imaginary section xx is passed such that the three bars 1, 2, and 3 are cut by it. The free-body diagram of one of the segments is shown in Fig. 4.4b, where

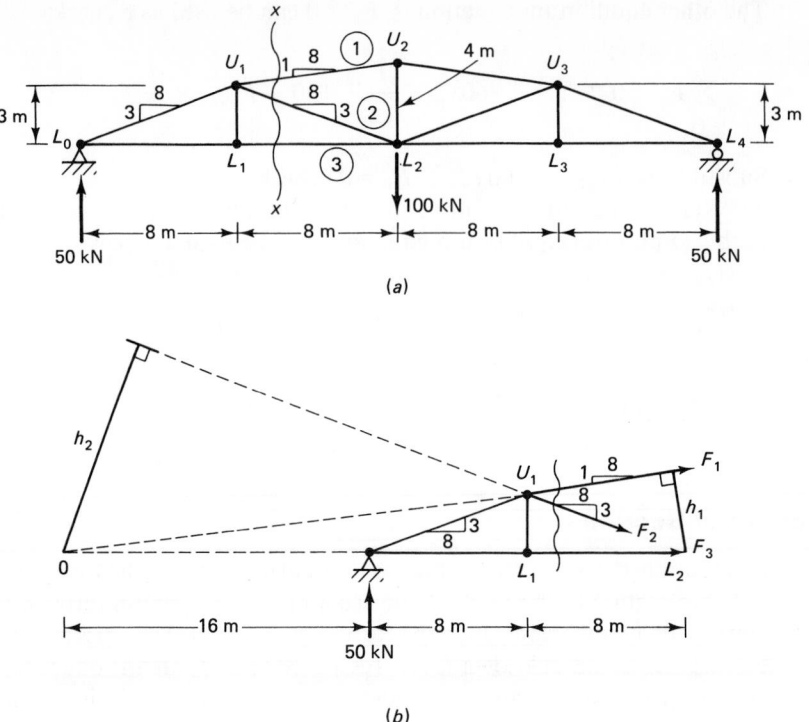

Figure 4.4 Example of the application of the method of sections. (*b*) Free-body diagram.

tensile forces of unknown magnitude are introduced in the cut members. If the three equilibrium equations are applied to the free-body diagram shown in Fig. 4.4*b*:

$$\Sigma F_x = 0 \rightarrow^+ : \qquad F_1 \times \frac{8}{\sqrt{65}} + F_2 \times \frac{8}{\sqrt{73}} + F_3 = 0 \tag{a}$$

$$\Sigma F_y = 0 \uparrow^+ : \qquad 50 + F_1 \times \frac{1}{\sqrt{65}} - F_2 \times \frac{3}{\sqrt{73}} = 0 \tag{b}$$

For the moment equation, it is convenient to take moments either about U_1 (where unknown forces F_1 and F_2 meet) or about L_2 (where F_2 and F_3 meet) or about O (where F_1 and F_3 meet). If $\Sigma M = 0 +$ ↺ about U_1 is considered:

$$(50)(8) - (F_3)(3) = 0 \tag{c}$$

$$\therefore F_3 = +133.3 \text{ kN tension}$$

Substituting this value in (a) and solving equations (a) and (b) simultaneously:

$$F_1 = -201.6 \text{ kN} \qquad \text{(compression)} \qquad F_2 = 71.2 \text{ kN} \qquad \text{(tension)}$$

Alternatively, after evaluating F_3 (by taking moments about U_1), F_1 can be directly computed by taking moments of all forces about L_2 (where F_2 and F_3 meet). The lever arm of F_1 is shown as h_1 in Fig. 4.4*b* and is rather difficult to compute. In order to avoid the need to compute the lever arm h_1, the moment of F_1 about L_2 is obtained by algebraically adding the moments of the horizontal and vertical components of force F_1 (whose lever arms about L_2 are 3 m and 8 m, respectively).

$$\Sigma\,M \text{ about } L_2 = 0 + \; \text{)}: \qquad (50)(16) + \left(F_1 \times \frac{8}{\sqrt{65}} \times 3 + F_1 \times \frac{1}{\sqrt{65}} \times 8 \right) = 0$$

$\therefore F_1 = -201.6$ kN, as before (compression).

To evaluate the only other remaining unknown force F_2, *any one* of the following three equations applied to the free-body diagram in Fig. 4.4*b* can be used:

$$\Sigma\,F_x = 0 \longrightarrow +: \qquad -201.6 \times \frac{8}{\sqrt{65}} + F_2 \times \frac{8}{\sqrt{73}} + 133.3 = 0 \qquad\qquad \text{(d)}$$

Solving, $F_2 = +71.2$ kN (tensile).

$$\Sigma\,F_y = 0 \uparrow +: \qquad 50 - 201.6 \times \frac{1}{\sqrt{65}} - F_2 \times \frac{3}{\sqrt{73}} = 0 \qquad\qquad \text{(e)}$$

$F_2 = +71.2$ kN (tensile).

$$\Sigma\,M \text{ about point } O = 0 + \text{)}\,.$$

$$-(50)(16) + \left(F_2 \times \frac{8}{\sqrt{73}} \times 3 + F_2 \times \frac{3}{\sqrt{73}} \times 24 \right) = 0 \qquad \text{(f)}$$

$F_2 = +71.2$ kN (tensile).

(Again, instead of computing the lever arm h_2 for force F_2 about point O, the moment is indirectly obtained by adding the moments of the horizontal and vertical components of force F_2.)

4.4.3 Combination of Method of Joints and Method of Sections

Whereas the application of method of joints alone is sufficient for the analysis of any statically determinate simple truss, in the case of compound trusses, a combination of method of joints and method of sections will be found expedient. (Refer to Chapter 3 for the definition of simple and compound trusses.) Consider, for example, the Fink roof truss shown in Fig. 4.5*a*. This is a compound truss consisting of two simple trusses *A* and *B* connected together by a common hinge at the top (two reaction components) and a member ③ − ④ at the bottom (one reaction component). The loads are as shown. The left and right reactions, computed by considering the equilibrium of the whole structure, are 30 and 10 kN, respectively.

If the method of joints is used, analysis can proceed in the following order:

1. Consider equilibrium of joint 1. Solve for forces in members 1-2 and 1-7.
2. Consider equilibrium of joint 7. Solve for forces in members 7-2 and 7-8.
3. Consider equilibrium of joint 2. Solve for forces in members 2-3 and 2-8.

At this stage any further application of the method of joints will fail since there are more than two unknown bars meeting at joint 3 or joint 8. (The same is the case even

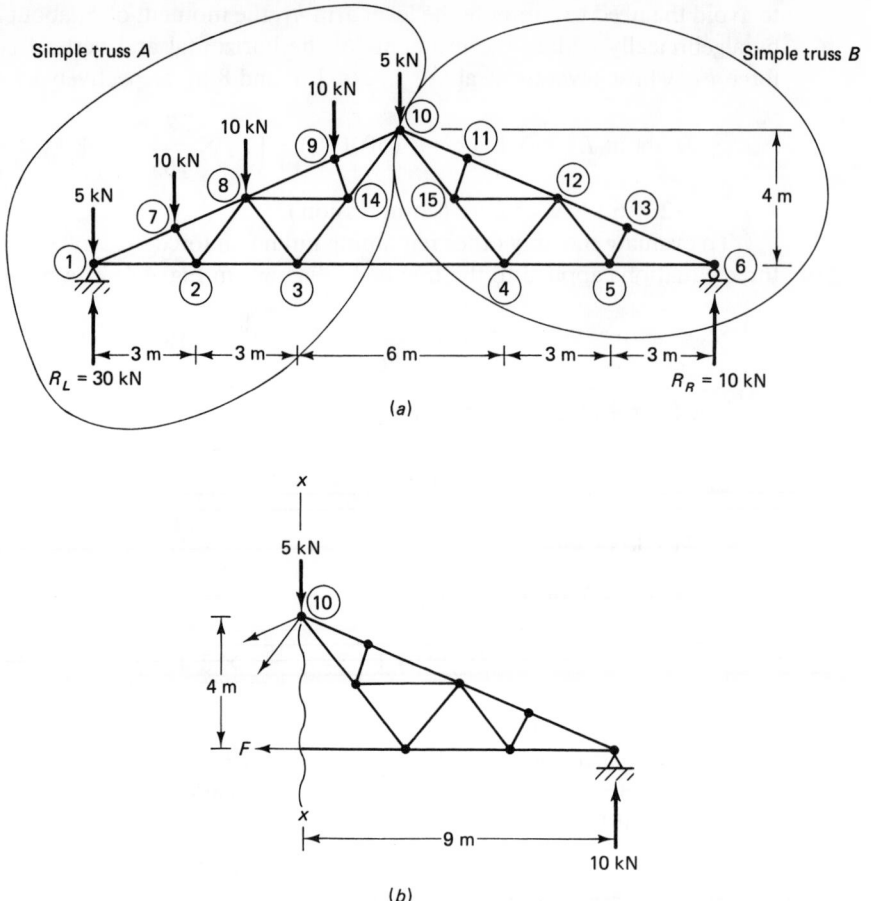

Figure 4.5 Illustration of the analysis of one type of compound truss.

if one starts from the right support. After analyzing joints 6, 13, and 5, it will be found that there are more than two unknowns at joints 4 and 12.) A very easy solution to the problem is to find the force in the member (or members) that connects the two simple trusses A and B. For the example problem, a section xx is passed through joint 10 to cut member 3-4. The free-body diagram of the right portion of the given truss is shown in Fig. 4.5b. The force in member 3-4 is readily computed by taking the moments of the forces about joint 10.

$$\Sigma M = 0 + \text{\large(} : \qquad F \times 4 - 10 \times 9 = 0$$

$F = +22.5$ kN (tensile).

Having thus determined the force in member 3-4, the analysis using the method of joints can continue from joint 3, where now the only two unknowns are the forces in members 3-8 and 3-14.

As another example on the application of the combination of the method of joints and the method of sections, consider the compound truss shown in Fig. 4.6a. The

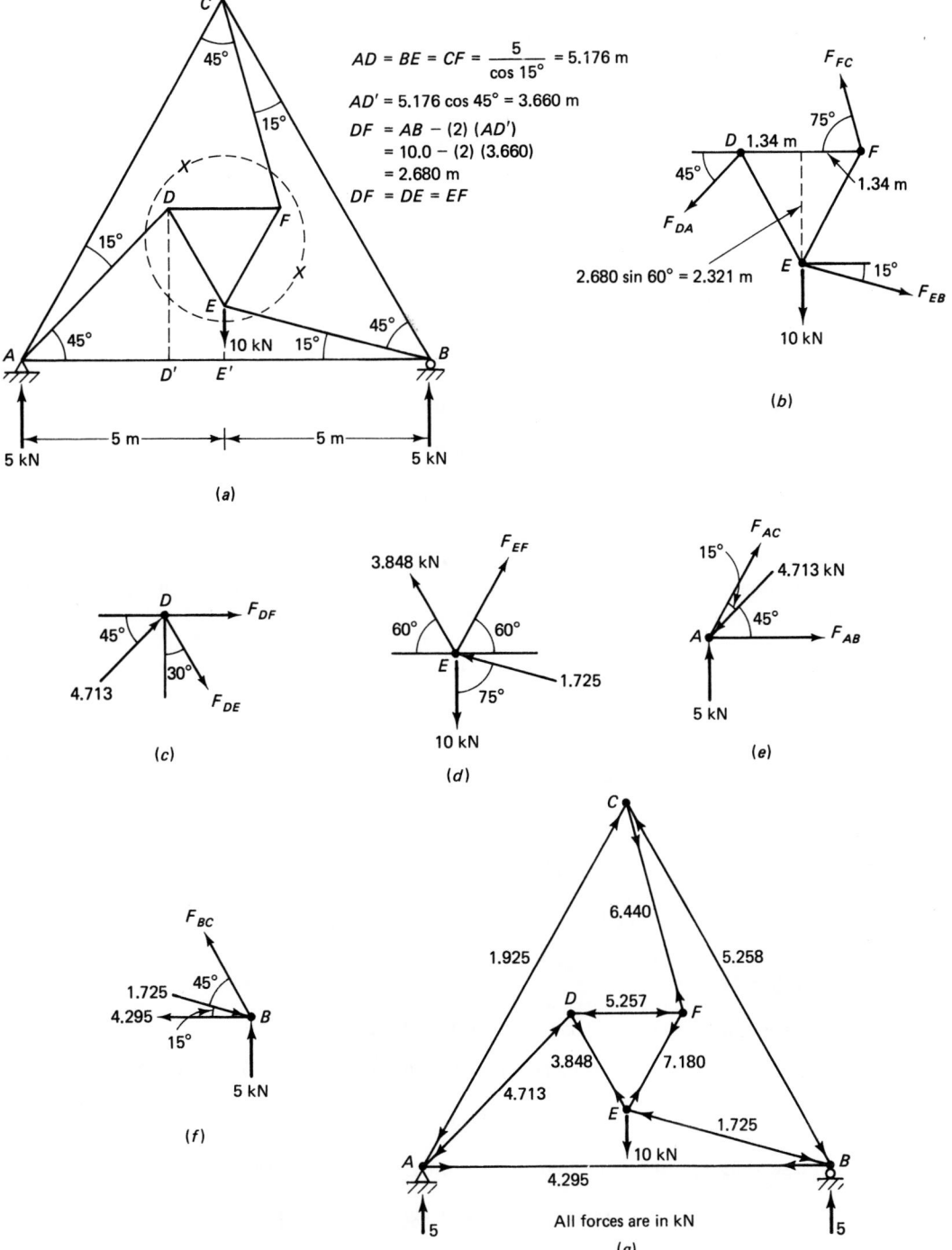

$$AD = BE = CF = \frac{5}{\cos 15°} = 5.176 \text{ m}$$

$$AD' = 5.176 \cos 45° = 3.660 \text{ m}$$

$$DF = AB - (2)(AD')$$
$$= 10.0 - (2)(3.660)$$
$$= 2.680 \text{ m}$$

$$DF = DE = EF$$

(a)

(b)

(c)

(d)

(e)

(f)

All forces are in kN

(g)

Figure 4.6 Illustration of the analysis of another type of compound truss. (*b*) Free-body diagram. (*c*) FBD of joint *D*. (*d*) FBD of joint *E*. (*e*) FBD of joint *A*. (*f*) FBD of joint *B*.

truss consists of two simple trusses *ABC* and *DEF* connected together by three non-parallel nonconcurrent bars *AD, BE,* and *CF.* The reactions at *A* and *B* are computed in the usual manner, by considering the equilibrium of the whole truss. Pass a section *xx* as shown in Fig. 4.6*a* cutting the three connecting members *AD, BE,* and *CF.* The free-body diagram of the simple truss *DEF* is shown in Fig. 4.6*b.* Applying the three equations of equilibrium to the free-body diagram:

$$\sum F_x = 0 \xrightarrow{+}: \qquad -F_{DA}\cos 45° - F_{FC}\cos 75° + F_{EB}\cos 15° = 0 \qquad (a)$$

$$\sum F_y = 0\uparrow{+}: \qquad -F_{DA}\sin 45° + F_{FC}\sin 75° - F_{EB}\sin 15° - 10 = 0 \qquad (b)$$

$$\sum M \text{ about } E = 0 + \Big\rangle: \qquad -(F_{DA}\cos 45°)(2.321) - (F_{DA}\sin 45°)(1.340)$$
$$- (F_{FC}\cos 75°)(2.321) - (F_{FC}\sin 75°)(1.340) = 0 \qquad (c)$$

Solving equations (a), (b), and (c) yields

$$F_{DA} = -4.713 \text{ kN} \qquad \text{(compression)}$$

$$F_{EB} = -1.725 \text{ kN} \qquad \text{(compression)}$$

$$F_{FC} = +6.440 \text{ kN} \qquad \text{(tension)}$$

The other forces in the members of the two simple trusses *ABC* and *DEF* are found in the usual manner by the method of joints as shown below.

Members *DE* and *DF*

Refer to the FBD of joint *D* shown in Fig. 4.6*c*:

$$\sum F_y = 0\uparrow{+}: \qquad 4.713\sin 45° - F_{DE}\cos 30° = 0$$

$$F_{DE} = +3.848 \text{ kN} \qquad \text{(tension)}$$

$$F_x = 0 \xrightarrow{+}: \qquad 4.713\cos 45° + F_{DF} + F_{DE}\sin 30° = 0$$

Substituting $F_{DE} = 3.848$, $F_{DF} = -5.257$ kN (compression).

Member *EF*

Refer to the FBD of joint *E* shown in Fig. 4.6*d*:

$$\sum F_x = 0 \xrightarrow{+}: \qquad -3.848\cos 60° + F_{EF}\cos 60° - 1.725\cos 15° = 0$$

$$F_{EF} = +7.180 \text{ kN} \qquad \text{(tension)}$$

The second equilibrium equation at joint *E* can be used as a check.

$$\sum F_y = 0\uparrow{+}: \qquad 3.848\sin 60° + 7.180\sin 60° + 1.725\sin 15° - 10$$

$$= -0.003 \approx 0 \qquad \text{(check)}$$

Members AB and AC

Refer to FBD of joint A shown in Fig. 4.6 e:

$$\Sigma\, F_y = 0\!\uparrow^+: \qquad 5 + F_{AC} \sin 60° - 4.713 \sin 45° = 0$$

$$F_{AC} = -1.925 \text{ kN} \qquad \text{(compression)}$$

$$\Sigma\, F_x = 0\!\rightarrow^+: \qquad F_{AC} \cos 60° - 4.713 \cos 45° + F_{AB} = 0$$

Substituting $F_{AC} = -1.925$, $F_{AB} = +4.295$ kN (tension).

Member BC

Refer to FBD of joint B shown in Fig. 4.6 f:

$$\Sigma\, F_y = 0\!\uparrow^+: \qquad -1.725 \sin 15° + F_{BC} \sin 60° + 5 = 0$$

$$F_{BC} = -5.258 \text{ kN} \qquad \text{(compression)}$$

The other equation $\Sigma\, F_x = 0$ can be used as a check.

$$\Sigma\, F_x = 0\!\rightarrow^+: \qquad -4.295 + 1.725 \cos 15° - F_{BC} \cos 60° = 0 \qquad \text{(check)}$$

All forces acting at joints C and F are known. Therefore, equilibrium of joints C and F need be considered only if a check on the computation is desired. The results of the analysis are summarized in Fig. 4.6 g.

4.5 ANALYSIS OF SPACE TRUSSES—METHOD OF TENSION COEFFICIENTS

So far, the discussion has been limited to plane trusses. The majority of trusses encountered in structural engineering are either plane trusses or, if space trusses, can be resolved into component plane trusses. However, some exceptions, like domes, cannot be resolved into plane trusses and have to be analyzed as space trusses only. For the analysis of statically determinate space trusses, an elegant method called "the method of tension coefficients" will be found very useful. This method does not involve new concepts but is a systematic application of the method of joints and is equally applicable to the analysis of plane trusses; however, its potential is best utilized in the analysis of space trusses.

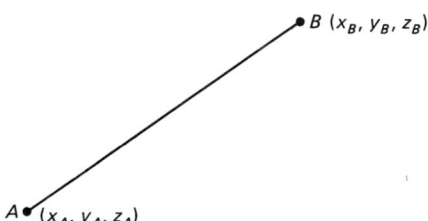

Figure 4.7 Tension coefficient of a member AB.

The tension coefficient of any member AB (Fig. 4.7) is defined as the force in the member AB divided by the length of the member AB, or

$$t_{AB} = \frac{F_{AB}}{L_{AB}}$$

A negative tension coefficient means a compressive force in the member. Referring to Fig. 4.7:

At Joint A

$$x \text{ component of force in } AB = \frac{F_{AB}(x_B - x_A)}{L_{AB}} = t_{AB}(x_B - x_A)$$

At Joint B

$$x \text{ component of force in } AB = t_{AB}(x_A - x_B)$$

At Joint A

$$y \text{ component of force in } AB = \frac{F_{AB}(y_B - y_A)}{L_{AB}} = t_{AB}(y_B - y_A)$$

At Joint B

$$y \text{ component of force in } AB = t_{AB}(y_A - y_B)$$

At Joint A

$$z \text{ component of force in } AB = \frac{F_{AB}(z_B - z_A)}{L_{AB}} = t_{AB}(z_B - z_A)$$

At Joint B

$$z \text{ component of force in } AB = t_{AB}(z_A - z_B)$$

At Any Joint

Force component in x direction = (tension coefficient)

$$\times \, (x \text{ coordinate of farther end} - x \text{ coordinate of nearer end})$$

Similarly for the y and z components.

Procedure for the Tension-Coefficient Method

1. Take the origin at the left-hand bottommost joint such that the coordinates of all the joints are positive.
2. Choose x and y axes as for plane trusses (i.e., x axis horizontal and y axis vertical). The z axis is horizontal and perpendicular to the plane containing x and y axes.

3. Number the bars 1, 2, 3, etc. The tension coefficients of bars 1, 2, 3, etc., are denoted by t_1, t_2, t_3, etc.
4. Write down the coordinates of all joints with reference to the axes chosen.
5. Compute the lengths of all members.
6. Start with a joint at which three or fewer than three unknowns (including reaction components) exist. (It is generally difficult to determine the reactions of space trusses, and it is also not necessary to compute them beforehand.) Write down the three equations of equilibrium $\Sigma \, F_x = 0$, $\Sigma \, F_y = 0$, $\Sigma \, F_z = 0$ and solve for the unknowns. Repeat the procedure for all other joints. (If there is no joint with fewer than four unknowns, the equations have to be solved simultaneously with the equations of equilibrium for the other joints.)
7. Multiply the tension coefficients by their corresponding lengths to get the forces in the members. If the tension coefficient of any member is negative, it means that the force in that member is compressive.
8. Check the equilibrium of the entire structure under the action of the external applied loads and internal developed reactions.

The method is illustrated by means of the example truss shown in Fig. 4.8.

Step 1. Take the origin at B.
Step 2. Choose the x, y, z axes as shown in Fig. 4.8.
Step 3. Number the members as shown in Fig. 4.8b.
Step 4. Coordinates of joint A $(4, 12, 5)$
    Coordinates of joint B $(0, 0, 0)$
    Coordinates of joint C $(3, 0, 15)$
    Coordinates of joint D $(15, 0, 0)$
Step 5.

$$L_1 = \sqrt{(x_D - x_B)^2 + (y_D - y_B)^2 + (z_D - z_B)^2}$$
$$= \sqrt{15^2 + 0^2 + 0^2} = 15 \text{ m}$$
$$L_2 = \sqrt{(x_C - x_B)^2 + (y_C - y_B)^2 + (z_C - z_B)^2}$$
$$= \sqrt{3^2 + 0^2 + 15^2} = 15.30 \text{ m}$$
$$L_3 = \sqrt{(x_C - x_D)^2 + (y_C - y_D)^2 + (z_C - z_D)^2}$$
$$= \sqrt{12^2 + 0^2 + 15^2} = 19.21 \text{ m}$$
$$L_4 = \sqrt{(x_A - x_B)^2 + (y_A - y_B)^2 + (z_A - z_B)^2}$$
$$= \sqrt{4^2 + 12^2 + 5^2} = 13.60 \text{ m}$$
$$L_5 = \sqrt{(x_C - x_A)^2 + (y_C - y_A)^2 + (z_C - z_A)^2}$$
$$= \sqrt{1^2 + 12^2 + 10^2} = 15.65 \text{ m}$$
$$L_6 = \sqrt{(x_A - x_D)^2 + (y_A - y_D)^2 + (z_A - z_D)^2}$$
$$= \sqrt{11^2 + 12^2 + 5^2} = 17.03 \text{ m}$$

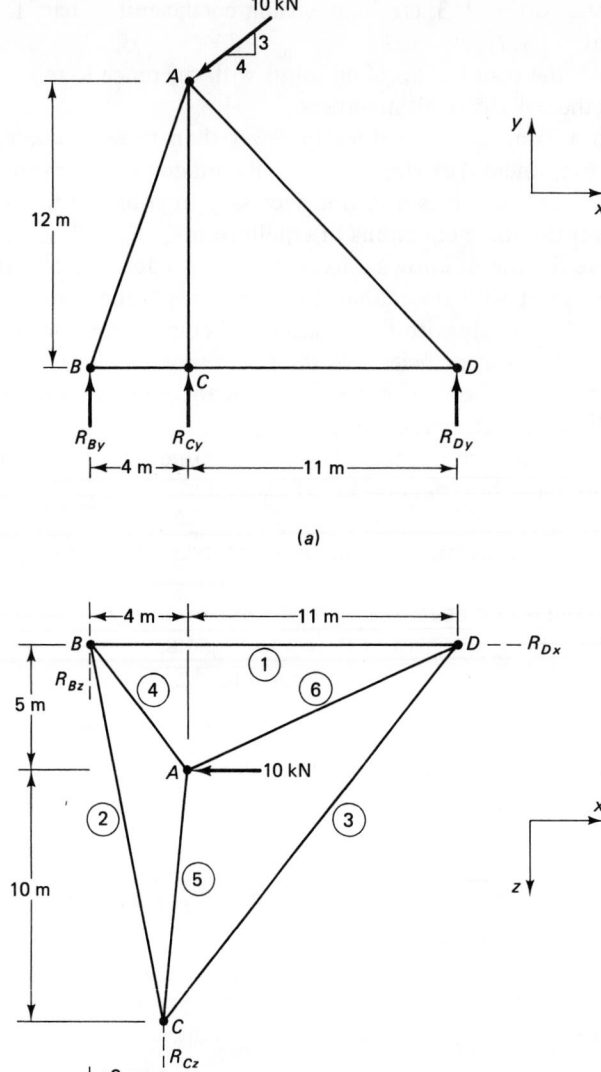

Figure 4.8 Analysis of a space truss by the tension-coefficient method. (*a*) Front view (elevation). (*b*) Top view (plan).

Step 6. *Joint A:* There are three unknowns t_4, t_5, and t_6.

$$\sum F_x = 0: \quad t_4(0 - 4) + t_6(15 - 4) + t_5(3 - 4) - 10 \times \tfrac{4}{5} = 0 \quad \text{(a)}$$

$$\sum F_y = 0: \quad t_4(0 - 12) + t_6(0 - 12) + t_5(0 - 12) - 10 \times \tfrac{3}{5} = 0 \quad \text{(b)}$$

$$\sum F_z = 0: \quad t_4(0 - 5) + t_6(0 - 5) + t_5(15 - 5) = 0 \quad \text{(c)}$$

Solving

$$t_4 = -\tfrac{23}{30} = -0.767 \text{ kN/m}$$

$$t_5 = -\tfrac{1}{6} = -0.167 \text{ kN/m}$$

$$t_6 = +\tfrac{13}{30} = +0.433 \text{ kN/m}$$

Next either joint B, C, or D can be chosen. At each joint there are four unknowns and the equations have to be solved simultaneously considering all the joints together. At joint B, the unknowns are t_1, t_2, R_{B_y}, and R_{B_z}. At joint C, the unknowns are t_2, t_3, R_{C_y}, and R_{C_z}. At joint D, the unknowns are t_1, t_3, R_{D_x}, and R_{D_y}.

Joint B

$$\sum F_x = 0: \qquad t_1(15 - 0) + t_2(3 - 0) + t_4(4 - 0) = 0 \qquad \text{(d)}$$

$$\sum F_y = 0: \qquad t_1(0 - 0) + t_2(0 - 0) + t_4(12 - 0) + R_{B_y} = 0 \qquad \text{(e)}$$

$$\sum F_z = 0: \qquad t_1(0 - 0) + t_2(15 - 0) + t_4(5 - 0) + R_{B_z} = 0 \qquad \text{(f)}$$

From equation (e), $R_{B_y} = +9.2$ kN. t_1, t_2, and R_{B_z} cannot be solved at this stage.

Joint D

$$\sum F_x = 0: \qquad t_1(0 - 15) + t_3(3 - 15) + t_6(4 - 15) + R_{D_x} = 0 \qquad \text{(g)}$$

$$\sum F_y = 0: \qquad t_1(0 - 0) + t_3(0 - 0) + t_6(12 - 0) + R_{D_y} = 0 \qquad \text{(h)}$$

$$\sum F_z = 0: \qquad t_1(0 - 0) + t_6(5 - 0) + t_3(15 - 0) = 0 \qquad \text{(i)}$$

Solving equation (h):

$$R_{D_y} = -12t_6 = -5.2 \text{ kN}$$

From equation (i):

$$t_3 = -\tfrac{1}{3} t_6 = -\tfrac{13}{90} \text{ kN/m}$$

Joint C

$$\sum F_x = 0: \qquad t_2(0 - 3) + t_3(15 - 3) + t_5(4 - 3) = 0 \qquad \text{(j)}$$

$$\sum F_y = 0: \qquad t_2(0 - 0) + t_3(0 - 0) + t_5(12 - 0) + R_{C_y} = 0 \qquad \text{(k)}$$

$$\sum F_z = 0: \qquad t_2(0 - 15) + t_3(0 - 15) + t_5(5 - 15) + R_{C_z} = 0 \qquad \text{(l)}$$

From equation (j):

$$t_2 = -\tfrac{19}{30} \text{ kN/m}$$

From equation (k):

$$R_{C_y} = +2.0 \text{ kN}$$

From equation (f):

$$R_{B_z} = +\tfrac{40}{3} \text{ kN}$$

From equation (d):

$$t_1 = \tfrac{149}{450} \text{ kN/m}$$

From equation (l):

$$R_{C_z} = -\tfrac{40}{3} \text{ kN}$$

From equation (g):

$$R_{D_x} = 8 \text{ kN}$$

Step 7.

$$F_1 = +\tfrac{149}{450} \times 15 = +4.97 \text{ kN}$$

$$F_2 = -\tfrac{19}{30} \times 15.30 = -9.69 \text{ kN}$$

$$F_3 = -\tfrac{13}{90} \times 19.21 = -2.78 \text{ kN}$$

$$F_4 = -\tfrac{23}{30} \times 13.60 = -10.4 \text{ kN}$$

$$F_5 = -\tfrac{1}{6} \times 15.65 = -2.61 \text{ kN}$$

$$F_6 = +\tfrac{13}{30} \times 17.03 = +7.38 \text{ kN}$$

Step 8. Check the equilibrium of the whole structure.

$$\sum F_x = R_{D_x} + P_x = 8 - 8 = 0 \qquad \text{(check)}$$

$$\sum F_y = R_{B_y} + R_{C_y} + R_{D_y} + P_y$$

$$= +9.2 + 2.0 - 5.2 - 6 = 0 \qquad \text{(check)}$$

$$\sum F_z = R_{B_z} + R_{C_z} = +\tfrac{40}{3} - \tfrac{40}{3} = 0 \qquad \text{(check)}$$

4.6 SOLVED PROBLEM

■ SOLVED PROBLEM 4.1

Determine the forces in the members of the compound truss shown in Fig. 4.9a.

Solution. The compound truss shown in Fig. 4.9a is subdivided into a primary truss (Fig. 4.9b) and six secondary trusses (three of which are shown in Fig. 4.9c to e). The secondary trusses are first analyzed by the method of joints, and the forces in the bars are shown in Table 4.1a, and Fig. 4.9c to e

The secondary trusses are supported by the primary truss. Secondary truss reactions are shown in Fig. 4.9c to e. Loads that are equal and opposite to the reactions must be applied at the corresponding panel points of the primary truss. For example,

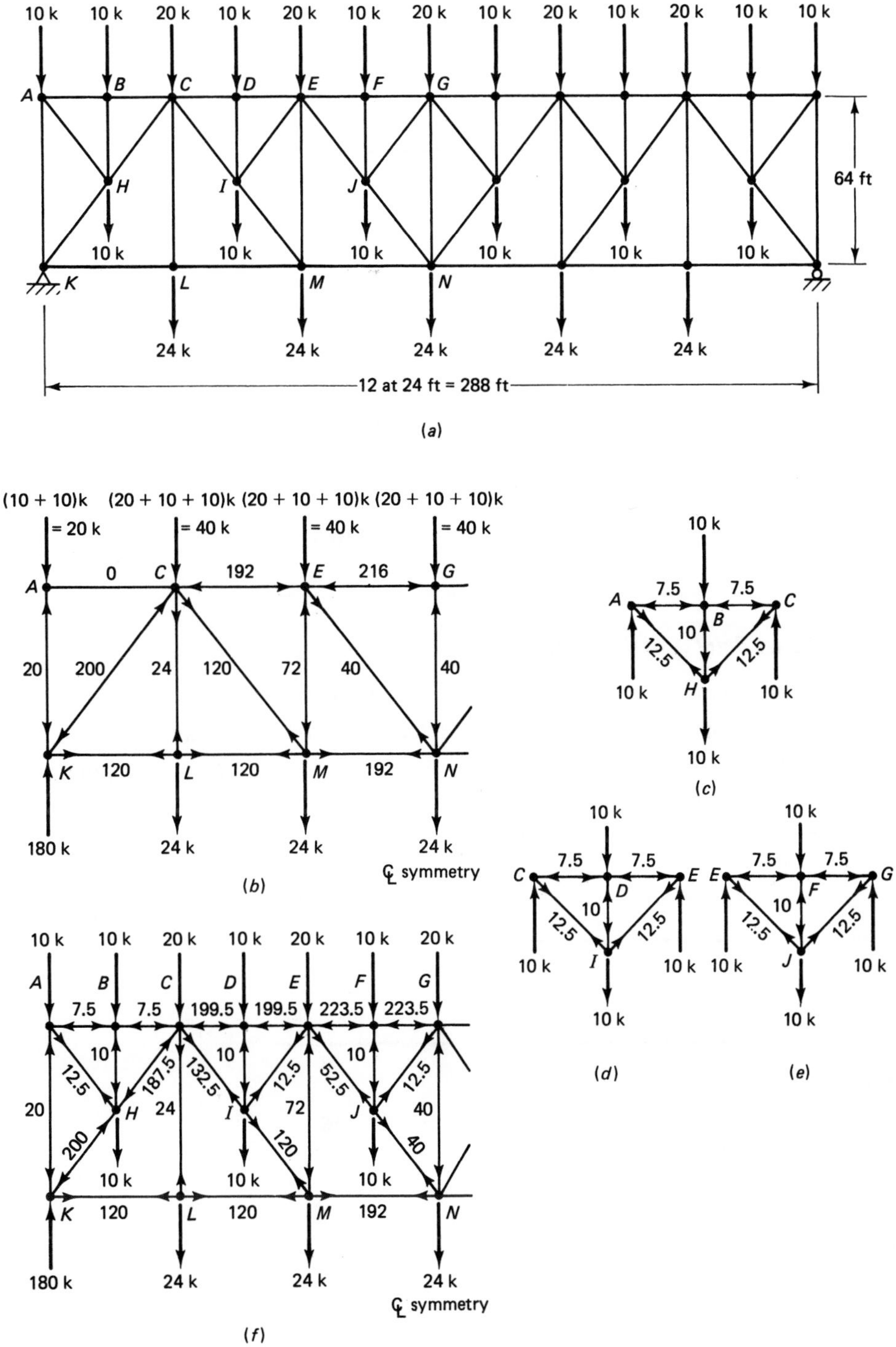

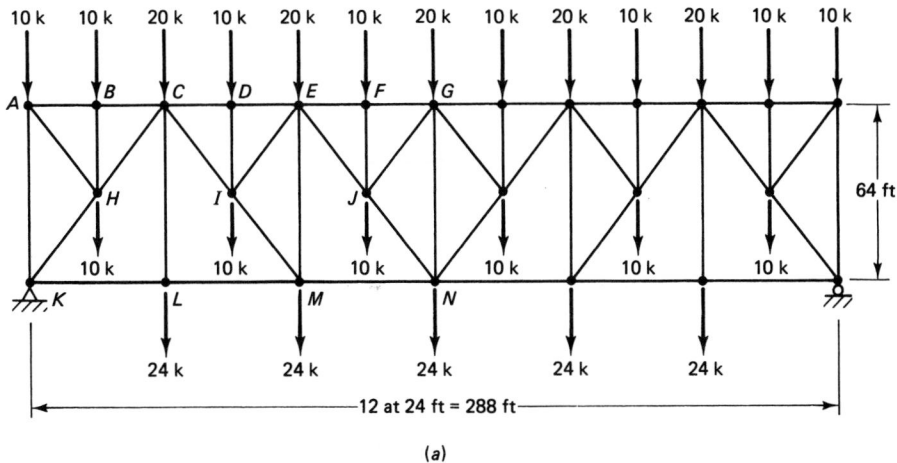

Figure 4.9 Structure for Solved Problem 4.1.

TABLE 4.1a FORCES IN THE MEMBERS OF
SECONDARY TRUSSES

Member	Force (k)	
	+ Tension	− Compression
AB		−7.5
BC		−7.5
AH	+12.5	
BH		−10.0
CH	+12.5	
CD		−7.5
DE		−7.5
CI	+12.5	
DI		−10.0
EI	+12.5	
EF		−7.5
FG		−7.5
EJ	+12.5	
FJ		−10.0
GJ	+12.5	

At Panel Point A of the Primary Truss

Load acting on the compound truss (Fig. 4.9a) at A = 10 k↓

Reaction at A on the secondary truss $ABCH$ (Fig. 4.9c) = 10 k↑

Therefore, load transferred from the secondary truss $ABCH$ to panel point A of primary truss (Fig. 4.9b) = 10 k↓

Total load acting at panel point A of the primary truss = 10 + 10 = 20 k↓

TABLE 4.1b FORCES IN THE MEMBERS OF THE
PRIMARY TRUSS

Member	Force (k)	
	+ Tension	− Compression
AC		0
CE		−192
EG		−216
KL	+120	
LM	+120	
MN	+192	
AK		−20
CL	+24	
EM		−72
GN		−40
CK		−200
CM	+120	
EN	+40	

At Panel Point C of the Primary Truss

Load acting on the compound truss (Fig. 4.9a) at C = 20 k↓

Reaction at C on the secondary truss *ABCH* (Fig. 4.9c) = 10 k↑

Reaction at C on the secondary truss *CDEI* (Fig. 4.9d) = 10 k↑

Therefore, loads transferred from the secondary trusses *ABCH* and *CDEI* to panel point C of primary truss (Fig. 4.9b) = 10 + 10 = 20 k↓

Total load acting at panel point C of the primary truss = 20 + 20 = 40 k↓

After all the panel-point loads are determined, the primary truss is analyzed by the method of joints, and the results are given in Table 4.1b and Fig. 4.9b.

The forces in the members of the given compound truss are obtained by algebraically adding the forces in the corresponding members of the primary and secondary trusses as shown in Table 4.2. The final results are given in Fig. 4.9f. ■

TABLE 4.2   FORCES IN THE MEMBERS OF THE COMPOUND TRUSS

Member (1)	Force as a primary truss member (k) (from Table 4.1b) (2)	Force as a secondary truss member (k) (from Table 4.1a) (3)	Resultant force (k) col. (2) + (3) (4)
AB	0	−7.5	−7.5
BC	0	−7.5	−7.5
CD	−192	−7.5	−199.5
DE	−192	−7.5	−199.5
EF	−216	−7.5	−223.5
FG	−216	−7.5	−223.5
AH	—	+12.5	+12.5
BH	—	−10.0	−10.0
CH	−200	+12.5	−187.5
CI	+120	+12.5	+132.5
DI	—	−10.0	−10.0
EI	—	+12.5	+12.5
EJ	+40	+12.5	+52.5
FJ	—	−10.0	−10.0
GJ	—	+12.5	+12.5
HK	−200	—	−200
IM	+120	—	+120
JN	+40	—	+40
AK	−20	—	−20
CL	+24	—	+24
EM	−72	—	−72
GN	−40	—	−40
KL	+120	—	+120
LM	+120	—	+120
MN	+192	—	+192

PROBLEMS

4.1 to 4.36 For trusses shown in Figs. P4.1 to P4.36 determine the forces in all members using the method of joints.

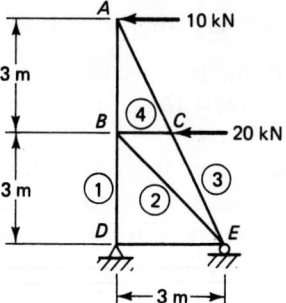

Figure P4.1

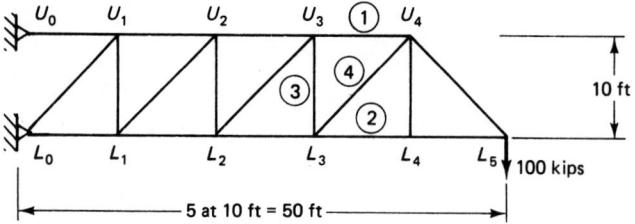

Figure P4.2

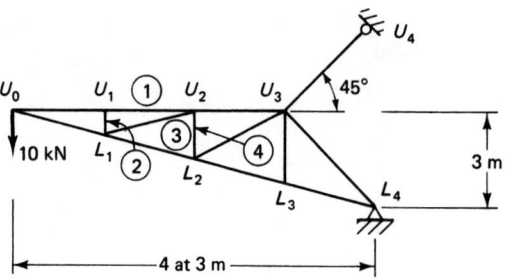

Figure P4.3

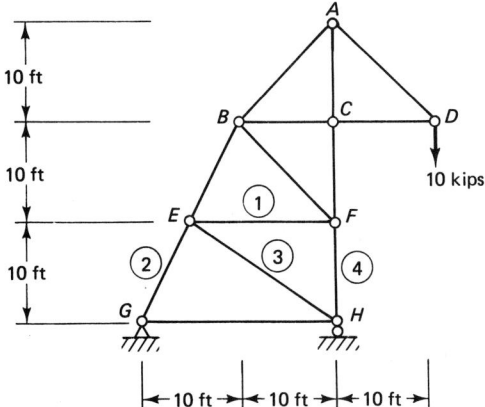

Figure P4.4

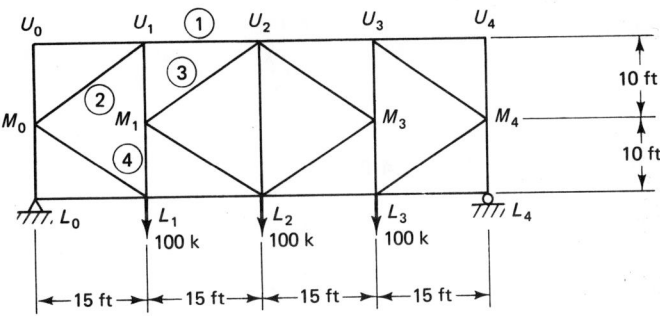

Figure P4.5

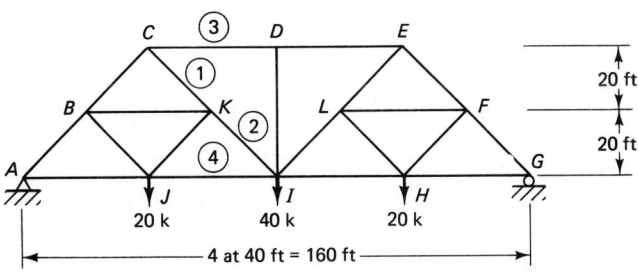

Figure P4.6

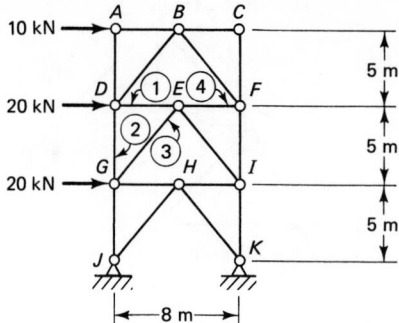

Figure P4.7

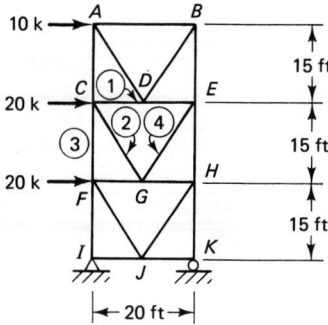

Figure P4.8

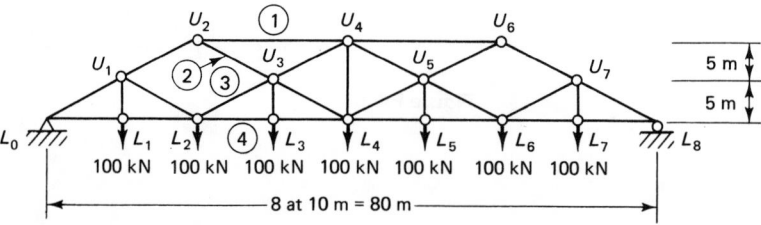

Figure P4.9

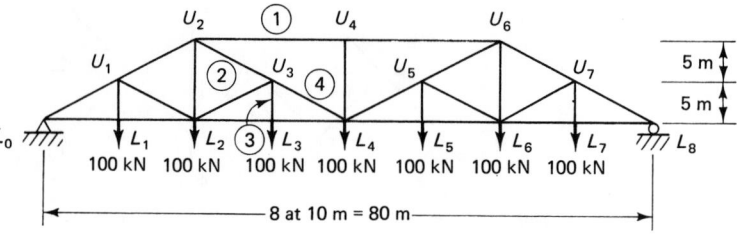

Figure P4.10

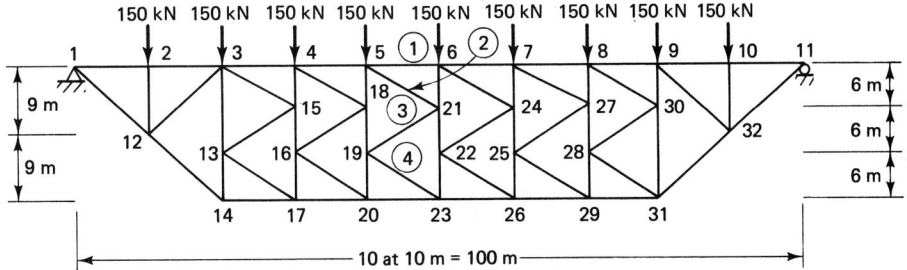

Figure P4.11

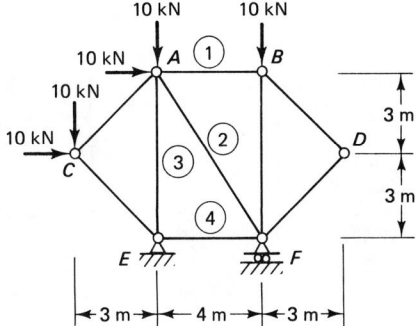

Figure P4.12

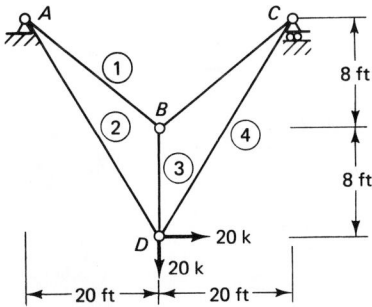

Figure P4.13

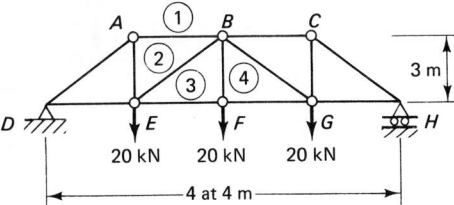

Figure P4.14

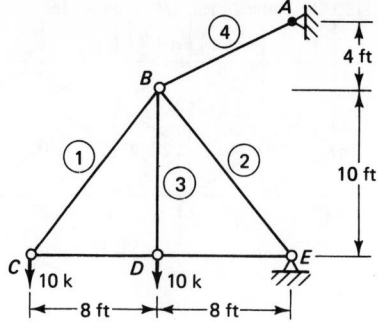

Figure P4.15

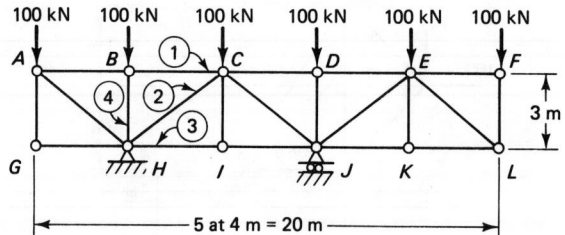

Figure P4.16

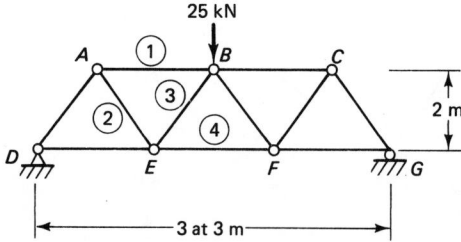

Figure P4.17

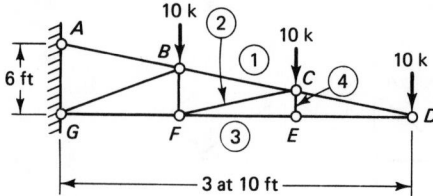

Figure P4.18

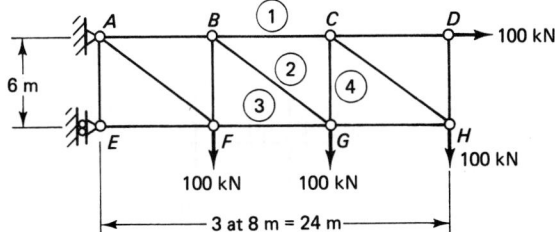

Figure P4.19

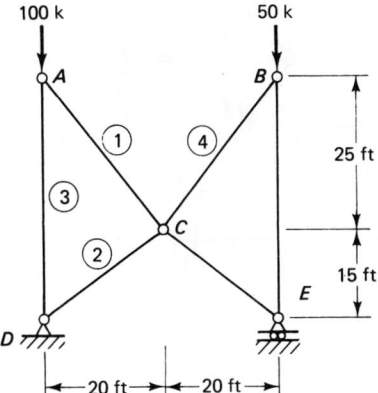

Figure P4.20

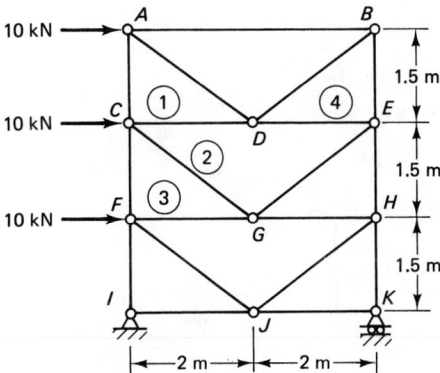

Figure P4.21

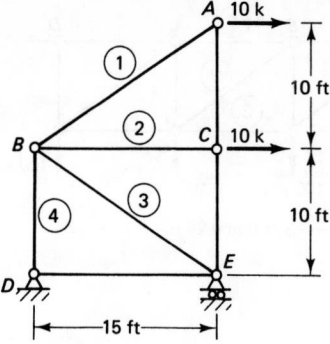

Figure P4.22

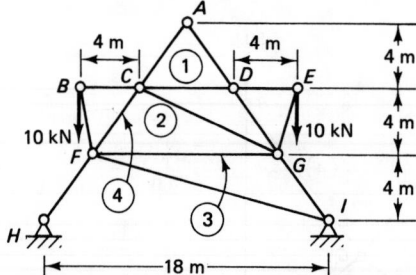

Figure P4.23

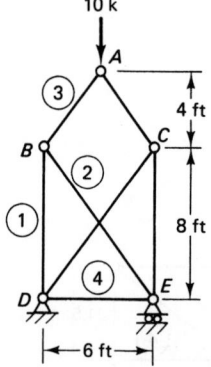

Figure P4.24

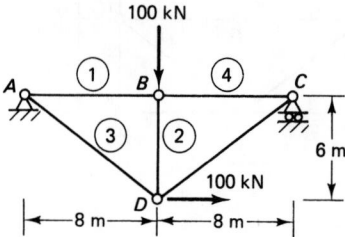

Figure P4.25

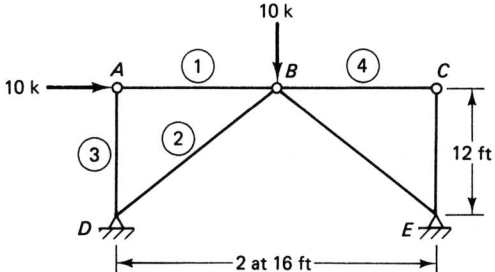

Figure P4.26

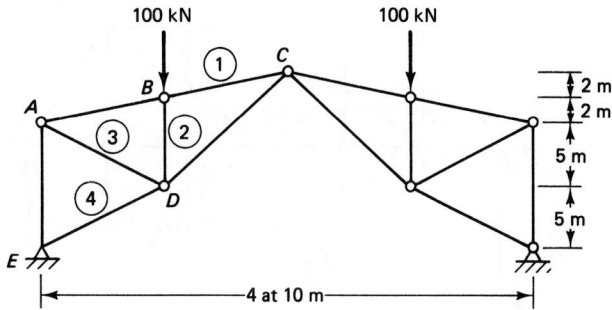

Figure P4.27

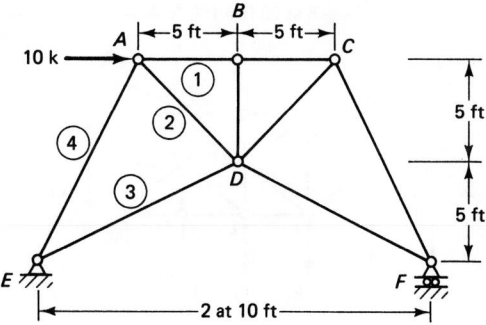

Figure P4.28

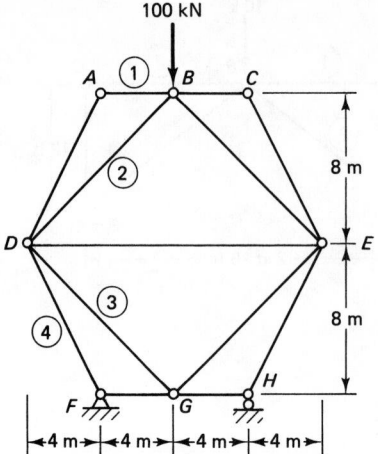

Figure P4.29

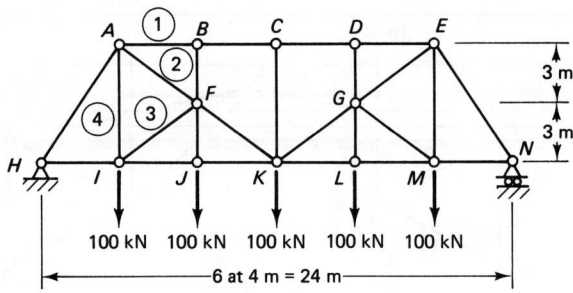

Figure P4.30

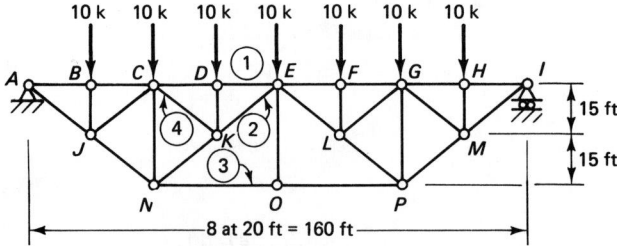

Figure P4.31

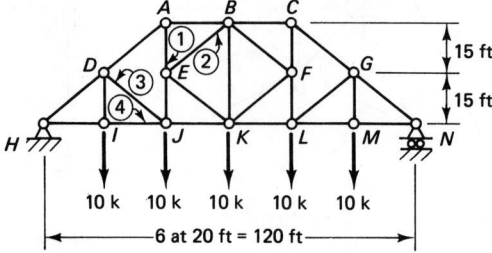

Figure P4.32

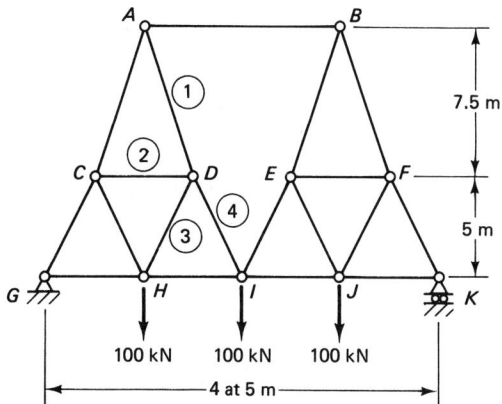

Figure P4.33

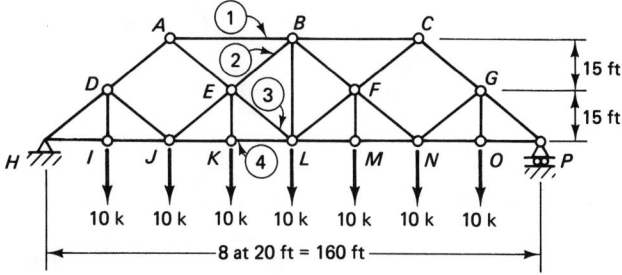

Figure P4.34

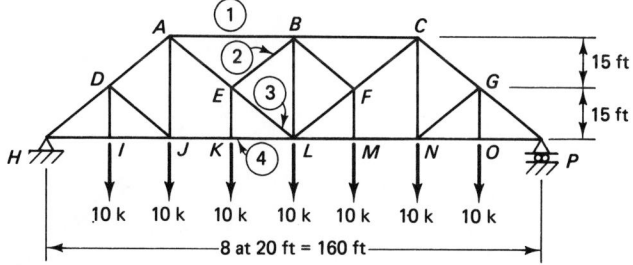

Figure P4.35

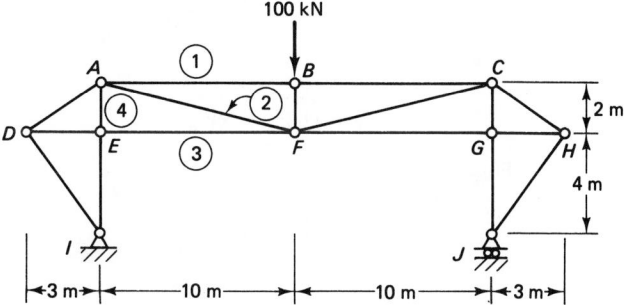

Figure P4.36

4.37 to 4.72 Using the method of sections, determine the forces in members ① to ④ of trusses shown in Figs. P4.1 to P4.36. (If necessary, use a combination of the method of joints and method of sections.)

4.73 Using the method of sections, determine the forces in members ① and ② of the complex truss shown. (*Hint:* Take sections *xx* and *yy* as shown; take the moment of the forces about the point of intersection of the other two members.)

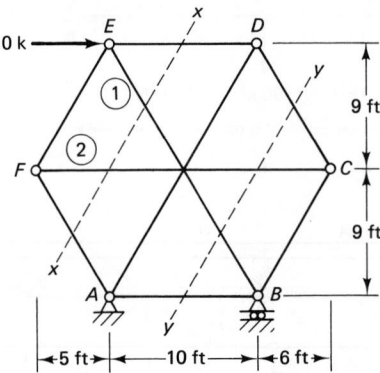

4.74 to 4.79 For the space trusses shown in Figs. P4.74 to P4.79, determine the forces in all members using the tension-coefficient method.

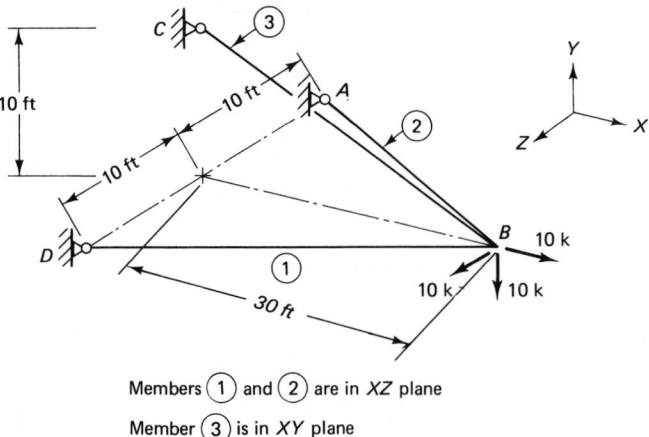

Members ① and ② are in *XZ* plane

Member ③ is in *XY* plane

Figure P4.74

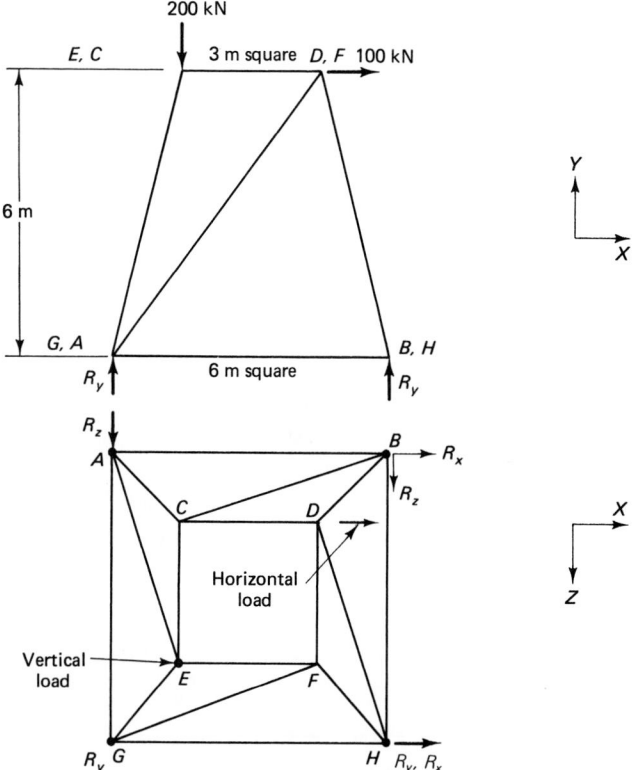

200 kN

E, C   3 m square D, F 100 kN

6 m

G, A
R_y 6 m square   B, H
                             R_y

R_z
A   B → R_x
                             ↓R_z
   C   D
          Horizontal
          load

Vertical
load   E   F

R_y G   H R_y, R_x

Figure P4.75

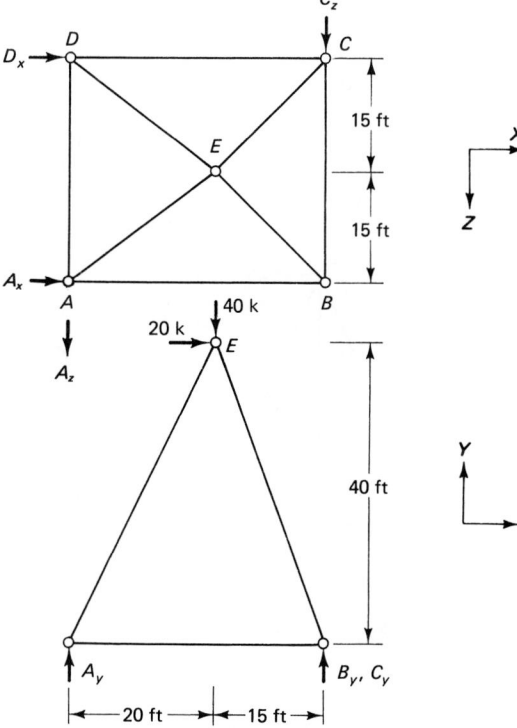

C_z

D_x → D   C

15 ft

       E

15 ft

A_x →
      A   B

          ↓40 k
20 k → E

A
↓
A_z

40 ft

↑A_y ↑B_y, C_y

|← 20 ft →|← 15 ft →|

Figure P4.76   125

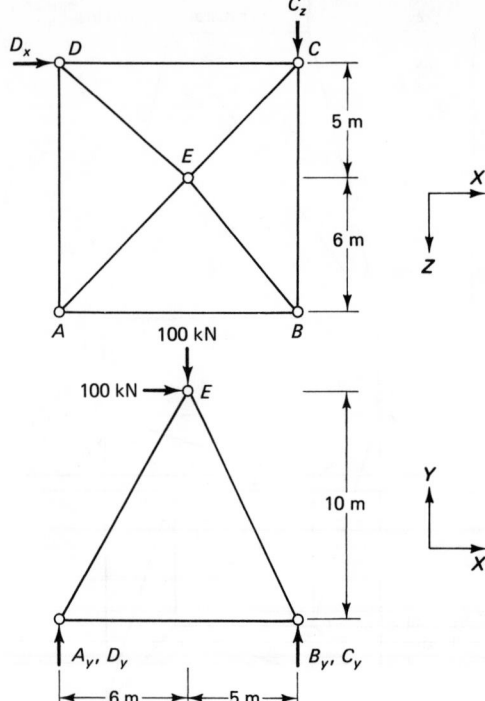

Figure P4.77

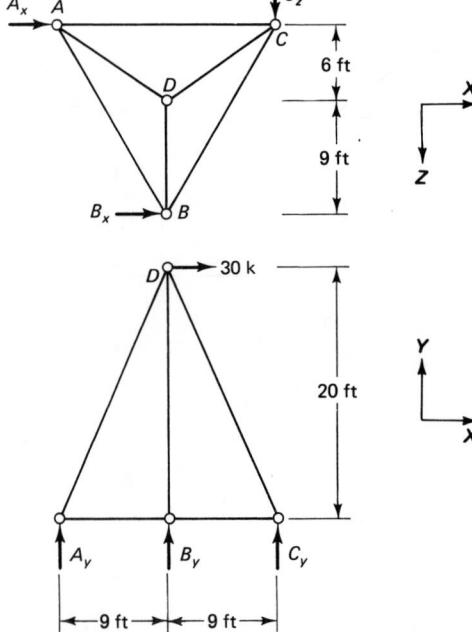

Figure P4.78

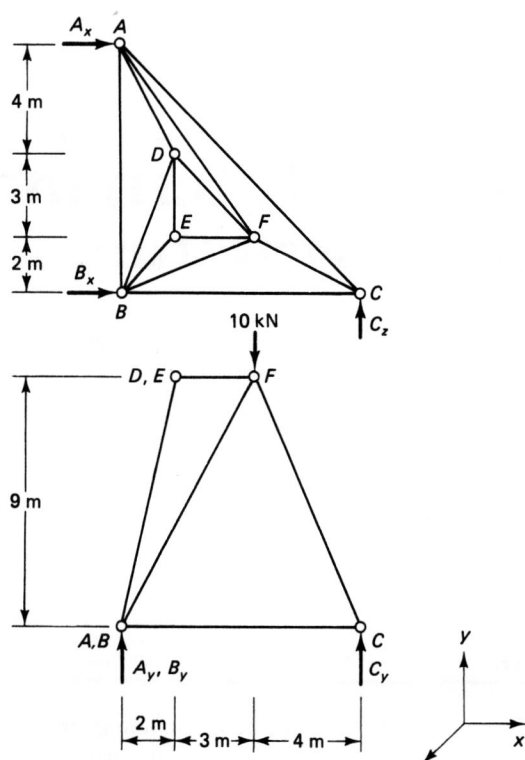

Figure P4.79

Chapter 5

Statically Determinate Flexural Structures

Bridges of half a mile span, for common or railway travel, may be built, using iron for the cables, with entire safety. But by substituting the best quality of steel wire, we may nearly double the span and afford the same degree of security.

John A. Roebling, 1855

5.1 INTRODUCTION

A flexural structure is defined as one consisting of members that can carry external loads through their internal resistance against shears and moments (bending and twisting). A pin-jointed structure, on the other hand, resists only axial forces. An intermediate type of structure is one whose members resist shears and moments as well as axial forces.

The first step in the analysis of all structural systems is the calculation of the support forces *and moments* that are *external* to the structural system. The next step in the analysis is to determine the *internal* forces *and moments* at desired locations of the structure; the maximum values of such internal forces and moments are required by the engineer in order to design the structural member of suitable cross section and material; furthermore, the *variation of moment* in a flexural structure *is also required* since it influences its behavior, for example, deflection of the structure.

If a loaded structure is in equilibrium, every portion of that structure is also in equilibrium. To determine the forces and moments acting across any section of a beam subjected to some random loads (Fig. 5.1a), let us imagine that the beam is severed at that section. Considering either part of the beam as a free body, the internal force R and moment M *acting on the section to maintain equilibrium* (Fig. 5.1b) can be found by applying the conditions of equilibrium ($\Sigma F = 0$, $\Sigma M = 0$). For a general type of loading the vectors R and M each will have three components, that is, $R = (F_x^2 + F_y^2 + F_z^2)^{1/2}$ and $M = (M_1^2 + M_2^2 + M_3^2)^{1/2}$. If we are dealing with a coplanar structure and loadings, only internal resisting forces F_x, F_y, and internal resisting moment M_3 can develop. This is demonstrated in Fig. 5.2a, where the beam AB is subjected to coplanar loading; the internal resisting forces and moment at a

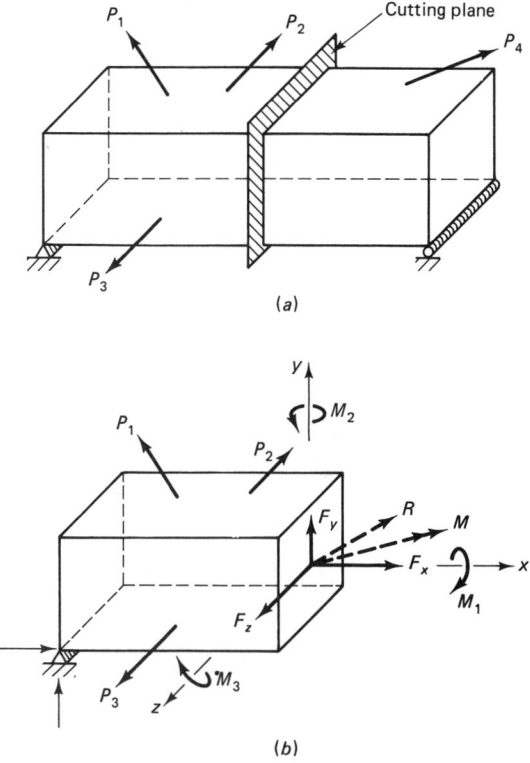

Figure 5.1 (*a*) Flexural structure acted upon by forces. (*b*) Free-body diagram of a portion of structure in *a*.

cross section[1] C are shown in Fig. 5.2*b* for the two free-body diagrams; note that V, N, and M are oppositely directed in these diagrams in accordance with Newton's third law.

Shear is defined as an internal reaction force acting in the plane *and through the centroid* of the section; it is a bound vector since it has magnitude, sense, and location; *it has units of a force. Bending moment* is defined as an internal reaction couple acting about a cross-sectional centroidal axis; it acts in a plane normal to the section and has units of force and distance; *it is a bound vector. Axial force* is defined as an internal reaction force acting *at the centroid of the section and* along an axis normal to the section; it is also a bound vector. The sign conventions for shear, moment, and axial force, *as applied to beams,* are shown in Fig. 5.3; the sign conventions are further illustrated in Fig. 5.4. Quite often, we have to deal also with a *twisting moment* that is defined as an internal reaction couple acting about a centroidal axis normal to the plane of the section; for example, M_1 in Fig. 5.1*b* is a twisting moment.

A graph showing a plot of shear force versus the distance x along a slender structural

---

[1] The section need not be a right one, but it is preferred for analytical convenience.

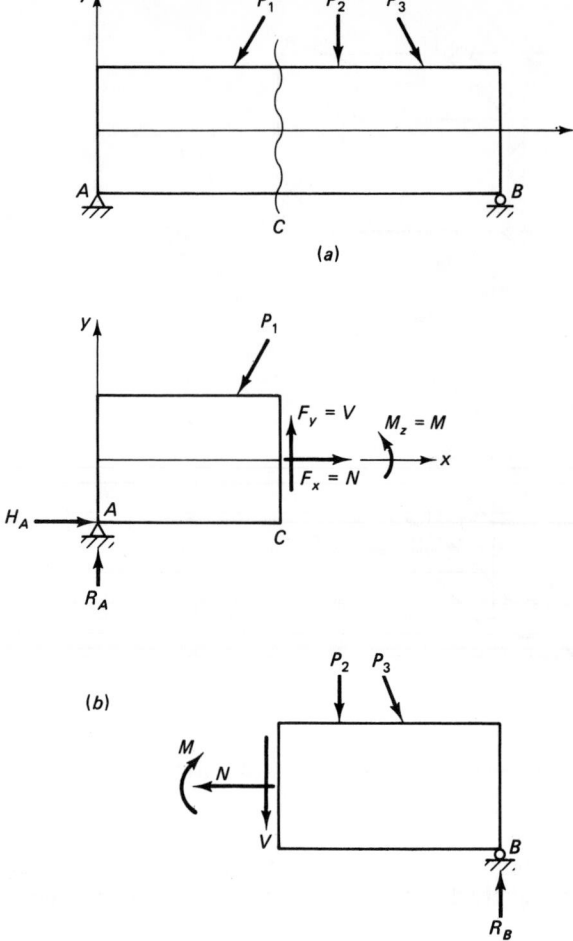

Figure 5.2 (*a*) Beam acted upon by loads. (*b*) Free-body diagram showing internal forces on the cut section.

member, for example, a beam, is called a *shear-force diagram* (SFD); similarly, a graph showing the variation of bending moment along the length of a member is called a *bending-moment diagram* (BMD). The same definition applies to axial-force diagram (AFD) and twisting-moment diagram (TMD).

5.2 AXIAL-FORCE, SHEAR-FORCE, AND BENDING-MOMENT DIAGRAMS

The *internal* axial and shear forces as well as the bending moment are determined by the following procedure:

1. The actual structure *and loading* are idealized by a model, supported by un-known reaction components.

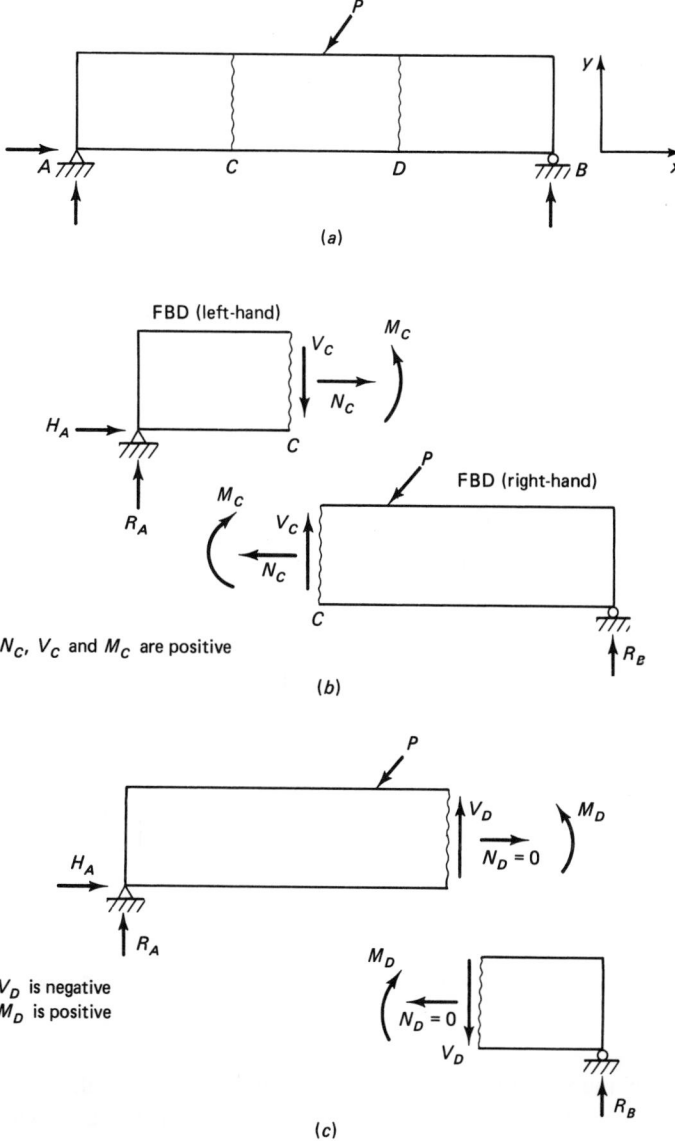

Figure 5.3 (*a*) Beam acted upon by loads. (*b*) and (*c*) Free-body diagrams showing sign convention for shear force and bending moment and axial force.

2. With all the external loads (including the unknown reaction components) acting on the model structure (*free-body diagram*) the equations of equilibrium ($\Sigma\ F = 0$ and $\Sigma\ M = 0$) are applied to determine the unknown reaction components.

3. *Cut the member* at the location where the internal forces are required and isolate one of the segments on either side of the cut section. Usually a cut is required whenever the load and/or geometry change along the structure.

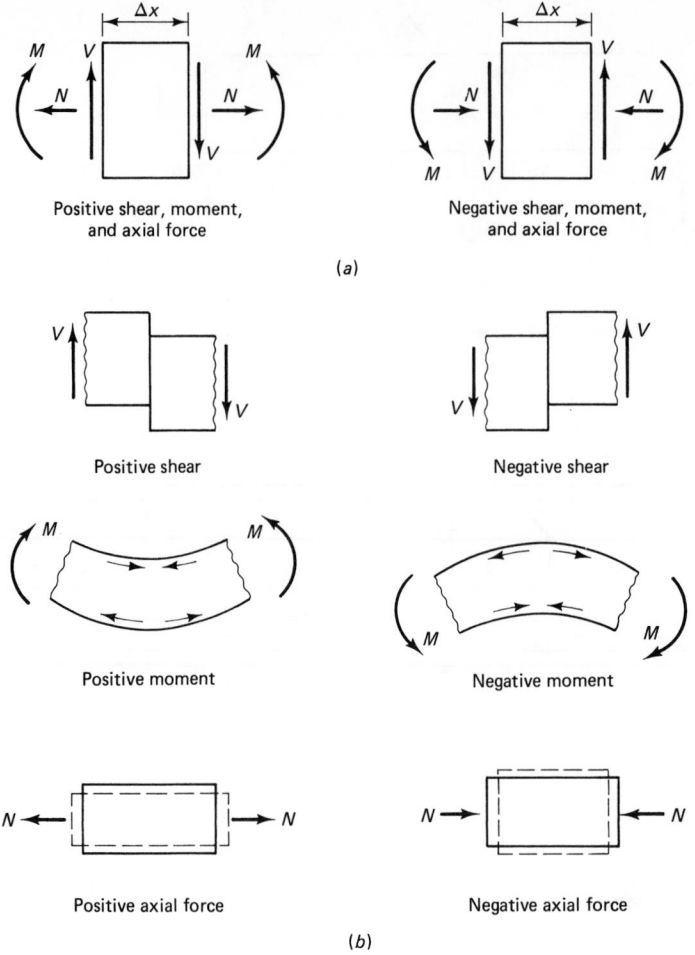

Figure 5.4 (*a*) Elements of a beam. (*b*) Physical analogy for shear force, bending moment, and axial force.

4. Draw the free-body diagram of the segment showing all the applied loads, the external reactions (now known from step 2), and the unknown positive internal forces V, N and bending moment M (Fig. 5.3*b*) at the cut section.
5. Apply the equations of equilibrium ($\Sigma F = 0$, $\Sigma M = 0$) to calculate V, N, and M in terms of the applied loads and the known external reactions.

Plotting V, N, and M versus the axial length of the slender member yields the shear-force, axial-force, and bending-moment diagrams. The above procedure is demonstrated in applying it to several already idealized structures; that is, we will start with step 2. To provide a physical interpretation of the structural behavior and a link to the qualitative deflected shapes, a curl (∪ for concave upward and ∩ for concave downward) is shown on the bending-moment diagrams.

■ EXAMPLE 5.1

Draw the axial-force, shear-force, and bending-moment diagrams for the beam shown loaded in Fig. 5.5a.

Solution

Step ①. Find reactions at A and B (Fig. 5.5b). Use equilibrium equations

$$\Sigma F_x = 0 \quad \Sigma F_y = 0 \quad \Sigma M = 0$$

$\Sigma F_x = 0$: $H_A - 60 = 0 \quad H_A = 60 \text{ kN} \rightarrow$

ΣM about $A = 0$: $(R_B)(8) - (20)(4 + 2)(12 - 3) - (80)(2) = 0$

$8R_B = 1080 + 160 = 1240$

$R_B = 155 \text{ kN} \uparrow$

$\Sigma F_y = 0$: $R_A - 80 + 155 - (20)(4 + 2) = 0$

$R_A = 80 - 155 + 120 = 45 \text{ kN} \uparrow$

Check.

ΣM about $B = 0$: $-(R_A)(8) + (80)(6) - (20)(4 + 2)(1) = 0$

$-(45)(8) + (80)(6) - (20)(6) = 0$

$-360 + 480 - 120 = 0 \quad$ (checks)

Step ②. Since the loading conditions change along four portions of the beam, we require four cuts, made at ①, ②, ③, and ④ as shown in Fig. 5.5b, and the internal axial force, shear force, and bending moment, calculated at these cuts. For cut ①: From the FBD of Fig. 5.5c: $x \leqslant 2$ m

$\Sigma F_x = 0$: $60 + N = 0 \quad N = -60 \text{ kN} = 60 \text{ kN} \quad$ compression

$\Sigma F_y = 0 \quad 45 - V = 0 \quad V = 45 \text{ kN} \downarrow$

ΣM about ① $= 0$: $-(45)(x) + M = 0 \quad M = 45x \quad$ kN·m

For cut ②: From the FBD of Fig. 5.5d: 2 m $\leqslant x \leqslant$ 6 m

$\Sigma F_x = 0$: $60 - 60 + N = 0 \quad N = 0$

$\Sigma F_y = 0$: $45 - 80 - V = 0 \quad V = -35 \text{ kN} = 35 \text{ kN} \uparrow$

ΣM about ② $= 0$: $-(45)(x) + (80)(x - 2) + M = 0$
$M = 160 - 35x \quad$ kN·m

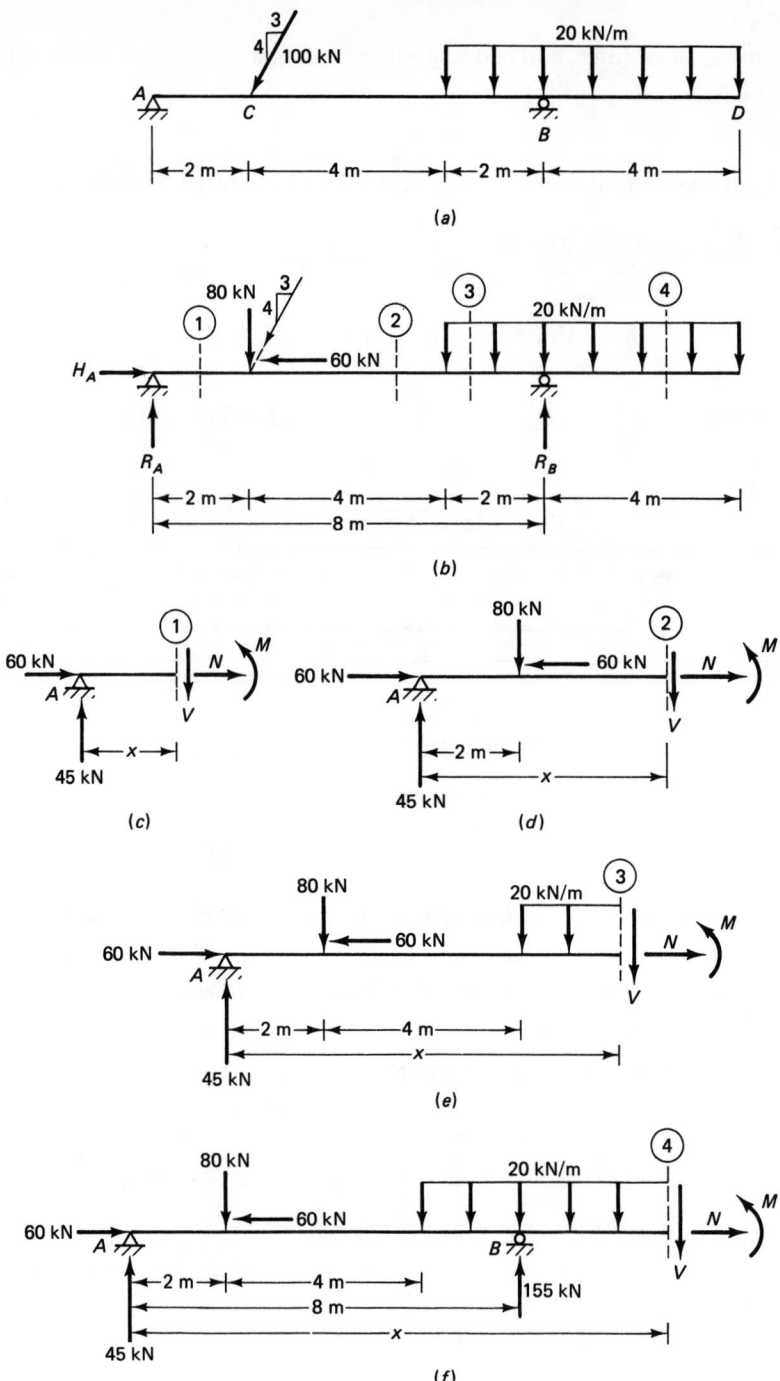

Figure 5.5 Structure for Example 5.1.

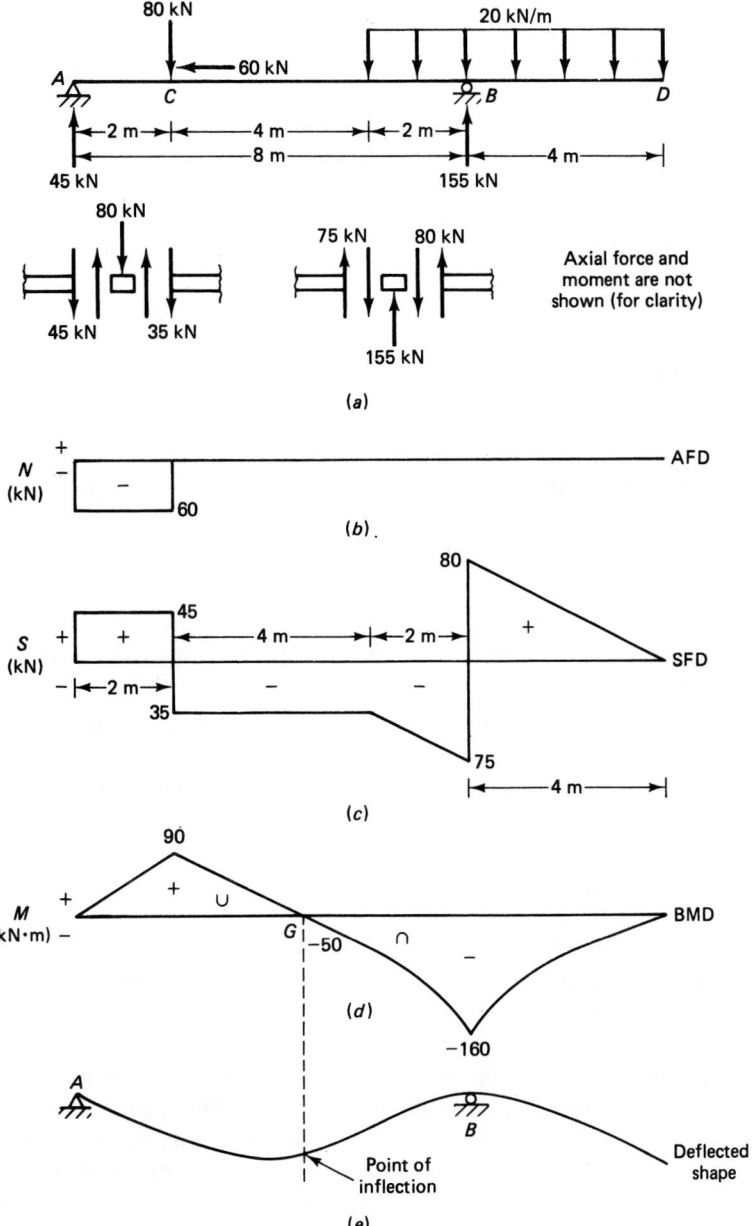

Figure 5.6 AFD, SFD, BMD, and deflected shape for structure in Example 5.1.

For cut ③: From the FBD in Fig. 5.5*e*: 6 m ⩽ *x* ⩽ 8 m

$$\Sigma F_x = 0: \qquad 60 - 60 + N = 0 \qquad N = 0$$

$$\Sigma F_y = 0: \qquad 45 - 80 - (20)(x - 6) - V = 0$$

$$V = 85 - 20x \qquad \text{kN}$$

$$\Sigma M \text{ about } ③ = 0: \quad -(45)(x) + (80)(x - 2) + (20)(x - 6)\left(\frac{x - 6}{2}\right) + M = 0$$

$$M = -200 + 85x - 10x^2 \qquad \text{kN} \cdot \text{m}$$

For cut ④: From the FBD in Fig. 5.5*f*: 8 m ⩽ *x* ⩽ 12 m

$$\Sigma F_x = 0: \qquad 60 - 60 + N = 0 \qquad N = 0$$

$$\Sigma F_y = 0: \qquad 45 - 80 - 20(x - 6) + 155 - V = 0$$

$$V = 240 - 20x \qquad \text{kN}$$

$$\Sigma M \text{ about } ④ = 0: \qquad -(45)(x) + (80)(x - 2)$$

$$+ (20)(x - 6)\left(\frac{x - 6}{2}\right) - (155)(x - 8) + M = 0$$

$$M = -1440 + 240x - 10x^2 \qquad \text{kN} \cdot \text{m} \quad \blacksquare$$

Having determined the internal axial (N), shear (V), and bending moment (M) at the four imaginary cuts, in terms of the distance x from the origin at end A, plots of N, V, and M versus x for the intervals will yield the AFD, SFD, and BMD shown in Fig. 5.6*b*, *c*, and *d*, respectively. It will be shown later that the deflected shape of the loaded beam is related to the beam curvature, which in turn is related to the bending moment. Thus, in Fig. 5.6*d* the moment at point G is zero, which corresponds to a point of inflection in the deflected shape of the beam (Fig. 5.6*e*). A point of inflection is one where the slope of the deflected shape changes sign and therefore the curvature is zero (or the radius of curvature = 1/curvature = ∞).

It is instructive to note that in the neighborhood of a concentrated load such as the 80 kN at point C in Fig. 5.6*a*, the *internal resisting* shears to the left and right of the load must add up to 80 kN; furthermore, the maximum *internal resisting* shear in the neighborhood of the 80-kN load is 45 kN. Similar argument applies to the reaction at B; the two internal shears add up to the value of the vertical reaction at B, that is, 75 + 80 = 155 kN; however, the maximum internal shear in the vicinity of the reaction at B is 80 kN just to the right of B.

■ **EXAMPLE 5.2**

Draw the axial-force, shear-force, and bending-moment diagrams for the frame shown in Fig. 5.7*a*. Note that a concentrated couple of 90 kN · m acts at D.

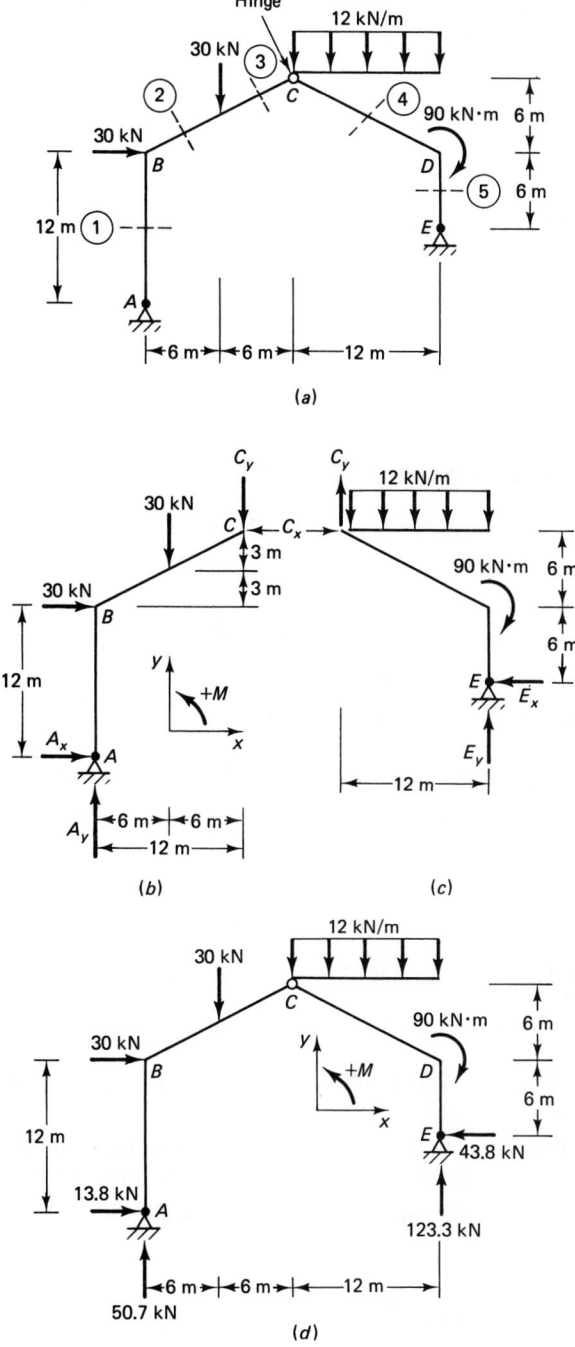

Figure 5.7 Structure for Example 5.2.

Solution

Step 1. Find the reactions at A and E. From the FBD in Fig. 5.7b, we find by applying the equilibrium equations:

$\sum F_x = 0$: $A_x + 30 - C_x = 0$; $A_x - C_x = -30$ (a)

$\sum F_y = 0$: $A_y - 30 - C_y = 0$; $A_y - C_y = 30$ (b)

$\sum M$ about $A = 0$: $(C_x)(18) - (C_y)(12) - (30)(6) - (30)(12) = 0$

or

$$18C_x - 12C_y = 540;   3C_x - 2C_y = 90$$ (c)

From the FBD in Fig. 5.7c:

$\sum F_x = 0$: $-E_x + C_x = 0$ (d)

$\sum F_y = 0$: $E_y + C_y - (12)(12) = 0$; $E_y + C_y = 144$ (e)

$\sum M$ about $E = 0$: $-(C_x)(12) - (C_y)(12) + (12)(12)(\frac{12}{2}) - 90 = 0$

or

$$-12C_x - 12C_y + 774 = 0;   2C_x + 2C_y = 129$$ (f)

Solving the six simultaneous equations (a) to (f) yields

$A_x = 13.8$ kN; $A_y = 50.7$ kN; $C_x = 43.8$ kN;

$C_y = 20.7$ kN; $E_x = 43.8$ kN; $E_y = 123.3$ kN

It is good practice to check the solution for the reactions; thus using Fig. 5.7d:

$\sum F_x = 0$: $13.8 + 30 - 43.8 = 0$ (checks)

$\sum F_y = 0$: $50.7 - 30 - (12)(12) + 123.3 = 0$ (checks)

$\sum M$ about $E = 0$: $-(50.7)(24) + (13.8)(12 - 6) - (30)(6)$

$$+ (30)(18) + (12)(12)(\tfrac{12}{2}) - 90 = 0$$

$$-1216.8 + 82.8 - 180 + 540 + 864 - 90 = 0$$ (checks)

Step 2. *Changes* in the loading conditions and geometry require five cuts as shown in Fig. 5.7a. In this example, a new origin is chosen for each interval; this would simplify the arithmetic as well as the equations for the internal forces and moment.

For Cut ①: Using the FBD of Fig. 5.8a:

$\sum F_x = 0$: $13.8 + V = 0$; $V = -13.8$ kN

$\sum F_y = 0$: $50.7 + N = 0$; $N = -50.7$ kN

$\sum M$ about ① $= 0$: $(13.8)(x) + M = 0$; $M = -13.8x$ kN·m

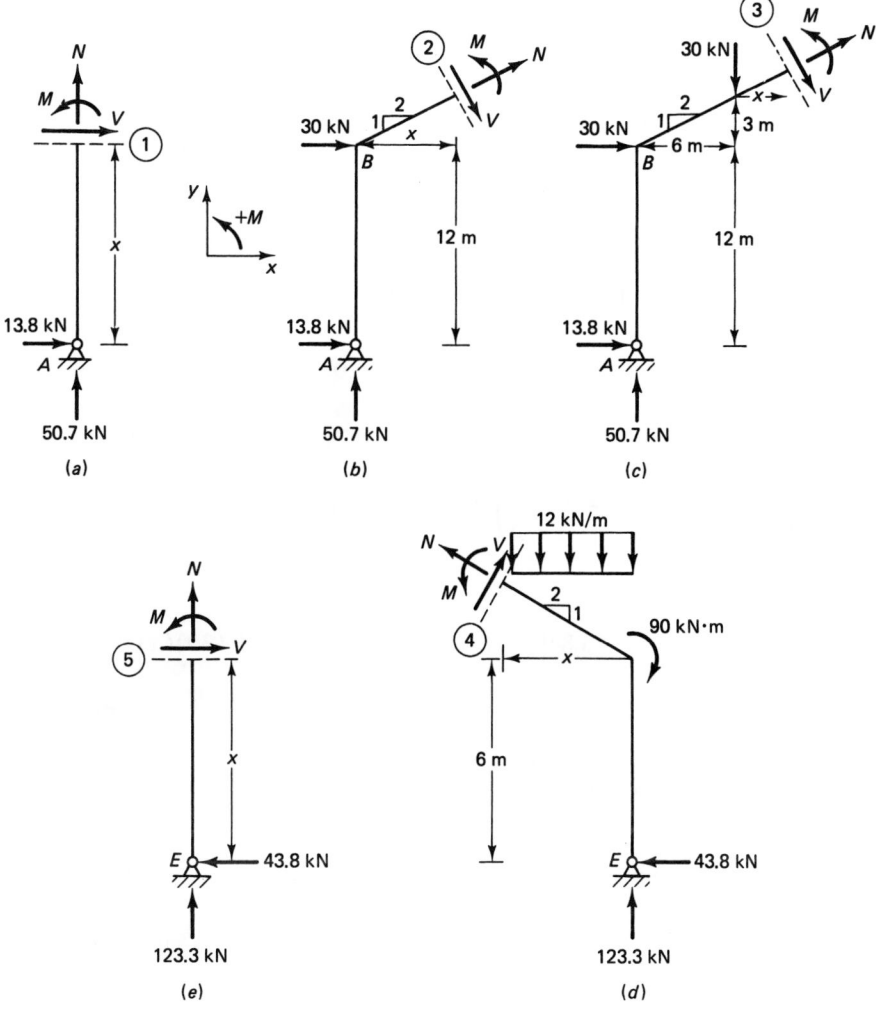

Figure 5.8 Free-body diagrams for structure in Example 5.2.

For Cut ②: From the FBD in Fig. 5.8*b*:

$$\sum F_x = 0: \qquad 13.8 + 30 + V\left(\frac{1}{\sqrt{5}}\right) + N\left(\frac{2}{\sqrt{5}}\right) = 0;$$

$$\text{or} \quad V + 2N = -(\sqrt{5})(43.8) \qquad \text{(a)}$$

$$\sum F_y = 0: \qquad 50.7 - V\left(\frac{2}{\sqrt{5}}\right) + N\left(\frac{1}{\sqrt{5}}\right) = 0;$$

$$\text{or} \quad 2V - N = (\sqrt{5})(50.7) \qquad \text{(b)}$$

$$\Sigma\,M \text{ about } ② = 0: \qquad (13.8)\left(12 + \frac{x}{2}\right)$$

$$+ M + (30)\left(\frac{x}{2}\right) - (50.7)(x) = 0; \qquad (c)$$

$$\text{or} \quad M = -165.6 + 28.8x \qquad \text{kN}\cdot\text{m}$$

From equations (a) and (b), we have

$$5V = (\sqrt{5})(101.4 - 43.8); \quad \text{or} \quad V = 25.8 \text{ kN} \qquad N = -61.8 \text{ kN}$$

For Cut ③: Using the FBD in Fig. 5.8*c*:

$$\Sigma\,F_x = 0: \qquad 13.8 + 30 + V\left(\frac{1}{\sqrt{5}}\right) + N\left(\frac{2}{\sqrt{5}}\right) = 0;$$

$$\text{or} \quad V + 2N = -(\sqrt{5})(43.8) \qquad (d)$$

$$\Sigma\,F_y = 0: \qquad 50.7 - 30 - V\left(\frac{2}{\sqrt{5}}\right) + N\left(\frac{1}{\sqrt{5}}\right) = 0;$$

$$\text{or} \quad 2V - N = +(\sqrt{5})(20.7) \qquad (e)$$

$$\Sigma\,M \text{ about } ③ = 0: \qquad (13.8)\left(15 + \frac{x}{2}\right) + (30)\left(3 + \frac{x}{2}\right) + (30)(x) + M$$

$$- (50.7)(6 + x) = 0 \quad \text{or} \quad M = (7.2 - 1.2x) \text{ kN}\cdot\text{m} \qquad (f)$$

From equations (d) and (e), $V = -1.1$ kN and $N = -48.4$ kN.
For Cut ④: From the FBD in Fig. 5.8*d*:

$$\Sigma\,F_x = 0: \qquad -43.8 + V\left(\frac{1}{\sqrt{5}}\right) - N\left(\frac{2}{\sqrt{5}}\right) = 0;$$

$$\text{or} \quad V - 2N = (\sqrt{5})(43.8) \qquad (g)$$

$$\Sigma\,F_y = 0: \qquad 123.3 - (12)(x) + V\left(\frac{2}{\sqrt{5}}\right) + N\left(\frac{1}{\sqrt{5}}\right) = 0;$$

$$\text{or} \quad 2V + N = (\sqrt{5})(-123.3 + 12x) \qquad (h)$$

$$\Sigma\,M \text{ about } ④ = 0: \qquad (123.3)(x) - (43.8)\left(6 + \frac{x}{2}\right) - 90 - (12)(x)\left(\frac{x}{2}\right)$$

$$+ M = 0; \quad \text{or} \quad M = (6x^2 - 101.4x + 352.8) \text{ kN}\cdot\text{m} \qquad (i)$$

From equations (g) and (h), $V = (-90.7 + 10.7x)$ kN, $N = (-94.3 + 5.4x)$ kN.
For Cut ⑤: Referring to the FBD of Fig. 5.8*e*:

$$\Sigma\,F_x = 0: \qquad -43.8 + V = 0; \quad \text{or} \quad V = 43.8 \text{ kN}$$

$$\Sigma\,F_y = 0: \qquad 123.3 + N = 0; \quad \text{or} \quad N = -123.3 \text{ kN}$$

$$\Sigma\,M \text{ about } ⑤ = 0: \qquad M - (43.8)(x) = 0; \quad \text{or} \quad M = 43.8x \qquad \text{kN}\cdot\text{m} \qquad \blacksquare$$

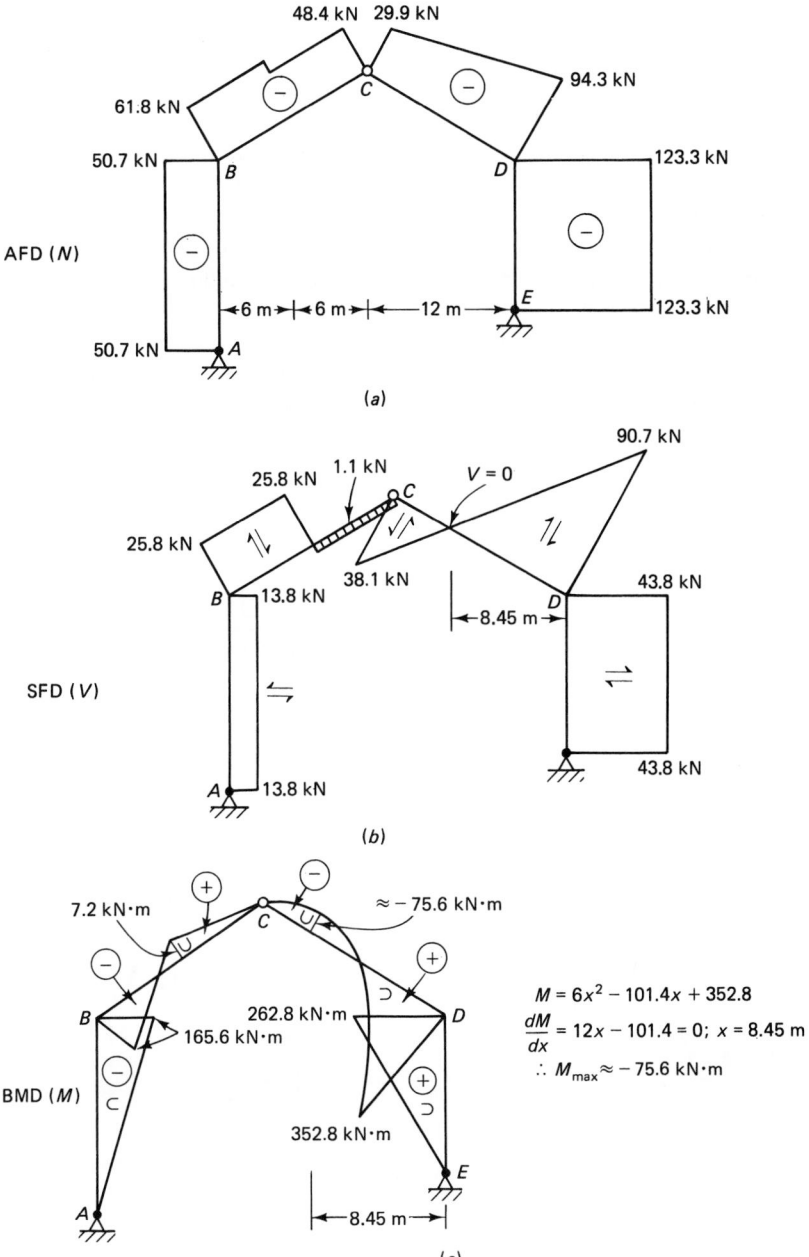

Figure 5.9 AFD, SFD, and BMD for structure in Example 5.2.

From the expressions for N, V, and M at each of the five cut sections, the axial-force, shear-force, and bending-moment diagrams are drawn as shown in Fig. 5.9. Examination of the SFD and BMD reveals that along CD where the shear is zero the bending moment has a maximum value; in fact, there are functional relationships between load, shear, and bending moment which are examined in the next section.

5.3 RELATIONSHIP BETWEEN LOAD, SHEAR, AND BENDING MOMENT

The determination of the shear and bending-moment diagrams can be much simplified if we use the functional relationships between load, shear, and bending moment. These relationships are derived as follows: Consider a beam with applied loads as shown in Fig. 5.10a. For a differential element of length Δx isolated from the beam by two parallel sections at x and $x + \Delta x$, and if all the forces and moments are shown acting on the element, then a FBD for the element will result as shown in Fig. 5.10b. Although the loading w is variable, it can be assumed as uniformly distributed over the length Δx since the variation of w with x [i.e., $(dw/dx)\Delta x$] leads to higher-order terms of Δx which are negligibly small compared with Δx; furthermore, an ideal weightless structure is assumed, with the dead loads being regarded as additional applied loads.

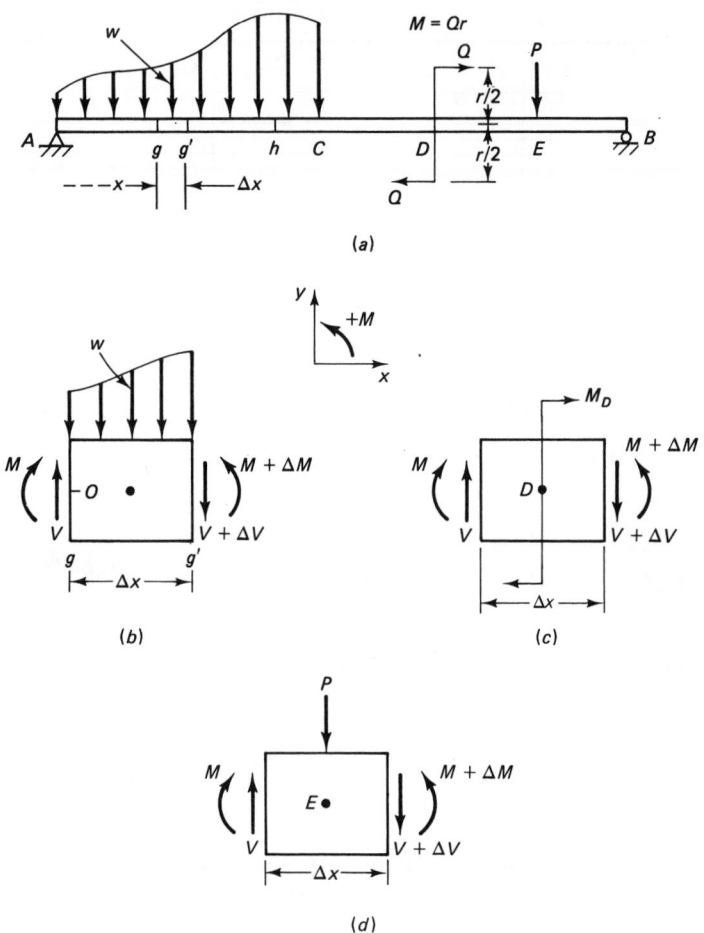

Figure 5.10 Relationships between load, shear, and bending moment.

Following the sign convention shown in Fig. 5.10 and applying the equilibrium condition that $\Sigma F_y = 0$ yield

$$V - (V + \Delta V) - w \, \Delta x = 0 \quad \text{or} \quad \Delta V = -w \, \Delta x$$

Dividing by Δx and having $\Delta x \to 0$, we find

$$\lim_{\Delta x \to 0} \frac{\Delta V}{\Delta x} = \frac{dV}{dx} = -w \quad \text{or} \quad \frac{dV}{dx} = -w \qquad (5.1)$$

Integrating Eq. (5.1) between g and h, we have

$$\int_{x_g}^{x_h} \frac{dV}{dx} \, dx = \int_{x_g}^{x_h} -w \, dx$$

or

$$V_h - V_g = -\int_{x_g}^{x_h} w \, dx \qquad (5.2)$$

Equations (5.1) and (5.2) can be translated, respectively, into the following statements of principle:

1. The rate of change of shear (i.e., slope of the shear curve) at a point is equal to the numerical value of the transverse distributed load at that point [Eq. (5.1)]. Thus the shear is constant when $w = 0$ and the shear diagram is linear where w is constant. In general, the shear equation will always be one degree higher than the equation for the transverse distributed load, i.e., loads of zero, constant, variable, and infinite intensities produce flat, linear, curved, and stepped shear diagrams, respectively. It should also be noted that if V is a continuous function of x in an interval of length, then a local maximum or minimum shear occurs at a point (in that interval) where $w = |dV/dx| = 0$.

2. The difference in the ordinates of the shear curve between any two sections is equal numerically to the total transverse loads applied between the two sections [Eq. (5.2)]. For equilibrium the resultant transverse load (equals the area under the transverse distributed load diagram in addition to the transverse concentrated loads) is zero and the difference in shear for the structure as a whole is zero. Thus the shear diagram must close (return to its original level) for equilibrium to exist.

It should be remarked that Eq. (5.1) does not apply at points of discontinuity in loading. For example, for the concentrated load P acting at point E in Fig. 5.10a, there is a sudden change in shear between a section just to the left and another section just to the right of the load P as shown in Fig. 5.10d. Therefore, the derivative (dV/dx) does not exist at point E and Eq. (5.1) does not apply at this point; however, Eq. (5.1) can be applied at a section just to either side of the point E.

Next, applying the equilibrium condition that ΣM at $O = 0$ (Fig. 5.10b) yields

$$M + \Delta M - M - (V + \Delta V)(\Delta x) - (w \, \Delta x) \frac{\Delta x}{2} = 0$$

(Again the intensity of the load w is assumed to be uniformly distributed over the length Δx since the variation of w leads to higher-order terms of Δx which are negligibly small compared with Δx itself.) Thus

$$\Delta M = V \, \Delta x + (\Delta V)(\Delta x) + w \frac{(\Delta x)^2}{2}$$

As Δx approaches 0 in the limit, the higher-order terms $w(\Delta x)^2/2$ and $(\Delta V)(\Delta x)$ become negligible, and therefore we can write the above equation as

$$\Delta M = V \, \Delta x$$

or

$$\frac{dM}{dx} = V \tag{5.3}$$

Integrating Eq. (5.3) between g and h, we have

$$\int_{x_g}^{x_h} \frac{dM}{dx} \, dx = \int_{x_g}^{x_h} V \, dx$$

or

$$M_h - M_g = \int_{x_g}^{x_h} V \, dx \tag{5.4}$$

Equations (5.3) and (5.4) lead to the following statements of principle, respectively:

3. The slope of the moment curve at any point is equal to the intensity of the shear force at that point. Here the moment equation will be one degree higher than the shear equation. If the moment is a continuous function of x in an interval of length in which the derivative (dM/dx) exists, then a local maximum or minimum moment occurs at a point in the interval where $V = dM/dx = 0$. It should be mentioned that when the moment is discontinuous as a result of applied loads, such as the concentrated couple M_D at point D of the beam in Fig. 5.10a, then the derivative (dM/dx) does not exist at point D and Eq. (5.3) cannot be used at that point; see Fig. 5.10c. However, local maximum or minimum moment can also occur where the moment diagram is discontinuous (even though $V \neq 0$) at points where concentrated moments act, for example, along the beam or at the fixed end of a beam. By this principle we can also deduce that shears of zero, constant, variable, and infinite intensities produce flat, linear, curved, and stepped bending-moment diagrams, respectively.

4. The difference in the ordinates of the moment curve between any two sections is equal to the area under the shear curve between the two sections, provided no concentrated moments (couples) act between those sections.

It should be noted that a transverse concentrated load gives rise to a step in the shear diagram equal to the transverse concentrated load P. At a concentrated load,

$$w = \lim_{\Delta x \to 0} \frac{P}{\Delta x} = \infty \qquad \Delta V = -\int_x w \, dx = -P \quad \text{and} \quad \Delta \frac{dM}{dx} = \Delta V = -P$$

The shear is assumed to pass through zero value over a very small distance and hence a local maximum moment occurs there. Also, the bending-moment curve undergoes an abrupt change in slope at, and equal to, the transverse concentrated load P.

Equations (5.1) to (5.4) are very helpful in the construction of shear-force and bending-moment diagrams from visual inspection of the load diagram. This is demonstrated in the examples below.

■ **EXAMPLE 5.3**

Draw the shear and bending-moment diagrams for the loaded beam shown in Fig. 5.11a.

Solution. The first step in the solution is to determine the external reactions. Assuming the positive sign convention shown in Fig. 5.11, we take moments of all forces about point E. Thus ΣM at $E = 0$,

$$(32)(10) + (40)(5) + (48)(4) - 8R_B = 0 \qquad R_B = 89 \text{ kN}\uparrow$$

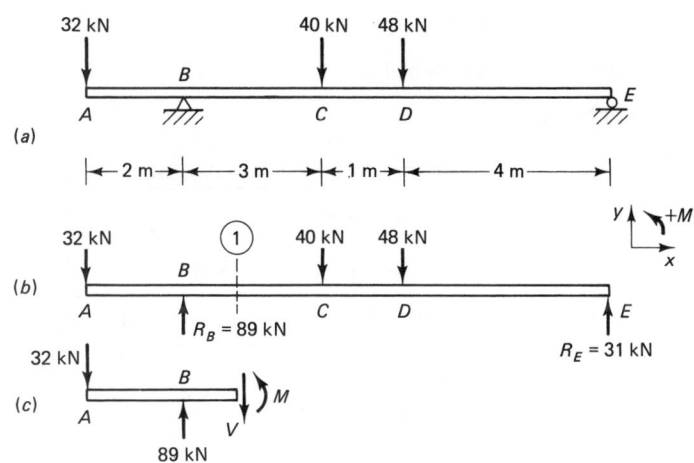

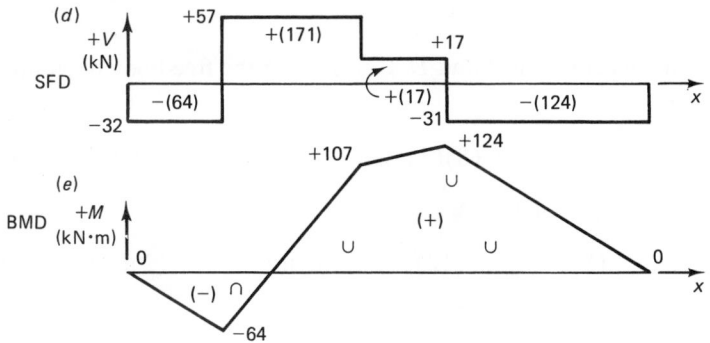

Figure 5.11 Structure for Example 5.3.

Taking moments about B: $\sum M$ at $B = 0$,

$$-[(48)(4) + (40)(3)] + (32)(2) + 8R_E = 0 \qquad R_E = 31 \text{ kN}\uparrow$$

Check $\sum F_y = 0$: $\qquad -32 - 40 - 48 + 89 + 31 = 0 \qquad (\therefore \text{ checks})$

It should be noted that bending moment $M_A = 0$ (since end A is free) and bending moment $M_E = 0$ (since end E is simply supported).

Shear-Force Diagram. Since $dV/dx = -w$ [Eq. (5.1)] and $w = 0$, the slope of shear diagram is zero between loads, that is, shear is constant. The shear at any point is obtained by dividing the beam into two parts and considering either part as a free body. For example, inducing a cut at 1 in Fig. 5.11b and applying the equilibrium condition $\sum F_y = 0$ to the free-body diagram in Fig. 5.11c yield

$$-32 + 89 - V = 0 \qquad V = 57 \text{ kN} \qquad \text{and so on}$$

The SFD can thus be completed as shown in Fig. 5.11d.

Bending-Moment Diagram. From Eq. (5.4):

$$M_B - M_A = \text{area of shear force between } A \text{ and } B = (-32)(2) = -64$$

Since $M_A = 0$, $M_B = -64$ kN $\cdot$ m

$$M_C - M_B = (57)(3) = 171 \quad \text{or} \quad M_C = +107 \text{ kN} \cdot \text{m}$$

$$M_D - M_C = (17)(1) = 17 \quad \text{or} \quad M_D = 124 \text{ kN} \cdot \text{m}$$

$$M_E - M_D = (-31)(4) = -124 \quad \text{or} \quad M_E = 0 \qquad \text{(checks)}$$

The change in slope of the BMD at B is $[107 - (-64)]/3 - (-64 - 0)/2 = 57 + 32 = 89$ kN, which is equal to the concentrated reaction (force) at B. Notice also that since the loads on the beam have zero intensity (Fig. 5.11a or b), the shear diagram is flat (Fig. 5.11d) and hence the moment diagram is linear (Fig. 5.11e). ∎

■ EXAMPLE 5.4

Draw the shear and bending-moment diagrams for the beam shown loaded in Fig. 5.12a.

Solution. We first find the reactions at A and D. Considering the free-body diagram in Fig. 5.12b:

$$\sum M \text{ about } D = 0: \qquad -(R_A)(5) + (8)(3)(\tfrac{3}{2} + 2) + (12)(1) - (12)(2)(\tfrac{2}{2}) = 0$$

$$R_A = 14.4 \text{ kN}\uparrow$$

$$\sum M \text{ about } A = 0: \qquad -(8)(3)(\tfrac{3}{2}) - (12)(4) + (R_D)(5) - (12)(2)(\tfrac{2}{2} + 5) = 0$$

$$R_D = 45.6 \text{ kN}\uparrow$$

Check $\sum F_y = 0$: $\qquad 14.4 - (8)(3) - 12 + 45.6 - (12)(2) = 0 \qquad (\therefore \text{ checks})$

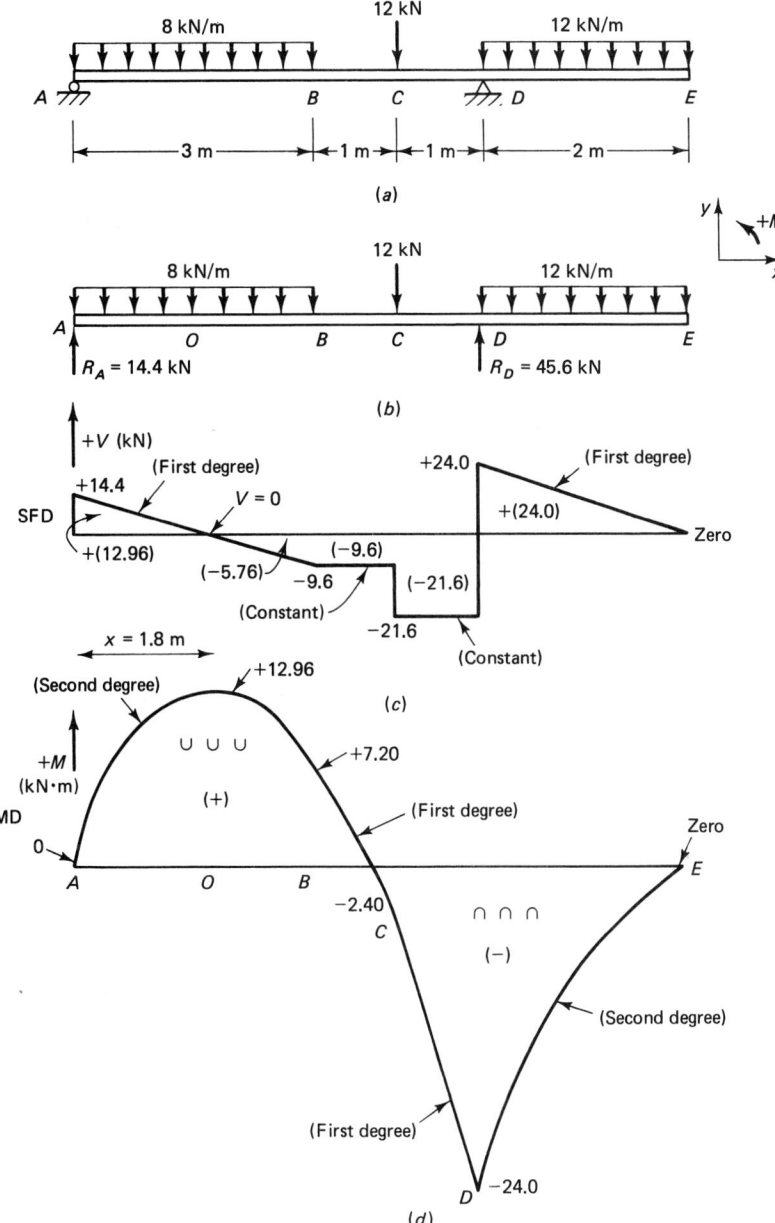

Figure 5.12 Structure for Example 5.4.

Shear-Force Diagram. At the left end of the beam, point A is a vertical reactive force of 14.4 kN, and the shear at a section close to A will be also 14.4 kN (from $\Sigma F_y = 0$) and in a positive sense. Hence, the initial value of the shear diagram is +14.4 kN. Figure 5.11b shows a constant uniform load between points A and B; thus, from Eq. (5.1),

$$\frac{dV}{dx} = -w = -8 \quad \text{or} \quad [V]_A^B = (-8)(3) = -24$$

$$\therefore V_B - V_A = -24 \quad \text{or} \quad V_B = -24 + 14.4 = -9.6 \text{ kN}$$

The shape of the shear diagram between points A and B is linear with a negative slope since $dV/dx = -8 = $ constant. The shear is constant between B and C and also between C and D. However, because of the concentrated load at C of 12 kN the shear will undergo an abrupt change just to the right of point C as shown in Fig. 5.12c. A similar situation occurs at D where the shear force changes abruptly just to the right of D owing to the presence of the reactive force of 45.6 kN. For portion DE we find from Eq. (5.2)

$$V_E - V_{\text{right of } D} = -(12)(2) = -24$$

or

$$V_E = -24 + V_{\text{right of } D} = -24 + 24 = 0$$

which closes the shear curve. Again the slope of the shear curve is negative and constant since, from Eq. (5.1), $dV/dx = -w = -12$ and therefore the shear is linear (first degree) between D and E.

The point of zero shear between A and B is found by cutting the beam at an arbitrary cross section between A and B, and at a distance x from A ($0 < x < 3$). Then from the equation $\Sigma F_y = 0$, we find

$$14.4 - (8)(x) = 0 \quad \text{or} \quad x = 1.8 \text{ m}$$

See Fig. 5.12c.

Bending-Moment Diagram. We should first note that $M_A = M_E = 0$ from the boundary conditions of the beam. From Eq. (5.4),

$$M_O - M_A = \text{area of shear force between } A \text{ and } O$$

$$= \frac{+(14.4)(1.8)}{2} = +12.96$$

$$M_O = 12.96 + M_A = 12.96 + 0 = 12.96 \text{ kN} \cdot \text{m}$$

From Eq. (5.3), $dM/dx = V = 0$ at point O, and therefore the moment at O has a stationary value as shown in Fig. 5.12d; furthermore, the variation of the moment between A and B will be a second-degree curve since the shear is a first-degree curve. For B:

$$M_B - M_O = \text{area of shear force between } O \text{ and } B$$

$$= \frac{(-9.6)(3 - 1.8)}{2} = -5.76$$

or

$$M_B = M_O - 5.76 = 12.96 - 5.76$$

$$= 7.20 \text{ kN} \cdot \text{m}$$

For C:

$$M_C - M_B = -9.6 \qquad M_C = -9.6 + M_B$$

$$M_C = -9.60 + 7.20 = -2.40 \text{ kN} \cdot \text{m}$$

For D:

$$M_D - M_C = -21.6 \qquad M_D = -24.0 \text{ kN} \cdot \text{m}$$

For portions BC and CD, the moment is a first-degree curve since the shear is constant in the two portions (Fig. 5.12c and d).

For E:

$$M_E - M_D = +24.0 \qquad M_E = 0 \qquad (\therefore \text{ checks})$$

For portion DE, the moment is a second-degree curve since the shear is a first-degree curve; since the shear V is positive and decreasing from D to E, the slope of the moment curve $dM/dx = V$ must also be positive and decrease as shown in Fig. 5.12d. ∎

■ EXAMPLE 5.5

A plate girder cantilever bridge in Fig. 5.13a is hinged at A and simply supported at supports B, C, and D. The suspended span EF is hinged to cantilever spans BE and CF. Draw the SFD and BMD for the bridge.

Solution. The first step is to calculate the reactions R_A, R_B, R_C, and R_D (Fig. 5.13b). This can be readily accomplished by considering the equilibrium of portions ABE, EF, and FCD (Fig. 5.13c). Since R_C and R_D are vertical, R_F must be vertical and hence R_E must be vertical. Thus $H_A = 0$ from $\Sigma F_x = 0$. From the equilibrium of portion EF we can deduce (by symmetry)

$$R_E = R_F = 160 \text{ kN}$$

Next consider the equilibrium of portion ABE:

$$\Sigma M \text{ about } A = 0: \qquad -(30)(20)(\tfrac{20}{2}) + R_B(12) - R_E(20) = 0$$

$$R_B \approx 766.67 \text{ kN}\uparrow$$

$$\Sigma M \text{ about } B = 0: \qquad -(R_A)(12) + (30)(20)(12 - \tfrac{20}{2}) - (R_E)(8) = 0$$

$$R_A \approx -6.67 \text{ kN} \approx 6.67 \text{ kN}\downarrow$$

Check $\Sigma F_y = 0$: $-(30)(20) + R_A + R_B - R_E = 0$

$$-600 + (-6.67) + (766.67) - 160 = 0 \qquad (\text{checks})$$

Dealing with the equilibrium of portion FCD:

$$\Sigma M \text{ about } D = 0: \qquad (160)(6) + (R_F)(20) - (R_C)(12) = 0$$

$$R_C = 346.67 \uparrow \text{kN}$$

$$\Sigma M \text{ about } C = 0: \qquad (R_D)(12) - (160)(6) + (R_F)(8) = 0$$

$$R_D = -26.67 \text{ kN} = 26.67 \text{ kN}\downarrow$$

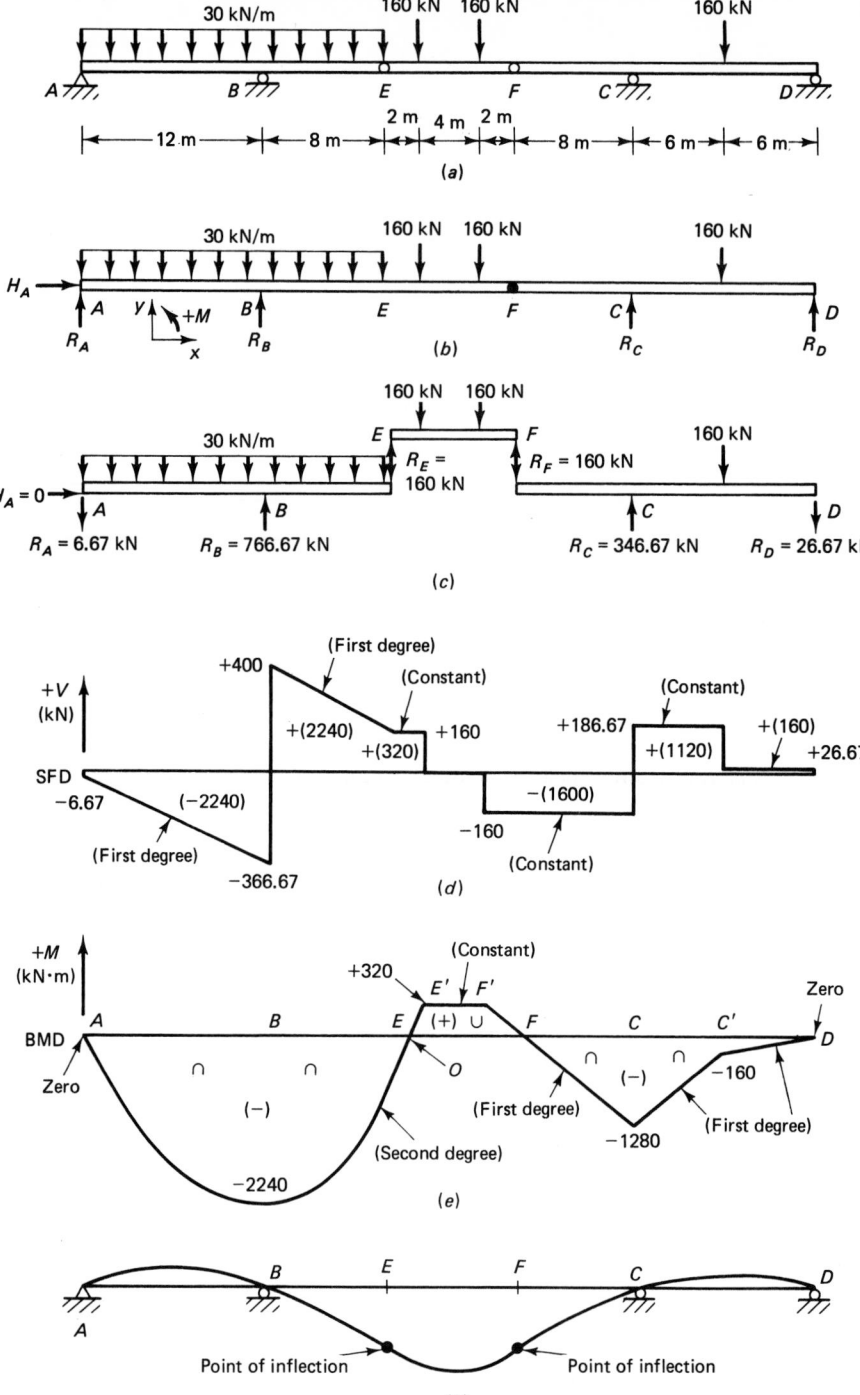

Figure 5.13 Structure for Example 5.5.

Check $\Sigma F_y = 0$: $R_D - 160 + R_C - R_F = 0$

$-26.67 - 160 + 346.67 - 160 = 0$ (checks)

Shear-Force Diagram.   By using Eqs. (5.1) and (5.2) the shear-force diagram can be constructed quite readily (see Fig. 5.13 d). Notice that the shear-force diagram is not affected by the internal vertical forces in the hinges at E and F.

Bending-Moment Diagram.   Making use of Eqs. (5.3) and (5.4) one can construct the bending-moment diagram. Notice that $M_A = M_D = 0$ (from the boundary conditions). Thus

$M_B - M_A$ = area of shear force diagram between AB

$$= \frac{(12)(-366.67 - 6.67)}{2} = -2240$$

Therefore, $M_B = -2240$ kN $\cdot$ m. It is a local maximum since the shear force changes abruptly. The moment curve in the portion ABE is a second-degree curve since the shear is a first-degree curve. Following this procedure will yield the bending-moment diagram shown in Fig. 5.13 e. Notice how the diagram closes to zero at D, as it should since $M_D = 0$ from the boundary condition of the bridge girder. Also, the bending moment at hinges E and F is zero, as it should be. ■

Quite often the structural engineer is faced with the problem of deducing the load diagram from a shear-force diagram and/or a bending-moment diagram. For example, in prestressing a continuous beam, the prestressing force will bend the beam, creating a tendency of the structure to deflect from its supports. However, restraint offered by the supports against such deflection will generate bending moments. To evaluate these moments, the following steps are taken:

1. Remove all supports and plot the bending moment for the beam due to the eccentricity of the prestressing force.
2. Deduce the loading on the beam corresponding to the bending-moment diagram in step 1.
3. With this loading on the continuous beam as it is actually supported, compute the moment in the beam by any one of the methods used in dealing with indeterminate structures (see Chapters 9 through 13).

For the present we are interested in step 2 only, and the following examples will illustrate the procedure.

■ EXAMPLE 5.6

The two-span continuous prestressed-concrete beam shown in Fig. 5.14a is prestressed in a way that results in a bending moment due to prestress assuming that the three supports at A, B, and C are removed (Fig. 5.14b). Deduce the SFD and the loading diagram from such a bending-moment diagram.

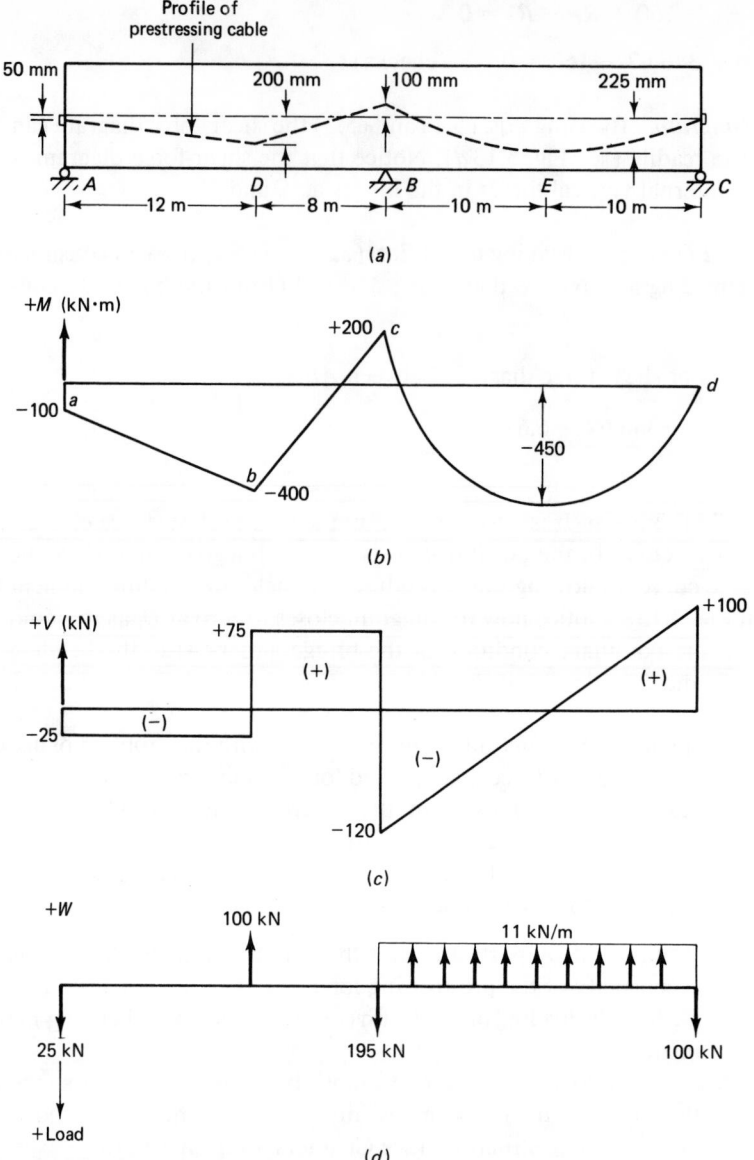

Figure 5.14 Structure for Example 5.6. (*a*) Elevation of prestressed beam. (*b*) Moment due to prestress (assuming that supports *A*, *B*, and *C* are removed).

Solution. We apply Eqs. (5.3) and (5.1) to generate the SFD and the loading diagram; since these involve calculating slopes we need to derive the equation for the parabolic curve *cd* in Fig. 5.14*b*. Referring to the parabolic curve in Fig. 5.15, it can be shown that

$$y = \frac{4D}{L^2} x(L - x) \tag{5.5}$$

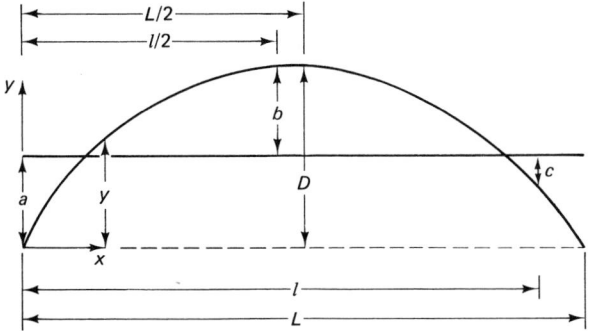

Figure 5.15 Parabolic shape of prestressing cable for Example 5.6.

Expressions for D and L can be generated in terms of a, b, c, and l, that is, known coordinate points on the curve. Thus

$$L = \frac{3a + 4b + c}{2a + 4b + 2c} l$$

and   (5.6)

$$D = \frac{(a + b)L^2}{2l(L - l/2)}$$

Now to obtain the shear V just to the left of support B due to the moment diagram in Fig. 5.14b, we have, from Eq. (5.3),

$$\frac{dM}{dx} = V$$

Or, from Eq. (5.5),

$$\frac{dM}{dx} = \left[\frac{4D}{L^2} (L - 2x) \right]_{\text{at } x=0} = \frac{4D}{L}$$

Using $a = -200$, $b = -450$, $c = 0$, $l = 20$, we find from the first of Eqs. (5.6), $L = \frac{240}{11}$. From the second of Eqs. (5.6), $D = (-650)L^2/[40(L - 10)] = -654.55$. Thus $V = 4D/L = -120$ kN.

Similarly at support C, $V = dM/dx = (4D/L^2)(L - 2x) = (4D/L^2)(L - 40) = +100$ kN. These shears are shown plotted in Fig. 5.14c. Since the bending moment is a second-degree curve, the shear force is a first-degree curve between supports B and C. For the portion of the beam between A and D, $dM/dx = V = [-400 - (-100)]/12 = -25$ kN. In the portion DB, $dM/dx = V = [+200 - (-400)]/8 = 75$ kN. In both portions the shear is a flat curve; at D the SFD is discontinuous since the slope of the BMD changes abruptly (Fig. 5.14c).

The loading diagram in Fig. 5.14d shows a concentrated load at A of $+25$ kN, at D of -100 kN, at B of $+195$ kN, and at C of $+100$ kN. Between B and C, since $dV/dx = -w = [100 - (-120)]/20 = 11$ kN/m or $w = -11$ kN/m, the loading is uniformly distributed with an intensity of 11 kN/m upward (negative sense!). ■

■ **EXAMPLE 5.7**

Draw the moment and load diagrams corresponding to the given shear diagram in Fig. 5.16a. Specify values at all change of load positions and at all points of zero shear.

Solution. The first step is to determine the points of zero shear. Thus for portion *BC* (Fig. 5.16a), by proportion

$$\frac{a}{2} = \frac{40}{40 + 37} \qquad \therefore a = 1.04 \text{ m} \qquad \text{hence } b = 0.96 \text{ m}$$

Similarly for portion *EF*,

$$\frac{c}{8} = \frac{17}{17 + 31} \qquad \therefore c = 2.83 \text{ m} \quad \text{and} \quad d = 5.17 \text{ m}$$

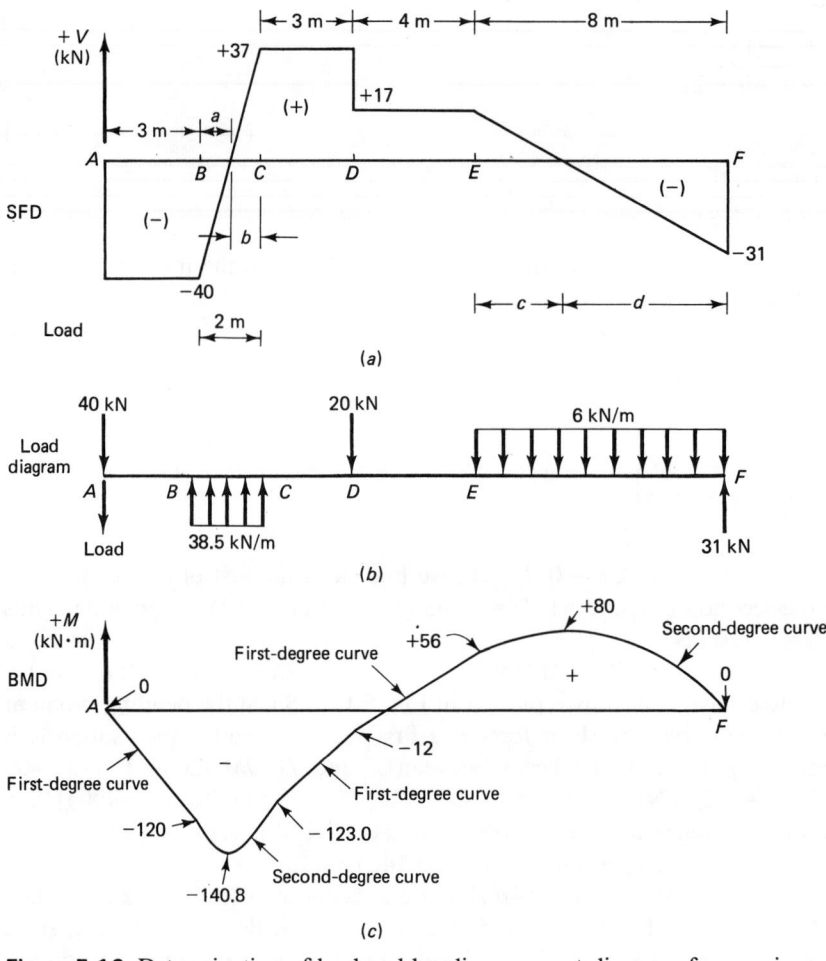

Figure 5.16 Determination of load and bending-moment diagrams from a given shear-force diagram (Example 5.7).

A shear of -40 kN at A with a constant distribution implies a concentrated load of 40 kN↓ at A. For portion BC, $dV/dx = [37 - (-40)]/2 = -w$; or $w = -38.5$ kN/m $= 38.5$ kN/m↑. At point D the shear-force diagram changes abruptly, pointing to a concentrated load of 20 kN↓; for portion EF, $dV/dx = [-31 - (17)]/8 = -6 = -w$; or $w = 6$ kN/m↓ as shown in Fig. 5.16b.

To generate the moment diagram (Fig. 5.16c), we make use of Eq. (5.4). Thus for portion AB, noting that $M_A = 0$,

$$M_B - M_A = \text{area of shear-force diagram between } A \text{ and } B$$

$$= -(40)(3) = -120 \qquad \therefore M_B = -120 \text{ kN} \cdot \text{m}$$

The moment between A and B is a first-degree curve since the shear is constant. At the point of zero shear in portion BC, we have

$$M_{max} - M_B = \frac{-(40)(a)}{2} = \frac{-40}{2}(1.04) = -20.8$$

$$\therefore M_{max} = -120 - 20.8 = -140.8 \text{ kN} \cdot \text{m}$$

The moment between B and C is a second-degree curve since the shear is a first-degree curve. At point of zero shear along portion EF, we have

$$M_{max} - M_E = \text{area of shear-force diagram between } E \text{ and point of zero shear}$$

Thus,

$$M_{max} - M_E = \frac{+(17)(c)}{2} = \frac{(17)(2.83)}{2} = +24$$

Hence

$$M_{max} = M_E + 24 = 56 + 24 = +80 \text{ kN} \cdot \text{m}$$

$$M_F - M_{max} = \text{area of shear-force diagram between point of zero shear and } F$$

Therefore

$$M_F = M_{max} + \frac{(-31)(5.17)}{2} = 80 - 80 = 0$$

which serves as a check, since M_F should be zero. ∎

5.4 SINGULARITY FUNCTIONS AND MACAULAY'S NOTATION

In the preceding section we emphasized that Eqs. (5.1) to (5.4) are applicable only if the loading functions are continuous; concentrated loads and distributed load with sudden changes along its length give rise to discontinuities or singularities. In 1862, A. Clebsch introduced a concise notation in treating such singularity functions; however, in 1919, Macaulay suggested using such notation in solving problems on the deflection of beams, and for this reason it is known as Macaulay's notation. Therefore, to handle discontinuity in loading without the necessity of writing several sets of equations connecting the piecewise-continuous functions, we introduce a family of singularity functions, defined by

$$F_n(x) = \{x - a\}^n \tag{5.7}$$

where n is an integer.

For $n \geqslant 0$,

$$\{x - a\}^n = (x - a)^n \qquad \text{when } x \geqslant a$$
$$= 0 \qquad \text{when } x < a$$

For $n < 0$,

$$\{x - a\}^n = \infty \qquad \text{when } x = a$$
$$= 0 \qquad \text{when } x \neq a$$

The function $F_n(x)$ has the following additional properties with respect to differentiation and integration:

$$\frac{d}{dx}\{x - a\}^n = \{x - a\}^{n-1} \qquad \text{for } n \leqslant 0$$
$$= n\{x - a\}^{n-1} \qquad \text{for } n > 0$$
$$\int \{x - a\}^n \, dx = \{x - a\}^{n+1} \qquad \text{for } n \leqslant 0$$
$$= \frac{1}{n+1}\{x - a\}^{n+1} \qquad \text{for } n > 0$$

While there is no limit to the order n, the lowest order for which the mathematical idealization represents physical loading is -2. Figure 5.17 shows an array of singularity functions that are susceptible of physical interpretation. Figure 5.18 presents some discontinuous loadings that can be represented by the singularity functions in Fig. 5.17. Thus, we can write

$$w = M_1\{x - a\}^{-2} \qquad \text{for loading in Fig. 5.18}a$$

$$w = P_1\{x - a\}^{-1} \qquad \text{for loading in Fig. 5.18}b$$

$$w = w_1[\{x - a\}^0 - \{x - b\}^0] \qquad \text{for loading in Fig. 5.18}c$$

$$w = \frac{w_1}{b - a}\{x - a\}^1 - \frac{w_1}{b - a}\{x - b\}^1 - w_1\{x - b\}^0$$

for loading in Fig. 5.18d

and

$$w = \frac{w_1}{(b - a)^2}\{x - a\}^2 - \frac{w_1}{(b - a)^2}\{x - b\}^2 - w_1\{x - b\}^0$$

for loading in Fig. 5.18e

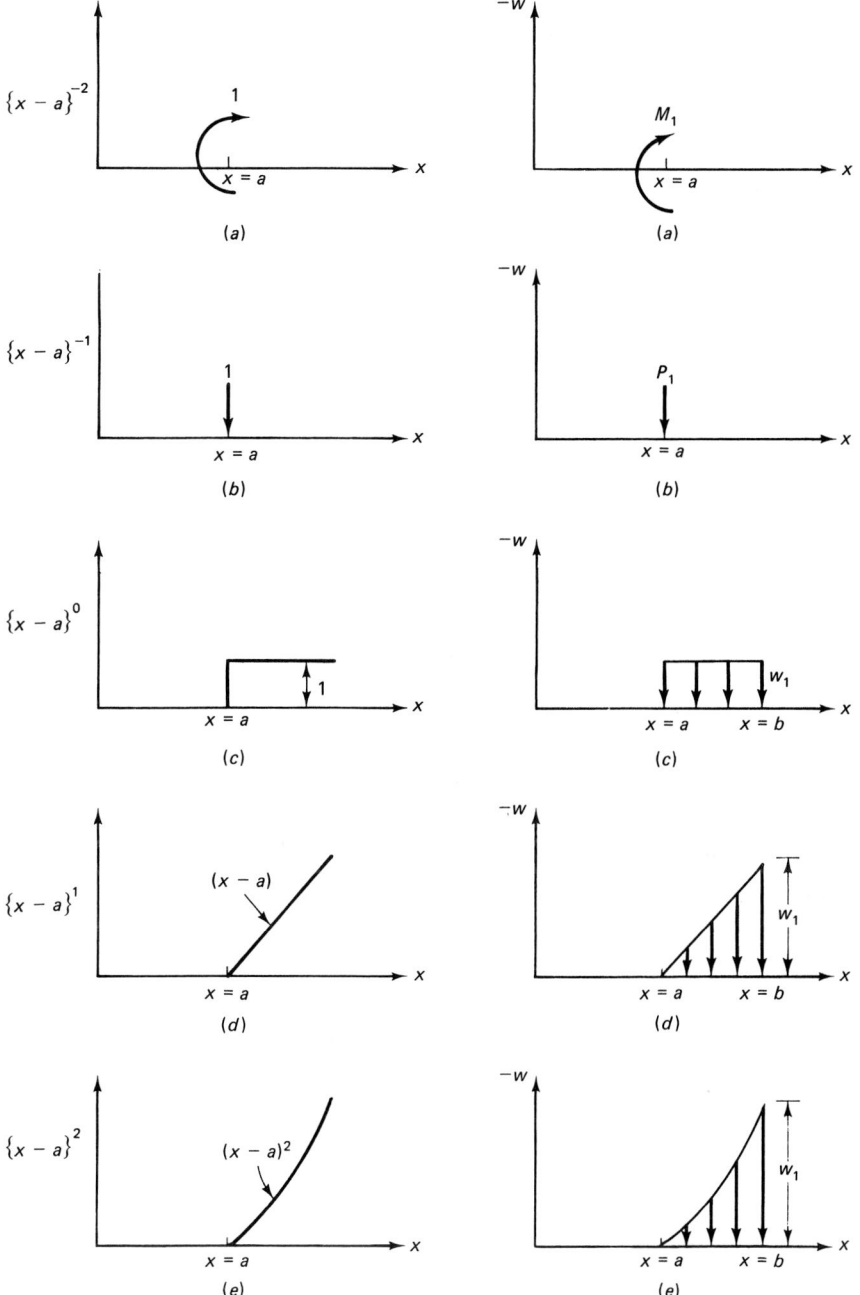

Figure 5.17 A few singularity functions. (*a*) Doublet. (*b*) Impulse. (*c*) Unit step. (*d*) Unit ramp. (*e*) Half parabola.

Figure 5.18 Discontinuous loadings represented by the singularity functions in Fig. 5.17.

It should be noted that the third, fourth, and fifth loading functions in Fig. 5.18c, d, and e, respectively, start at $x = a$ and terminate at $x = b$; in order to confine these functions between the limits of $x = a$ to $x = b$, we must subtract the portion of the area under the loading curve beyond $x = b$. This does not apply to the first and second loading functions (Fig. 5.18a and b), since they are nonzero only at $x = a$.

■ EXAMPLE 5.8

Using singularity functions, deduce the shear and bending-moment functions for the beam loaded as shown in Fig. 5.19.

Solution. The first step in the solution is to calculate the external reactions at A and B. Let us assume that these are found to be R_A and R_B, both acting upward as shown in Fig. 5.19. Following our convention of signs that a positive vertical load acts downward, opposite to the positive y direction, we can write an expression for

$$w(x) = -R_A\{x - 0\}^{-1} + \frac{w_1}{r}\{x - 0\}^1 - \frac{w_1}{r}\{x - r\}^1 - w_1\{x - r\}^0$$
$$- M_1\{x - (r + s)\}^{-2} + P_1\{x - (r + s + t)\}^{-1} - R_B\{x - L\}^{-1}$$

Since $dV/dx = -w(x)$ and hence $V = -\int w(x)dx$, the above expression is integrated to yield an expression for the shear force. Thus,

$$V = R_A\{x\}^0 - \frac{w_1}{2r}\{x\}^2 + \frac{w_1}{2r}\{x - r\}^2 + w_1\{x - r\}^1$$
$$+ M_1\{x - (r + s)\}^{-1} - P_1\{x - (r + s + t)\}^0 + R_B\{x - L\}^0 + C_1$$

where C_1 is a constant of integration.

Using the relationship $dM/dx = V$ and hence $M = \int V dx$, we deduce the following expression for the bending moment:

$$M = R_A\{x\}^1 - \frac{w_1}{6r}\{x\}^3 + \frac{w_1}{6r}\{x - r\}^3 + \frac{w_1}{2}\{x - r\}^2$$
$$+ M_1\{x - (r + s)\}^0 - P_1\{x - (r + s + t)\}^1 + R_B\{x - L\}^1 + C_1x + C_2$$

where C_2 is a constant of integration. The constants C_1 and C_2 are found by satisfying the boundary conditions: $M = 0$ at A and B, that is, at $x = 0$ and $x = L$. Thus from $M = 0$ at $x = 0$

$$C_2 = 0$$

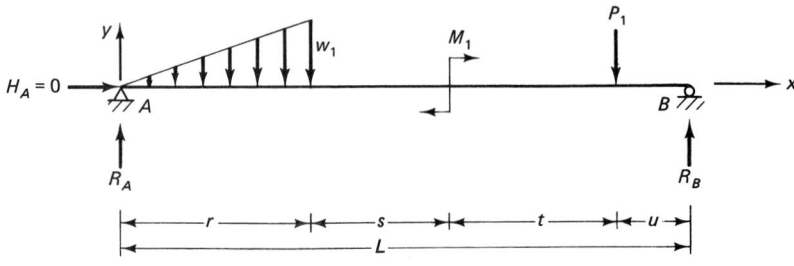

Figure 5.19 Structure for Example 5.8.

The condition that $M = 0$ at $x = L$ yields

$$C_1 L = -R_A L + \frac{w_1 L^3}{6r} - \frac{w_1}{6r}(L - r)^3 - \frac{w_1}{2}(L - r)^2 - M_1 + P_1[L + (r + s + t)]$$

hence the final expression for the bending moment M.   ■

Singularity functions are quite often used to simplify finding the deflection curve of beams; this will be demonstrated in Chapter 7. As a final note we should remark that errors can be avoided if it is remembered that the term inside the "braces" must be omitted when, on substituting for x, it has a negative value.

5.5 SOLVED PROBLEMS

■ SOLVED PROBLEM 5.1

Draw the shear and moment diagrams for the beam shown in Fig. 5.20*a*.

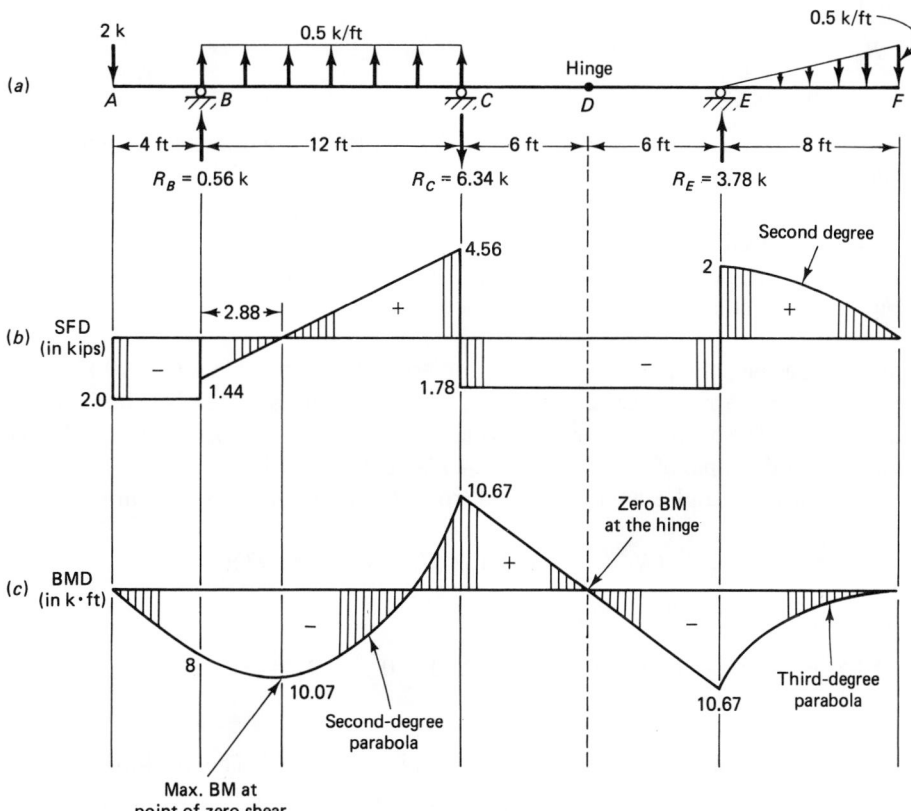

Figure 5.20 Structure for Solved Problem 5.1.

Solution. We must first find the reactions. Assume R_B, R_C, and R_E all acting upward:
From $\sum F_y = 0$,

$$\therefore R_B + R_C + R_E = 2 - 0.5 \times 12 + 0.5 \times 8 \times \tfrac{1}{2} = -2.0 \qquad (a)$$

From

$$+ \left(\curvearrowright \right. \sum M_B = 0,$$

$$2 \times 4 + 12 R_C + 0.5 \times 12 \times 6 + 24 R_E - 0.5 \times \tfrac{8}{2} \times (\tfrac{2}{3} \times 8 + 24) = 0$$

$$\therefore R_C + 2 R_E = 1.22 \qquad (b)$$

Moment at $D = 0$; from the sum of moments of all forces to the right of hinge D,

$$+ \left(\curvearrowright \right. \sum M_D(\text{right}) = 0$$

$$6 R_E = 0.5 \times \tfrac{8}{2} (\tfrac{2}{3} \times 8 + 6) = 0$$

$$\therefore R_E = 3.78 \text{ k} \uparrow$$

From (b),

$$R_C = 1.22 - 2 R_E = -6.34 \text{ k} \downarrow$$

From (a),

$$R_B = -2.0 - R_C - R_E = -2 + 6.34 - 3.78$$

$$= +0.56 \text{ k} \uparrow$$

After the reactions are computed, the SFD and BMD are drawn as shown in Fig. 5.20*b* and *c*. ∎

■ SOLVED PROBLEM 5.2

Draw axial-force, shear-force, and bending-moment diagrams for the beam *ABC* shown in Fig. 5.21*a*. Assume the pulley at *C* is frictionless.

Solution. Because the pulley at *C* is frictionless, tension in cable *CD* = 10 kN. The forces acting on beam *ABC* are shown in Fig. 5.21*b*. The forces are resolved horizontally and vertically. The horizontal force acting at *D* is transferred to *B* as a horizontal force of 7.07 kN and couple of $7.07 \times 2 = 14.14$ kN · m (Fig. 5.21*c*).

Reactions at *A* and *B* are determined from the equations of equilibrium.

$$\overset{+}{\curvearrowright} \sum M_A = 0: \qquad -(R_B)(2) + 14.14 + (7.07)(2) + (2.93)(4) = 0$$

$$R_B = 20 \text{ kN} \uparrow$$

$$\uparrow^+ \sum F_y = 0: \qquad R_A + 20 - 7.07 - 2.93 = 0 \qquad R_A = -10 \text{ kN}$$

$$R_A = 10 \text{ kN} \downarrow$$

The free-body diagram for beam *ABC* is shown in Fig. 5.21*d*. The axial-force, shear-force, and bending-moment diagrams are shown in Fig. 5.21*e, f,* and *g,* respectively. The qualitative deflected shape is shown in Fig. 5.21*h*.

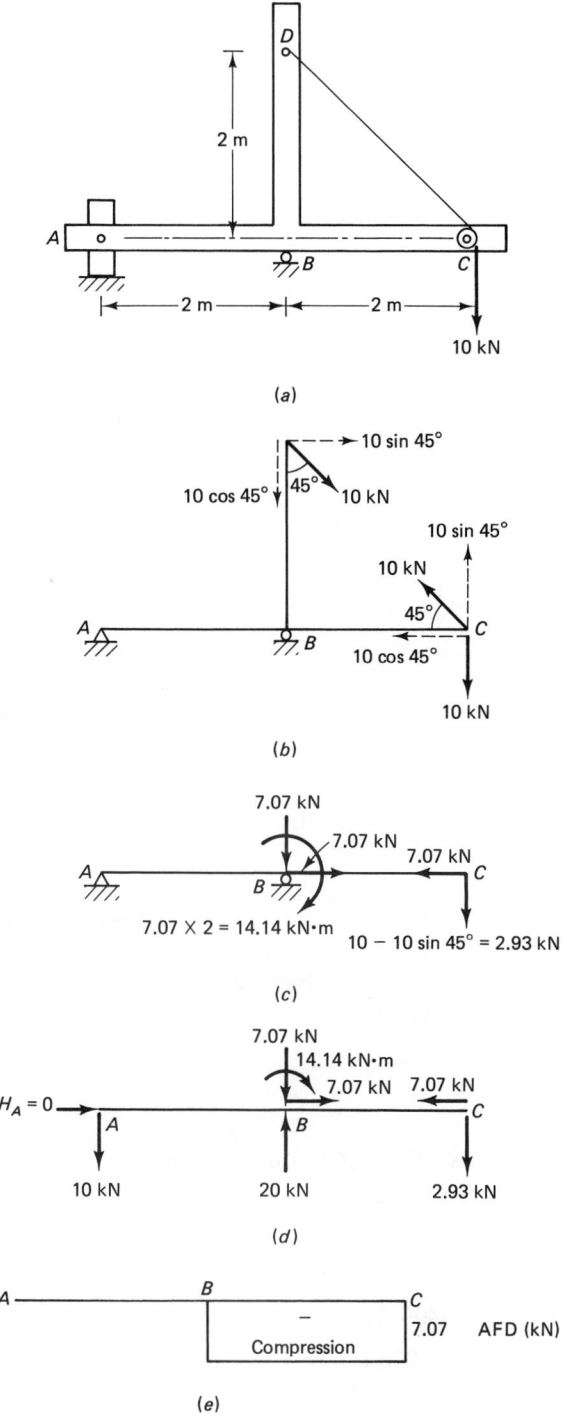

Figure 5.21 Structure for Solved Problem 5.2.

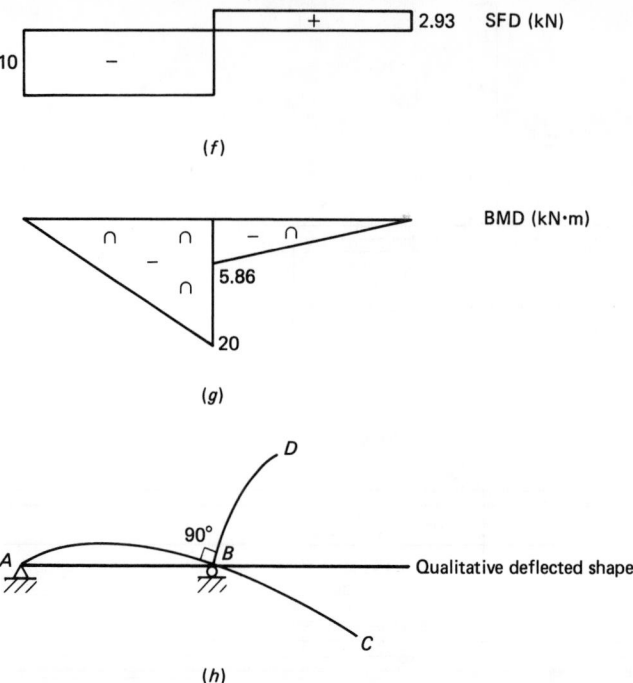

(f)

(g)

(h)

Figure 5.21 (*Continued*) ∎

■ SOLVED PROBLEM 5.3

Assuming the pulley at D to be smooth (frictionless), draw axial-force, shear-force, and bending-moment diagrams for beam ACB shown in Fig. 5.22a.

Solution. Cut the cable DE and replace it by a tension force of 5 k. The forces acting at D can be calculated from the FBD of the pulley shown in Fig. 5.22b. Equal and opposite forces will act at D on beam ACB as shown in Fig. 5.22c.

    Reaction V_A is determined by taking moments of all forces about B.

$$\sum M_B = 0 + \circlearrowright : \qquad -(V_A)(10) - (3.536)(6) + (8.536)(5) = 0$$

$$V_A = 2.146 \text{ k}\uparrow$$

Reaction V_B is determined by taking moments of all forces about A.

$$\sum M_A = 0 + \circlearrowright : \qquad +(V_B)(10) - (8.536)(5) - (3.536)(6) = 0$$

$$V_B = 6.390 \text{ k}\uparrow$$

Since the roller at B is at 45° to the horizontal,

$$H_B = 6.390 \leftarrow \quad \text{and} \quad R_B = 6.390\sqrt{2} = 9.036 \text{ k}\nwarrow$$

Reaction H_A is determined from $\sum F_x = 0$:

$$H_A + 3.536 - 6.390 = 0 \quad \text{or} \quad H_A = 2.854 \text{ k}\rightarrow$$

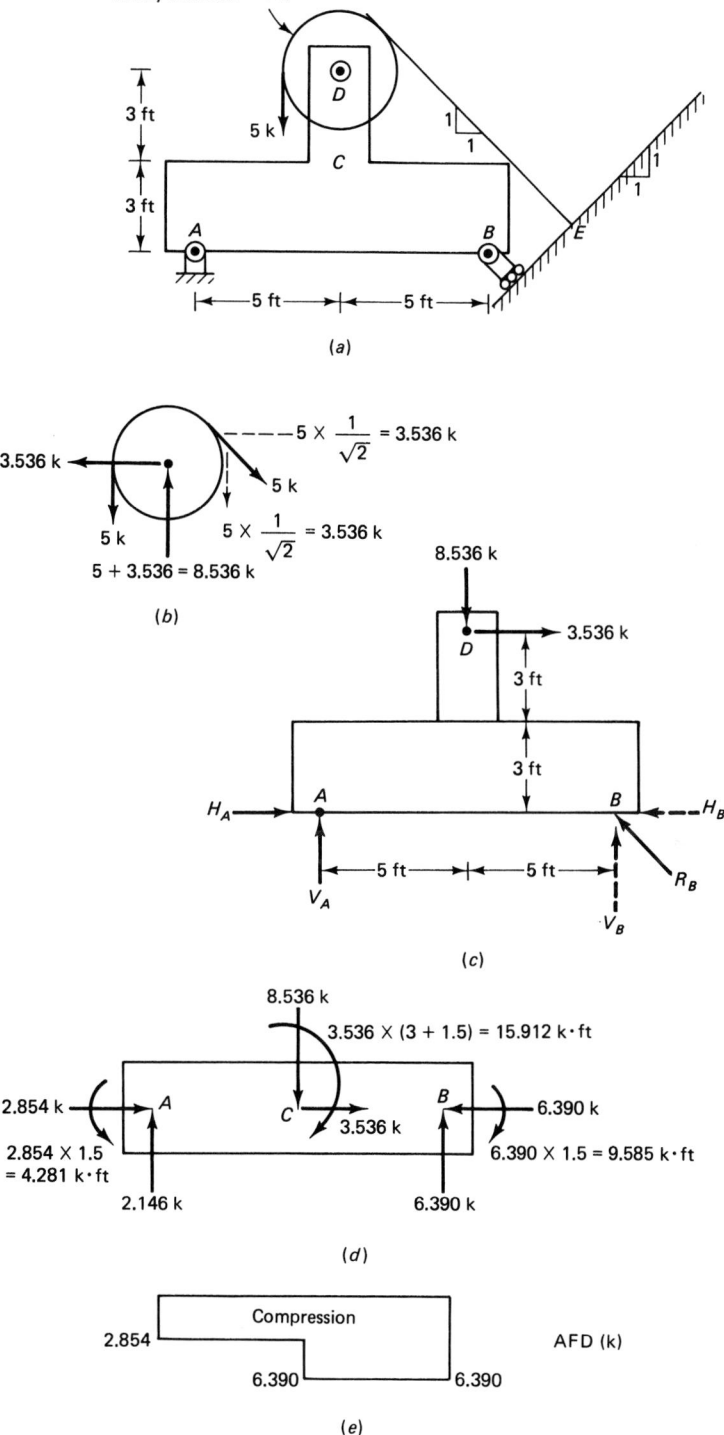

Figure 5.22 Structure for Solved Problem 5.3.

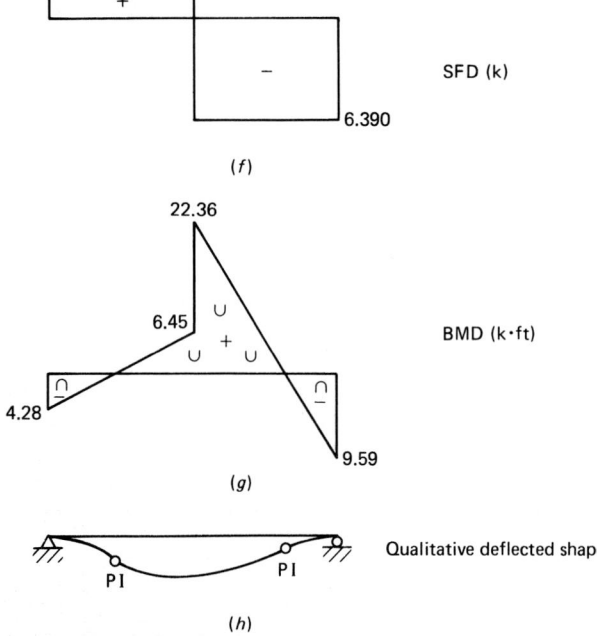

Figure 5.22 (*Continued*)

To draw AFD, SFD, and BMD, the forces and reactions are transferred to the centerline of beam *ACB* as shown in Fig. 5.22*d*. It should be noted that when H_A, H_B, and H_D are transferred to the centerline of beam *ACB*, additional moments are induced as shown in Fig. 5.22*d*. The AFD, SFD, and BMD are shown in Fig. 5.22*e*, *f*, and *g*, respectively. A qualitative deflected shape is shown in Fig. 5.22*h*. ∎

PROBLEMS

5.1 to 5.14 Draw shear-force and bending-moment diagrams for the beams shown. Determine location and magnitude of maximum shear and maximum moment.

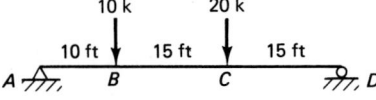

Figure P5.1

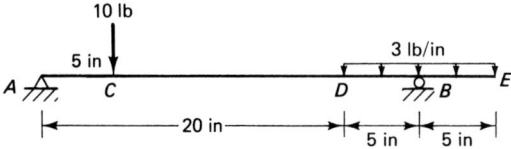

Figure P5.2

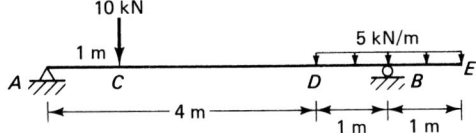

Figure P5.3

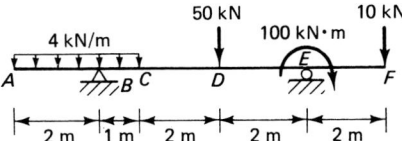

Figure P5.4

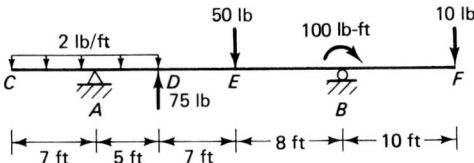

Figure P5.5

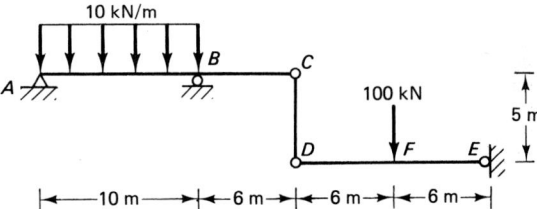

Figure P5.6

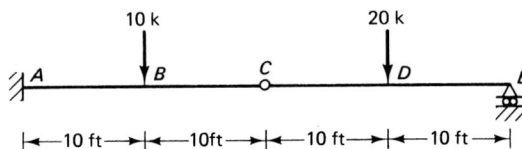

Figure P5.7

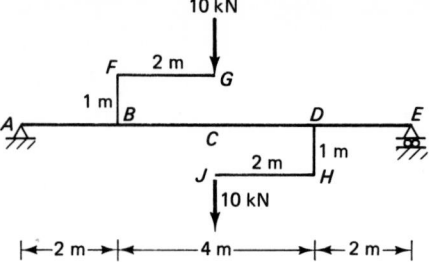

Figure P5.8

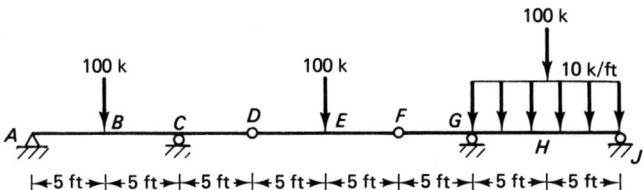

Figure P5.9

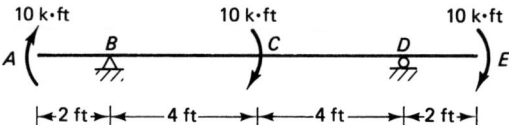

Figure P5.10

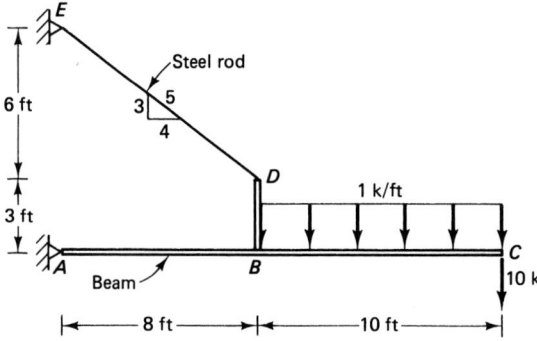

Figure P5.11

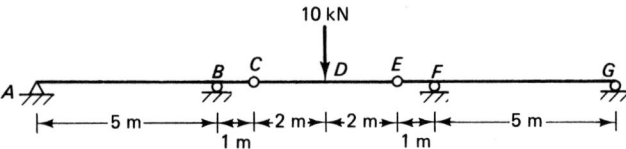

Figure P5.12

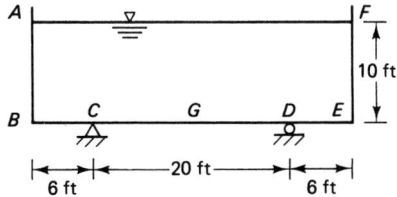

Figure P5.13

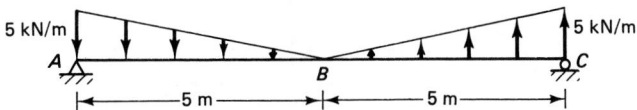

Figure P5.14

5.15 to 5.34 Draw axial-force, shear-force, and bending-moment diagrams for the structures shown.

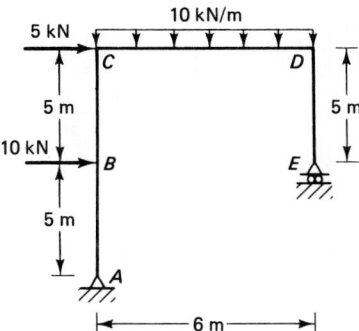

Figure P5.15

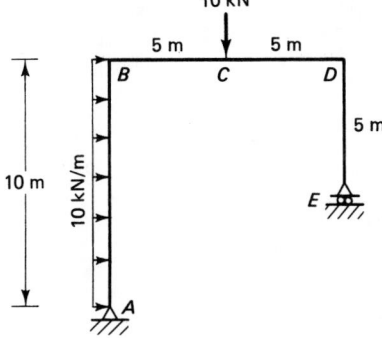

Figure P5.16

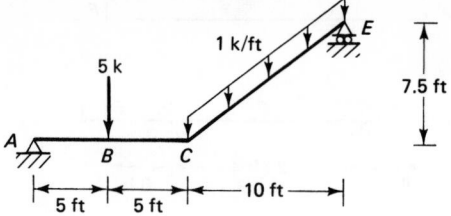

Figure P5.17

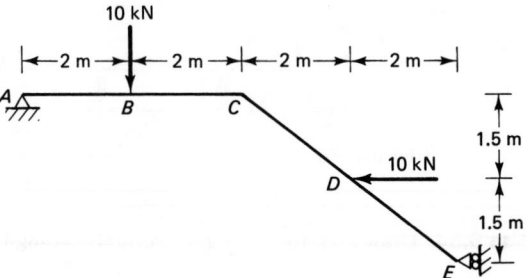

Figure P5.18

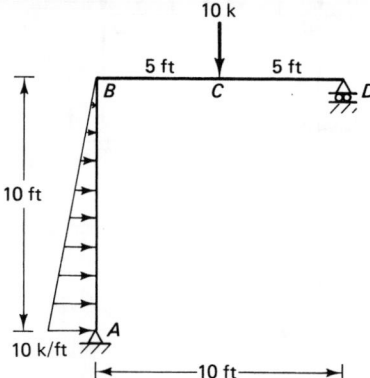

Figure P5.19

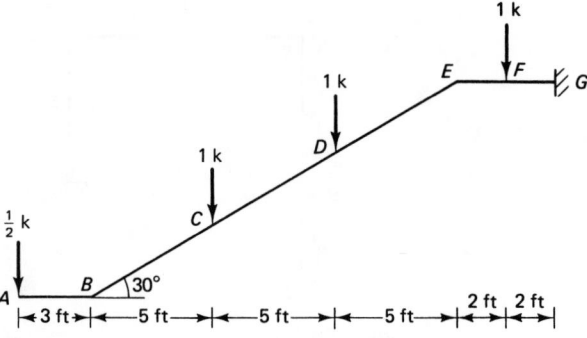

Figure P5.20

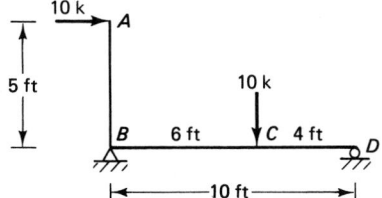

Figure P5.21

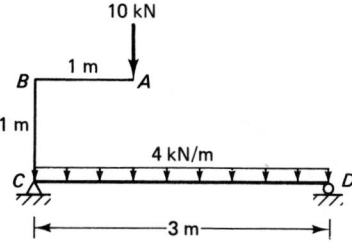

Figure P5.22

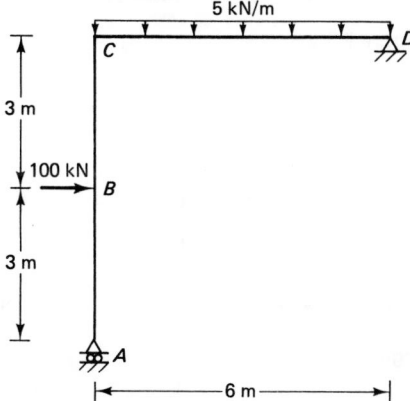

Figure P5.23

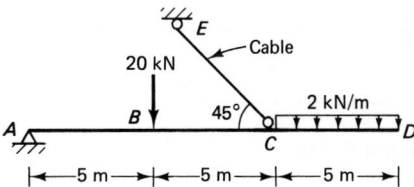

Figure P5.24

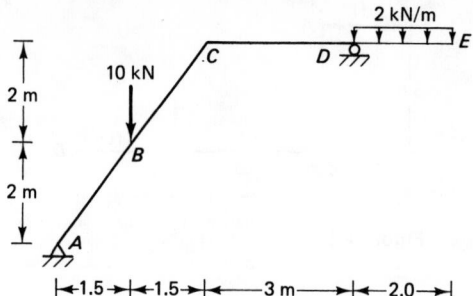

Figure P5.25

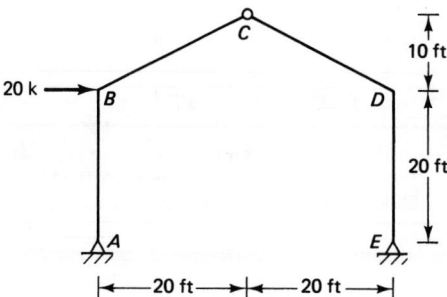

Figure P5.26

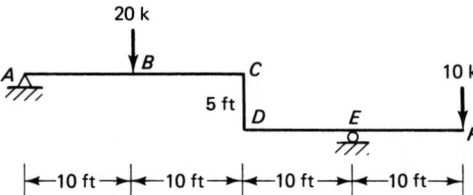

Figure P5.27

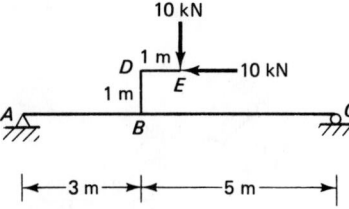

Figure P5.28

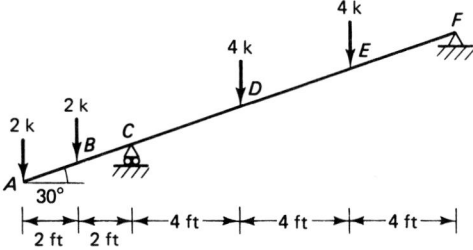

Figure P5.29

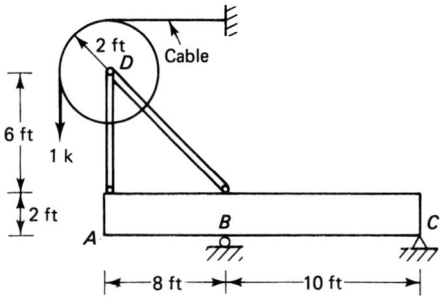

Figure P5.30

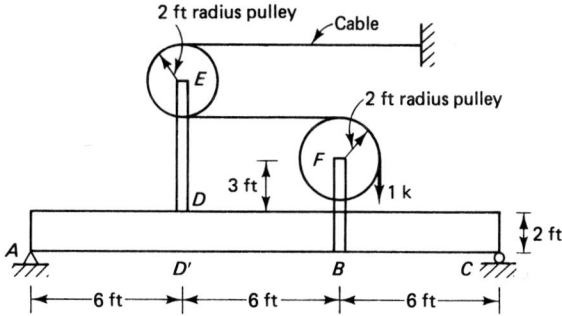

Figure P5.31

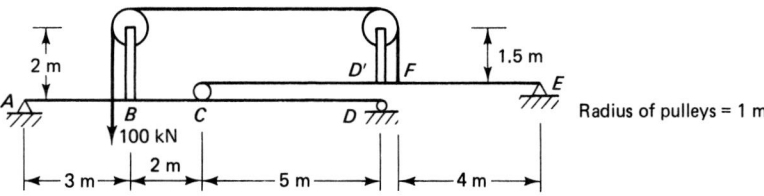

Figure P5.32

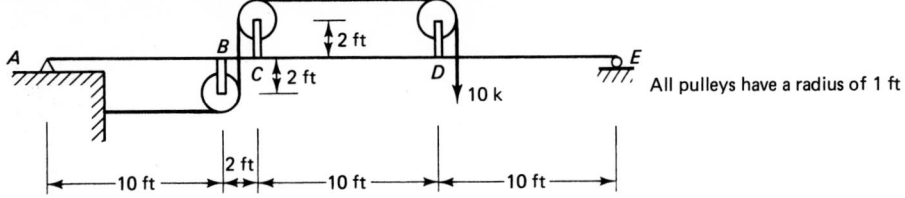

Figure P5.33

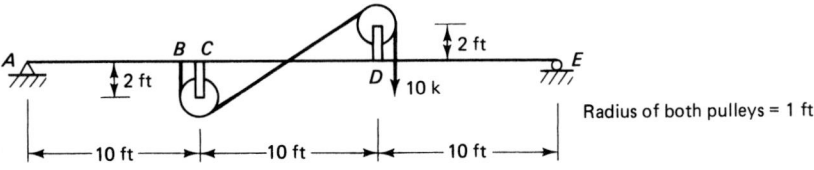

Figure P5.34

5.35 For the foundation-grade beam shown, assuming uniform soil pressure, determine the value of x and draw shear-force and bending-moment diagrams.

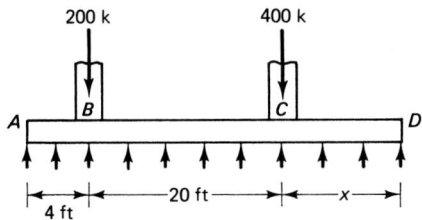

Figure P5.35

5.36 (a) and (b) If maximum negative moment is numerically equal to the maximum positive moment, determine the value of x and draw shear-force and bending-moment diagrams.

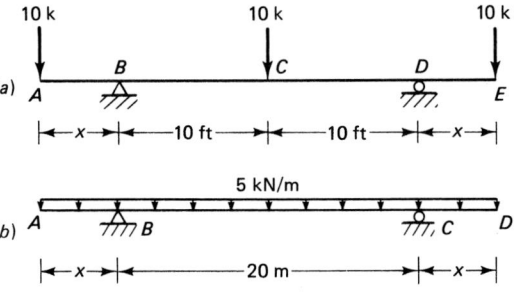

Figure P5.36

5.37 to 5.39 Draw the load and shear-force diagrams for the beams whose bending-moment diagrams are given.

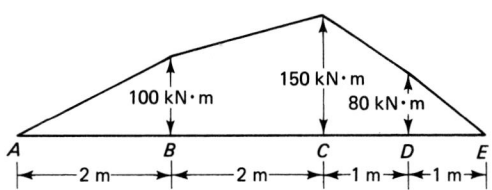

Figure P5.37

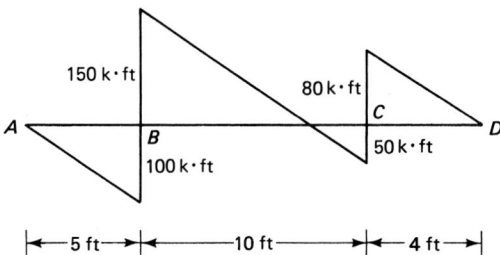

Figure P5.38

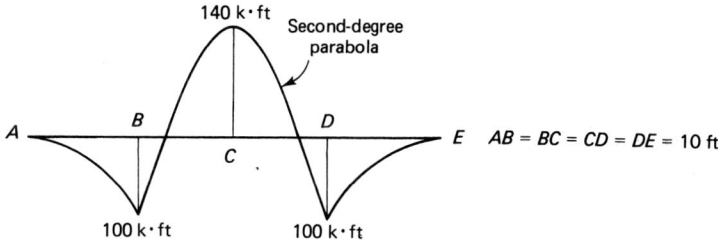

Figure P5.39

5.40 to 5.42 For the shear diagrams shown, draw the load and bending-moment diagrams for the beams. No couples act on the beams except at the fixed supports.

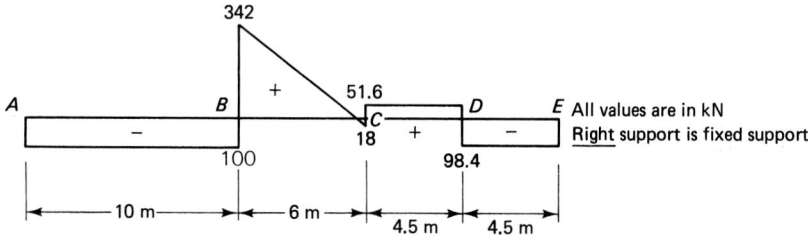

Figure P5.40

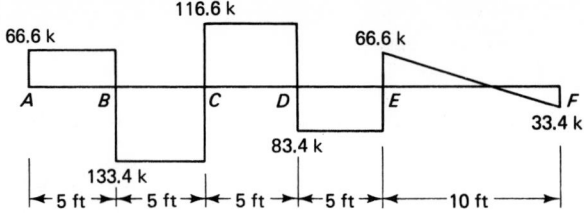

Figure P5.41

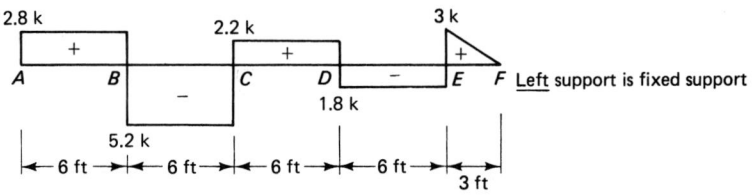

Figure P5.42

Chapter **6**

Cable Structures and Three-Hinged Arches

The modern bridge engineer has to be an artist and a poet as well as mathematician, scientist, financier and contractor.

D. B. Steinman

6.1 INTRODUCTION

Cables are used in many important types of structures such as suspension bridges and guyed electrical-transmission and antenna-supporting structures. When a cable supports a load that is uniform per unit length of cable, such as its self-weight, the shape of the cable is a catenary. However, the catenary can be approximated by a parabola, as shown in Fig. 6.1, without any significant loss of accuracy, for most engineering problems. Such an approximation results in a significant simplification of the analysis.

The cable is a flexible structure; that is, it has no moment-carrying capacity; thus every section of the cable is in pure tension. Because of its lack of flexural rigidity, a cable acted upon by external loads deforms in such a way that there is no bending moment at any section of the cable. When either the position or the nature of the external load changes, the deformed shape of cable also changes to suit the new load configuration, as shown in Fig. 6.2a to e. The deformed shape of the cable is called a "funicular" shape. If the deformed shapes of the cables in Fig. 6.2a to e are inverted, the resulting configurations will be as shown in Fig. 6.3a to e; any one of these new structures is called a "funicular arch," with the only internal forces acting being pure compressive forces as shown. Theoretically, such an arch can be constructed by simply stacking individual blocks (either concrete, brick, or wooden blocks) without any need for rigid interconnection of the blocks. The structure will be stable for that particular loading but will collapse under a new loading. A real arch, however, has to resist more than one loading condition, for example,

1. Dead load only
2. Dead load + live load

175

Pasco-Kennewick Intercity Bridge over the Columbia River, Washington. (Photo courtesy of Portland Cement Association.)

In such a case, the shape of the arch will be chosen such that the internal forces are mostly compressive under this primary loading condition. For other loading conditions, some bending moment will be induced in the arch, producing bending stresses that should also be taken into account in the design of the arch. If only the span and loading are defined in a given case, a family of funicular shapes (funicular cables and funicular arches) having different sags (for cables) and rises (for arches) can be found,

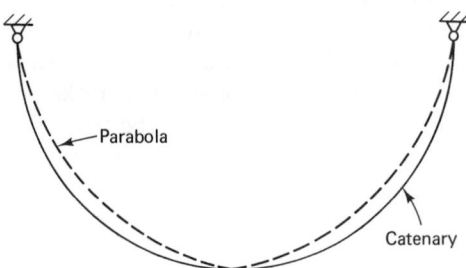

Figure 6.1 Comparison between a parabola and a catenary.

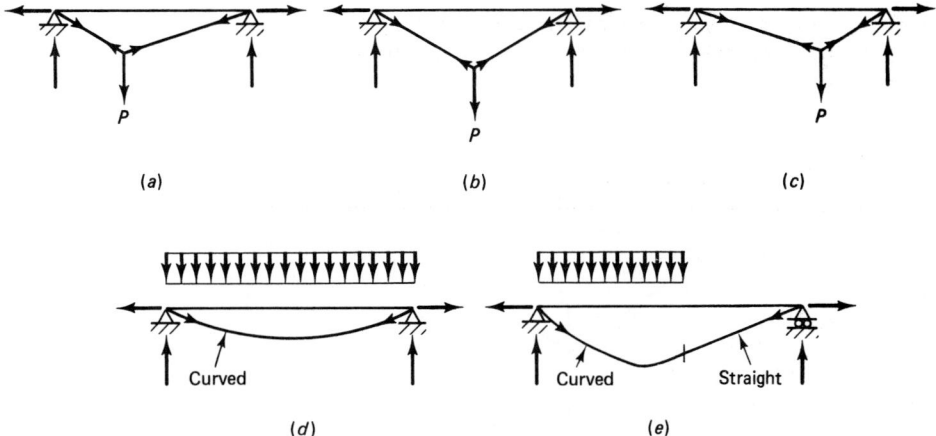

Figure 6.2 Change of cable shape with loading.

as shown in Fig. 6.4; here if the sag (or rise, for the arch) is increased, **the internal** force decreases and vice versa.

6.2 ANALYSIS OF CABLE SUBJECTED TO CONCENTRATED VERTICAL LOADS

Consider the case of a cable supported at A and B and acted upon by a system of concentrated vertical loads $W_1, W_2, W_3, \ldots, W_n$, as shown in Fig. 6.5a. Since all the external loads are vertical, the horizontal components of the two reactions, that is,

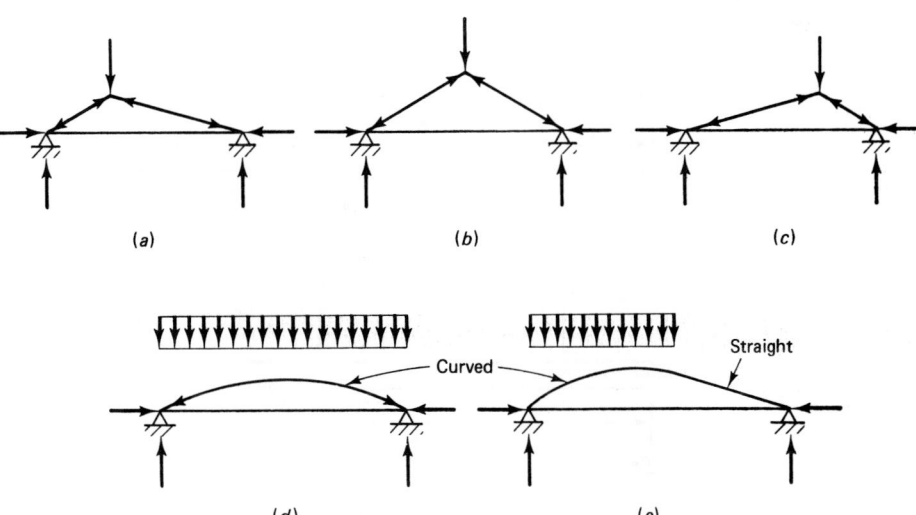

Figure 6.3 The development of "funicular arch" from the cables shown in Fig. 6.2.

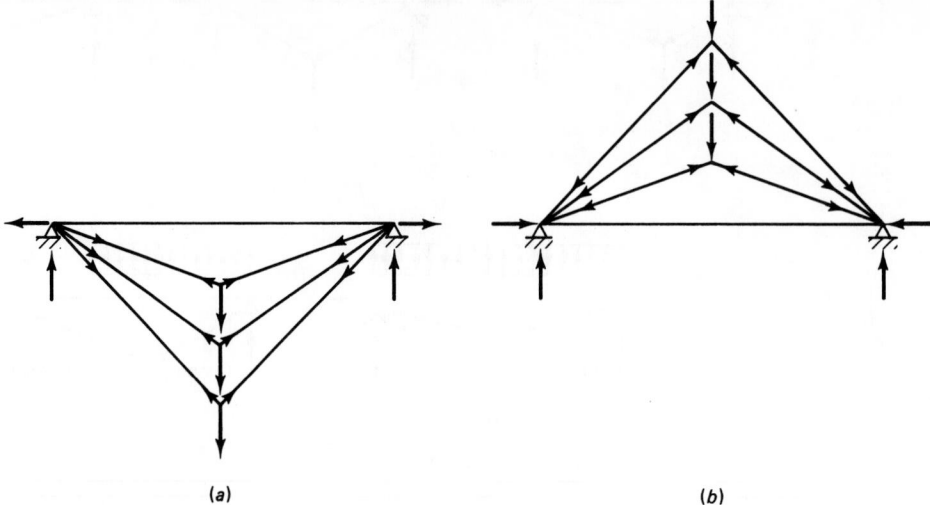

(a)   *(b)*

Figure 6.4 Family of "funicular cables" and "funicular arches."

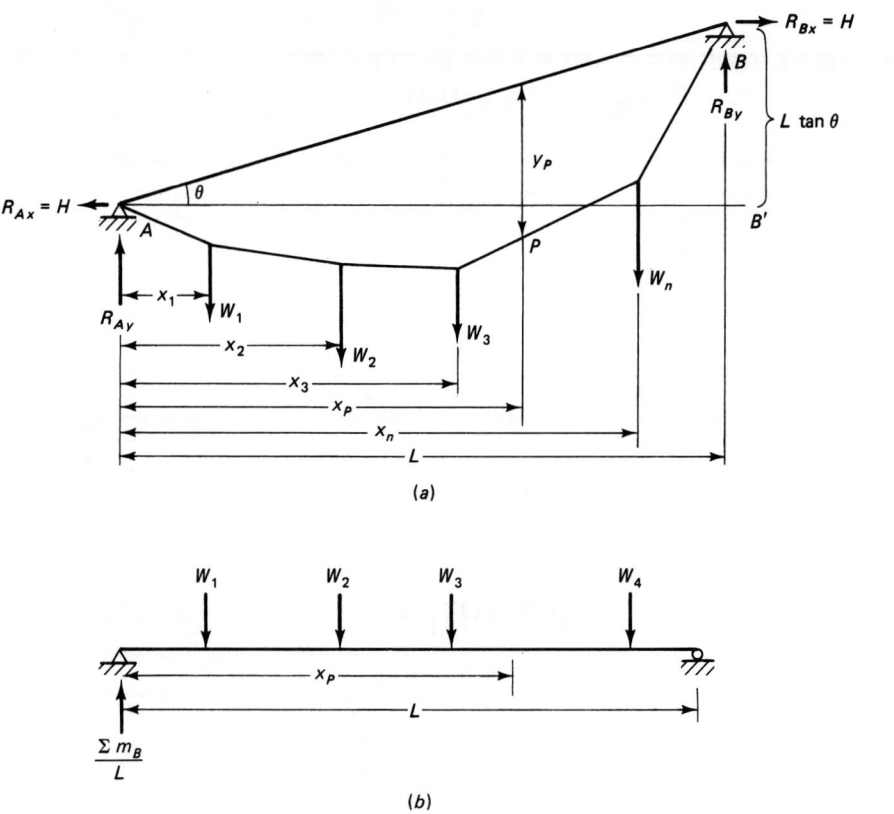

(a)

(b)

Figure 6.5 Cable subjected to concentrated loads.

R_{Ax} and R_{Bx}, must be equal and opposite to each other. These are denoted by H in Fig. 6.5a. From force equilibrium, the horizontal component of cable tension at any point in the cable is constant and is equal to H. If the distance from the cable chord (the line AB obtained by joining the two supports A and B) to the cable is known at any point P (shown as y_P at point P at distance x_P from A in Fig. 6.5a), the reactions can be computed as follows: Let $\Sigma\ m_B$ = sum of moments about B of all external loads $W_1, W_2, W_3, \ldots, W_n$

$\Sigma\ m_P$ = sum of moments about P of those loads W_1, W_2, W_3
acting on cable to left of point P

For moment equilibrium, moment of all the forces acting on the structure about point $B = 0$.

Assuming clockwise moments to be positive $(+\ \curvearrowright\ )$

$$(R_{Ay})(L) + (H)(L\tan\theta) - \Sigma\ m_B = 0 \tag{6.1}$$

$$\therefore R_{Ay} = \frac{\Sigma\ m_B}{L} - H\tan\theta \tag{6.1a}$$

Since the cable is assumed to be perfectly flexible, the bending moment is zero at all points. Applying that condition to point P,

$$(R_{Ay})(x_P) - H(y_P - x_P\tan\theta) - \Sigma\ m_P = 0 \tag{6.2}$$

Substituting for R_{Ay} from Eq. (6.1a) and simplifying, Eq. (6.2) becomes

$$H \cdot y_P = \frac{x_P}{L}\Sigma\ m_B - \Sigma\ m_P \tag{6.3}$$

Figure 6.5b shows a horizontal simply supported beam subjected to the same loading system acting on the cable. The left reaction for the beam, from statics, is $\Sigma\ m_B/L$. The bending moment at $P = (\Sigma\ m_B/L)x_P - \Sigma\ m_P$, which is the right-hand side of Eq. (6.3). Therefore, Eq. (6.3) leads to a theorem that can be stated as follows: Product of horizontal component of cable tension and vertical distance between the cable chord and cable at any point = bending moment at that point on a horizontal simply supported beam subjected to the same external loading as that on the cable.

Equation (6.3) will enable us to compute H and is thus of fundamental importance in the analysis of cables. As can be readily seen, this theorem is also applicable to the case of a cable acted upon by uniformly distributed vertical loading. After H is computed, the vertical reactions can be determined by applying the equations of statics to the cable. The shape of the cable can be determined from the condition that there should be no bending moment at any point in the cable.

As a numerical example, consider the cable loaded as shown in Fig. 6.6a. Let the maximum vertical distance between the cable chord and cable be 15 m. It is required to determine the reactions and the shape of the cable.

Let us consider the corresponding simply supported beam shown in Fig. 6.6b. Reactions at A' and B' are determined in the usual manner and are as shown in Fig. 6.6b. For the simply supported beam (Fig. 6.6b):

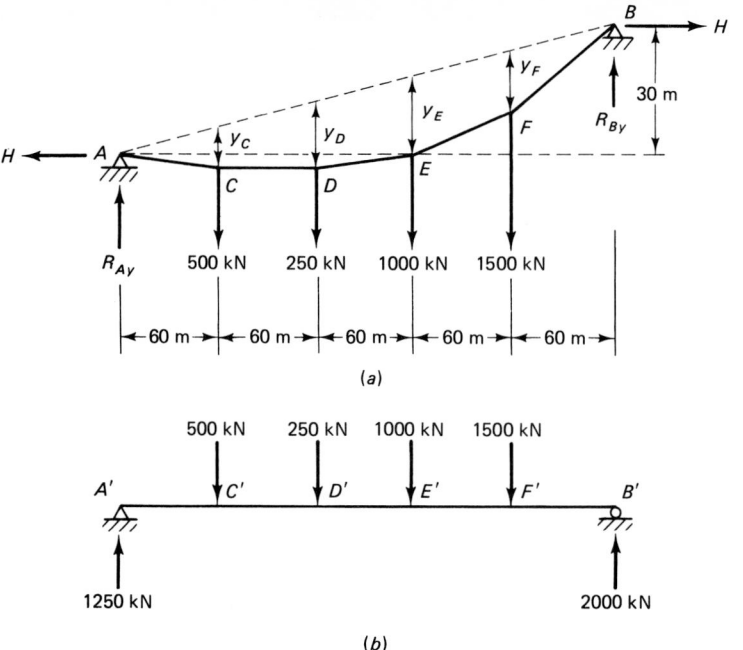

Figure 6.6 Analysis of a loaded cable.

Bending moment at $C' = (1250)(60) = 75{,}000$ kN $\cdot$ m

Bending moment at $D' = (1250)(120) - (500)(60) = 120{,}000$ kN $\cdot$ m

Bending moment at $E' = (1250)(180) - (500)(120) - (250)(60)$

$$= 150{,}000 \text{ kN} \cdot \text{m}$$

Bending moment at $F' = (2000)(60) = 120{,}000$ kN $\cdot$ m

Since for the case of a simply supported beam acted upon by vertical concentrated loads, maximum bending moment occurs under one of the point loads,

Maximum bending moment $=$ bending moment at $E' = 150{,}000$ kN $\cdot$ m

Applying Eq. (6.3),

$(H)(y_{\max}) =$ maximum bending moment of 150,000

$(H)(15) = 150{,}000$

$H = 10{,}000$ kN

Referring to Fig. 6.6a,

$\Sigma F_y = 0\uparrow +:$ $R_{Ay} + R_{By} - 500 - 250 - 1000 - 1500 = 0$ (a)

$\Sigma M_B = 0 \,\rotatebox[origin=c]{-45}{$\circlearrowright$}\, +:$ $(R_{Ay})(300) + (H)(30) - (500)(240) - (250)(180)$

$$- (1000)(120) - (1500)(60) = 0 \qquad \text{(b)}$$

From equation (b), $R_{Ay} = 250$ kN, whence $R_{By} = 3000$ kN from equation (a). Repeated application of Eq. (6.3) will yield the shape of the cable:

$$y_C = \frac{M'_C}{H} = \frac{75,000}{10,000} = 7.5 \text{ m}$$

$$y_D = \frac{M'_D}{H} = \frac{120,000}{10,000} = 12 \text{ m}$$

$$y_E = \frac{M'_E}{H} = \frac{150,000}{10,000} = 15 \text{ m}$$

$$y_F = \frac{M'_F}{H} = \frac{120,000}{10,000} = 12 \text{ m}$$

(y_C, y_D, y_E, and y_F are the vertical distances of points C, D, E, and F from the inclined cable chord AB, as shown in Fig. 6.6a.)

If it is desired to compute the tension in any particular segment of the cable, the horizontal component of tension H, which is constant throughout the cable, is vectorially added to the vertical component of tension in that particular cable segment. (The vertical component of tension is equal to the algebraic sum of vertical forces acting either to the left or to the right of the cable segment.) Continuing with the example problem (Fig. 6.6a), the computations for cable tension in various segments are conveniently shown in Table 6.1.

TABLE 6.1

Segment	Horizontal component (kN)	Vertical component (kN)	Resultant tension (kN)
AC	10,000	250	$\sqrt{10,000^2 + 250^2} = 10,003$
CD	10,000	$250 - 500 = -250$	$\sqrt{10,000^2 + (-250)^2} = 10,003$
DE	10,000	$250 - 500 - 250 = -500$	$\sqrt{10,000^2 + (-500)^2} = 10,012$
EF	10,000	$250 - 500 - 250 - 1000 = -1500$	$\sqrt{10,000^2 + (-1500)^2} = 10,112$
FB	10,000	$250 - 500 - 250 - 1000 - 1500 = -3000$	$\sqrt{10,000^2 + (-3000)^2} = 10,440$

Maximum tension in the cable occurs in segment *FB* and is equal to 10,440 kN.

6.3 ANALYSIS OF CABLE SUBJECTED TO UNIFORMLY DISTRIBUTED VERTICAL LOADS

Cables in structures like suspension bridges are acted upon by uniformly distributed vertical loads, that is, loading that is uniform per horizontal meter. If the sag is not significant, that is, has a shallow profile, a cable loaded by its self-weight can also be analyzed by considering the loading to be uniform per horizontal meter. Such a cable loading is shown in Fig. 6.7a.

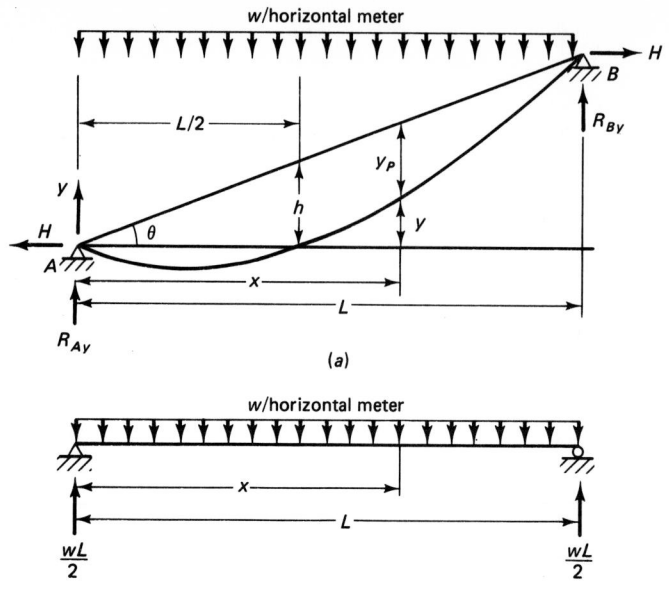

Figure 6.7 Analysis of cable subjected to uniformly distributed load.

Applying Eq. (6.3) to Fig. 6.7a and b,

$$(H)(y_P) = \left(\frac{wL}{2}\right)(x) - (wx)\left(\frac{x}{2}\right) = \frac{wx}{2}(L - x) \tag{6.4}$$

Let the sag of the cable at midspan be h; that is, when $x = L/2$, $y_P = h$. Substitution of these values in Eq. (6.4) yields

$$H \cdot h = \frac{w}{2} \cdot \frac{L}{2}\left(L - \frac{L}{2}\right)$$

or

$$H = \frac{wL^2}{8h} \tag{6.5}$$

For the case of a cable loaded with a uniformly distributed load, H can be determined by the application of Eq. (6.4) or Eq. (6.5). The vertical reactions R_{Ay} and R_{By} (Fig. 6.7a) can then be evaluated from statics ($\Sigma F_y = 0$, $\Sigma M = 0$).

The tension in the cable at any point can be computed by vectorially adding the horizontal and vertical components, similar to the case of the cable loaded with concentrated vertical loads. Maximum tension occurs at the higher end of the cable. (If the two supports are at the same level, that is, if the cable chord is horizontal, then the tensions at the two end supports are equal.) Minimum tension occurs at the point where the magnitude of the algebraic sum of the vertical component of forces is minimum. At any point, the slope of the cable to the horizontal is equal to the slope of the resultant tension.

The shape of the cable acted upon by uniformly distributed vertical loads is obtained by substituting in Eq. (6.4) the value of H from Eq. (6.5):

$$\left(\frac{wL^2}{8h}\right)(y_P) = \frac{wx}{2}(L-x)$$

or

$$y_P = \frac{4hx}{L^2}(L-x) \tag{6.6}$$

Equation (6.6) defines the shape of the cable, when x is measured horizontally from the left support and y_P is measured vertically from the cable chord (the line joining the two end supports). Sometimes it will be necessary to measure y from the horizontal x axis instead of from the inclined cable chord.

Referring to Fig. 6.7a, since $y = x \tan \theta - y_P$ (y is positive upward), Eq. (6.6) with respect to orthogonal x and y axes becomes

$$x \tan \theta - y = \frac{4hx}{L^2}(L-x)$$

$$y = x \tan \theta - \frac{4hx}{L^2}(L-x) \tag{6.7}$$

As a numerical example, consider the case of a cable shown in Fig. 6.8. The horizontal component of cable tension can be calculated from Eq. (6.5).

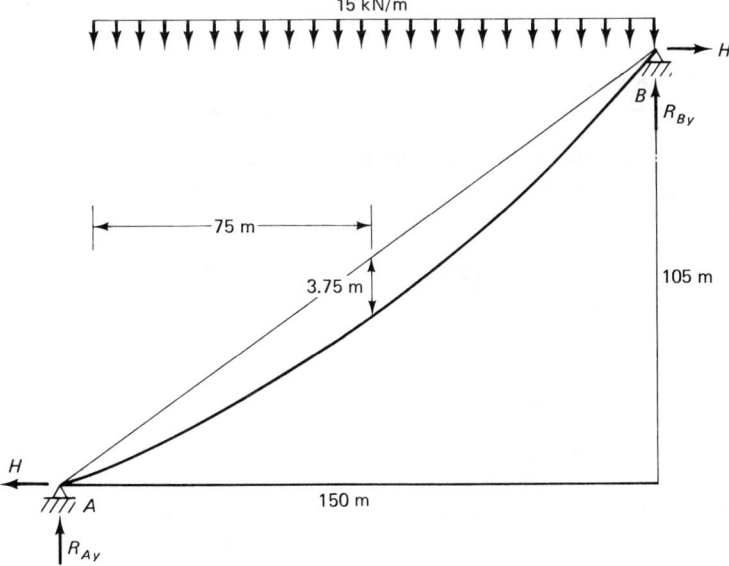

Figure 6.8 Analysis of cable with inclined cable chord.

$$H = \frac{wL^2}{8h} = \frac{(15)(150)^2}{(8)(3.75)} = 11{,}250 \text{ kN}$$

Vertical reaction at A is determined from $\Sigma M_B = 0$:

$$[\mathcal{C}+] \qquad (11{,}250)(105) + (R_{Ay})(150) - (15)(150)(75) = 0$$

$R_{Ay} = -6750$ kN, that is, downward. Vertical reaction at B is obtained from $\Sigma F_y = 0$:

$$[\uparrow+] \qquad -6750 - (15)(150) + R_{By} = 0 \qquad R_{By} = 9000 \text{ kN}$$

Maximum cable tension occurs at the higher end, that is, at B

$$T_{max} = \sqrt{H^2 + R_{By}^2} = \sqrt{11{,}250^2 + 9000^2} = 14{,}407 \text{ kN}$$

By inspection of Fig. 6.8, since $R_{Ay} = 6750$ kN downward, the magnitude of the algebraic sum of the vertical components of forces is minimum at left support A and is equal to

$$T_{min} = \sqrt{11{,}250^2 + 6750^2} = 13{,}120 \text{ kN}$$

The slope of the cable is maximum at point of maximum tension, that is, at right support B.

Max slope of cable = inclination of line of action of maximum tension

to horizontal

$$= \tan^{-1}\left(\frac{R_{By}}{H}\right) = \tan^{-1}\left(\frac{9000}{11{,}250}\right)$$

$$= 38.7° \text{ to the horizontal axis}$$

6.4 LENGTH OF CABLE

The total length of a cable acted upon by concentrated vertical loads can be obtained by adding the length of the various straight inclined segments computed from the formula

$$\Delta s = \Delta x \sqrt{1 + \left(\frac{\Delta y}{\Delta x}\right)^2} \tag{6.8a}$$

or

$$\Delta s = \Delta x \sqrt{1 + \left(\frac{V}{H}\right)^2} \tag{6.8b}$$

where V and H are vertical and horizontal components of cable tension in the segment. Suppose it is desired to determine the total length of cable shown in Fig. 6.6a. The computations can be conveniently tabulated as shown.

Segment	Δx (meters)	Vertical component of tension V (kN)	Horizontal component of tension H (kN)	$\Delta s = \Delta x \sqrt{1 + \left(\dfrac{V}{H}\right)^2}$ (meters)
AC	60	250	10,000	60.019
CD	60	$250 - 500 = -250$	10,000	60.019
DE	60	$250 - 500 - 250 = -500$	10,000	60.075
EF	60	$250 - 500 - 250 - 1000 = -1500$	10,000	60.671
FB	60	$250 - 500 - 250 - 1000 - 1500 = -3000$	10,000	62.642
Total length of cable =				303.426

The length of a uniformly loaded cable can only be obtained by integrating the expression

$$ds = dx \sqrt{1 + \left(\frac{dy}{dx}\right)^2} \tag{6.8a}$$

The quantity dy/dx can be obtained by differentiating Eq. (6.7). As an example, if it is desired to compute the length of cable shown in Fig. 6.8,

$$y = x \tan \theta - \frac{4hx}{L^2}(L - x)$$

$$= (x)\left(\frac{105}{150}\right) - \frac{(4)(3.75)x}{150^2}(150 - x)$$

$$= 0.7x - \frac{x(150 - x)}{1500}$$

Differentiating with respect to x,

$$\frac{dy}{dx} = 0.7 - 0.1 + \frac{x}{750} = 0.6 + \frac{x}{750} = \left(\frac{450 + x}{750}\right)$$

Applying Eq. (6.8a),

$$ds = dx \sqrt{1 + \left(\frac{dy}{dx}\right)^2}$$

$$s = \int_0^{150} \sqrt{1 + \left(\frac{450 + x}{750}\right)^2}\, dx$$

Expanding binomially and taking only the first few terms,

$$s = \int_0^{150} \left[(1)^{1/2} + \frac{1}{2} (1)^{-1/2} \frac{(450+x)^2}{750^2} + \frac{(\frac{1}{2})(-\frac{1}{2})}{2!} (1)^{-3/2} \frac{(450+x)^4}{750^4} \right.$$

$$\left. + \frac{(\frac{1}{2})(-\frac{1}{2})(-\frac{3}{2})}{3!} (1)^{-5/2} \frac{(450+x)^6}{750^6} + \frac{(\frac{1}{2})(-\frac{1}{2})(-\frac{3}{2})(-\frac{5}{2})}{4!} (1)^{-7/2} \frac{(450+x)^8}{750^8} \right] dx$$

$$s = \int_0^{150} \left[1 + \frac{1}{2} \frac{(450+x)^2}{750^2} - \frac{1}{8} \frac{(450+x)^4}{750^4} \right.$$

$$\left. + \frac{1}{16} \frac{(450+x)^6}{750^6} - \frac{5}{128} \frac{(450+x)^8}{750^8} \right] dx$$

$$= \left[x + \frac{1}{2} \frac{(450+x)^3}{(3)(750)^2} - \frac{1}{8} \frac{(450+x)^5}{(5)(750)^4} + \frac{1}{16} \frac{(450+x)^7}{(7)(750)^6} - \frac{5}{128} \frac{(450+x)^9}{(9)(750)^8} \right]_0^{150}$$

$$= (150 - 0) + \frac{600^3 - 450^3}{(2)(3)(750)^2} - \frac{600^5 - 450^5}{(8)(5)(750)^4}$$

$$+ \frac{600^7 - 450^7}{(16)(7)(750)^6} - \frac{5(600^9 - 450^9)}{(128)(9)(750)^8}$$

$$= 150 + 37 - 4.686 + 1.217 - 0.404$$

$$= 183.127 \text{ m}$$

6.5 ELASTIC STRETCH OF CABLE

When a load acts on a cable, the cable undergoes elastic stretch (elongation), which is often of importance in determining cable sags. From mechanics of materials,

$$\text{Elastic stretch (elongation)} = \delta = \frac{TL}{AE} \tag{6.9a}$$

where T = cable tension
L = cable length
A = cross-sectional area of cable
E = modulus of elasticity of cable material

For the case of a cable acted upon by concentrated vertical loads, the tension being constant in each straight inclined segment, the elastic stretch for each segment can be computed by Eq. (6.9a), and the total elastic stretch for the whole cable is obtained by adding the elastic stretches of individual cable segments.

On the other hand, for the case of a cable acted upon by uniformly distributed vertical loads, the tension continuously varies along the length of the cable. For such cases, Eq. (6.9a) should be modified as shown:

$$\text{Elastic stretch} = \delta = \int_0^s \frac{T_x \cdot ds}{AE} \tag{6.9b}$$

Substituting

$$ds = dx \sqrt{1 + \left(\frac{dy}{dx}\right)^2} \tag{6.8a}$$

and

$$T_x = \sqrt{H^2 + V^2} = H\sqrt{1 + \left(\frac{V}{H}\right)^2} = H\sqrt{1 + \left(\frac{dy}{dx}\right)^2} \tag{6.10}$$

Eq. (6.9b) becomes

$$\delta = \frac{H}{AE} \int_0^L \left[1 + \left(\frac{dy}{dx}\right)^2\right] dx \tag{6.9c}$$

As an example of the application of Eq. (6.9c), let it be required to find the elastic stretch of cable shown in Fig. 6.8. As earlier calculated,

$$\frac{dy}{dx} = \frac{450 + x}{750}$$

and

$$H = 11{,}250 \text{ kN}$$

From Eq. (6.9c)

$$\delta = \frac{11{,}250}{AE} \int_0^{150} \left[1 + \left(\frac{450 + x}{750}\right)^2\right] dx$$

$$= \frac{11{,}250}{AE} \left[x + \frac{(450 + x)^3}{(3)(750)^2}\right]_0^{150}$$

$$= \frac{11{,}250}{AE} \left[150 + \frac{600^3 - 450^3}{(3)(750)^2}\right]$$

$$= \frac{11{,}250}{AE} (150 + 74) = \frac{(11{,}250)(224)}{AE}$$

If A is in m^2, E is in kN/m^2, elastic stretch δ will be in meters.

6.6 GUYED STRUCTURES

As mentioned in the introduction, cables are commonly used in guyed structures. Figure 6.9a shows a simple guyed structure; guy AB holds the mast BC in equilibrium under the given loading. Taking moments of all forces acting on BC about C (Fig. 6.9b),

$$(H)(50) = (0.5)(50)(25)$$

$$H = 12.5 \text{ kN} = 12{,}500 \text{ N}$$

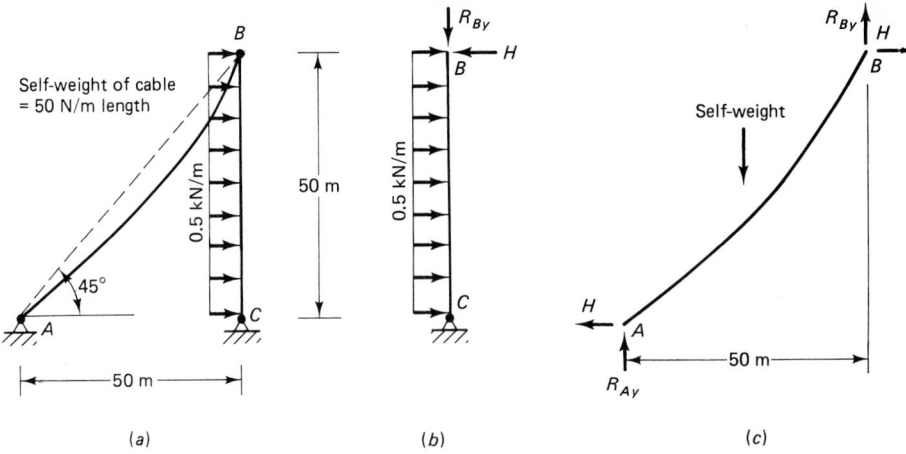

Figure 6.9 Example of a guyed structure.

To compute the vertical components of reaction R_{Ay} and R_{By} of the cable, it is first necessary to find the vertical load acting per horizontal meter length of cable.

The chord length $AB = \sqrt{50^2 + 50^2} = 70.711$ m. The actual length of cable can be taken as approximately equal to the chord length.

$$\text{Total weight of cable} = (70.711)(50)$$

$$= 3535.55 \text{ N}$$

$$\text{Horizontal projected length} = 50 \text{ m}$$

$$\text{Weight of cable per horizontal meter} = \frac{3535.55}{50}$$

$$= 70.711 \text{ N/m}$$

The cable sag must conform to Eq. (6.5), that is,

$$H = \frac{wL^2}{8h}$$

Thus,

$$h = \frac{wL^2}{8H} = \frac{(70.711)(50)^2}{(8)(12,500)} = 1.768 \text{ m}$$

Isolating cable AB, and taking moments of all forces about A (Fig. 6.9c),

$$(R_{By})(50) - (H)(50) - (70.711)(50)(25) = 0 \qquad R_{By} = 14{,}268 \text{ N}$$

Since $\Sigma F_y = 0$, $R_{Ay} = -14{,}268 + 3536 = -10{,}732$ N. From the equation of the shape of the cable [Eq. (6.7)]:

$$y = x \tan \theta - \frac{4hx}{L^2}(L - x)$$

$$= (x)(1) - \frac{(4)(1.768)x}{(50)^2}(50 - x)$$

$$\frac{dy}{dx} = 1 - \frac{(4)(1.768)}{50^2}(50 - 2x) = 1 + \frac{(4)(1.768)}{50^2}(2x - 50)$$

The slope is maximum at $x = 50$ m

$$\left(\frac{dy}{dx}\right)_{max} = 1 + \frac{(4)(1.768)}{50^2}(100 - 50) = 1.1414$$

$$T_{max} = H\sqrt{1 + \left(\frac{dy}{dx}\right)^2_{max}} = 12,500\sqrt{1 + (1.1414)^2}$$

$$= (12,500)(1.5175)$$

$$= 18,969 \text{ N}$$

Maximum tension can also be calculated as $\sqrt{H^2 + R^2_{By}} = \sqrt{12,500^2 + 14,268^2} = 18,969$ N. Sometimes, as an approximation, the sag of the cable is neglected and tension is assumed constant throughout the cable and acting along the cable chord AB:

$$(T_{max})_{approx} = \frac{H}{\cos 45°} = \frac{12,500}{\cos 45°} = 17,678 \text{ N}$$

6.7 SUSPENSION BRIDGES

Another very important class of cable structures is a suspension bridge, shown in Fig. 6.10a. The load from the bridge deck is transferred to the main cable by means of hangers, which are in direct tension. Since the live load can act anywhere on the bridge, the cable changes its shape (and hence the bridge deck as well) with the position of the live load unless the bridge deck is suitably stiffened either by means of a stiffening girder or by means of a stiffening truss. The stiffening girder or stiffening truss can be either three-hinged or two-hinged. Since a three-hinged stiffening girder or truss is statically determinate, the discussion here is limited to the three-hinged stiffening girder or truss. The purpose of the stiffening girder or truss is to supply sufficient rigidity to the bridge deck such that all the hangers carry equal load for all positions of the live load. The main cable thus retains its parabolic shape for all load positions.

The analysis of the three-hinged stiffening girder can be carried out as shown below. Referring to Fig. 6.10b, let P be the concentrated live load acting at a distance x from support A, where $x \leqslant L/2$. The stiffening girder transfers the load to the main cable as a uniformly distributed load of w per unit length, thus enabling the main cable to retain its parabolic shape. (The magnitude of w varies with the position of load P.)

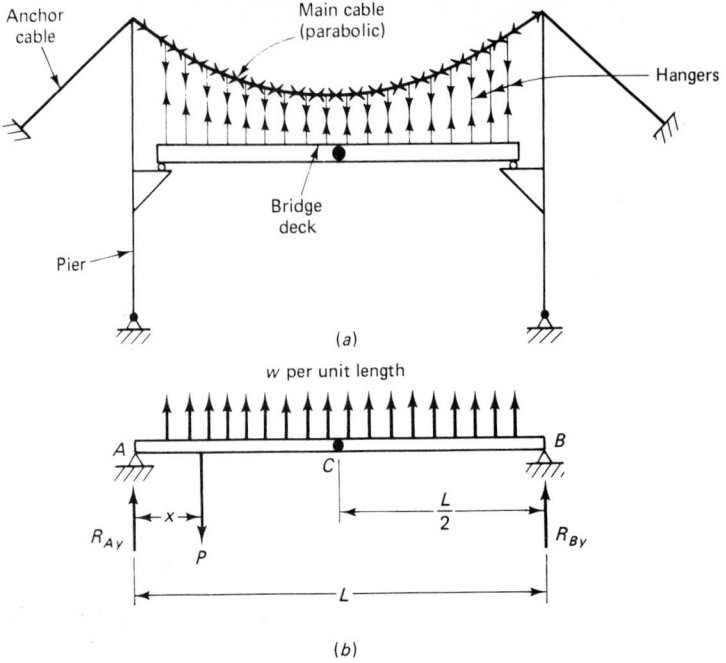

Figure 6.10 Three-hinged stiffening girder.

Since the bending moment at hinge C is zero, considering the moment of forces to the right of C,

$$(R_{By})\left(\frac{L}{2}\right) + (w)\left(\frac{L}{2}\right)\left(\frac{L}{4}\right) = 0$$

or

$$R_{By} = -\frac{wL}{4}$$

Since $\Sigma F_y = 0$,

$$R_{Ay} + R_{By} + wL - P = 0$$

$$R_{Ay} - \frac{wL}{4} + wL - P = 0$$

$$R_{Ay} = P - \tfrac{3}{4}wL$$

Taking moments of all the forces about B,

$$(R_{Ay})(L) + wL \cdot \frac{L}{2} - P(L - x) = 0$$

$$(P - \tfrac{3}{4}wL)(L) + \frac{wL^2}{2} - P(L - x) = 0$$

that is,

$$w = \frac{4Px}{L^2} \tag{6.11}$$

Knowing P, w, R_{Ay}, and R_{By}, the bending moment and shear force at any section of the three-hinged girder can be computed in the usual manner. As another example, consider the suspension bridge shown in Fig. 6.11a. Let it be required to calculate the maximum tension in the main cable and the anchor cable.

Referring to Fig. 6.11a, from Eq. (6.5):

$$H = \frac{wL^2}{8h} = \frac{(16)(120)^2}{(8)(12)} = 2400 \text{ kN}$$

Applying Eq. (6.7),

$$y = x \tan \theta - \frac{4hx}{L^2}(L - x)$$

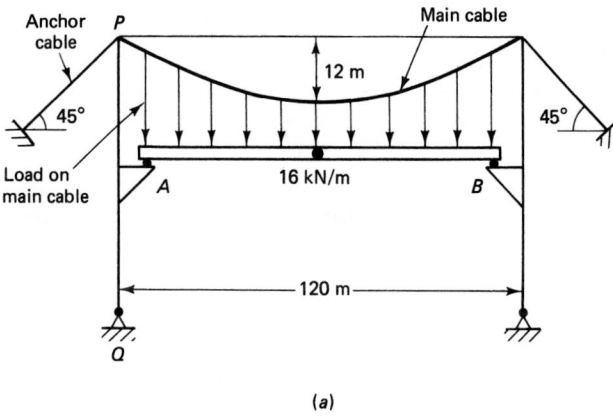

(a)

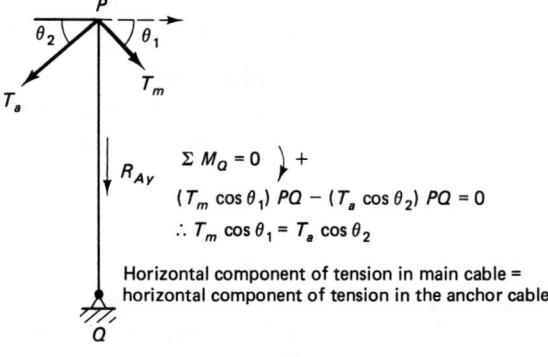

$\Sigma M_Q = 0 \quad \rangle\, +$

$(T_m \cos \theta_1)\, PQ - (T_a \cos \theta_2)\, PQ = 0$

$\therefore T_m \cos \theta_1 = T_a \cos \theta_2$

Horizontal component of tension in main cable = horizontal component of tension in the anchor cable

(b)

Figure 6.11 Analysis of a three-hinged stiffening girder.

$$= 0 - \frac{(4)(12)x}{120^2}(120 - x)$$

$$\frac{dy}{dx} = -\frac{(4)(12)}{120^2}(120 - 2x)$$

Maximum slope (numerical value) occurs at either support ($x = 0$ or 120).

$$\left(\frac{dy}{dx}\right)_{max} = -\frac{(4)(12)}{120^2}(120 - 240) = 0.4$$

Applying Eq. (6.10),

$$T_{max} = H\sqrt{1 + \left(\frac{dy}{dx}\right)_{max}^2} = H\sqrt{1 + (0.4)^2} = 2400\sqrt{1 + 0.16} = 2585 \text{ kN}$$

The horizontal component of tension in the anchor cable is equal to the horizontal component of tension in the main cable. (This can be verified by isolating pier PQ and taking moments of forces about Q, as shown in Fig. 6.11b.)

Neglecting the sag of the anchor cable, the tension in the anchor cable (constant throughout the length) is

$$\frac{H}{\cos 45°} = \frac{2400}{\cos 45°} = 3394 \text{ kN}$$

It should be mentioned that the sag of the cable greatly influences the tensions in the cable segments. For all other structures covered in this book, the deflection of a structure has negligible effect on its structural analysis; for example, for rigid structures, the change in geometry is ignored. However, for a cable structure, the change in geometry has a pronounced effect on the tension in the cable; in this case the final sag will not be known a priori and the problem is essentially iterative in nature. For a detailed discussion of the subject, the student is referred to the bibliography at the end of the chapter; only a simple example is presented in Solved Problem 6.3.

6.8 THREE-HINGED ARCHES

An arch can be visualized as an inversion of a cable and is thus an essentially compressive structure. Whereas a cable cannot resist any bending moment, an arch is sufficiently rigid to be able to resist bending moments. There are three types of arches:

1. Three-hinged arches
2. Two-hinged arches
3. Hingeless (or fixed) arches

These are shown in Fig. 6.12a to c. The two end supports for an arch can be either at the same level or at different levels. Of the three types of arches, only the three-hinged arch is a statically determinate structure. (The four unknown reactions can be determined from the three equations of static equilibrium plus the condition equation

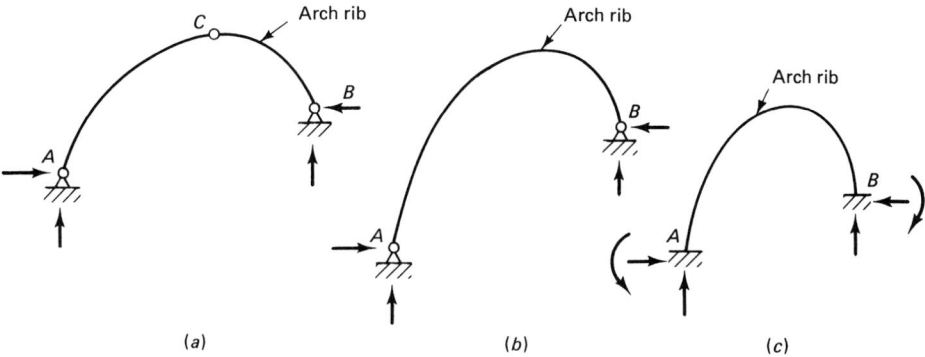

Figure 6.12 Different types of arches. (*a*) Three-hinged arch. (*b*) Two-hinged arch. (*c*) Hingeless (fixed) arch.

that the bending moment at hinge *C* is zero.) The two-hinged arch is statically inde-terminate to the first degree (four unknown reaction components and three equations of static equilibrium) and the hingeless (fixed) arch is statically indeterminate to the third degree (six unknown reaction components and three equations of equilibrium). The discussion in the remainder of this chapter is limited to three-hinged arches; two-hinged arches are treated in Chapters 13 and 14.

Consider a three-hinged arch loaded with a uniformly distributed load as shown in Fig. 6.13. From symmetry, the vertical reactions at the two end supports are each equal to one-half the total, that is, $R_{Ay} = R_{By} = wL/2$. From consideration of horizon-tal equilibrium ($\Sigma F_x = 0$), the horizontal components of reaction at the two end supports must be equal and opposite to each other as shown. Since the bending moment at hinge *C* is zero:

$$M_C = 0 \; \rangle \; +: \qquad (R_{Ay})\left(\frac{L}{2}\right) - (H)(y_C) - (w)\left(\frac{L}{2}\right)\left(\frac{L}{4}\right) = 0$$

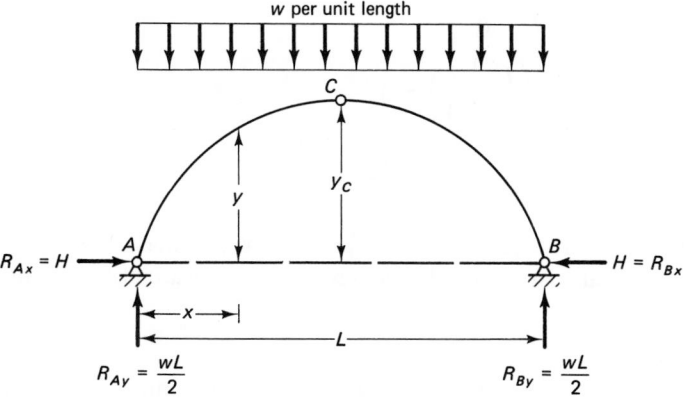

Figure 6.13 Analysis of a three-hinged arch.

that is,

$$\left(\frac{wL}{2}\right)\left(\frac{L}{2}\right) - (H)(y_C) - (w)\left(\frac{L}{2}\right)\left(\frac{L}{4}\right) = 0$$

$$\therefore H = \frac{wL^2}{8\,y_C}$$

the horizontal component of reaction depends on y_C, the rise of the arch at the interior hinge.

The bending moment at any section distant x from A is

$$M_x = \left(\frac{wL}{2}\right)(x) - (H)(y) - (w)(x)\left(\frac{x}{2}\right)$$

$$= \frac{wL}{2}x - \frac{wx^2}{2} - Hy$$

For the special case when $H \cdot y = (wL/2)x - (wx^2/2)$, the bending moment is zero at all sections of the arch and the arch is thus in pure compression; such a shape of the arch, as already mentioned, is called the "funicular" shape.

For a given span, location of hinges, and loading, there is only one funicular shape of the arch. When the loading changes, the shape is no longer funicular and there will be bending moment at different sections of the arch rib. In actual practice, since an arch rib has to resist different loading conditions, it is, in general, subjected to moment and shear in addition to axial thrust.

Suppose it is required to find the funicular shape for the three-hinged arch with extremities shown in Fig. 6.14a:

Take moments of all the forces about A

$$\Sigma\, M_A = 0 \text{) } +: \qquad (R_{By})(L) + (H)\left(\frac{L}{4}\right) - (P)\left(\frac{L}{4}\right) = 0$$

$$R_{By} + \frac{H}{4} = \frac{P}{4}$$

Bending moment at $C = 0$:

$$(R_{By})\left(\frac{L}{2}\right) - (H)\left(\frac{L}{2} - \frac{L}{4}\right) = 0$$

$$R_{BY} = \frac{H}{2}$$

Solving the two equations, $H = \frac{1}{3}P$, $R_{By} = \frac{1}{6}P$. Since $\Sigma\, F_y = 0$, $R_{Ay} = \frac{5}{6}P$. To find the funicular shape, consider each segment of the arch rib separately (Fig. 6.14b):

Segment AD

Let the rise of arch be y at a distance x from A (y is measured from the horizontal through A).

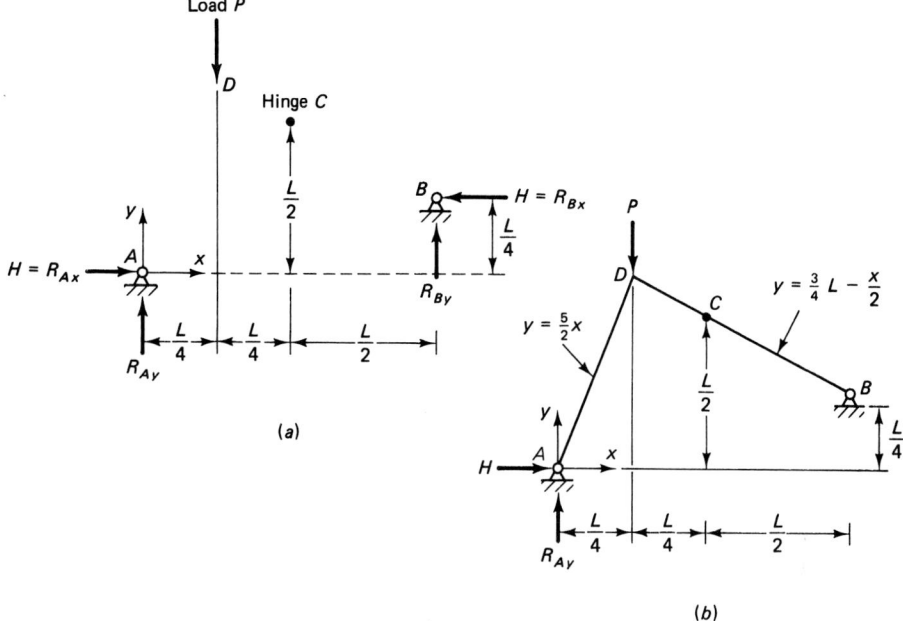

Figure 6.14 Determination of funicular shape for a three-hinged arch.

Bending moment at $x\left(0 \leqslant x \leqslant \dfrac{L}{4}\right) = R_{Ay} \cdot x - H \cdot y$

$$= (\tfrac{5}{6}P)(x) - (\tfrac{1}{3}P)(y)$$

$$= 0 \qquad \text{for funicular shape}$$

Therefore,

Equation of arch rib is $y = \tfrac{5}{2}x$

Rise of arch at $D\left(x = \dfrac{L}{4}\right) = y_D = \dfrac{5L}{8}$

Segment DC

Again measuring x from A and y perpendicular to x,

Bending moment at $x\left(\dfrac{L}{4} \leqslant x \leqslant \dfrac{L}{2}\right) = (R_{Ay})(x) - H \cdot y - P\left(x - \dfrac{L}{4}\right)$

$$= \frac{5}{6}P \cdot x - \frac{1}{3}P \cdot y - P\left(x - \frac{L}{4}\right)$$

$$= 0 \text{ for funicular shape}$$

that is,

$$y = \frac{3}{4}L - \frac{x}{2}$$

Rise of arch at $D\left(x = \frac{L}{4}\right) = y_D = \frac{5}{8}L$

same as before.

Rise of arch at $C\left(x = \frac{L}{2}\right) = y_C = \frac{L}{2}$ given

Segment CB

With reference to the same set of coordinate axes,

$$\text{Bending moment at } x\left(\frac{L}{2} \leqslant x \leqslant L\right) = (R_{Ay})(x) - (H)(y) - P\left(x - \frac{L}{4}\right)$$

$$= \left(\frac{5}{6}P\right)(x) - \left(\frac{1}{3}P\right)(y) - P\left(x - \frac{L}{4}\right)$$

$$= 0 \qquad \text{for funicular shape}$$

that is,

$$y = \frac{3}{4}L - \frac{x}{2}$$

same equation as for segment DC.

Rise of arch at $C\left(x = \frac{L}{2}\right) = y_C = \frac{L}{2}$ given

Rise of arch at $B\ (x = L) = y_B = \frac{L}{4}$ given

The funicular shape of the arch is shown in Fig. 6.14b.

If the shape of the arch is not funicular, there will be bending moment and also shear force V (since $V = dM/dx$) at different sections in the arch rib, in addition to axial compression. As an example of the computation of these forces and moments, consider the three-hinged arch shown in Fig. 6.15a. Suppose it is required to find the axial force, shear force, and bending moment at section D. First, compute the reactions in the usual manner:

$$\Sigma M_A = 0 \curvearrowright +: \qquad (R_{By})(100) - (100)(20) = 0$$

$$\therefore R_{By} = 20 \text{ kN}$$

Bending moment at $C = 0$: $(R_{By})(50) - (H)(30) = 0$

$$(20)(50) - (H)(30) = 0$$

$$H = \tfrac{100}{3}\text{kN}$$

$$\Sigma F_y = 0: \qquad R_{Ay} = 100 - 20 = 80 \text{ kN}$$

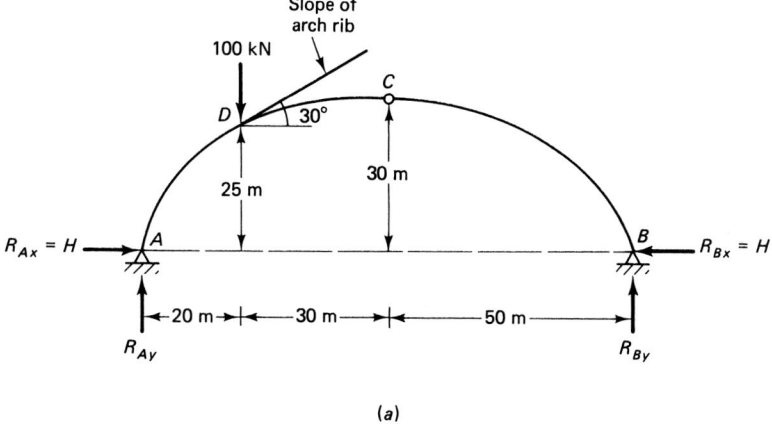

(a)

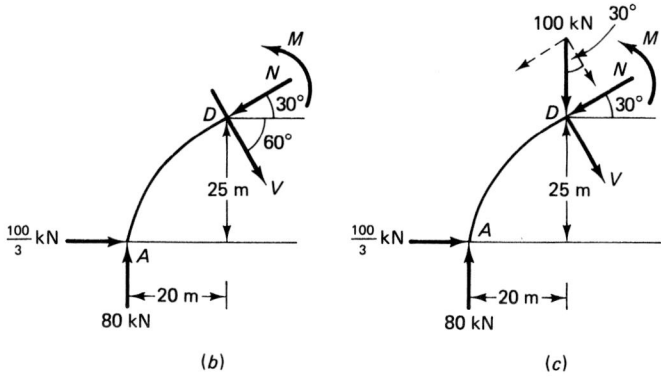

(b)   (c)

Figure 6.15 Numerical example of a three-hinged arch.

Since at D there is a point load acting, the axial force and shear force just to the left of D will be different from the axial force and shear force just to the right of D. So two different sections should be taken and two different sets of forces computed.

Section Just to the Left of D

The free-body diagram is shown in Fig. 6.15b. Resolving the forces in the direction of N,

$$80 \sin 30° + \tfrac{100}{3} \cos 30° - N = 0$$

$$N = 68.87 \text{ kN} \qquad (\text{compressive})$$

Resolving the forces in the direction of V,

$$80 \cos 30° - \tfrac{100}{3} \sin 30° - V = 0$$

$$V = 52.61 \text{ kN in the direction shown}$$

Taking moments of all forces about D,

$$(80)(20) - (\tfrac{100}{3})(25) - M = 0$$

$M = 766.7$ kN·m in the direction shown

Section Just to the Right of D

The free-body diagram for this case is shown in Fig. 6.15c. Resolving the forces in the direction of N,

$$80 \sin 30° + \tfrac{100}{3} \cos 30° - 100 \sin 30° - N = 0$$

$N = 18.87$ kN   (compressive)

Resolving the forces in the direction of V,

$$80 \cos 30° - \tfrac{100}{3} \sin 30° - 100 \cos 30° - V = 0$$

$V = -33.99$ kN

that is, opposite to the direction assumed, that is in the upward direction.
Taking moments of all forces about D,

$$(80)(20) - \left(\frac{100}{3}\right)(25) - M = 0$$

$M = 766.7$ kN·m

same as for the section just to the left of D.

6.9 THREE-HINGED TRUSSED ARCH

Sometimes instead of an arch rib, a truss system can be used as shown in Fig. 6.16; such a structure is called a three-hinged trussed arch. Once the end reactions are determined, the forces in the truss members can be computed in the usual manner either by the method of joints or by the method of sections. Referring to Fig. 6.16, it is to be noted that the top member NO is made inactive to facilitate hinge action about the panel point E. (Member MN could have been made inactive instead of member NO.) Suppose it is required to compute the forces in bars LM, CD, LC, and LD. We first determine the reactions at the two end hinges A and I in the usual manner. From symmetry,

$$R_{Ay} = R_{Iy} = 1200 \text{ kN}$$

Since the bending moment at hinge E is zero,

$$(R_{Ay})(120) - (H)(48) - (150)(120) - (300)(90)$$

$$- (300)(60) - (300)(30) = 0$$

$H = 1500$ kN

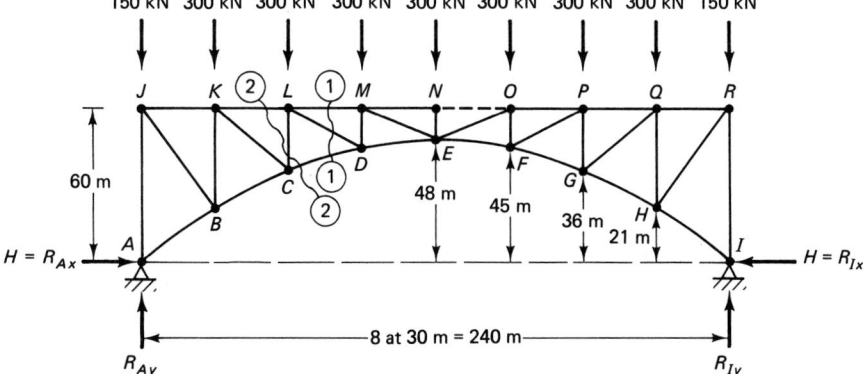

Figure 6.16 Three-hinged trussed arch.

Force in *LM*

To compute the force in member *LM*, consider section ① – ① as shown in Fig. 6.16 and take moments of all forces acting to the left of ① – ① about joint *D*:

$$(1200)(90) - (1500)(45) - (150)(90)$$
$$- (300)(60) - (300)(30) + (F_{LM})(15) = 0$$

$$\therefore F_{LM} = 0$$

Force in *CD*

Again, consider section ① – ① as shown in Fig. 6.16 and take moments of all forces acting to the left of ① – ① about joint *L*:

$$(1200)(60) - (1500)(60) - (150)(60)$$
$$- (300)(30) - \left(F_{CD} \times \frac{30}{\sqrt{981}}\right)(24) = 0$$

$$F_{CD} = -50\sqrt{981} \quad \text{that is} \quad 1566 \text{ kN} \quad \text{(compression)}$$

Force in *LC*

Consider section ② – ② as shown in Fig. 6.16. For the vertical equilibrium of all forces acting to the left of section ② – ② we have

$$1200 - 150 - 300 + F_{LC} - (1566) \times \frac{9}{\sqrt{981}} = 0$$

$$F_{LC} = -300 \quad \text{that is} \quad 300 \text{ kN} \quad \text{(compression)}$$

Force in *LD*

Consider section ① − ① and the vertical equilibrium of all the forces acting to the left of ① − ①:

$$1200 - 150 - 300 - 300 - (F_{LD})\frac{15}{\sqrt{1125}} - (1566) \times \frac{9}{\sqrt{981}} = 0$$

$$F_{LD} = 0$$

6.10 SOLVED PROBLEMS

■ **SOLVED PROBLEM 6.1**

A steel suspension bridge cable, having a central sag equal to one-tenth of its span, carries a superimposed uniformly distributed load of 2 k/ft. The density of the steel cable is 490 lb/ft^3. The self-weight of the cable can be considered as uniformly distributed per foot of horizontal span.

(a) If the permissible stress in the cable is 32 ksi, determine the maximum possible span of the cable.
(b) If the permissible stress is 32 ksi and the anchorage difficulties limit the maximum tension in the cable to 16,000 k, what is the maximum possible span?

Solution

(a) From Eq. (6.8a),

$$\text{Length of cable} = S = \int_0^L \sqrt{1 + \left(\frac{dy}{dx}\right)^2}\, dx$$

From Eq. (6.7), substituting $\theta = 0$,

$$y = -\frac{4hx}{L^2}(L - x)$$

$$\frac{dy}{dx} = \frac{4h}{L^2}(2x - L)$$

$$\therefore S = \int_0^L \sqrt{1 + \left[\frac{4h}{L^2}(2x - L)\right]^2}\, dx$$

$$= \int_0^L \sqrt{1 + \frac{16h^2}{L^4}(2x - L)^2}\, dx$$

Expanding binomially and neglecting higher-order small quantities,

$$S = \int_0^L \left[1 + \frac{1}{2}\frac{16h^2}{L^4}(2x - L)^2\right] dx$$

$$= \left[x + \frac{8h^2}{L^4}\frac{(2x - L)^3}{(3)(2)}\right]_0^L$$

$$= L + \frac{8}{3}\frac{h^2}{L}$$

Total weight of cable $= (A)\left(L + \dfrac{8}{3}\dfrac{h^2}{L}\right)\dfrac{490}{1000}$ k

$$= (A)\left[L + \dfrac{8}{3}\dfrac{(0.1L)^2}{L}\right]\dfrac{490}{1000}\text{ k}$$

Weight/foot $= \dfrac{\text{total weight}}{L} = A(1 + 0.0267) \times 0.490 = 0.503A$ k

Total uniformly distributed load $= w = (2 + 0.503A)$ k

From Eq. (6.5),

$$H = \dfrac{wL^2}{8h} = \dfrac{wL^2}{8(0.1L)} = 1.25wL$$

Vertical reaction $= \dfrac{wL}{2} = 0.5wL$

Maximum tension $= \sqrt{(1.25wL)^2 + (0.5wL)^2} = 1.3463wL$

Substituting $w = (2 + 0.503A)$ k

$T_{\max} = 1.3463(2 + 0.503A)L$ k

Max stress $= \dfrac{T_{\max}}{A} = \dfrac{1.3463(2 + 0.503A)L}{A}$ k/ft^2

This should not exceed 32 k/in^2 or 4608 k/ft^2.

$$\therefore \dfrac{1.3463(2 + 0.503A)L}{A} \not> 4608$$

$$\left(\dfrac{2.6926}{A} + 0.6773\right)L \not> 4608$$

For maximum possible span, A has to be maximum. When $A \rightarrow \infty$,

$0.6773L = 4608$ $L = 6803$ ft

(b) If the maximum tension is limited to 16,000 k,

$1.3463(2 + 0.503A)L \not> 16,000$

Area of cross section of cable $A = \dfrac{16,000}{4608} = 3.472$ ft^2

$\therefore 1.3463(2 + 0.503 \times 3.472)L \not> 16,000$ $L \not> 3172$ ft   ∎

■ SOLVED PROBLEM 6.2

A suspension cable of 160-ft span and 16-ft sag carries a load of 1 k per lineal horizontal foot. Calculate

(a) The maximum and minimum tensions in the cable
(b) The horizontal and vertical forces in each pier if the cable passes over fric-

tionless rollers on the tops of the piers and the backstays are inclined at 30°
to the horizontal

(c) The horizontal and vertical forces in each pier if the cable is connected to
saddles carried on frictionless rollers on the tops of piers and the backstays
are inclined at 30° to the horizontal.

Solution

(a) From Eq. (6.5), and referring to Fig. 6.17*a*,

$$T_{min} = H = \frac{wL^2}{8h} = \frac{(1)(160)^2}{(8)(16)} = 200 \text{ k}$$

Vertical reaction on each pier due to suspension cable is

$$\frac{wL}{2} = \frac{(1)(160)}{2} = 80 \text{ k}$$

$$T_{max} = \sqrt{200^2 + 80^2} = 215.4 \text{ k}$$

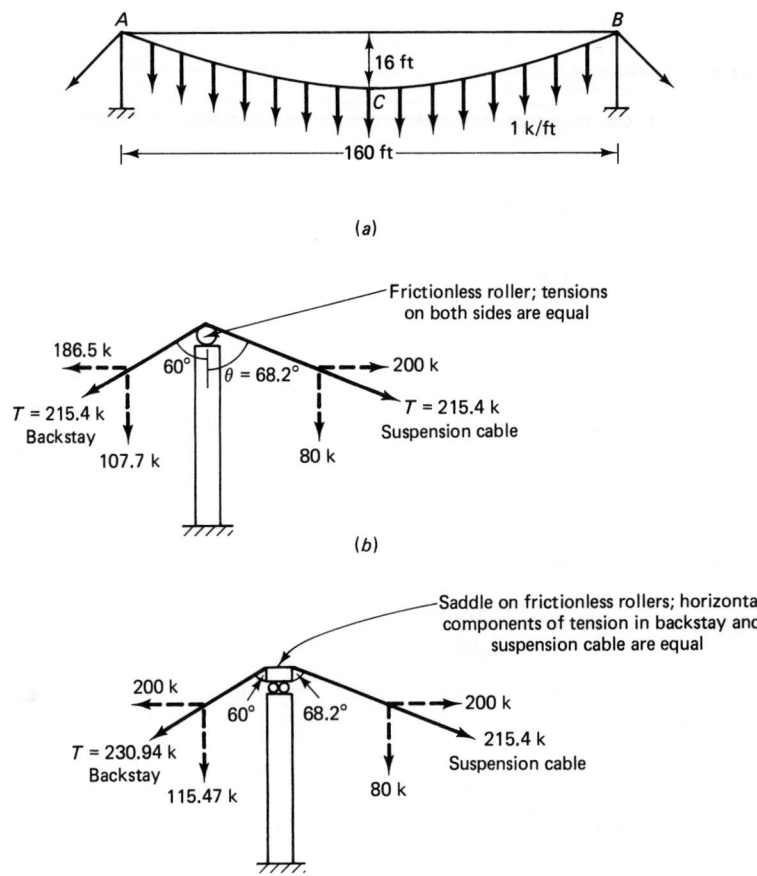

(a)

(b)

(c)

Figure 6.17 Structure for Solved Problem 6.2.

(b) Refer to Fig. 6.17b.

$$\tan \theta = \frac{H}{\text{vertical reaction due to suspension cable}} = \frac{200}{80} = 2.5$$

$$\theta = 68.2°$$

The tension in the backstay = tension in suspension cable at support = 215.4 k.

From Fig. 6.17b

Due to suspension cable:

Vertical component of tension = 80 k↓

Horizontal component of tension = 200 k→

Due to backstay:

Vertical component of tension = 215.4 cos 60° = 107.7 k↓

Horizontal component of tension = 215.4 sin 60° = 186.5 k←

∴ Total vertical force on the pier = 80 + 107.7 = 187.7 k↓

Total horizontal force on the pier = 200 − 186.5 = 13.5 k→

(c) Refer to Fig. 6.17c. The horizontal components of tension in the suspension cable and backstay must be equal for horizontal equilibrium.

$$T \sin 60° = 200 \qquad T = 230.94 \text{ k}$$

Due to backstay:

$$\text{Vertical component of tension} = T \cos 60°$$
$$= 230.94 \cos 60°$$
$$= 115.47 \text{ k↓}$$

Horizontal component of tension = $T \sin 60°$ = 200 k←

The components of tension due to suspension cable are as calculated in (b) above.

∴ Total vertical force on the pier = 80 + 115.47 = 195.47 k↓

Total horizontal force on the pier = 200 − 200 = 0 ∎

■ **SOLVED PROBLEM 6.3**

A steel cable of cross-sectional area 2000 mm^2 (E = 180 GPa) with an initial sag of 5 m is acted upon by a single concentrated load of 100 kN as shown in Fig. 6.18a. Determine the final sag and maximum cable tension, taking into account the elastic stretch of cable.

Solution

Unstretched length $AC = \sqrt{150^2 + 5^2} = 150.0833$ m

Unstretched length $BC = \sqrt{50^2 + 5^2} = 50.2494$ m

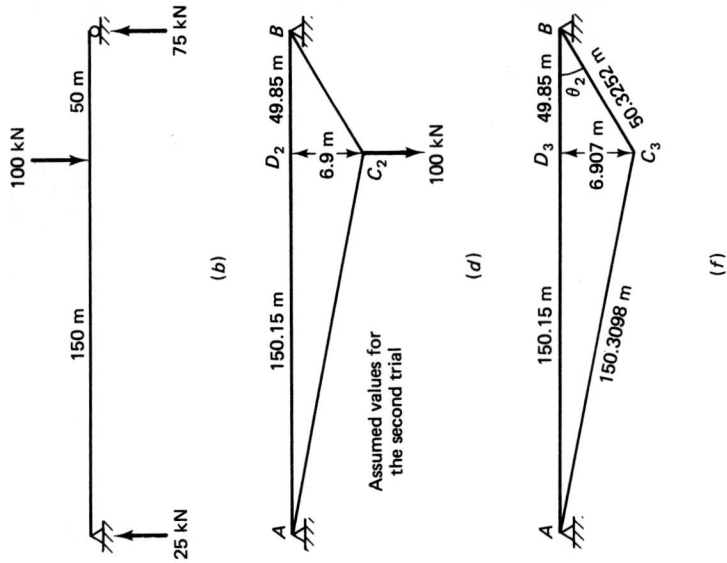

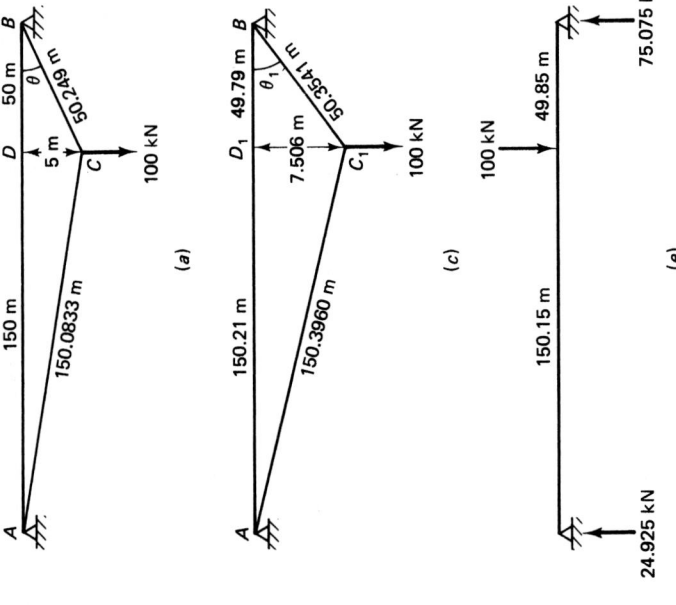

Figure 6.18 Cable for Solved Problem 6.3.

Referring to Fig. 6.18b, applying Eq. (6.3),

$$H = \frac{25 \times 150}{5} = 750 \text{ kN}$$

Elastic stretch of AC [from Eq. (6.9c)]

$$\delta_{AC} = \frac{750,000}{2000 \times 180,000} \int_0^{150} \sqrt{1 + \left(\frac{5}{150}\right)^2} \, dx = 0.3127 \text{ m}$$

Elastic stretch of CB [from Eq. (6.9c)]

$$\delta_{CB} = \frac{750,000}{2000 \times 180,000} \int_0^{50} \sqrt{1 + \left(\frac{5}{150}\right)^2} \, dx = 0.1047 \text{ m}$$

Length of AC after elastic stretch = unstretched length + elastic stretch

$$= 150.0833 + 0.3127 = 150.3960 \text{ m}$$

Length of CB after elastic stretch = $50.2494 + 0.1047 = 50.3541$ m

The position of the cable after the elastic stretch is shown in Fig. 6.18c. Angle θ_1 can be obtained from the cosine rule of the triangle thus:

$$AC_1^2 = AB^2 + BC_1^2 - 2(AB)(BC_1)\cos \theta_1$$

Substituting $AC_1 = 150.3960$ m, $BC_1 = 50.3541$ m, and $AB = 200$ m, we get

$$\cos \theta_1 = 0.9888$$

$$\theta_1 = 8.57°$$

and

Sag $C_1 D_1 = 50.3541 \sin \theta_1 = 7.506$ m

Since the sag is one and a half times the original value for unstretched cable, the horizontal component of tension will be about two-thirds of the earlier calculated value, and thus the computations have to be revised. The actual value will be intermediate between these two values; let us assume the sag to be 6.9 m. Also assume that $AD = 150.15$ m and $DB = 49.85$ m corresponding to the assumed sag of 6.9 m as shown in Fig. 6.18d, and repeat the calculations.

From Fig. 6.18e, applying Eq. (6.3),

$$H = \frac{24.925 \times 150.15}{6.9} = 542.4 \text{ kN}$$

Elastic stretch of AC [Eq. (6.9c)]

$$\delta_{AC} = \frac{542,400}{2000 \times 180,000} \int_0^{150.15} \sqrt{1 + \left(\frac{6.9}{150.15}\right)^2} \, dx = 0.2265 \text{ m}$$

Similarly

$$\delta_{CB} = \frac{542,400}{2000 \times 180,000} \int_0^{49.85} \sqrt{1 + \left(\frac{6.9}{49.85}\right)^2} \, dx = 0.0758 \text{ m}$$

Length of AC after elastic stretch = $150.0833 + 0.2265 = 150.3098$ m

Length of CB after elastic stretch = $50.2494 + 0.0758 = 50.3252$ m

The position of the cable after the elastic stretch is shown in Fig. $6.18f$. Applying the cosine rule as before, θ_2 is calculated to be $7.89°$ and

Sag $C_3 D_3 = 50.3252 \sin 7.89° = 6.907$ m

which is close to the assumed value of 6.9 m

Maximum tension = $(H)(\sec 7.89°) = 547.6$ kN

This example clearly demonstrates that the final sag is not known a priori and the problem is essentially iterative in nature. ■

PROBLEMS

6.1 and 6.2 Determine **(a)** the reactions, and **(b)** the location and magnitude of maximum tension in the cable.

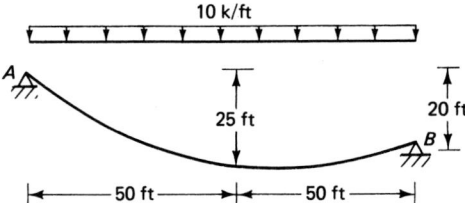

Figure P6.1

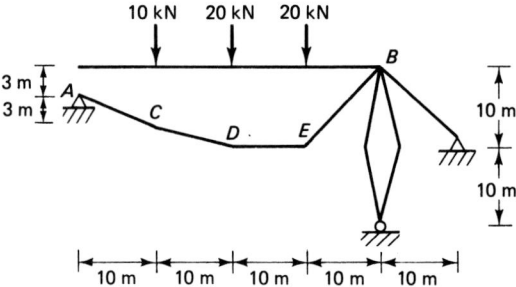

Figure P6.2

6.3 and 6.4 Determine the maximum cable tension for the cable structures shown.

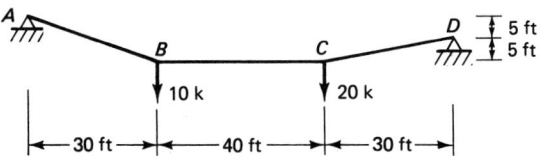

Figure P6.3

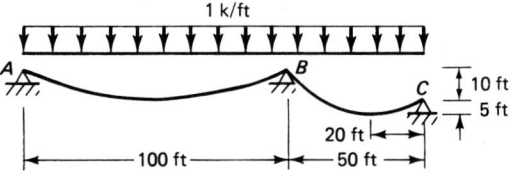

Figure P6.4

6.5 and 6.6 For the structures shown, determine (**a**) maximum tension in the main cable, (**b**) maximum tension in the anchor cable, and (**c**) the compressive force in the mast.

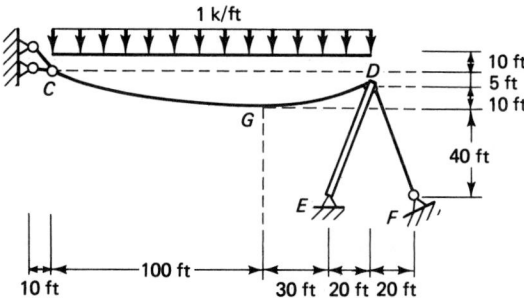

Figure P6.5

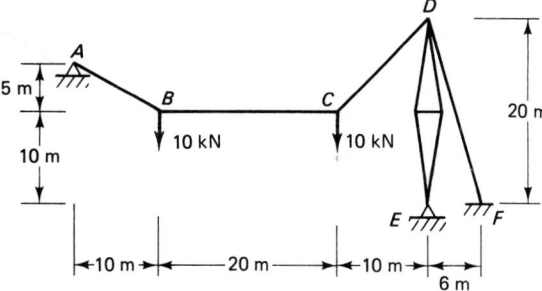

Figure P6.6

6.7 and 6.8 For the cables shown, determine the **(a)** maximum tension, **(b)** location of lowest point, **(c)** magnitude of sag.

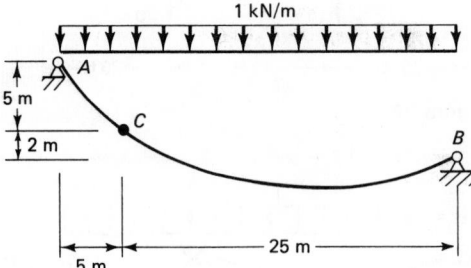

Figure P6.7

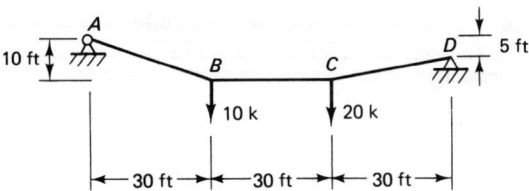

Figure P6.8

6.9 A cable with supports at different levels as shown is used to carry a pipeline weighing 5 kN/m. Determine the maximum tension in the cable.

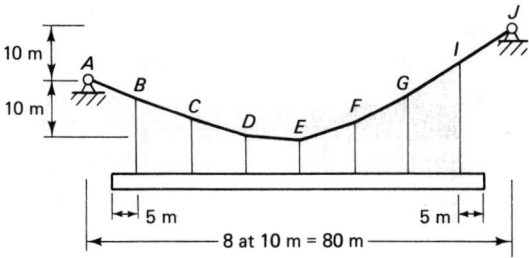

Figure P6.9

6.10 Determine the cable sag and cable tension under a central load of 200 kN for a steel cable with an unstretched length of 52 m supported between two supports at a horizontal distance of 50 m. Assume $E = 180$ GPa, area of cross section = 500 mm^2.

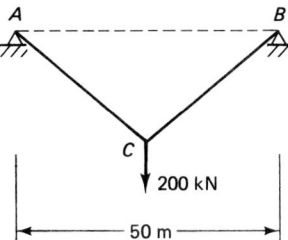

Figure P6.10

6.11 A wire 105 ft long, weighing 0.2 lb/ft is suspended from two points 100 ft apart in the same horizontal plane. Determine **(a)** the sag, **(b)** the tension at center, and **(c)** the tension at the ends.

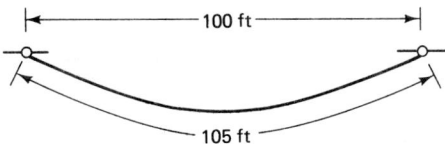

Figure P6.11

6.12 A fully extended cable (unstretched) weighing 0.5 lb/ft is nailed to two points 18 ft apart. Determine the sag below the level of the nails, if the stiffness of the cable is 2000 lb/ft.

6.13 A cable that weighs 0.4 lb/horizontal foot swings from two supports at a horizontal distance of 1900 ft and at a vertical distance of 300 ft as shown. Find the tension at the upper support, if the sag below the lower support is 25 ft.

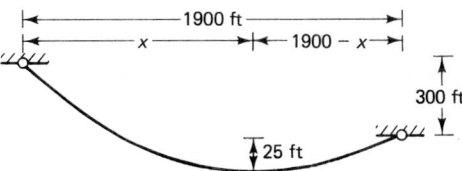

Figure P6.13

6.14 A uniform cable is suspended from two points in a ceiling 50 m apart; the sag in the middle is 15 m. Calculate the length of the cable.

6.15 A cable that weighs 2 lb/ft is to be suspended across a river 1000 ft wide. If the minimum overhead clearance for shipping is 25 ft and the maximum tension in the cable is limited to 2000 lb, calculate the minimum height of supporting towers above the water level.

6.16 A uniform wire, 3 N/m, was stretched between two points 50 m apart until the sag was only 2 m. How much wire would be in the span? What is the maximum tension?

6.17 A wire weighing 1 N/m is stretched between two points 100 m apart. What must be the sag in order that the maximum tension is 1 kN?

6.18 Draw axial-force, shear-force, and bending-moment diagrams for the member *AB* supporting cable *BC*.

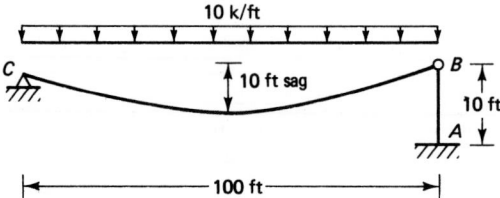

Figure P6.18

6.19 Define the shape of a two-hinged arch with no bending moment for the loads applied.

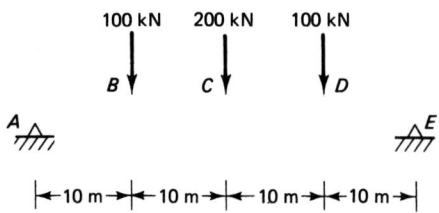

Figure P6.19

6.20 A three-hinged semicircular arch of 100 m span has a rise of 50 m and is loaded as shown in the figure. Determine the reactions at the supports and draw the bending-moment diagram.

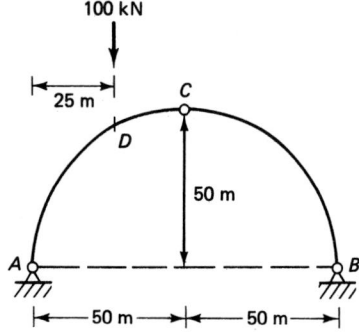

Figure P6.20

6.21 A three-hinged parabolic arch of span 100 m and rise 20 m carries a uniformly distributed load of 1 kN/m length on the right half as shown in the figure. Determine the maximum bending moment in the arch.

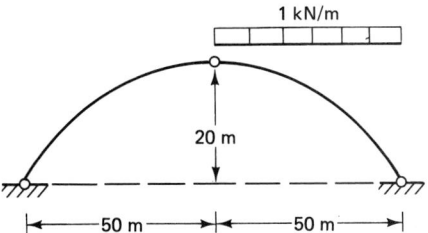

Figure P6.21

6.22 The cables of a suspension bridge have a horizontal span of 750 ft and a sag of 50 ft. Each cable carries a total uniformly distributed load of 500 k. If the allowable stress for the cable material is 20 ksi, determine the required cross-sectional area of each cable.

6.23 A suspension bridge with two three-hinged stiffening girders (one on each side) has a span of 350 ft and a sag of 35 ft. The bridge carries a total uniformly distributed dead load of 700 k. If the tensile stress in the cables cannot exceed 28 ksi, determine the required cross-sectional area of each cable. Compute also the maximum bending moment in the stiffening girder.

6.24 A steel suspension-bridge cable having a sag equal to one-tenth the span carries a superimposed uniformly distributed load of 30 kN/m. The unit weight of steel is 77.0 kN/m^3. If the permissible stress in the cable is 200 MPa, determine the maximum span of the cable. The self-weight of the cable can be considered as uniformly distributed per meter of horizontal span.

6.25 If the anchorage difficulties in Problem 6.24 limit the maximum tension in the cable to 80 MN, determine the maximum possible span.

6.26 A suspension cable of 160 m span and 16 m sag carries a load of 15 kN per lineal horizontal meter. Calculate **(a)** the maximum and minimum tensions in the cable, **(b)** the horizontal and vertical forces in each pier if the cable passes over frictionless rollers on the tops of the piers and the backstays are inclined at 30° to the horizontal, **(c)** the horizontal and vertical forces in each pier if the cable is firmly clamped to saddles carried on frictionless rollers on the tops of the piers.

6.27 A three-hinged trussed arch has a span of 100 m and a central rise of 20 m. It is hinged at the springing and crown. The load on the arch consists of a dead load of 10 kN/m. Determine the forces in the members ①, ②, and ③.

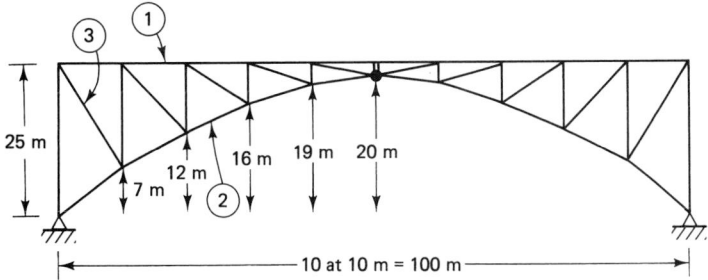

Figure P6.27

6.28 to 6.32   For the three-hinged arches shown determine **(a)** the support reactions and **(b)** axial force, shear force, and bending moment at *D*.

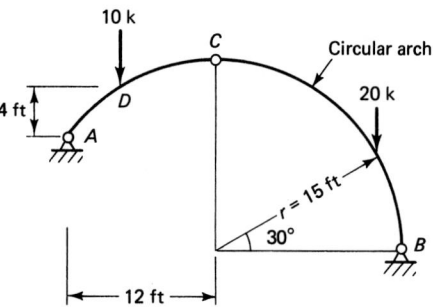

Figure P6.28

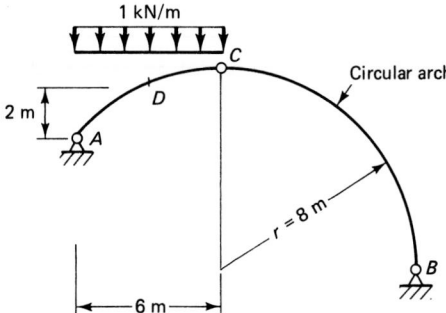

Figure P6.29

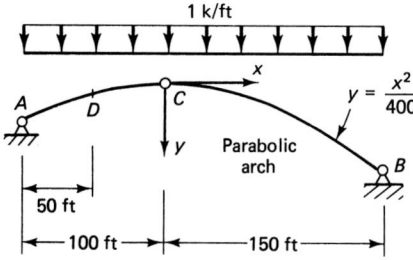

Figure P6.30

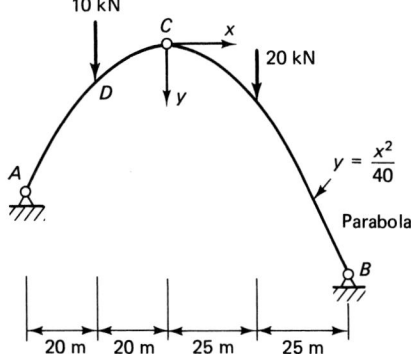

Figure P6.31

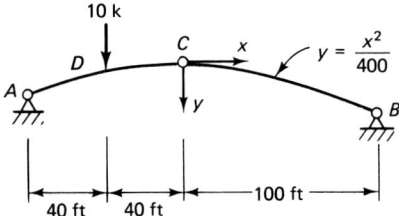

Figure P6.32

6.33 For the three-hinged gabled frame shown, determine (**a**) support reactions, (**b**) axial force, shear force, and bending moment at B.

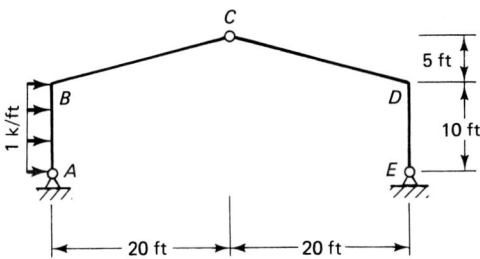

Figure P6.33

6.34 For the three-hinged truss arch shown, determine the reactions at the supports and forces in members ①, ②, ③, ④, and ⑤.

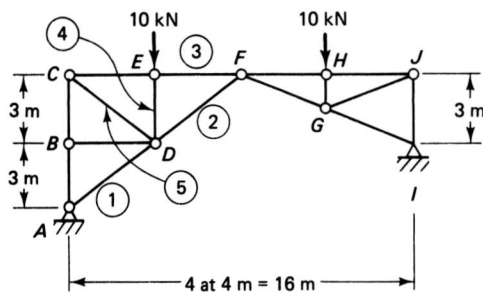

Figure P6.34

BIBLIOGRAPHY

D. Allan Firmage, "Fundamental Theory of Structures," Wiley, New York, 1963, pp. 247–265.

Frei Otto, "Tensile Structures," vol. 2, "Cable Structures," The M.I.T. Press, Cambridge, Mass., 1969, pp. 99–107.

W. Podolny and J. B. Scalzi, "Construction and Design of Cable-Stayed Bridges," Wiley, New York, 1976, pp. 337–353.

Prem Krishna, "Cable-Suspended Roofs," McGraw-Hill, New York, 1978, pp. 27–57.

John W. Leonard, "Tension Structures," McGraw-Hill, New York, 1988, pp. 17–71.

Chapter 7

Deflections

It is probably obvious by this time that these deflections, be they large or small, generate the forces of resistance which make a solid hard and stiff and resistant to external loads. In other words, a solid deflects exactly far enough to build up forces which just counter the external load applied to it. This is the automatic process at the basis of all structures.''

J. E. Gordon, 1973

7.1 INTRODUCTION

All engineering structures are made of materials that undergo some deformation on application of load or a change in temperature, resulting in displacement or deflection of the structure. If the structure returns to its original shape on removal of load or on return to the original temperature, the deflection is said to be *elastic.* The following are some of the other causes of deflection of structures:

1. Settlement of supports
2. Shrinkage of concrete (for concrete structures)
3. Slack in hinged joints (for trusses)

Owing to the above causes, the structure is permanently displaced and the associated deflections are called *nonelastic,* to distinguish them from elastic deflections. Deflection is a measure of the stiffness of the structure.

It is important for a structure to be not only *strong enough* to carry the super-imposed loads but also *stiff enough* to serve its intended function satisfactorily. For example, excessive deflection of (1) ceiling beams in a building may cause plaster to crack, even though the beams may be structurally strong; (2) long-span bridges may result in undesirable vibrations under dynamic loads; (3) an airplane wing may seriously affect the aerodynamic characteristics of the airplane; (4) a machine shaft may seriously affect the proper functioning of the bearings of the shaft; (5) rollers in a steel plate rolling mill may result in a rolled plate having uneven thickness.

Therefore, it is important to be able to compute the deflections of structures and ensure that they do not exceed the permissible values, which are usually specified in standard codes of practice.

Computation of deflections is also important for the analysis of indeterminate structures. In the case of statically indeterminate structures the available number of equations of equilibrium is less than the number of unknowns. Therefore, additional equations—called equations of compatibility—based on slopes and deflections at various points of the structure are required to determine the additional unknowns.

Computation of deflections is required not only in the analysis and design of structures but also in determining the erection procedures for some structures, for example, cantilever and continuous bridges, and in the amount of camber required for long-span structures (see Section 7.17).

7.2 METHODS OF DETERMINING DEFLECTIONS

Many methods are available for computing deflections of structures. They can be broadly classified into two groups:

1. Geometric methods
2. Energy methods

With geometric methods it is possible to determine deflections at *several locations* of the structure, while with energy methods, it is possible to compute only one component of deflection at *one location* of the structure. Each method has some advantages and disadvantages. It is important for the student to know several methods so that he or she can apply the one method that is particularly suitable for the given situation. In some cases, there may not be much to choose from among the various methods. Under those circumstances, engineers probably use that method with which they are most familiar. Only a few methods that have wide application and are considered fundamental are included in this chapter. These are

1. *Geometric methods* (applicable to beams and rigid frames)
    (a) Double integration
    (b) Moment area
    (c) Conjugate beam
2. *Energy methods* (applicable to beams, rigid frames, trusses, etc.)
    (a) Castigliano's theorem—Part II
    (b) Virtual work

7.3 THE DOUBLE-INTEGRATION METHOD

This is a geometric method and, as the name implies, involves integrating twice the differential equation to the elastic curve ($d^2y/dx^2 = M/EI$). If there is discontinuity of slope at any point of the elastic curve (e.g., at locations of hinges in the beam), *two separate expressions* must be written and constants of integration must be evaluated from compatibility conditions (at the hinge) and boundary conditions. Obviously, the expression for M/EI must be an integrable function of x, for the method to be applicable. By the use of singularity functions mentioned in Chapter 5 as shown in

Examples 7.1 and 7.2, a single expression can be written for the whole beam and constants of integration evaluated from the known boundary conditions. For the method to be applicable, the beam should be divided into segments whose ends coincide with points of application of load or changes in the expression for the flexural rigidity *EI* of the beam.

■ EXAMPLE 7.1

For the beam shown in Fig. 7.1*a* find the expression for the elastic curve. What is the maximum deflection of the beam and where does it occur?

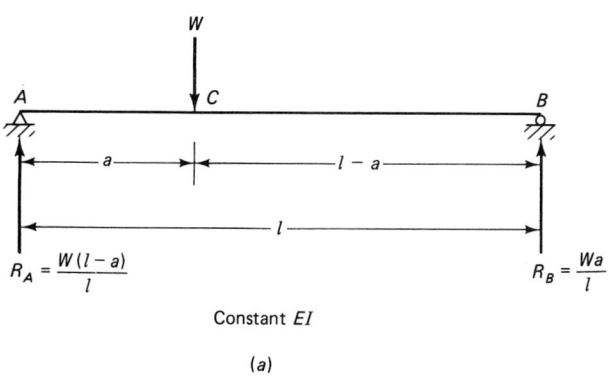

(*a*)

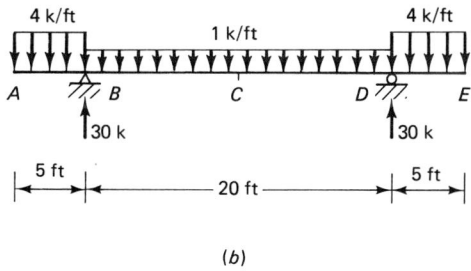

(*b*)

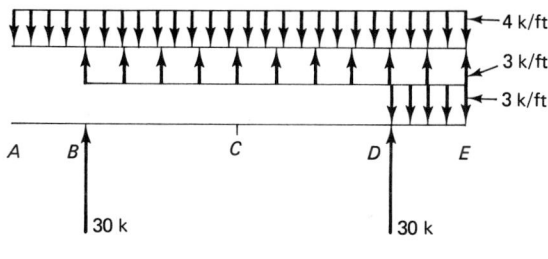

(*c*)

Figure 7.1 (*a*) Beam for Example 7.1. (*b*) and (*c*) Beam for Example 7.2.

Solution. Since there is a concentrated load at *C,* one expression for bending moment is valid for the portion *AC* of the beam and another expression for the portion *CB* of the beam. When using singularity functions, the bending-moment expression is developed by considering moments of all the forces acting to one side (generally the left) of the section.

For the portion *AC*:

$$M_x = R_A x = \frac{W(l-a)}{l} x \qquad 0 \leqslant x \leqslant a$$

For the portion *CB*:

$$M_x = R_A x - W(x-a)$$

$$= \frac{W(l-a)}{l} x - W(x-a) \qquad a \leqslant x \leqslant l$$

These two expressions can be combined into a single expression

$$M_x = \frac{W(l-a)}{l} x - W\{x-a\}$$

where $\{x-a\} = 0 \qquad$ if $x \leqslant a$
$= (x-a) \qquad$ if $a \leqslant x \leqslant l$

In using singularity functions, it is essential to keep each term separate and to write the integral of $\{x-a\}$ as $\{x-a\}^2/2$. Remembering this, we write

$$M_x = EI\frac{d^2y}{dx^2} = \frac{W(l-a)}{l} x - W\{x-a\} \tag{7.1}$$

Integrating once with respect to *x,*

$$EI\frac{dy}{dx} = \frac{W(l-a)}{l}\frac{x^2}{2} - \frac{W\{x-a\}^2}{2} + C_1 \tag{7.2}$$

where C_1 is a constant of integration. Integrating once again with respect to *x* yields

$$EIy = \frac{W(l-a)}{l}\frac{x^3}{6} - \frac{W\{x-a\}^3}{6} + C_1 x + C_2 \tag{7.3}$$

where C_2 is another constant of integration. Integration constants C_1 and C_2 are evaluated by the two boundary conditions:

1. At the left support:

$$x = 0 \qquad y = 0 \tag{7.4}$$

2. At the right support:

$$x = l \qquad y = 0 \tag{7.5}$$

Substituting the left support boundary condition [Eq. (7.4)] in Eq. (7.3), and remembering that $\{x-a\} = 0$ if $x \leqslant a$, we get

$$0 = 0 - 0 + 0 + C_2 \quad \text{or} \quad C_2 = 0 \tag{7.6}$$

Substituting the right support boundary condition [Eq. (7.5)] in Eq. (7.3), we get

$$0 = \frac{W(l-a)}{l}\frac{l^3}{6} - \frac{W(l-a)^3}{6} + C_1 l$$

or

$$C_1 = \frac{W(l-a)^3}{6l} - \frac{W(l-a)}{6}l \tag{7.7}$$

Substituting Eq. (7.6) and Eq. (7.7) in Eq. (7.3), we find

$$EIy = \frac{W(l-a)}{l}\frac{x^3}{6} - \frac{W\{x-a\}^3}{6} + \frac{W(l-a)^3}{6l}x - \frac{W(l-a)}{6}lx \tag{7.8}$$

where $\{x-a\} = 0$ if $x \leq a$
              $= x - a$ if $a \leq x \leq l$

Equation (7.8) gives the equation of the elastic curve (centerline of the deflected beam). To find the position of maximum deflection, we equate the slope dy/dx to zero and solve for x. Thus,

$$\frac{dy}{dx} = \frac{1}{EI}\left[\frac{W(l-a)}{l}\frac{x^2}{2} - \frac{W\{x-a\}^2}{2} + \frac{W(l-a)^3}{6l} - \frac{W(l-a)}{6}l\right] \tag{7.9}$$

If we assume that the load W is to the right of the center of the beam, that is, $a \geq l/2$, then maximum deflection occurs in the left portion of the beam, that is, $x < a$ or $\{x-a\} = 0$. Therefore,

$$\frac{dy}{dx} = \frac{1}{EI}\left[\frac{W(l-a)}{l}\frac{x^2}{2} + \frac{W(l-a)^3}{6l} - \frac{W(l-a)}{6}l\right] = 0 \tag{7.10}$$

Simplifying Eq. (7.10),

$$x = \sqrt{\frac{a(2l-a)}{3}} \tag{7.11}$$

Substituting this value of x in Eq. (7.8), we get the magnitude of the maximum deflection. Examining Eq. (7.11):
When $a = l/2$, that is, load at center of beam,

$$x = \sqrt{\frac{(l/2)(3l/2)}{3}} = \frac{l}{2} = 0.5l \quad \text{(max deflection occurs at center)}$$

When $a \to L$ (load very close to the right support),

$$x \to \sqrt{\frac{l \cdot l}{3}} = \frac{l}{\sqrt{3}} = 0.577l$$

Therefore, as the load moves from position $l/2$ to l, the location of maximum deflection moves only a short distance (from $0.5l$ to $0.577l$). Therefore, the deflection at center can be considered to be the maximum for all practical purposes. When the load is applied at the center, $a = l/2$ and the maximum deflection will be found to be $WL^3/48EI$. ∎

■ EXAMPLE 7.2

Using the double-integration method, calculate the deflection at the center of a beam of length 30 ft shown in Fig. 7.1b. Take $E = 30,000$ ksi and $I = 400$ in^4.

Solution. Singularity functions using Macaulay's notation mentioned in Chapter 5 will be used and a single expression for bending moment applicable to the entire beam will be written.

In Macaulay's method, any distributed load must extend to the right end on the beam. (It can start anywhere.) Therefore, the given loading is replaced by the equivalent loading shown in Fig. 7.1c. The reactions remain unaffected. Referring to Fig. 7.1c,

$$EI\frac{d^2y}{dx^2} = M_x = -\frac{4x^2}{2} + 30\{x - 5\} + \frac{3\{x - 5\}^2}{2}$$

$$+ 30\{x - 25\} - \frac{3\{x - 25\}^2}{2} \qquad (a)$$

$$EI\frac{dy}{dx} = -\frac{4x^3}{6} + \frac{30\{x - 5\}^2}{2} + \frac{3\{x - 5\}^3}{6}$$

$$+ \frac{30\{x - 25\}^2}{2} - \frac{3\{x - 25\}^3}{6} + c_1 \qquad (b)$$

$$EIy = -\frac{4x^4}{24} + \frac{30\{x - 5\}^3}{6} + \frac{3\{x - 5\}^4}{24}$$

$$+ \frac{30\{x - 25\}^3}{6} - \frac{3\{x - 25\}^4}{24} + c_1x + c_2 \qquad (c)$$

When $x = 5$ ft, $y = 0$ (at support B).

$$0 = \frac{-4(5)^4}{24} + 0 + 0 + 0 - 0 + 5c_1 + c_2$$

or

$$5c_1 + c_2 = 104.17 \qquad (d)$$

When $x = 25$ ft, $y = 0$ (at support D).

$$0 = \frac{-4(25)^4}{24} + \frac{30(20)^3}{6} + \frac{3(20)^4}{24} + 0 - 0 + 25c_1 + c_2$$

or

$$25c_1 + c_2 = 5104.17 \qquad (e)$$

Solving equations (d) and (e),

$$c_1 = 250 \qquad c_2 = -1145.83$$

Deflection at center is obtained by substituting $x = 15$ ft in equation (c).

$$EIy = -\frac{4(15)^4}{24} + \frac{30(10)^3}{6} + \frac{3(10)^4}{24} + 0 - 0 + (250)(15) - 1145.83$$

$$= 416.67$$

Therefore,

$$y = \frac{416.67}{EI} = \frac{416.67}{(30{,}000 \times 144)[400/(144 \times 144)]} \text{ ft}$$

$$= 0.005 \text{ ft}$$

$$= 0.06 \text{ in upward}$$ ∎

■ EXAMPLE 7.3

Calculate the slope and deflection at the free end of the beam shown in Fig. 7.2a. $EI = 10{,}000 \text{ k} \cdot \text{ft}^2$.

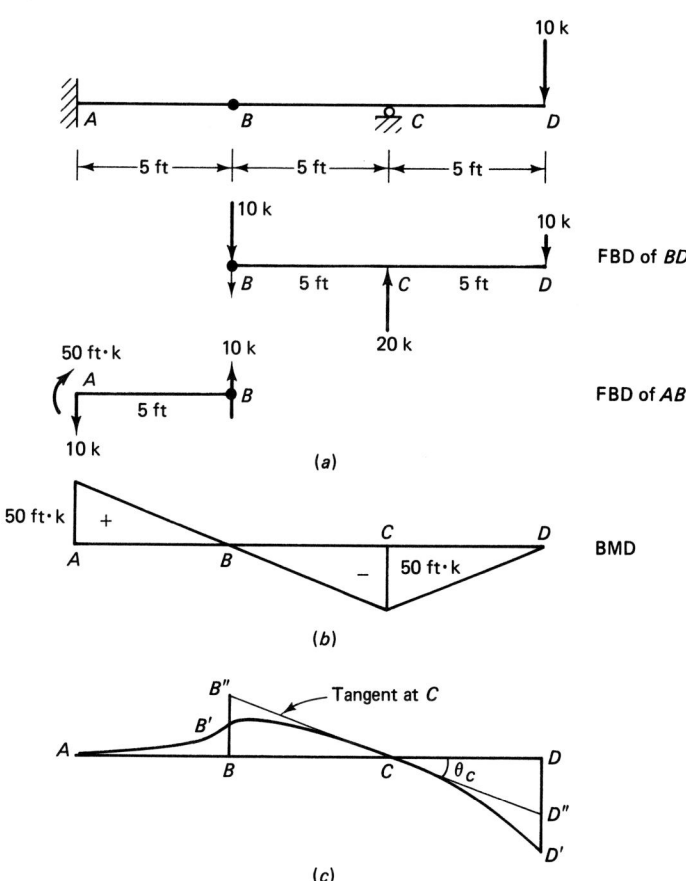

Figure 7.2 Beam for Examples 7.3 and 7.7. (c) Elastic curve.

Solution. Since there is an internal hinge at B, the elastic curve is not continuous at B and the double-integration method can only be applied to segments AB and BD separately.

The reactions are determined from statics by considering AB and BD separately as shown in Fig. 7.2b.

For Beam AB

$$EI\frac{d^2y}{dx^2} = 50 - 10x \qquad\qquad\text{(a)}$$

$$EI\frac{dy}{dx} = 50x - 5x^2 + C_1 \qquad\qquad\text{(b)}$$

$$EIy = 25x^2 - \frac{5}{3}x^3 + C_1x + C_2 \qquad\qquad\text{(c)}$$

Substituting the boundary conditions (support A is fixed),
when $x = 0$,

$$\frac{dy}{dx} = 0$$

when $x = 0$,

$$y = 0$$

we get $C_1 = 0$, $C_2 = 0$.

Slope at B just to the left of the hinge and deflection at B can be calculated from equations (b) and (c) by substituting $x = 5$ ft.
From equation (b):

$$\theta_{B\text{ left}} = \frac{1}{EI}(50 \times 5 - 5 \times 5^2)$$

$$= \frac{125}{10{,}000} = 0.0125 \text{ radian (counterclockwise)}$$

From equation (c):

$$\Delta_B = \frac{1}{EI}\left[25(5)^2 - \frac{5}{3}(5)^3\right]$$

$$= \frac{1250}{3 \times 10{,}000} = +0.04167 \text{ ft} = 0.5 \text{ in upward}$$

For Beam BCD (origin at B, x positive to right)

$$EI\frac{d^2y}{dx^2} = -10x + 20\{x - 5\} \qquad\qquad\text{(d)}$$

$$EI\frac{dy}{dx} = -5x^2 + 10\{x - 5\}^2 + C_3 \qquad\qquad (e)$$

$$EIy = -\tfrac{5}{3}x^3 + \tfrac{10}{3}\{x - 5\}^3 + C_3x + C_4 \qquad\qquad (f)$$

C_3 and C_4 are determined from the following boundary conditions: when $x = 0$, $y = +0.04167$ ft; when $x = 5$ ft, $y = 0$ (support C). (Note that the *deflection* to the left and right of B is the same, but the slopes to the left and right are different.)

Substituting these two conditions in equation (f), we get $C_3 = -41.67$ and $C_4 = 416.7$.

Substituting these values in equations (e) and (f),

$$EI\frac{dy}{dx} = -5x^2 + 10\{x - 5\}^2 - 41.67 \qquad\qquad (g)$$

$$EIy = -\tfrac{5}{3}x^3 + \tfrac{10}{3}\{x - 5\}^3 - 41.67x + 416.7 \qquad\qquad (h)$$

Slope at D and deflection at D are obtained by substituting $x = 10$ ft in equations (g) and (h).

$$\theta_D = \frac{1}{EI}[-5(10)^2 + 10(10 - 5)^2 - 41.67]$$

$$= -\frac{291.67}{10,000} = -0.029167 \text{ radian}$$

that is, 0.029167 radian clockwise.

$$\Delta_D = \frac{1}{EI}\left[-\frac{5}{3}(10)^3 + \frac{10}{3}(10 - 5)^3 - (41.67)(10) + 416.7\right]$$

$$= -0.125 \text{ ft}$$

that is, 0.125 ft downward.

To calculate the slope just to the right of hinge B, substitute $x = 0$ in equation (g):

$$\theta_{B \text{ right}} = \frac{1}{EI}(-0 + 0 - 41.67)$$

$$= -0.004167 \text{ radian}$$

that is, 0.004167 radian clockwise. ∎

7.4 THE MOMENT-AREA METHOD

The moment-area method is based on a consideration of the geometry of the elastic curve and the relation between the rate of change of slope and the bending moment at a point on the elastic curve. This method was developed by Professor C. E. Greene of the University of Michigan in 1873. Two theorems are associated with the moment-area method.

THEOREM 1. Provided that the elastic curve is continuous between two points A and B (i.e., there are no hinges between A and B), the rate of change of slope of the elastic curve between points A and B is equal to the area of the M/EI diagram between A and B. The M/EI diagram is the bending-moment diagram whose ordinates are divided by the (EI) values for the beam at different sections.

Proof. Let ACB be a continuous portion of the elastic curve as shown in Fig. 7.3a. $A_0 B_0$ is the undeformed position of the beam. Let the corresponding bending-moment

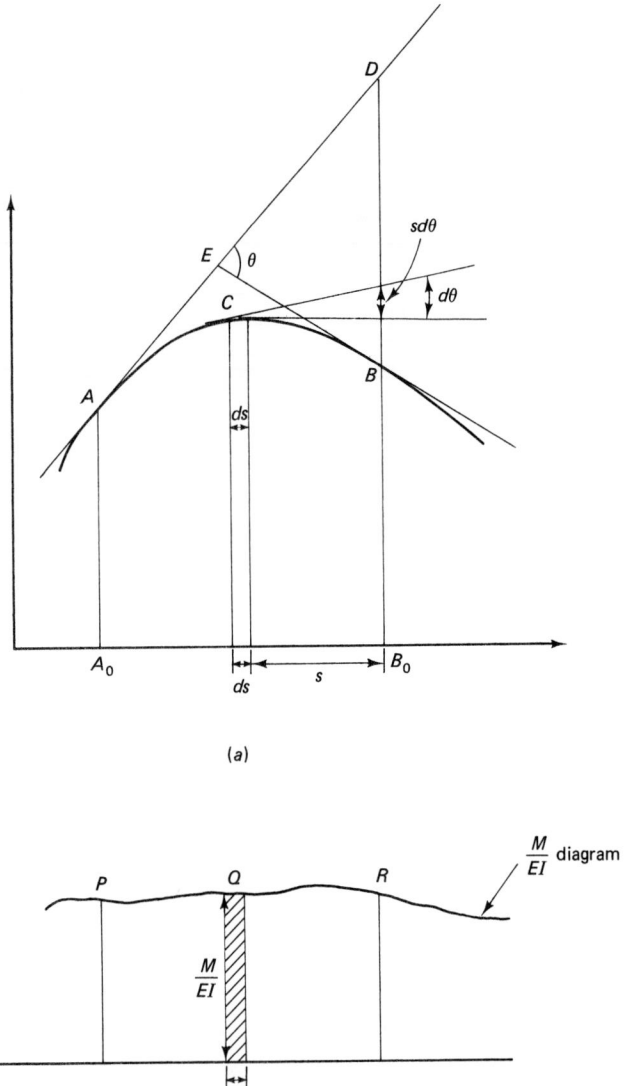

(a)

(b)

Figure 7.3 A portion of the elastic curve and the corresponding M/EI diagram.

diagram, whose ordinates are divided by EI, be PQR, as shown in Fig. 7.3b. For clarity, the deflected shape in Fig. 7.3a is shown greatly exaggerated. The slope of the deflection curve generally is so small that the sine and tangent of a slope are equal to the slope in radians, and the cosine of a slope is equal to unity.

The angle between the tangents at A and B, $\angle BED$ in Fig. 7.3a, is denoted as θ. If we consider a differential element of length ds at a distance s from B_0, the change of slope is $d\theta$. Referring to Fig. 7.4,

$$d\theta = \tan(d\theta) = \frac{uu'}{uO} = \frac{\text{shortening of fiber}}{\text{distance of fiber from neutral axis}}$$

or

$$d\theta = \frac{\text{unit strain} \times \text{original length}}{c}$$

$$= \frac{(\text{unit stress}/E)ds}{c}$$

$$= \frac{(\sigma/E)ds}{c}$$

From mechanics of materials, we know that

$$\sigma = \frac{M}{I}c$$

Therefore,

$$d\theta = \frac{(M/I)c}{E}\frac{ds}{c} = \frac{M\,ds}{EI} \tag{7.12}$$

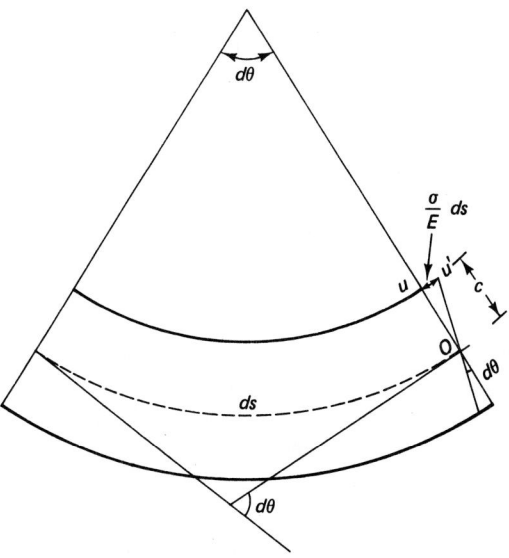

Figure 7.4 Elastic curve for a differential element.

Referring to Fig. 7.3b, $(M/EI)ds$ is the area of the M/EI diagram under the differential element ds. Integrating Eq. (7.12) yields

$$\theta = \int_A^B d\theta = \int_A^B \frac{M}{EI} ds$$

$$= \text{area of } \frac{M}{EI} \text{ diagram between } A \text{ and } B$$

Theorem 2.   Provided that the elastic curve is continuous between two points A and B, the deflection of point B on the elastic curve from the tangent at another point A on the elastic curve is equal to the moment of the area of the M/EI diagram between A and B about B.

Proof.   Consider again Fig. 7.3a. Draw line BD from point B, perpendicular to the original position A_0B_0 of the beam. The intercept on BD of the two tangents at the extremities of the differential element ds can be taken as equal to $s\,d\theta$. Substituting $d\theta = M\,ds/EI$ [Eq. (7.12)], the intercept $s\,d\theta = (M\,ds/EI)s \equiv$ moment about B of area of M/EI diagram under the differential element ds. Summing up all these intercepts, we find

$$BD = \int_A^B s\,d\theta = \int_A^B \frac{M\,ds}{EI} s \tag{7.13}$$

Referring to Fig. 7.3b, the integral on the right-hand side of Eq. (7.13) is, in fact, the statical moment of the area of the M/EI diagram between A and B about B.

It is to be clearly understood that the second moment-area theorem does not give the absolute value of deflection directly, but only the deflection of one point B *from the tangent at another point A,* measured perpendicular to the undeformed position.

The application of moment-area theorems for the determination of slopes and deflections is shown in Examples 7.4 to 7.8. For consistency, the following sign convention is adopted for the moment-area method:

1. Change in slope is positive if the area of the M/EI diagram is positive. The moment M is positive when it produces tension in the bottom fibers of the beam.
2. Point B on the elastic curve is above the tangent at A if the moment of the area of the M/EI diagram between A and B about B is positive.

Since the moment-area method involves computation of areas and moments of areas of the M/EI diagram, it will be found convenient sometimes to draw the M/EI diagram in parts (one diagram for each reaction, load, etc.) instead of a composite M/EI diagram showing the cumulative effect of all the loads. For example, let it be required to draw the M/EI diagram for the beam shown in Fig. 7.5a. The reactions R_A and R_B are determined from the equilibrium equations in the usual manner. The bending-moment diagrams due to R_A, P_1, P_2, and w are drawn separately as shown

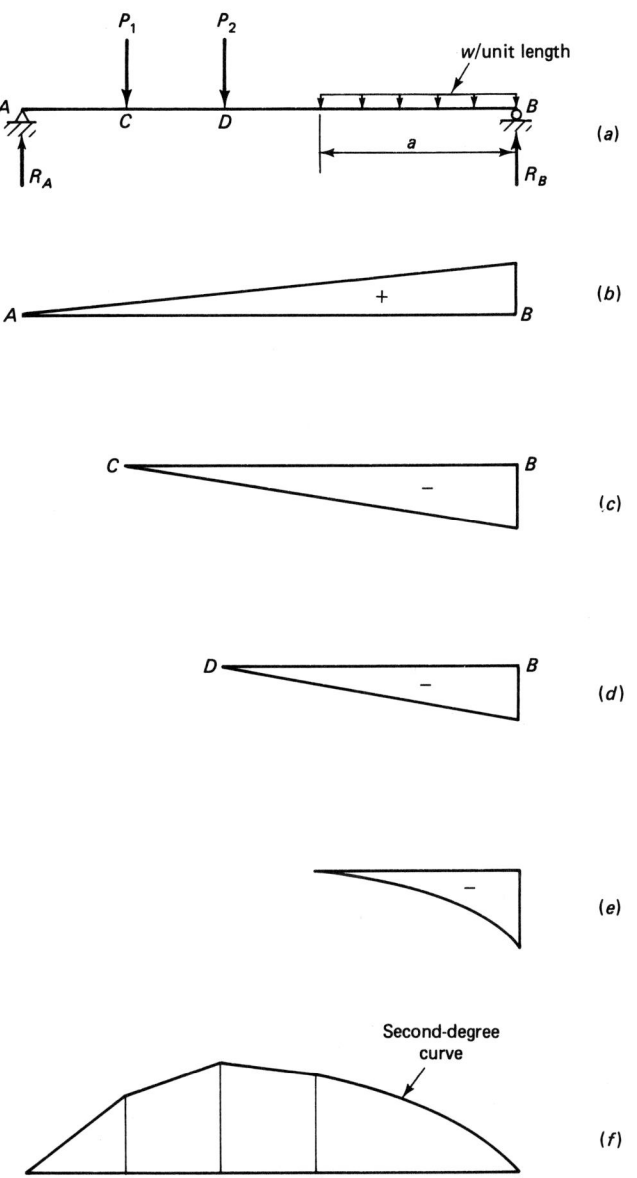

Figure 7.5 Bending-moment diagram by parts. (b) BMD due to R_A. (c) BMD due to P_1. (d) BMD due to P_2. (e) BMD due to w. (f) Composite BMD.

in Fig. 7.5b to e. Since these are simple geometrical figures, their areas and moments of the areas can be computed very easily, when compared with the composite bending-moment diagram shown in Fig. 7.5f. For ready reference, areas and moments of areas of simple geometrical figures that are frequently encountered in the application of the moment-area method are given in Fig. 7.6.

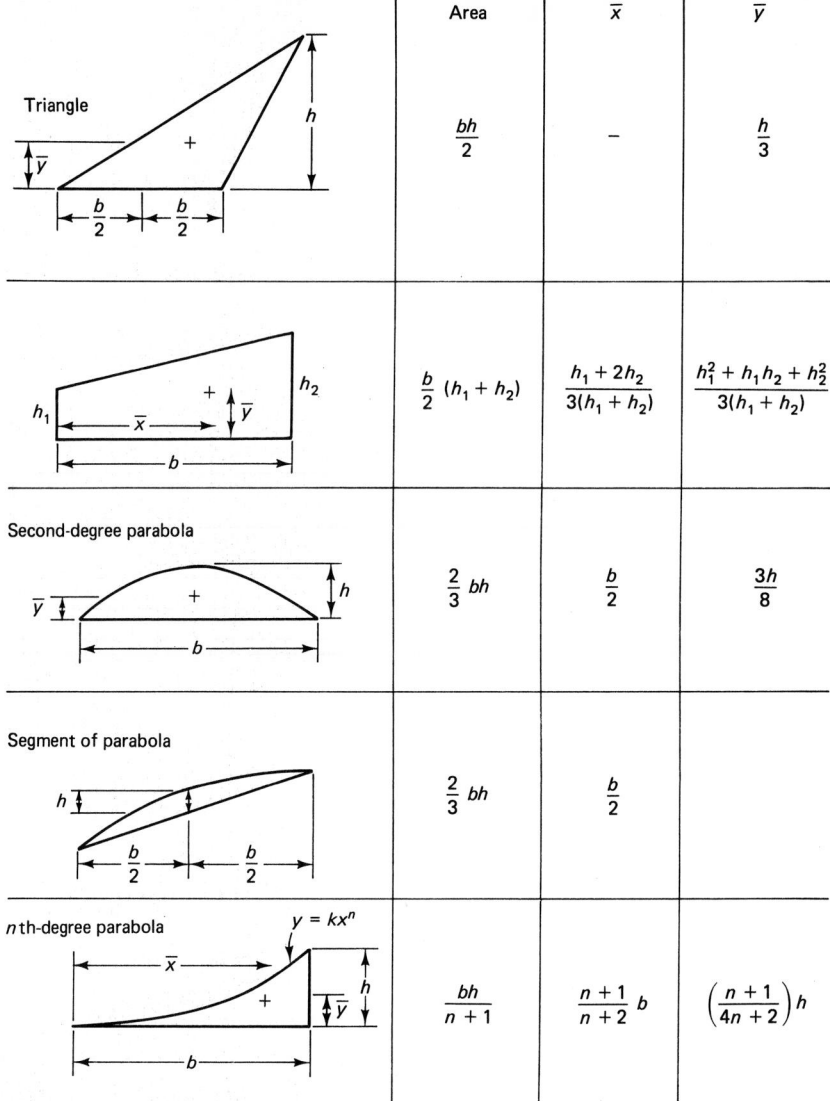

	Area	$\bar{x}$	$\bar{y}$
Triangle	$\dfrac{bh}{2}$	—	$\dfrac{h}{3}$
	$\dfrac{b}{2}(h_1 + h_2)$	$\dfrac{h_1 + 2h_2}{3(h_1 + h_2)}$	$\dfrac{h_1^2 + h_1 h_2 + h_2^2}{3(h_1 + h_2)}$
Second-degree parabola	$\dfrac{2}{3}\,bh$	$\dfrac{b}{2}$	$\dfrac{3h}{8}$
Segment of parabola	$\dfrac{2}{3}\,bh$	$\dfrac{b}{2}$	
nth-degree parabola	$\dfrac{bh}{n+1}$	$\dfrac{n+1}{n+2}\,b$	$\left(\dfrac{n+1}{4n+2}\right)h$

Figure 7.6 Properties of simple geometrical figures.

■ EXAMPLE 7.4

For the cantilever beam shown in Fig. 7.7a, using the moment-area method, determine the slope and deflection at B and C.

$$EI = 10,000,000 \ \text{k} \cdot \text{in}^2$$

Solution. Draw the bending-moment diagram in parts, as shown in Fig. 7.7b. Note that for a cantilever beam, the bending-moment diagram can be drawn without determining the reactions at the fixed support.

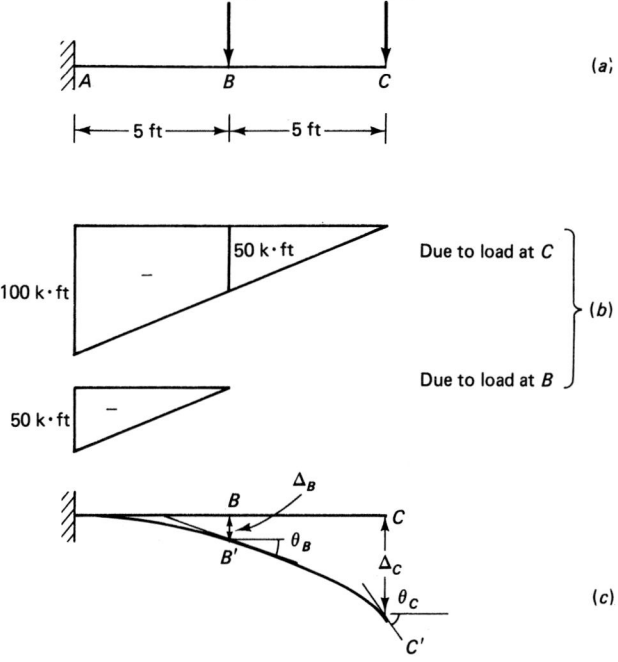

Figure 7.7 Beam for Example 7.4. (*b*) BMD. (*c*) Elastic curve.

(a) Determination of θ_B. Apply the first moment-area theorem to points A and B:

$$\theta_B - \theta_A = \text{area of } \frac{M}{EI} \text{ diagram between } A \text{ and } B$$

$$= \frac{1}{EI}\left(-\frac{50 + 100}{2} \times 5 - \frac{1}{2} \times 5 \times 50\right)$$

$$= -\frac{500}{EI}$$

Since θ_A is zero,

$$\theta_B = -\frac{500 \text{ k} \cdot \text{ft}^2}{(10,000,000/144) \text{ k} \cdot \text{ft}^2} = -0.0072 \text{ radian}$$

that is, 0.0072 radian clockwise.

(b) Determination of θ_C. Similarly, applying the first moment-area theorem to points A and C,

$$\theta_C - \theta_A = \text{area of } \frac{M}{EI} \text{ diagram between } A \text{ and } C$$

$$= \frac{1}{EI}\left(-\frac{1}{2} \times 10 \times 100 - \frac{1}{2} \times 5 \times 50\right)$$

$$= -\frac{625}{EI}$$

Since A is a fixed end, $\theta_A = 0$. Therefore,

$$\theta_C = -\frac{625}{10,000,000/144} = -0.0090 \text{ radian}$$

that is, 0.0090 radian clockwise.

(c) *Determination of Deflection at B.*   Apply the second moment-area theorem to points A and B. The deviation of B from the tangent at A

$t_{B/A}$ = moment about B of area of $\dfrac{M}{EI}$ diagram between A and B

$$= \frac{1}{EI}\left\{-\frac{1}{2}(50 + 100)(5)\left[\frac{50 + (2)(100)}{50 + 100}\right]\left(\frac{5}{3}\right) - \frac{1}{2} \times 50 \times 5 \times \left(\frac{2}{3} \times 5\right)\right\}$$

$$= -\frac{8750}{6EI}$$

Since the tangent at A is horizontal,

$$\Delta_B = t_{B/A} = -\frac{8750}{6(10,000,000/144)} \text{ ft} = -0.021 \text{ ft} = -0.252 \text{ in}$$

that is, 0.252 in downward.

(d) *Determination of Deflection at C.*   As before, apply the second moment-area theorem to points A and C. The deviation of C from the tangent at A

$t_{C/A}$ = moment about C of area of $\dfrac{M}{EI}$ diagram between A and C

$$= \frac{1}{EI}\left[-\frac{1}{2} \times 10 \times 100 \times \left(\frac{2}{3} \times 10\right) - \frac{1}{2} \times 5 \times 50\left(5 + \frac{10}{3}\right)\right]$$

$$= -\frac{26,250}{6EI}$$

Since the tangent at A is horizontal,

$$\Delta_C = t_{C/A} = -\frac{26,250}{6(10,000,000/144)} = -0.063 \text{ ft} = -0.756 \text{ in}$$

that is, 0.756 in downward.   ∎

■ **EXAMPLE 7.5**

For the cantilever beam of nonuniform cross section shown in Fig. 7.8a, determine the slope and deflection at B and C due to the distributed load shown.

Solution. The bending-moment diagram due to uniformly distributed load is shown in Fig. 7.8b. The M/EI diagram shown in Fig. 7.8c is obtained by dividing the bending-moment ordinates by the corresponding EI values. Since there is a sudden change in EI at B, the M/EI diagram has two values at B: one value corresponding to an EI of

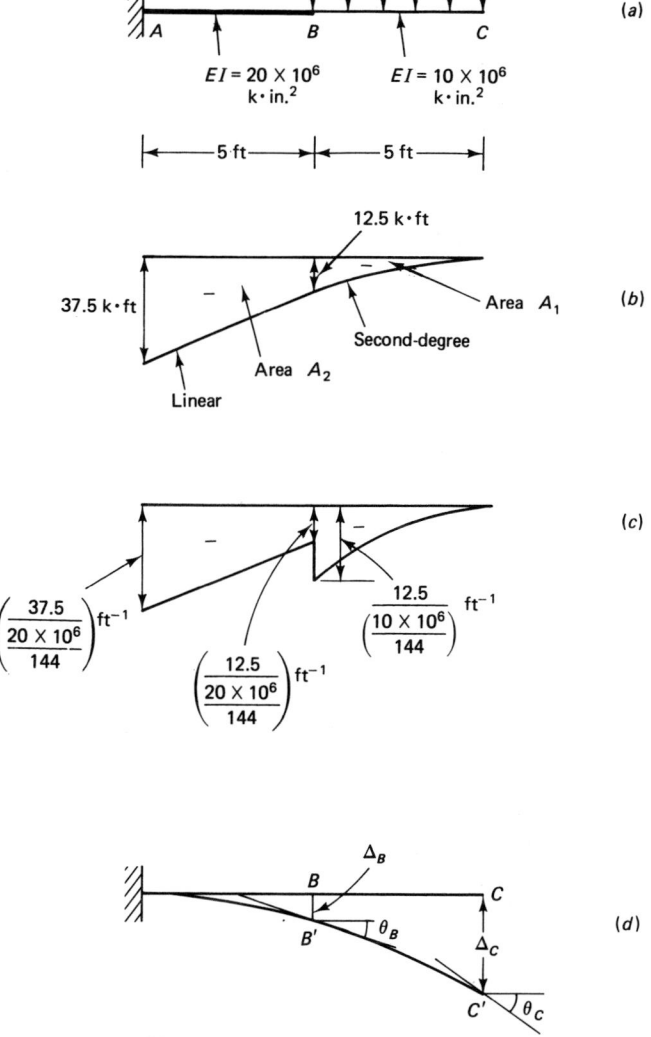

Figure 7.8 Beam for Example 7.5. (*b*) BMD. (*c*) M/EI diagram. (*d*) Elastic curve.

10×10^6 k·in² and the other value corresponding to an EI of 20×10^6 k·in². The elastic curve is sketched in Fig. 7.8d.

(a) Determination of θ_B. Applying the first moment-area theorem to points A and B on the elastic curve and realizing that the slope at A is zero,

$$\theta_B - \theta_A = \frac{1}{(20 \times 10^6)/144}\left(-\frac{12.5 + 37.5}{2} \times 5\right) = -0.0009 \text{ radian}$$

Therefore, $\theta_B = 0.0009$ radian clockwise.

(b) Determination of θ_C. Once again apply the first moment-area theorem to points A and C:

$$\theta_C - \theta_A = \frac{1}{(20 \times 10^6)/144}\left(-\frac{12.5 + 37.5}{2} \times 5\right)$$

$$+ \frac{1}{(10 \times 10^6)/144}\left(-\frac{1}{3} \times 12.5 \times 5\right)$$

Therefore, $\theta_C = -0.0012$ radian, that is, 0.0012 radian clockwise.

(c) Determination of Δ_B. Apply the second moment-area theorem to points A and B. Deviation of B from the tangent at A

$$t_{B/A} = \Delta_B = \text{moment about } B \text{ of area of } \frac{M}{EI} \text{ diagram between } A \text{ and } B$$

$$= \frac{1}{(20 \times 10^6)/144}\left(-\frac{12.5 + 37.5}{2} \times 5 \times \frac{5}{3} \times \frac{2 \times 37.5 + 12.5}{37.5 + 12.5}\right)$$

$$= -0.002625 \text{ ft} = -0.0315 \text{ in}$$

that is, 0.0315 in downward.

(d) Determination of Δ_C. Applying once again the second moment-area theorem to points A and C, deviation of C from the tangent at A

$$t_{C/A} = \Delta_C = \text{moment about } C \text{ of area of } \frac{M}{EI} \text{ diagram between } A \text{ and } C$$

Referring to Fig. 7.8b,

Area $A_1 = -\frac{1}{3} \times 12.5 \times 5 = -20.833$ k·ft²

Distance of centroid from $C = \left(\frac{3}{4} \times 5\right) = 3.75$ ft

Area $A_2 = -\frac{1}{2}(12.5 + 37.5) \times 5 = -125$ k·ft²

Distance of centroid from $C = 5 + \left(\frac{5}{3} \times \frac{12.5 + 2 \times 37.5}{12.5 + 37.5}\right)$

$$= 5 + 2.917 = 7.917 \text{ ft}$$

Therefore,

$$\Delta_C = -\frac{20.833 \times 3.75}{(10 \times 10^6)/144} - \frac{125 \times 7.917}{(20 \times 10^6)/144}$$

$$= -0.001125 - 0.007125 = -0.00825 \text{ ft} = -0.099 \text{ in}$$

that is, 0.099 in downward. ∎

 As can be seen from above, the moment-area method can be readily used to determine the slopes and deflections of beams having different EI values in different segments of the beam.

■ **EXAMPLE 7.6**
Determine the location and magnitude of the maximum upward deflection of the overhanging beam shown in Fig. 7.9a. Also determine the slope and deflection at A. $EI = 1000 \text{ kN} \cdot \text{m}^2$.

Solution. The reactions are determined and the bending-moment diagram is drawn in parts as shown in Fig. 7.9b.
 The maximum upward deflection obviously occurs in the supported span BC (Fig. 7.9c).
 The procedure for locating the position of maximum of deflection x is as follows:

1. Determine θ_B by first computing $t_{C/B}$ (see Fig. 7.9c).
2. Let maximum deflection occur at x, distance x meters to the right of B. Slope at x is zero.
3. Apply the first moment-area theorem to points B and x; solve for x.

When once the location of maximum deflection is found, the magnitude can be determined in the usual manner.

Step 1

$t_{C/B}$ = deviation of C from the tangent at B

 = moment about C of the area of $\frac{M}{EI}$ diagram between B and C

$$= \frac{1}{EI}\left[-\frac{1}{2} \times 4(10+50) \times \frac{4}{3} \times \frac{2 \times 10 + 50}{50 + 10} + \frac{1}{2} \times 4 \times 30 \times \left(\frac{1}{3} \times 4\right)\right]$$

$$= -\frac{106.67}{EI}$$

that is, point C is below the tangent at B as shown in Fig. 7.9c. From Fig. 7.9c, $\theta_B = \tan\theta_B = |t_{C/B}|/4 = 26.67/EI$ radian (counterclockwise, by inspection). Therefore,

$$\theta_B = +\frac{26.67}{EI} \text{ radian} = 0.02667 \text{ radian}$$

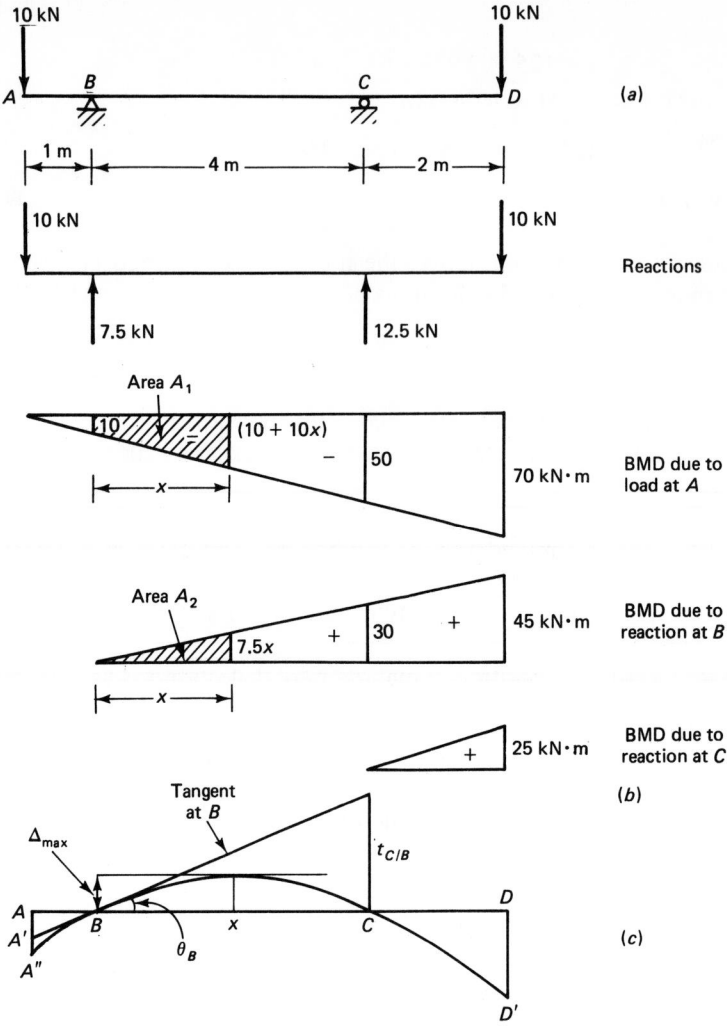

Figure 7.9 Beam for Example 7.6. (*c*) Elastic curve.

(It should be noted that the magnitude and direction of θ_B are obtained from the geometry and not by the moment-area theorems.)

Steps 2 and 3. Apply the first moment-area theorem to B and x. Slope at $x = 0$. Therefore,

$$\theta_x - \theta_B = \text{area of } \frac{M}{EI} \text{ diagram between } B \text{ and } x$$

$$= \frac{1}{EI}\left\{-\frac{1}{2}(x)[10 + (10 + 10x)] + \frac{1}{2}(x)(7.5x)\right\}$$

$$0 - \frac{26.67}{EI} = \frac{1}{EI}\left(-\frac{20x + 10x^2}{2} + \frac{7.5x^2}{2}\right)$$

Solving, $x = 2.11$ m. Therefore, maximum upward deflection occurs at a distance of 2.11 m to the right of B.

Magnitude of Maximum Upward Deflection. From Fig. 7.9c, it can be seen that maximum upward deflection is *numerically* equal to the deviation of B from the tangent at x.

$t_{B/x}$ = moment about B of area of $\dfrac{M}{EI}$ diagram between B and x

Referring to Fig. 7.9b,

$$A_1 = -\frac{2.11}{2}[10 + (10 + 21.1)] = -43.36$$

Distance of centroid from $B = \dfrac{2.11}{3}\left(\dfrac{2 \times 31.1 + 10}{31.1 + 10}\right) = 1.236$ m

$$A_2 = +\tfrac{1}{2} \times 2.11(7.5 \times 2.11) = +16.70$$

Distance of centroid from $B = \tfrac{2}{3} \times 2.11 = 1.407$ m

Therefore,

$$t_{B/x} = \frac{1}{EI}[(-43.36)(1.236) + (16.70)(1.407)]$$

$$= -\frac{30.1}{1000} = -0.0301 \text{ m} = -30.1 \text{ mm}$$

Therefore, Δ_{max} in span BC = 30.1 mm *upward*.

Slope at A. Since the slope at B is known, the slope at A can be readily computed by applying the first moment-area theorem to points A and B.

$\theta_B - \theta_A$ = area of $\dfrac{M}{EI}$ diagram between A and B

$$= \frac{1}{EI}\left(-\frac{1}{2} \times 1 \times 10\right) = -\frac{5}{1000} = -0.005 \text{ radian}$$

$$\theta_A = \theta_B + 0.005 = 0.03167 \text{ radian counterclockwise}$$

Deflection at A. From Fig. 7.9c, point A deflects downward; magnitude of Δ_A is given by

$$\Delta_A = |AA'| + |A'A''| = |\theta_B \cdot AB| + |t_{A/B}|$$

$t_{A/B}$ = moment about A of area of $\dfrac{M}{EI}$ diagram between A and B

$$= \frac{1}{EI}\left[-\frac{1}{2} \times 1 \times 10 \times \left(\frac{2}{3} \times 1\right)\right] = -\frac{3.33}{1000} = -0.00333 \text{ m}$$

that is, A is below the tangent drawn at B. Therefore,

$$\Delta_A = |(0.02667)(1)| + |0.00333| = |0.03| \text{ m} = |30| \text{ mm}$$

Point A deflects *downward* by 30 mm.   ∎

■ EXAMPLE 7.7

Find the slope and deflection at the free end D of the beam shown in Fig. 7.2a; $EI =$ 10,000 k·ft^2.

Solution. This problem was solved earlier by the double-integration method. Since there is an internal hinge at B, the elastic curve is not continuous at B and moment-area theorems can only be applied to segments AB and BD separately.

From Fig. 7.2c, it can be easily seen that deflection at D is downward and is given by

$$\Delta_D = DD' = DD'' + D''D' = |(5)\theta_C| + |t_{D/C}|$$

θ_C is calculated as follows:

$$\tan \theta_C = \theta_C = \frac{BB''}{BC} = \frac{BB' + B'B''}{5} = \frac{|\Delta_B| + |t_{B/C}|}{5}$$

Δ_B is readily calculated by applying the second moment-area theorem to points A and B

$$\Delta_B = BB' = \text{moment about } B \text{ of area of } \frac{M}{EI} \text{ diagram between } A \text{ and } B$$

$$= \frac{1}{EI}\left[\frac{1}{2} \times 5 \times 50 \times \left(\frac{2}{3} \times 5\right)\right] = \frac{1250}{3EI} = \frac{1250}{3 \times 10,000} = 0.0417 \text{ ft}$$

$$t_{B/C} = B'B'' = \text{moment about } B \text{ of area of } \frac{M}{EI} \text{ diagram between } B \text{ and } C$$

$$= \frac{1}{EI}\left[-\frac{1}{2} \times 5 \times 50 \times \left(\frac{2}{3} \times 5\right)\right] = -0.0417 \text{ ft}$$

As shown in Fig. 7.2c, point B' is below the tangent drawn at C.

$$BB'' = BB' + B'B'' = 0.0417 + 0.0417 = 0.08333 \text{ ft}$$

Therefore,

$$\theta_C = \frac{0.08333}{5} = 0.0167 \text{ radian} \qquad \text{(clockwise, by examining Fig. 7.2c)}$$

$$t_{D/C} = D'D'' = \text{moment about } D \text{ of area of } \frac{M}{EI} \text{ diagram between } C \text{ and } D$$

$$= \frac{1}{EI}\left[-\frac{1}{2} \times 50 \times 5 \times \left(\frac{2}{3} \times 5\right)\right] = -0.0417 \text{ ft}$$

that is, point D' is 0.0417 ft below the tangent drawn at C, as shown in Fig. 7.2c.

Therefore,

$$\Delta_D = DD' = DD'' + D''D' = (5)(0.0167) + 0.0417$$

$$= 0.125 \text{ ft} = 1.5 \text{ in downward}$$

Slope at D can be readily determined by the application of the first moment-area theorem between C and D.

$$\theta_D - \theta_C = \text{area of } \frac{M}{EI} \text{ diagram between } C \text{ and } D$$

Since $\theta_C = 0.0167$ radian clockwise,

$$\theta_C = -0.0167$$

Therefore,

$$\theta_D - (-0.0167) = \frac{1}{EI}\left(-\frac{1}{2} \times 5 \times 50\right) = -\frac{125}{10,000} = -0.0125$$

Therefore,

$$\theta_D = -0.02917 \text{ radian}$$

that is, 0.02917 radian clockwise. ∎

■ EXAMPLE 7.8
Calculate the vertical deflection at point C for the structure shown in Fig. 7.10a using the moment-area method.

Solution. Refer to Fig. 7.10b.

$$\theta_B - \theta_A = \text{area of } \frac{M}{EI} \text{ diagram between } A \text{ and } B$$

$$\theta_B - 0 = \left(\frac{-WL}{EI}\right)(2L) = \frac{-2WL^2}{EI} \qquad \text{radians}$$

that is, clockwise.

$$\text{Vertical deflection at } C = CC'' = CC' + C'C''$$

$$= (\theta_B)2L + \text{deviation of } C \text{ from tangent at } B$$

$$= \left(\frac{-2WL^2}{EI}\right)2L + \text{moment about } C \text{ of area}$$

$$\text{of } \frac{M}{EI} \text{ diagram between } B \text{ and } C$$

$$= \frac{-4WL^3}{EI} + \left[\frac{1}{2}L\left(\frac{-WL}{EI}\right)\left(L + \frac{2}{3}L\right)\right] = \frac{-29}{6}\frac{WL^3}{EI}$$

that is, downward deflection. ∎

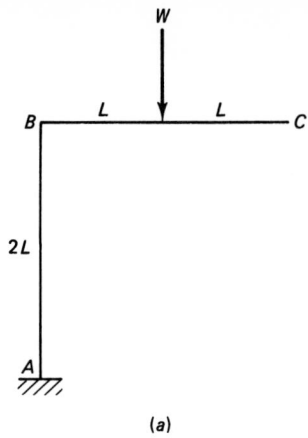

(a)

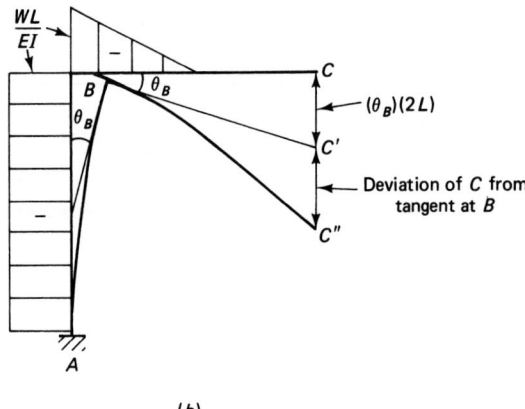

(b)

Figure 7.10 Structure for Example 7.8. Deflected shape is grossly exaggerated for clarity.

7.5 THE CONJUGATE-BEAM METHOD

This method was developed by Otto Mohr in a paper published in 1868. It is based on the similarity between the mathematical expressions for load, shear, and moment in an analogous beam and M/EI, slope and deflection of the real beam.

In Chapter 5, the relationships between load, shear, and moment were derived in the following form:

$$\frac{dV}{dx} = -w \tag{7.14}$$

and

$$\frac{dM}{dx} = V \tag{7.15}$$

where w, V, and M refer to load, shear, and moment, respectively, in a given beam.

In the development of the moment-area method, it was shown that

$$d\theta = \frac{M \, ds}{EI} \qquad (7.12)$$

Replacing ds by dx, for straight members, we have

$$d\theta = \frac{M \, dx}{EI}$$

or

$$\frac{d\theta}{dx} = \frac{M}{EI} \qquad (7.16)$$

We also know that the slope $dy/dx = \tan \theta$, by definition. Or, since θ is small, we have

$$\frac{dy}{dx} = \theta \qquad (7.17)$$

where θ and y refer to the slope (in radians) and deflection at any section of the real beam. Comparing Eqs. (7.14) and (7.15) with (7.16) and (7.17) shows that they are mathematically similar and an analogy exists; that is, the relationship between load, shear, and moment is similar to the relationship between M/EI, slope, and deflection. Therefore, if an analogous beam is loaded with the M/EI of the original beam, the slope and deflection of the original beam are given by the shear and bending moment at the corresponding section of the analogous beam. The analogous beam referred to is called the conjugate beam, subjected to an elastic load M/EI. This method of determining slopes and deflections by computing the shears and moments in the conjugate beam is called the conjugate-beam method. It should be noted that the conjugate beam has the same span as the original beam and is supported and constrained in such a manner that the interior support conditions and external boundary conditions of the actual beam are satisfied. This will be expounded below.

Conjugate-Beam Supports   The supports for the conjugate beam can be readily determined from the following two fundamental relationships:

Slope in actual beam $\equiv$ shear in conjugate beam $\qquad$ (a)

and

Deflection in actual beam $\equiv$ bending moment in conjugate beam $\qquad$ (b)

Following these relationships, we have

1. At a *hinged or rollered end support* of an actual beam, there is slope and no deflection. The corresponding conjugate-beam support must have shear and no moment. Therefore, the conjugate-beam support must also be a *hinged or rollered end support.*
2. At a *free end of an actual beam,* there is slope as well as deflection. Thus, the

corresponding conjugate-beam end must have both shear and moment; that is, the *conjugate-beam support must be a fixed-end support.*

3. At a *fixed-end support* of a real beam, there is neither slope nor deflection. Thus, there must be neither shear nor moment at the corresponding end of the conjugate beam; that is, the conjugate-beam end must be a *free end.*

4. At an *interior support* of a real beam, there is slope and no deflection; therefore, the corresponding point in the conjugate beam must have shear but no moment; that is, the *conjugate beam* must have an *internal hinge at that location.* If the support of the real beam settles by a given amount, the internal hinge is considered to be a "rusty" hinge capable of developing a moment equal in magnitude to the settlement of support of the actual beam.

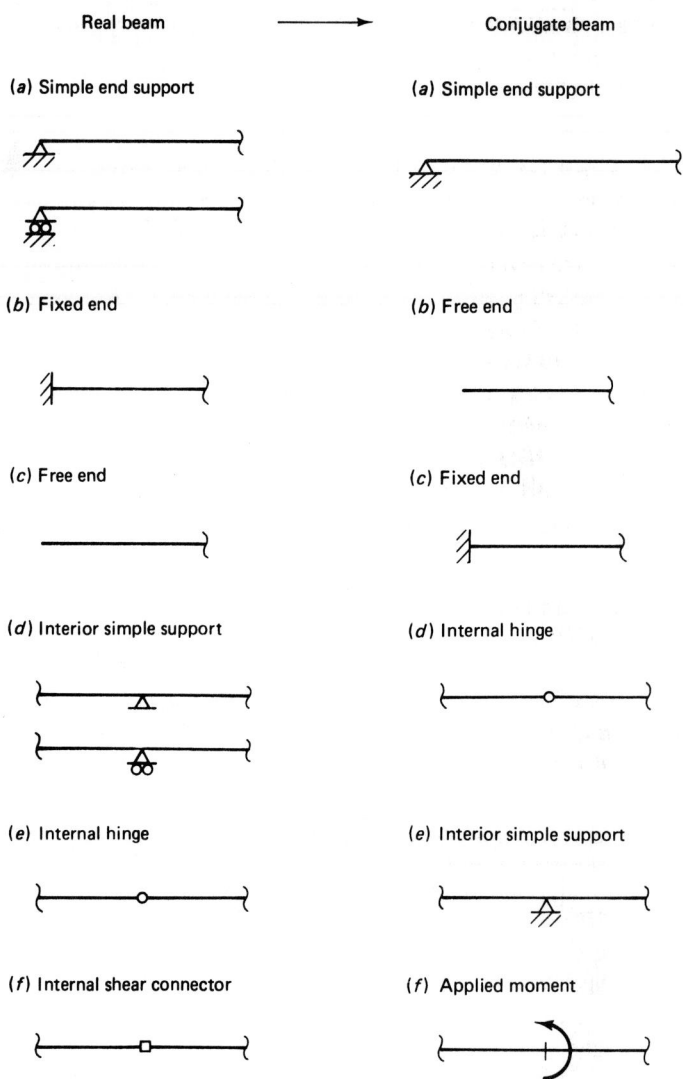

Figure 7.11 Some real-beam supports and the corresponding conjugate-beam supports.

5. At an *internal hinge* of the real beam, there is sudden change of slope but no sudden change in deflection. Therefore, at the corresponding section in the conjugate beam, there must be a sudden change in shear but no abrupt change in bending moment. These conditions are satisfied by an *interior support in the conjugate beam* corresponding to an internal hinge in the real beam.

6. At an *interior shear connector* in a real beam, there is no sudden change in slope but there can be discontinuity in deflection. Therefore, in the conjugate beam, there cannot be any sudden change in shear but only abrupt change in moment. This is accomplished by applying an external couple to the conjugate beam at the location of the interior shear connector in the real beam.

Figure 7.11 shows the real-beam supports with some construction features and the corresponding conjugate-beam supports and releases required for the application of the conjugate-beam method. Figure 7.12 shows some examples of real beams and the corresponding conjugate beams, based on the above principles. By examining the conjugate beams, which sometimes are referred to as fictitious beams, the following conclusions can be drawn:

1. The conjugate beams corresponding to statically determinate real beams are themselves statically determinate.

2. The conjugate beams corresponding to statically indeterminate real beams are in unstable equilibrium. However, the elastic load M/EI stabilizes such conjugate beams, and the usual methods of determining reactions, shearing forces, and bending moments can be applied to them in the usual manner to yield consistent results. A special case in point is a fixed-ended real beam, whose conjugate beam has no supports at all (a fictitious beam, indeed). However, the M/EI diagram on the conjugate beam is self-balancing, requiring no external supports for equilibrium. The slope and deflection at any section can be readily obtained by computing the shear and bending moment at the corresponding section of the conjugate beam.

■ **EXAMPLE 7.9**

Using the conjugate-beam method, determine the slopes and deflections at B and C of the cantilever beam shown in Fig. 7.13a. Take $EI = 10,000,000 \text{ k} \cdot \text{in}^2$.

Solution. This is the same beam analyzed in Example 7.4 by the moment-area method. The conjugate beam $A'C'$ corresponding to the given beam AC is shown in Fig. 7.13b. End A is fixed in the real beam; A' is therefore free. End C is free in the real beam; C' is fixed. The loading on the conjugate beam is the M/EI diagram of the given beam and is shown in Fig. 7.13c. The shears and moments in the conjugate beam give the slopes and deflections in the real beam.

Slope at B in real beam = shear at B' in conjugate beam

$$= -\frac{1}{2} \times 5 \times \frac{50}{EI} - \frac{1}{2} \times 5 \times \frac{100}{EI} - \frac{1}{2} \times 5 \times \frac{50}{EI}$$

$$= -\frac{500}{EI} = -\frac{500}{10,000,000/144} = -0.0072 \text{ radian}$$

that is, 0.0072 radian clockwise.

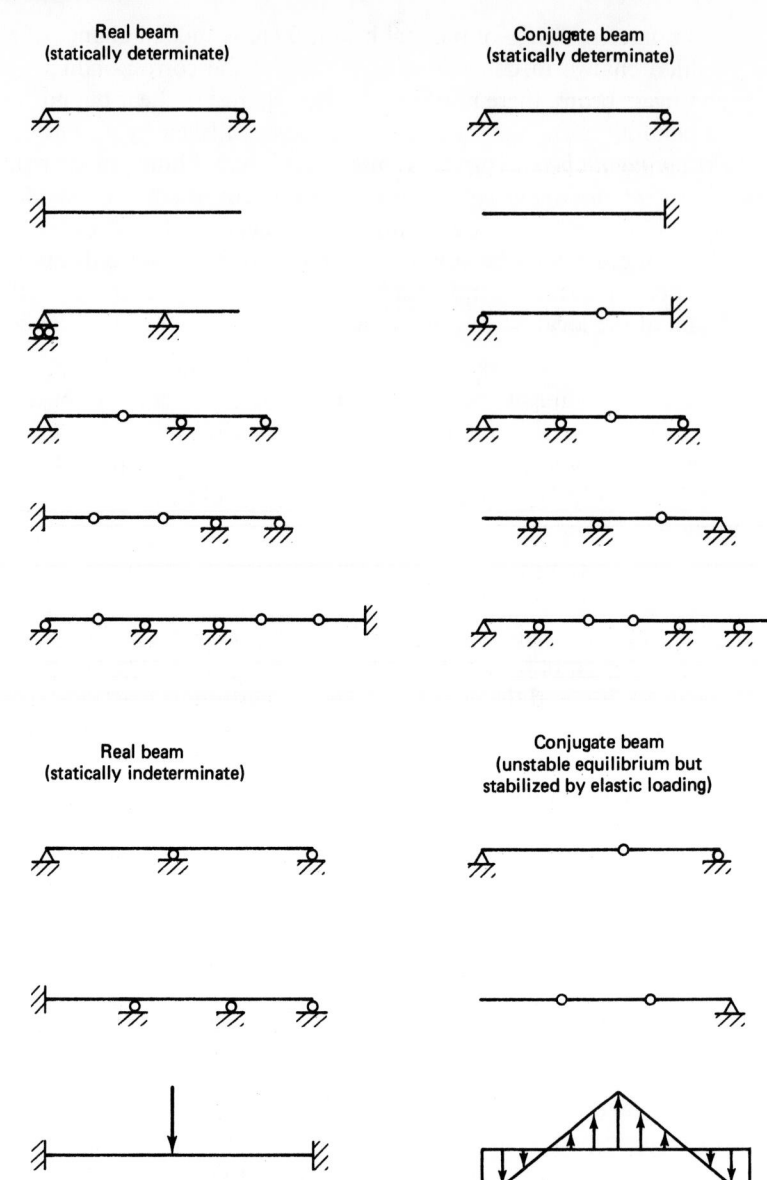

Real beam
(statically determinate)

Conjugate beam
(statically determinate)

Real beam
(statically indeterminate)

Conjugate beam
(unstable equilibrium but
stabilized by elastic loading)

Figure 7.12 Some real beams and the corresponding conjugate beams.

Deflection at B in real beam = bending moment at B' in conjugate beam

$$= -\frac{1}{2} \times 5 \times \frac{50}{EI} \times \left(\frac{1}{3} \times 5\right) - \frac{1}{2} \times 5 \times \frac{100}{EI}$$

$$\times \left(\frac{2}{3} \times 5\right) - \frac{1}{2} \times 5 \times \frac{50}{EI}\left(\frac{2}{3} \times 5\right)$$

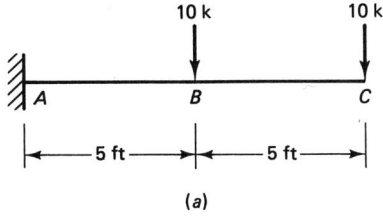

(a)

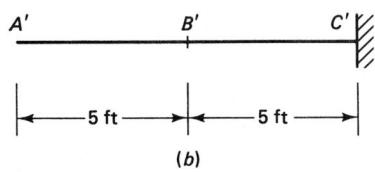

(b)

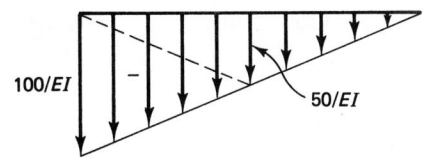

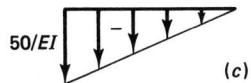

(c)

Figure 7.13 Beam for Example 7.9. (*a*) Given beam and loading. (*b*) Conjugate beam. (*c*) Load on the conjugate beam.

$$= -\frac{8750}{6EI} = -\frac{8750}{6(10,000,000/144)} = -0.021 \text{ ft}$$

$$= -0.252 \text{ in}$$

that is, 0.252 in downward.

Slope at C in real beam = shear in C' in conjugate beam

$$= -\frac{1}{2} \times 10 \times \frac{100}{EI} - \frac{1}{2} \times 5 \times \frac{50}{EI}$$

$$= -\frac{625}{EI} = -\frac{625}{10,000,000/144} = -0.009 \text{ radian}$$

that is, 0.009 radian clockwise.

Deflection at C in real beam = bending moment at C' in conjugate beam

$$= -\frac{1}{2} \times 10 \times \frac{100}{EI}\left(\frac{2}{3} \times 10\right) - \frac{1}{2} \times 5$$

$$\times \frac{50}{EI}\left(5 + \frac{2}{3} \times 5\right)$$

$$= -\frac{26,250}{6EI} = -\frac{26,250}{6(10,000,000/144)} = -0.063 \text{ ft}$$

$$= -0.756 \text{ in}$$

that is, 0.756 in downward.   ∎

■ EXAMPLE 7.10

Using the conjugate-beam method, determine the slope and deflection at B and C of the cantilever beam of nonuniform cross section shown in Fig. 7.14a.

Solution.   This beam was analyzed by the moment-area method in Example 7.5. The conjugate beam $A'C'$ corresponding to the real beam AC is shown in Fig. 7.14b. The bending-moment diagram for the real beam is shown in Fig. 7.14c. The load on the conjugate beam is obtained by dividing the ordinates of the bending-moment diagram by the EI values. Since segment AB has a flexural rigidity EI of 20×10^6

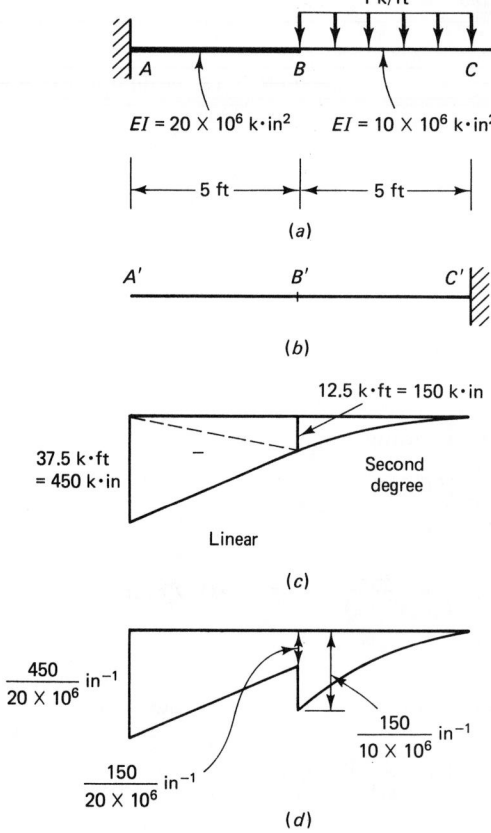

Figure 7.14 Beam for Example 7.10. (a) Given beam and loading, (b) Conjugate beam. (c) BMD for the real beam. (d) Loading on the conjugate beam.

$k \cdot in^2$, the ordinates of the bending-moment diagram in segment AB are divided by 20×10^6; segment BC has a flexural rigidity EI of only 10×10^6 $k \cdot in^2$; therefore, the ordinates of the moment diagram in BC are divided by 10×10^6 $k \cdot in^2$ to get the load on the conjugate beam.

Slope at B in real beam = shear at B' in conjugate beam

$$= \frac{1}{20 \times 10^6}\left(-\frac{1}{2} \times 150 \times 60 - \frac{1}{2} \times 450 \times 60\right)$$

$$= -0.0009 \text{ radian}$$

0.0009 radian clockwise.

Deflection at B = moment at B' in conjugate beam

$$= \frac{1}{20 \times 10^6}\left[-\frac{1}{2} \times 150 \times 60 \times \left(\frac{1}{3} \times 60\right)\right.$$

$$\left. -\frac{1}{2} \times 450 \times 60 \times \left(\frac{2}{3} \times 60\right)\right]$$

$$= -0.0315 \text{ in}$$

that is, 0.0315 in downward.

Slope at C = shear at C' in conjugate beam

$$= \frac{1}{20 \times 10^6}\left(-\frac{1}{2} \times 150 \times 60 - \frac{1}{2} \times 450 \times 60\right)$$

$$+ \frac{1}{10 \times 10^6}\left(-\frac{1}{3} \times 150 \times 60\right)$$

$$= -0.0009 - 0.0003 = -0.0012 \text{ radian}$$

that is, 0.0012 radian clockwise.

Deflection at C = bending moment at C' in conjugate beam

$$= \frac{1}{20 \times 10^6}\left[-\frac{1}{2} \times 150 \times 60 \times \left(\frac{1}{3} \times 60 + 60\right)\right.$$

$$\left. -\frac{1}{2} \times 450 \times 60 \times \left(\frac{2}{3} \times 60 + 60\right)\right]$$

$$+ \frac{1}{10 \times 10^6}\left[-\frac{1}{3} \times 150 \times 60 \times \left(\frac{3}{4} \times 60\right)\right]$$

$$= -0.099 \text{ in}$$

that is, 0.099 in downward. ∎

■ EXAMPLE 7.11

Using the conjugate-beam method, determine the location and magnitude of the maximum upward deflection of the overhanging beam shown in Fig. 7.15a. Also determine the slope and deflection at A. $EI = 1000 \text{ kN} \cdot \text{m}^2$.

Solution. This overhanging beam was analyzed by the moment-area method in Example 7.6. Since there is slope and deflection at the overhanging ends A and D of the real beam, the corresponding ends A' and D' of the conjugate beam are fixed as shown in Fig. 7.15b. Since there is only slope and no deflection at the interior supports B and C of the given beam, hinges are inserted in the conjugate beam at B' and C' (to make moment in the conjugate beam zero at those locations).

The loading on the conjugate beam is shown in Fig. 7.15c. Let the maximum upward deflection in span BC occur at a distance x to the right of support B. Since the slope at that point is zero, equate the shear in the conjugate beam at that section to zero.

Referring to Fig. 7.15d,

$$\text{Shear at } x = \frac{80}{3EI} - \frac{10}{EI}x - \frac{1}{2}x\left(\frac{40}{EI}\right)\left(\frac{x}{4}\right) + \frac{1}{2}x\left(\frac{30}{EI}\right)\left(\frac{x}{4}\right) = 0$$

Solving,

$$x = 2.11 \text{ m}$$

Therefore, maximum upward deflection in span BC occurs 2.11 m to the right of B.

Magnitude of maximum deflection

$$= \text{bending moment at } x \text{ in conjugate beam}$$

$$= +\left(\frac{80}{3EI}\right)(2.11) + \left(\frac{1}{2}\right)(2.11)\left(\frac{30}{EI} \times \frac{2.11}{4}\right)\left(\frac{1}{3} \times 2.11\right)$$

$$- \left(\frac{10}{EI}\right)(2.11)\left(\frac{2.11}{2}\right) - \frac{1}{2}(2.11)\left(\frac{40}{EI} \times \frac{2.11}{4}\right)\left(\frac{1}{3} \times 2.11\right)$$

$$= \frac{30.1}{EI} = \frac{30.1}{1000} = 0.0301 \text{ m} = 30.1 \text{ mm upward}$$

Slope and deflection at A can be readily computed from the free-body diagram of segment $A'B'$ of the conjugate beam. Referring to Fig. 7.15e,

$$\text{Slope at } A = +\frac{31.67}{EI} = \frac{31.67}{1000} = 0.03167 \text{ radian counterclockwise}$$

$$\text{Deflection at } A = -\frac{30}{EI} = -\frac{30}{1000} = -0.03 \text{ m} = 30 \text{ mm downward}$$ ■

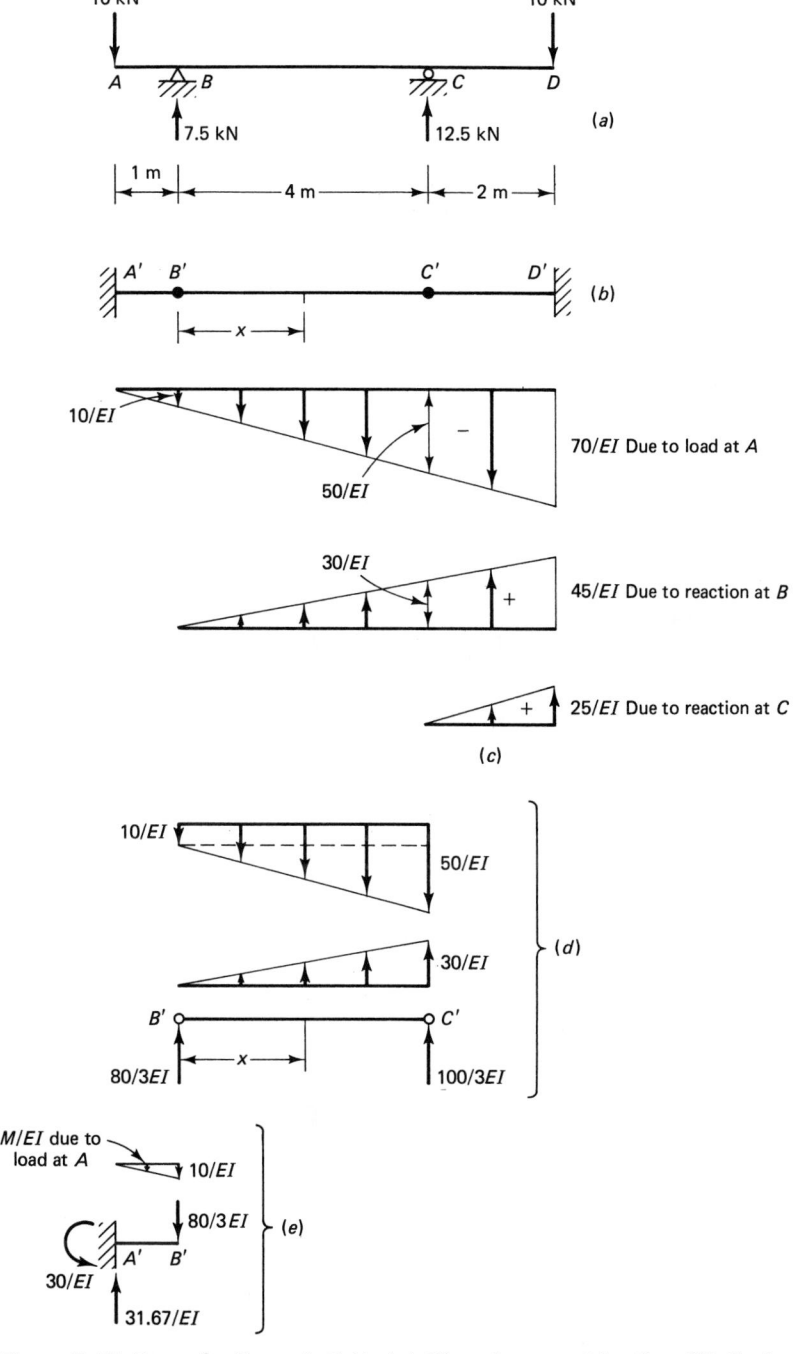

Figure 7.15 Beam for Example 7.11. (*a*) Given beam and loading. (*b*) Conjugate beam. (*c*) Loading on the conjugate beam. (*d*) Loading and reactions on segment *B'C'* of the conjugate beam. (*e*) Loading and reactions on segment *A'B'* of the conjugate beam.

■ EXAMPLE 7.12

Using the conjugate-beam method, find the slope and deflection at the free end D of the beam shown in Fig. 7.16a. $EI = 10,000 \text{ k} \cdot \text{ft}^2$.

Solution. This problem was solved earlier by the double-integration method and the moment-area method.

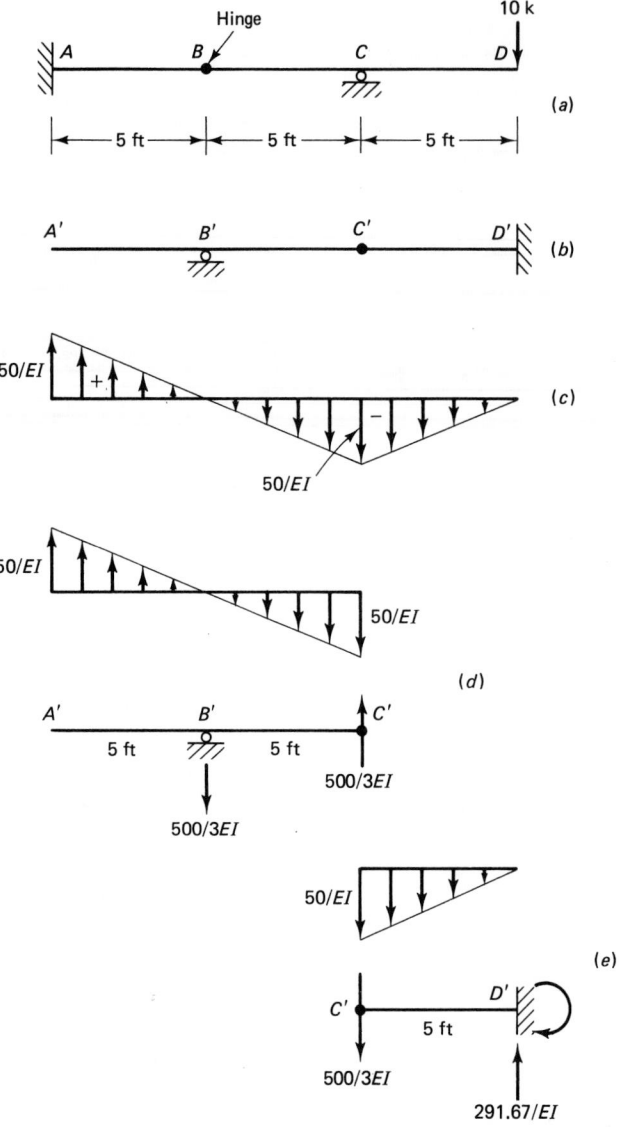

Figure 7.16 Beam for Example 7.12. (*a*) Given beam and loading. (*b*) Conjugate beam. (*c*) Loading on the conjugate beam. (*d*) FBD of segment $A'B'C'$ of the conjugate beam. (*e*) FBD of segment $C'D'$ of the conjugate beam.

The conjugate beam $A'B'C'D$ shown in Fig. 7.16b is drawn as follows:

Real beam	Conjugate beam
End A fixed	End A' free
Internal hinge at B	Interior support at B'
Interior support at C	Internal hinge at C'
End D free	End D' fixed

The loading on the conjugate beam shown in Fig. 7.16c is obtained by dividing the ordinates of the bending-moment diagram of Fig. 7.2b. Segment $A'B'C'$ is isolated, and the free-body diagram is shown in Fig. 7.16d. Segment $C'D'$ is then isolated, and the free-body diagram is shown in Fig. 7.16e.

$$\text{Slope at } D = \text{shear at } D' = -\frac{500}{3EI} - \frac{1}{2} \times \frac{50}{EI} \times 5 = -\frac{291.67}{EI}$$

$$= 0.029167 \text{ radian clockwise}$$

$$\text{Deflection at } D = \text{bending moment at } D' = -\left(\frac{500}{3EI} \times 5\right)$$

$$-\frac{1}{2} \times \frac{50}{EI} \times 5 \times \left(\frac{2}{3} \times 5\right)$$

$$= -0.125 \text{ ft} = 1.5 \text{ in downward.} \qquad \blacksquare$$

7.6 ENERGY METHODS—INTRODUCTION

The concepts of virtual work and energy are extremely important in the determination of deflection of structures. Before embarking on a discussion of these concepts, it is important that we review the ideas of work and energy. When a force acting on a body moves its point of application a certain distance, the product of force and that distance is defined as *work;* this capacity to do work is defined as energy; as a result of this work, the *kinetic energy* of the body changes if the body is in motion, or its *potential energy* changes if the body is at rest.

Consider the nonrigid body in Fig. 7.17a subjected to a force of magnitude P; the displacement of the body along the line of action of P is Δ. If the force on the body increases from P to $P + dP$ and the deformation from Δ to $\Delta + d\Delta$, then the increment in the *external work* will be approximately

$$dW_e = \left(P + \frac{dP}{2}\right)d\Delta \approx P \cdot d\Delta \tag{7.18}$$

which is the small shaded area in Fig. 7.17b, depicting the load-displacement curve. The total external work done over the full displacement Δ_1 is

$$W_e = \int_0^{\Delta_1} P \cdot d\Delta \tag{7.19}$$

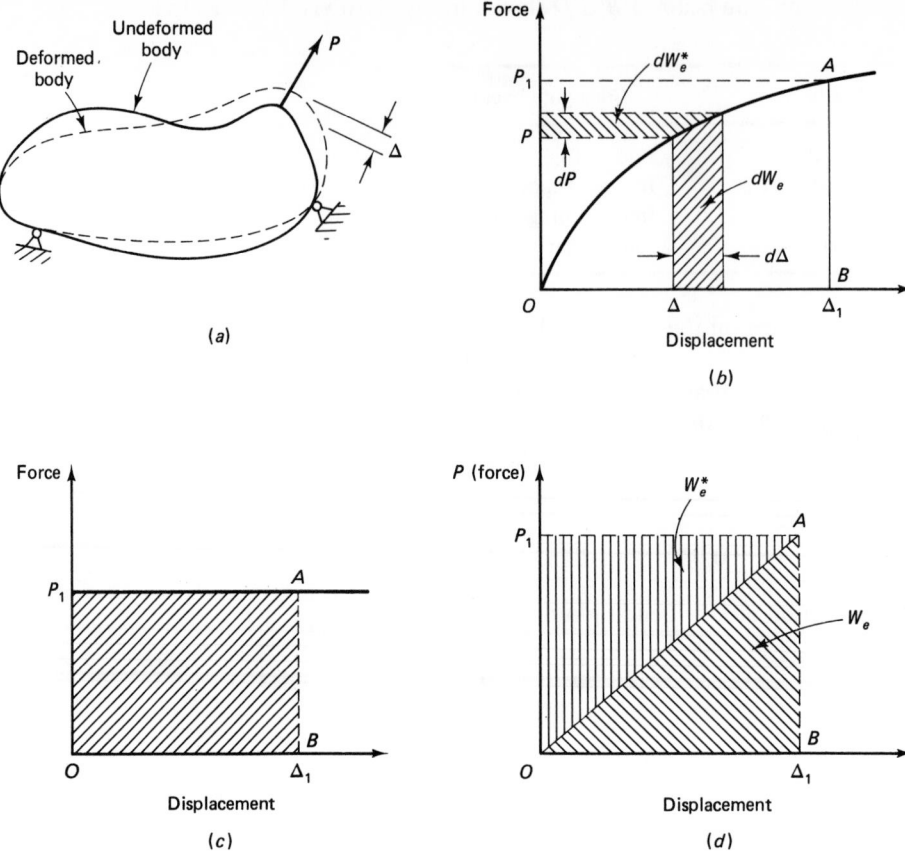

Figure 7.17 Nonrigid body subjected to a force P.

This is the area OAB under the curve OA, shown in Fig. 7.17b. If the force P_1 is constant on the body while the displacement Δ is caused by an action independent of P, the external work done by P is

$$W_e = P_1 \cdot \Delta_1 \qquad (7.20)$$

as shown in Fig. 7.17c. When the body is linearly elastic, that is, the load displacement is linear, or $P = (P_1/\Delta_1)\Delta$, then the external work becomes, from Eq. (7.19),

$$W_e = \tfrac{1}{2}P_1 \cdot \Delta_1 \qquad (7.21)$$

If the body is subjected to a system of n external forces, the total external work becomes, from Eq. (7.19),

$$W_e = \sum_{i=1}^{n} \int_0^{\Delta_1} P_i \cdot d\Delta_i \qquad (7.22)$$

It should be noted that W_e is a scalar quantity.

As a result of the external work, strain energy U is stored in the deformable body and the body may be set into motion, acquiring some kinetic energy T. The strain energy W_i, or the internal work, is one-half the sum of the product of the internal forces and the corresponding deformations in the body. Ignoring energy losses, by the law of conservation of energy we can write

$$W_e = W_i + T \qquad (7.23)$$

For a body in statical equilibrium, we have $T = 0$; hence Eq. (7.23) reduces to

$$W_e = W_i \qquad (7.24)$$

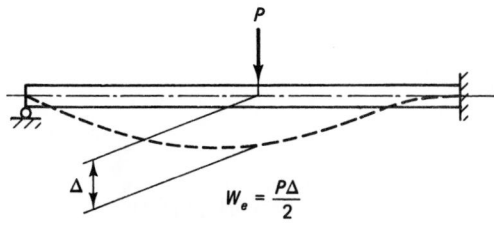

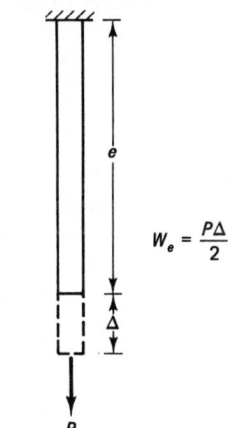

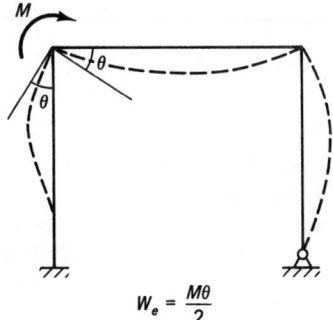

Figure 7.18 Examples of external work.

It should be noted that the expression W_e will also include the work done by the supporting reactions of the body if these reactions displace. Furthermore, the term P represents generalized forces, that is, forces and moments, while Δ represents the corresponding generalized displacements, that is, deflections and rotations. Some examples of external work W_e are shown in Fig. 7.18.

7.7 COMPLEMENTARY ENERGY OR WORK

Referring to the relationship between the force P and the displacement Δ in its line of action (Fig. 7.17b), the shaded area above the curve OA can be expressed as

$$dW_e^* = (\Delta)dP \tag{7.24a}$$

Thus the total area above the curve OA is

$$W_e^* = \int_0^{P_1} (\Delta)dP \tag{7.24b}$$

This is denoted as the external complementary work performed on the body; it should be mentioned that unlike the external work W_e, the complementary work or energy has no physical meaning. Correspondingly the expression for the internal complementary work, for example, in one of the members of an axially loaded pin-jointed structure, can be written as

$$W_i^* = \int_0^{P_1} (e)dP \tag{7.24c}$$

where e is the extension in the member and P is the internal force in the member.

For nonlinear elastic structures, $W_i^* \neq W_i$ and $W_e^* \neq W_e$. However, for linear elastic structures, $W_i^* = W_i$ and $W_e^* = W_e$, as can be observed from Fig. 7.17d. In the absence of gross distortion (as is the case in real structures) it can be shown from the law of conservation of complementary work (or energy) that

$$W_e^* = W_i^* \tag{7.24d}$$

7.8 FORMS OF ELASTIC STRAIN ENERGY (INTERNAL WORK)

As mentioned earlier, the internal work in a deformable body is the product of the force (equal to stress multiplied by the respective area) and the distance that the force moves (this distance being a function of the strain and hence the deformation in the line of force). There are four types of forces that a structural member can be subjected to: axial load, shear, moment, and twist. Each of these forces causes different deformation in the body. Consider a beam AB, with circular cross section, shown in Fig. 7.19a and loaded arbitrarily so that an elemental length dx is subjected to an axial force P (Fig. 7.19b), moment M (Fig. 7.19c), twist T (Fig. 7.19d), and shear force V (Fig. 7.19e). Let us assume that these forces, which may vary along the beam AB,

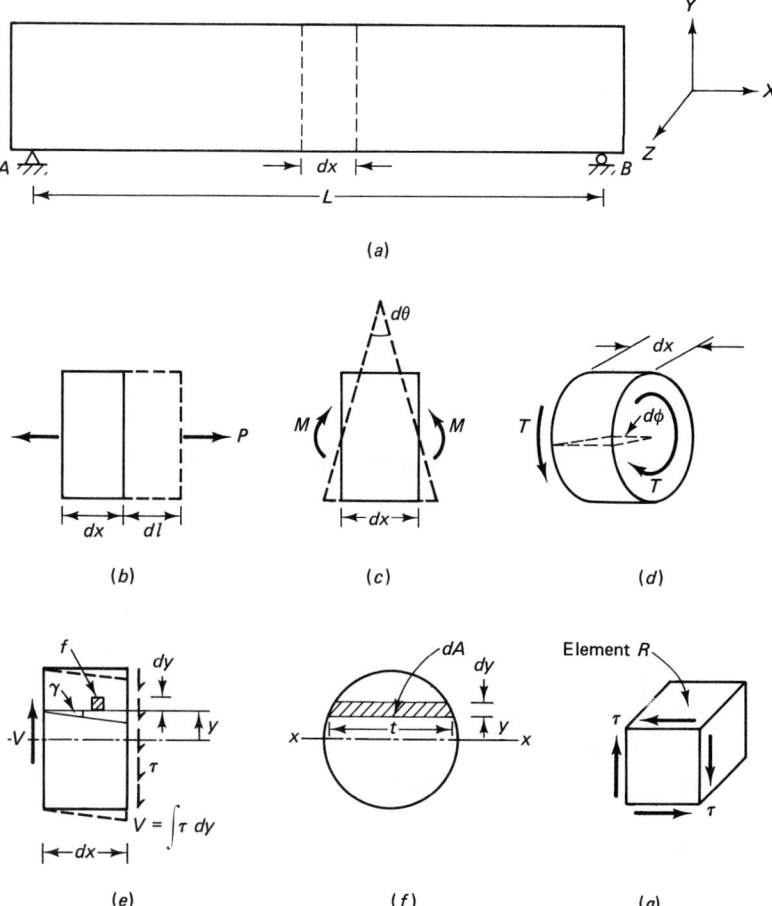

Figure 7.19 Forms of elastic strain energy (internal work).

produce stresses within the elastic limit of the material; that is, the structure will return to its original position when the forces are removed.

Elastic Strain Energy in Tension (or Compression) For an elemental length dx subjected to an axial force P, the displacement de is given by

$$de = \frac{P\,dx}{AE} \tag{7.25}$$

in which A is the cross-sectional area of the elemental length and E is the modulus of elasticity of the material. For a linearly elastic material, the internal work performed on the elemental length dx is given by

$$dW_i = \tfrac{1}{2}P \cdot de \tag{7.26}$$

Substituting Eq. (7.25) into Eq. (7.26) yields

$$dW_i = \frac{P^2}{2AE} dx \tag{7.27}$$

Therefore, the internal work or strain energy stored in the beam AB of length L becomes

$$W_i = \int_0^L \frac{P^2}{2AE} dx \tag{7.28}$$

For a uniform member, this becomes

$$W_i = \frac{P^2L}{2AE} \tag{7.28a}$$

For a structure with several members, say a truss, the internal work done becomes

$$W_i = \sum \frac{P^2L}{2AE} \tag{7.28b}$$

We can write W_i in terms of the axial strain ε_a as

$$W_i = \int_0^L \frac{AE}{2} \varepsilon_a^2 dx \tag{7.29}$$

where $\varepsilon_a = de/dx$.

For a uniform axial strain $\varepsilon_a = e/L$, Eq. (7.29) reduces to

$$W_i = \frac{AE}{2} \varepsilon_a^2 L = \frac{AE}{2L} e^2 \tag{7.29a}$$

For a truss,

$$W_i = \sum \frac{AEe^2}{2L} \tag{7.29b}$$

Elastic Strain Energy in Bending   Referring to Fig. 7.19c one can write the strain energy due to bending of an elemental length dx as

$$dW_i = \frac{M \, d\theta}{2} \tag{7.30}$$

But from Eq. (7.12)

$$d\theta = \frac{M}{EI} dx$$

Therefore

$$dW_i = \frac{M^2}{2EI} dx \tag{7.31}$$

which when integrated for the entire beam AB becomes

$$W_i = \int_0^L \frac{M^2}{2EI} dx \tag{7.32}$$

Or, in terms of displacement,

$$W_i = \int_0^L \frac{EI}{2} \left(\frac{d\theta}{dx}\right)^2 dx \tag{7.33}$$

If v is the displacement component in the y direction, $d\theta/dx =$ curvature $= d^2v/dx^2$ and Eq. (7.33) becomes, in terms of curvature,

$$W_i = \int_0^L \frac{EI}{2} \left(\frac{d^2v}{dx^2}\right)^2 dx \tag{7.34}$$

Strain Energy in Twisting   The elemental length dx is shown twisted by a torque T, producing an angle of twist $d\phi$ as shown in Fig. 7.19d. For simplicity we confine our discussion to circular sections. The strain energy stored in the elemental length becomes

$$dW_i = \frac{T\, d\phi}{2} \tag{7.35}$$

But from strength of materials, the relationship between angle of twist and torque is

$$d\phi = \frac{T\, dx}{GJ} \tag{7.36}$$

for a circular shaft in which G is the shear modulus $= E/2(1 + v)$; $v =$ Poisson's ratio, and J is the polar moment of inertia of a circular cross section.

Thus putting Eq. (7.36) into Eq. (7.35) yields

$$dW_i = \frac{T^2\, dx}{2GJ} \tag{7.37}$$

Therefore, the total strain energy stored in the beam AB due to twisting becomes

$$W_i = \int_0^L \frac{T^2}{2GJ} dx \tag{7.38}$$

In terms of the angle of twist ϕ

$$W_i = \int_0^L \frac{GJ}{2} \left(\frac{d\phi}{dx}\right)^2 dx \tag{7.39}$$

Strain Energy in Shear   Owing to the shear force V applied to the vertical sides of the element dx (Fig. 7.19e), and with a cross section shown in Fig. 7.19f, the stored strain energy in the element of volume $(A)(dx)$ is

$$dW_i = \int_A \frac{1}{2}\, \tau(dA)\, dV \tag{7.40}$$

From Fig. 7.19e, the displacement $dV = \gamma\, dx$, where τ is the shear stress on the element of area and γ is the shear strain given by $\gamma = \tau/G$. Therefore, with $dA = t \cdot dy$ we can rewrite dW_i as

$$dW_i = \left(\int \frac{\tau^2}{2G} t\, dy \right) dx \tag{7.41}$$

But from strength of materials, $\tau = VQ/It$, where V = shear force on the section; Q = statical moment of area above y about the neutral axis xx (Fig. 7.19f). Thus,

$$dW_i = \frac{V^2}{2I^2 G} \left(\int \frac{Q^2}{t}\, dy \right) dx \tag{7.42}$$

Hence the strain energy stored in beam AB becomes

$$W_i = \int_0^L \frac{V^2}{2I^2 G} \left(\int \frac{Q^2}{t}\, dy \right) dx \tag{7.43}$$

In general we can rewrite the above expression as

$$W_i = K_s \int_0^L \frac{V^2}{2GA}\, dx \tag{7.44}$$

where K_s is a constant whose value is dependent on the shape of the cross section. For a circular cross section, $K_s = \frac{10}{9}$; for a rectangular cross section, $K_s = \frac{6}{5}$. In the case of an I-beam cross section, where the web carries most of the shearing force and the shearing stress is nearly uniformly distributed over the web, $K_s \approx 1$, with A taken equal to the area of the web. In this derivation we have neglected the warping of the cross section (if it is noncircular) as well as the error arising from the assumption that the shearing stress τ is uniformly distributed across a horizontal interface.

Therefore, in terms of the resultant forces P, M, T, and V on a section of beam AB we can write the total internal work or strain energy stored as

$$W_i = \int_0^L \frac{P^2\, dx}{2AE} + \int_0^L \frac{M^2\, dx}{2EI} + \int_0^L \frac{T^2\, dx}{2GJ} + K_s \int_0^L \frac{V^2}{2GA}\, dx \tag{7.45}$$

The internal work for a stressed body can be also derived in terms of the internal stresses. Figure 7.20 shows an infinitesimal element from the stressed member, with sides dx, dy, and dz; during gradual loading of the body the possible stresses on the element increase from zero to their final values shown in Fig. 7.20; these stresses (σ and τ) cause strains (ε and γ). The internal work done by σ_x is

$$dW_i = \frac{(\text{force})(\text{displacement})}{2} = \frac{(\sigma_x\, dy\, dz)(\varepsilon_x\, dx)}{2} = \sigma_x \varepsilon_x \frac{dV}{2}$$

where $dV = dx\, dy\, dz$. When all the stresses in Fig. 7.20 are considered, it can be shown that

$$dW_i = \tfrac{1}{2}(\sigma_x \varepsilon_x + \sigma_y \varepsilon_y + \sigma_z \varepsilon_z + \tau_{xy} \gamma_{xy} + \tau_{xz} \gamma_{xz} + \tau_{yz} \gamma_{yz})\, dV \tag{7.46}$$

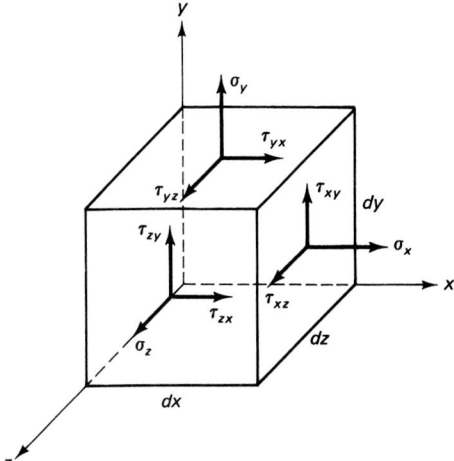

Figure 7.20 An element of a body subjected to stresses in three dimensions.

For a three-dimensional elastic material, the general stress-strain relationships are

$$\varepsilon_x = \frac{\sigma_x}{E} - \frac{\nu}{E}(\sigma_y + \sigma_z)$$

$$\varepsilon_y = \frac{\sigma_y}{E} - \frac{\nu}{E}(\sigma_x + \sigma_z) \qquad (7.47)$$

$$\varepsilon_z = \frac{\sigma_z}{E} - \frac{\nu}{E}(\sigma_x + \sigma_y)$$

$$\gamma_{xy} = \frac{\tau_{xy}}{G} \qquad \gamma_{xz} = \frac{\tau_{xz}}{G} \qquad \gamma_{yz} = \frac{\tau_{yz}}{G}$$

Substituting these relationships in Eq. (7.46) and integrating over the volume of the structural member yields

$$w_i = \int_V \left[\frac{1}{2E}(\sigma_x^2 + \sigma_y^2 + \sigma_z^2) - \frac{\nu}{E}(\sigma_x\sigma_y + \sigma_y\sigma_z + \sigma_z\sigma_x) \right.$$

$$\left. + \frac{1}{2G}(\tau_{xy}^2 + \tau_{yz}^2 + \tau_{zx}^2) \right] dV \qquad (7.48)$$

7.9 REAL WORK AND APPLICATIONS

Consider the cantilever beam of circular cross section loaded as shown in Fig. 7.21. The load is gradually applied from zero to a value P. The external work done by the load P is

$$W_e = \tfrac{1}{2}P \cdot \Delta_A$$

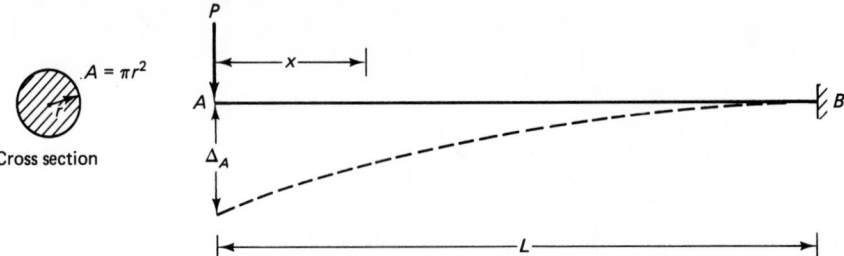

Figure 7.21 Cantilever beam of circular cross section with an end-concentrated load.

and the internal work or strain energy stored is

$$W_i = \int_0^L \frac{1}{2} \frac{M^2 \, dx}{EI} + K_s \int_0^L \frac{V^2}{2GA} \, dx$$

$$= (W_i)_{\text{bending}} + (W_i)_{\text{shear}}$$

Now at x, $M = -P \cdot x$ and $V = -P$ (M is negative since it produces tension at the top and V is negative since it is a downward force to the left of the section. Substituting in the above equation for W_i and integrating yields

$$W_i = \int_0^L \frac{1}{2EI} (-Px)^2 \, dx + K_s \int_0^L \frac{(-P)^2}{2GA} \, dx$$

$$= \frac{P^2 L^3}{6EI} + \frac{K_s P^2 L}{2GA}$$

where $K_s = \frac{10}{9}$ for a circular cross section.
From the law of conservation of energy

$$W_e = W_i$$

and hence the deflection Δ_A becomes

$$\Delta_A = \frac{PL^3}{3EI} + \left(\frac{10}{9}\right) \frac{PL}{GA}$$

By taking ν for steel $= 0.3$, I for a circle $= \pi r^4/4$, and $A = \pi r^2$,

$$\Delta_A = \frac{PL^3}{3EI} \left[1 + 2.17 \left(\frac{r}{L}\right)^2\right]$$

The second term inside the brackets is caused by shear, and we observe that for relatively short deep beams it can be important; however, it is quite small in ordinary slender beams, say, with $L/r \geqslant 10$.

■ EXAMPLE 7.13
Determine the horizontal deflection of joint C in the pin-jointed truss in Fig. 7.22a.
Use the method of real work.

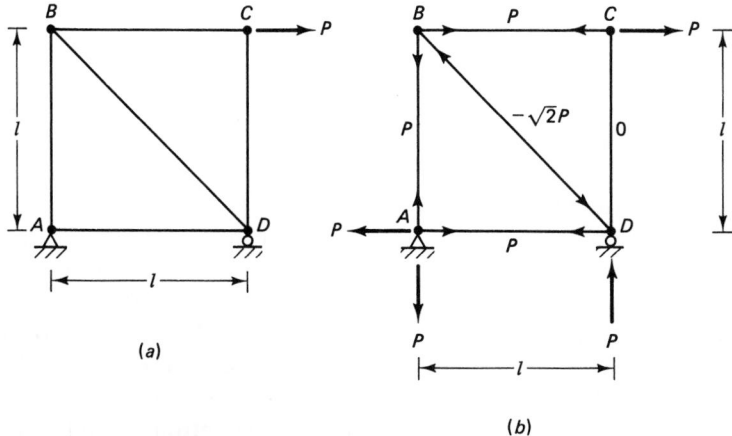

Figure 7.22 Truss for Example 7.13.

Solution. From statics, one can readily find the forces in this truss *ABCD* as shown in Fig. 7.22*b*. Since this is a pin-jointed truss, the strain energy stored is the sum of the strain energies due to internal forces in the members, that is,

$$W_i = \sum_{\text{all members}} \frac{P^2 L}{2AE} \tag{7.28b}$$

In tabular form, we have

Member	Length L	Area A	Force P	$W_i = P^2L/2AE$
AB	l	A	P	$P^2l/2AE$
BC	l	A	P	$P^2l/2AE$
CD	l	A	0	0
DA	l	A	P	$P^2l/2AE$
BD	$\sqrt{2}l$	A	$-\sqrt{2}P$	$2\sqrt{2}P^2l/2AE$

$$\Sigma W_i = \frac{P^2l}{2AE}(3 + 2\sqrt{2})$$

The external work performed by *P* is

$$W_e = \tfrac{1}{2}P\Delta$$

where Δ = required horizontal deflection. From $W_e = W_i$,

$$\frac{1}{2}P\Delta = \frac{P^2l}{2AE}(3 + 2\sqrt{2})$$

or

$$\Delta = 5.83 \, \frac{Pl}{AE}$$ ∎

This method can be applied also to find the angle of rotation θ (or angle of twist ϕ) at any point A in the structure where a moment M_A (or twist T_A) is applied. In this case the external work W_e will be equal to $\frac{1}{2} M_A \theta_A$ (or $\frac{1}{2} T_A \phi_A$). However, if the displacement is required at any other points where no external forces are applied, or in a different direction than the line of action of the applied loads, it is not possible to apply the method of real work; furthermore, if more than one external force is applied simultaneously on the structure, then more than one unknown displacement will be present in the expression for W_e and hence the one equation ($W_i = W_e$) cannot solve for more than one unknown. For these reasons, the method of real work has some limitations and is not widely used. We now introduce and concentrate on the method of virtual work, which is the foundation of the other energy methods.

7.10 VIRTUAL WORK

The principle of virtual work forms the foundation for other energy principles and theorems; the principle was developed by Johann Bernoulli in 1717.

Let us consider a rigid particle A shown in Fig. 7.23, subjected to a system of concurrent coplanar forces P_1, P_2, ... having a resultant R; the particle is in static equilibrium and therefore satisfies the equilibrium conditions

$$\sum F_x = 0 \qquad \sum F_y = 0$$

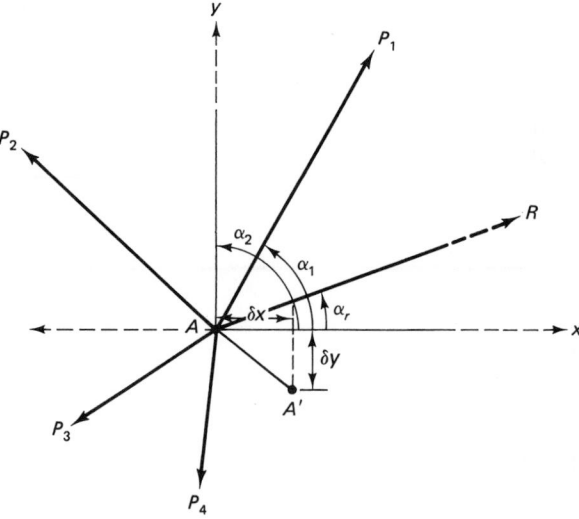

Figure 7.23 Rigid particle subjected to a system of concurrent coplanar forces.

Suppose the particle is given arbitrary imaginary or virtual displacements δx and δy along the x and y axes as shown. (The word "virtual" signifies that the displacement is independent of the P forces.) Then the virtual work W_r done (product of real forces and virtual displacements) is

$$W_r = R \, \delta x \cos \alpha_r + R \, \delta y \sin \alpha_r$$

$$= \delta x (\sum F_x) + \delta y (\sum F_y)$$

Since δx and δy are arbitrary and $\sum F_x = \sum F_y = 0$, the virtual work

$$W_r = 0 \tag{7.49}$$

This is true for any rigid body subjected to forces and in equilibrium. Therefore, the *principle of virtual work* for a rigid body can be stated as follows:

Given a rigid body held in equilibrium by a system of forces (or couples), the virtual work done by this system of forces (or couples) during a virtual displacement, causing rigid-body motion, is zero.

The converse of this statement is also true: If the work done by a system of forces (or couples) on a rigid body during a virtual displacement vanishes, the system of forces (or couples) is in equilibrium. It should be emphasized that virtual work is also defined as the product of virtual forces and real displacements, and it is then referred to as complementary virtual work; furthermore, a displacement can be either linear or rotational.

7.10.1 Virtual Work on an Elastic Body

Let us consider the elastic body shown in Fig. 7.24a, which is in equilibrium under a set of P external forces; the P-induced stresses σ_P on an element are also shown. The body is perturbed and thus given a virtual distortion δD as shown by the dashed line, while the P-force system is present; the virtual distortion must be consistent with the geometrical constraints of the elastic body; for example, in Fig. 7.24a we must have $\delta D = 0$ at the supports A and B. Furthermore, the magnitudes of δD must be vanishingly small such that the lines of action of the P forces and the P-induced internal stresses can be assumed to be unaffected by the distortion; thus δD does not affect the equilibrium of the P-force system. During this displacement, any element will displace and the stresses σ_P at the boundaries of such an element will do work; part of this work will be due to rigid-body movement of the element and the other part is due to the set of virtual strains $\delta \varepsilon_D$. Thus the total external virtual work W_e is composed of two parts:

1. W_r that couples the P-induced stresses with rigid-body displacements of all elements in the body
2. W_i that couples the P-induced stresses with virtual strains $\delta \varepsilon_D$

Thus, we can write

$$W_e = W_r + W_i \tag{7.50}$$

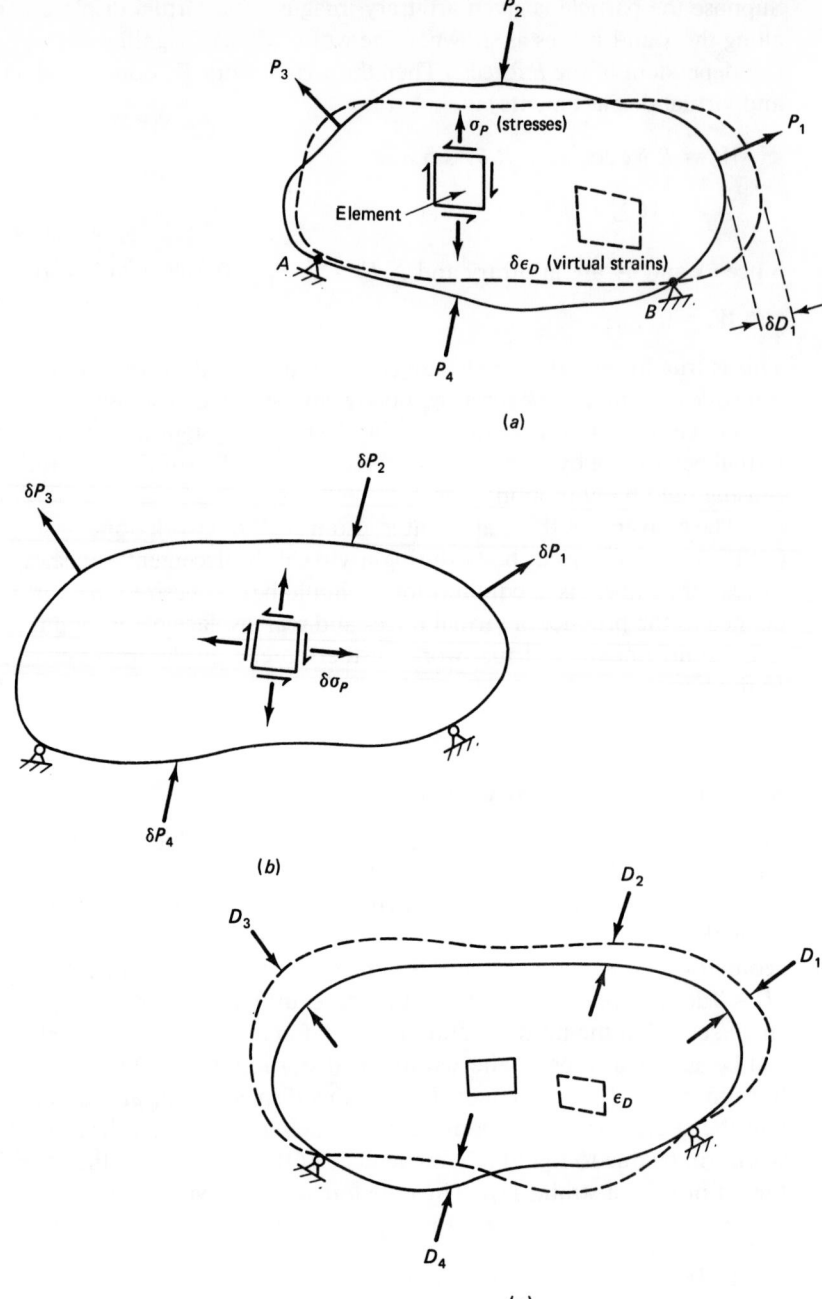

Figure 7.24 Virtual work on an elastic body.

But, from the principle of virtual displacement, $W_r = 0$ [Eq. (7.49)]; hence

$$W_e = W_i \tag{7.51}$$

where

$$W_i = \int_{\text{volume}} \sigma_P(\delta \varepsilon_D) d(\text{volume}) \tag{7.52}$$

Thus the principle of virtual work for an elastic body may be stated as follows: If an elastic body is in equilibrium and remains in equilibrium while it is subjected to a virtual distortion compatible with the geometrical constraints, the virtual work done by the external P forces is equal to the virtual work done by the P-induced internal stresses.

It should be noted that the expressions for the external and internal virtual work in Eq. (7.51) do not contain the factor of $\frac{1}{2}$ as was found in the expression for real work [e.g., Eq. (7.21)]; the reason for this is that the virtual distortion δD (and $\delta \varepsilon_D$) is independent of the P forces (and P-induced stresses); hence the virtual work done will be the area of the rectangle shown in Fig. 7.17c. In effect the virtual distortion gives a ride to the P-force system (and the same applies to the P-induced stresses).

Thus, from Eqs. (7.22), (7.51), and (7.52) we can write

$$\sum_{i=1}^{n} P_i(\delta D)_i = \int_{\text{vol}} \sigma_P(\delta \varepsilon_D) d\text{ vol} \tag{7.53}$$

In words:

Virtual displacement (compatible)

Real force $\times$ virtual displacement = real internal forces $\times$ virtual internal displacements

P system (actual) in equilibrium $\tag{7.53a}$

For all possible types of stresses on an element the expression for the internal virtual work [right-hand side of Eq. (7.53)] can be written as

$$W_i = \int_{\text{vol}} (\sigma_x \, \delta \varepsilon_x + \sigma_y \, \delta \varepsilon_y + \sigma_z \, \delta \varepsilon_z + \tau_{xy} \, \delta \gamma_{xy} + \tau_{xz} \, \delta \gamma_{xz} + \tau_{yz} \, \delta \gamma_{yz}) d\text{ vol}$$

After selecting a compatible set of δD virtual displacements, one can deduce the P forces from Eq. (7.53).

An analogous proof can be carried out for the principle of virtual complementary work showing that

$$W_e^* = W_i^* \tag{7.54}$$

where W_e^* is the virtual external complementary work and W_i^* is the virtual internal complementary work.

Thus, if the displacement is required, the virtual-work method can be used by applying a small virtual force δP to an elastic body (Fig. 7.24b) with real deformation pattern D and ε_D (Fig. 7.24c) that is geometrically compatible; the virtual force δP, producing internal virtual stresses $\delta\sigma_P$, must be in equilibrium. Thus, applying Eq. (7.54), one can write

$$\sum_{i=1}^{n}(\delta P)_i D_i = \int_{\text{vol}} \delta\sigma_P \varepsilon_D \, d\text{ vol} \tag{7.54a}$$

where $W_e^* = \sum_{i=1}^{n}(\delta P)_i D_i$

$$W_i^* = \int_{\text{vol}}(\delta\sigma_P)\varepsilon_D \, d\text{ vol}$$

In words:

Virtual forces in equilibrium

Virtual force × real displacement = virtual internal forces × real internal displacements

Real displacement (compatible)   (7.54b)

After a set of δP virtual forces are selected, the displacements D of the actual structure can be obtained from Eq. (7.54). For a general system of strains on an element, the virtual internal complementary work [right-hand side of Eq. (7.54a and b) can be written as

$$W_i = \int_{\text{vol}}(\delta\sigma_x \varepsilon_x + \delta\sigma_y \varepsilon_y + \delta\sigma_z \varepsilon_z + \delta\tau_{xy}\gamma_{xy} + \delta\tau_{xz}\gamma_{xz} + \delta\tau_{yz}\gamma_{yz})d\text{ vol} \tag{7.54c}$$

Quite often, the virtual-force system comprises a unit load (or unit couple) at the point and in the direction of the required displacement; in such cases *the virtual-force method* becomes known as the *dummy-unit-load method*. We illustrate the method in the subsequent sections of this chapter.

7.10.2 Forms of Virtual Internal Complementary Work Due to Virtual Forces

To find the displacement in a structure, we require expressions for the virtual internal complementary work, given by the right-hand side of Eq. (7.54). Consider again the element dx of the beam in Fig. 7.19a subjected to a real force system P, M, T, and V as shown in Fig. 7.19b, c, d, and e. Denoting virtual forces by lowercase letters, let us subject the element dx to virtual forces p, m, t, and u as shown in Fig. 7.25. The internal virtual work developed due to each of the virtual forces is as follows:

Due to Virtual Force p

Since the element dx experiences a real displacement de due to P (Fig. 7.25b), the virtual work performed by the virtual force p on the element dx is

$$dW_i^* = p \cdot de \tag{a}$$

But

$$de = \frac{P\,dx}{AE}$$

in which A = cross-sectional area and E = Young's modulus. Or

$$dW_i^* = p\left(\frac{P\,dx}{AE}\right)$$

For the entire member AB (Fig. 7.19a), the virtual internal complementary work becomes

$$W_i^* = \int dW_i^* = \int_0^L p\left(\frac{P\,dx}{AE}\right) \tag{7.55}$$

For the analysis of trusses where p and P forces are constant along the member of a truss, Eq. (7.55) can be written as

$$W_i^* = \sum_{\text{all members}} p\left(\frac{Pl}{AE}\right) \tag{7.56}$$

We should be reminded that p represents the internal virtual forces in a member while (Pl/AE) represents the internal real displacement of a member.

Due to Virtual Moment m

Figure 7.25c shows a virtual moment m acting on the element dx with a real displacement $d\theta$ caused by the applied external real moment M; the virtual work performed by m on the element dx is

$$dW_i^* = m \cdot d\theta$$

The real displacement $d\theta$ is given by [see Eq. (7.12)]

$$d\theta = \frac{M\,dx}{EI}$$

Therefore

$$dW_i^* = m\,\frac{M\,dx}{EI}$$

For the entire beam, the internal virtual work becomes

$$W_i^* = \int_0^L m\left(\frac{M\,dx}{EI}\right) \tag{7.57}$$

Here again, the term m represents the internal virtual forces in a bent member while $(M\,dx/EI)$ represents the internal real displacement (rotation) in a bent member.

Applying the above procedure for twist and shear yields the following expressions for the internal virtual work for twist T:

$$W_i^* = \int_0^L t\left(\frac{T\,dx}{GJ}\right) \tag{7.58}$$

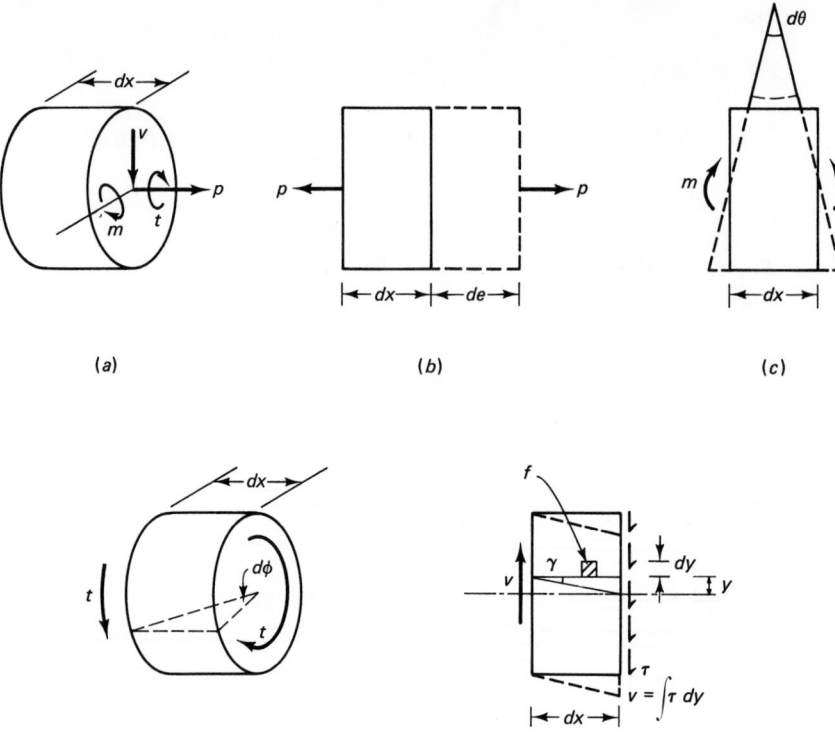

Figure 7.25 Forms of virtual internal complementary work due to virtual forces.

applicable only to members with circular cross section; and for shear V,

$$W_i^* = K_s \int_0^L v \frac{V \, dx}{GA} \tag{7.59}$$

where K_s is the same constant as that in Eq. (7.44).

For convenience, values of $\int_0^L f_1(x)f_2(x)\,dx$ in Eqs. (7.57) to (7.59) for common loading cases are presented in Table 7.1.

Sometimes it is advantageous to express the internal virtual stresses $\delta\sigma_x$, etc., in terms of the internal virtual forces p, m, etc., on the element dx in Fig. 7.25. From strength of materials, it can be shown that

$$\delta\sigma_x = \frac{p}{A} - \frac{my}{I}$$

$$\delta\sigma_y = 0$$

$$\delta\sigma_z = 0$$

$$\delta\tau_{xy} = -\frac{vQ}{It} + \frac{t \cdot z}{J} \qquad \delta\tau_{xz} = -\frac{t \cdot y}{J} \quad \text{and} \quad \delta\tau_{yz} = 0$$

TABLE 7.1 VALUES OF $\int_0^L f_1(x)f_2(x)\,dx$

$f_1(x)$ \\ $f_2(x)$	rectangle (c, L)	triangle rising (d, L)	trapezoid (c, d, L)	parabola* (c, L)
triangle rising (b, L)	$\dfrac{1}{2}Lbc$	$\dfrac{1}{3}Lbd$	$\dfrac{Lb}{6}(c+2d)$	$\dfrac{1}{3}Lbc$
triangle falling (a, L)	$\dfrac{1}{2}Lac$	$\dfrac{1}{6}Lad$	$\dfrac{La}{6}(2c+d)$	$\dfrac{1}{3}Lac$
trapezoid (a, b, L)	$\dfrac{L}{2}(a+b)c$	$\dfrac{Ld}{6}(a+2b)$	$\dfrac{L}{6}(2ac+ad+2bd+bc)$	$\dfrac{Lc}{3}(a+b)$
parabola* (a, L)	$\dfrac{2}{3}Lac$	$\dfrac{1}{3}Lad$	$\dfrac{La}{3}(c+d)$	$\dfrac{8}{15}Lac$
Tangent (b, L)*	$\dfrac{2}{3}Lbc$	$\dfrac{5}{12}Lbd$	$\dfrac{Lb}{12}(3c+5d)$	$\dfrac{7}{15}Lbc$
Tangent (a, L)*	$\dfrac{2}{3}Lac$	$\dfrac{1}{4}Lad$	$\dfrac{La}{12}(5c+3d)$	$\dfrac{7}{15}Lac$
Tangent (b, L)*	$\dfrac{1}{3}Lbc$	$\dfrac{1}{4}Lbd$	$\dfrac{Lb}{12}(c+3d)$	$\dfrac{1}{5}Lbc$
Tangent (a, L)*	$\dfrac{1}{3}Lac$	$\dfrac{1}{12}Lad$	$\dfrac{La}{12}(3c+d)$	$\dfrac{1}{5}Lac$
$\int f_2^2(x)\,dx$	Lc^2	$\dfrac{1}{3}Ld^2$	$\dfrac{L}{3}(c^2+cd+d^2)$	$\dfrac{8}{15}Lc^2$

* Second-degree parabola.

The signs attached to the above terms are necessary for consistency with the positive directions of stresses in Fig. 7.20. Thus, in matrix form we can write

$$
\begin{Bmatrix} \delta\sigma_x \\ \delta\sigma_y \\ \delta\sigma_z \\ \delta\tau_{xy} \\ \delta\tau_{xz} \\ \delta\tau_{yz} \end{Bmatrix} = \begin{bmatrix} 1/A & -y/I & 0 & 0 \\ 0 & 0 & 0 & 0 \\ 0 & 0 & 0 & 0 \\ 0 & 0 & z/J & -Q/It \\ 0 & 0 & -y/J & 0 \\ 0 & 0 & 0 & 0 \end{bmatrix} \begin{Bmatrix} p \\ m \\ t \\ v \end{Bmatrix}
\tag{7.60}
$$

7.10.3 Displacements by Virtual Work

To determine the real displacements in a structure by the method of virtual work, we require the expression for the virtual external complementary work given by the left-hand side of Eq. (7.54), namely,

$$
W_e^* = \sum_{i=1}^{n} (\delta P)_i D_i
\tag{7.61}
$$

If a displacement D at only one point in a bent structure is required, we can use a unit value for the external virtual force δP and hence Eq. (7.61) reduces to

$$
W_e^* = 1 \cdot D
\tag{7.62}
$$

Based on Eq. (7.57), Eq. (7.54b) then yields

$$
1 \cdot D = \int m \frac{M \, dx}{EI}
\tag{7.63}
$$

In this case the internal virtual moment m in the structure is that due to a unit external virtual force applied to the structure at the point where the displacement is required. As mentioned earlier, the required displacement can be either linear or rotational; in the case of linear displacement, we apply a unit virtual load, whereas in the case of rotational displacement we apply a unit virtual moment. In such a case Eq. (7.54b) becomes

$$
1 \cdot \theta = \int_0^L m \frac{M \, dx}{EI}
\tag{7.63a}
$$

The concept of applying a unit virtual force can be extended to structures subjected to axial loads, torsion, and shear. Thus, for axially loaded pin-connected structures, we can write

$$
1 \cdot D = \sum_{\text{all members}} p\left(\frac{Pl}{AE}\right)
$$

or

$$
D = \sum_{\text{all members}} \frac{pPL}{AE}
\tag{7.64}
$$

and so on. We now illustrate the use of the virtual-work method.

■ **EXAMPLE 7.14**

Determine the vertical deflection and slope at A for the cantilever beam AB loaded as shown in Fig. 7.26a. Use the virtual-work method.

Solution. The bending moment due to the external load P is shown in Fig. 7.26b. For the vertical deflection at A we apply a unit load ($P = 1$) at A as shown in Fig. 7.26c; the resulting moment m is given by Fig. 7.26d. Thus, applying Eq. (7.63),

$$(1)(D) = \int_0^{L/2} \frac{(-x)(0)}{EI}\, dx + \int_{L/2}^{L} (-x) \frac{(-P)(x - L/2)}{EI}\, dx$$

or

$$D = \frac{5}{48EI} PL^3$$

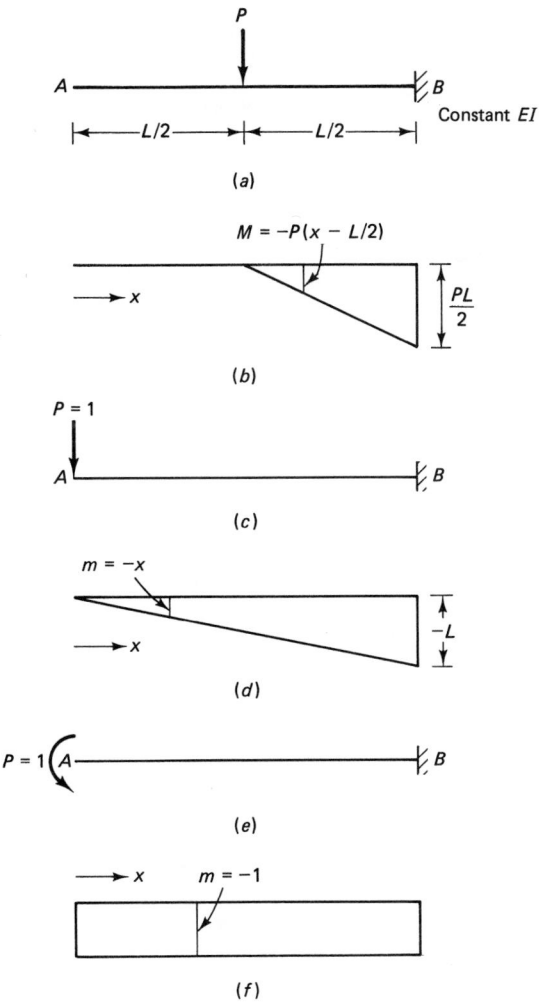

(a)

(b)

(c)

(d)

(e)

(f)

Figure 7.26 Beam for Example 7.14.

The positive sign indicates that the deflection is in the same direction as the applied unit virtual load, that is, downward. For slope at A we apply a unit moment at A as shown in Fig. 7.26e, and the resulting moment diagram is given in Fig. 7.26f. Again from Eq. (7.63a), we have

$$(1)(\theta_A) = \int_0^{L/2} \frac{(-1)(0)}{EI}\,dx + \int_{L/2}^L \frac{(-1)[-P(x - L/2)]}{EI}\,dx$$

or

$$\theta_A = \frac{PL^2}{8EI}$$

∎

∎ EXAMPLE 7.15

Determine the slope at A and B of the beam AB subjected to a moment M_A as shown in Fig. 7.27a. Use the virtual-work method.

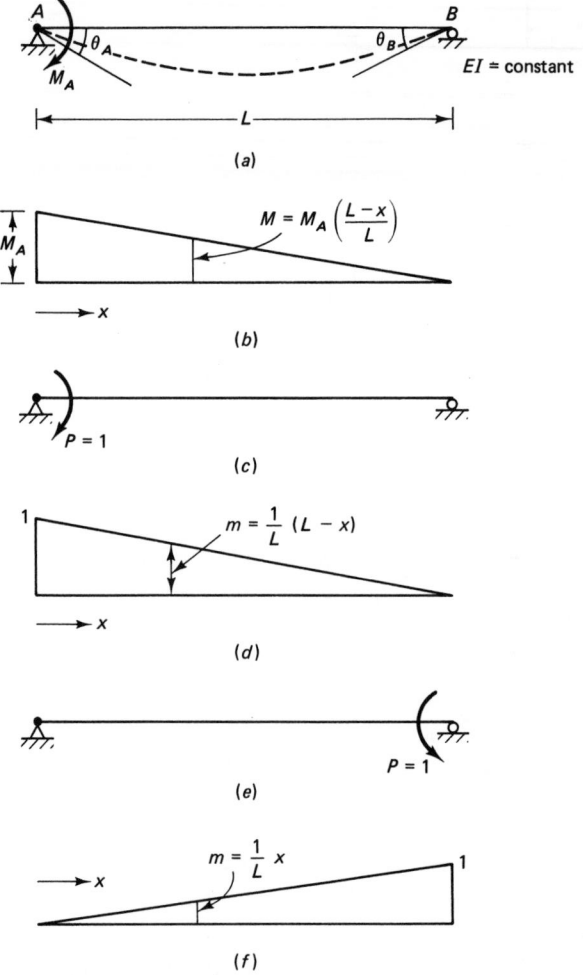

Figure 7.27 Beam for Example 7.15.

Solution. The bending-moment diagram due to the moment M_A is given in Fig. 7.27b. To deduce the slope at A we apply a unit moment at A (Fig. 7.27c), and the corresponding m diagram is presented in Fig. 7.27d. From Eq. (7.63a)

$$(1)(\theta_A) = \int_0^L \frac{1}{EI}\frac{L-x}{L}\left[\frac{M_A}{L}(L-x)\right]dx = \frac{M_A L}{3EI}$$

or

$$\theta_A = \frac{M_A L}{3EI}$$

To obtain the slope at B, apply a unit moment at B as shown in Fig. 7.27e with the corresponding m diagram given in Fig. 7.27f. Thus, from Eq. (7.63a)

$$(1)(\theta_B) = \int_0^L \frac{1}{EI}\left(\frac{x}{L}\right)\left[\frac{M_A}{L}(L-x)\right]dx = \frac{M_A L}{6EI}$$

Therefore,

$$\theta_B = \frac{M_A L}{6EI}$$ ∎

■ EXAMPLE 7.16
Determine the (a) horizontal deflection and (b) vertical deflection of joint C in the pin-jointed truss of Fig. 7.28a, using the virtual-work method.

Solution
   (a) The forces in the members due to the external load P are shown in Fig. 7.28b. To find the horizontal deflection of joint C, we apply a unit load to joint C, and the resulting p system of forces is given in Fig. 7.28c. Therefore, from Eq. (7.64),

$$1 \cdot D^H = \sum_{\text{all members}} p\frac{PL}{AE}$$

In tabular form:

Member	P	L/AE	PL/AE	p	$p(PL/AE)$
AB	P	l/AE	Pl/AE	1	Pl/AE
BC	P	l/AE	Pl/AE	1	Pl/AE
CD	0	l/AE	0	0	0
DA	P	l/AE	Pl/AE	1	Pl/AE
BD	$-\sqrt{2}P$	$\sqrt{2}l/AE$	$-2Pl/AE$	$-\sqrt{2}$	$2\sqrt{2}Pl/AE$

$$\sum p\frac{PL}{AE} = \frac{Pl}{AE}(3+2\sqrt{2})$$

Therefore, D^H = horizontal deflection of joint C = 5.83 Pl/AE as before.
   (b) For the vertical deflection of joint C, apply a unit vertical load at C, and the corresponding internal virtual forces in the members are shown in Fig. 7.28d. By

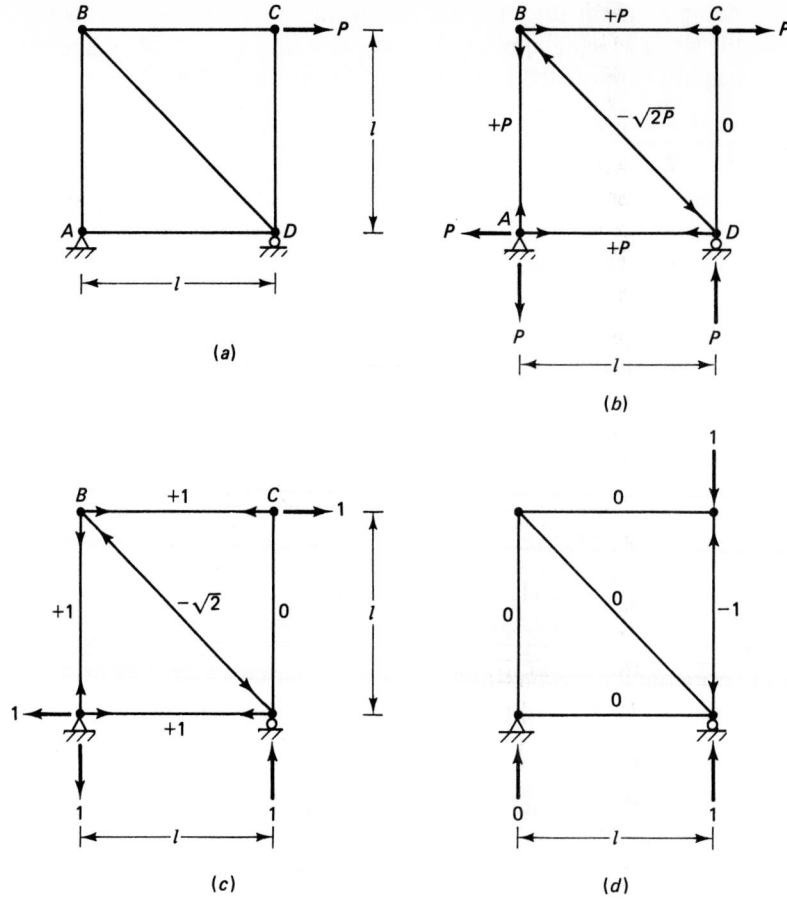

Figure 7.28 Truss for Example 7.16. (*a*) Area of all members = *A*. (*b*) *P*-force system. (*c*) *p* virtual-force system.

inspection, since the force in *CD* due to *P* is zero and since the *p*-force system shows that all *p* forces are zero except *CD*, the product $\Sigma\, p(PL/AE) = 0$; therefore,

$$1 \cdot D^V = \Sigma\, p\, \frac{PL}{AE} = 0$$

or

D^V = vertical deflection of joint *C* = 0 (as expected)   ■

■ EXAMPLE 7.17

Figure 7.29 shows a hollow circular bar in a horizontal plane loaded by a vertical force *P* at *A*. Determine the vertical component of deflection at *A*. Take *L* = 4 m, area $A = 1200$ mm^2, $E = 200$ GPa, $G = 80$ GPa, $I = 250 \times 10^6$ mm^4, $J = 500 \times 10^6$ mm^4, and *P* = 20 kN.

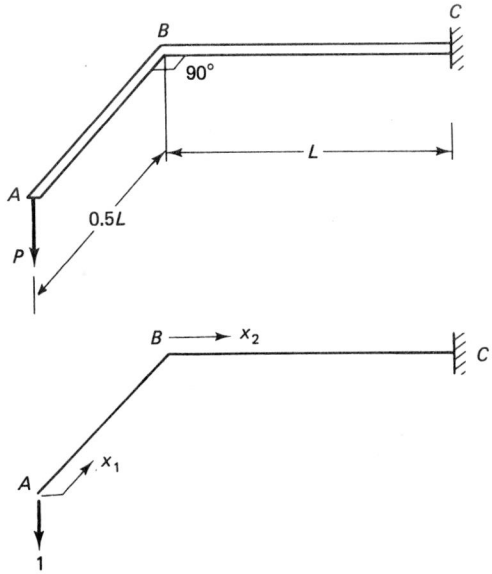

Figure 7.29 Structure for Example 7.17.

Solution. The virtual internal complementary work here is due to bending, shear, and torsion. Thus, referring to the geometry of Fig. 7.29,

From A to B: Limits 0 to 0.5L

$$M = -Px_1 \qquad V = -P \qquad T = 0$$

$$m = -x_1 \qquad\quad v = -1 \qquad t = 0$$

From B to C: Limits 0 to L

$$M = -Px_2 \qquad V = -P \qquad T = 0.5\,PL$$

$$m = -x_2 \qquad\quad v = -1 \qquad t = 0.5\,L$$

From Eq. (7.51), we can write using Eqs. (7.57), (7.58), and (7.59),

$$1 \cdot D_A^V = \int_{\text{frame}} m\!\left(\frac{M}{EI}\,dx\right) + K_s \int_{\text{frame}} v\!\left(\frac{V}{GA}\,dx\right) + \int_{\text{frame}} t\!\left(\frac{T}{GJ}\,dx\right)$$

or

$$D_A^V = \frac{1}{EI}\int_0^{0.5L} (-x_1)(-Px_1)\,dx_1 + \frac{K_s}{GA}\int_0^{0.5L} (-1)(-P)\,dx_1 + 0$$

$$+ \frac{1}{EI}\int_0^{L} (-x_2)(-Px_2)\,dx_2 + \frac{K_s}{GA}\int_0^{L} (-1)(-P)\,dx_2$$

$$+ \frac{1}{GJ}\int_0^{L} (0.5L)(0.5PL)\,dx_2$$

$$= \frac{1}{EI}\left(\frac{PL^3}{24}\right) + \frac{K_s}{2GA}PL + \frac{1}{EI}\left(\frac{PL^3}{3}\right) + \frac{K_sPL}{GA} + \frac{1}{GJ}\left(\frac{PL^3}{4}\right)$$

$$D_A^V = \frac{3}{8}\frac{PL^3}{EI} + \frac{3K_sPL}{2GA} + \frac{PL^3}{4GJ}$$

Using $K_s = \frac{10}{9}$,

$$D_A^V = \frac{(3)(20 \text{ kN})(4 \text{ m})^3}{8(200 \times 10^6 \text{ kN/m}^2)(250 \times 10^{-6} \text{ m}^4)}$$

$$+ \frac{(3)(10/9)(20 \text{ kN})(4 \text{ m})}{(2)(80 \times 10^6 \text{ kN/m}^2)(1200 \times 10^{-6} \text{ m}^2)}$$

$$+ \frac{(20 \text{ kN})(4 \text{ m})^3}{(4)(80 \times 10^6 \text{ kN/m}^2)(500 \times 10^{-6} \text{ m}^4)}$$

$$D_A^V = 0.0096 + 0.0014 + 0.0080$$

$$= 0.019 \text{ m} = 19 \text{ mm}$$

This value is positive, and therefore the deflection is downward in the direction of the unit load. It should be noted that the contribution of shear to the deflection D_A^V is approximately 7 percent, which can be considered to be negligibly small. ∎

∎ EXAMPLE 7.18

Using the virtual-work method calculate the vertical and horizontal displacements of joint d in the truss shown loaded in Fig. 7.30a, and accounting for the following support movements:

Support a: Horizontal $= u_a \leftarrow$

Support b: Vertical $= v_b\downarrow$ and horizontal $u_b \rightarrow$

(AE) is same for all members.

Solution

For the vertical displacement of joint d (v_d): First the forces in the members of the pin-jointed truss (Fig. 7.30a) are determined from statics. We now apply a unit vertical load at d, and the corresponding forces are shown in Fig. 7.30b. Now the external virtual complementary work W_e^* must account also for the support movements. Therefore,

$$W_e^* = (1)(v_d) + (2)(-u_a) + (2)(-u_b) + (1)(-v_b)$$

The negative sign attached to some of the terms in W_e^* is due to the fact that the virtual reactions are performing negative external work since they are opposite to the support movements. For the internal virtual complementary work

$$W_i^* = \sum_{\text{all bars}} p\left(\frac{PL}{AE}\right)$$

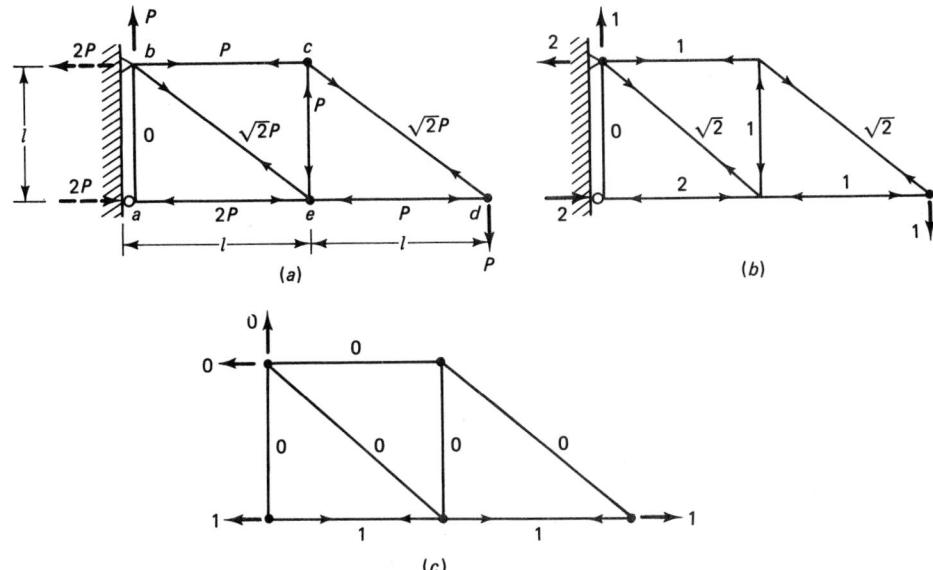

Figure 7.30 Truss for Examples 7.18, 7.19, and 7.20. *P*-force systems.

$$= \frac{l}{AE}[(-1)(-P) + (-2)(-2P) + (1)(P) + (\sqrt{2})(\sqrt{2}P)(\sqrt{2})$$

$$+ (-1)(-P) + (\sqrt{2})(\sqrt{2}P)(\sqrt{2})]$$

$$= 12.657 \frac{Pl}{AE}$$

Equating $W_i^* = W_e^*$ yields

$$v_d = 12.657 \frac{Pl}{AE} + 2(u_a + u_b) + v_b$$

For the horizontal displacement of joint d(u_d): Applying the unit load horizontally at d (Fig. 7.30c) yields W_e^* as

$$W_e^* = (1)(u_d) + (1)(u_a)$$

The internal virtual work is

$$W_i^* = \frac{l}{AE}[(1)(-P) + (1)(-2P)] = \frac{-3Pl}{AE}$$

From $W_e^* = W_i^*$

$$u_d = \frac{-3Pl}{AE} - u_a$$

$$= -\left(\frac{3Pl}{AE} + u_a\right)$$

The negative sign indicates that the horizontal deflection is to the left, opposite to the direction of the unit virtual horizontal load applied. ∎

■ **EXAMPLE 7.19**

Calculate the length change in members *be, ae,* and *ed* of the truss in Fig. 7.30a in order to induce an initial vertical upward deflection at *d* equal to 50 mm. This is accomplished by shortening *bc* and lengthening *ae* and *ed* by the same amount.

Solution. Let us denote the length change by Δ. Applying a unit virtual load vertically upward at *d* will result in the same internal virtual forces shown in Fig. 7.30b except that the sign is changed. Thus

$$W_e^* = (1)(50)$$

and the internal virtual forces ride along the real displacements to yield

$$W_i^* = (+1)(+\Delta) + (+2)(+\Delta) + (-1)(-\Delta) = 4\Delta$$

From $W_i^* = W_e^*$,

$$4\Delta = 50 \qquad \Delta = 12.5 \text{ mm}$$ ■

■ **EXAMPLE 7.20**

If member *bc* of the truss in Fig. 7.30a is subjected to a temperature increase of 40°C, calculate the resulting vertical deflection at joint *d* (v_d). Assume that the coefficient of thermal expansion for the material is $12 \times 10^{-6}/°C$ and $l = 4$ m.

Solution. Since there is no external load on the truss, the only real displacement is the change in length in member *bc* due to a temperature increase of 40°C. This is given by $\Delta = \alpha L(\Delta T) = (12 \times 10^{-6})(4)(40) = 0.00192$ m. The virtual internal forces due to a unit virtual load applied vertically at *d* are shown in Fig. 7.30b. Thus the external virtual complementary work

$$W_e^* = (1)(v_d)$$

and the internal virtual complementary work is the product of the virtual member forces and the real displacements, or

$$W_i^* = (1)(\Delta) = (1)(0.00192)$$

From $W_e^* = W_i^*$,

$$(1)(v_d) = 0.00192$$

or

$$v_d = 0.00192 \text{ m} = 1.92 \text{ mm}$$

A positive sign means that the displacement is downward, coinciding with the direction of the unit vertical load in Fig. 7.30b. ■

■ **EXAMPLE 7.21**

The upper and lower surfaces of beam *ABC* are uniformly heated to 140°F and 70°F, respectively, as shown in Fig. 7.31a. Take $\alpha = 6.5 \times 10^{-6}$ in/in/°F. Using the method of virtual work, calculate (a) the vertical deflection of point *A*, and (b) the slope at point *A*.

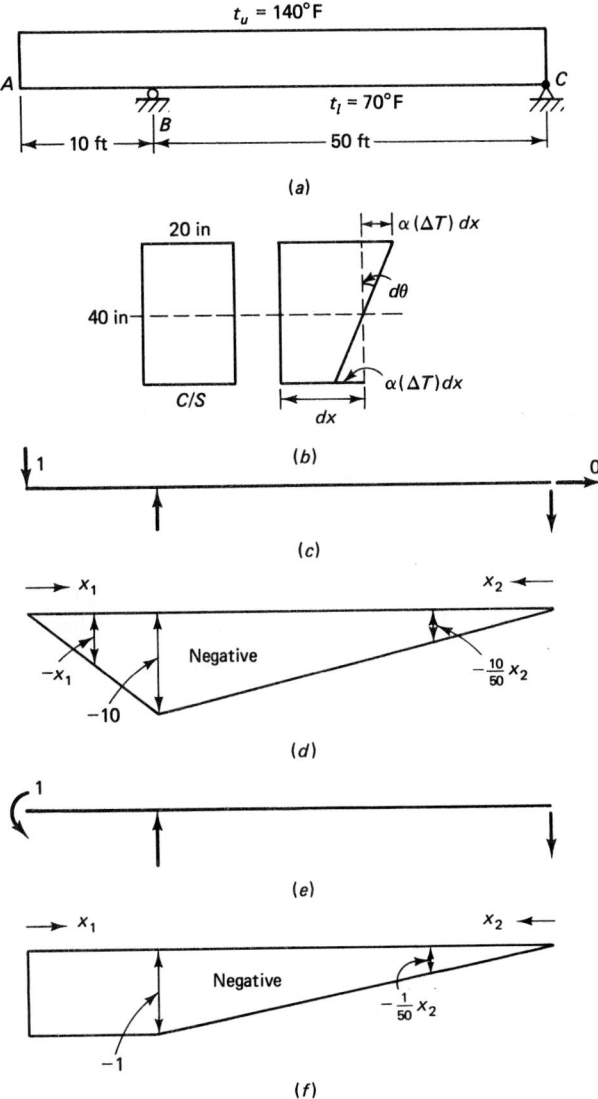

Figure 7.31 Beam for Example 7.21.

Solution

(a) The real displacement is the rotation caused by the temperature gradient; the rotation at any one section is (Fig. 7.31b)

$$d\theta = -\frac{\alpha(\Delta T)dx}{h/2} \qquad \text{where } \Delta T = 35°F$$

$$= -\frac{(70)(\alpha)dx}{h} = \frac{-(70)(6.5 \times 10^{-6})dx}{\frac{40}{12}}$$

$$= -136.5 \times 10^{-6} \, dx$$

(Since the upper surface expands more than the lower surface, the beam bends concave downward; hence the negative sign.)

The other real displacement is the axial extension of an element of beam dx, $\Delta L = (105)(\alpha)dx$. Now apply a unit virtual load at A (Fig. 7.31c) and the corresponding bending-moment diagram (m) is shown in Fig. 7.31d.

$$W_e^* = (1)(v_A)\downarrow$$

The internal complementary virtual work W_i^* is given by

$$W_i^* = \int_0^{10} (-x_1)d\theta + \int_0^{50} \left(\frac{-10}{50} x_2\right)d\theta + \int (0)(\Delta L)$$

The zero in the last integral term means that there is no virtual force in the direction of the real displacement along the horizontal axis of the beam. Therefore,

$$W_i^* = \int_0^{10} (-x_1)(-136.5 \times 10^{-6})dx_1 + \int_0^{50} \left(\frac{-x_2}{5}\right)(-136.5 \times 10^{-6})dx_2$$

$$= 0.491 \text{ in}\downarrow$$

(b) To calculate the resulting slope at A (θ_A) due to the temperature differential, we apply a unit virtual moment as shown in Fig. 7.31e; the corresponding m diagram is given in Fig. 7.31f. Here

$$W_e^* = (1)(\theta_A)$$

$$W_i^* = \int_0^{10} (-1)(d\theta) + \int_0^{50} \left(\frac{-1}{50} x_2\right)d\theta$$

$$= (+136.5 \times 10^{-6})\left(\int_0^{10} dx_1 + \frac{1}{50}\int_0^{50} x_2\, dx_2\right)$$

$$= (136.5 \times 10^{-6})\left[10 + \frac{50^2}{(50)(2)}\right]$$

$$= (136.5 \times 10^{-6})(35) = 4.78 \times 10^{-3} \text{ radian} \qquad \blacksquare$$

7.11   BETTI'S LAW AND MAXWELL'S LAW

Betti's law states that in any structure that is linearly elastic, the virtual work done by the Q forces under a distortion caused by a system of P forces is equal to the virtual work by the system of P forces under a distortion caused by the system of Q forces.

Proof. Consider the linearly elastic body, shown in Fig. 7.32 and loaded *in turn* by two different systems of generalized forces, P and Q systems (Fig. 7.32a and b). Let the displacements of the points of application of the P forces caused by the Q forces be denoted as Δ_i^Q; similarly let us denote the displacements of the points of application of the Q forces caused by the P forces as Δ_k^P. If σ^Q are the internal stresses due to the

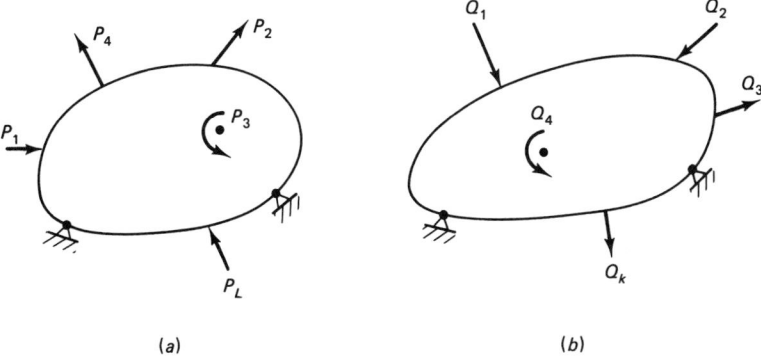

Figure 7.32 Linear elastic body subjected to two different force systems P and Q.

displacements caused by the P forces, then the virtual work done by the Q system of forces is given by

$$\sum_{k=1}^{K} Q_k \, \Delta_k^P = \int_{\text{vol } V} \sigma^Q \, \delta\varepsilon^P \, dV \qquad (7.65)$$

in which $\delta\varepsilon^P$ are the internal virtual strains caused by the P forces. Similarly the virtual work performed by the P system of forces and the internal stresses σ^P under the displacements caused by the Q-force system is

$$\sum_{l=1}^{L} P_l \, \Delta_l^Q = \int_{\text{vol } V} \sigma^P \, \delta\varepsilon^Q \, dV \qquad (7.66)$$

in which $\delta\varepsilon^Q$ are the internal virtual strains caused by the Q-induced stresses σ^Q. For a linearly elastic material, we have $\delta\varepsilon^Q = e \cdot \sigma^Q$ and $\delta\varepsilon^P = e \cdot \sigma^P$, where e is the constant of proportionality. Putting these expressions in Eqs. (7.65) and (7.66), we find the right-hand sides of Eqs. (7.65) and (7.66) are equal. Hence

$$\sum_{k=1}^{K} Q_k \, \Delta_k^P = \sum_{l=1}^{L} P_l \, \Delta_l^Q \qquad (7.67)$$

which is the mathematical statement of *Betti's law.*

Maxwell's law of reciprocal deflections is a special case of Betti's law: In any linearly elastic structure, the generalized deflection at point m caused by a generalized force Q at point n is numerically equal to the generalized deflection at point n caused by the same magnitude of generalized force Q at point m.

Proof. If we consider just one Q force and one P force which are numerically equal applied to an elastic body as shown in Fig. 7.33, then according to Betti's law we find from Eq. (7.67)

$$Q \cdot \Delta_1^P = P \cdot \Delta_2^Q = Q \cdot \Delta_2^Q \qquad (7.68)$$

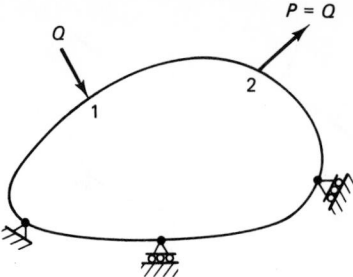

Figure 7.33 Elastic body subjected to two numerically equal forces.

Hence

$$\Delta_1^P = \Delta_2^Q \tag{7.69}$$

We can generalize this result as follows: Let us denote the deflection at point m caused by a load at point n by Δ_{mn} and the deflection at point n caused by the same magnitude of load applied at point m by Δ_{nm}. Then by Betti's law, Maxwell's law of reciprocal deflections can be stated mathematically as

$$\Delta_{mn} = \Delta_{nm} \tag{7.70}$$

If we define f_{mn} as the displacement at point m due to a unit load at point n, and f_{nm} as the displacement at n due to a unit load at m, then, based on Eq. (7.70), we can write

$$f_{mn} = f_{nm} \tag{7.71}$$

The displacements in Eq. (7.71) are denoted as the *flexibility coefficients*. For a linear elastic structure with several coordinates, the resulting flexibility coefficient matrix can be shown to be symmetrical.

7.12 THEOREM OF STATIONARY COMPLEMENTARY ENERGY

The theorem states that for a body in equilibrium, the first partial derivative of the internal complementary energy U^* with respect to any particular force (or moment) is equal to the displacement (or rotation) of the point of action of that force (or moment) in the direction of its line of action. This theorem is applicable to both linear and nonlinear structures.

Proof. Let us consider a body in equilibrium under a system of forces. The difference between the internal complementary energy and the external complementary work is the total potential complementary energy denoted by π^*. Thus

$$\pi^* = U^* - W_e^* \tag{7.72}$$

Now if the body is subjected to a virtual change in the ith force, the change in π^* according to Eq. (7.72) is

$$\delta\pi^* = \delta U^* - \delta W_e^* = \frac{\partial U^*}{\partial P_i}\,\delta P_i - \Delta_i\,\delta P_i = 0$$

in which P_i is a virtual force corresponding to a displacement Δ_i; $\partial U^*/\partial P_i$ is the change in the internal complementary energy with respect to a change in P_i, and U^* is in terms of the internal forces.

Since $\delta P_i \neq 0$, it readily follows that

$$\frac{\partial U^*}{\partial P_i} = \Delta_i \tag{7.73}$$

This is the *theorem of stationary complementary energy.*

$$\text{If } \Delta_i = 0, \quad \text{then} \quad \frac{\partial U^*}{\partial P_i} = 0 \tag{7.73a}$$

■ **EXAMPLE 7.22**

Determine the horizontal deflection at C for the loaded truss shown in Fig. 7.34. Member AC has material nonlinearity; the relationship between the force F and its extension e is $F = (1/\beta)e^2$. All other members have a linear $F - e$ relationship. Use the theorem of stationary complementary energy.

Solution. The forces in the members due to P applied at C are shown in Fig. 7.34. Thus for *member AC*:

$$F = \frac{1}{\beta}e^2 \quad \text{or} \quad e = (\beta F)^{1/2}$$

Hence

$$U^* = \int_0^F e\, dF = \int_0^F (\beta F)^{1/2}\, dF = \frac{2}{3}\beta^{1/2}F^{3/2}$$

But

$$F = \sqrt{2}P \quad \text{therefore} \quad U^* = \tfrac{2}{3}\beta^{1/2}(\sqrt{2}P)^{3/2}$$

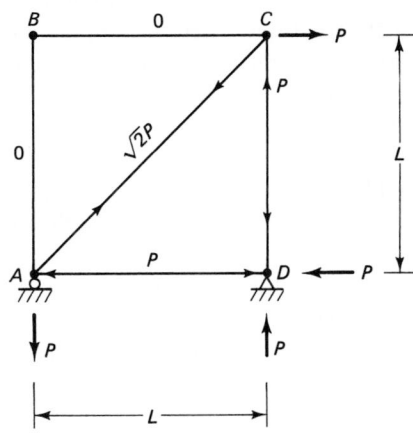

Figure 7.34 Truss for Example 7.22.

For members CD and AD: $F = AEe/L$ (forces in the other members are zero),

or

$$e = \frac{FL}{AE}$$

Hence

$$U^* = \int_0^F e \, dF = \frac{F^2 L}{2AE}$$

But

$$F = -P \quad \text{therefore} \quad U^* = \frac{P^2 L}{2AE}$$

$$\text{Total } U^* = \frac{2}{3}\beta^{1/2}(\sqrt{2}P)^{3/2} + \left(\frac{P^2 L}{2AE}\right) \times 2$$

From Eq. (7.73),

$$\Delta(\text{horizontal}) = \frac{\partial U^*}{\partial P} = 2^{3/4}\beta^{1/2}\sqrt{P} + \frac{2PL}{AE} \qquad\qquad \blacksquare$$

7.13 CASTIGLIANO'S THEOREM—PART II

In 1879, Alberto Castigliano published the results of research on statically indeterminate structures based on two theorems: "The Theorem of the Differential Coefficients of the Internal Work—Part I and Part II." There is considerable confusion in the literature about the designation of these two theorems. Some people refer to the theorems as Castigliano's first and second theorems. However, in this book, we refer to them as Castigliano's theorem—part I and part II, in conformity with Castigliano's original titles to the theorems. Castigliano's theorem—part II is an energy method of computation of a deflection component at any point of a truss, beam, or frame due to applied loads and is a special case of the theorem of stationary complementary energy for linearly elastic structures. For such structures, internal complementary energy U^* is equal to the internal strain energy U. The theorem *cannot* be used for finding deflections due to temperature changes or settlement of supports. For the theorem to be applicable, the following conditions must be satisfied:

1. The structure must be in stable equilibrium.
2. The material of the structure must be linearly elastic, that is, should obey Hooke's law.
3. The temperature of the structure must be constant.
4. The supports of the structure should not yield.
5. The deflections must be small so as not to change significantly the geometry of the structure.

Castigliano's theorem—part II states that "For a linearly elastic stable structure at constant temperature, resting on unyielding supports, and subjected to small deflections due to applied loads, the first partial derivative of the strain energy with respect to any particular force (or moment) is equal to the displacement (or rotation) of the point of application of that force (or moment) in the direction of its line of action."

Mathematically, Castigliano's theorem—part II can be expressed as

$$\frac{\partial U}{\partial P_n} = \Delta_n$$

or   (7.74)

$$\frac{\partial U}{\partial M_n} = \theta_n$$

where U = strain energy, which is equal to the internal work W_i
             Δ_n = displacement of force P_n in direction of its line of action
             θ_n = rotation of moment M_n in direction of its line of action

Proof. The theorem follows directly from Eq. (7.73) by substituting $U^* = U$. The theorem can also be derived from first principles as follows: Consider any structure, for example, a beam, loaded with two forces P and Q, as shown in Fig. 7.35. Let Δ and γ be the deflections under P and Q in the directions of their lines of action.

Strain energy stored in beam $= U$

External work done $= \frac{1}{2}P \cdot \Delta + \frac{1}{2}Q \cdot \gamma$ (7.75)

Now, let the force P be increased by a small amount δP and in the same direction as P. This will increase the deflections by $\delta\Delta$ and $\delta\gamma$ in the directions of P and Q. During the period the additional deflections take place, the force P is increased from P to P

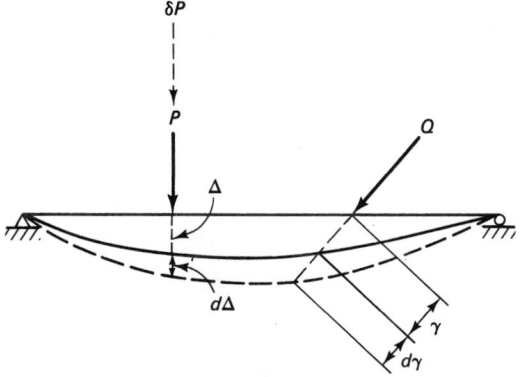

Figure 7.35 A simply supported beam to prove Castigliano's theorem—part II.

$+ \delta P$ (i.e., average force $= P + \delta P/2$) while Q remains constant (i.e., average force $= Q$). Additional strain energy stored in the beam

$$\delta U = (P + \tfrac{1}{2}\delta P) \cdot \delta\Delta + Q \cdot \delta\gamma$$

or   (7.76)

$$\delta U = P \cdot \delta\Delta + Q \cdot \delta\gamma$$

neglecting $\tfrac{1}{2} \cdot \delta P \cdot \delta\Delta$, which is a small quantity of second order. Adding Eqs. (7.75) and (7.76),

Total strain energy stored $= U + \Delta U$

$$= \tfrac{1}{2} \cdot P \cdot \Delta + \tfrac{1}{2}Q\gamma + P \cdot \delta\Delta + Q \cdot \delta\gamma \qquad (7.77)$$

Removing the forces $P + \delta P$ and Q, the beam will return to its original position (since its material is elastic).

Now apply the forces $(P + \delta P)$ and Q at the same time. Since the material of the structure is assumed to be linearly elastic, the principle of superposition is valid and the total deflection is independent of the loading history of the structure. Therefore, the deflections in this case will be $(\Delta + \delta\Delta)$ and $(\gamma + \delta\gamma)$ as before.

Total strain energy stored $=$ total external work done $= U' + \delta U'$

$$= \tfrac{1}{2}(P + \delta P)(\Delta + \delta\Delta) + \tfrac{1}{2}Q(\gamma + \delta\gamma)$$

$$= \tfrac{1}{2}P \cdot \Delta + \tfrac{1}{2}P \cdot \delta\Delta + \tfrac{1}{2}\delta P \cdot \Delta + \tfrac{1}{2}\delta P \, \delta\Delta$$

$$+ \tfrac{1}{2}Q \cdot \gamma + \tfrac{1}{2}Q \cdot \delta\gamma$$

or

$$U' + \delta U' = \tfrac{1}{2}P \cdot \Delta + \tfrac{1}{2}Q \cdot \gamma + \tfrac{1}{2}P \cdot \delta\Delta + \tfrac{1}{2}Q \cdot \delta\gamma + \tfrac{1}{2}\delta P \cdot \Delta \qquad (7.78)$$

neglecting second-order terms. Since the total strain energy for a linearly elastic structure is independent of the loading sequence, the strain energy given by Eqs. (7.77) and (7.78) must be equal. Equating these two quantities and simplifying,

$$\tfrac{1}{2}P \cdot \delta\Delta + \tfrac{1}{2}Q \cdot \delta\gamma = \tfrac{1}{2}\delta P \cdot \Delta$$

or

$$P \cdot \delta\Delta + Q \cdot \delta\gamma = \delta P \cdot \Delta$$

Substituting Eq. (7.76),

$$\delta U = \delta P \cdot \Delta$$

In the limit, when $\delta P \rightarrow 0$, we have

$$\frac{\partial U}{\partial P} = \Delta$$

In order to use Castigliano's theorem—part II, it is necessary to use expressions developed earlier [Eqs. (7.28a), (7.32), (7.38), and (7.44)] for the strain energy stored due to the action of different types of forces (axial forces, bending moments, twisting moments, and shearing forces).

Application of Castigliano's Theorem—Part II for Computation of Deflections In order to apply Castigliano's theorem there must be a force P (or moment M) acting in the desired direction at the section where the deflection component (rotation) is to be computed. If a load having a certain numerical value is already acting at the required location, it should be temporarily replaced by the symbol P (or M). After the partial derivative for the strain-energy expression is obtained, the symbol P must be replaced by the numerical value of the load, and the magnitude of the deflection component is evaluated. If there is no load (moment for the determination of slope) acting at the section, a fictitious force P (fictitious moment M) is introduced. After the partial derivative for the strain-energy expression is obtained, the fictitious force P (or M) is set equal to zero to get the deflection (rotation) due to the given load system. The following examples illustrate the application of the above principles.

■ **EXAMPLE 7.23**
Using Castigliano's theorem—part II, calculate the vertical deflection at point D for the structure shown in Fig. 7.36.

Solution. Apply a fictitious load Q at D in the downward direction (see Fig. 7.36).

From Eq. (7.32),

$$U = \int_{\text{Frame}} \frac{M_x^2 \, dx}{2EI}.$$

From Eq. (7.74),

$$\text{vertical deflection at } D = \frac{\partial U}{\partial Q} = \int_{\text{Frame}} \frac{1}{EI} M_x \frac{\partial M_x}{\partial Q} \, dx.$$

Portion DC. Origin at D; $0 \leqslant x \leqslant L$

$$M_x = -Qx \quad \text{and} \quad \frac{\partial M_x}{\partial Q} = -x$$

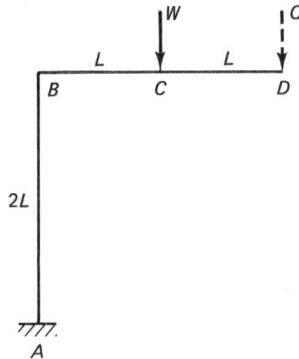

Figure 7.36 Structure for Example 7.23.

Portion CB. Origin at C; $0 \leqslant x \leqslant L$

$$M_x = -Wx - Q(x + L)$$

$$\frac{\partial M_x}{\partial Q} = -(x + L)$$

Portion BA. Origin at B; $0 \leqslant x \leqslant 2L$

$$M_x = -Q2L - WL$$

$$\frac{\partial M_x}{\partial Q} = -2L$$

Vertical deflection at $D = \displaystyle\int \frac{M_x(\partial M_x / \partial Q)\,dx}{EI}$

$$= \frac{1}{EI} \left(\int_D^C M_x \frac{\partial M_x}{\partial Q}\,dx + \int_C^B M_x \frac{\partial M_x}{\partial Q}\,dx \right.$$

$$\left. + \int_B^A M_x \frac{\partial M_x}{\partial Q}\,dx \right)$$

Putting $Q = 0$ in the expressions for M_x and substituting:

Vertical deflection at D

$$= \frac{1}{EI} \left\{ 0 + \int_0^L (-Wx)[-(x + L)]\,dx + \int_0^{2L} (-WL)(-2L)\,dx \right\}$$

$$= \frac{1}{EI} \left[W\left\{ \frac{x^3}{3} + \frac{Lx^2}{2} \right\}_0^L + 2WL^2 \{x\}_0^{2L} \right] = \frac{+29}{6} \frac{WL^3}{EI}$$

Positive sign indicates the deflection is in the assumed direction of Q, that is, downward. ∎

■ EXAMPLE 7.24

Find the deflection and slope at A of the loaded cantilever in Fig. 7.37, using Castigliano's Theorem—Part II.

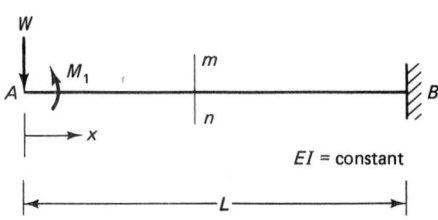

Figure 7.37 Beam for Example 7.24.

Solution. Here the stored strain energy is mainly due to bending moment (the effect of shear on the strain energy has been shown to be small and will not be considered). At a cross section *mn*,

$$M_x = -Wx - M_1$$

From Eq. (7.32), the stored strain energy $U\,(=W_i)$ is

$$U = \int_0^L \frac{M_x^2\,dx}{2EI}$$

From Eq. (7.74), the expression for deflection at A can be written as

$$\Delta_A = \frac{\partial}{\partial W}\left(\int_0^L \frac{M_x^2\,dx}{2EI}\right)$$

It is much easier to first differentiate the quantity under the integral sign and then evaluate the integral; thus

$$\Delta_A = \int_0^L M_x \frac{\partial M_x}{\partial W}\frac{dx}{EI} \tag{7.79}$$

Since $M_x = -Wx - M_1$, $\partial M_x/\partial W = -x$. Hence

$$\Delta_A = \int_0^L (-Wx - M_1)(-x)\frac{dx}{EI} = \frac{1}{EI}\left(\frac{WL^3}{3} + M_1\frac{L^2}{2}\right)$$

For the slope θ_A, we have

$$\theta_A = \frac{\partial}{\partial M_1}\left(\int_0^L \frac{M_x^2\,dx}{2EI}\right) = \int_0^L M_x \frac{\partial M_x}{\partial M_1}\frac{dx}{EI}$$

Since $\partial M_x/\partial M_1 = -1$,

$$\theta_A = \int_0^L (-Wx - M_1)(-1)\frac{dx}{EI}$$

or

$$\theta_A = \frac{1}{EI}\left(\frac{WL^2}{2} + M_1L\right)$$

The positive signs for Δ_A and θ_A indicate that the deflection and rotation of end A have the same directions respectively as the load W and the couple M_1.

If there is no load at the point where the displacement is desired, a fictitious force Q is added to the structure at the point in the direction of the desired displacement. Then, *after* the derivative has been taken, the force Q is set equal to zero. This added force is sometimes called a *dummy force*. ∎

■ EXAMPLE 7.25

Determine the vertical deflection at A of the cantilever beam loaded as shown in Fig. 7.38*a*, using Castigliano's Theorem—Part II.

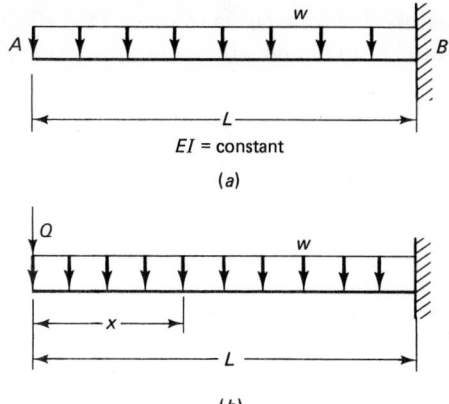

(a)

(b)

Figure 7.38 Beam for Example 7.25.

Solution. Applying a dummy force Q as shown in Fig. 7.38b, at a section x distance from A, $M_x = -Q \cdot x - wx^2/2$. Hence $\partial M_x/\partial Q = -x$. From Eq. (7.74),

$$\Delta_A = \frac{\partial}{\partial Q}\left(\int_0^L \frac{M_x^2\, dx}{2EI}\right) = \int_0^L M_x \frac{\partial M_x}{\partial Q}\frac{dx}{EI}$$

$$= \int_0^L \left(-Q\cdot x - \frac{wx^2}{2}\right)(-x)\frac{dx}{EI}$$

Putting $Q = 0$, and simplifying

$$\Delta_A = \frac{wL^4}{8EI}$$ ∎

■ EXAMPLE 7.26

Find the angular displacement of bar AB produced by the load W in Fig. 7.39a. Bars AB and BC have the same (AE/L).

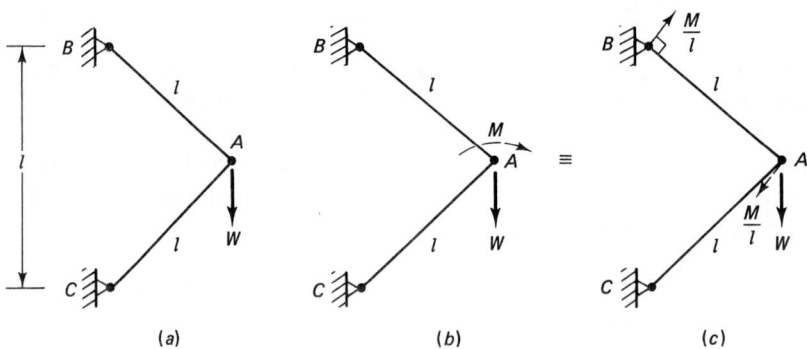

(a)   (b)   (c)

Figure 7.39 Structure for Example 7.26.

Solution. A fictitious dummy couple M is applied as shown in Fig. 7.39b. This is equivalent to a couple formed by two forces M/l at A and B shown dotted in Fig. 7.39c. From equilibrium at joint A, we have

$$P_{AB} = W + \frac{M}{\sqrt{3}l}$$

$$P_{AC} = -W - \frac{2M}{\sqrt{3}l}$$

Then the strain energy stored in the axially loaded structure

$$U = \Sigma \frac{P^2 L}{2AE}$$

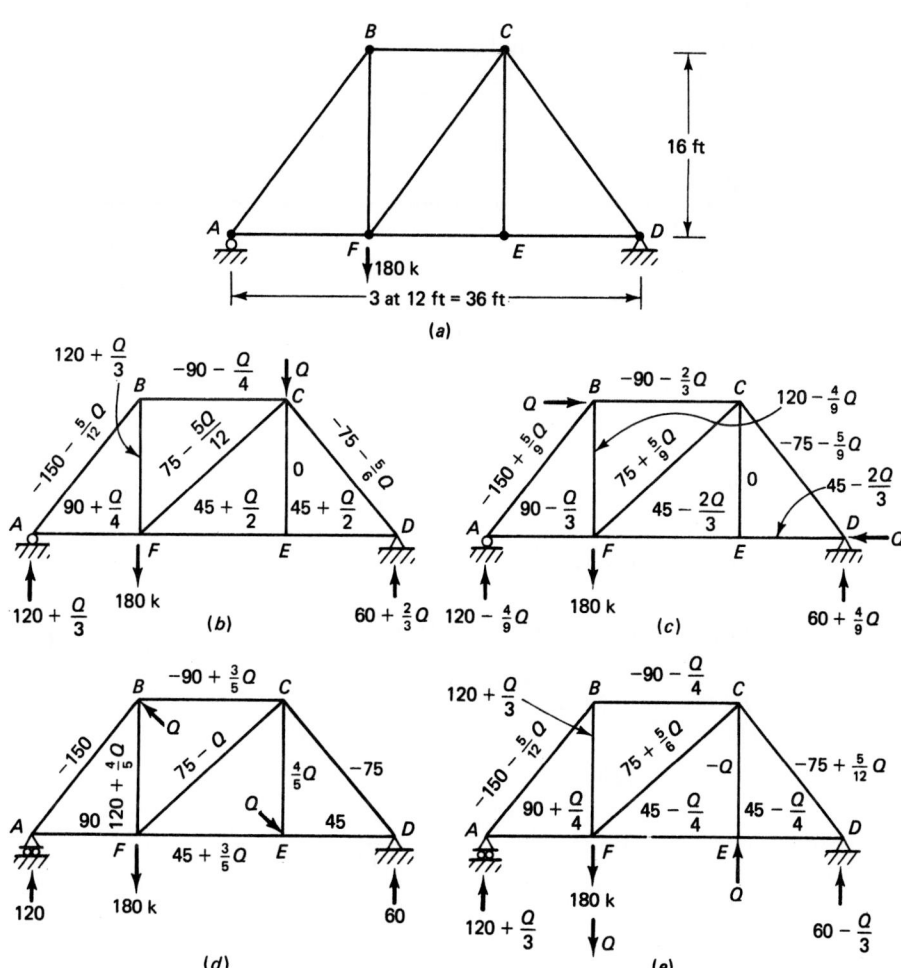

Figure 7.40 Truss for Example 7.27.

The rotation θ of bar AB due to load W is, by Eq. (7.74),

$$\theta = \left(\frac{\partial U}{\partial M}\right)_{M=0} = \Sigma\left(P\frac{\partial P}{\partial M}\frac{L}{AE}\right)$$

$$\frac{\partial P_{AB}}{\partial M} = \frac{1}{\sqrt{3}l} \qquad \frac{\partial P_{AC}}{\partial M} = \frac{-2}{\sqrt{3}l}$$

Thus, putting $M = 0$,

$$\theta = \frac{l}{AE}\left[(W)\left(\frac{1}{\sqrt{3}l}\right) + (-W)\left(\frac{-2}{\sqrt{3}l}\right)\right] = \frac{\sqrt{3}W}{AE} \qquad \blacksquare$$

■ **EXAMPLE 7.27**

For the loaded truss in Fig. 7.40a, determine by Castigliano's theorem—part II: (a) the vertical deflection at joint C, (b) the horizontal deflection at joint B, (c) the relative deflection between joints B and E along the line joining them, and (d) the rotation of member EF. (L/AE) of all members = $(0.8)(10)^{-3}$ in/k.

TABLE 7.2

Member	$\dfrac{L}{AE}$	P	$\dfrac{\partial P}{\partial Q}$	$\left(P\dfrac{\partial P}{\partial Q}\dfrac{L}{AE}\right)_{Q=0}$
AB	$(0.8)(10)^{-3}$	$-150 - \dfrac{5}{12}Q$	$-\dfrac{5}{12}$	$\left(\dfrac{750}{12}\right) \times (0.8)(10)^{-3}$
BC	$(0.8)(10)^{-3}$	$-90 - \dfrac{Q}{4}$	$-\dfrac{1}{4}$	$\left(\dfrac{90}{4}\right) \times (0.8)(10)^{-3}$
CD	$(0.8)(10)^{-3}$	$-75 - \dfrac{5}{6}Q$	$-\dfrac{5}{6}$	$\left(\dfrac{375}{6}\right) \times (0.8)(10)^{-3}$
ED	$(0.8)(10)^{-3}$	$45 + \dfrac{Q}{2}$	$+\dfrac{1}{2}$	$\left(\dfrac{45}{2}\right) \times (0.8)(10)^{-3}$
EF	$(0.8)(10)^{-3}$	$45 + \dfrac{Q}{2}$	$+\dfrac{1}{2}$	$\left(\dfrac{45}{2}\right) \times (0.8)(10)^{-3}$
FA	$(0.8)(10)^{-3}$	$90 + \dfrac{Q}{4}$	$+\dfrac{1}{4}$	$\left(\dfrac{90}{4}\right) \times (0.8)(10)^{-3}$
BF	$(0.8)(10)^{-3}$	$120 + \dfrac{Q}{3}$	$+\dfrac{1}{3}$	$(40) \times (0.8)(10)^{-3}$
CF	$(0.8)(10)^{-3}$	$75 - \dfrac{5Q}{12}$	$-\dfrac{5}{12}$	$\left(-\dfrac{375}{12}\right) \times (0.8)(10)^{-3}$
CE	$(0.8)(10)^{-3}$	0	0	0

$$\Delta = \Sigma P\frac{\partial P}{\partial Q}\frac{L}{AE} = 0.179 \text{ in}\downarrow$$

Solution

(a) Assume that joint C moves downward. To apply the theorem, we place an imaginary vertical force Q acting at C as shown in Fig. 7.40b. The resulting bar forces are as shown in Fig. 7.40b. The required terms such as L/AE, bar force P, $\partial P/\partial Q$ are given in Table 7.2. Notice that the last column containing the term $\left(P \dfrac{\partial P}{\partial Q} \dfrac{L}{AE} \right)$ is evaluated for $Q = 0$. The summation of the last column yields the required deflection, that is,

$$\Delta = \Sigma \left(P \frac{\partial P}{\partial Q} \frac{L}{AE} \right)_{Q=0} = 0.179 \text{ in} \downarrow$$

(b) Assume that joint B moves to the right. We apply an imaginary horizontal force Q at B as shown in Fig. 7.40c. The resulting bar forces are given in the same figure. The complete solution is shown in Table 7.3. Hence

$$\Delta = \Sigma \left(P \frac{\partial P}{\partial Q} \frac{L}{AE} \right)_{Q=0} = 0.0053 \text{ in} \rightarrow$$

TABLE 7.3

Member	$\dfrac{L}{AE}$	P	$\dfrac{\partial P}{\partial Q}$	$\left(P\dfrac{\partial P}{\partial Q}\dfrac{L}{AE} \right)_{Q=0}$
AB	$(0.8)(10)^{-3}$	$-150 + \dfrac{5}{9}Q$	$\dfrac{5}{9}$	$\left(-\dfrac{750}{9} \right) \times (0.8)(10)^{-3}$
BC	$(0.8)(10)^{-3}$	$-90 - \dfrac{2}{3}Q$	$-\dfrac{2}{3}$	$+60 \times (0.8)(10)^{-3}$
CD	$(0.8)(10)^{-3}$	$-75 - \dfrac{5}{9}Q$	$-\dfrac{5}{9}$	$\left(+\dfrac{375}{9} \right) \times (0.8)(10)^{-3}$
ED	$(0.8)(10)^{-3}$	$45 - 2\dfrac{Q}{3}$	$+\dfrac{1}{3}$	$(15) \times (0.8)(10)^{-3}$
EF	$(0.8)(10)^{-3}$	$45 - 2\dfrac{Q}{3}$	$+\dfrac{1}{3}$	$(15) \times (0.8)(10)^{-3}$
FA	$(0.8)(10)^{-3}$	$90 - \dfrac{Q}{3}$	$-\dfrac{1}{3}$	$(-30) \times (0.8)(10)^{-3}$
BF	$(0.8)(10)^{-3}$	$120 - \dfrac{4}{9}Q$	$-\dfrac{4}{9}$	$\left(-\dfrac{480}{9} \right) \times (0.8)(10)^{-3}$
CF	$(0.8)(10)^{-3}$	$75 + \dfrac{5}{9}Q$	$+\dfrac{5}{9}$	$\left(+\dfrac{375}{9} \right) \times (0.8)(10)^{3}$
CE	$(0.8)(10)^{-3}$	0	0	0

$$\Delta = \Sigma P \frac{\partial P}{\partial Q} \frac{L}{AE} = 0.0053 \text{ in} \rightarrow$$

TABLE 7.4

Member	$\dfrac{L}{AE}$	P	$\dfrac{\partial P}{\partial Q}$	$\left(P\dfrac{\partial P}{\partial Q}\dfrac{L}{AE}\right)_{Q=0}$
AB	$(0.8)(10)^{-3}$	-150	0	$0 \times (0.8)(10)^{-3}$
BC	$(0.8)(10)^{-3}$	$-90 + \dfrac{3}{5}Q$	$\dfrac{3}{5}$	$\left(-\dfrac{270}{5}\right) \times (0.8)(10)^{-3}$
CD	$(0.8)(10)^{-3}$	-75	0	$0 \times (0.8)(10)^{-3}$
ED	$(0.8)(10)^{-3}$	45	0	$0 \times (0.8)(10)^{-3}$
EF	$(0.8)(10)^{-3}$	$45 + \dfrac{3}{5}Q$	$\dfrac{3}{5}$	$(27) \times (0.8)(10)^{-3}$
FA	$(0.8)(10)^{-3}$	90	0	$0 \times (0.8)(10)^{-3}$
BF	$(0.8)(10)^{-3}$	$120 + \dfrac{4}{5}Q$	$\dfrac{4}{5}$	$\left(\dfrac{480}{5}\right) \times (0.8)(10)^{-3}$
CF	$(0.8)(10)^{-3}$	$75 - Q$	-1	$(-75) \times (0.8)(10)^{-3}$
CE	$(0.8)(10)^{-3}$	$\dfrac{4}{5}Q$	$\dfrac{4}{5}$	$0 \times (0.8)(10)^{-3}$

$$\Delta = \Sigma\, P\frac{\partial P}{\partial Q}\frac{L}{AE} = -0.0048 \text{ in} \quad \text{or} \quad 0.0048 \text{ in toward each other}$$

(c) Here we apply two equal and opposite imaginary forces at B and E along the line joining them as shown in Fig. 7.40d; the resulting bar forces are also shown. The solution is presented in Table 7.4, where we find that the relative deflection between joints B and E is

$$\Delta = \Sigma\left(P\frac{\partial P}{\partial Q}\frac{L}{AE}\right)_{Q=0} = -0.0048 \text{ in} \quad \text{or} \quad 0.0048 \text{ in } \textit{toward} \text{ each other}$$

(d) Finding the rotation of member EF is equivalent to determining the relative displacement between ends E and F, in a direction perpendicular to EF, divided by the length EF. We start by applying a pair of imaginary forces Q to joints E and F as shown in Fig. 7.40e; the resulting bar forces are also given. Following the solution presented in Table 7.5, we have the relative displacement between ends E and F in a direction perpendicular to EF

$$\Sigma\left(P\frac{\partial P}{\partial Q}\frac{L}{AE}\right)_{Q=0} = 0.125$$

Therefore, the required rotation

$$\theta = \frac{0.125}{\text{length of } EF} = \frac{0.125}{(12)(12)} = 0.00087 \text{ radian counterclockwise} \qquad \blacksquare$$

TABLE 7.5

Member	$\dfrac{L}{AE}$	P	$\dfrac{\partial P}{\partial Q}$	$\left(P\dfrac{\partial P}{\partial Q}\dfrac{L}{AE}\right)_{Q=0}$
AB	$(0.8)(10)^{-3}$	$-150-\dfrac{5}{12}Q$	$\left(-\dfrac{5}{12}\right)$	$\left(+\dfrac{750}{12}\right)\times(0.8)(10)^{-3}$
BC	$(0.8)(10)^{-3}$	$-90-\dfrac{Q}{4}$	$\left(-\dfrac{1}{4}\right)$	$\left(+\dfrac{90}{4}\right)\times(0.8)(10)^{-3}$
CD	$(0.8)(10)^{-3}$	$-75+\dfrac{5}{12}Q$	$\left(\dfrac{5}{12}\right)$	$\left(-\dfrac{375}{12}\right)\times(0.8)(10)^{-3}$
ED	$(0.8)(10)^{-3}$	$45-\dfrac{Q}{4}$	$\left(-\dfrac{1}{4}\right)$	$\left(-\dfrac{45}{4}\right)\times(0.8)(10)^{-3}$
EF	$(0.8)(10)^{-3}$	$45-\dfrac{Q}{4}$	$\left(-\dfrac{1}{4}\right)$	$\left(-\dfrac{45}{4}\right)\times(0.8)(10)^{-3}$
FA	$(0.8)(10)^{-3}$	$90+\dfrac{Q}{4}$	$\left(+\dfrac{1}{4}\right)$	$\left(\dfrac{90}{4}\right)\times(0.8)(10)^{-3}$
BF	$(0.8)(10)^{-3}$	$120+\dfrac{Q}{3}$	$\left(+\dfrac{1}{3}\right)$	$+40\times(0.8)(10)^{-3}$
CF	$(0.8)(10)^{-3}$	$75+\dfrac{5}{6}Q$	$\left(+\dfrac{5}{6}\right)$	$\left(+\dfrac{375}{6}\right)\times(0.8)(10)^{-3}$
CE	$(0.8)(10)^{-3}$	$-Q$	-1	$0\times(0.8)(10)^{-3}$

Relative displacement between ends F and E in a direction perpendicular to $FE =$

$$\Sigma\, P\frac{\partial P}{\partial Q}\frac{L}{AE} = 0.125.$$

7.14 THE UNIT-LOAD METHOD

This method can be shown to be derivable from the theorem of virtual work. In using this method to determine deflections, it becomes identical to the method of Castigliano's theorem—part II if instead of an imaginary force Q we use a unit force and the term $\partial P/\partial Q$ is changed to p, where p is the bar force due to the unit load. As an exercise, the student should re-solve Example 7.27 using the unit-load method. The method is also demonstrated in Chapter 13, page 663.

7.15 DEFLECTION DUE TO SHEAR

In most engineering applications, the deflection due to shear is small compared with the deflection due to bending moment and is therefore ignored in the calculations. Deflection due to shear is, of course, very important in the case of short deep beams. Sometimes, even for beams of common proportions, the magnitude of shear deflections

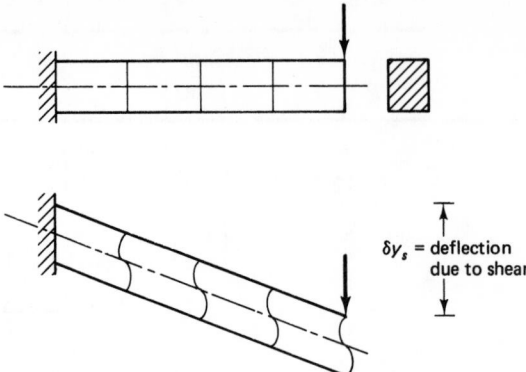

Figure 7.41 Beam with shear deflection due to concentrated load.

can be 10 to 15 percent of bending deflections and may have to be considered. Shearing deflections can be very easily computed by the method of virtual work presented in Section 7.10. Shearing forces produce warping of the transverse planes, and the theory of "pure" bending of beams assuming that plane cross sections remain plane is not strictly applicable. Fortunately, however, the error introduced by assuming that the theory of pure bending is also applicable to cases involving shear is negligible.

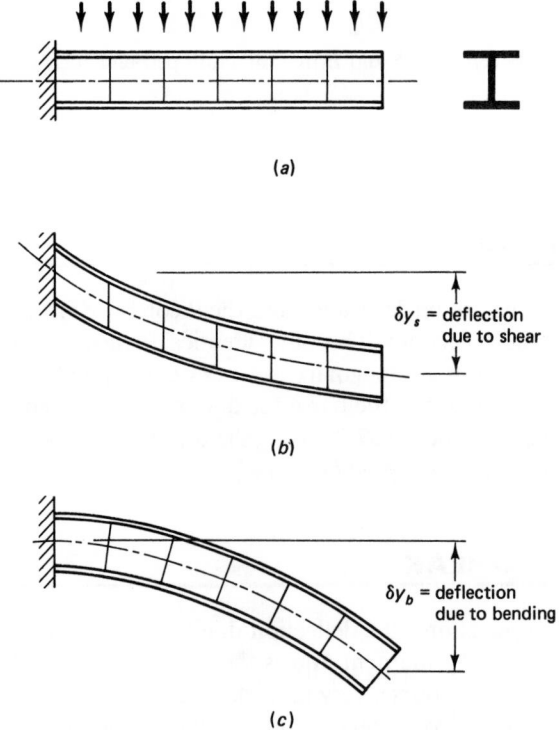

Figure 7.42 Beam with shear and bending deflections due to distributed load.

Figure 7.41 shows the effect of shear stresses on a cross section for a solid beam. On the other hand, for an ideal two-flange beam (with a very thin web), there is no warping of cross sections, as shown in Fig. 7.42. The deflection curve for the case of a cantilever subjected to uniformly distributed load is a parabola (Fig. 7.42b). The completely different nature of the bending deflection is illustrated in Fig. 7.42c. The shearing deflection at the free end of a cantilever for the loading case shown in Fig. 7.42a is computed as follows:

We use the method of virtual work to compute shearing deflections. Since the vertical deflection at the free end is required, using the method of virtual work, apply a unit vertical load in the downward direction at the free end, as shown in Fig. 7.43a.

$$\text{External virtual work done by the } Q \text{ forces} = \sum Q \cdot \Delta = 1 \cdot \Delta$$

$$\text{Internal virtual work of shearing deformation} = K_s \int v \frac{V\,dx}{GA} \qquad (7.59)$$

For an I section, approaching an ideal two-flange beam, $K_s = 1.0$. Substituting the expressions for v and V, (Figs. 7.43b and c.)

$$\text{Internal virtual work of shearing deformation} = (1.0) \int_0^l (+1) \frac{(+wx)}{AG}\,dx$$

$$= \frac{w}{AG}\left[\frac{x^2}{2}\right]_0^l = \frac{wL^2}{2AG}$$

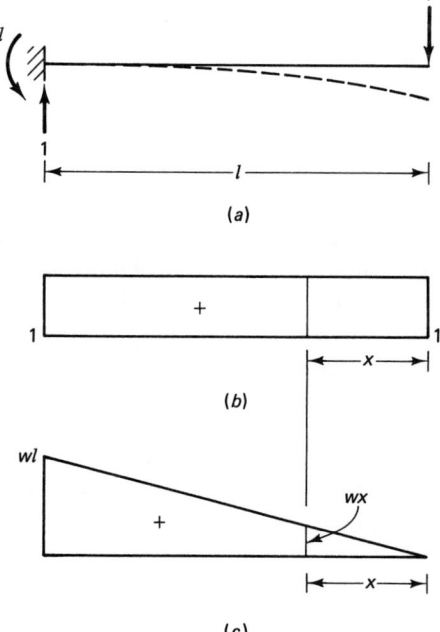

(a)

(b)

(c)

Figure 7.43 Beam with a unit load at end, the corresponding v diagram and V diagram due to loading in Fig. 7.42a. (a) Unit-load system. (b) v diagram. (c) V diagram due to applied uniformly distributed load.

Equating the external virtual work to internal virtual work,

$$1 \cdot \Delta = \frac{wl^2}{2AG} \quad \text{or} \quad \Delta = \frac{wl^2}{2AG}$$

The positive sign indicates the deflection is in the same direction as the applied fictitious unit load, that is, downward.

7.16 DEFLECTIONS OF STATICALLY INDETERMINATE STRUCTURES

All the examples in this chapter referred to computation of deflections of statically determinate structures only. However, the methods presented earlier are equally applicable for the computation of deflections of statically indeterminate structures. But before the methods can be applied, the magnitudes of the redundant forces must be computed. The determination of these forces will be dealt with when we analyze indeterminate (redundant) structures in Chapters 9 through 13.

7.17 CAMBERING OF STRUCTURES

In Section 7.1 it was mentioned that computation of deflections is required to determine the amount of camber required for long-span structures. Cambering is the process of introducing a deflection in the unloaded structure in a direction opposite to the deflection caused by the superimposed loads, such that under some specified camber loading, the structure attains its theoretical shape. The purpose of cambering is

1. To ensure that the geometry of the loaded structure corresponds to the shape assumed in the theoretical analysis
2. To improve the appearance of a loaded structure

This camber is visible even to the naked eye in all long-span bridges. As an example, to camber a truss, the truss is assumed loaded with some specified loading called "camber loading." The forces in all the members are computed and changes in their lengths under those forces are determined. At the time of fabricating the members of the truss, the tension members are fabricated too short by amounts equal to their extensions under the camber load, and the compression members are fabricated too long, by amounts equal to their shortening under the camber load. Then when the truss is assembled and loaded with the cambering load, it will deflect to the shape assumed in the theoretical analysis. The advantage of this method of cambering is that both statically determinate and indeterminate trusses so cambered can be assembled free from initial stresses. Since the deflection is primarily due to changes in lengths of the chord members, in one approximate method of cambering, only the chord members are fabricated either too long (compression chord members) or too short (tension chord members). In another approximate method, either the tension chord is fabricated too short or the compression chord is fabricated too long by an amount equal to the sum of the absolute change in length of compression and tension chords. These two approximate methods can be used for determinate trusses without causing initial stresses but should be used with caution for indeterminate trusses since initial stresses are introduced when some members have to be forced in place.

7.18 SOLVED PROBLEMS

■ SOLVED PROBLEM 7.1

Using the double-integration method, determine the deflection of the beam shown in Fig. 7.44a under the 10-k load. $E = 30,000$ ksi and $I = 120$ in ^{4}.

Solution. To enable the application of Macaulay's method, the distributed load of 4 k/ft is extended to the right end of the beam and an upward distributed load of 4 k/ft between C and D is superimposed, as shown in Fig. 7.44b.

The reactions (determined by taking moments about D and A) are 20.8 and 13.2 k respectively as shown in Fig. 7.44b.

$$EI \frac{d^2y}{dx^2} = M_x = 20.8x - 10\{x - 2\} - \frac{4\{x - 2\}^2}{2}$$

$$- 8\{x - 6\} + \frac{4\{x - 6\}^2}{2} \qquad (a)$$

$$EI \frac{dy}{dx} = \frac{20.8x^2}{2} - \frac{10\{x - 2\}^2}{2} - \frac{4\{x - 2\}^3}{6}$$

$$- \frac{8\{x - 6\}^2}{2} + \frac{4\{x - 6\}^3}{6} + C_1 \qquad (b)$$

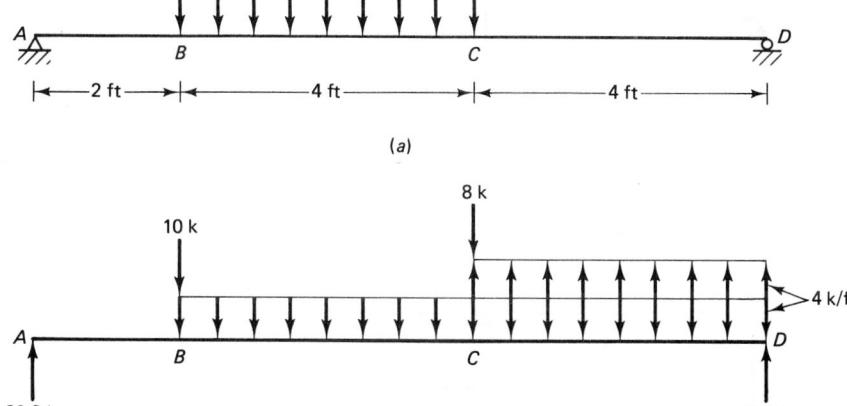

(a)

(b)

Figure 7.44 Beam for Solved Problem 7.1.

$$EIy = \frac{20.8x^3}{6} - \frac{10\{x-2\}^3}{6} - \frac{4\{x-2\}^4}{24}$$

$$- \frac{8\{x-6\}^3}{6} + \frac{4\{x-6\}^4}{24} + C_1x + C_2 \qquad\qquad \text{(c)}$$

The boundary conditions are $x = 0$, $y = 0$; $x = 10$, $y = 0$. Therefore,

$$0 = 0 - 0 - 0 - 0 + 0 + C_1(0) + C_2 \quad \text{or} \quad C_2 = 0$$

$$0 = \frac{20.8(10)^3}{6} - \frac{10(8)^3}{6} - \frac{4(8)^4}{24} - \frac{8(4)^3}{6} + \frac{4(4)^4}{24} + 10C_1 \quad \text{or} \quad C_1 = -188.8$$

Deflection under 10 k load is obtained by substituting $C_1 = -188.8$ and $x = 2$ in equation (c).

$$EIy = \frac{20.8(2)^3}{6} - 0 - 0 - 0 + 0 - 188.8(2)$$

$$y = \frac{-349.87}{EI} = \frac{-349.87 \times 144}{30,000 \times 120} = -0.014 \text{ ft} = 0.168 \text{ in downward} \qquad \blacksquare$$

■ SOLVED PROBLEM 7.2

Calculate the vertical deflection at point D for the beam shown in Fig. 7.45a using Castigliano's theorem.

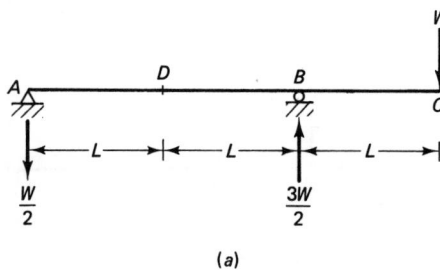

(a)

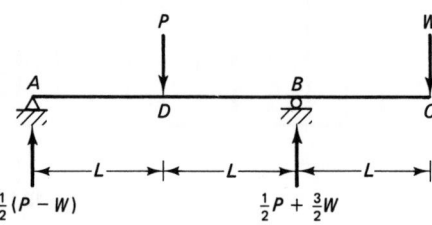

(b)

Figure 7.45 Beam for Solved Problem 7.2.

Solution. Since deflection at *D* is required, apply a fictitious load *P* at *D* as shown in Fig. 7.45*b*. The reactions are determined in the usual manner.

Segment A − D: 0 ⩽ x ⩽ L, Origin at A

$$M_x = R_A \cdot x = \tfrac{1}{2}(P - W)x$$

$$\frac{\partial M_x}{\partial P} = +\frac{x}{2}$$

Segment D − B: L ⩽ x ⩽ 2L, Origin at A

$$M_x = R_A x - P(x - L) = \tfrac{1}{2}(P - W)x - P(x - L)$$

$$\frac{\partial M_x}{\partial P} = +\frac{x}{2} - (x - L) = L - \frac{x}{2}$$

Segment B − C: 0 ⩽ x ⩽ L, Origin at C

$$M_x = -Wx$$

$$\frac{\partial M_x}{\partial P} = 0$$

$$\delta = \frac{\partial U}{\partial P} = \int \frac{M_x(\partial M_x/\partial P)dx}{EI}$$

It is convenient to carry out the integration after letting the fictitious load $P = 0$ in the expression for M_x.

Segment A − D: $M_x = \dfrac{-Wx}{2}$; *Segment D − B:* $M_x = \dfrac{-Wx}{2}$

$$\delta = \frac{1}{EI}\left[\int_0^L \left(\frac{-1}{2}Wx\right)\left(\frac{x}{2}\right)dx + \int_L^{2L}\left(\frac{-Wx}{2}\right)\left(L - \frac{x}{2}\right)dx + \int_0^L (-Wx)(0)dx \right]$$

$$= \frac{1}{EI}\left[\int_0^L \frac{-Wx^2}{4}dx + \int_L^{2L}\left(\frac{-WxL}{2} + \frac{Wx^2}{4}\right)dx \right]$$

$$= \frac{1}{EI}\left(\frac{-WL^3}{12} - \frac{WL^3}{6}\right) = \frac{-WL^3}{4EI}$$

The negative sign indicates that the deflection is opposite to the direction of the applied load, that is, upward. ∎

■ SOLVED PROBLEM 7.3

Determine the horizontal deflection of joint *C* of the truss shown in Fig. 7.46*a*. Use the virtual-work method. The cross-sectional areas in square inches are shown in parentheses. Take $E = 12{,}000$ ksi.

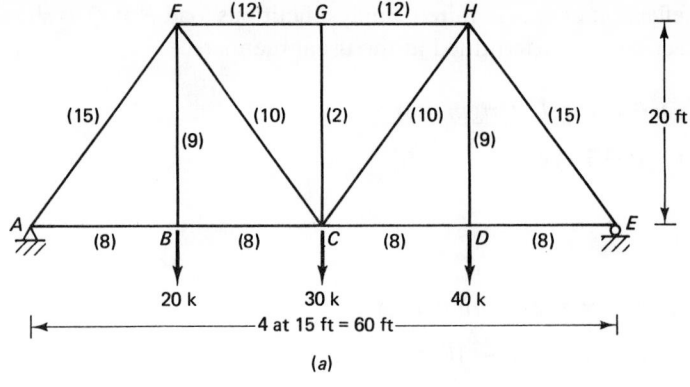

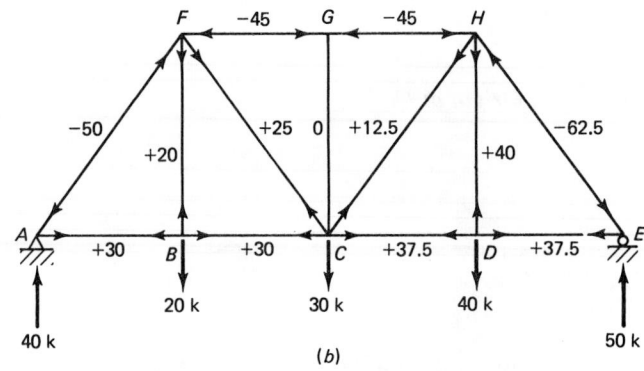

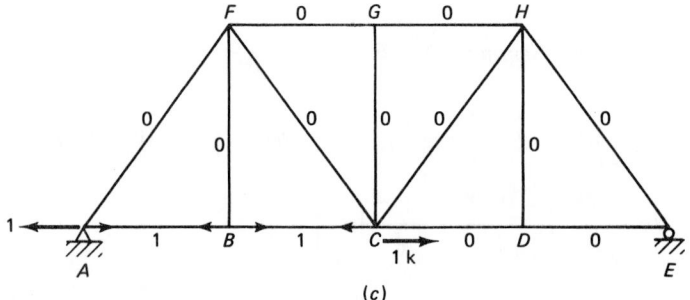

Figure 7.46 Truss for Solved Problem 7.3. (*b*) *P*-force system. (*c*) Unit-load system and *p* forces.

Solution. Let the fictitious unit-load system consist of a horizontal force at *C* as shown in Fig. 7.46*c*. The forces in the members of the truss due to the fictitious unit-load system are also shown in the same figure. δL is the change in the length of the member due to the given *P*-force system ($= PL/AE$ for each member). The forces in

TABLE 7.6

Member	P (k)	L (in)	A (in^2)	p (k)	$p\dfrac{PL}{A}$
AB	+30.0	180	8	+1	+675
BC	+30.0	180	8	+1	+675
CD	+37.5	180	8	0	0
DE	+37.5	180	8	0	0
FG	−45.0	180	12	0	0
GH	−45.0	180	12	0	0
AF	−50.0	300	15	0	0
EH	−62.5	300	15	0	0
CF	+25.0	300	10	0	0
CH	+12.5	300	10	0	0
BF	+20.0	240	9	0	0
CG	0	240	2	0	0
DH	+40.0	240	9	0	0

$$\Sigma\, p\frac{PL}{A} = +1350$$

the various members of the truss due to the given loading are shown in Fig. 7.46b. Then

$$1\cdot\Delta = \Sigma\, p\frac{PL}{AE}$$

The details of the calculations are shown in Table 7.6.

$$1\cdot\Delta = \frac{1350}{12,000} = 0.1125$$

Δ_c horizontal = 0.1125 in

The positive sign indicates that the deflection is in the same direction as the unit load, that is, to the right. The reader may have noticed from Table 7.6 that it is necessary to calculate P and p only for those members for which neither P nor p is zero. ■

■ SOLVED PROBLEM 7.4

Determine the horizontal displacement at B and the angle of rotation at A for the frame shown in Fig. 7.47a. Take $EI = 480,000$ k·ft^2. Consider only the effects of flexure. Use the virtual-work method.

Solution

(a) *Horizontal Displacement at B.* The bending-moment diagram due to the given loading (P-force system) is shown in Fig. 7.47b and the m diagram due to a unit horizontal load at B is given in Fig. 7.47c. Thus, making use of Table 7.1:

For Member AE. m and M diagrams are triangles (both negative).

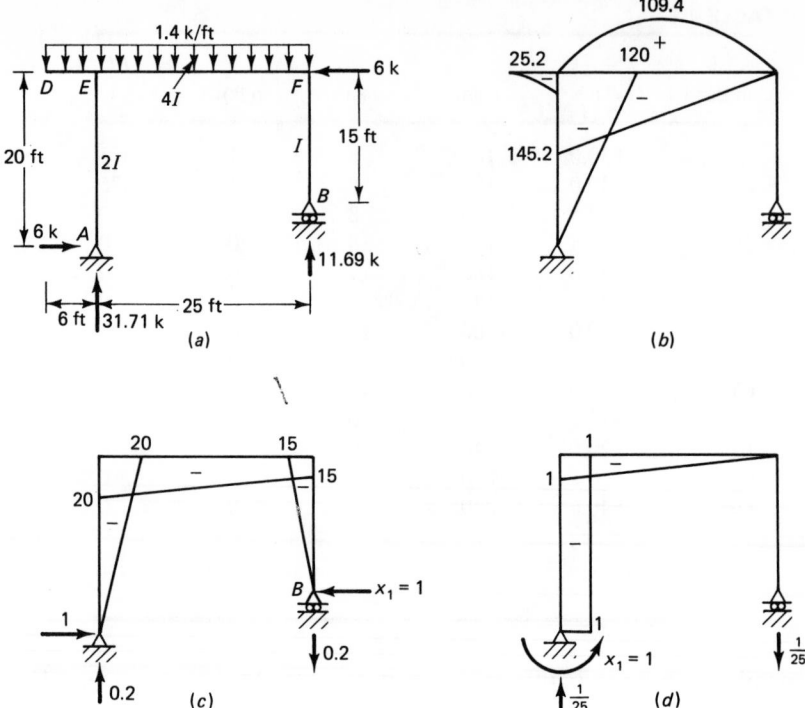

Figure 7.47 Frame for Solved Problem 7.4. (*b*) *M* diagram. (*c*) *m* diagram for Δ_{Bx}. (*d*) *m* diagram for θ_A.

For Member EF. m diagram is a trapezium (negative) and *M* diagram consists of a triangle (negative) and a parabola (positive).

For Member FB

$$M = 0; \therefore \int mM\, dx = 0$$

For Member DE

$$m = 0; \therefore \int mM\, dx = 0$$

Also note that the moments of inertia (I) of the members are different.

$$1 \times \Delta_{Bx} = \int_{\text{frame}} \frac{mM\, dx}{EI}$$

$$\frac{1}{(E)(2I)} \left[\tfrac{1}{3}(20)(-20)(-120)\right]$$

$$+ \frac{1}{(E)(4I)} \left[\left(\frac{25}{6} \right)(-145.2)((-15) + (2)(-20)) \right.$$

$$+ \frac{(25)(109.4)}{3} ((-20) + (-15)) \Big] + 0 + 0$$

$$\therefore \Delta_{Bx} = \frac{8341.67}{480,000} = 0.0174 \text{ ft} \leftarrow$$

(Direction is the same as that of the applied fictitious unit load.) ∎

(b) *Rotation at A.* The m diagram due to a fictitious unit moment at A is shown in Fig. 7.47d. Again, using Table 7.1,

$$1 \times \theta_A = \int \frac{mM \, dx}{EI}$$

$$= \frac{1}{E(2I)} \left[\frac{(20)(-120)}{6} ((-1) + 2(-1)) \right]$$

$$+ \frac{1}{E(4I)} \left[\frac{25}{3}(-1)(-145.2) + \frac{25}{3}(-1)(109.4) \right] + 0 + 0$$

$$\theta_A = \frac{674.58}{480,000} = +0.0014 \text{ radian}$$

The positive sign indicates that the rotation is in the same direction as the fictitious unit moment, that is, counterclockwise. ∎

■ SOLVED PROBLEM 7.5

Determine the vertical deflection at D, the horizontal deflection at B, and the angle of rotation at C for the structure shown in Fig. 7.48a. Consider the effects of bending moment M, axial force P, and shear force V. Use the virtual-work method.

Solution

(a) *Vertical Deflection at D.*

$$\text{Virtual work due to bending deformation} = \int \frac{mM \, dx}{EI} \tag{a}$$

$$\text{Virtual work due to axial deformation} = \int p \frac{P \, dx}{AE} \tag{b}$$

$$\text{Virtual work due to shear deformation} = K_s \int v \frac{V \, dx}{GA} \tag{c}$$

Equating the external virtual work and internal virtual work,

$$1 \cdot \Delta_{Dy} = \int m \frac{M \, dx}{EI} + \int p \frac{P \, dx}{AE} + K_s \int v \frac{V \, dx}{GA} \tag{d}$$

From the M, P, and V diagrams shown in Fig. 7.48b and the m, p, v diagrams due to a unit vertical load at D shown in Fig. 7.48c using Table 7.1:

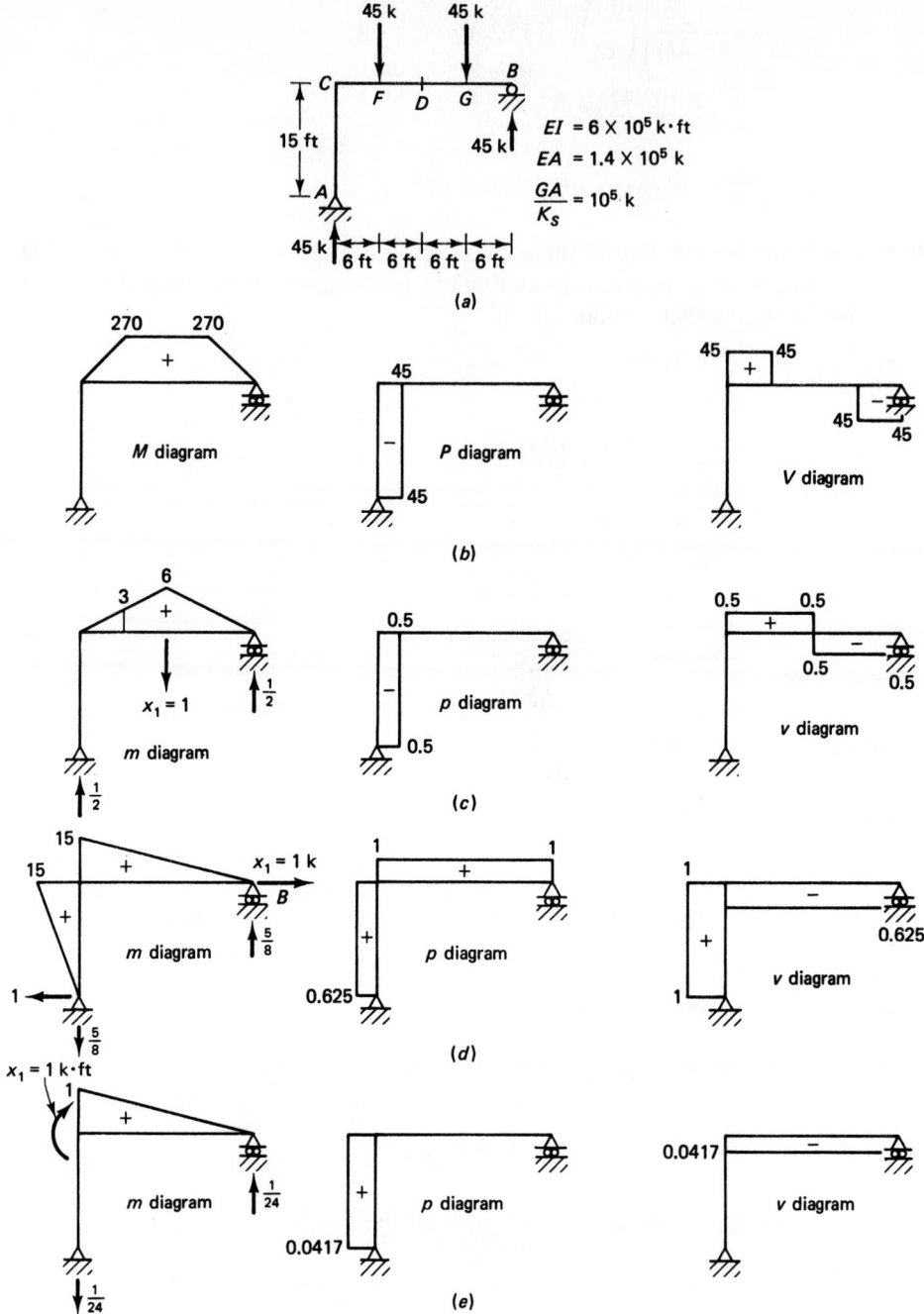

Figure 7.48 Structure for Solved Problem 7.5.

∫ *mM dx*

For *AC*: $m = 0$; $M = 0$.

For *CF*: *m* diagram is a triangle and *M* diagram is also a triangle. Therefore, from the first row and second column of Table 7.1, with $L = 6$, $b = 3$, and $d = 270$, we get $\int mM\,dx = \frac{6}{3} \times 3 \times 270 = 1620$.

For *FD*: *m* diagram is a trapezium, while *M* diagram is a rectangle. Therefore, from the third row and first column of Table 7.1, with $L = 6$, $a = 3$, $b = 6$, and $c = 270$, we get $\int mM\,dx = \frac{6}{2}(3 + 6)(270) = 7290$.

For *DG*: same as for *FD*: $\int mM\,dx = 7290$.

For *GB*: same as for *CF*: $\int mM\,dx = 1620$.

∫ *pP dx*

For *AC*: From the third row and first column of Table 7.1, with $L = 15$, $a = b = -0.5$, and $c = -45$, we get $\int pP\,dx = \frac{15}{2}(-0.5 - 0.5)(-45) = 337.5$.

For *CB*: $p = 0$, $P = 0$.

∫ *vV dx*

For *AC*: $v = 0$ and $V = 0$

For *CF*: From the third row and first column of Table 7.1, with $L = 6$, $a = b = +0.5$, and $c = +45$, we get $\int vV\,dx = \frac{6}{2}(0.5 + 0.5)(45) = 135$.

For *FD* and *DG*: $V = 0$

For *GB*: From the third row and first column of Table 7.1, with $L = 6$, $a = b = -0.5$, and $c = -45$, we get $\int vV\,dx = \frac{6}{2}(-0.5 - 0.5)(-45) = +135$.

From eq. (d):

$$1 \times \Delta_{Dy} = \frac{1}{6 \times 10^5}(1620 + 7290 + 7290 + 1620)$$

$$+ \frac{1}{1.4 \times 10^5}(337.5) + \frac{1}{10^5}(135 + 0 + 0 + 135)$$

$$= 0.0348\ \text{ft} = 0.418\ \text{in}\downarrow$$

(b) *Horizontal Displacement at B.* From the *M, P,* and *V* diagrams shown in Fig. 7.48*b* and the *m, p, v* diagrams due to a unit horizontal load at *B* shown in Fig. 7.48*d,* using Table 7.1:

∫ *mM dx*

For *AC*: $M = 0$.

For *CF*: From the third row and second column of Table 7.1, with $L = 6$, $d = 270$, $a = 15$, and $b = \frac{45}{4}$, we get $\int mM\,dx = \frac{6}{6} \times 270(15 + 2 \times \frac{45}{4}) = 10,125$.

For *FD*: From the third row and first column of Table 7.1, with $L = 6$, $a = \frac{45}{4}$, $b = 7.5$, and $c = 270$, we get $\int mM\,dx = \frac{6}{2}(\frac{45}{4} + 7.5)270 = 15{,}187.5$.

For DG: From the third row and first column of Table 7.1, with $L = 6$, $a = 7.5$, $b = \frac{15}{4}$ and $c = 270$, we get $\int mM\,dx = \frac{6}{2}(7.5 + \frac{15}{4})270 = 9112.5$.

For GB: From the first row and second column of Table 7.1, with $L = 6$, $b = \frac{15}{4}$, and $d = 270$, we get $\int mM\,dx = \frac{6}{3} \times \frac{15}{4} \times 270 = 2025$.

$\int pP\,dx$

For AC: From the third row and first column of Table 7.1, with $L = 15$, $a = b = +0.625$, and $c = -45$, we get $\int pP\,dx = \frac{15}{2}(0.625 + 0.625)(-45) = -421.875$.

For CB, $P = 0$.

$\int vV\,dx$

For AC: $V = 0$.

For CF: From the third row and first column of Table 7.1, with $L = 6$, $a = -0.625$, $b = -0.625$, and $c = +45$, we get $\int vV\,dx = \frac{6}{2}[-0.625 + (-0.625)] \times 45 = -168.75$.

For FD and DG, $V = 0$.

For GB: From the third row and first column of Table 7.1, with $L = 6$, $a = -0.625$, $b = -0.625$, and $c = -45$, we get $\int vV\,dx = \frac{6}{2}[-0.625 - 0.625] \times (-45) = +168.75$.

$$1 \times \Delta_{Bx} = \int \frac{mM\,dx}{EI} + \int \frac{pP\,dx}{AE} + \int k_s \frac{vV\,dx}{GA}$$

$$= \frac{(10{,}125 + 15{,}187.5 + 9{,}112.5 + 2025)}{6 \times 10^5}$$

$$+ \frac{(-421.875)}{1.4 \times 10^5} + \frac{(-168.75 + 168.75)}{10^5}$$

$$= 0.0577 \text{ ft} = 0.692 \text{ in} \rightarrow$$

(c) *Angle of Rotation at C.* From the M, P, and V diagrams shown in Fig. 7.48*b* and the m, p, and v diagrams due to a unit clockwise moment at C shown in Fig. 7.48*e*, using Table 7.1:

$\int mM\,dx$

For AC: $m = 0$, $M = 0$.

For *CB*: The *m* diagram is a triangle with an ordinate (=1) equal to $\frac{1}{15}$ of the ordinate of the *m* diagram of Fig. 7.48*d*. Therefore, $\int mM\ dx = \frac{1}{15}(10{,}125 + 15{,}187.5 + 9{,}112.5 + 2025) = 2430$

$\int pP\ dx$

For *AC*: The *p* diagram is a rectangle with ordinate $(+0.0417)/(-0.5)$ of the *p* diagram of Fig. 7.48*c*. Therefore,

$$\int pP\ dx = 337.5\ \frac{(0.0417)}{(-0.5)} = -28.1475$$

For *CB*: $p = 0$.

$\int vV\ dx$

For *AC*: $v = 0$.

For *CB*: *v* diagram is a rectangle with an ordinate equal to $(-0.0417)/(-0.625)$ of the *v* diagram in Fig. 7.48*d*. Therefore, $\int vV\ dx = (-0.0417)/(-0.625)(0) = 0$.

$$\therefore \int \frac{mM\ dx}{EI} + \int \frac{pP\ dx}{AE} + \int k_s \frac{vV\ dx}{AG}$$

$$= \frac{2430}{6 \times 10^5} + \frac{(-28.1475)}{1.4 \times 10^5} + 0$$

$$= 0.0038\ \text{radian} \,\rangle \qquad \blacksquare$$

PROBLEMS

Solve Problems 7.1 to 7.7 using the (**a**) double-integration method, (**b**) moment-area method, (**c**) conjugate-beam method, (**d**) Castigliano's theorem—part II, and (**e**) virtual-work method.

7.1 Find the deflection and slope at *A* of the loaded cantilever shown. *EI* is constant.

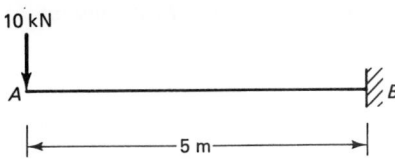

Figure P7.1

7.2 Determine the vertical deflection and slope at *A* for the cantilever beam shown. *EI* is constant.

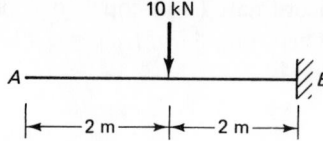

Figure P7.2

7.3 Calculate the vertical deflection and slope at D. EI is constant.

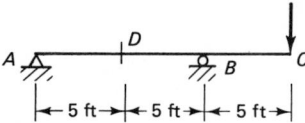

Figure P7.3

7.4 Calculate the vertical deflection and slope at points C and D of the beam shown. EI is constant.

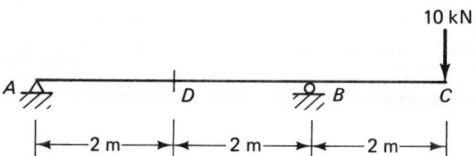

Figure P7.4

7.5 Determine the vertical deflection and slope at A of the cantilever beam loaded as shown. EI is constant.

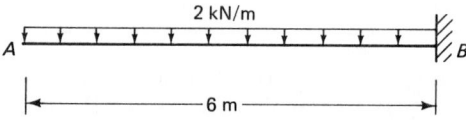

Figure P7.5

7.6 Calculate the vertical deflection and slope at A. EI is constant.

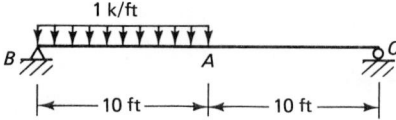

Figure P7.6

7.7 Determine **(a)** the vertical deflection and slope at C, **(b)** the vertical deflection and slope at B, of the cantilever beam. $E = 200$ GPa, $I = 50 \times 10^6$ mm^4.

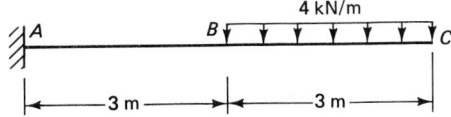

Figure P7.7

7.8 Determine the slopes at A and B of the simple beam AB subjected to a moment of 10 kN·m at A as shown in the figure. Use the virtual-work method.

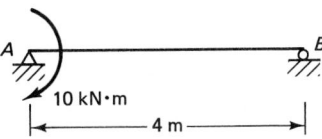

Figure P7.8

7.9 The propped cantilever beam AB is subjected to a clockwise moment of 100 kN·m. Assume that the moment reaction at A is 50 kN·m clockwise. Determine the position and magnitude of the maximum vertical deflection. Use **(a)** double-integration method, **(b)** moment-area method, **(c)** conjugate-beam method. $E = 200$ GPa, $I = 250 \times 10^6$ mm^4.

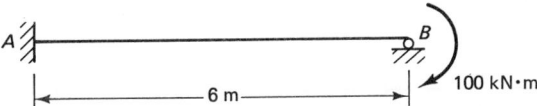

Figure P7.9

7.10 Using the conjugate-beam method, determine the vertical deflection of A for the structure shown. $E = 200$ GPa, $I = 700 \times 10^6$ mm^4.

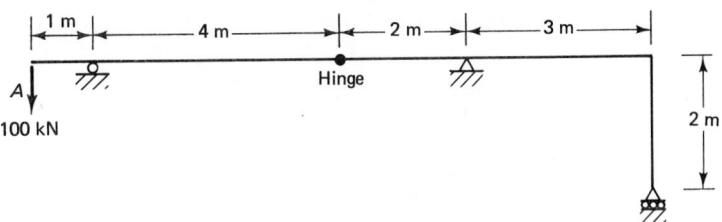

Figure P7.10

7.11 Determine the slope and vertical deflection at B and C of the cantilever beam using **(a)** moment-area method, **(b)** conjugate-beam method, **(c)** Castigliano's theorem—part II, and **(d)** virtual-work method. $E = 26,800$ ksi.

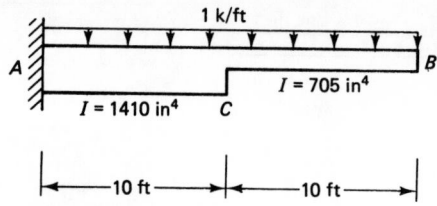

Figure P7.11

7.12 Determine the maximum upward deflection in the portion AB of the beam shown. Support B settles down 3 mm. Use the double-integration method. $EI = 10,000 \text{ KN} \cdot \text{m}^2$.

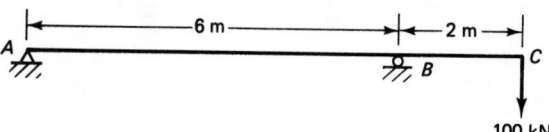

Figure P7.12

7.13 Using the conjugate-beam method, determine the slope and deflection at hinge C of the structure shown.

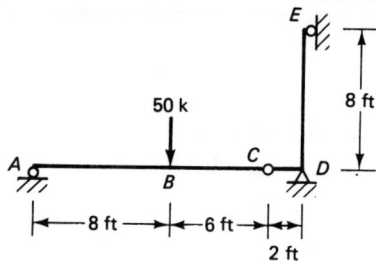

Figure P7.13

7.14 Calculate the vertical deflection of point D using **(a)** moment-area method, **(b)** conjugate-beam method, and **(c)** Castigliano's theorem—part II. Constant EI.

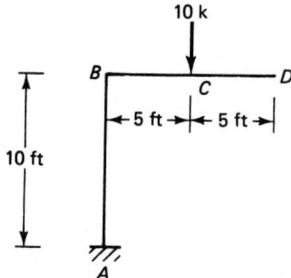

Figure P7.14

7.15 Calculate the vertical and horizontal deflection of point C using (a) Castigliano's theorem—part II, and (b) virtual-work method. Neglect the effect of axial loads on deformation. Constant EI.

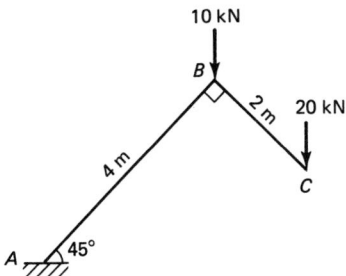

Figure P7.15

7.16 The upper and lower surfaces are uniformly heated along the length of the beam ABC, having a rectangular cross section. Calculate (a) the vertical deflection of point A, (b) the slope at point A. Take $\alpha = 12 \times 10^{-6}$ m/m/°C. Use the virtual-work method.

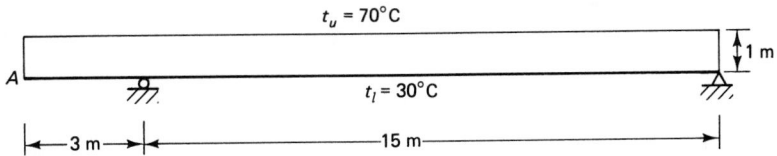

Figure P7.16

7.17 The figure shows a hollow circular bar in a horizontal plane loaded by a vertical load of 20 kN at A. Determine the vertical component of deflection at A, using the virtual-work method. Cross-sectional area = 1200 mm², $E = 200$ GPa, $G = 80$ GPa, $I = 250 \times 10^{6}$ mm⁴, $J = 500 \times 10^{6}$ mm⁴.

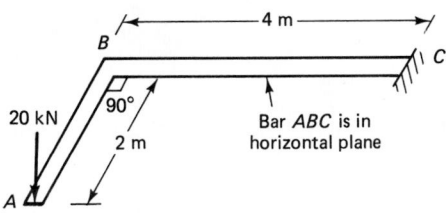

Figure P7.17

7.18 Calculate horizontal and vertical deflection of A using (a) Castigliano's theorem—part II, and (b) virtual-work method.

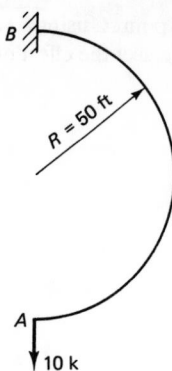

Figure P7.18

7.19 Determine vertical and horizontal deflection of C using **(a)** Castigliano's theorem—part II, and **(b)** virtual-work method.

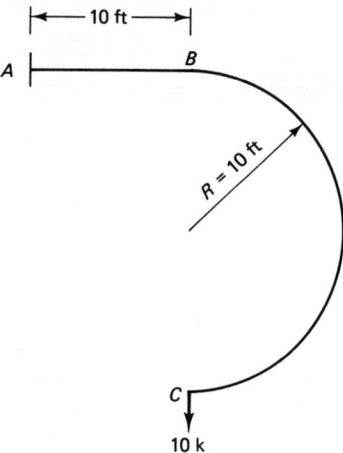

Figure P7.19

7.20 Using Castigliano's theorem—part II, determine the vertical deflection of point C.

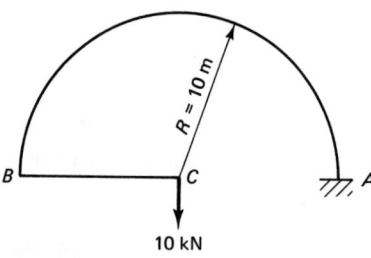

Figure P7.20

7.21 Find the angular displacement of bar AB of the structure shown. A for each member = 1000 mm². E = 200 GPa. Use the virtual-work method.

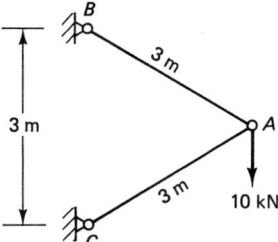

Figure P7.21

7.22 If the member BC of the truss is subjected to a temperature rise of 70°F, calculate the resulting vertical deflection at joint D. Assume that the coefficient of thermal expansion for the material is 6.5×10^{-6}/°F. Use the virtual-work method.

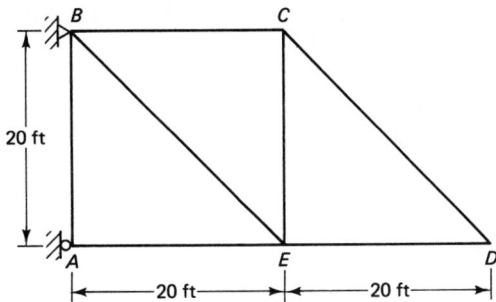

Figure P7.22

7.23 Calculate the length change in members AE, BC, and ED of the truss shown in Fig. P7.22 in order to induce an initial vertical upward deflection at D equal to 2 in. This is accomplished by shortening BC and lengthening AE and ED by the same amount. Use the virtual-work method.

7.24 Calculate the vertical and horizontal displacement of joint D in the truss shown. Support A moves horizontally 3 mm to the left; support B moves 3 mm vertically downward and 3 mm to the right. Assume A = 1000 mm², E = 200 GPa. Use the virtual-work method.

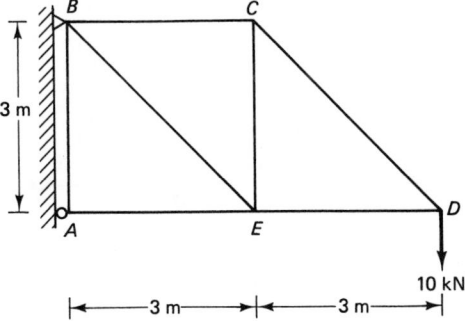

Figure P7.24

7.25 Determine   (a) the horizontal deflection, and   (b) the vertical deflection of joint C in the pin-jointed truss shown, using the virtual-work method. Each member has a cross-sectional area of 1000 mm^2; $E = 200$ GPa.

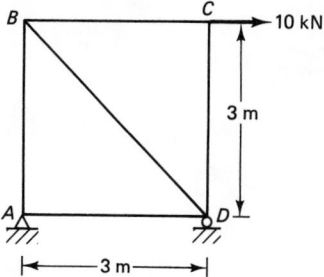

Figure P7.25

7.26 For the truss shown, determine   (a) the vertical deflection at joint C,   (b) the horizontal deflection at joint B,   (c) the relative deflection between joints B and E along the line joining them, and   (d) the rotation of member EF. L/AE of all members $= 0.8 \times 10^{-3}$ in/k. Use the virtual-work method.

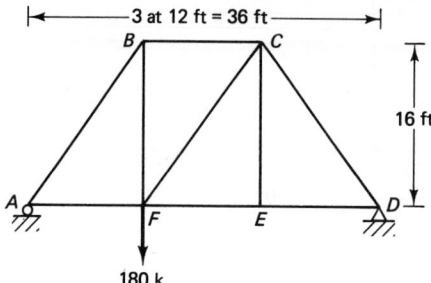

Figure P7.26

7.27 Determine the horizontal and vertical deflection of joint C in the truss shown using (a) Castigliano's theorem—part II, and   (b) virtual-work method. Cross-sectional areas in mm^2 are shown circled. $E = 200$ GPa.

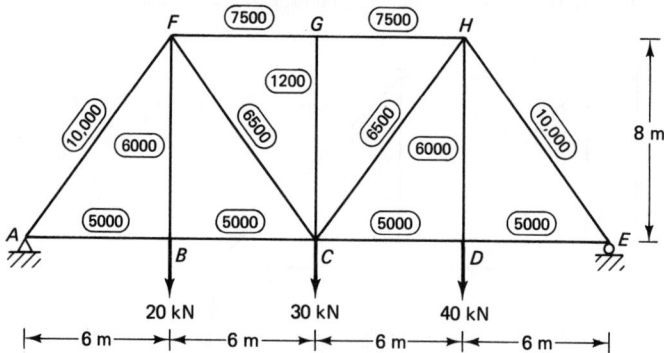

Figure P7.27

7.28 Compute the vertical deflection of point E taking into account axial and bending effects. I for columns $= 200 \times 10^6$ mm^4. I for beam $= 400 \times 10^6$ mm^4. Cross-sectional area of each column $= 9000$ mm^2. Cross-sectional area of beam $= 15{,}000$ mm^2. $E = 200$ GPa. Use the virtual-work method.

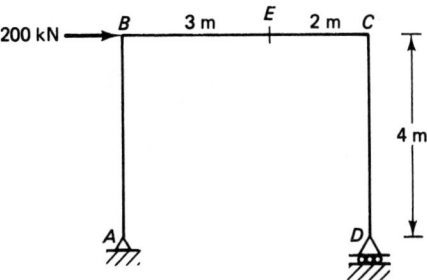

Figure P7.28

7.29 Using the method of virtual work, compute the vertical component of deflection of point B of the structure shown. I for $ABCD = 4000$ in^4.

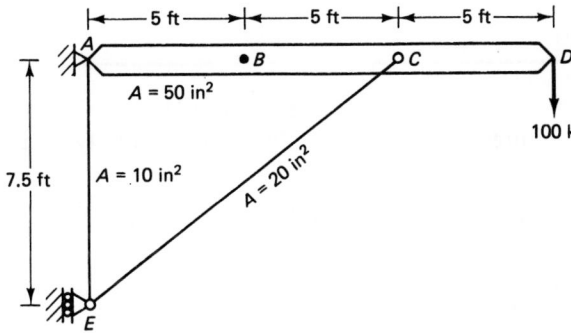

Figure P7.29

Chapter **8**

Approximate Methods of Analysis of Statically Indeterminate Structures

Ah, but a man's reach should exceed his grasp, or what's a heaven for?

Robert Browning

8.1 INTRODUCTION

We should realize by now that there is no "exact" analysis of an engineering structure. Indeed, all structures are analyzed on the basis of simplifying assumptions; they may include the following:

1. Riveted, bolted, or welded connections in an actual truss are generally assumed to be frictional pins.
2. Loads and reactions acting over a small area on an actual structure are considered to be concentrated loads.
3. The material properties of a structure, although nonhomogeneous on the microscopic scale, are assumed to be homogeneous from the macroscopic point of view. This renders the actual behavior of the structure different from the idealized behavior; for example, it is assumed that the behavior of concrete, a nonhomogeneous material, is linearly elastic and isotropic (i.e., having the same properties in all directions) if it is subjected to a compressive stress not greater than 50 percent of its ultimate compressive strength.

Because of the above and other assumptions the so-called exact methods of analysis are, in fact, not "exact"; however, experimentation shows that the results obtained by using these methods are quite close to the exact values. The approximate methods discussed in this chapter are based on intuition leading to further assumptions beyond those used in conventional "exact" methods, to analyze indeterminate structures by statics alone. Even with all the advanced computer programs available in analyzing highly redundant structures, there is still a need for the use of the approximate methods of analysis for the following reasons: The structure is so highly redundant that it requires a large computer storage which may not be available; the cost of preparation of the input, execution of the program, and interpretation of the output may be prohibitively expensive. Since the analysis of indeterminate structures depends upon the elastic properties of the members, an approximate method of analysis using statics can generate a preliminary design of members which can then serve as input for a complete indeterminate structural analysis; in fact, this is the most important application of the approximate methods since it calls for an instinctive understanding of the behavior of a redundant structure.

The approximate methods of analysis discussed below are based on intuition and the deflected shape of the loaded indeterminate structure; this approach leads to additional assumptions that generate sufficient equations to analyze the structure by statics only. Because of limited space only some of the more commonly used approximate methods of analysis will be illustrated; it is hoped that they will give insight to students to make their own assumptions suitable to new structural forms that they may encounter in their professional careers. We now apply these approximate methods to the solution of two kinds of indeterminate structures: one in which direct stress predominates and the second in which flexural stress predominates.

8.2 INDETERMINATE STRUCTURES IN WHICH AXIAL FORCE DOMINATES

Figure 8.1a shows a loaded truss, statically indeterminate to the fourth degree. To analyze this structure by means of statics alone we require four assumptions. Based on the properties of the diagonal members, the following two types of assumptions are available to us:

1. If the diagonal members are slender (i.e., long with a small radius of gyration), their compressive load-carrying capacity is relatively small; thus, such diagonal members in compression in any panel are assumed *not to carry* any load and the shear in the panel is then resisted entirely by the tension diagonals.
2. If the diagonal members in any panel are stocky (i.e., not slender) and with equal cross-sectional area, the shear in the panel is assumed to be carried equally by the two diagonal members in that panel.

Solution Based on Assumption 1 That the Diagonals Are Not Effective in Carrying Compressive Forces   The steps leading to the solution are as follows:

Step 1. Determine the external reactions by statics. Thus the vertical reactions are 15 kN each and the horizontal reaction $H = 0$.

Step 2. Determine the shears in the panels. Thus shear in panel $L_0 L_1' = 15$ kN; shear in panel $L_1' L_2' = 15 - 10 = 5$ kN; shear in panel $L_2' L_3' = 5 - 10 = -5$ kN; shear in panel $L_3' L_4' = -5 - 10 = -15$ kN.

Step 3. By inspection, determine which diagonal members are in compression for the given loading case. For example, the action of the shear 15 kN↑ ↓15 kN in panel $L_0 L_1'$ will tend to elongate the diagonal $L_0 L_1'$ (hence this diagonal is in tension) and will compress the diagonal $L_0' L_1$ (hence this diagonal is in compression). Similar reasoning applied to the re-

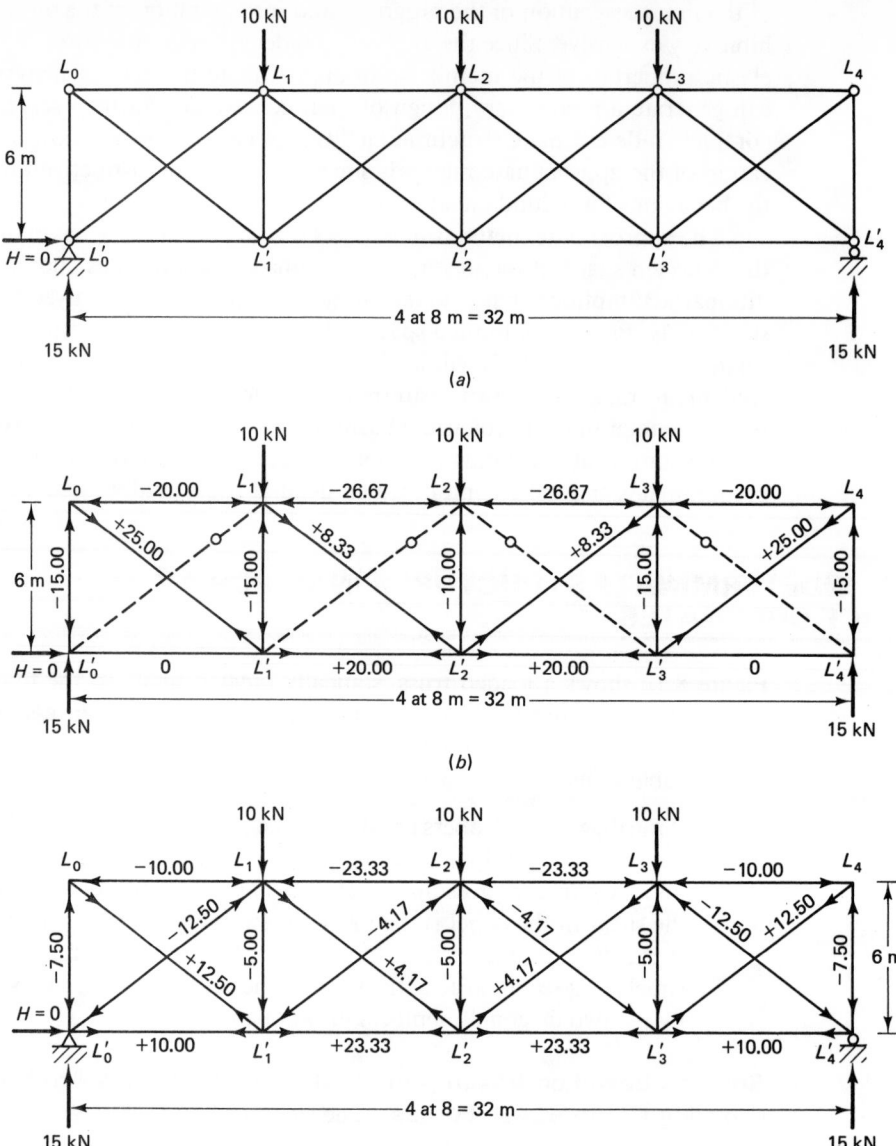

Figure 8.1

Joints are marked by open circles
Number of unknown reactions = 3
Number of unknown bar forces = 21
Number of joints = 10
∴ Number of available equations from static equilibrium = 10 × 2 = 20
Degree of indeterminacy = (21 + 3) − 20 = 4
Number of assumptions required = 4

(a) An internally indeterminate truss; determination of forces in indeterminate truss of Fig. 8.1a assuming (b) compressive diagonals do not carry any load, (c) shear in any panel is equally divided between the compressive and tensile diagonals.

maining panels will reveal that diagonals L'_1L_2, $L_2L'_3$ and $L_3L'_4$ are in compression also.

Step 4. Eliminate the diagonal members in compression and draw the statically determinate truss formed by the remaining members as shown in Fig. 8.1*b*. (It should be clearly understood that under a different load condition, it is quite possible that the diagonals that are now in compression can be in tension and therefore fully effective.)

Step 5. Compute the bar forces in the determinate truss, resulting from step 4, by the method of joints or sections (or some other suitable method discussed in Chapter 4); of course, the forces in the compressive diagonals are zero. The bar forces are shown in Fig. 8.1*b*.

Solution Based on Assumption 2 That the Shear in Any Panel Is Equally Divided between the Two Diagonals in That Panel   The steps here are as follows:

Step 1. Determine the reactions by statics; these are 15 kN each as before.

Step 2. Determine the shears in the panels; as shown earlier, these are 15, 5, −5, and −15 kN in panels $L'_0L'_1$, $L'_1L'_2$, $L'_2L'_3$, and $L'_3L'_4$, respectively.

City Hall, Toronto, Ontario, Canada. (Photo courtesy of Canadian Institute of Steel Construction.)

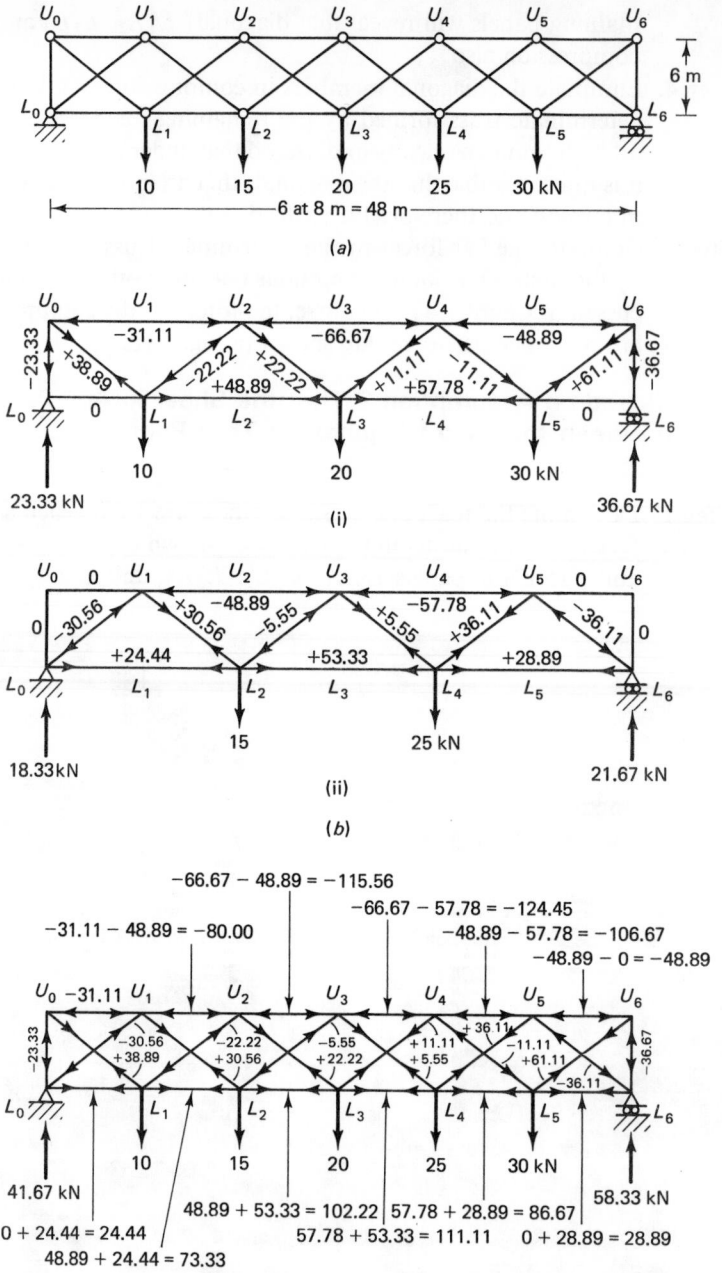

Figure 8.2 Determination of forces in a multiple-web truss. (*a*) Multiple web-system truss.

Number of joints = 14
Number of members = 26
Number of reactions = 3
Degree of indeterminacy = $(26r^3) - 2 \times 14 = 1$

(*b*) Component trusses i and ii of truss in *a*. (*c*) Member forces (in kN) of truss in *a*.

Step 3. Divide the shear equally between the two diagonals in each panel; therefore, the shear resisted by each diagonal in any one panel is equal to one-half the panel shear.

Step 4. From the direction of the external shear in the panel, determine the nature of the forces in the diagonals (whether compressive or tensile). See step 3 for the solution based on assumption 1. The type of force in the various diagonals is shown in Fig. 8.1c.

Step 5. Determine the magnitude of the forces in the diagonals. Consider the diagonal L_0L_1'; it resists a shear of $\frac{1}{2} \times 15 = 7.5$ kN; therefore, force in diagonal $L_0L_1' = (7.5)(\frac{10}{6}) = 12.5$ kN (tension); similarly force in diagonal $L_0'L_1 = 12.5$ kN (compression); and so on (see Fig. 8.1c).

Step 6. Knowing the forces in the diagonal members, compute the forces in the other members of the truss by the method of joints or any other applicable method as discussed in Chapter 4. The results obtained by the method of joints are shown in Fig. 8.1c.

It should be noted that if the cross-sectional areas of the diagonals are known a priori and are not equal to each other in a panel, the assumption of equal shear distribution may not be justified. Under such circumstances, distributing the panel shear to the two diagonals in proportion to their axial stiffness EA/L values will yield results closer to the actual distribution.

Multiple Web-System Truss This type of indeterminate truss can be analyzed by decomposing it into two or more statically determinate trusses; the load applied at any panel point of the original truss is assumed acting on the appropriate component truss. For the method to be applicable, it should be possible to split the given truss into a number of component trusses equal to the degree of indeterminacy plus one. Consider, for example, the truss shown in Fig. 8.2a; the degree of indeterminacy is one, and the truss can be decomposed into two component trusses as shown in Fig. 8.2b. The loads acting at joints L_1, L_3, and L_5 of the original truss are assumed resisted by the truss of Fig. 8.2b (i), while the loads acting at joints L_2 and L_4 are resisted by the truss of Fig. 8.2b (ii). Determining the forces in each of the two determinate trusses of Fig. 8.2b, adding them algebraically, will yield the net forces in the original statically indeterminate truss as shown in Fig. 8.2c. It should be noted that this method can only be applied when it is possible to apportion the external loads on the redundant truss to the component trusses; for example, this method will fail in the case of the truss shown in Fig. 8.1a.

8.3 INDETERMINATE STRUCTURES IN WHICH FLEXURE DOMINATES

8.3.1 Continuous Beams under Uniformly Distributed Vertical Loading

Figure 8.3a shows the bending-moment diagram for a simply supported beam of span L under a uniformly distributed load of w per unit length. The support moments are zero, while the maximum positive moment is $wL^2/8$. For the fixed-ended beam shown

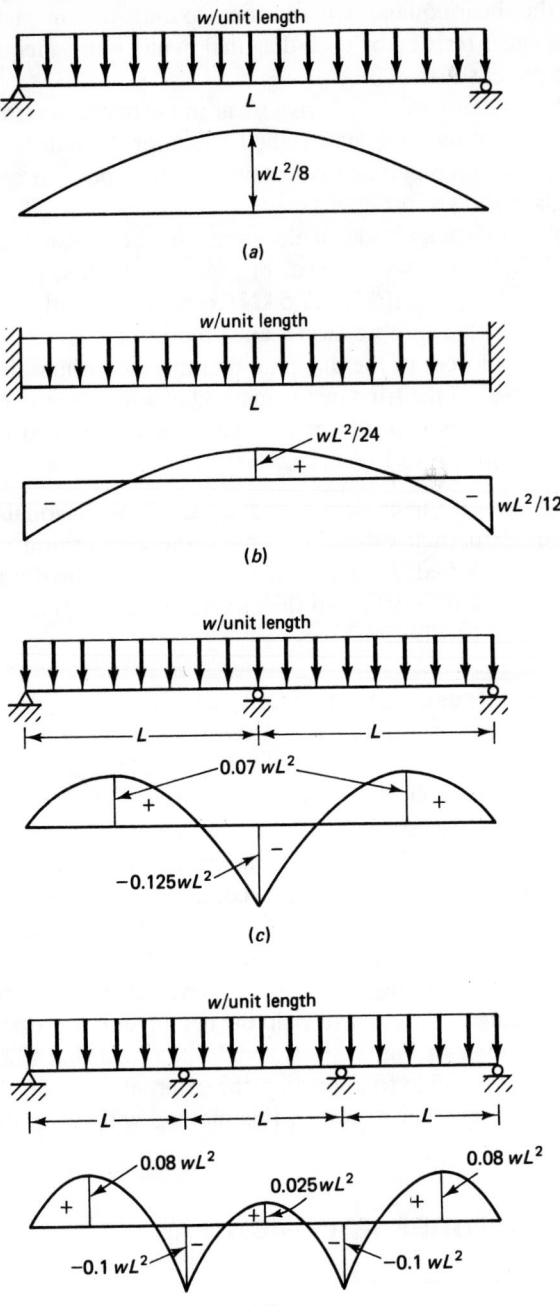

Figure 8.3 Continuous beams under uniformly distributed vertical loading.

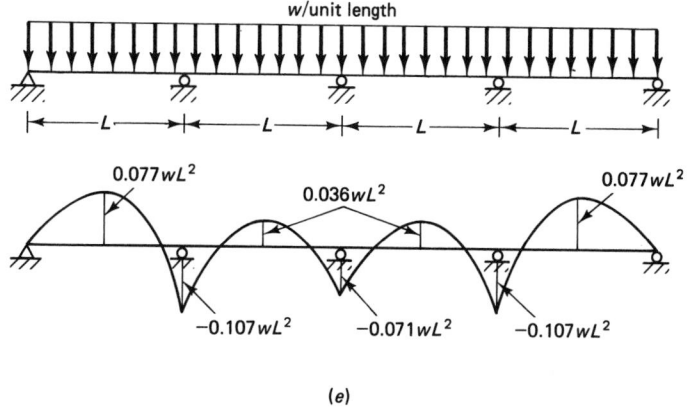

(e)

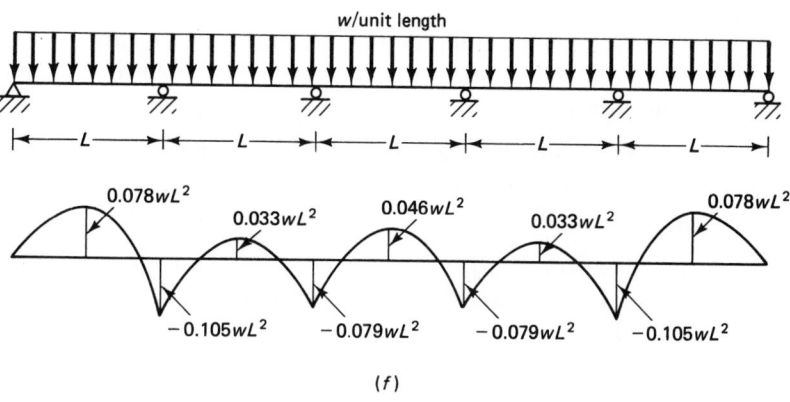

(f)

Figure 8.3 (*Continued*)

in Fig. 8.3b, the support moments are $wL^2/12$ and the maximum positive moment is $wL^2/24$. For a continuous beam with two, three, four, and five equal spans, the bending-moment diagrams are shown in Fig. 8.3c to f. An examination of these diagrams suggests that the following approximate formulas for support moments can be used for continuous beams of three or more spans under uniformly distributed loads:

$$\text{Support moment at first interior support} = -\frac{wL^2}{10}$$

$$\text{Support moment at other interior supports} = -\frac{wL^2}{12}$$

When the support moments are known, the support reactions and maximum positive moments can be readily determined by statics.

8.3.2 Multistory Rigid Frames Subjected to Horizontal Loads

Multistory rigid frame structures are highly indeterminate. Therefore, it would be advantageous if we could analyze such structures by means of approximate methods using statics alone in order to arrive at a preliminary design of the various members in the structure. The approximate analysis chosen for a given structure depends on the type of load applied to that structure. For example, if a multistory rigid frame structure is acted upon by lateral loads (i.e., wind and/or earthquake) two methods of analysis are suggested, the portal method and the cantilever method, whereas vertical loads on a rigid frame structure call for different assumptions in the approximate analysis than the ones used in these two methods.

The Portal Method   The results given by this method are best when the width of the multistory frame structure is greater than or at least equal to the total height of the structure. The assumptions made in this method are

1. There is a point of inflection at the center of each girder.
2. There is a point of inflection at the center of each rigidly connected column. This does not apply to columns with pinned bases where the moment is zero.

The above assumptions are sufficient for the analysis of one-bay frames of any number of stories. However, for multibay frames, the following additional assumption is required:

3. The shear resisted by an exterior column of any story is equal to $1/2n$ of the total story shear, and the shear resisted by each of the $(n-2)$ *adjacent* interior columns is $1/n$ of the total story shear, where n is the number of bays in any story. (For two-bay frames, since $n-2=0$, no assumption regarding interior column shear need be made.)

As an example, consider a four-bay one-story frame acted upon by a 10-k story shear (Fig. 8.4). The degree of indeterminacy of the frame is $4 \times 3 = 12$. There are four girders and five columns. Therefore, nine assumptions regarding the locations of hinges are made. Three more assumptions are required to make the frame determinate. Thus, it is assumed that

1. The shear resisted by the exterior column $I = [1/(2 \times 4)] \times 10 = 1.25$ k.
2. The shear resisted by each of the adjacent $n - 2 (= 2)$ interior columns II and III $= \frac{1}{4} \times 10 = 2.5$ k.

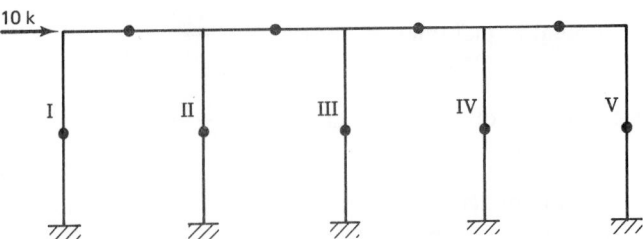

Figure 8.4 Four-bay building bent.

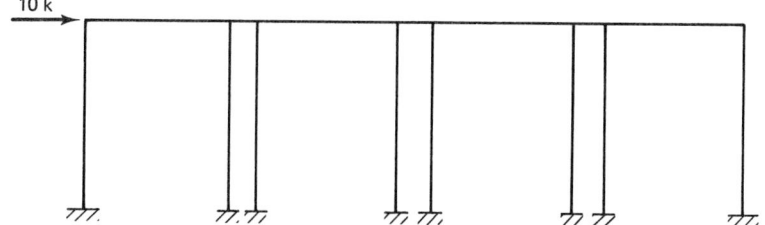

Figure 8.5 Building bent assumed as consisting of a series of portal frames.

With these assumptions it is possible to analyze the frame completely by statics. It will be found (this is left as an exercise for the student) that irrespective of the dimensions of the bays and story height, the interior column IV shear is also 2.5 k and the shear in the exterior column V is 1.25 k. To reduce the computation effort, it can thus be assumed that all interior columns carry twice the shear resisted by exterior columns. Since this is analogous to a building acting as a series of portal frames connected together as shown in Fig. 8.5, this method is called the "portal method."

■ EXAMPLE 8.1

Carry out an approximate analysis of the structure in Fig. 8.6a using the portal method.

Solution. This structure is indeterminate to the eighteenth degree. By means of imaginary hinges in the middle of the columns (total of eight hinges) and in the middle of the girders (total of six hinges) as well as the assumptions on the distribution of the horizontal column shears, one is able to analyze the structure by statics only. The analysis requires the following steps:

Step 1. Insert imaginary hinges in the middle of all columns and girders, as shown in Fig. 8.6b.

Step 2. Determine *column shears* in all stories considering that the interior column carries twice the shear of the exterior column. Referring to Fig. 8.6b, for horizontal equilibrium, we have

$$x_1 + 2x_1 + 2x_1 + x_1 = 3 + 6 \quad \text{or} \quad x_1 = 1.5 \text{ k}$$

Referring to Fig. 8.6c,

$$x_2 + 2x_2 + 2x_2 + x_2 = 3 \quad \text{or} \quad x_2 = 0.5 \text{ k}$$

Step 3. Determine *column end moments acting on the joints* by multiplying the column shears by one-half of the story heights as shown in Fig. 8.6d and e. (If a column base is pinned, the moment at the top end of that column is simply the product of the column shear and the *full* story height.) Thus, referring to Fig. 8.6d, $M_{CB} = 0.5 \times 6 = 3.0 \text{ k} \cdot \text{ft} \, \rangle = M_{BC}; M_{FE} = 1.0 \times 6 = 6.0 \text{ k} \cdot \text{ft} \, \rangle = M_{EF}; M_{IH} = 1.0 \times 6 = 6.0 \text{ k} \cdot \text{ft} \, \rangle = M_{HI}; M_{LK} = 0.5 \times 6 = 3.0 \text{ k} \cdot \text{ft} \, \rangle = M_{KL}$. Note that these are the moments exerted *on the joints* by the ends of columns. Referring to Fig. 8.6e,

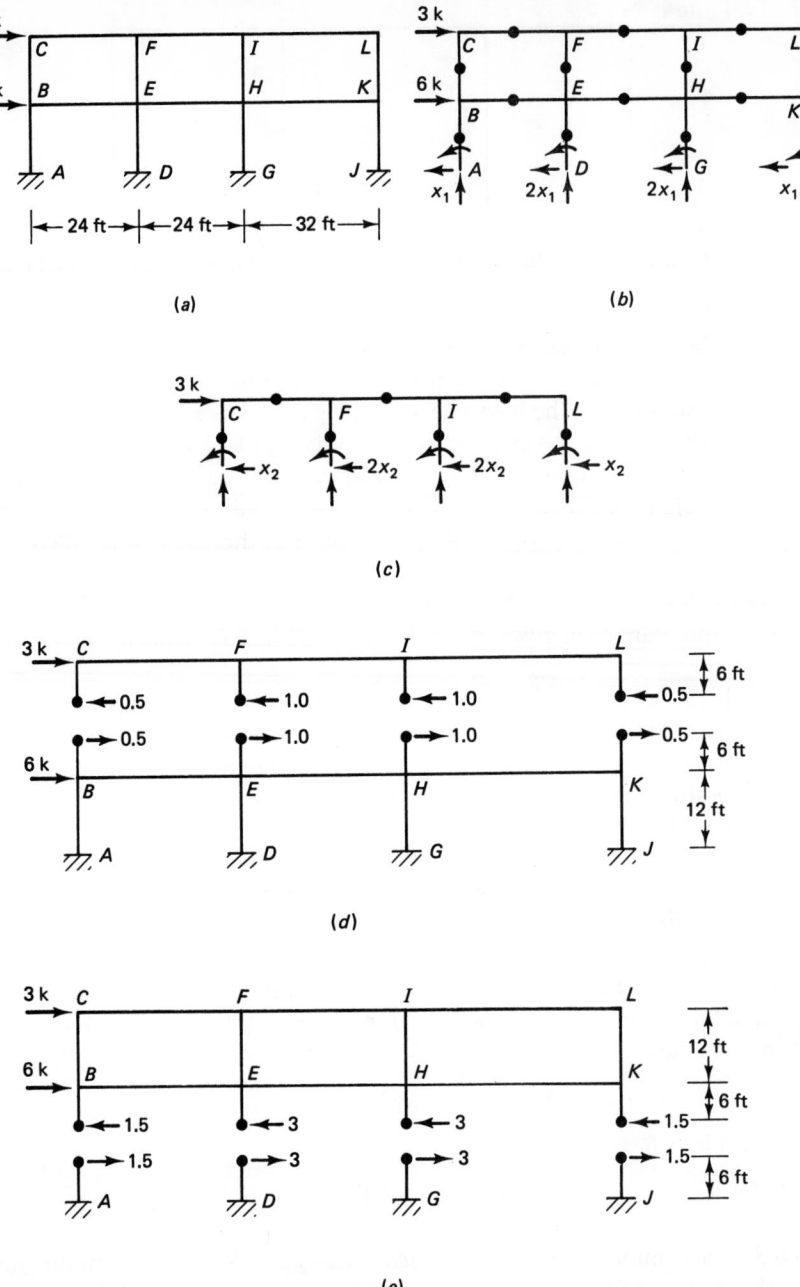

Figure 8.6 Structure for Example 8.1 analyzed by the portal method. (*a*) Three-bay two-story building bent. (*b*) Bent with 10 imaginary hinges and bottom-story column shears. (*c*) Top-story column shears. (*d*) Top-story column end moments. (*e*) Bottom-story column end moments. (*f*) Girder end moments. (*g*) Girder shears. (*h*) Column axial forces. (*i*) Girder axial forces. (*j*) Exaggerated deflected shape of the rigid frame (note points of inflection). (*k*) Bending-moment diagram drawn on compression side (values in ft·k).

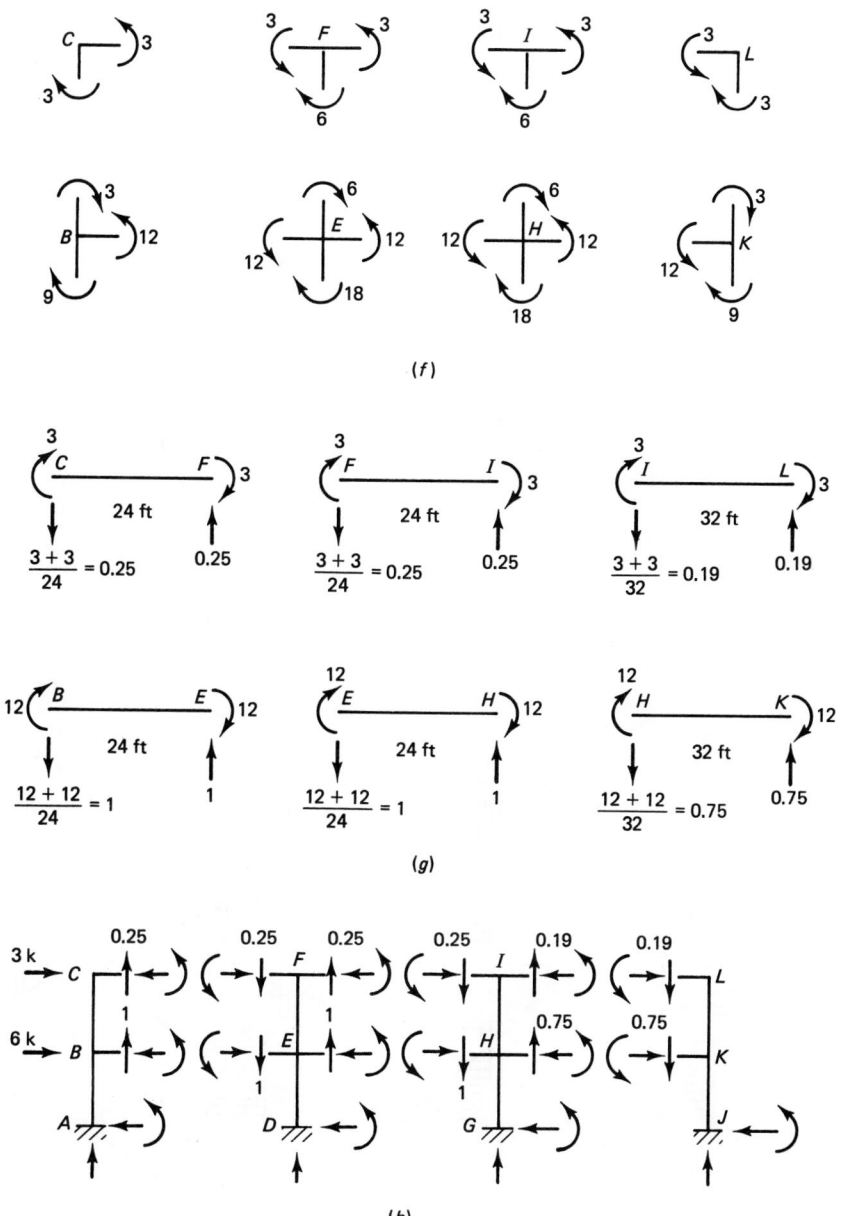

Figure 8.6 (*Continued*)

we have $M_{BA} = 1.5 \times 6 = 9.0 \text{ k} \cdot \text{ft} \; \downcurvearrowright \; = M_{AB}$; $M_{ED} = 3 \times 6 = 18.0 \text{ k} \cdot \text{ft} \; \downcurvearrowright \; = M_{DE}$; $M_{HG} = 3 \times 6 = 18.0 \text{ k} \cdot \text{ft} \; \downcurvearrowright \; = M_{GH}$; $M_{KJ} = 1.5 \times 6 = 9.0 \text{ k} \cdot \text{ft} \; \downcurvearrowright \; = M_{JK}$.

Step 4. Determine *girder end moments* (i.e., moments exerted on the joints by the ends of girders) by starting at a joint where there is only one unknown girder

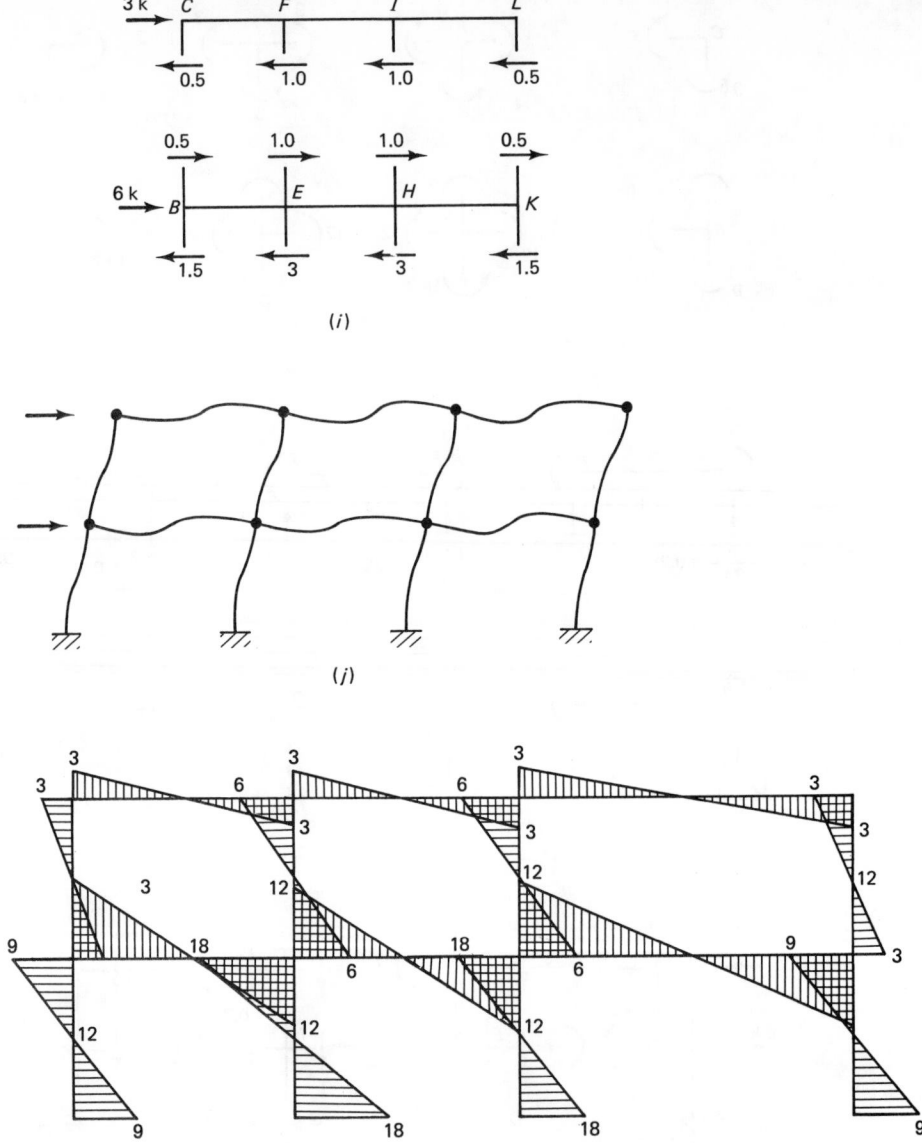

Figure 8.6 (*Continued*)

moment such as joint C (Fig. 8.6f) and considering moment equilibrium of the joint. Since M_{CB} = 3.0 k · ft, M_{CF} = 3.0 k · ft. Referring to Fig. 8.6f, since it is assumed that there is an imaginary hinge or a point of inflection at the center of each girder, M_{FC} = 3.0 k · ft also. Following this procedure, the remaining girder end moments can be found as shown in Fig. 8.6f.

Step 5. Calculate the *girder shears* from the moment equilibrium of each girder as shown in Fig. 8.6g.

Step 6. Determine the *axial forces* on the columns from the equilibrium of the vertical forces resisting the girder shears and acting on the columns, isolated as free bodies, as shown in Fig. 8.6h. Thus, axial force:

On column $BC = 0.25$ k

On column $BA = 1 + 0.25 = 1.25$ k

On columns EF and $ED = 0$

On column $HI = 0.25 - 0.19 = 0.06$ k (compression)

On column $GH = 0.25 + 1.0 - 0.19 - 0.75 = 0.31$ k (compression)

On column $KL = 0.19$ k (compression)

On column $KJ = 0.75 + 0.19 = 0.94$ k (compression)

Step 7. From equilibrium of the horizontal forces acting above and below each floor level, calculate the *axial forces on the girders*. Thus, based on the results shown in Fig. 8.6d and e, equilibrium of the forces shown in Fig. 8.6i yields axial force:

On girder $CF = 3 - 0.5 = 2.5$ k (compression)

On girder $FI = 2.5 - 1.0 = 1.5$ k (compression)

On girder $IL = 1.5 - 1.0 = 0.5$ k (compression)

On girder $BE = 6 + 0.5 - 1.5 = 5$ k (compression)

On girder $EH = 5 + 1.0 - 3 = 3$ k (compression)

On girder $HK = 3 + 1.0 - 3 = 1$ k (compression)

The deflected shape of the rigid frame is shown in Fig. 8.6 j, drawn to an exaggerated scale. From the moment results obtained, the bending-moment diagram for the entire frame is shown in Fig. 8.6k. ■

The Cantilever Method   This method yields best results when applied to tall slender frames subjected to lateral loads. The following assumptions are made in applying the method:

1. There is a point of inflection at the center of each girder and at the center of each rigidly connected column. (For a column with a hinged base, the point of zero moment is at the base of the column.) The above assumptions are sufficient for the analysis of one-bay frames having any number of stories. However, for frames with n number of bays ($n > 1$), the following additional assumptions are required.
2. The axial stresses (and not forces) in *one* of the exterior columns and in the adjacent ($n - 2$) interior columns of any story are the same as the axial stresses computed on the assumption that the building frame is a vertical cantilever subjected to transverse horizontal loads and that the elementary flexural beam formula ($\sigma = My/I$) is applicable. Obviously, in a two-bay rigid frame where

$n - 2 = 0$ the assumption with respect to the axial stress in the interior column is not required.

With the above assumptions a statical analysis will indicate that the axial stresses in the remaining two columns in each story are the same as those found by considering the rigid frame as a vertical cantilever subjected to horizontal loads. Thus, to reduce the computational effort, one can assume at the outset that the axial stresses in *all* the columns of any story are the same as those obtained on the basis of a horizontally loaded vertical cantilever. We now illustrate the method by an example.

■ EXAMPLE 8.2

Analyze the loaded rigid frame in Fig. 8.7a by the *cantilever method.* Assume all columns have the same cross-sectional area A.

Solution. The following steps are used to yield a solution.

Step 1. Determine the location of the centroid of the column group in each story. Thus, taking moments of the areas of the columns of the first bay about line $ABCD$ (Fig. 8.7a) we find

$$\text{Distance of centroid from line } ABCD = \frac{(A)(0) + (A)(6) + (A)(14)}{A + A + A} = 6.67 \text{ m}$$

The centroid of the column group for the other stories is also located at 6.67 m from line $ABCD$ since all columns have the same cross-sectional area A.

Step 2. Compute moment of inertia I_0 of the column group about its centroid. For the first story,

$$I_0 = \sum_{\substack{\text{all columns} \\ \text{in first story}}} (\text{cross-sectional area of column})$$

$$\times (\text{distance of column from centroidal axis})^2$$

$$= (A)(6.67)^2 + (A)(0.67)^2 + (A)(7.33)^2$$

$$= 98.67A$$

(This will be the same for all the other stories in this particular example.)

Step 3. Determine the column axial stresses (and hence *column axial forces*) for each story using the flexural beam equation $\sigma = My/I$. With the imaginary hinges in the middle of the three columns of the first story (Fig. 8.7b), calculate the moment M of the external forces about the horizontal line $B'F'J'$ (Fig. 8.7b). Thus

$$M = (12)(7.5) + (24)(4.5) + (24)(1.5) = 234 \text{ kN} \cdot \text{m}$$

If we denote the stress in the windward column AB as σ_1 (tensile), in column EF as σ_2 (tensile), and in the leeward column IJ as σ_3 (compressive), then

$$\sigma_1 = \frac{My}{I} = \frac{(234)(6.67)}{98.67A}$$

similarly

$$\sigma_2 = \frac{(234)(0.67)}{98.67A} \quad \text{and} \quad \sigma_3 = \frac{(234)(7.33)}{98.67A}$$

Then the axial forces in the columns, equal to stress $\sigma \times$ area A, are

Force in column $AB = (\sigma_1)(A) = [(234)(6.67)/98.67A] \times A = 15.81$ kN (tensile)

Force in column $EF = (\sigma_2)(A) = 1.58$ kN (tensile)

Force in column $IJ = (\sigma_3)(A) = 17.39$ kN (compressive)

This procedure is repeated for all the columns in the second and third stories, and the results are shown in Fig. 8.7c.

Step 4. By considering the vertical-force equilibrium compute the girder shears from the free-body diagrams of each girder, as shown in Fig. 8.7d. Thus, the shear in $DH = -1.22$ kN; in $HL = -1.22 - 0.12 = -1.34$ kN; in $CG = -6.08 + 1.22 = -4.86$ kN; in $GK = -4.86 - 0.61 + 0.12 = -5.35$ kN; in $BF = 6.08 - 15.81 = -9.73$ kN; and in $FJ = -9.73 - 1.58 + 0.61 = -10.70$ kN.

Step 5. Compute the girder end moments from the girder shears assuming that there is a point of inflection (imaginary hinge) at the center of each girder (Fig. 8.7e). The girder end moment = girder shear $\times \frac{1}{2}$ span of girder. Thus $M_{DH} = (1.22)(\frac{6}{2}) = 3.66$ kN·m $\mathbf{)} = M_{HD}$; and so on (see Fig. 8.7e).

Step 6. Compute the column end moments acting on the joints by considering the moment equilibrium of each rigid joint and assuming an imaginary hinge at the center of each rigidly connected column. Starting at a joint with one unknown column moment, such as D (Fig. 8.7f), we find from $\Sigma M = 0$, $M_{DC} - 3.66 = 0$; or $M_{DC} = 3.66$ kN·m $= M_{CD}$ [since a point of inflection (a hinge) is assumed at the center of column DC]; and so on for the other column end moments (Fig. 8.7f).

Step 7. Calculate the column shears from the moment equilibrium of each column, considered as a free-body diagram (Fig. 8.7g). For example, the shear at the ends of column $DC = (3.66 + 3.66)/3 = 2.44$ kN, etc.

Step 8. Determine the girder axial forces by isolating each floor and considering the horizontal force equilibrium of the external forces and the column shears at each floor level (Fig. 8.7h). For example,

Axial force in $DH = 12 - 2.44 = 9.56$ kN (compression)

Axial force in $HL = 9.56 - 6.01 = 3.55$ kN (compression)

Axial force in $CG = 24.0 + 2.44 - 7.32 = 19.12$ kN (compression)

Axial force in $GK = 19.12 + 6.01 - 17.97 = 7.16$ kN (compression)

Axial force in $BF = 24 + 7.32 - 12.14 = 19.18$ kN (compression)

Axial force in $FJ = 19.18 + 17.97 - 30.02 = 7.13$ kN (compression)

The bending-moment diagram for the rigid frame can now be drawn like the one in Fig. 8.6k. ∎

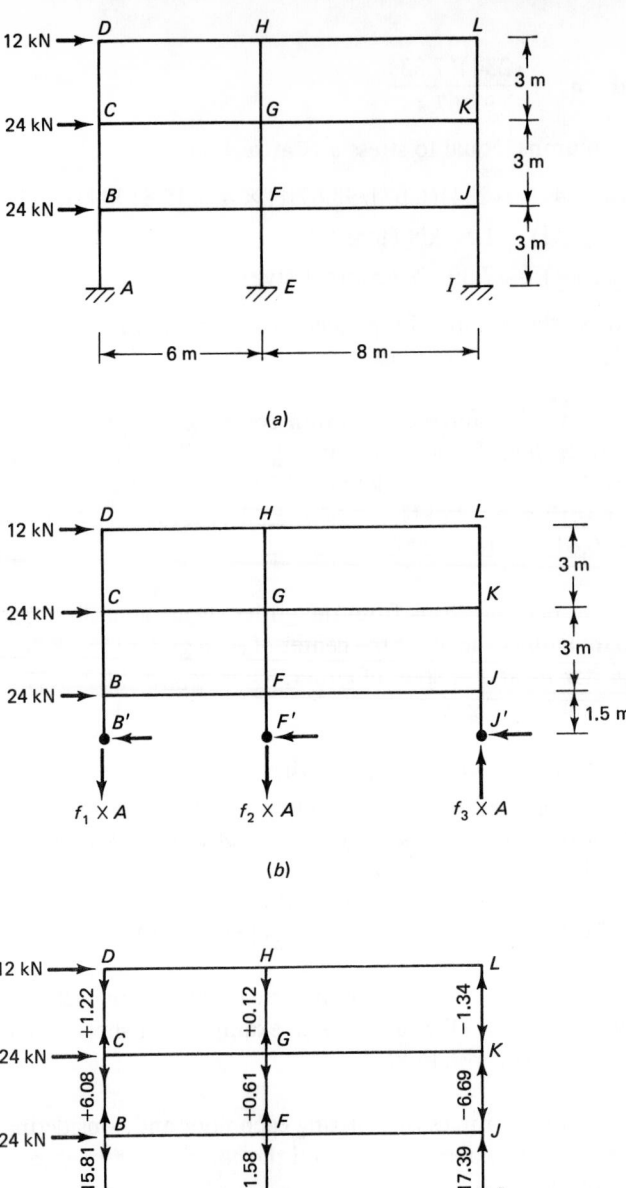

Figure 8.7 Structure for Example 8.2 analyzed by the cantilever method. All joints are rigid; cross-sectional areas of all columns are A. (a) Two-bay three-story building bent. (b) Column axial forces in first story. (c) Column axial forces. (d) Girder shears. (e) Girder end moments. (f) Column end moments (showing moments only). (g) Column shears (showing end moments and reactive shears only). (h) Girder axial forces (column shears and external forces only).

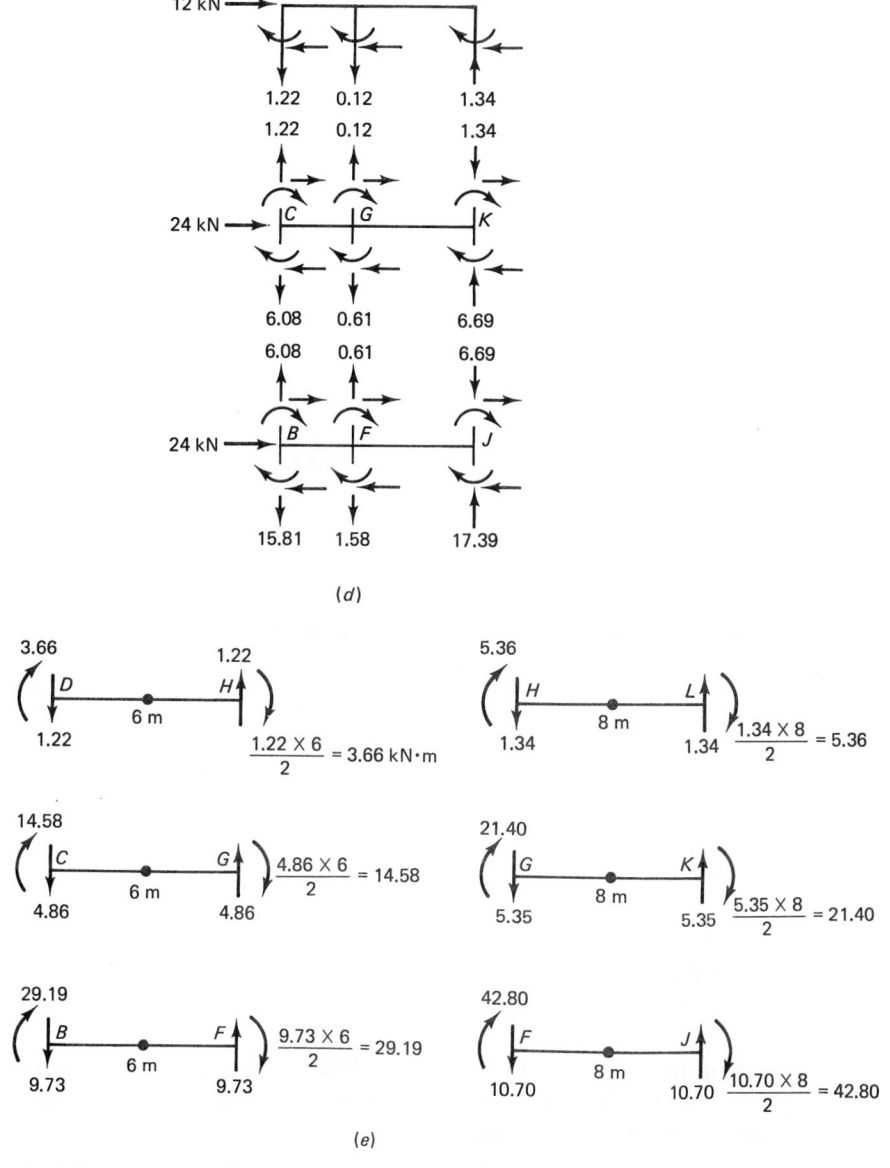

(d)

(e)

Figure 8.7 (*Continued*)

8.3.3 Rigid Frames Subjected to Vertical Loads

The behavior of rigid frames subjected to vertical loads is different from their behavior under the action of horizontal loads. This necessitates the use of different assumptions than the ones made earlier in the portal and cantilever methods. In an actual rigid frame under vertical load, the girders will act as if their ends were partially fixed and

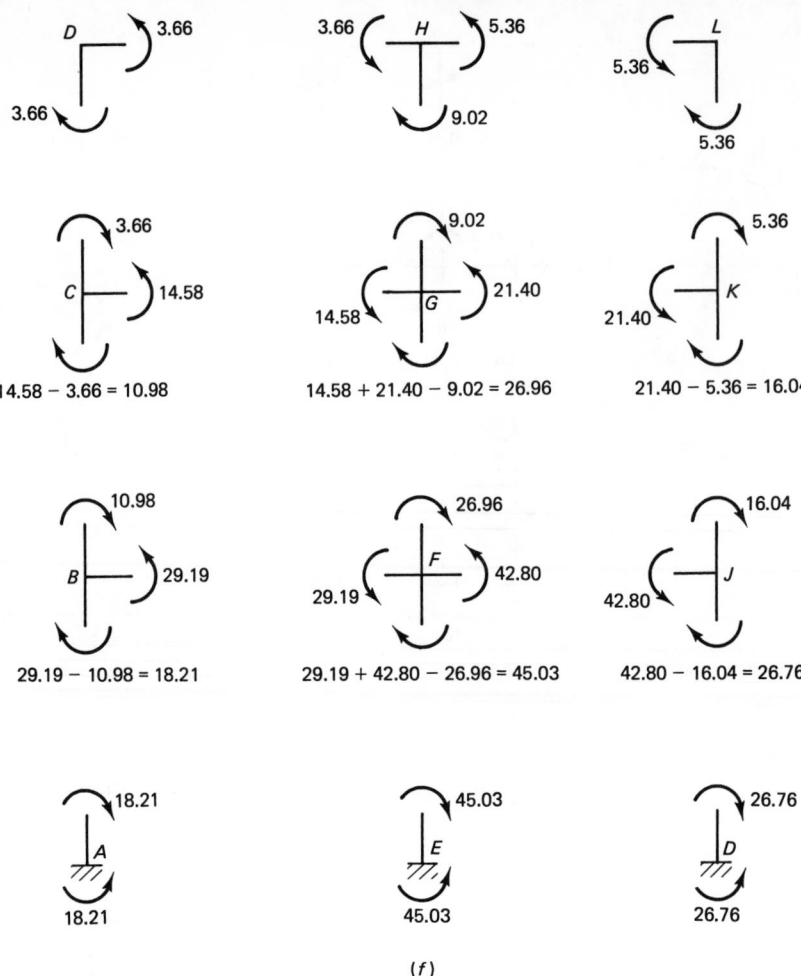

(f)

Figure 8.7 (*Continued*)

points of inflection can be located at about one-tenth of the span from either end. Furthermore, it is assumed that the axial force in the girder is negligible and can be taken as zero. Thus, three assumptions are made for each girder, with a total number of assumptions equal to the degree of indeterminacy. We illustrate this analysis by an example.

■ EXAMPLE 8.3

Analyze the rigid frame shown in Fig. 8.8a.

Solution

Step 1. We begin the solution by locating the points of inflection for the rigid frame (at $0.1 L$, where L = girder span), as shown in Fig. 8.8b; the simply supported

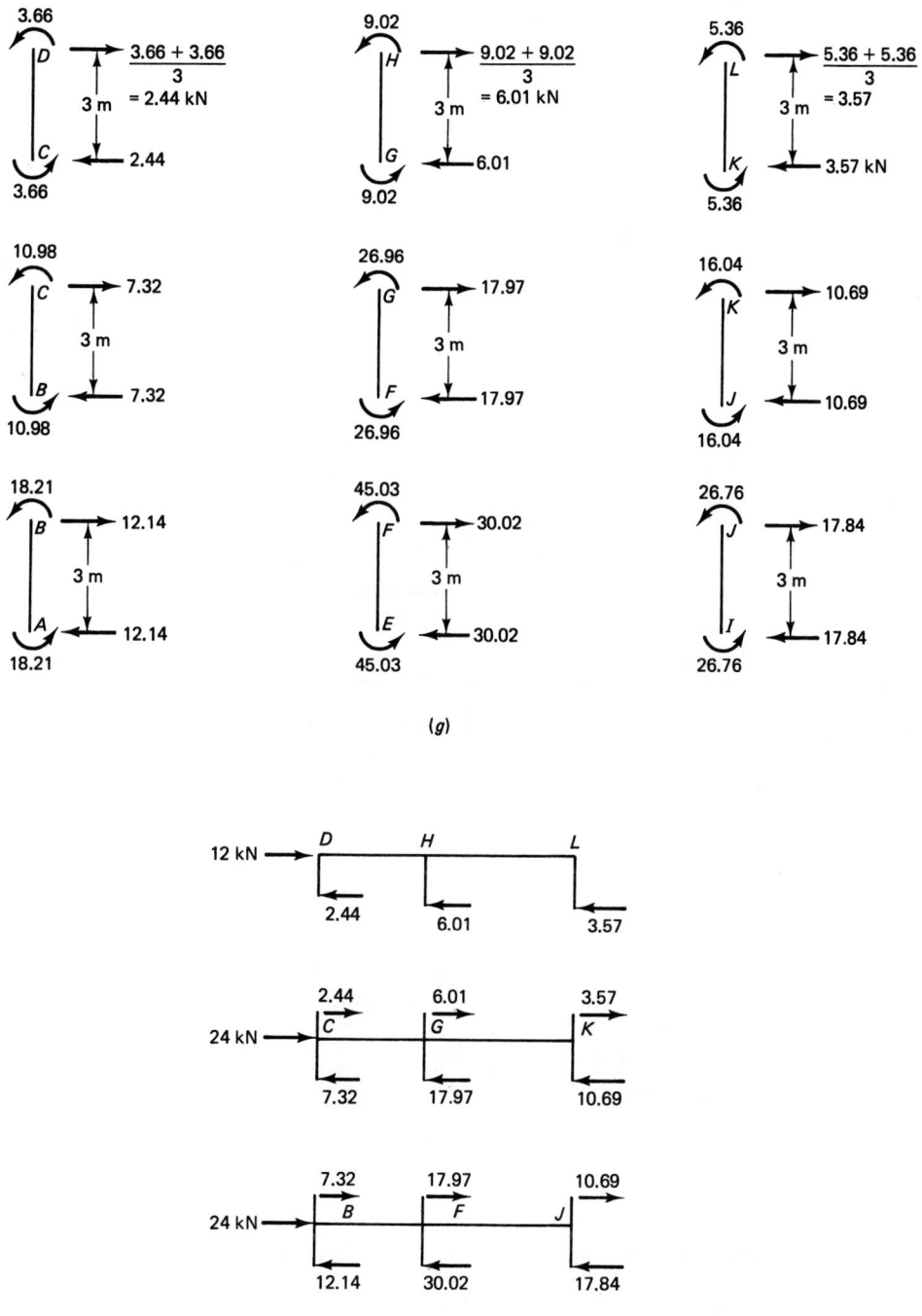

(g)

(h)

Figure 8.7 (*Continued*)

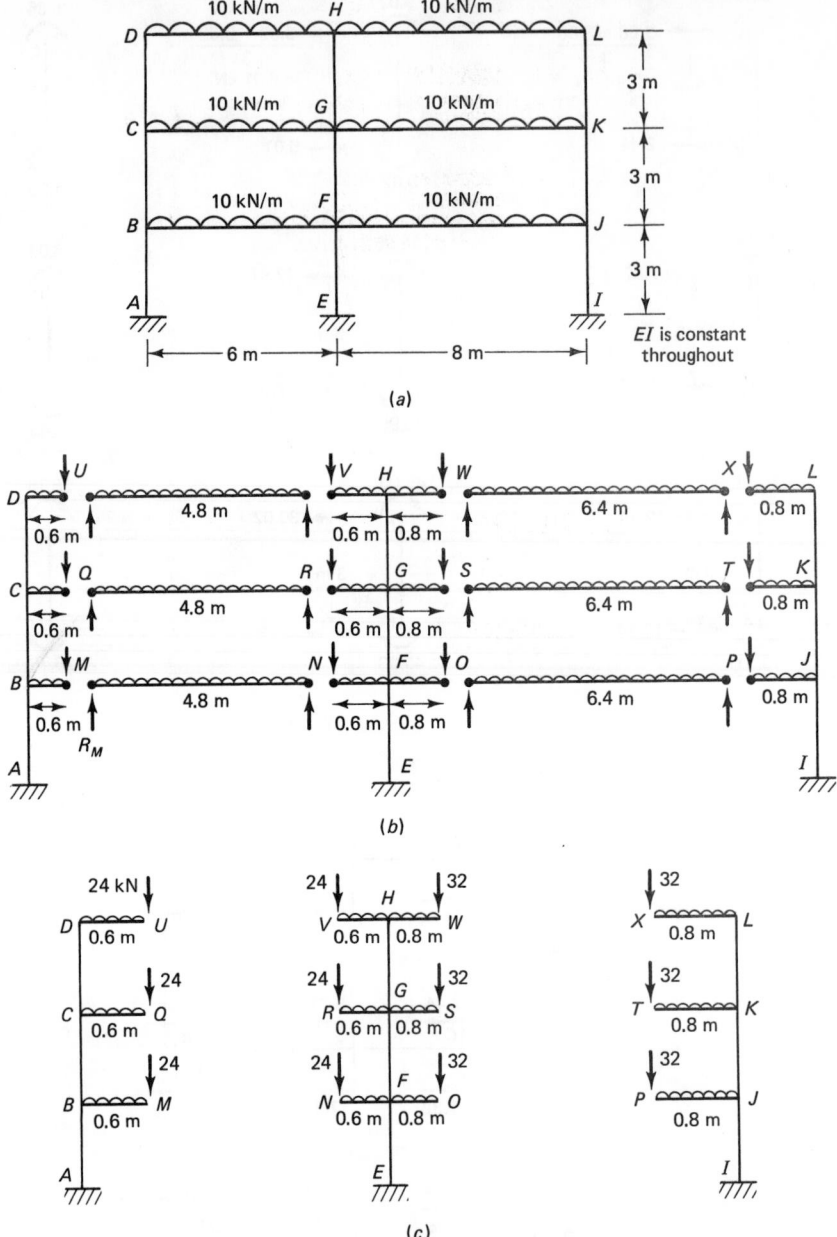

Figure 8.8 Structure for Example 8.3 subjected to vertical loads. (*a*) Building bent subjected to uniformly distributed vertical loads. (*b*) Assumed points of inflection. (*c*) Analysis of girder stubs. (*d*) Column end moments. (*e*) Column shears. (*f*) Bending-moment diagram drawn on compression side (values in kN·m).

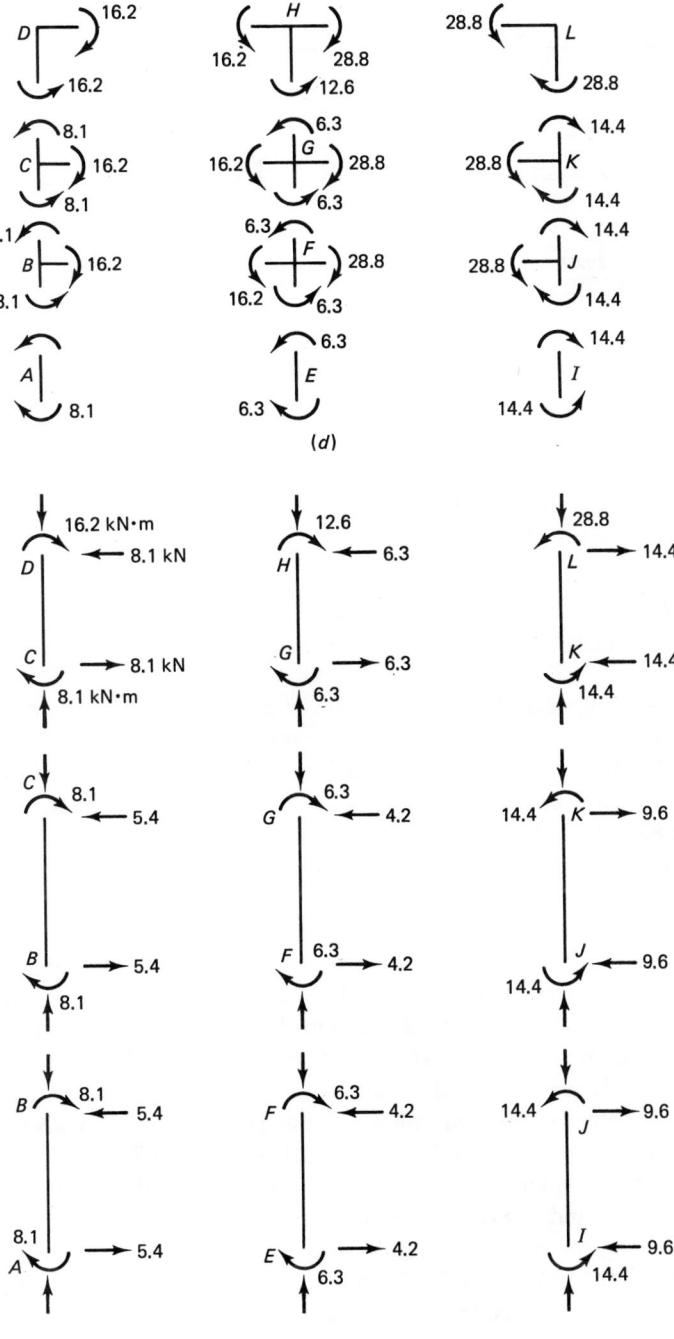

Figure 8.8 (*Continued*)

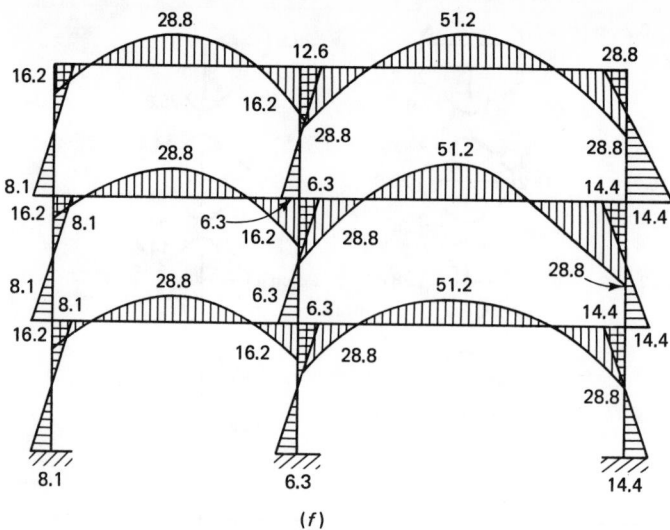

Figure 8.8 (*Continued*)

girders *MN, OP, QR, ST, UV,* and *W X* are analyzed in the usual manner. For example, the reaction $R_M = wL/2 = (10)(4.8)/2 = 24$ kN; and so on.

Step 2.   Determine the shears and moments at the ends of each girder. From an analysis of the girder stubs (Fig. 8.8*c*), we find maximum shears at ends of girders *BF, CG,* and *DH* = 24 + (10)(0.6) = 30 kN; the maximum negative moments at the same ends = (24)(0.6) + (10)(0.6)(0.6)/2 = 16.2 kN · m. Similarly, the maximum shears at ends of girders *FJ, GK,* and *HL* = 32 + (10)(0.8) = 40 kN, and the maximum negative moments at the same ends = (32)(0.8) + (10)(0.8)(0.8)/2 = 28.8 kN · m.

Step 3.   Determine the column end moments, column shears, and axial forces. At each joint, the girder end moments are resisted by the columns meeting at that joint in proportion to their stiffness, that is, *EI/L* values. Referring to joint *G* in Fig. 8.8*d*. Σ girder end moments = 28.8 − 16.2 = 12.6 kN · m clockwise. This net moment must be resisted by the columns *GH* and *GF* in proportion to their *EI/L* values. Since, in this case, both their length *L* and *EI* are the same, the moment of 12.6 kN · m will be resisted equally by the two columns, that is, 12.6/2 = 6.3 kN · m counterclockwise as shown in Fig. 8.8*d*. At the bases of columns *AB, EF,* and *IJ,* that is, at *A, E,* and *I,* the column moments are assumed to be the same as those acting at their top ends, that is, at *B, F,* and *J.* From the column moments, the column shears are calculated by considering the moment equilibrium of the columns as shown in Fig. 8.8*e*. The column axial forces are readily determined by considering the vertical force equilibrium on the columns; thus, from Fig. 8.8c, we have

Axial force in column *CD* = 24 + (10)(0.6) = 30 kN (compression)

Axial force in column *BC* = 30 + 24 + (10)(0.6) = 60 kN (compression)

Axial force in column *AB* = 60 + 24 + (10)(0.6) = 90 kN (compression)

Axial force in column *GH* = 24 + 32 + (10)(0.6 + 0.8) = 70 kN (compression)

Axial force in column FG = 140 kN (compression)

Axial force in column EF = 210 kN (compression)

Axial force in column KL = 40 kN (compression)

Axial force in column JK = 80 kN (compression)

Axial force in column IJ = 120 kN (compression)

Finally, the bending-moment diagram for the rigid frame is drawn as shown in Fig. 8.8f. *Note:* The existence of shears in the columns invalidates the assumption made regarding zero axial forces in the girders. ∎

8.4 BRIDGE PORTALS

A bridge portal (Figs. 8.9 and 8.10) is used to transfer the horizontal loads applied at the top of a bridge truss to its foundation; all members of such a rigid frame are designed to carry bending-moment, shearing, and axial forces. The portals in Figs. 8.9 and 8.10 are indeterminate to the first and third degree, respectively. It is assumed that the column shears are distributed in proportion to EI/L^3 values of the columns; while this assumption is sufficient to analyze the portal in Fig. 8.9, the analysis of the portal in Fig. 8.10 requires two additional assumptions: Each column has a point of inflection at its center. As an example, let us analyze the portal shown in Fig. 8.11. Since the horizontal external force = 10 kN, the horizontal reaction at each hinge = $(\frac{1}{2})(10)$ = 5 kN. (Note that EI/L^3 is the same for both columns.) Vertical reactions are such as to produce the balancing couple, that is, $(10)(9)/12$ = 7.5 kN, acting as shown in Fig. 8.11a. The moment M_{BA} = 5 × 9 = 45 kN·m = M_{CD}. The bending-moment diagram drawn on the compression side is shown in Fig. 8.11b.

Figure 8.12a shows a portal with fixed bases. Again since (EI/L^3) is the same for both columns, the external force of 10 kN is divided equally between the two columns as shown in Fig. 8.12a; next, points of inflection are assumed at the center of columns AB and CD. Isolating $EBCF$ as a free-body diagram and for moment equilibrium, we find the vertical reaction = $(10)(4.5)/12$ = 3.75 kN, acting as shown in Fig. 8.12b. Thus the moments at column ends A and D can be found from the moment equilibrium of columns AE and DF as shown in Fig. 8.12c. Finally, the BMD, SFD, and AFD are drawn as shown in Fig. 8.12d to f. Comparing Figs. 8.11b and 8.12d, it is seen that by fixing the bases, the maximum moments in the columns and beam are reduced

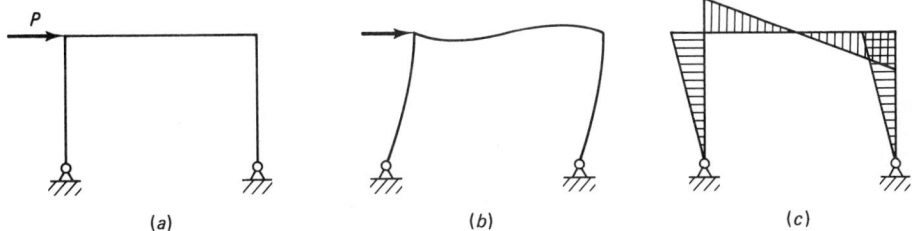

Figure 8.9 A bridge portal (hinged base). (b) Deflected shape. (c) Bending-moment diagram drawn on compression side.

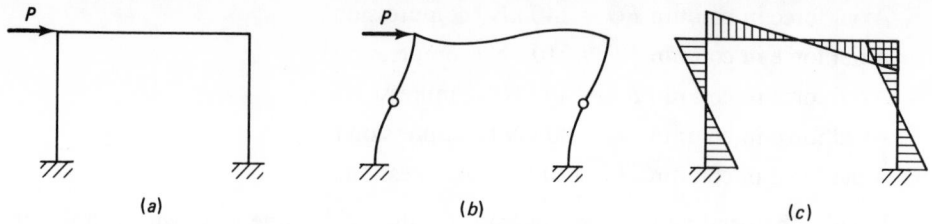

Figure 8.10 A bridge portal (fixed base). (b) Deflected shape. (c) BMD drawn on compression side.

by 50 percent. It should be mentioned, however, that fixed-ended column base connections are more elaborate and must be designed properly to transfer the moment to the supporting foundation.

Quite often the bridge portals have a structural form shown in Fig. 8.13a. Columns ABC and DEF are continuous members designed to resist axial and shearing forces as well as moments. Since the structure is indeterminate to the third degree, the following three assumptions are used to analyze it by statics alone: The horizontal force is resisted equally by the two columns; there is a point of inflection at the midpoints of AB and DE. The following steps are used to analyze the portal frame of Fig. 8.13a.

Step 1. Determine the horizontal reactions at A and D (5 kN in this case).

Step 2. Locate points of inflection J and K in columns ABC and DEF as shown.

Step 3. By considering the moment equilibrium of the structure above JK, determine the vertical reactions at J and K, that is, $V_J = -V_K = (10)(6)/12 = 5$ kN, as shown in Fig. 8.13b.

Step 4. From the free-body diagrams of AJ and DK (Fig. 8.13c), compute the moments at A and D.

Step 5. From the free-body diagrams of columns ABC and DEF (Fig. 8.13d), determine the horizontal reactions at the truss connections, that is, at B, C, E, and F. Thus, for *column ABC*: From $\sum M$ about $C = 0$:

$$(5)(9) - 15 + (H_B)(3) = 0 \quad \text{or} \quad H_B = -10 \text{ kN}$$

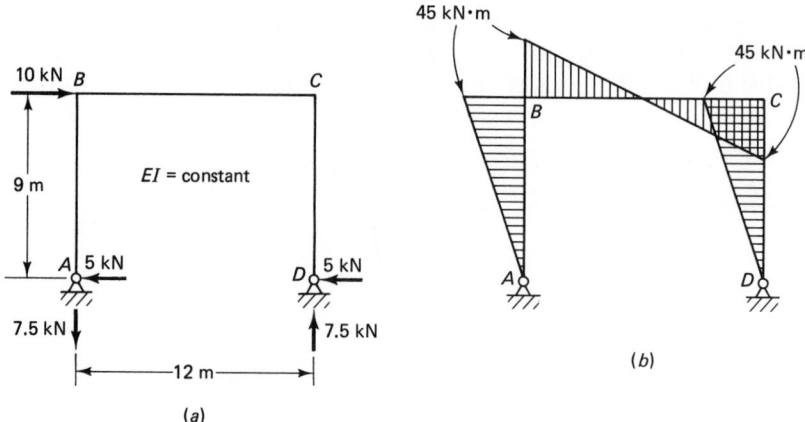

Figure 8.11 A portal frame (hinged-base) subjected to horizontal load.

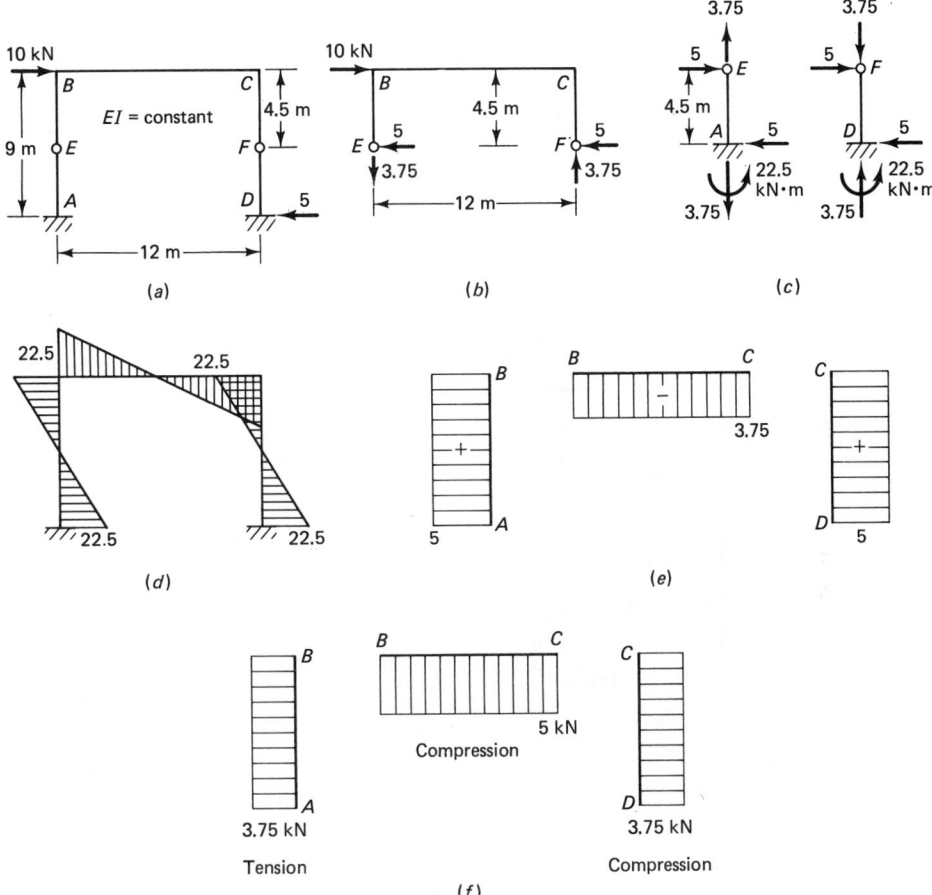

Figure 8.12 A portal frame (fixed base) subjected to horizontal load. (b) Vertical reactions. (c) End moments at A and D. (d) BMD drawn on compression side. (e) SFD. (f) AFD.

From $\Sigma H = 0$:

$$10 - H_C - H_B - 5 = 0 \quad \text{or} \quad 10 - H_C - (-10) - 5 = 0$$

hence $H_C = 15$ kN

For *column DEF*: From ΣM about $F = 0$, $H_E = 10$ kN, and from $\Sigma H = 0$, $H_F = -5$ kN. The correct directions of the horizontal reactions at B, C, E, and F are shown in Fig. 8.13e.

Step 6. Considering the geometry of the truss, force equilibrium of joints B and E of the truss will yield the vertical force component at $B = (10)/2 = 5$ kN↓, and that at $E = (10)/2 = 5$ kN↑. The reactions on the columns ABC and DEF will be equal and opposite as shown in Fig. 8.13f.

Step 7. From the vertical force equilibrium of columns ABC and DEF (Fig. 8.13f), determine $V_C = 0$ and $V_F = 0$.

Step 8. Compute the bar forces in the truss in the usual manner, knowing the reactions at B, C, D, and E (Fig. 8.13g).

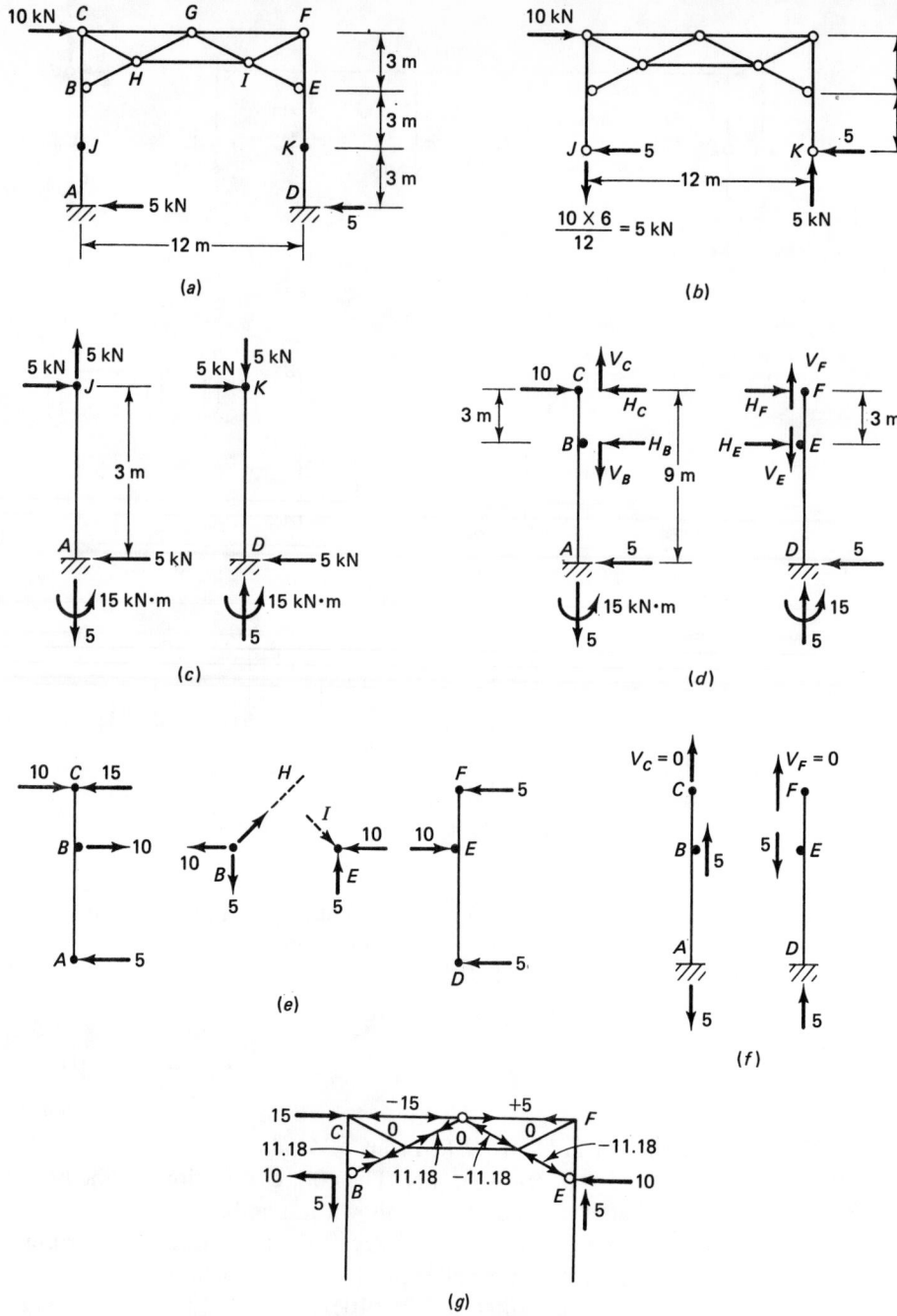

Figure 8.13 A trussed portal. (*a*) Bridge portal truss. (*b*) Vertical reactions. (*c*) Moments at ends *A* and *D*. (*d*) Assumed horizontal and vertical reactions at truss connection points. (*e*) Horizontal reactions at truss connection points (vertical reactions are not shown for clarity). (*f*) Vertical reactions at truss connection points (horizontal reactions are omitted for clarity). (*g*) Forces in truss members. (*h*) Axial-force, shear-force, and bending-moment diagrams: (i) axial-force diagram; (ii) shear-force diagram; (iii) bending-moment diagram.

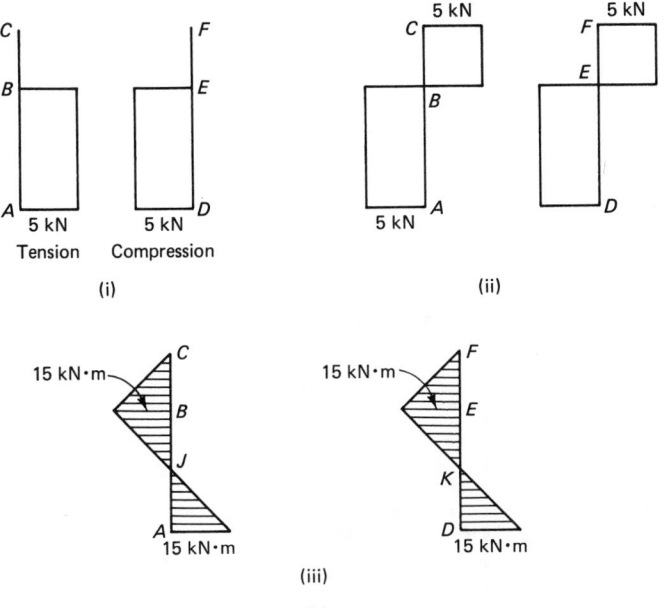

Figure 8.13 (*Continued*)

Step 9. Draw the axial-force, shear-force, and bending-moment diagrams for the columns *ABC* and *DEF* (Fig. 8.13*h*).

Mill bents with knee braces, such as the one shown in Fig. 8.14, are analyzed in a manner similar to the bridge portal truss in Fig. 8.13*a*.

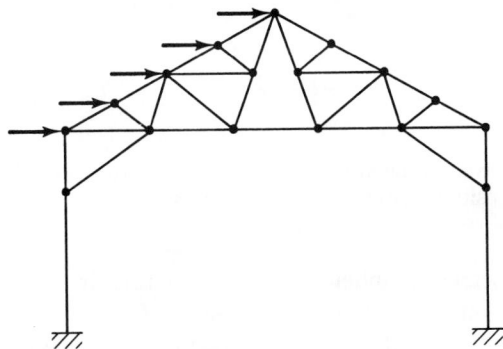

Figure 8.14 Mill bent with knee braces.

8.5 SOLVED PROBLEM

■ SOLVED PROBLEM 8.1

Determine the horizontal shears in all the columns of the four-bay one-story portal rigid frame shown in Fig. 8.15*a*.

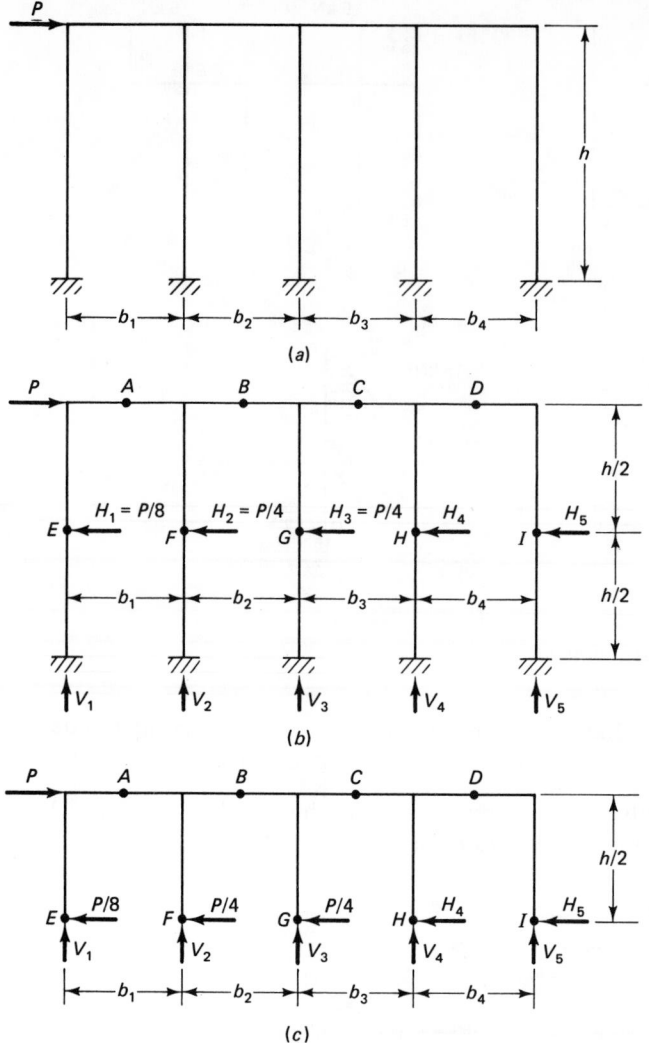

Figure 8.15 Structure for Solved Problem 8.1. (*a*) Four-bay one-story portal rigid frame. (*b*) Frame with assumed hinge positions. (*c*) Column axial forces.

Solution. The frame has $(4 \times 3) = 12$ degrees of indeterminacy. Therefore, in order to reduce this structure to a determinate one, we assume the following: (1) points of inflection in the middle of the four beams and the five columns, as shown in Fig. 8.15*b*; (2) shear in exterior column $= P/2n = P/8$, where $n =$ number of bays; (3) shear in each of the two adjacent interior columns $= P/n = P/4$. The above assumptions are $(4 + 5 + 1 + 2) = 12$ in number and equal to the degree of indeterminacy of the structure. Using Fig. 8.15*c*, bending moment at $A = 0$ (considering forces to the left of A):

$$(V_1)\left(\frac{b_1}{2}\right) + \left(\frac{P}{8}\right)\left(\frac{h}{2}\right) = 0$$

or

$$V_1 = \frac{-Ph}{8b_1}$$
(a)

Bending moment at $B = 0$ (considering forces to the left of B):

$$V_1\left(b_1 + \frac{b_2}{2}\right) + V_2\frac{b_2}{2} + \left(\frac{P}{8} + \frac{P}{4}\right)\frac{h}{2} = 0$$

Substituting for V_1 from equation (a) yields

$$V_2 = \frac{Ph}{8b_2}\frac{b_2 - b_1}{b_1}$$
(b)

Bending moment at $C = 0$ (considering forces to the left of C):

$$V_1\left(b_1 + b_2 + \frac{b_3}{2}\right) + V_2\left(b_2 + \frac{b_3}{2}\right) + V_3\left(\frac{b_3}{2}\right) + \left(\frac{P}{8} + \frac{P}{4} + \frac{P}{4}\right)\frac{h}{2} = 0$$

Using equations (a) and (b) and simplifying yields

$$V_3 = \frac{Ph}{8b_3}\frac{b_3 - b_2}{b_2}$$
(c)

Bending moment at $D = 0$ (considering forces to the right of D):

$$(V_5)\left(\frac{b_4}{2}\right) - H_5\frac{h}{2} = 0$$

or

$$V_5 = H_5\frac{h}{b_4}$$
(d)

Σ moments of all forces at $I = 0$:

$$\frac{Ph}{2} + V_1(b_1 + b_2 + b_3 + b_4) + V_2(b_2 + b_3 + b_4) + V_3(b_3 + b_4) + V_4(b_4) = 0$$

Substituting for V_1, V_2, and V_3 and simplifying lead to

$$V_4 = \frac{Ph}{8b_4}\frac{b_4 - b_3}{b_3}$$
(e)

Now Σ vertical forces $= 0$, or

$$V_1 + V_2 + V_3 + V_4 + V_5 = 0$$
(f)

Using equations (a), (b), (c), (d), and (e) in equation (f) we find

$$H_5 = \frac{P}{8}$$
(g)

Furthermore, Σ horizontal forces $= 0$ (Fig. 8.15c), that is,

$$P - \left(\frac{P}{8} + \frac{P}{4} + \frac{P}{4} + H_4 + H_5\right) = 0$$
(h)

Knowing H_5, equation (h) yields

$$H_4 = \frac{P}{4} \tag{i}$$

This solved problem shows that the horizontal shears carried by the remaining interior and exterior columns are also P/n and $P/2n$, respectively, where $n \, (= 4$ in this case) is the number of bays. ∎

PROBLEMS

8.1 to 8.5 Assume suitable points of inflection and draw the approximate bending-moment diagram for the statically indeterminate beams shown.

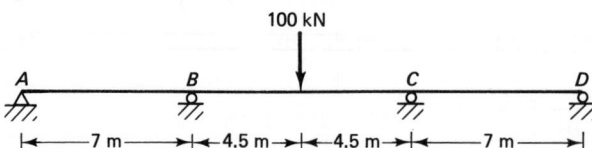

Figure P8.1

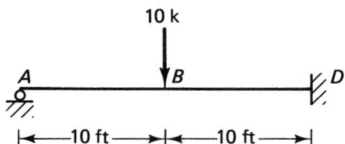

Figure P8.2

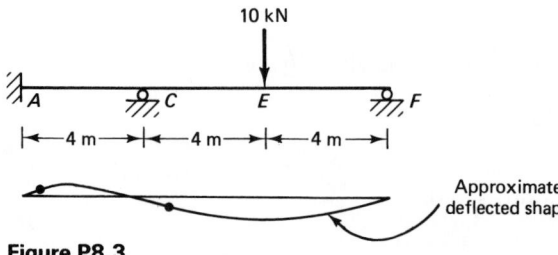

Figure P8.3

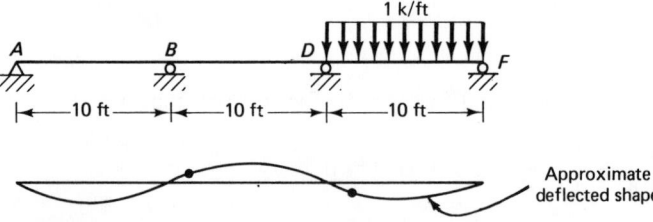

Figure P8.4

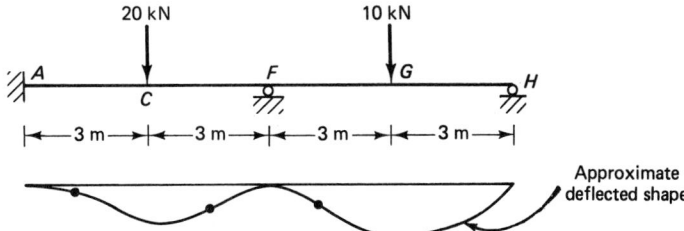

Figure P8.5

8.6 Determine the forces in all the members of the indeterminate truss assuming **(a)** the diagonals share the wind load equally, and **(b)** diagonals in tension only carry the wind load.

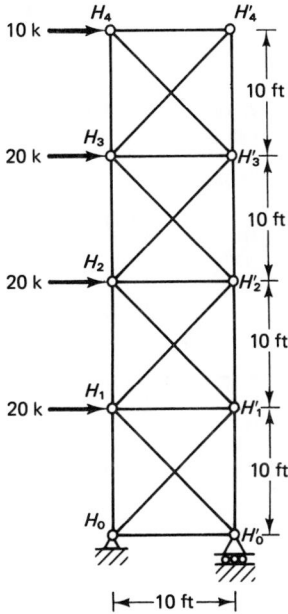

Figure P8.6

8.7 Assuming that the arch rib carries no bending moment (therefore, it retains its parabolic shape), and hangers are inextensible, determine the horizontal thrust in the two-hinged parabolic stiffened arch. Draw also the shear-force and bending-moment diagrams for the girder *ABC*. (*Hint:* Hanger loads are equal; $H = wl^2/8h$.)

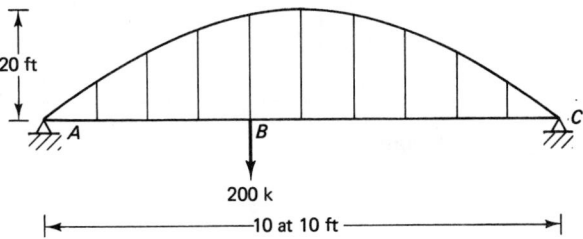

Figure P8.7

8.8 Assuming that the horizontal reactions are one-half of those of a three-hinged arch, draw axial-force, shear-force, and bending-moment diagrams for the two-hinged arch shown.

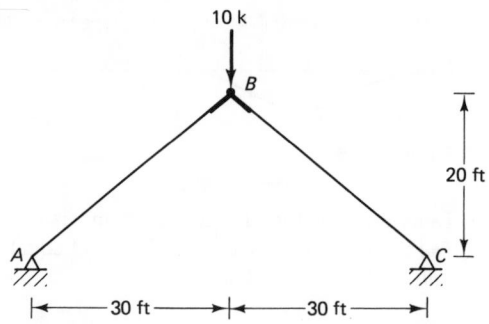

Figure P8.8

8.9 Draw the axial-force, shear-force, and bending-moment diagrams for the two-hinged portal frame shown.

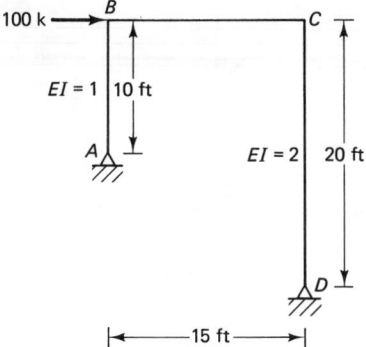

Figure P8.9

8.10 Draw axial-force, shear-force, and bending-moment diagrams for the columns of the mill bent shown.

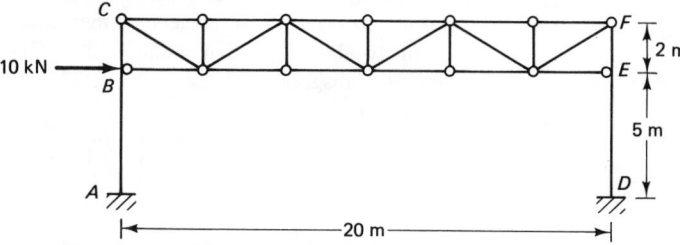

Figure P8.10

8.11 Use the portal method to determine axial forces, shear forces, and bending moments in the vierendeel truss shown.

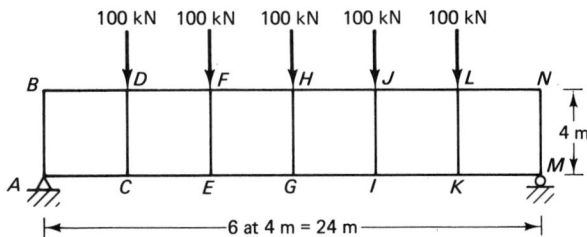

Figure P8.11

8.12 and 8.13 Determine the shears, bending moments, and axial forces in the rigid frames shown using the portal method.

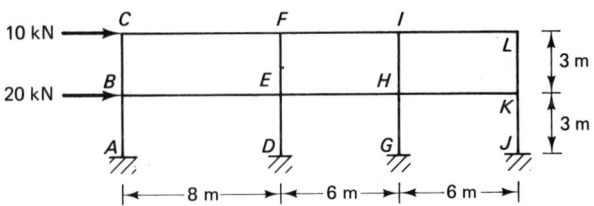

Figure P8.12

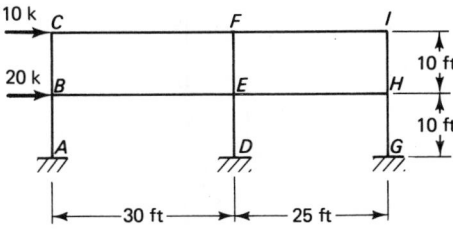

Figure P8.13

8.14 to 8.16 Determine the shears, bending moments, and axial forces in the rigid frames shown using the cantilever method.

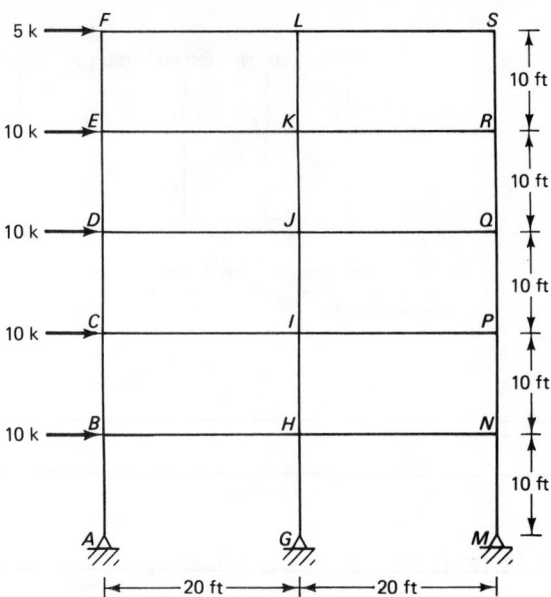

Figure P8.14

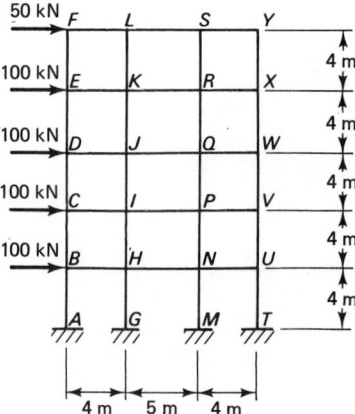

Figure P8.15

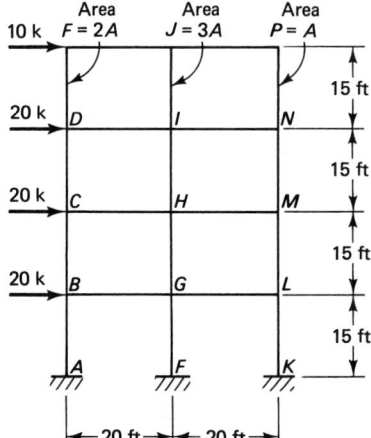

Figure P8.16

8.17 and 8.18 Determine the shears, bending moments, and axial forces in the rigid frames shown using **(a)** the portal method and **(b)** the cantilever method.

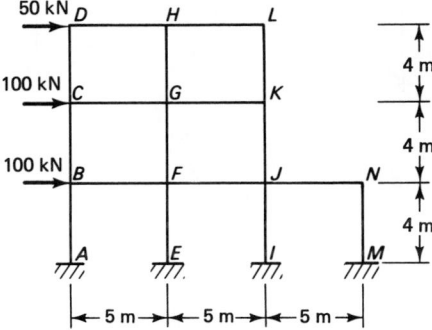

Figure P8.17

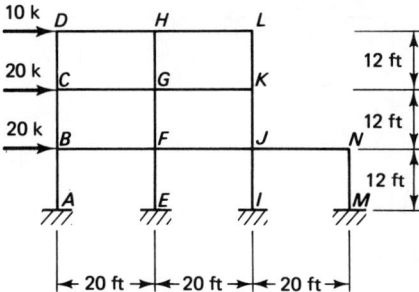

Figure P8.18

8.19 to 8.21 Making suitable assumptions, analyze the industrial mill bents shown.

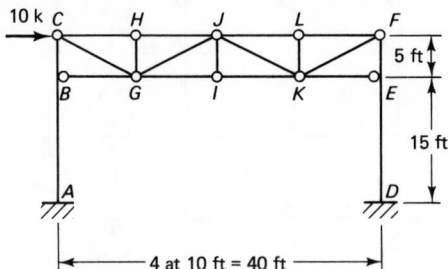

Figure P8.19

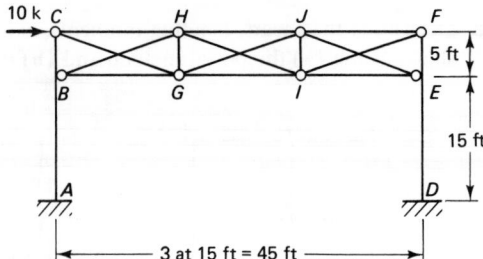

Figure P8.20

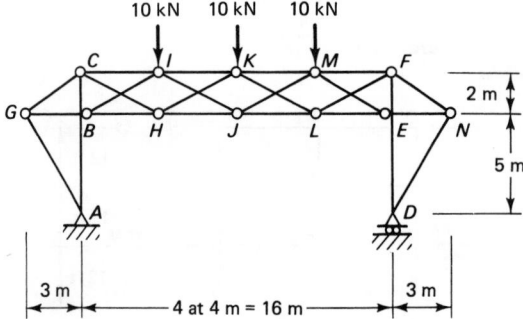

Figure P8.21

8.22 and 8.23 Make an approximate analysis of the vierendeel girders shown.

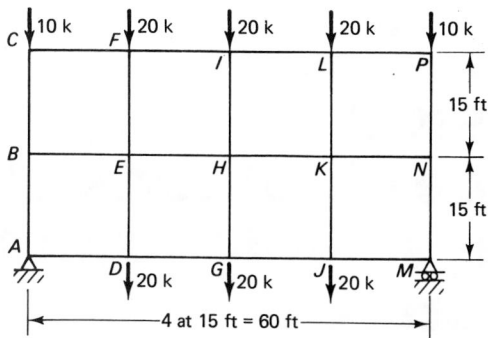

Figure P8.22

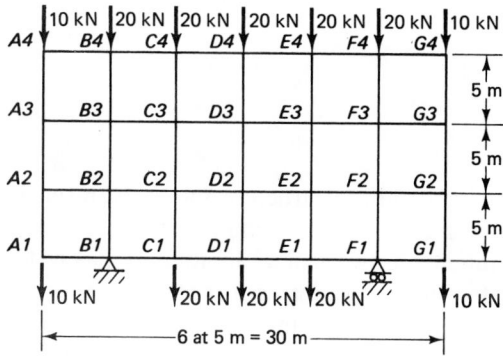

Figure P8.23

Chapter 9

Classical Compatibility (Force) Methods of Analysis—Indeterminate Structures

To know the road ahead ask those coming back.

Chinese Proverb

In analyzing an indeterminate structure, the structural engineer must arrive at a solution that satisfies the requirements of equilibrium and compatibility (or geometry). To meet these requirements two general approaches may be followed. In the first approach, a statically determinate solution is first assumed satisfying the equilibrium requirements; the geometrical errors incurred by such an assumption are then corrected without disturbing the equilibrium conditions by using compatibility equations—hence the name *compatibility approach.* Furthermore, the relations between forces and deformations are expressed such that deformations are in terms of forces and hence forces become the independent variables or unknowns to be determined. This is why such a method of analysis is referred to also as the *force (or flexibility) method* in which the relations between the forces and deformations are expressed through *flexibility coefficients;* the methods of three-moment equation, of consistent deformation, and of least work are classified as force methods. In the second approach, a solution is assumed satisfying the geometrical requirements; the resulting errors in statical equilibrium are then corrected using equilibrium equations without disturbing the geometrical conditions—hence the name *equilibrium approach.* Here forces are expressed in terms of deformations by means of *stiffness coefficients* and the deformations become the independent variables or unknowns in the problem. This method of analysis is also denoted as the *displacement (or stiffness) approach;* the methods of slope deflection and moment distribution are classified as displacement methods. In this chapter we discuss the force methods—the methods of three-moment equation and of consistent deformation.

9.1 THE THREE-MOMENT EQUATION

9.1.1 General

The basis of the three-moment method, developed by Clapeyron in 1849, is the following continuity requirement over a support: The slopes of the elastic curve at a common support of two spans of a continuous beam must be equal. This is shown in Fig. 9.1a. The dashed line shows an exaggerated assumed elastic curve; because the beam is continuous, the elastic curve over an interior support is a smooth curve, that is, with no discontinuity. For example, for support C in Fig. 9.1,

$$\theta_1 = \theta_2 \tag{9.1}$$

If we cut a continuous beam at the supports (Fig. 9.1b), we obtain a series of simply supported beams, which are loaded with the transverse loads as well as subjected to support moments as shown in Fig. 9.1b. The magnitude and direction of these support moments must be such that the continuity of the elastic curve is preserved. Thus, the moments at the supports are the unknowns, or redundants, in this case. Once these are determined, the reactions can be calculated and the shear-force and bending-moment diagrams can be drawn.

To develop the three-moment equation, let us consider two adjacent spans from a continuous beam shown in Fig. 9.2a. First, we cut the beam at supports A, B, and C, as shown in Fig. 9.2b, and then apply the continuity condition for the slope at B. This gives us a relationship between the moments M_A, M_B, and M_C. By considering two adjacent spans at a time, and applying the same procedure, we are able to develop

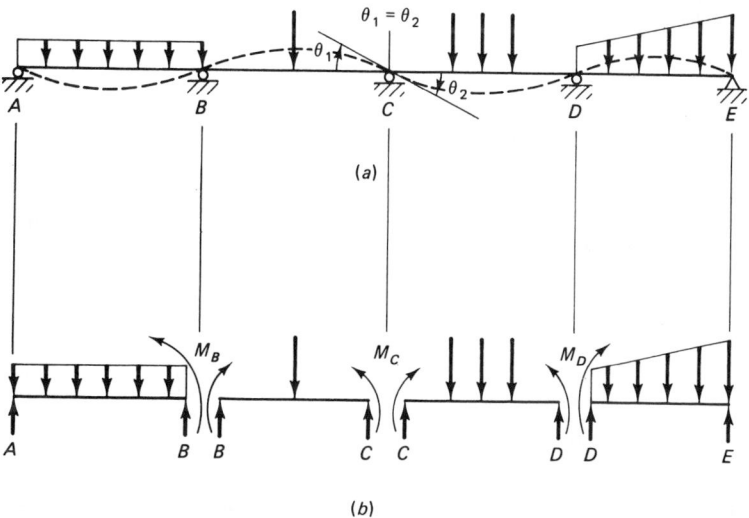

(a)

(b)

Figure 9.1 A four-span continuous beam showing (a) elastic curve; (b) support moments and shears.

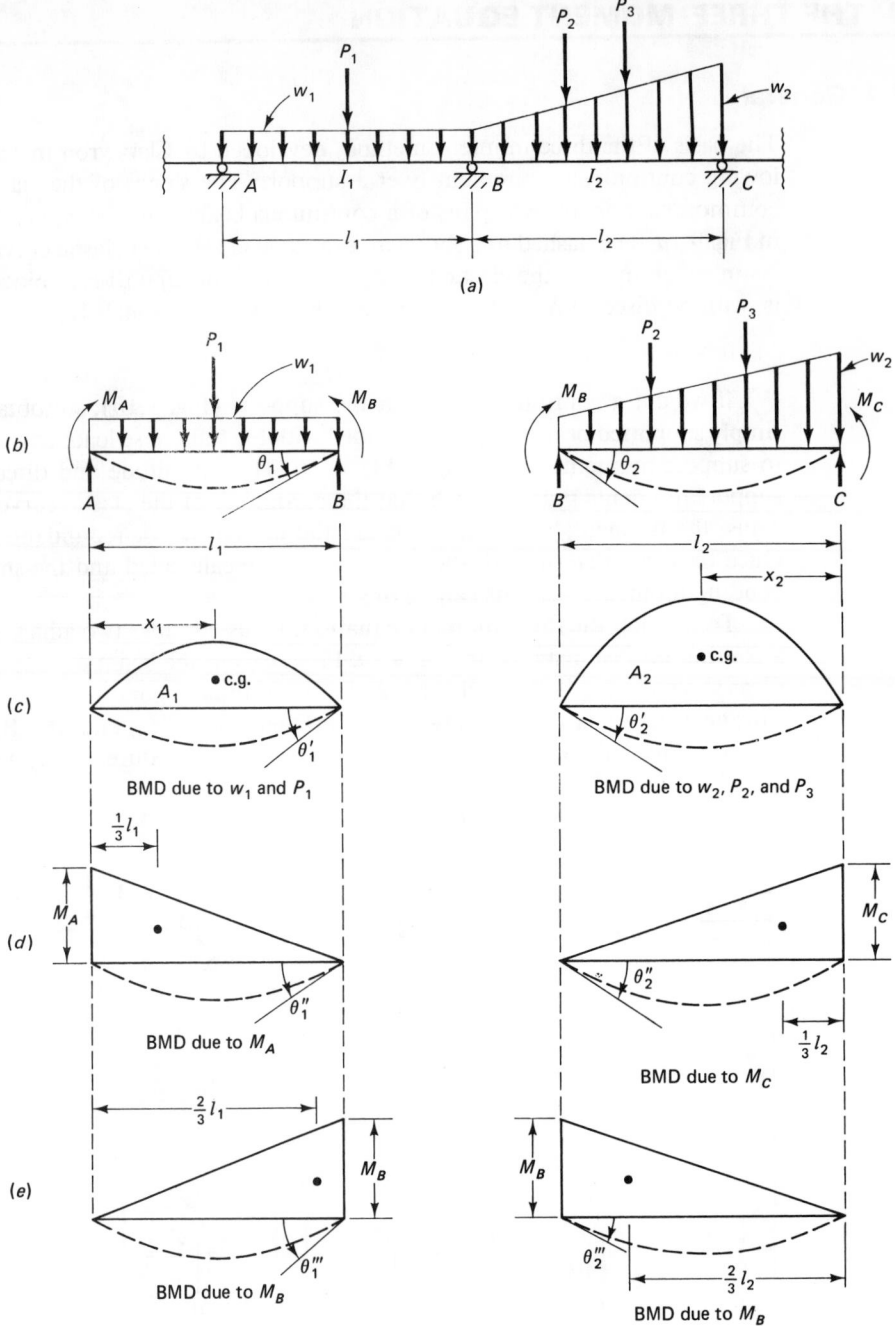

Figure 9.2 Development of three-moment equation considering two adjacent spans.

the necessary number of equations to determine the unknown moments at the supports. It should be mentioned that support moments are assumed positive in the sense shown in Fig. 9.2b; if a negative sign is obtained for one of the moments, it simply means a reversal of direction and that this moment is producing a hogging effect on the beam.

We begin first by finding the slopes θ_1 and θ_2 (Fig. 9.2b) by the method of superposition. Using the conjugate-beam method for calculating slopes, for the bending-moment diagrams shown in Fig. 9.2, we obtain

For Span *AB*

$$\theta_1 = \theta_1' + \theta_1'' + \theta_1''' \tag{9.2}$$

Using the conjugate-beam method (Chapter 7):
From Fig. 9.2c

$$\theta_1' = \frac{1}{l_1}\frac{A_1 x_1}{EI_1}$$

From Fig. 9.2d

$$\theta_1'' = \frac{1}{l_1}\left[\left(\frac{M_A}{EI_1}\right)\left(\frac{l_1}{2}\right)\left(\frac{1}{3}l_1\right)\right] = \frac{M_A l_1}{6EI_1} \tag{9.3}$$

From Fig. 9.2e

$$\theta_1''' = \frac{1}{l_1}\left[\left(\frac{M_B}{EI_1}\right)\left(\frac{l_1}{2}\right)\left(\frac{2}{3}l_1\right)\right] = \frac{M_B l_1}{3EI_1}$$

Referring to Fig. 9.2c, it should be noted that A_1 and A_2 represent the areas of the bending-moment diagrams drawn based on the assumption that each of the spans of the beam is *simply supported,* and x_1 and x_2 designate the distances of the centroids of each of these moment diagrams from A and C, respectively.

Substituting Eq. (9.3) in Eq. (9.2) yields

$$\theta_1 = \frac{A_1 x_1}{EI_1 l_1} + \frac{M_A l_1}{6EI_1} + \frac{M_B l_1}{3EI_1} \quad \left. \right) \quad = -\left(\frac{A_1 x_1}{EI_1 l_1} + \frac{M_A l_1}{6EI_1} + \frac{M_B l_1}{3EI_1}\right) \left. \right) \tag{9.4}$$

For Span BC

$$\theta_2 = \theta_2' + \theta_2'' + \theta_2''' \tag{9.5}$$

Following the same procedure as above, we can write

$$\theta_2' = \frac{1}{l_2}\frac{A_2 x_2}{EI_2}$$

$$\theta_2'' = \frac{M_C l_2}{6EI_2} \tag{9.6}$$

$$\theta_2''' = \frac{M_B l_2}{3EI_2}$$

Thus the slope θ_2 becomes

$$\theta_2 = \frac{A_2 x_2}{EI_2 l_2} + \frac{M_C l_2}{6EI_2} + \frac{M_B l_2}{3EI_2} \text{ }\Big)$$ (9.7)

Substituting for θ_1 and θ_2 in Eq. (9.1) yields, after rearranging,

$$M_A \frac{l_1}{I_1} + 2M_B \left(\frac{l_1}{I_1} + \frac{l_2}{I_2} \right) + M_C \frac{l_2}{I_2} = - \frac{6A_1 x_1}{I_1 l_1} - \frac{6A_2 x_2}{I_2 l_2}$$ (9.8)

which is the *three-moment equation;* this equation will be adjusted later to account for support settlements. When the moment of inertia I is constant for the entire continuous beam, that is, $I_1 = I_2 = \cdots = I_n$, then Eq. (9.8) reduces to

$$M_A l_1 + 2M_B (l_1 + l_2) + M_C l_2 = - \frac{6A_1 x_1}{l_1} - \frac{6A_2 x_2}{l_2}$$ (9.9)

The procedure is to take two adjacent spans and write a three-moment equation for each pair until we have as many three-moment equations as the number of redundant (unknown) support moments. The solution of these equations will yield the unknown support moments. The terms $6A_1 x_1 / l_1$ and $6A_2 x_2 / l_2$ for some typically loaded simply supported beams are given in Table 9.1. We demonstrate the method by applying the three-moment equation to the loaded continuous beam shown in Fig. 9.3a.

■ **EXAMPLE 9.1**

Find the support moments for the continuous beam $ABCD$ loaded as shown in Fig. 9.3a. Hence sketch the shear-force and bending-moment diagrams.

Solution. The beam is statically indeterminate to the second degree. Therefore, we must generate two three-moment equations. The first equation is generated using spans AB and BC (Fig. 9.3b, c, and d). Since EI is constant, we use Eq. (9.9).

$$M_A l + 2M_B (l + l) + M_C l = - \frac{6A_1 x_1}{l_1} - \frac{6A_2 x_2}{l_2}$$

From Fig. 9.3d:

$$\frac{6A_1 x_1}{l_1} = \frac{wl^3}{4} \quad \text{and} \quad \frac{6A_2 x_2}{l_2} = \frac{wl^3}{4}$$

(See also Table 9.1.) Since $M_A = 0$, the above equation becomes

$$2M_B (2l) + M_C l = - \frac{wl^3}{4} - \frac{wl^3}{4}$$

or

$$4M_B + M_C = - \frac{wl^2}{2}$$ (a)

The second equation is generated by considering spans BC and CD (Fig. 9.3e, f, and g).

TABLE 9.1 TERMS $\dfrac{6Ax_1}{l}$ AND $\dfrac{6Ax_2}{l}$ FOR SOME TYPICALLY LOADED SIMPLE SPANS

Loaded simply supported beam	$\dfrac{6Ax_1}{l}$	$\dfrac{6Ax_2}{l}$
w/unit length	$\dfrac{wl^3}{4}$	$\dfrac{wl^3}{4}$
	$\dfrac{Pa(l^2 - a^2)}{l}$	$\dfrac{Pb(l^2 - b^2)}{l}$
	$\dfrac{8}{60}\,wl^3$	$\dfrac{7}{60}\,wl^3$
	$-\dfrac{M}{l}(3a^2 - l^2)$	$\dfrac{M}{l}(3b^2 - l^2)$
	$\dfrac{5}{32}\,wl^3$	$\dfrac{5}{32}\,wl^3$

(Note that when applying successively the three-moment equation, we keep one span common as we proceed from one step to the next; in this case *BC* is the common span.)

$$M_B(l) + 2M_C(l + l) + M_D(l) = -\frac{6A_1x_1}{l_1} - \frac{6A_2x_2}{l_2}$$

Since $M_D = 0$, and from Fig. 9.3 *e, f,* and *g,*

$$\frac{6A_1x_1}{l_1} = \frac{6A_2x_2}{l_2} = \frac{wl^3}{4}$$

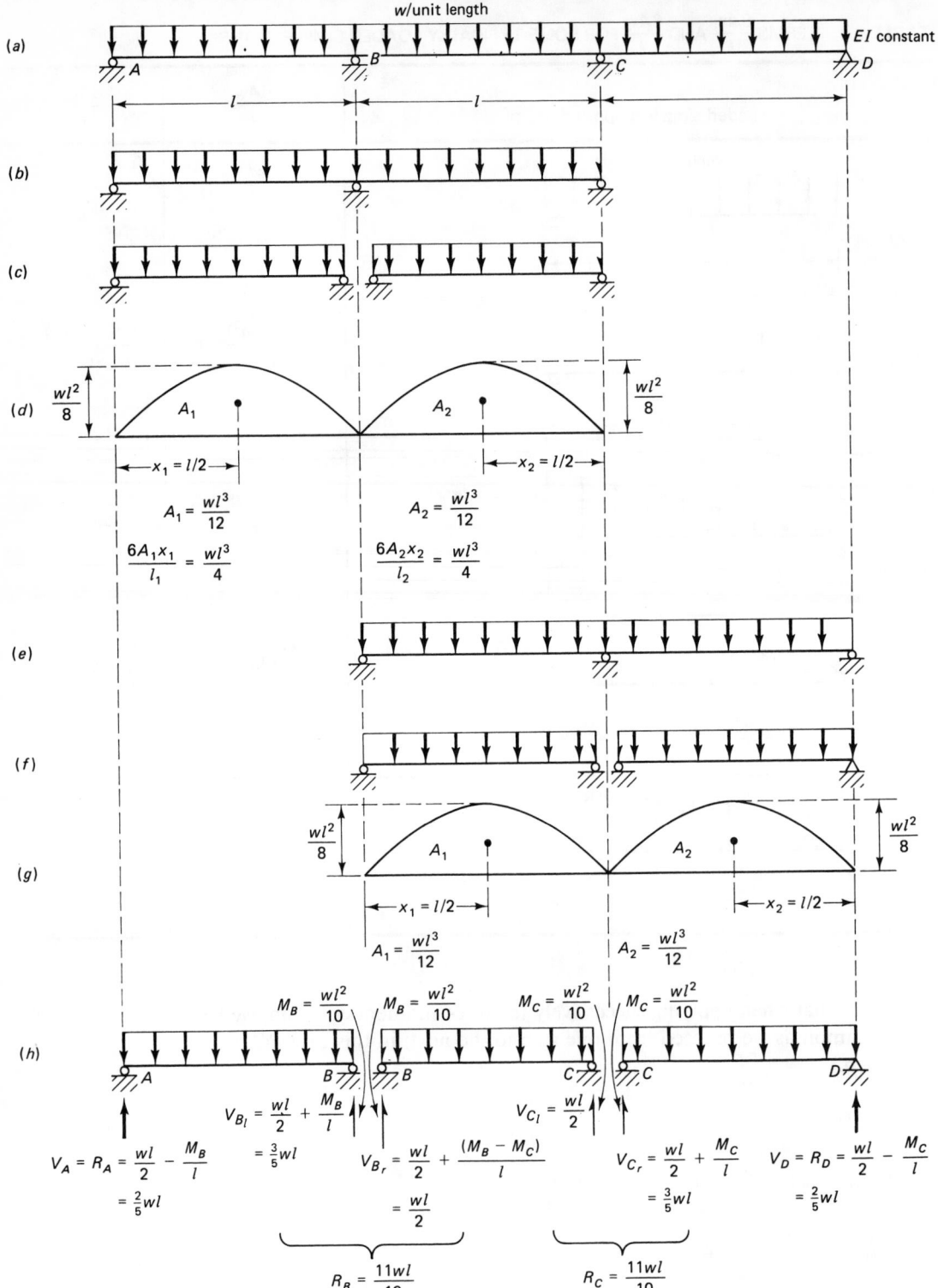

Figure 9.3 Structure for Example 9.1.

the above equation reduces to

$$M_B(l) + 2M_C(2l) = -\frac{wl^3}{4} - \frac{wl^3}{4}$$

or

$$M_B + 4M_C = -\frac{wl^2}{2} \tag{b}$$

Solving equations (a) and (b), we get

$$M_B = M_C = -\frac{wl^2}{10}, \text{ that is negative moment producing tension at top.}$$

These support moments are shown on the free-body diagrams in Fig. 9.3h. The reactions are

$$R_A = \tfrac{2}{5}wl\uparrow \qquad R_B = \tfrac{11}{10}wl\uparrow \qquad R_C = \tfrac{11}{10}wl\uparrow \quad \text{and} \quad R_D = \tfrac{2}{5}wl\uparrow$$

Check.

$$\sum Fy = 0\downarrow^{+}: \qquad (w)(3l) - \tfrac{2}{5}wl - \tfrac{11}{10}wl - \tfrac{11}{10}wl - \tfrac{2}{5}wl = 0$$

Therefore, the solution of the reactions is correct. With the redundant support moments being known, the shear-force and bending-moment diagrams can be drawn by super-position as shown in Fig. 9.4. Figure 9.4b shows the reactions for simple spans loaded as shown (with no end moments), while Fig. 9.4c gives the reactions due to the support moments applied at the ends of the simple spans. Superimposing Fig. 9.4b on Fig. 9.4c results in free-body diagrams of the three individual spans, shown in Fig. 9.4d and hence the SFD shown in Fig. 9.4e. The bending-moment diagrams for the loaded simple spans (Fig. 9.4f) are superimposed on the bending-moment diagrams due to support moments (Fig. 9.4g), to yield the BMD for the loaded continuous beam as shown in Fig. 9.4h. Drawing the BMD from a horizontal baseline yields Fig. 9.4i. ∎

■ EXAMPLE 9.2
(a) Determine the term $6Ax/l$ for the simply supported beam loaded as shown in Fig. 9.5a. (b) Determine the fixed-end moments for the fixed-ended beam loaded as shown in Fig. 9.5d.

Solution

(a) For the beam in Fig. 9.5a, the reaction at A is found by taking moments about B. Therefore, from $\big(\ +\sum M_B = 0$,

$$\frac{wl}{2}\left(\frac{l}{2}\right) - R_A l = 0$$

Therefore,

$$R_A = \frac{wl}{4}$$

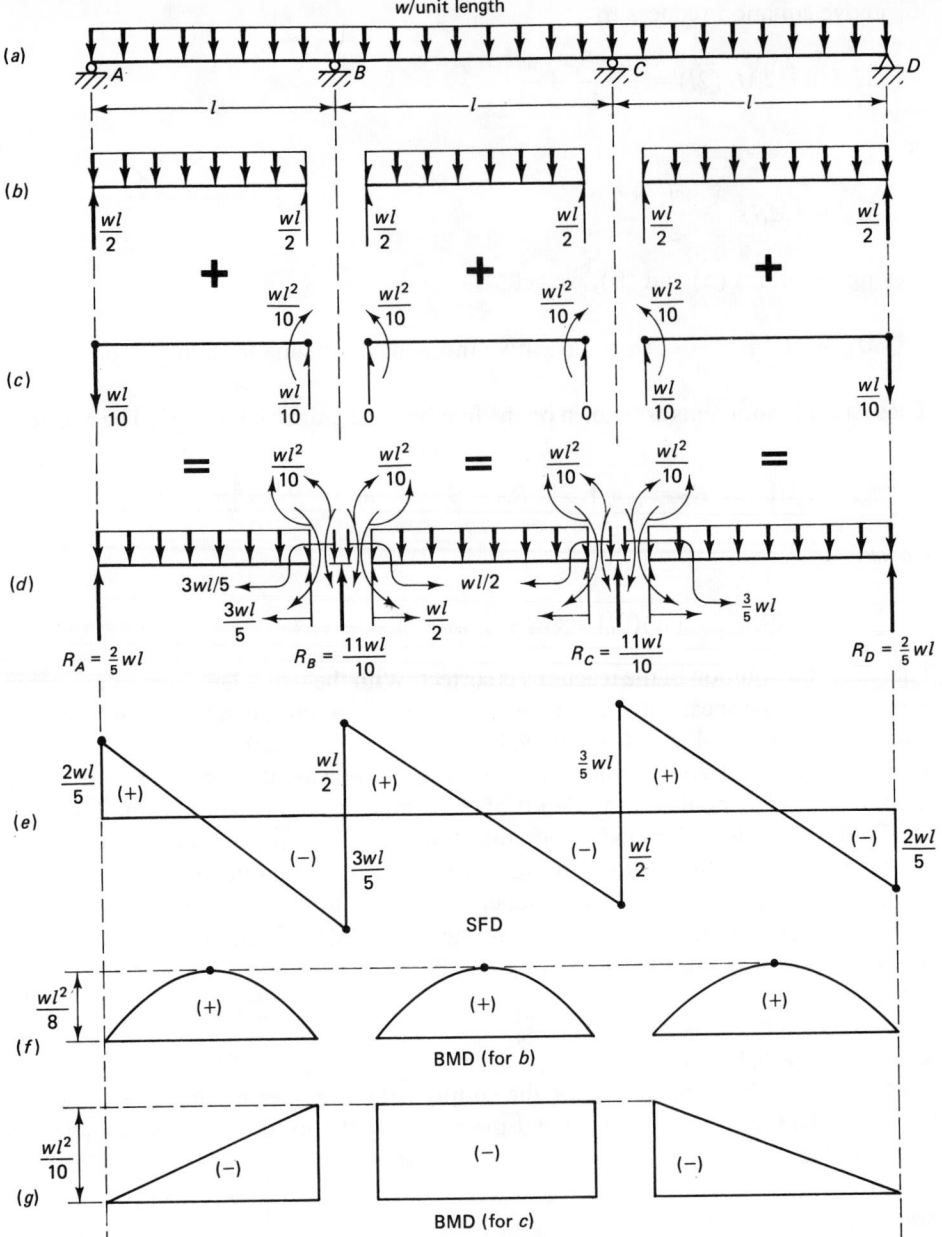

Figure 9.4 Development of shear-force and bending-moment diagrams for structure in Example 9.1.

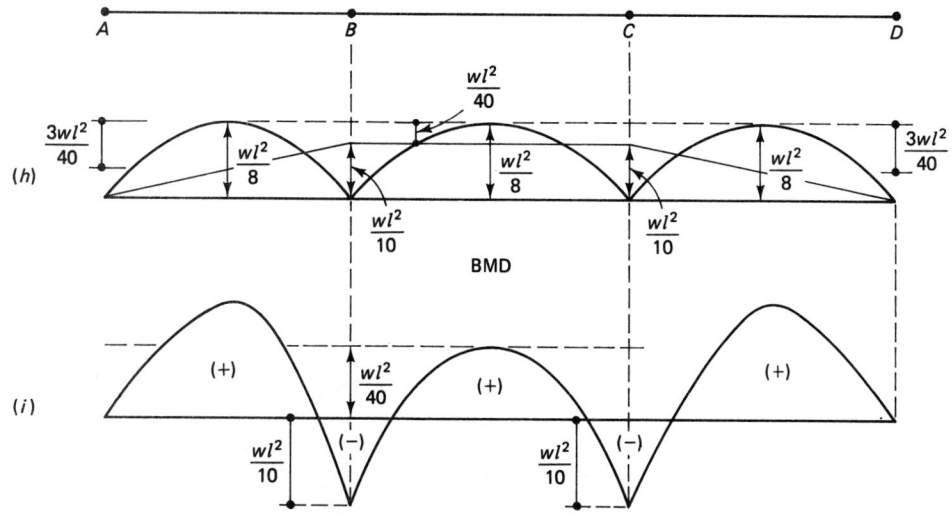

Figure 9.4 (*Continued*)

From symmetry,

$$R_B = \frac{wl}{4}$$

The bending moment at any section x from A is

$$M_x = R_A(x) - \left(\frac{wx}{l/2}\right)\left(\frac{x}{2}\right)\left(\frac{x}{3}\right)$$

or

$$M_x = wx\left(\frac{l}{4} - \frac{x^2}{3l}\right)$$

This is shown plotted in Fig. 9.5c. Since the load on the beam is discontinuous at $x = l/2$, the BMD, which is a third-degree parabola, will be discontinuous at $x = l/2$. Therefore,

Ax_1 = moment of area of BMD about left support

$$= \int_0^{l/2}\left[wx\left(\frac{l}{4} - \frac{x^2}{3l}\right)\right]dx(x) + \int_0^{l/2}\left[wx\left(\frac{l}{4} - \frac{x^2}{3l}\right)\right]dx(l - x) = \frac{5}{192}\,wl^4$$

The above integration limits were necessary because of the discontinuity of the moment diagram at $x = l/2$. Because of symmetry, $6Ax_1/l = 6Ax_2/l$. Therefore,

$$\frac{6Ax_1}{l} = \frac{6Ax_2}{l} = \frac{5}{32}\,wl^3$$

(b) The beam in Fig. 9.5d is statically indeterminate to the second degree and therefore requires two three-moment equations. To apply the three-moment equation for a beam with a fixed end, an imaginary span to the left of the fixed support A, having an arbitrary length l' and moment of inertia $I' = \infty$ is assumed; the same is

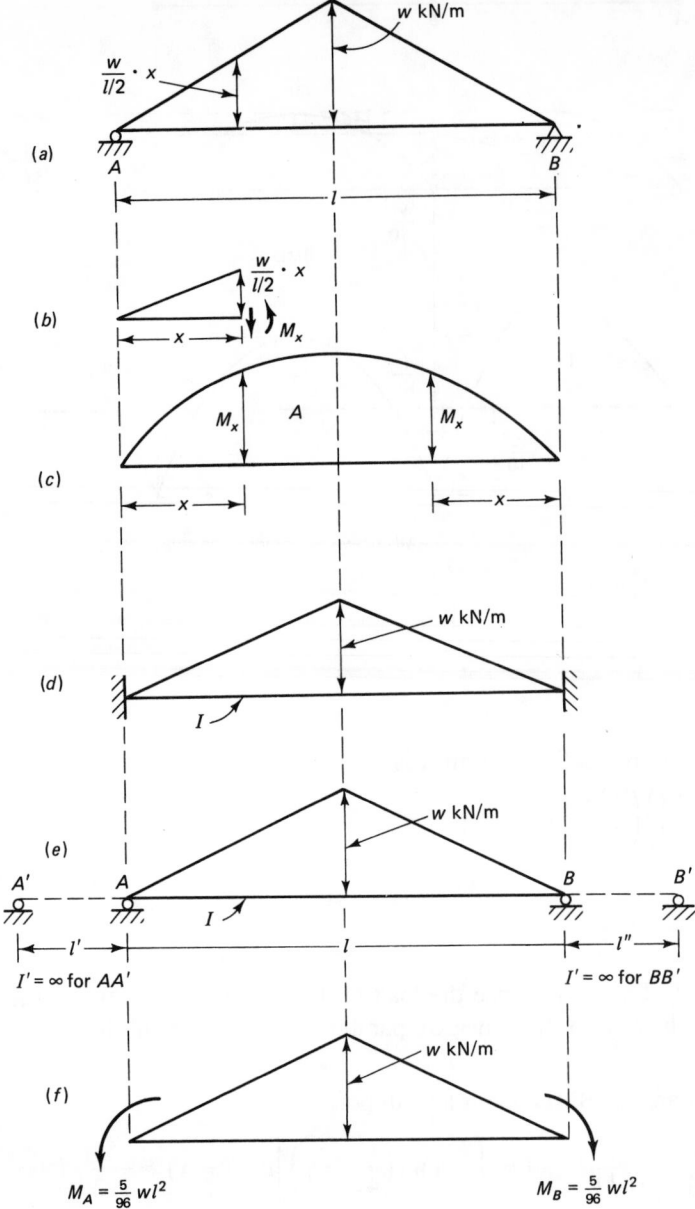

Figure 9.5 Structure for Example 9.2.

applied to the right-hand side of the beam, as shown in Fig. 9.5 e. Thus, for spans $A'A$ and AB, applying Eq. (9.8) we find

$$M_{A'} \cancelto{0}{\frac{l'}{\infty}} + 2M_A \left(\cancelto{0}{\frac{l'}{\infty}} + \frac{l}{I} \right) + M_B \left(\frac{l}{I} \right) = -\frac{6A_1 \cancelto{0}{x_1}}{(\infty) l'} - \frac{6A_2 x_2}{Il}$$

or

$$2M_A l + M_B l = -\frac{6A_2 x_2}{l} = -\frac{5}{32} w l^3 \qquad \text{(from Table 9.1)} \qquad \text{(a)}$$

Similarly, for spans AB and BB', we obtain

$$M_A l + 2M_B l = -\tfrac{5}{32} w l^3 \qquad \text{(b)}$$

Solving (a) and (b) yields

$$M_A = M_B = -\tfrac{5}{96} w l^2$$

The negative sign means that the direction of these support moments should be reversed as shown in Fig. 9.5f, causing tension at the top fibers of the beam.

It should be noted that the above equations (a) and (b) could have been directly deduced from Eqs. (9.4) and (9.7). Thus, for Eq. (9.7), putting $\theta_2 = 0$ at A and writing M_B for M_C and M_A for M_B will reduce to

$$0 = \frac{A_2 x_2}{EIl} + \frac{M_B l}{6EI} + \frac{M_A l}{3EI}$$

or

$$M_B l + 2M_A l = -\frac{6A_2 x_2}{l}$$

and from Eq. (9.4), putting $\theta_1 = 0$ at B will yield

$$0 = \frac{A_1 x_1}{EIl} + \frac{M_A l}{6EI} + \frac{M_B l}{3EI}$$

or

$$M_A l + 2M_B l = -\frac{6A_1 x_1}{l}$$

∎

9.1.2 Settlement of Supports

If the supports A, B, and C of the two-span continuous beam in Fig. 9.2a settled downward as shown in Fig. 9.6a, no internal support moments are induced since such settlement involves rigid body motion and the slopes of lines AB and BC remain unchanged. However, if the same beam settles differentially as shown in Fig. 9.6b, discontinuity of the slope at B occurs which must then be corrected by internal support moments.

From Fig. 9.6b, due to downward settlements, Δ_A, Δ_B, and Δ_C, the slopes to the left and to the right of support B, are given by, respectively,

$$\theta_1'''' = \frac{\Delta_A - \Delta_B}{l_1} \qquad (9.10a)$$

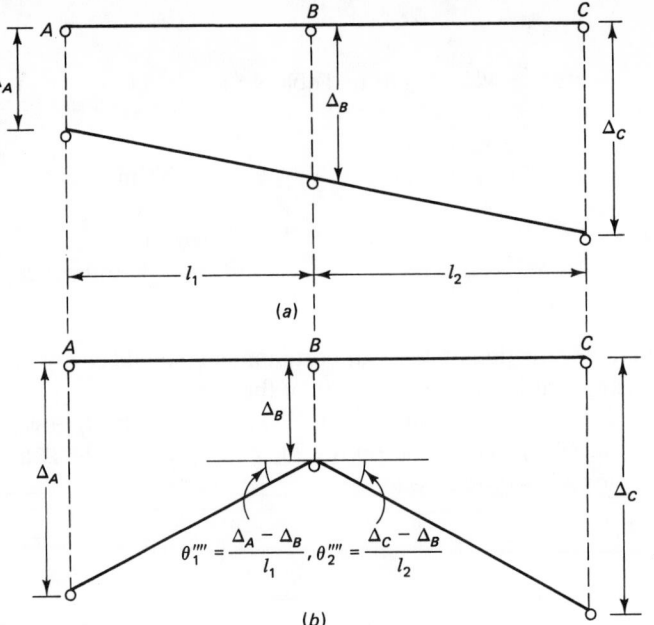

Figure 9.6 Treatment of differential settlement of supports.

and

$$\theta_2''''' = \frac{\Delta_C - \Delta_B}{l_2} \tag{9.10b}$$

The slope θ_1, given by Eq. (9.4), can now be adjusted as follows to include the influence of differential settlement:

$$\theta_1 = \frac{A_1 x_1}{EI_1 l_1} + \frac{M_A l_1}{6 EI_1} + \frac{M_B l_1}{3 EI_1}$$

$$+ \frac{\Delta_A - \Delta_B}{l_1} \qquad \text{counterclockwise, i.e., } -\theta_1 \text{ clockwise} \tag{9.11}$$

Similarly, for θ_2, given by Eq. (9.7):

$$\theta_2 = \frac{A_2 x_2}{EI_2 l_2} + \frac{M_C l_2}{6 EI_2} + \frac{M_B l_2}{3 EI_2} + \frac{\Delta_C - \Delta_B}{l_2} \qquad \text{clockwise} \tag{9.12}$$

Equating the above slopes θ_1 and θ_2 will yield the following modified three-moment equation to account for differential settlement of supports:

$$M_A \frac{l_1}{I_1} + 2 M_B \left(\frac{l_1}{I_1} + \frac{l_2}{I_2} \right) + M_C \frac{l_2}{I_2}$$

$$= - \frac{6 A_1 x_1}{I_1 l_1} - \frac{6 A_2 x_2}{I_2 l_2} - \frac{6 E (\Delta_A - \Delta_B)}{l_1} - \frac{6 E (\Delta_C - \Delta_B)}{l_2} \tag{9.13}$$

■ **EXAMPLE 9.3**

For the continuous beam $A'ABC$ loaded as shown in Fig. 9.7a, deduce the support moments and draw the SFD and BMD, assuming that support A sinks by 60 mm under the given loading. Given $EI = $ constant $= 20,000$ kN $\cdot$ m^2.

Solution. From statics of Fig. 9.7b, $M_{AA'} = -(100)(10) = -1000$ kN $\cdot$ m. Therefore, from moment equilibrium at support A, we have $M_A = M_{AA'} = -1000$ kN $\cdot$ m (i.e., producing tension at the top fiber of span AB). The beam is indeterminate to the second degree. Applying Eq. (9.13) to spans AB and BC and using the information in Table 9.1 for the terms $6A_1x_1/l_1$ and $6A_2x_2/l_2$ yield:

$$M_A(6) + 2M_B(6 + 9) + M_C(9)$$

$$= -\frac{wl^3}{4} - \frac{1}{l}P\left(\frac{l}{2}\right)\left(l^2 - \frac{l^2}{4}\right) - 6EI\left(\frac{60}{6000} - 0\right) - 0$$

that is,

$$(6)(-1000) + 30M_B + 9M_C = -3240 - 4556.25 - 1200$$

or

$$30M_B + 9M_C = -2996.25 \tag{a}$$

Applying Eq. (9.13) to spans BC and CC' (having $I = \infty$) and with no differential settlement yields

$$M_B\frac{9}{I} + 2M_C\left(\frac{9}{I} + \frac{\overset{0}{\cancel{l'}}}{\cancel{\infty}}\right) + M_C\frac{\overset{0}{\cancel{l'}}}{\cancel{\infty}} = -\frac{P(l/2)(l^2 - l^2/4)}{Il} - \frac{6A\overset{0}{\cancel{x_2}}}{\cancel{\infty}l'} - 0 - 0$$

that is,

$$9M_B + 18M_C = -4556.25 \tag{b}$$

Solving (a) and (b) we find

$$M_B = -28 \text{ kN} \cdot \text{m} \quad \text{and} \quad M_C = -239 \text{ kN} \cdot \text{m}$$

The negative sign means that the two moments act on the ends of the beams producing tension at the top fibers. From the free-body diagrams of the three spans, shown in Fig. 9.7b, we obtain the shears to the left and to the right of the supports as well as the reactions. To check on the reactions:

$$\sum Fy = 0\uparrow^+: \quad 442 + 69.6 + 98.4 - 100 - (60)(6) - 150 = 0$$

Therefore, it checks. The shear-force and bending-moment diagrams are shown in Fig. 9.7c and g.

If there was no settlement, equation (a) will be

$$30M_B + 9M_C = -1796.25$$

and equation (b) will remain the same. Solving,

$$M_B = +19 \text{ kN} \cdot \text{m} \quad \text{and} \quad M_C = -263 \text{ kN} \cdot \text{m} \quad ■$$

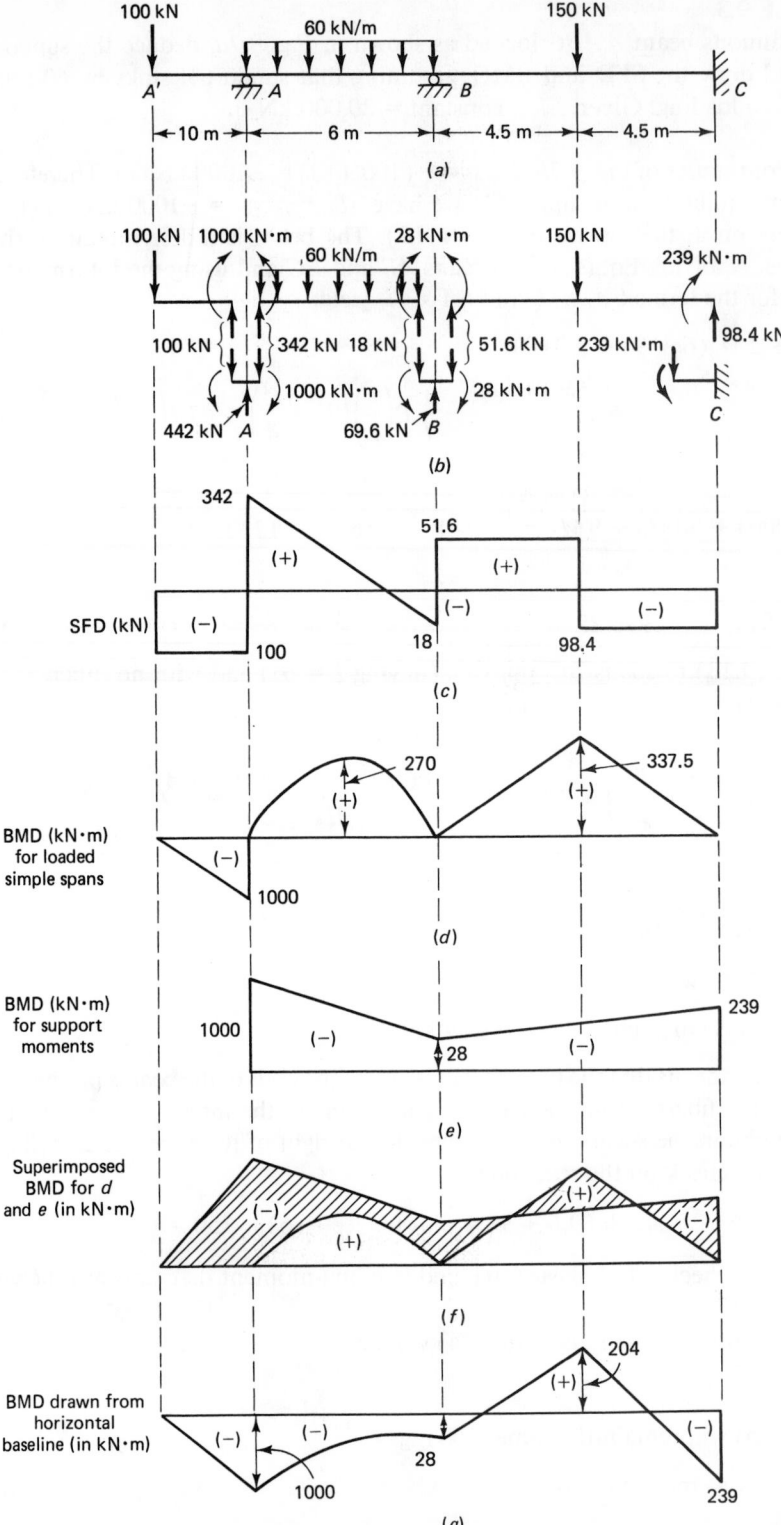

Figure 9.7 Structure for Example 9.3

9.1.3 Three-Moment Equation in Terms of Flexibility Coefficients

The three-moment equation can also be derived using *flexibility coefficients*. The portion of a continuous beam is shown in Fig. 9.8*a*, and Fig. 9.8*b* shows the free-body diagrams of two adjoining spans *AB* and *BC*. Figure 9.8*c* shows the rotation of the beam just to the right and to the left of support *B* due to external loads on the simple

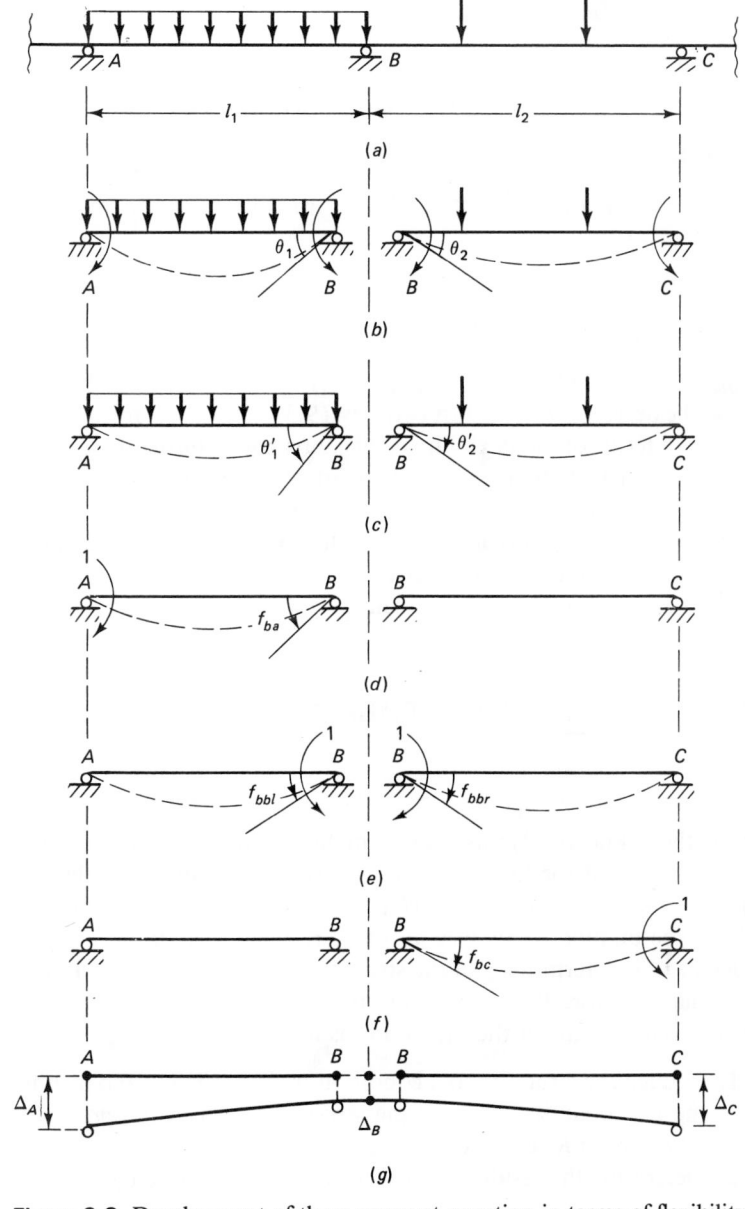

Figure 9.8 Development of three-moment equation in terms of flexibility coefficients.

spans AB and BC (without the end moments). A unit moment at A is applied as shown in Fig. 9.8d producing a rotation f_{ba} just to the left of support B. Similarly, unit moments at B and then at C are applied separately as shown in Fig. 9.8e and f; these produce rotations to the left and right of B as shown. Therefore, considering the redundant moments M_A, M_B, and M_C we can write that

$$\theta_1 = \theta'_1 + M_A f_{ba} + M_b f_{bbl} \quad \text{(counterclockwise)} \tag{9.14}$$

and

$$\theta_2 = \theta'_2 + M_B f_{bbr} + M_C f_{bc} \quad \text{(clockwise)} \tag{9.15}$$

But

$$\theta_1 = \theta_2 \quad \text{(in magnitude and direction)}$$

Therefore,

$$M_A f_{ba} + M_B(f_{bbl} + f_{bbr}) + M_C f_{bc} = -\theta'_1 - \theta'_2 \tag{9.16}$$

where

$$f_{ba} = \frac{l_1}{6EI_1} \qquad f_{bbl} = \frac{l_1}{3EI_1} \qquad f_{bbr} = \frac{l_2}{3EI_2}$$

$f_{bc} = l_2/6EI_2$, $\theta'_1 = A_1 x_1/EI_1 l_1$, $\theta'_2 = A_2 x_2/EI_2 l_2$. The flexibility coefficients, f_{ba}, etc., can be derived from the last two Eqs. (9.3). This is the form of the three-moment equation in terms of the flexibility coefficients. Substituting these values in Eq. (9.16) will yield Eq. (9.8). If there is settlement of supports as shown in Fig. 9.8g, then adding the terms $(\Delta_A - \Delta_B)/l_1$ to θ_1 [Eq. (9.14)] and $(\Delta_C - \Delta_B)/l_2$ to θ_2 [Eq. (9.15)] will modify Eq. (9.16) to yield Eq. (9.13). It should be noted that a flexibility coefficient, such as f_{ba}, is defined as the displacement (rotation here) at b due to a unit force (moment here) at a. More will be said on this in Section 9.2.

9.2 METHOD OF CONSISTENT DEFORMATIONS

9.2.1 General

This method, known also as a force method, can be applied to any indeterminate structure analyzed for the effects of applied loads, support settlement, temperature changes, or any other effect, *provided* that the principle of superposition holds. In applying the method to analyze an indeterminate structure, we must also recognize the degree of indeterminacy of the structure or the number of redundants, that is, the forces that are more than the minimum necessary for the static equilibrium of the structure. The outline of the method is as follows:

1. Reduce the structure to a condition of determinacy and stability by removing the redundants. The resulting structure is called the *primary structure* (Fig. 9.9b, where R_2 is removed).
2. Determine the resulting errors in geometry incurred by subjecting the primary

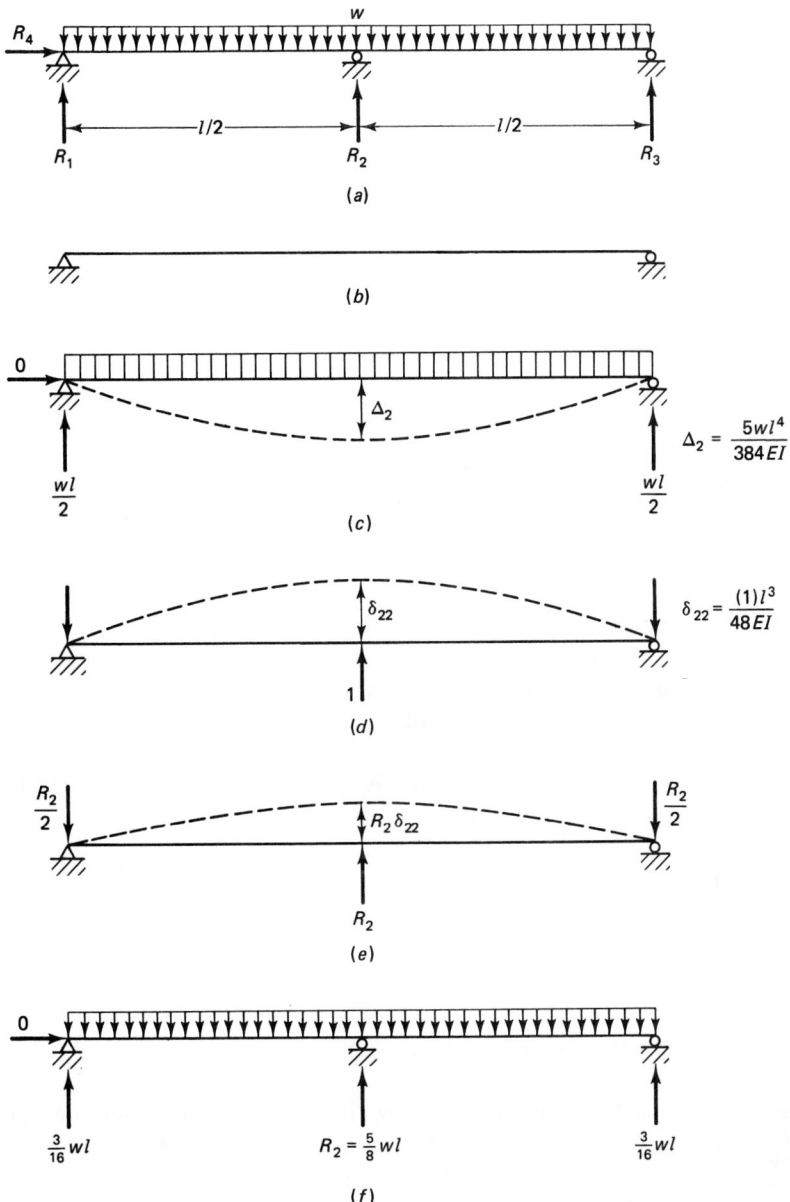

Figure 9.9 Determination of a redundant reaction by the method of consistent deformation.

structure to loads on the original indeterminate structure (Fig. 9.9c; the displacement Δ_2 is the error in geometry).

3. Apply the redundants to the primary structure and determine their effects on the displacements where the geometry has to be corrected (R_2 is the redundant applied as shown in Fig. 9.9e).

4. Determine the redundants to eliminate the original errors in geometry (by equating Δ_2 to $R_2\delta_{22}$, Fig. 9.9 c and e).
5. Obtain the final forces by adding the redundant forces to the corresponding forces of the primary structure (Fig. 9.9 f).

Thus for the continuous beam in Fig. 9.9, the reaction R_2 is removed to render the structure determinate and stable. (Note: We could have removed the reaction R_1 or R_3 or the internal moment M_2 at support 2.) Since in the actual structure (Fig. 9.9 a, the deflection at support 2 is zero, we can write (assuming downward deflections to be positive)

$$\Delta_2 - R_2\delta_{22} = 0 \tag{9.17}$$

or

$$\frac{5}{384}\frac{wl^4}{EI} - R_2\left(\frac{l^3}{48\,EI}\right) = 0$$

Thus

$$R_2 = \tfrac{5}{8}wl$$

(Deflections can be determined by any of the methods discussed in Chapter 7.) We can now determine the reactions from Fig. 9.9 c and e as shown in Fig. 9.9 f. It should be observed that δ_{22} is a *flexibility coefficient* which in this case is the displacement at point 2 due to a unit action at joint 2, all other points being assumed unloaded, as shown in Fig. 9.9 d.

If support 2 of the structure in Fig. 9.9 a had settled downward a known distance Δ_2', then Eq. (9.17) becomes

$$\Delta_2 - R_2\delta_{22} = \Delta_2'$$

from which R_2 is obtained.

■ EXAMPLE 9.4

Determine the reactions for the continuous beam in Fig. 9.10 a.

Solution. The primary structure is chosen as shown in Fig. 9.10 b. With the external loads applied to the primary structure (Fig. 9.10 c), the deflections Δ_2 and Δ_3 at points 2 and 3 can be calculated. Using the conjugate-beam method, the calculations for Δ_2 and Δ_3 are shown in Fig. 9.11 for each of the loads applied separately. Therefore, by superposition, we have

$$\Delta_2 = \Delta_2' + \Delta_2'' + \Delta_2''' = \frac{1}{EI}(52{,}778 + 95{,}833 + 20{,}833) = \frac{169{,}444}{EI}$$

$$\Delta_3 = \Delta_3' + \Delta_3'' + \Delta_3''' = \frac{1}{EI}(43{,}056 + 95{,}833 + 25{,}000) = \frac{163{,}889}{EI}$$

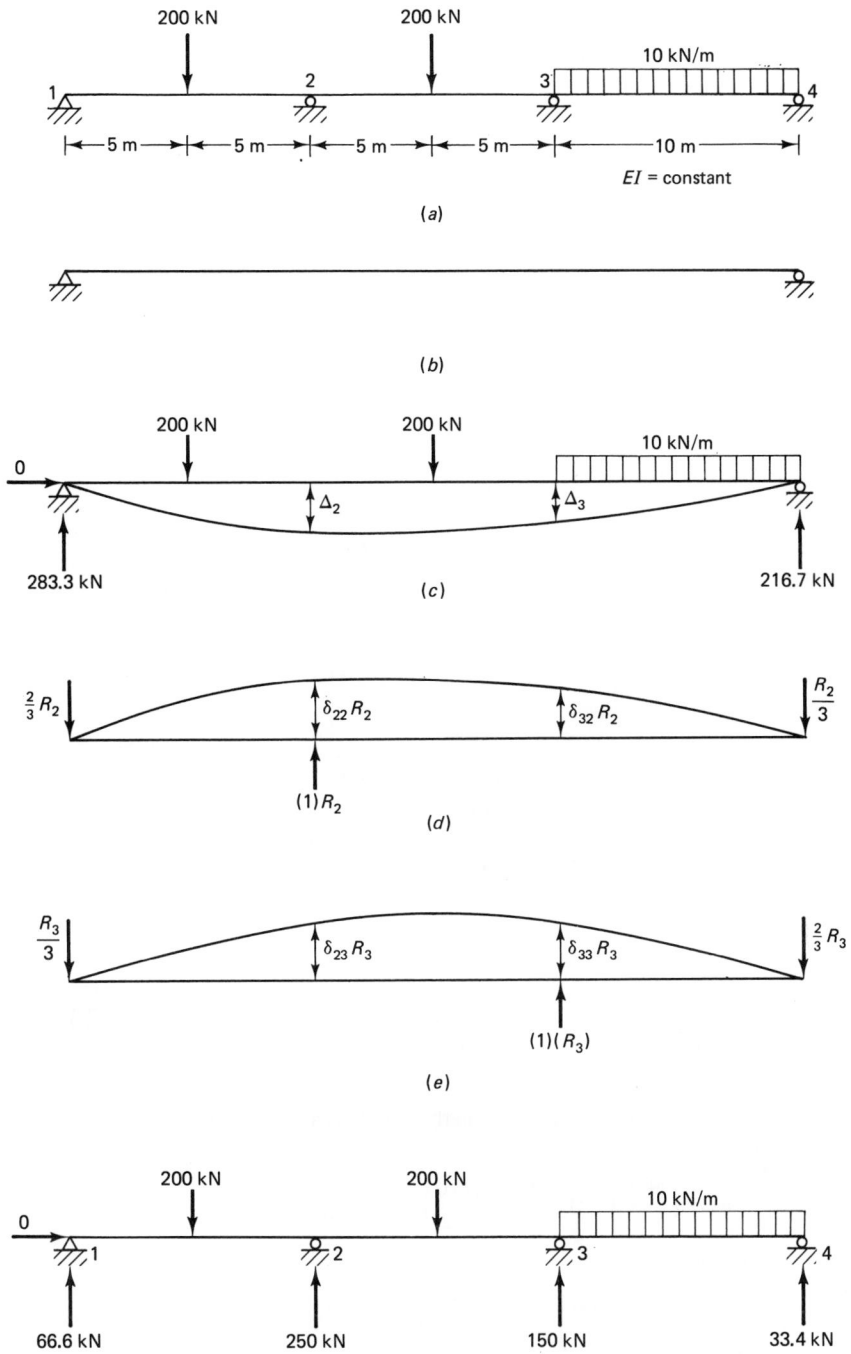

Figure 9.10 Structure for Example 9.4.

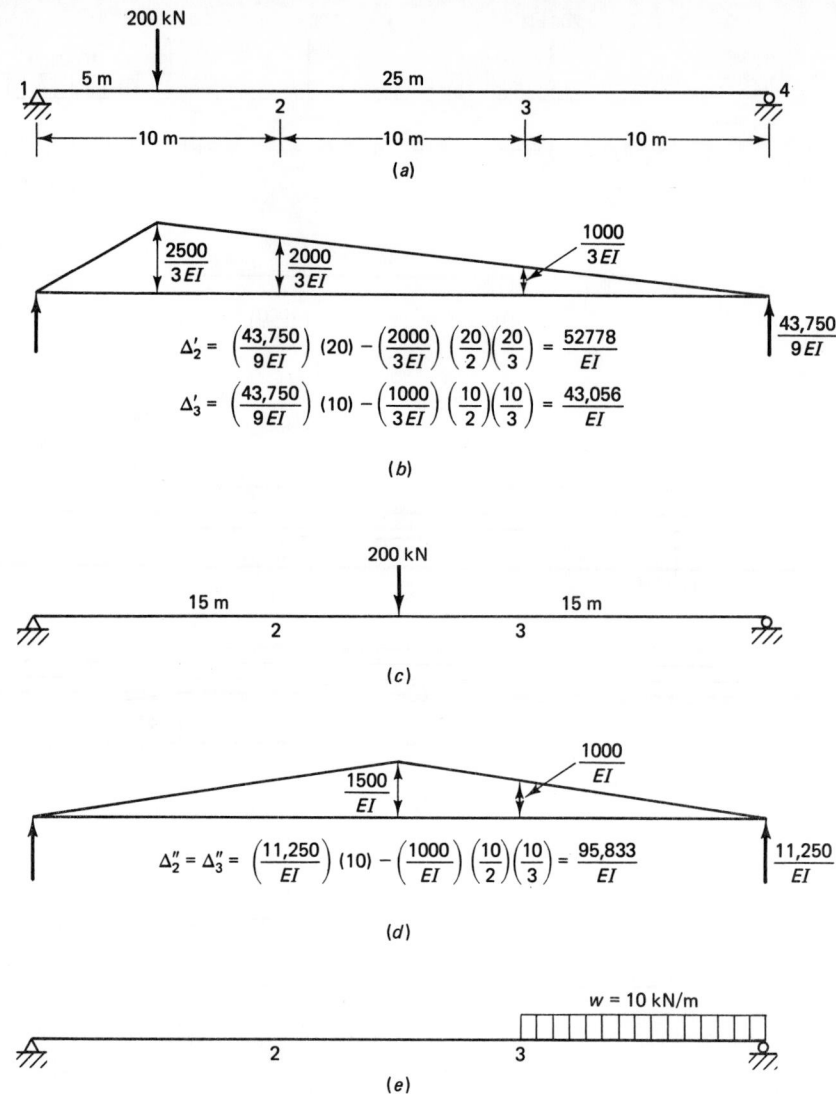

Figure 9.11 Conjugate-beam method of determining deflections for the structure in Example 9.4.

$$\delta_{22} = \left(\frac{500}{9EI}\right)(10) - \left(\frac{20}{3EI}\right)\left(\frac{10}{2}\right)\left(\frac{10}{3}\right) = \frac{4000}{9EI}$$

From symmetry $\delta_{22} = \delta_{23}$

$$\delta_{23} = \delta_{32} = \left(\frac{400}{9EI}\right)(10) - \left(\frac{10}{3EI}\right)\left(\frac{10}{2}\right)\left(\frac{10}{3}\right) = \frac{3500}{9EI}$$

(by Maxwell's reciprocal theorem).

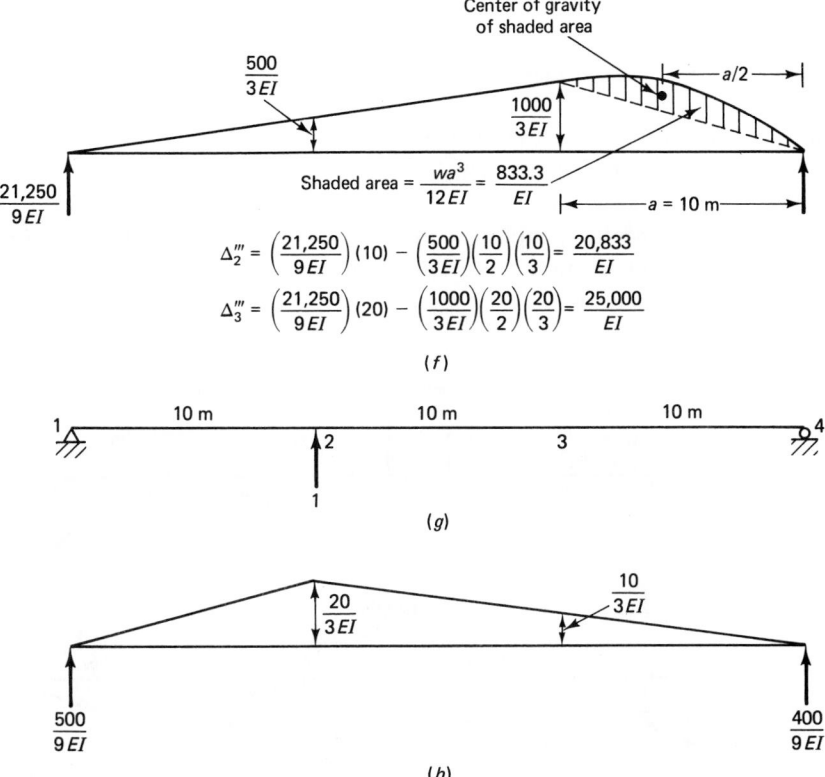

$$\Delta_2''' = \left(\frac{21,250}{9EI}\right)(10) - \left(\frac{500}{3EI}\right)\left(\frac{10}{2}\right)\left(\frac{10}{3}\right) = \frac{20,833}{EI}$$

$$\Delta_3''' = \left(\frac{21,250}{9EI}\right)(20) - \left(\frac{1000}{3EI}\right)\left(\frac{20}{2}\right)\left(\frac{20}{3}\right) = \frac{25,000}{EI}$$

(f)

(g)

(h)

Figure 9.11 (*Continued*)

Also the calculations of the flexibility coefficients δ_{22}, δ_{33}, δ_{23}, and δ_{32} are given in Fig. 9.11h. Since the deflections at supports 2 and 3 of the continuous beam in Fig. 9.10a are zero, we can write, using Fig. 9.10c, d, and e:

$$\Delta_2 - \delta_{22}R_2 - \delta_{23}R_3 = 0 \qquad (a)$$

$$\Delta_3 - \delta_{32}R_2 - \delta_{33}R_3 = 0 \qquad (b)$$

or

$$\frac{169,444}{EI} - \left(\frac{4000}{9EI}\right)R_2 - \left(\frac{3500}{9EI}\right)R_3 = 0$$

$$\frac{163,889}{EI} - \left(\frac{3500}{9EI}\right)R_2 - \left(\frac{4000}{9EI}\right)R_3 = 0$$

Solving the above equations will yield

$$R_2 = 250 \text{ kN} \quad \text{and} \quad R_3 = 150 \text{ kN}$$

Therefore,

$$R_1 = 283.3 - \frac{2}{3}R_2 - \frac{R_3}{3} = 66.6 \text{ kN}$$

and

$$R_4 = 216.7 - \frac{R_2}{3} - \frac{2}{3}R_3 = 33.4 \text{ kN}$$

The reactions are shown in Fig. 9.10f. From statics one can find $M_2 = (66.6)(10) - (200)(5) = -335$ kN·m (producing tension at top). ∎

It is generally simpler to compute rotations than deflections; in such cases we consider support moments rather than reactions as unknowns. This is illustrated in the following example.

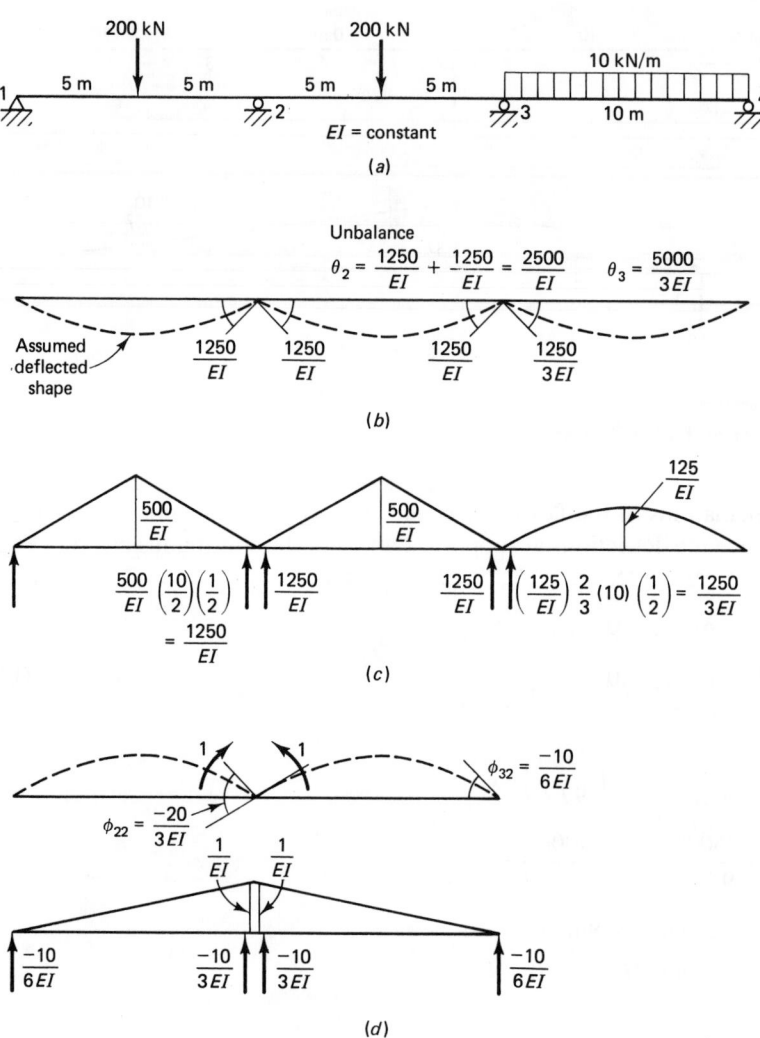

Figure 9.12 Structure for Example 9.5.

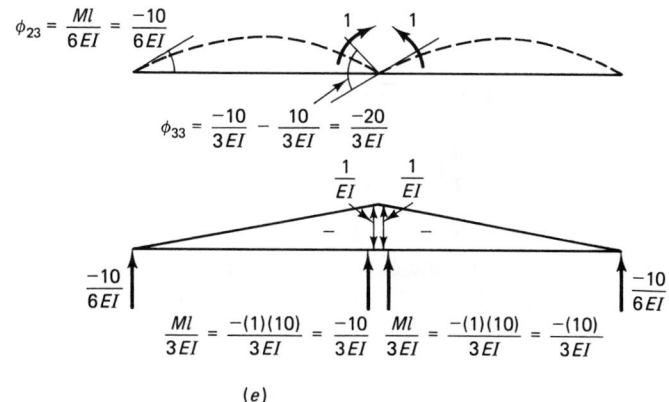

$$\phi_{23} = \frac{Ml}{6EI} = \frac{-10}{6EI}$$

$$\phi_{33} = \frac{-10}{3EI} - \frac{10}{3EI} = \frac{-20}{3EI}$$

$$\frac{-10}{6EI} \qquad \frac{-10}{6EI}$$

$$\frac{Ml}{3EI} = \frac{-(1)(10)}{3EI} = \frac{-10}{3EI} \qquad \frac{Ml}{3EI} = \frac{-(1)(10)}{3EI} = \frac{-(10)}{3EI}$$

(e)

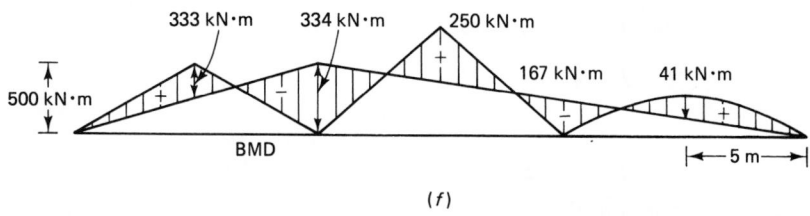

333 kN·m   334 kN·m   250 kN·m

500 kN·m   167 kN·m   41 kN·m

BMD |←——5 m——→|

(f)

Figure 9.12 (*Continued*)

■ EXAMPLE 9.5

For the structure in Fig. 9.12a, determine the moments M_2 and M_3, using the compatibility of slopes at points 2 and 3.

Solution. Let us assume that the beam is cut at each interior support; then the assumed deflected shape is shown in Fig. 9.12b. The slopes at joints 2 and 3 can be calculated from the conjugate beams loaded by the M/EI diagrams as shown in Fig. 9.12c. The errors in the geometry are the rotations at the joints; thus from Fig. 9.12b and c, $\theta_2 = 2500/EI$ and $\theta_3 = 5000/3EI$, obtained from the algebraic sum of the end rotations of the intersecting members at each joint. Rotations are assumed positive when produced by moments producing tension at bottom. To correct for these errors, unit negative moments (producing tension at top) are applied to the cut ends, as shown in Fig. 9.12d; the resulting end rotations are considered negative.

Thus, from Fig. 9.12d and e we have

$$\phi_{22} = \frac{-20}{3EI} \qquad \phi_{32} = \frac{-10}{6EI} \qquad \phi_{33} = \frac{-20}{3EI} \qquad \text{and} \qquad \phi_{23} = \frac{-10}{6EI}$$

The equations of geometry are as follows:

$$\phi_{22}M_2 + \phi_{23}M_3 + \theta_2 = 0 \tag{c}$$

$$\phi_{32}M_2 + \phi_{33}M_3 + \theta_3 = 0 \tag{d}$$

or

$$M_2\left(\frac{-20}{3EI}\right) + M_3\left(\frac{-10}{6EI}\right) + \frac{2500}{EI} = 0 \tag{c}$$

$$M_2\left(\frac{-10}{6EI}\right) + M_3\left(\frac{-20}{3EI}\right) + \frac{5000}{3EI} = 0 \tag{d}$$

Solving (c) and (d) yields $M_2 = 334$ kN·m and $M_3 = 167$ kN·m as before; the positive sign indicates that the correction moments have the same sense as the unit moments in Fig. 9.12d and e. The bending-moment diagram is given in Fig. 9.12f.

∎

As illustrated by Examples 9.4 and 9.5, this method can be used to analyze any indeterminate structure by solving a set of simultaneous equations of the type:

$$
\begin{aligned}
f_{11}x_1 + f_{12}x_2 + \cdots + f_{1n}x_n + \Delta_1 &= 0 \\
f_{21}x_1 + f_{22}x_2 + \cdots + f_{2n}x_n + \Delta_2 &= 0 \\
\vdots \qquad\qquad\qquad \vdots \\
f_{n1}x_1 + f_{n2}x_2 + \cdots + f_{nn}x_n + \Delta_n &= 0
\end{aligned}
\tag{9.18}
$$

in which the terms f represent flexibility coefficients (i.e., displacements due to unit force or rotations due to unit moment); the terms x represent the required correction forces or moments; and the terms Δ represent the errors in geometry (i.e., either displacements or rotations).

In case of prescribed settlement (or rotation) of supports, say Δ_{i0}, then Eq. (9.18) can be written as

$$
\begin{aligned}
f_{11}x_1 + f_{12}x_2 + \cdots + f_{1n}x_n + \Delta_1 &= \Delta_{10} \\
\vdots \\
f_{i1}x_1 + f_{i2}x_2 + \cdots + f_{in}x_n + \Delta_i &= \Delta_{i0} \\
\vdots \\
f_{n1}x_1 + f_{n2}x_2 + \cdots + f_{nn}x_n + \Delta_n &= \Delta_{n0}
\end{aligned}
\tag{9.19}
$$

in which Δ_{i0} are the actual given displacements (settlement or rotation) at the redundant reaction points of the original structure.

∎

∎ EXAMPLE 9.6

Calculate the reactions and moment at joint A of the rigid frame loaded as shown in Fig. 9.13a.

Solution. First we remove the three redundant reaction components at A as shown in Fig. 9.13b and calculate the displacements and rotation at A. This is followed by calculating the displacement and rotation at A due to the three redundants x_1, x_2, and x_3. Since joint A is fixed, we can apply the compatibility equations (9.18), to find x_1, x_2, and x_3. To find the displacements and rotation at A for the frame in Fig. 9.13b,

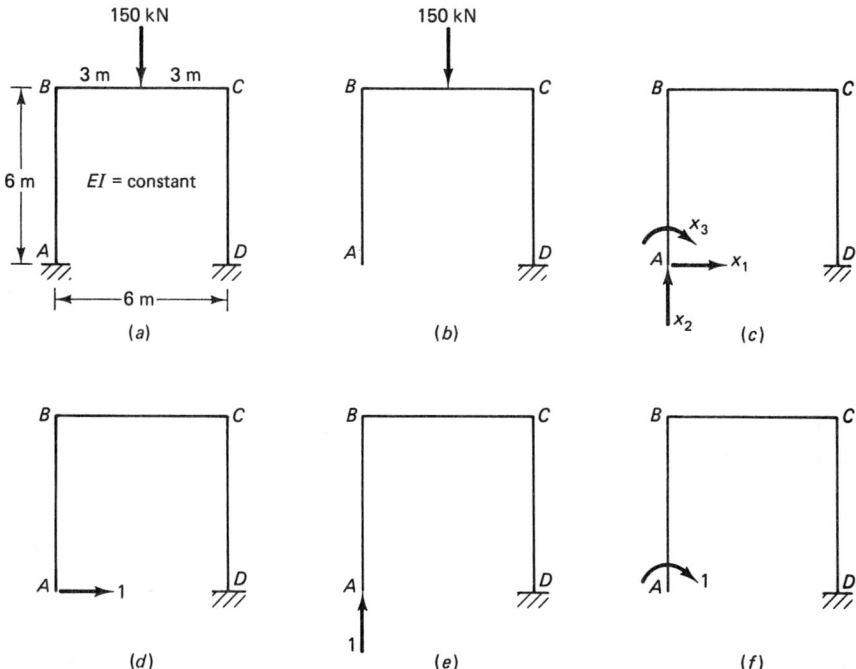

Figure 9.13 Structure for Example 9.6.

we can apply the virtual-work expression $\int_{\text{frame}} Mm\, dx/EI$ (see Chapter 7). Taking the centroidal axis of each member as the x axis, we can write

$$\int_{\text{frame}} \frac{Mm\, dx}{EI} = \int_A^B \frac{Mm\, dx}{EI} + \int_B^C \frac{Mm\, dx}{EI} + \int_C^D \frac{Mm\, dx}{EI}$$

Table 9.2 shows the expressions for M and m, where M denotes the moment in frame of Fig. 9.13b due to the load of 150 kN and m_1 denotes the moment due to a unit horizontal force at A (Fig. 9.13d), m_2 that due to a unit vertical force at A (Fig. 9.13e), and m_3 that due to a unit couple at A (Fig. 9.13f). We assume that moment producing tension on the inside of the frame is considered positive. Thus, the horizontal deflection at A, Δ, for the frame in Fig. 9.13b is, using the expressions in Table 9.2,

TABLE 9.2

Member	Origin	Limits (m)	M (Fig. 9.13b)	m_1 (Fig. 9.13d)	m_2 (Fig. 9.13e)	m_3 (Fig. 9.13f)
AB	A	0 to 6	0	$-x$	0	1
BC	B	0 to 3	0	-6	x	1
	B	3 to 6	$-150(x-3)$	-6	x	1
CD	C	0 to 6	-450	$x-6$	6	1

$$\Delta_1 = \int_{\text{frame}} \frac{Mm_1 \, dx}{EI} = \frac{1}{EI}\left[0 + \int_3^6 (900)(x-3)\,dx + \int_0^6 (450)(6-x)\,dx \right]$$

$$= \frac{12{,}150}{EI} \text{ m } (\rightarrow)$$

Since Δ_1 is positive, it is in the same direction as the unit horizontal force at A (Fig. 9.13d), and therefore to the right.

$$\Delta_2 = \int_{\text{frame}} \frac{Mm_2 \, dx}{EI} = \frac{1}{EI}\left[0 + \int_3^6 (-150x)(x-3)\,dx + \int_0^6 (-450)(6)\,dx \right]$$

$$= -\frac{19{,}575}{EI} \text{ m } (\downarrow)$$

This is a downward displacement since it is negative and therefore opposite to the direction of the vertical unit force shown in Fig. 9.13e.

$$\Delta_3 = \int_{\text{frame}} \frac{Mm_3 \, dx}{EI} = \frac{1}{EI}\left[0 + \int_3^6 (-150)(x-3)(1)\,dx + \int_0^6 (-450)(1)\,dx \right]$$

$$= -\frac{3375}{EI} \quad \text{radians (counterclockwise)}$$

To find the flexibility coefficients at joint A:

$$f_{11} = \delta_{11} = \int_{\text{frame}} \frac{(m_1)^2 \, dx}{EI} \qquad f_{21} = \delta_{21} = \int_{\text{frame}} \frac{m_1 m_2 \, dx}{EI}$$

$$f_{31} = \delta_{31} = \int_{\text{frame}} \frac{m_1 m_3 \, dx}{EI}$$

since $M = m_1$ in this case. Similarly, with $M = m_2$:

$$f_{12} = \delta_{12} = \int_{\text{frame}} \frac{m_1 m_2 \, dx}{EI} \qquad f_{22} = \delta_{22} = \int_{\text{frame}} \frac{(m_2)^2 \, dx}{EI}$$

$$f_{32} = \delta_{32} = \int_{\text{frame}} \frac{m_2 m_3 \, dx}{EI}$$

Also, with $M = m_3$:

$$f_{13} = \delta_{13} = \int_{\text{frame}} \frac{m_1 m_3 \, dx}{EI} \qquad f_{23} = \delta_{23} = \int_{\text{frame}} \frac{m_2 m_3 \, dx}{EI}$$

$$f_{33} = \int_{\text{frame}} \frac{(m_3)^2 \, dx}{EI}$$

Using the expressions in Table 9.2:

$$f_{11} = \delta_{11} = \frac{1}{EI}\left[\int_0^6 (-x)^2 \, dx + \int_0^3 (-6)^2 \, dx + \int_3^6 (-6)^2 \, dx + \int_0^6 (x-6)^2 \, dx \right]$$

$$= \frac{360}{EI} \quad \text{m } (\rightarrow)$$

$$f_{21} = \delta_{21} = \frac{1}{EI}\left[0 + \int_0^3 (-6)(x)dx + \int_3^6 (-6)(x)dx + \int_0^6 (6)(x-6)dx\right]$$

$$= \frac{-216}{EI} \quad \text{m } (\downarrow)$$

$$f_{31} = \delta_{31} = \frac{1}{EI}\left[\int_0^6 (-x)dx + \int_0^3 (-6)(1)dx + \int_3^6 (-6)(1)dx\right.$$

$$\left. + \int_0^6 (1)(x-6)dx\right] = \frac{-72}{EI} \quad \text{radians } (\,\rangle\,)$$

$$f_{12} = \delta_{12} = \delta_{21} = \frac{-216}{EI} \quad \text{m } (\leftarrow)$$

$$f_{22} = \delta_{22} = \frac{1}{EI}\left[0 + \int_0^3 x^2\,dx + \int_3^6 x^2\,dx + \int_0^6 (6)^2\,dx\right] = \frac{288}{EI} \quad \text{m } (\uparrow)$$

$$f_{32} = \delta_{32} = \frac{1}{EI}\left[0 + \int_0^3 (x)(1)dx + \int_3^6 (x)(1)dx + \int_0^6 (6)(1)dx\right]$$

$$= \frac{54}{EI} \quad \text{radians } (\,\rangle\,)$$

$$f_{13} = \delta_{13} = \delta_{31} = \frac{-72}{EI} \quad \text{m } (\leftarrow)$$

$$f_{23} = \delta_{23} = \delta_{32} = \frac{54}{EI} \quad \text{m } (\uparrow)$$

$$f_{33} = \delta_{33} = \frac{1}{EI}\left[\int_0^6 (1)dx + \int_0^3 (1)dx + \int_3^6 (1)dx + \int_0^6 (1)dx\right]$$

$$= \frac{18}{EI} \quad \text{radians } (\,\rangle\,)$$

Thus the flexibility coefficients in matrix form become

$$\begin{bmatrix} f_{11} & f_{12} & f_{13} \\ f_{21} & f_{22} & f_{23} \\ f_{31} & f_{32} & f_{33} \end{bmatrix} = \begin{bmatrix} \delta_{11} & \delta_{12} & \delta_{13} \\ \delta_{21} & \delta_{22} & \delta_{23} \\ \delta_{31} & \delta_{32} & \delta_{33} \end{bmatrix} = \begin{bmatrix} \int \dfrac{m_1^2\,dx}{EI} & \int \dfrac{m_1 m_2\,dx}{EI} & \int \dfrac{m_1 m_3\,dx}{EI} \\[2mm] \int \dfrac{m_1 m_2\,dx}{EI} & \int \dfrac{m_2^2\,dx}{EI} & \int \dfrac{m_2 m_3\,dx}{EI} \\[2mm] \int \dfrac{m_1 m_3\,dx}{EI} & \int \dfrac{m_2 m_3\,dx}{EI} & \int \dfrac{m_3^2\,dx}{EI} \end{bmatrix}$$

$$= \frac{1}{EI}\begin{bmatrix} 360 & -216 & -72 \\ -216 & 288 & 54 \\ -72 & 54 & 18 \end{bmatrix}$$

Since A is fixed, we can apply the compatibility conditions; thus using Eqs. (9.18), we can write

$$
\begin{bmatrix} f_{11} & f_{12} & f_{13} \\ f_{21} & f_{22} & f_{23} \\ f_{31} & f_{32} & f_{33} \end{bmatrix} \begin{Bmatrix} x_1 \\ x_2 \\ x_3 \end{Bmatrix} + \begin{Bmatrix} \Delta_1 \\ \Delta_2 \\ \Delta_3 \end{Bmatrix} = 0
$$

Substituting for the flexibility coefficients and the Δ's, we find

$$
\begin{bmatrix} 360 & -216 & -72 \\ -216 & 288 & 54 \\ -72 & 54 & 18 \end{bmatrix} \begin{Bmatrix} x_1 \\ x_2 \\ x_3 \end{Bmatrix} + \begin{Bmatrix} 12{,}150 \\ -19{,}575 \\ -3{,}375 \end{Bmatrix} = \begin{Bmatrix} 0 \\ 0 \\ 0 \end{Bmatrix}
$$

The solution of this matrix equation yields

$$
\begin{Bmatrix} x_1 \\ x_2 \\ x_3 \end{Bmatrix} = \begin{Bmatrix} 18.75 \\ 75.0 \\ 37.5 \end{Bmatrix} \begin{array}{l} \text{kN} \\ \text{kN} \\ \text{kN} \cdot \text{m} \end{array}
$$ ■

9.2.2 Elastic Supports

If the support points in a structure are provided with elastic supports, one can write

$$R_i = k_{si}\Delta_i \tag{9.20}$$

in which k_{si} is the elastic or spring constant. Now the unit load, such as the one shown in Fig. 9.14b, is applied at the bottom of the spring at point 2; as such the unit load, say at support i, shortens the spring at i by an amount equal to the unit load divided by the spring constant k_{si}. This amount must be included to obtain the total movement of the support; thus the term δ_{ii} (or f_{ii}) changes to $(f_{ii} + 1/k_{si})$ and so on. As a result, Eq. (9.18) takes the form

$$\left(f_{11} + \frac{1}{k_{si}}\right)x_1 + f_{12}x_2 + \cdots + f_{1n}x_n + \Delta_1 = 0$$

$$\vdots$$

$$f_{i1}x_1 + f_{i2}x_2 + \cdots + \left(f_{ii} + \frac{1}{k_{si}}\right)x_i + \cdots f_{in}x_n + \Delta_i = 0 \tag{9.21}$$

$$\vdots$$

$$f_{n1}x_1 + f_{n2}x_2 + \cdots + \left(f_{nn} + \frac{1}{k_{sn}}\right)x_n + \Delta_n = 0$$

■ EXAMPLE 9.7

For the continuous beam in Fig. 9.14a assume that supports 2 and 3 are elastically supported by springs whose constants are 4500 and 3000 kN/m. Determine the support reactions R_2 and R_3. Assume $EI = 138{,}000$ kN $\cdot$ m^2.

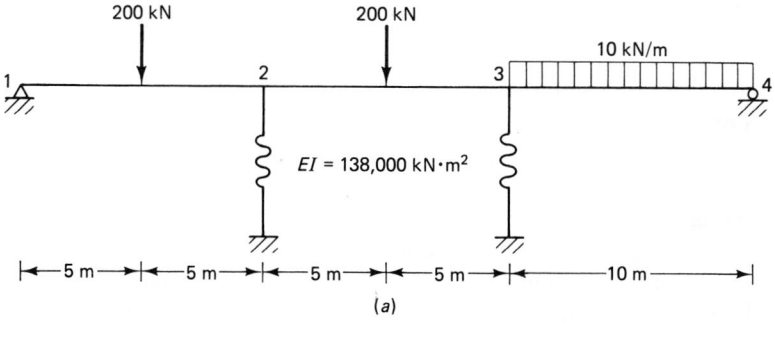

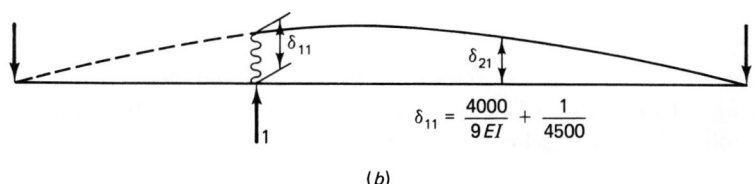

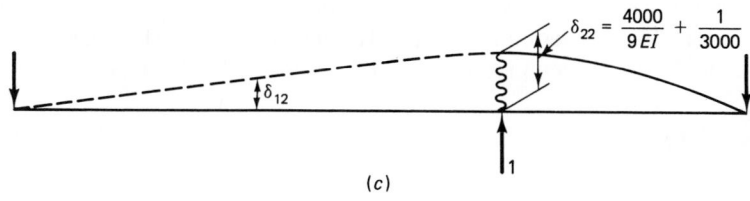

Figure 9.14 Structure for Example 9.7.

Solution. Because of the influence of elastic support, equations (a) and (b) of Section 9.2.1 will become

$$\Delta_2 - \left(\delta_{22} + \tfrac{1}{4500}\right) R_2 - \delta_{23} R_3 = 0 \tag{e}$$

$$\Delta_3 - \delta_{32} R_2 - \left(\delta_{33} + \tfrac{1}{3000}\right) R_3 = 0 \tag{f}$$

or

$$\frac{169{,}444}{EI} - \left(\frac{4000}{9EI} + \frac{1}{4500}\right) R_2 - \frac{3500}{9EI} R_3 = 0$$

$$\frac{163{,}889}{EI} - \frac{3500}{9EI} R_2 - \left(\frac{4000}{9EI} + \frac{1}{3000}\right) R_3 = 0$$

Solving the above two equations yields

$$R_2 = 236.8 \text{ kN} \quad \text{and} \quad R_3 = 146.4 \text{ kN} \qquad \blacksquare$$

9.2.3 Method of Consistent Deformations for the Analysis of Statically Indeterminate Trusses

A truss may be statically indeterminate owing to the presence of redundant supports, redundant members, or any combination thereof. The procedure for analysis of trusses with redundant supports (Fig. 9.15a) is the same as the one described for continuous beams and frames. For an internal redundant member(s), the member is first cut and the relative movement of the two cut ends due to the loading on the truss is calculated. The required continuity is restored by applying two equal and opposite forces to the cut section of such intensity as to nullify the relative movement between the cut ends. The following examples will demonstrate the cases of external and internal redundancies in trusses.

■ **EXAMPLE 9.8**

For the Pratt truss shown in Fig. 9.15a, determine the forces in the members. Assume AE = constant for all members. (This assumption, which may not be realistic in practice, will simplify the numerical calculations.)

Solution. First we must recognize that this truss is indeterminate externally to the first degree. Let us select the center support at c to be the redundant quantity and remove it. The truss is thus converted to a statically determinate one (Fig. 9.15b); by virtual work (Chapter 7) the deflection Δ_c due to the external load at b can be calculated as

$$\Delta_c = \sum_{\text{all members}} p \, \frac{PL}{AE} \qquad (9.22)$$

in which P is the force in any member of the truss in Fig. 9.15b and p is the force in that member of the truss due to a unit load applied vertically at c as shown in Fig. 9.15c. For the statically determinate truss loaded by a unit load in Fig. 9.15c, the vertical deflection δ_{cc} is given again by Eq. (9.22) with P becoming p, as

$$\delta_{cc} = \sum_{\text{all members}} \frac{p^2 L}{AE}$$

By superposition, it follows that the vertical deflection at c due to force R_c is $\delta_{cc} R_c$. If support c of the truss in Fig. 9.15a does not settle, the compatibility equation for no settlement is

$$\Delta_c + \delta_{cc} R_c = 0 \qquad (9.23)$$

or

$$R_c = - \frac{\sum (pPL/AE)}{\sum (p^2 L/AE)} \qquad (9.24)$$

where $\sum$ is for all members of the truss. The various terms for Eq. (9.24) are shown in Table 9.3. Thus from Eq. (9.24)

$$R_c = - \frac{391.4d/AE}{7.32d/AE} = -53.4 \text{ k}$$

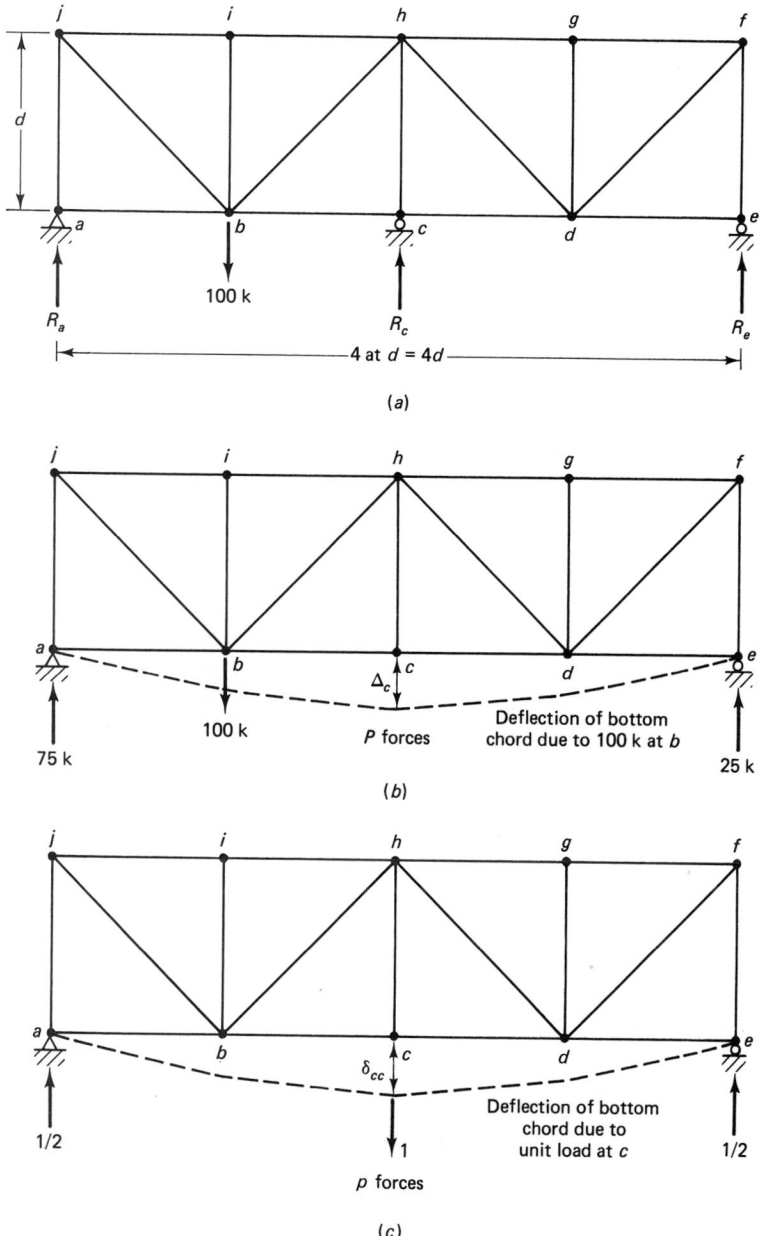

Figure 9.15 Structure for Example 9.8.

The negative sign indicates that this reaction is in the opposite direction to the applied unit load in Fig. 9.15c. The bar forces F for the truss in Fig. 9.15a can now be determined from

$$F = P + pR_c \qquad (9.25)$$

as shown in Table 9.3. The other reactions are

$$R_a = 75 + \tfrac{1}{2}R_c = 75 + \tfrac{1}{2}(-53.4) = 48.3 \text{ k}$$

and

$$R_b = 25 + \tfrac{1}{2}R_c = 25 + \tfrac{1}{2}(-53.4) = -1.7 \text{ k}$$

(the negative sign indicates that it is downward instead of upward as assumed). ∎

TABLE 9.3

Member	l	P	p	pPl	p^2l	$F = P + R_cp$
ab	d	0	0	0	0	0
bc	d	$+50$	$+1$	$+50d$	d	-3.4
cd	d	$+50$	$+1$	$+50d$	d	-3.4
de	d	0	0	0	0	0
ef	d	-25	$-\dfrac{1}{2}$	$\dfrac{25}{2}d$	$\dfrac{1}{4}d$	$+1.7$
fg	d	-25	$-\dfrac{1}{2}$	$\dfrac{25}{2}d$	$\dfrac{1}{4}d$	$+1.7$
gh	d	-25	$-\dfrac{1}{2}$	$\dfrac{25}{2}d$	$\dfrac{1}{4}d$	$+1.7$
hi	d	-75	$-\dfrac{1}{2}$	$\dfrac{75}{2}d$	$\dfrac{1}{4}d$	-48.3
ij	d	-75	$-\dfrac{1}{2}$	$\dfrac{75}{2}d$	$\dfrac{1}{4}d$	-48.3
ja	d	-75	$-\dfrac{1}{2}$	$\dfrac{75}{2}d$	$\dfrac{1}{4}d$	-48.3
jb	$d\sqrt{2}$	$+75\sqrt{2}$	$+\dfrac{\sqrt{2}}{2}$	$75d\sqrt{2}$	$d\dfrac{\sqrt{2}}{2}$	$+68.3$
ib	d	0	0	0	0	0
bh	$d\sqrt{2}$	$+25\sqrt{2}$	$-\dfrac{\sqrt{2}}{2}$	$-25d\sqrt{2}$	$d\dfrac{\sqrt{2}}{2}$	$+73.1$
hc	d	0	$+1$	0	d	-53.4
hd	$d\sqrt{2}$	$-25\sqrt{2}$	$-\dfrac{\sqrt{2}}{2}$	$+25d\sqrt{2}$	$d\dfrac{\sqrt{2}}{2}$	$+2.4$
dg	d	0	0	0	0	0
df	$d\sqrt{2}$	$+25\sqrt{2}$	$+\dfrac{\sqrt{2}}{2}$	$+25d\sqrt{2}$	$d\dfrac{\sqrt{2}}{2}$	-2.4
		Σ		$391.4d$	$7.32d$	

■ **EXAMPLE 9.9**

Analyze the Pratt truss shown in Fig. 9.16a. Assume AE = constant for all members.

Solution. This truss is indeterminate externally to the first degree and internally to the first degree also. Let us choose the vertical reaction at c (i.e., x_1), and the force x_2 in the member ci to be the two redundant quantities. Removing x_1 and cutting the member ci anywhere along its length and then loading the truss with the 100-k load will result in the deflected shape of the basic determinate truss shown in Fig. 9.16b. Let the vertical deflection at c be denoted as Δ_c and the gap between the ends of the member ci be Δ_{ci}. By virtual work and using Eq. (9.22), one can write

$$\Delta_c = \sum p_1 \frac{PL}{AE} \quad \text{and} \quad \Delta_{ci} = \sum p_2 \frac{PL}{AE}$$

in which $\sum$ includes all truss members in Fig. 9.16b, p_1 = internal force in any member of the basic determinate structure due to a unit force at c (Fig. 9.16c), and p_2 = internal force in any member of the basic determinate structure due to a pair of unit axial forces acting at the cut ends of member ci (Fig. 9.16d). Loading the basic determinate structure with a unit load at c, as shown in Fig. 9.16c, will produce deflections at c as well as at the gap in the cut member ic; again from virtual work these deflections are given by

$$\delta_{cc} = \sum p_1 \frac{p_1 l}{AE} \quad \text{and} \quad \delta_{(ci)c} = \sum p_1 \frac{p_2 l}{AE}$$

Furthermore, subjecting the determinate truss to a pair of unit axial forces acting along the cut ends of member ci (Fig. 9.16d) will cause deflections at c as well as at the gap in the cut member ic given by

$$\delta_{c(ci)} = \sum p_2 \frac{p_1 L}{AE} \quad \text{and} \quad \delta_{(ci)ci} = \sum p_2 \frac{p_2 L}{AE}$$

where $\sum$ denotes all truss members in Fig. 9.16b. Applying the compatibility conditions, namely, (1) the vertical deflection at joint c is zero, and (2) the overlap at the cut ends of member ci is zero, one can write

$$\Delta_c + x_1 \delta_{cc} + x_2 \delta_{c(ci)} = 0$$

$$\Delta_{ci} + x_1 \delta_{(ci)c} + x_2 \delta_{(ci)ci} = 0 \qquad (9.26)$$

In matrix form this becomes

$$\begin{bmatrix} \delta_{cc} & \delta_{c(ci)} \\ \delta_{(ci)c} & \delta_{(ci)ci} \end{bmatrix} \begin{Bmatrix} x_1 \\ x_2 \end{Bmatrix} + \begin{Bmatrix} \Delta_c \\ \Delta_{ci} \end{Bmatrix} = 0$$

in which the left-hand side matrix is the flexibility matrix, which is always symmetric by virtue of Maxwell's theorem of reciprocal deflections. Table 9.4 presents the calculations for the δ's and Δ's.

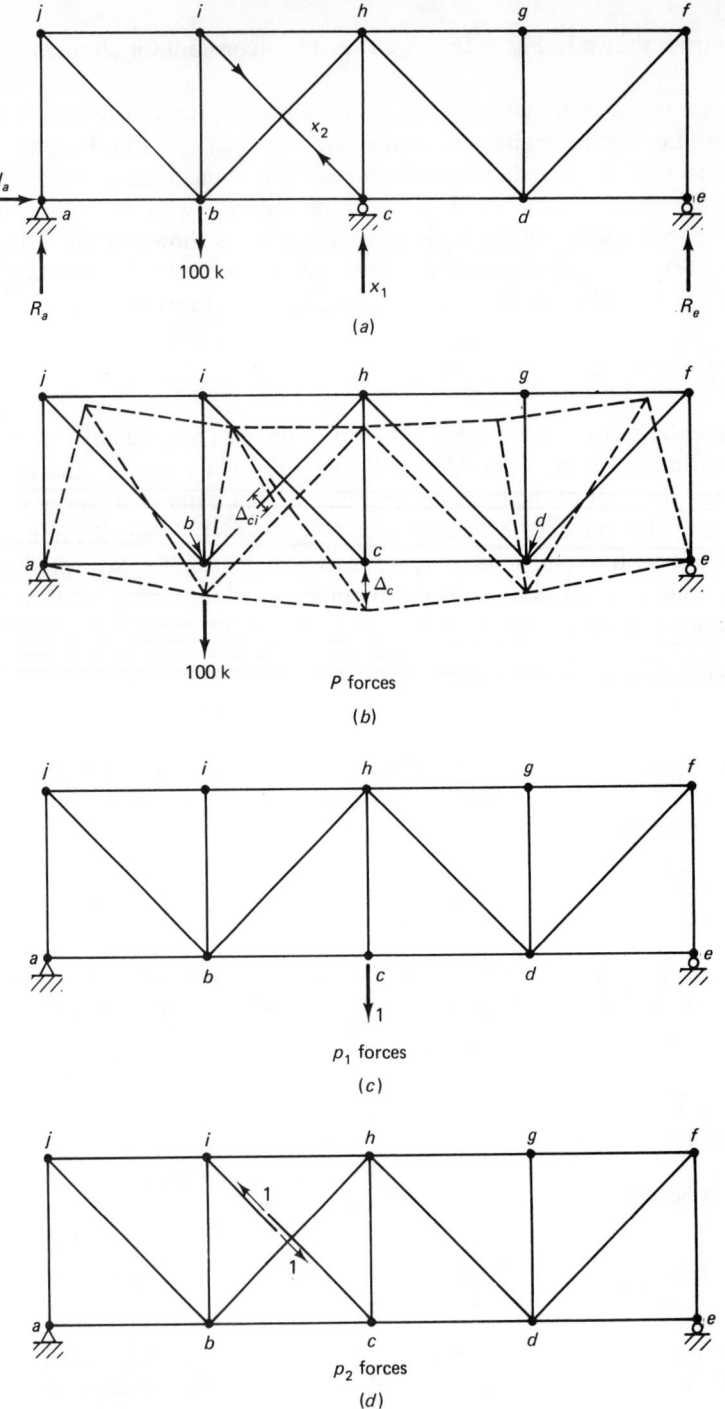

Figure 9.16 Structure for Example 9.9.

TABLE 9.4

Member	l	P	p_1	p_2	p_1Pl	p_2Pl	p_1^2l	p_2^2l	p_1p_2l	$F = P + X_1p_1 + X_2p_2$
ab	d	0	0	0	0	0	0	0	0	0
bc	d	$+50$	$+1$	$-\dfrac{\sqrt{2}}{2}$	$+50d$	$-25\sqrt{2}\,d$	d	$\dfrac{d}{2}$	$-\dfrac{d\sqrt{2}}{2}$	$+14.3$ k
cd	d	$+50$	$+1$	0	$+50d$	0	d	0	0	-15.2 k
de	d	0	0	0	0	0	0	0	0	0
ef	d	-25	$-\dfrac{1}{2}$	0	$+\dfrac{25}{2}d$	0	$\dfrac{1}{4}d$	0	0	$+7.6$
fg	d	-25	$-\dfrac{1}{2}$	0	$+\dfrac{25}{2}d$	0	$\dfrac{1}{4}d$	0	0	$+7.6$
gh	d	-25	$-\dfrac{1}{2}$	0	$+\dfrac{25}{2}d$	0	$\dfrac{1}{4}d$	0	0	$+7.6$
hi	d	-75	$-\dfrac{1}{2}$	$-\dfrac{\sqrt{2}}{2}$	$\dfrac{75}{2}d$	$+\dfrac{75}{2}\sqrt{2}\,d$	$\dfrac{1}{4}d$	$\dfrac{d}{2}$	$\dfrac{d\sqrt{2}}{4}$	-12.8
ij	d	-75	$-\dfrac{1}{2}$	0	$\dfrac{75}{2}d$	0	$\dfrac{1}{4}d$	0	0	-42.4
ja	d	-75	$-\dfrac{1}{2}$	0	$\dfrac{75}{2}d$	0	$\dfrac{1}{4}d$	0	0	-42.4
jb	$d\sqrt{2}$	$+75\sqrt{2}$	$+\dfrac{\sqrt{2}}{2}$	0	$75d\sqrt{2}$	0	$\dfrac{d\sqrt{2}}{2}$	0	0	$+59.9$
ib	d	0	0	$-\dfrac{\sqrt{2}}{2}$	0	0	0	$\dfrac{d}{2}$	0	$+29.5$
bh	$d\sqrt{2}$	$+25\sqrt{2}$	$-\dfrac{\sqrt{2}}{2}$	$+1$	$-25d\sqrt{2}$	$+50d$	$\dfrac{d\sqrt{2}}{2}$	$d\sqrt{2}$	$-d$	$+39.6$
hc	d	0	$+1$	$-\dfrac{\sqrt{2}}{2}$	0	0	d	$\dfrac{d}{2}$	$-\dfrac{d\sqrt{2}}{2}$	-35.6
hd	$d\sqrt{2}$	$-25\sqrt{2}$	$-\dfrac{\sqrt{2}}{2}$	0	$+25d\sqrt{2}$	0	$\dfrac{d\sqrt{2}}{2}$	0	0	$+10.7$
dg	d	0	0	0	0	0	0	0	0	0
df	$d\sqrt{2}$	$+25\sqrt{2}$	$+\dfrac{\sqrt{2}}{2}$	0	$+25d\sqrt{2}$	0	$\dfrac{d\sqrt{2}}{2}$	0	0	-10.7
ci	$d\sqrt{2}$	0	0	$+1$	0	0	0	$d\sqrt{2}$	0	-41.8
				Σ	$391.4d$	$67.68d$	$7.32d$	$4.83d$	$-2.06d$	

$$\begin{bmatrix} 7.32\dfrac{d}{AE} & -2.06\dfrac{d}{AE} \\[2mm] -2.06\dfrac{d}{AE} & 4.83\dfrac{d}{AE} \end{bmatrix} \begin{Bmatrix} x_1 \\ x_2 \end{Bmatrix} + \begin{Bmatrix} 391.4\dfrac{d}{AE} \\[2mm] 67.68\dfrac{d}{AE} \end{Bmatrix} = 0$$

Solving for x_1 and x_2 gives $x_1 = -65.2$ k and $x_2 = -41.8$ k. The negative sign means that the reaction at c is upward and that the force in member ci is compressive. ∎

■ EXAMPLE 9.10

The top chords of the Pratt truss in Fig. 9.17a are subjected to a temperature rise of 60°F. Determine the forces in the truss, assuming $\alpha = 6.5 \times 10^{-6}$ in/in/°F; $E = 30,000$ ksi and area = 10 in^2 for all members.

Solution. The Pratt truss is internally indeterminate to the second degree. Choose bars ci and cg as the two redundant members; these two members are first cut, and the temperature rise in the upper chords will create gaps Δ_{ci} and Δ_{cg} in the truss. By virtual work

$$\Delta_{ci} = \sum p_1(\alpha t° l) \quad \text{and} \quad \Delta_{cg} = \sum p_2(\alpha t° l)$$

in which p_1 and p_2 are the internal forces in any member of the basic determinate structure due to a pair of unit axial forces acting at the cut ends of members ci and cg, respectively, as shown in Fig. 9.17c and d. Subjecting the determinate truss to a pair of unit axial forces acting along the cut ends of member ci (Fig. 9.17c) will cause gaps in members ci and cg, determined by

$$\delta_{ci(ci)} = \sum p_1^2 \frac{l}{AE} \quad \text{and} \quad \delta_{cg(ci)} = \sum p_1 \frac{p_i l}{AE}$$

Similarly, the pair of unit axial forces shown in Fig. 9.17d will cause gaps in members ci and cg given by

$$\delta_{cg(cg)} = \sum \frac{p_2^2 l}{AE} \quad \text{and} \quad \delta_{ci(cg)} = \sum p_2 \frac{p_i l}{AE}$$

Applying the compatibility conditions that the overlap at the cut ends of member ci and cg must be zero, we can write

$$\Delta_{ci} + X_1 \delta_{ci(ci)} + X_2 \delta_{ci(cg)} = 0$$

$$\Delta_{cg} + X_1 \delta_{cg(ci)} + X_2 \delta_{cg(cg)} = 0$$

The calculations for the Δ's and δ's are presented in Table 9.5. For the top chord members

$$\alpha t° l = +(6.5)(10^{-6})(60)(d) = (390)(10^{-6})(d)$$

Thus from Table 9.5:

$$(-195)(10^{-6})(d\sqrt{2}) + X_1 \frac{4.828 d}{30 \times 10^4} + X_2 \frac{d/2}{30 \times 10^4} = 0$$

$$(-195)(10^{-6})(d\sqrt{2}) + X_1 \frac{(d/2)}{30 \times 10^4} + X_2 \frac{4.828 d}{30 \times 10^4} = 0$$

From symmetry, $X_1 = X_2$; hence $X_1 = X_2 = 15.5$ k in tension.   ■

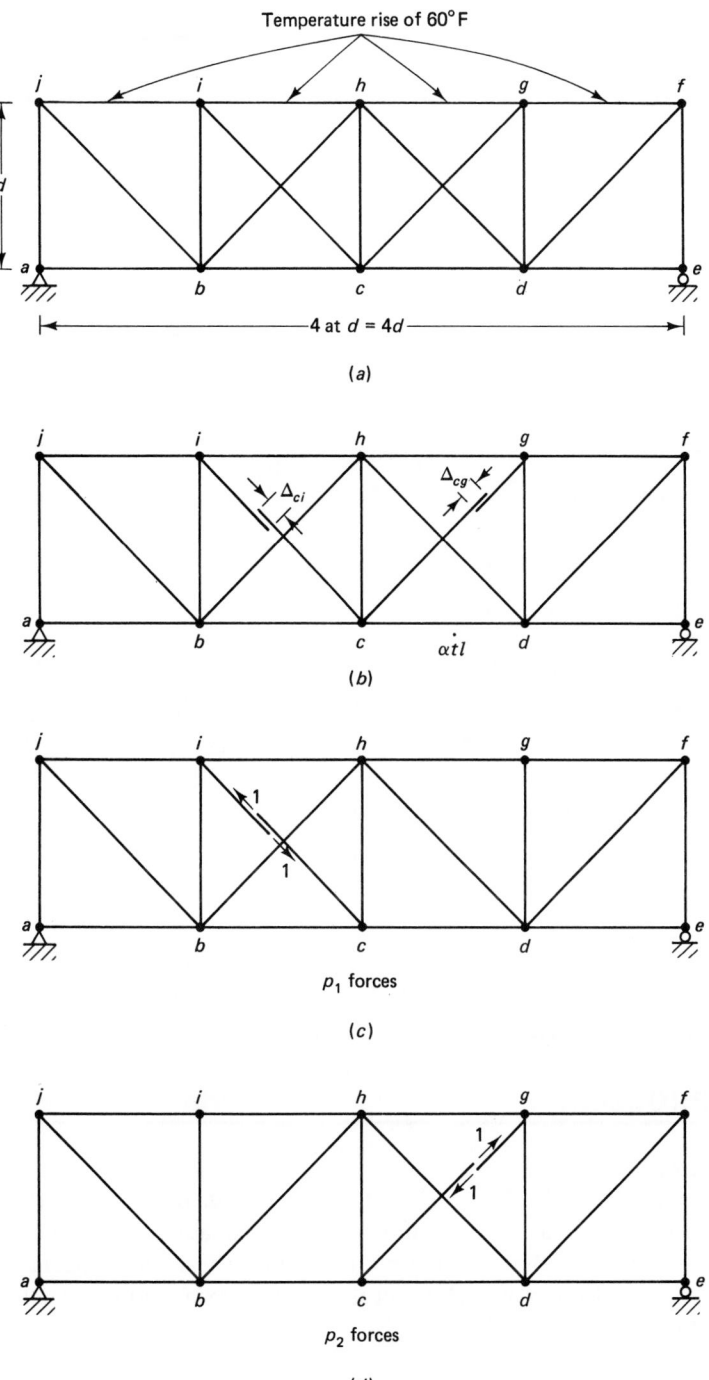

Figure 9.17 Structure for Example 9.10.

TABLE 9.5

Member	l	$\alpha t°l$	p_1	p_2	$p_1(\alpha t°l)$	$p_2(\alpha t°l)$	$p_1^2 l$	$p_2^2 l$	$p_1 p_2 l$
ab	d	0	0	0	0	0	0	0	0
bc	d	0	$-\dfrac{\sqrt{2}}{2}$	0	0	0	$\dfrac{d}{2}$	0	0
cd	d	0	0	$-\dfrac{\sqrt{2}}{2}$	0	0	0	$\dfrac{d}{2}$	0
de	d	0	0	0	0	0	0	0	0
ef	d	0	0	0	0	0	0	0	0
fg	d	$(390)(10)^{-6}d$	0	0	0	0	0	0	0
gh	d	$(390)(10)^{-6}d$	0	$-\dfrac{\sqrt{2}}{2}$	0	$(-195)(10)^{-6}d\sqrt{2}$	0	$\dfrac{d}{2}$	0
hi	d	$(390)(10)^{-6}d$	$-\dfrac{\sqrt{2}}{2}$	0	$(-195)(10)^{-6}d\sqrt{2}$	0	$\dfrac{d}{2}$	0	0
ij	d	$(390)(10)^{-6}d$	0	0	0	0	0	0	0
ja	d	0	0	0	0	0	0	0	0
jb	$d\sqrt{2}$	0	0	0	0	0	0	0	0
ib	d	0	$-\dfrac{\sqrt{2}}{2}$	0	0	0	$\dfrac{d}{2}$	0	0
bh	$d\sqrt{2}$	0	$+1$	0	0	0	$d\sqrt{2}$	0	0
hc	d	0	$-\dfrac{\sqrt{2}}{2}$	$-\dfrac{\sqrt{2}}{2}$	0	0	$\dfrac{d}{2}$	$\dfrac{d}{2}$	$\dfrac{d}{2}$
hd	$d\sqrt{2}$	0	0	$+1$	0	0	0	$d\sqrt{2}$	0
dg	d	0	0	$-\dfrac{\sqrt{2}}{2}$	0	0	0	$\dfrac{d}{2}$	0
df	$d\sqrt{2}$	0	0	0	0	0	0	0	0
ci	$d\sqrt{2}$	0	$+1$	0	0	0	$d\sqrt{2}$	0	0
cg	$d\sqrt{2}$	0	0	$+1$	0	0	0	$d\sqrt{2}$	0
				Σ	$(-195)(10)^{-6}d\sqrt{2}$	$(-195)(10)^{-6}d\sqrt{2}$	$4.828d$	$4.828d$	$\dfrac{d}{2}$

9.3 SOLVED PROBLEM

■ SOLVED PROBLEM 9.1

The continuous beam $ABCD$ is loaded as shown in Fig. 9.18a. It is also assumed support A settles down 0.1 in. **(a)** Use the three-moment equation to deduce the support moments. **(b)** Sketch the bending-moment and shear-force diagrams. **(c)** Calculate the vertical deflection at G. Neglect the effect of axial and shear deformations on flexure. Given: $EI = 50{,}000 \text{ k} \cdot \text{ft}^2$.

Solution.

$$M_C = \frac{wL^2}{2} = \frac{1(3)^2}{2} = 4.5 \text{ k} \cdot \text{ft}$$

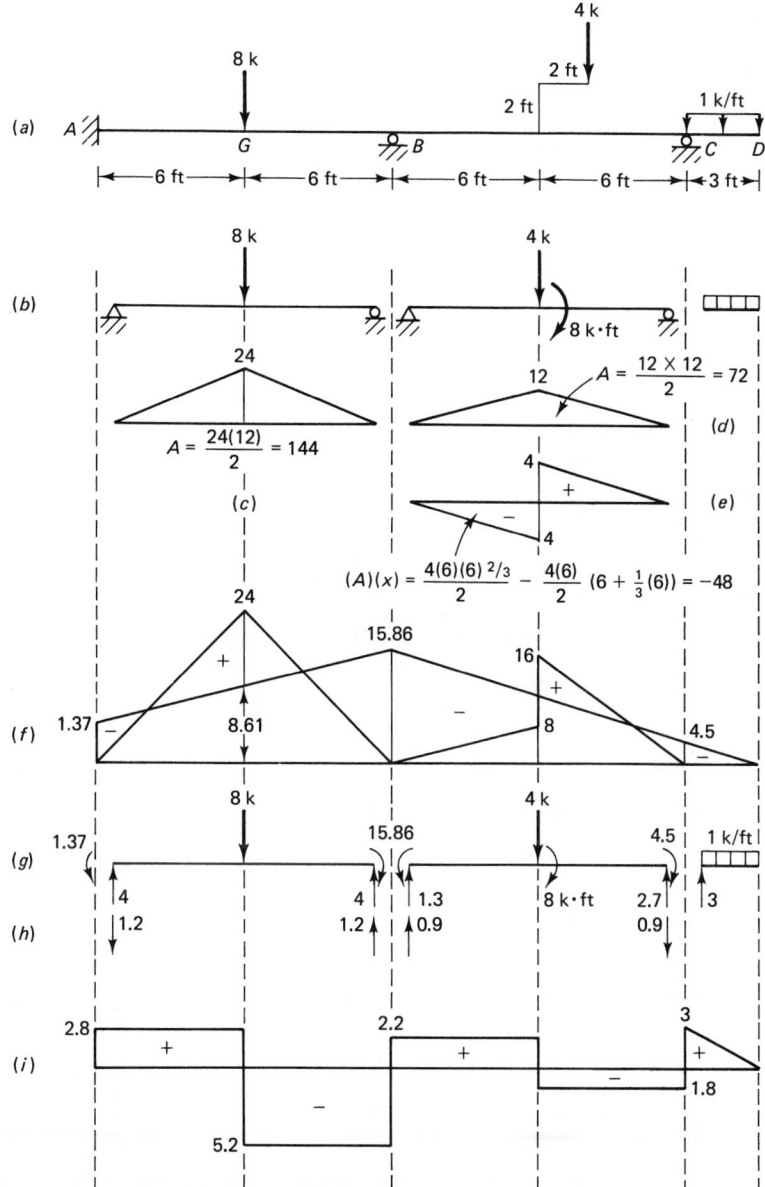

Figure 9.18 Structure for Solved Problem 9.1.

Span AA' at AB:

$$2M_A(12) + M_B(12) = -\frac{6(144)(6)}{12} - \frac{6EI}{12}(0 - 0.1/12)$$

$$2M_A + M_B = 36 - 17.4$$

$$2M_A + M_B = 18.6 \tag{a}$$

Spans AB and BC:

$$M_A(12) + 2M_B(12 + 12) + 12M_C$$

$$= -\frac{6(144)(6)}{12} - \frac{6(72)(6)}{12} + \frac{6(48)}{12} - 6EI\frac{(0.1/12 - 0)}{12}$$

$$M_A + 4M_B + M_C = -36 - 18 + 2 - 17.4 = -69.4$$

But

$$M_C = -4.5 \text{ k} \cdot \text{ft} \qquad\qquad\qquad (b)$$

Therefore,

$$M_A + 4M_B = -64.9$$

Solving (a) and (b),

$$M_A = -1.37 \text{ k} \cdot \text{ft}$$

$$M_B = -15.86 \text{ k} \cdot \text{ft}$$

SFD and BMD are shown in Fig. 9.18i and f, respectively.

Vertical deflection at G = moment of area of BMD between A and G about point G divided by EI

Therefore,

$$\Delta_G = \text{deflection due to bending} + \text{deflection due to settlement}$$

$$= \frac{1}{EI}\left\{\frac{24(6)(6)}{2 \times 3} - \left[\frac{(1.37)(6)}{2}\left(\frac{2}{3}6\right) + \frac{(8.61)(6)(\frac{6}{3})}{2}\right]\right\} + \frac{0.1}{2}$$

$$= \frac{1}{EI}(144 - 68.1) + 0.05$$

$$= \frac{75.9}{50,000} \times 12 + 0.05 = 0.068 \text{ in} \qquad\qquad\blacksquare$$

PROBLEMS

9.1 to 9.16 Use the three-moment equation method to determine the support moments for the continuous beams shown in Figs. P9.1 to P9.16. Draw also shear-force and bending-moment diagrams.

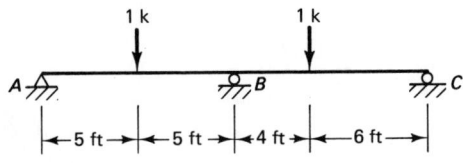

Figure P9.1

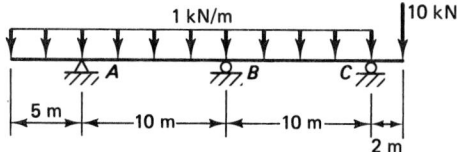

Figure P9.2

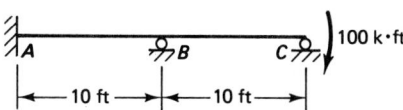

Figure P9.3

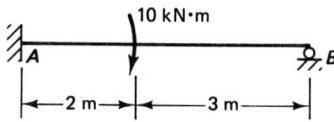

Figure P9.4

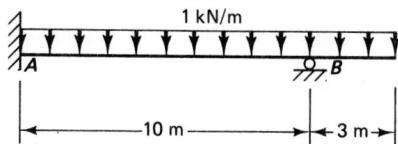

Figure P9.5

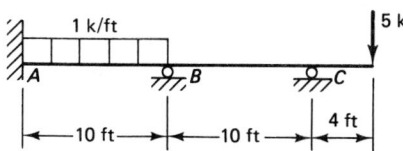

Figure P9.6

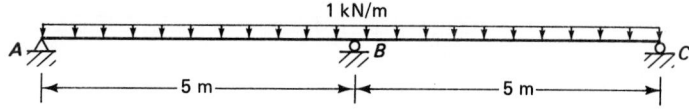

Figure P9.7

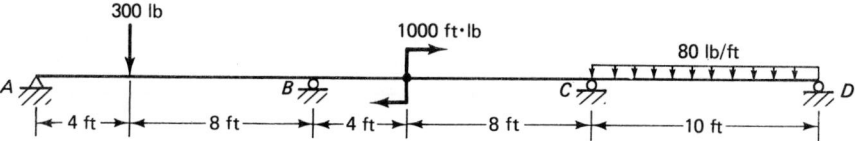

Figure P9.8

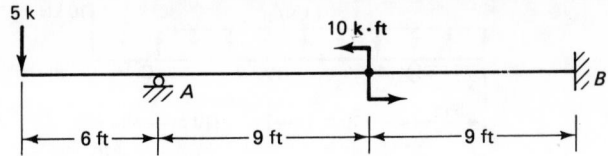

Figure P9.9

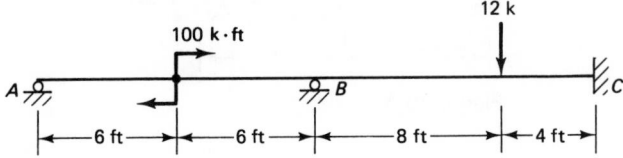

Figure P9.10

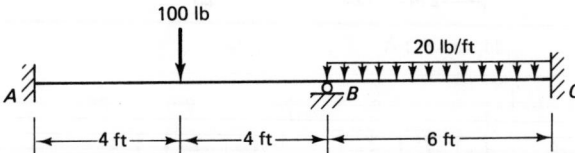

Figure P9.11

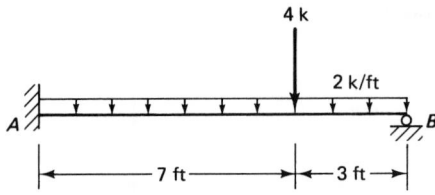

Figure P9.12

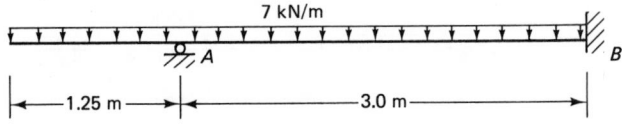

Figure P9.13

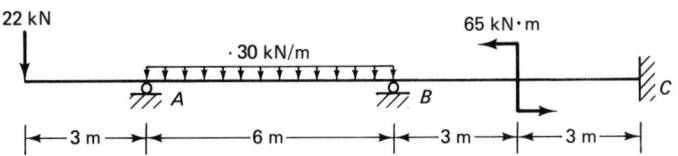

Figure P9.14

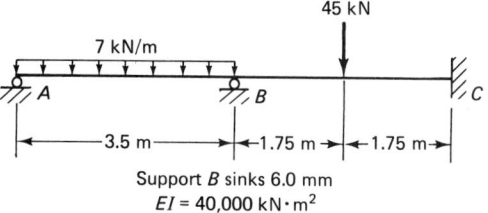

Figure P9.15

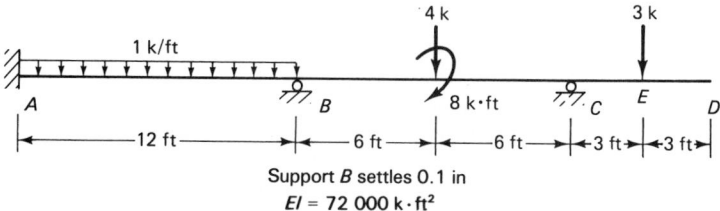

Figure P9.16

9.17 Using the three-moment equation method, determine the support moments for the two-span continuous beam shown. Draw shear-force and bending-moment diagrams. Using the conjugate-beam method, determine the vertical deflection at G. $EI = 80,000$ k·ft².

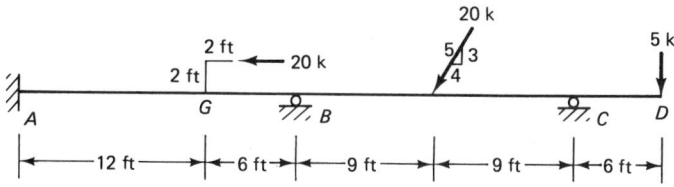

Figure P9.17

9.18 The continuous beam $ABCD$ is loaded as shown; it is also assumed that support B settles down 0.1 in. Assume axial deformations are negligible. Use the three-moment equation to deduce the support moments. Sketch the bending-moment and shear-force diagrams. Also calculate the vertical deflection at E. $EI = 50,000$ k·ft².

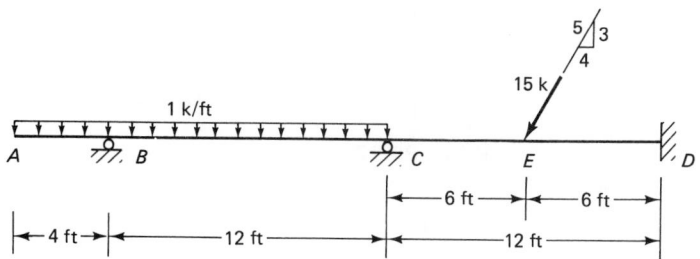

Figure P9.18

9.19 to 9.25 Use the consistent-deformation method to solve the structures shown in Figs. P9.7, P9.9, P9.10, P9.11, P9.12, P9.13, and P9.15.
Use the consistent-deformation method to solve Problems 9.26 to 9.39.

9.26 Determine the stress in the steel cable.

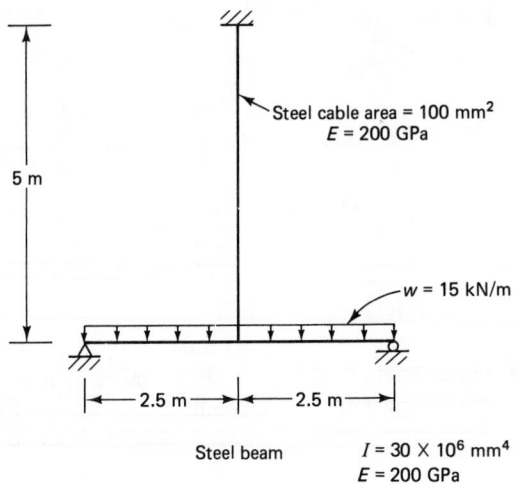

Figure P9.26

9.27 Determine the horizontal reaction at A.

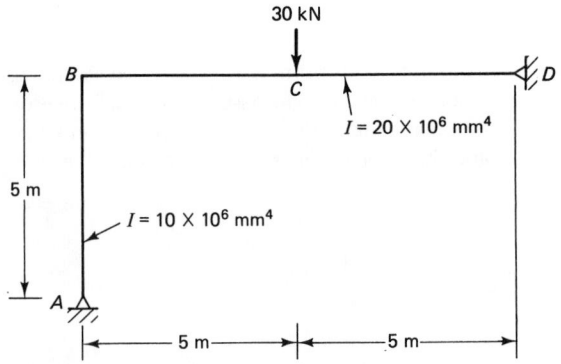

Figure P9.27

9.28 Determine the stress in the steel tie.

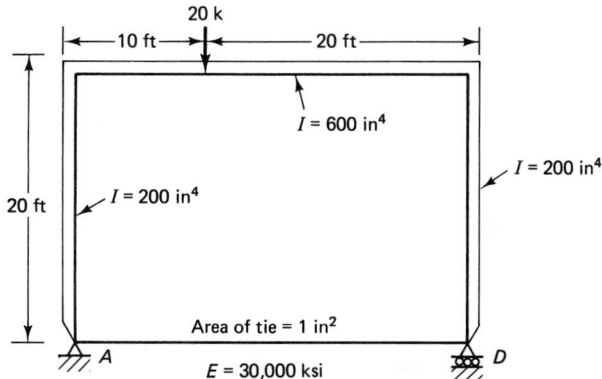

Figure P9.28

9.29 to 9.32 Determine the forces in all members; cross-sectional area of each member is 1000 mm^2; $E = 200$ GPa.

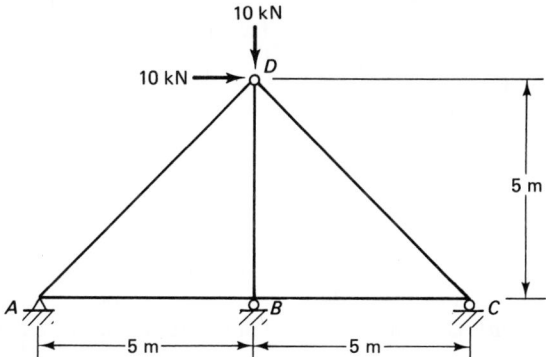

Figure P9.29

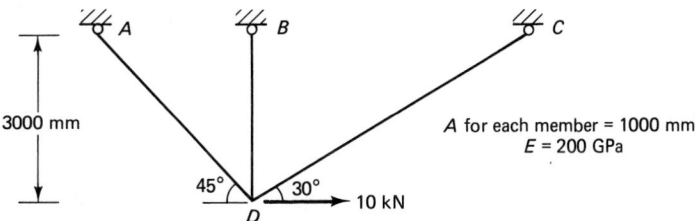

Figure P9.30

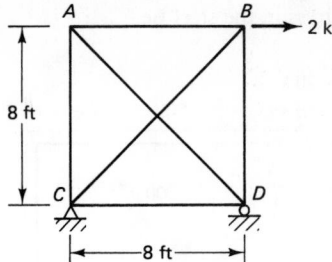

Figure P9.31

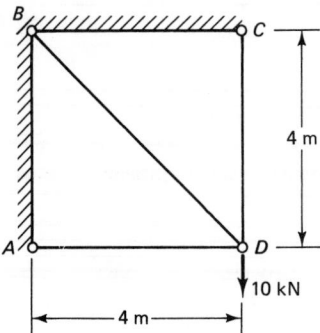

Figure P9.32

9.33 In the truss shown, AB is a tie bar. The members in tension have a cross-sectional area of 1 in^2, and the members in compression have a cross-sectional area of 2 in^2. Determine the force in the tie, and hence the final loads in the members.

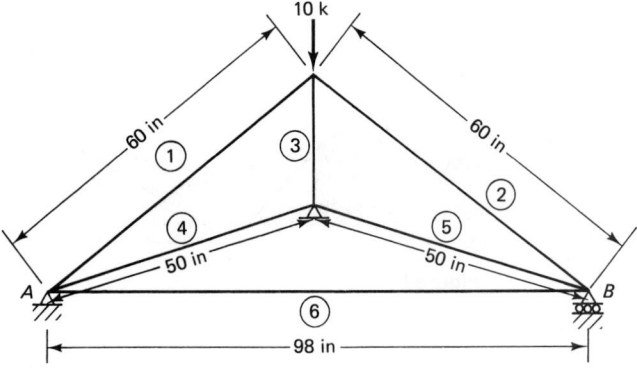

Figure P9.33

9.34 Determine the reaction at B for the pin-connected frame shown, when (a) support B does not settle, and (b) support B settles downward a distance of 2.5 mm. $E = 200$ GPa, A (for each member) $= 1300$ mm^2.

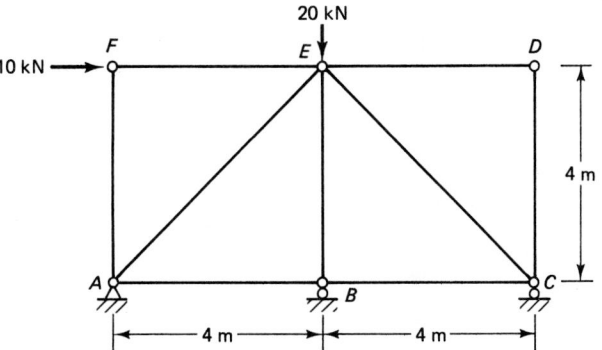

Figure P9.34

9.35 to 9.39 Determine the reactions for the trusses shown. Assume the cross-sectional areas to be the same for all members.

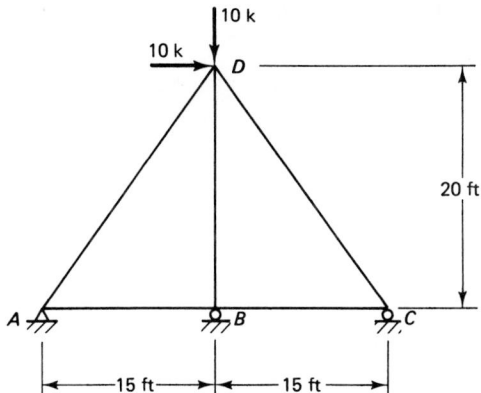

Figure P9.35

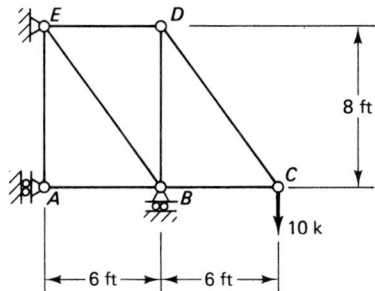

Figure P9.36

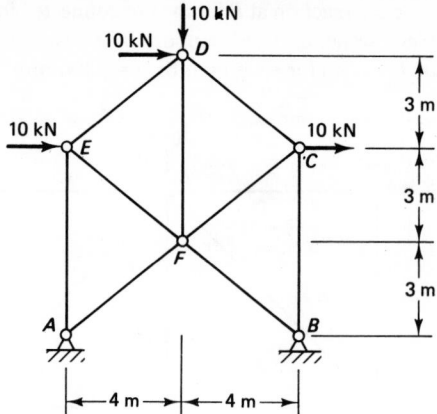

Figure P9.37

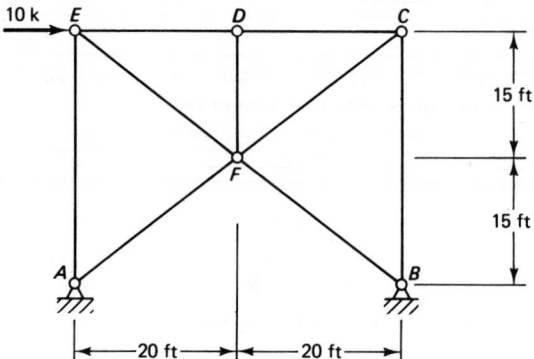

Figure P9.38

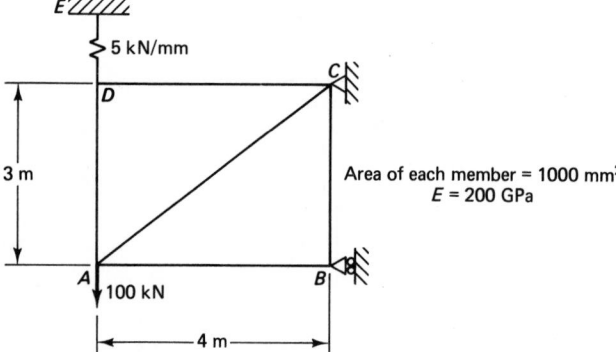

Figure P9.39

Chapter **10**

Classical Equilibrium (Displacement) Methods of Analysis— Indeterminate Structures

There is nothing more frightful than ignorance in action.

Goethe

10.1 THE SLOPE-DEFLECTION METHOD

10.1.1 General

We indicated that the three-moment equation method, using support moments as unknowns, is a force method. The governing equations were derived by enforcing compatibility of displacements (deflection Δ and slope θ) at specified supports. These equations implicitly satisfy force-displacement relations, equilibrium, as well as the boundary conditions. In contrast, the slope-deflection method uses displacements as unknowns, and because of that, it is known as a "displacement method." Here the compatibility of displacements of adjacent members at a support or joint is automatically satisfied since the members are rigidly connected to one another. After the force-displacement relations are established for each member in the structure (i.e., the relations between the moments at the ends of a member and the displacements at its ends as well as any transverse loads on the member), equilibrium conditions are applied at points of connectivity. This procedure leads to a set of simultaneous equations with displacements as unknowns. After these displacements are evaluated, the end moments can be computed from the force-displacement relation for each member of the structure.

The slope-deflection method is fairly easy to apply even when the structure becomes relatively complex since the individual force-displacement equations are readily deduced, regardless of the number of unknown joint displacements. The method shows to best advantage when the number of joint displacements is small in comparison with the number of members.

10.1.2 Sign Convention

Up to now we have followed the convention that a moment with a plus sign causes tension in the bottom fibers and one with a negative sign causes tension in the top

Allied Bank Plaza, Houston, Texas. (Photo courtesy of American Institute of Steel Construction.)

fibers of a member. This distinction between the tension and compression side of a member is not possible for the analysis of structures when the rotational equilibrium of the joints must be considered as is the case in the slope-deflection method. Therefore, we must affix the same sign to a moment at the end of the member that rotates the

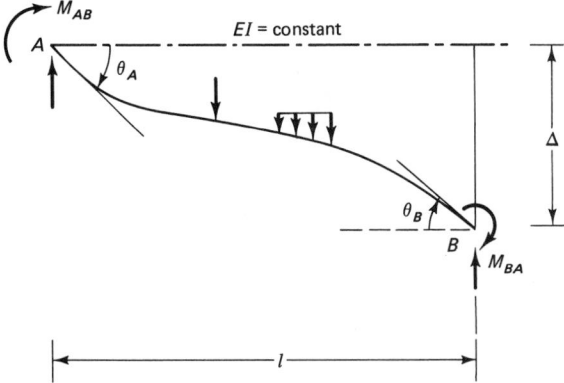

Figure 10.1 Beam with positive end moments, rotations, and relative displacement.

member in a certain direction regardless of the type of bending it imposes on the member. Accordingly, the following sign convention is adopted: Moments and rotations at the ends of a member and displacements of one end relative to the other are all considered positive if they are *clockwise*. Figure 10.1 shows a beam AB with the end moments, rotations, and relative displacements being positive. After all the unknown moments are determined, they are given the correct signs following the usual convention for constructing bending-moment diagrams.

10.1.3 Derivation of the Slope-Deflection Equation

Assume that the beam AB in Fig. 10.1 has been isolated from a loaded statically indeterminate beam or rigid frame. Owing to the end rotations, relative deflection Δ, and the external loads on the beam, end moments M_{AB} and M_{BA} are induced because of continuity or support requirements. Therefore, we can say that M_{AB} and M_{BA} are functions of θ_A, θ_B, Δ, and the external load on the span. Let us consider the effects of these four variables separately.

End Moments Due to Rotation θ_A While θ_B = Δ = Load = 0   This condition is shown in Fig. 10.2a. From the conjugate-beam method (Chapter 7), the slope at A, θ_A, is the reaction at A of the conjugate beam:

$$\theta_A = \frac{1}{2}\left(\frac{M'_{AB}}{EI}\right)(l) - \frac{1}{2}\left(\frac{M'_{BA}}{EI}\right)(l)$$

$$= \frac{l}{2EI}(M'_{AB} - M'_{BA})$$

Also deflection at A is the moment of the load on the conjugate beam about end A:

$$\text{Deflection at } A = 0 = \frac{1}{2}\left(\frac{M'_{AB}}{EI}\right)(l)\left(\frac{1}{3}l\right) - \frac{1}{2}\left(\frac{M'_{BA}}{EI}\right)(l)\left(\frac{2}{3}l\right)$$

$$0 = M'_{AB} - 2M'_{BA}$$

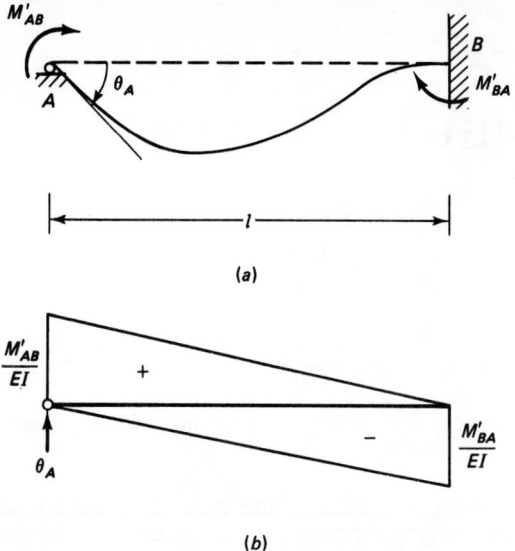

Figure 10.2 End moments due to rotation θ_A.

or

$$M'_{BA} = \frac{1}{2} M'_{AB}$$

$$\theta_A = \frac{l}{2EI} \left(M'_{AB} - \frac{1}{2} M'_{AB} \right) = \frac{l}{4EI} \cdot M'_{AB}$$

Hence,

$$M'_{AB} = \frac{4EI}{l} \cdot \theta_A$$

and   (10.1)

$$M'_{BA} = \frac{2EI}{l} \cdot \theta_A$$

End Moments Due to Rotation θ_B While $\theta_A = \Delta = $ Load $= 0$ For the beam
AB in Fig. 10.3 satisfying the above conditions, the end moments can be deduced in
a similar fashion to the previous case. Thus we can write

$$M''_{BA} = \frac{4EI}{l} \cdot \theta_B$$

                                                                                              (10.2)

$$M''_{AB} = \frac{2EI}{l} \cdot \theta_B$$

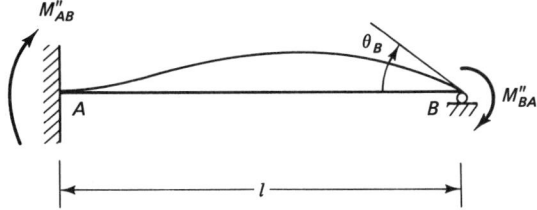

Figure 10.3 End moments due to rotation θ_B.

Moments Developed at the Ends Due to a Relative Joint Displacement Δ, with $\theta_A = \theta_B =$ Load = 0   This condition is shown in Fig. 10.4a. Using the conjugate-beam method, we can show by using Fig. 10.4b that the change in the slope between A and B is equal to the area of the load diagram of the conjugate beam (M/EI diagram); thus

$$\theta_A - \theta_B = 0 = \left(\frac{1}{2}\right)\left(\frac{M'''_{BA}}{EI}\right)(l) - \left(\frac{1}{2}\right)\left(\frac{M'''_{AB}}{EI}\right)(l)$$

or

$$M'''_{AB} = M'''_{BA}$$

Since there are no reactions for the conjugate beam (Fig. 10.4b), the moment of the M/EI diagram about B must be equal to the displacement Δ. Thus,

$$\left(\frac{1}{2}\right)\left(\frac{M'''_{BA}}{EI}\right)(l)\left(\frac{l}{3}\right) - \left(\frac{1}{2}\right)\left(\frac{M'''_{AB}}{EI}\right)(l)\left(\frac{2}{3}l\right) = \Delta$$

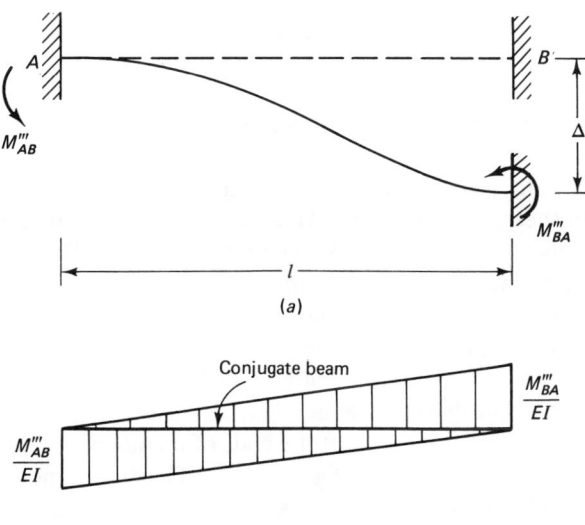

(a)

Conjugate beam

(b)

Figure 10.4 End moments due to relative support displacement Δ.

Therefore,

$$\left(\frac{1}{2}\right)\left(\frac{M'''_{AB}}{EI}\right)(l^2)\left(\frac{1}{3} - \frac{2}{3}\right) = \Delta$$

$$-\frac{M'''_{AB}l^2}{6EI} = \Delta$$

or

$$M'''_{AB} = -\frac{6EI\Delta}{l^2} = M'''_{BA} \tag{10.3}$$

End Moments When External Loads Act on the Member, and $\theta_A = \theta_B = \Delta = 0$ End moments FEM_{AB} and FEM_{BA} are produced; such end moments are denoted as fixed-end moments (FEM). The Appendix gives expressions for fixed-end moments for different loading conditions.

Combining the end moments due to θ_A, θ_B, Δ, and load from above, we obtain

$$M_{AB} = \frac{2EI}{l}\left(2\theta_A + \theta_B - \frac{3\Delta}{l}\right) + \text{FEM}_{AB}$$

and $\tag{10.4}$

$$M_{BA} = \frac{2EI}{l}\left(2\theta_B + \theta_A - \frac{3\Delta}{l}\right) + \text{FEM}_{BA}$$

Equations (10.4) are the basic *slope-deflection equations.* It should be noted that in the derivation of the above equations, only transverse bending deformations are accounted for; this is because shear and axial-load deformations in bending members are in most cases generally very small compared with bending deformations and are therefore ignored.

Equations (10.4) are written for each member end in terms of the unknown displacements on which compatibility conditions have been imposed. Then an equation of equilibrium is written at each joint with unknown rotational displacements. Similarly, in the case of frames with freedom to sway, an equation of equilibrium is written in terms of column shears and moments. This means that there are the same number of equilibrium equations as there are unknown displacements. Thus one can say that the slope-deflection method is an *equilibrium method.*

It is obvious from Eqs. (10.4) that the *final* moments at the ends of a member are made up of two parts: the fixed-end moments due to any loading on the member as well as changes in these fixed-end moments created by the displacements (rotational and/or translational) at the ends of the member. For example, for the two structures shown in Fig. 10.5a, the final moments at the ends of members AB and BC are fixed-end moments only; owing to symmetry of geometry and loads, the ends of the members do not incur any displacement, that is, $\theta_B = 0$ (and $\theta_C = 0$) and remain so, before and after loading. In Fig. 10.5b (i) $\theta_B \neq 0$, and in Fig. 10.5b (ii) $\theta_B \neq 0$, $\theta_C \neq 0$, and $\theta_E \neq 0$ and joints B, C, and E translate horizontally an amount Δ.

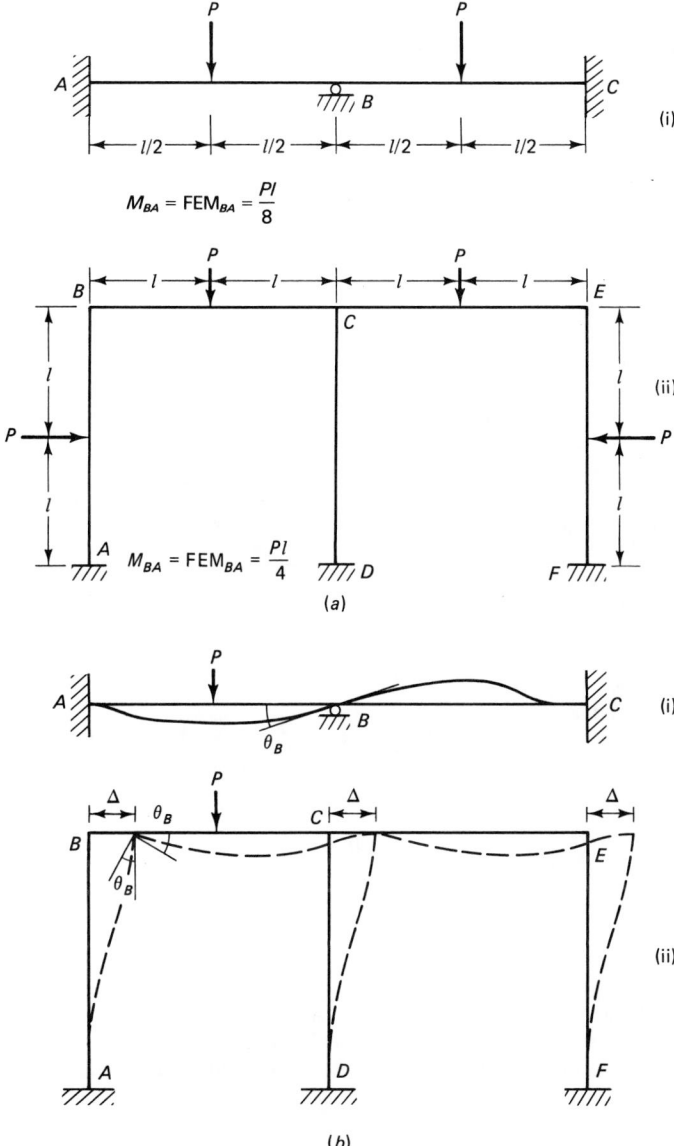

Figure 10.5 Symmetric structures subjected to (a) symmetric loading and (b) unsymmetric loading.

■ EXAMPLE 10.1

Apply the slope-deflection method to determine the support moments for the continuous beam loaded as shown in Fig. 10.6a.

Solution. The *assumed* deformed shape of the beam and the end moments of each member are shown in Fig. 10.6b and c; the joint rotations and end moments are

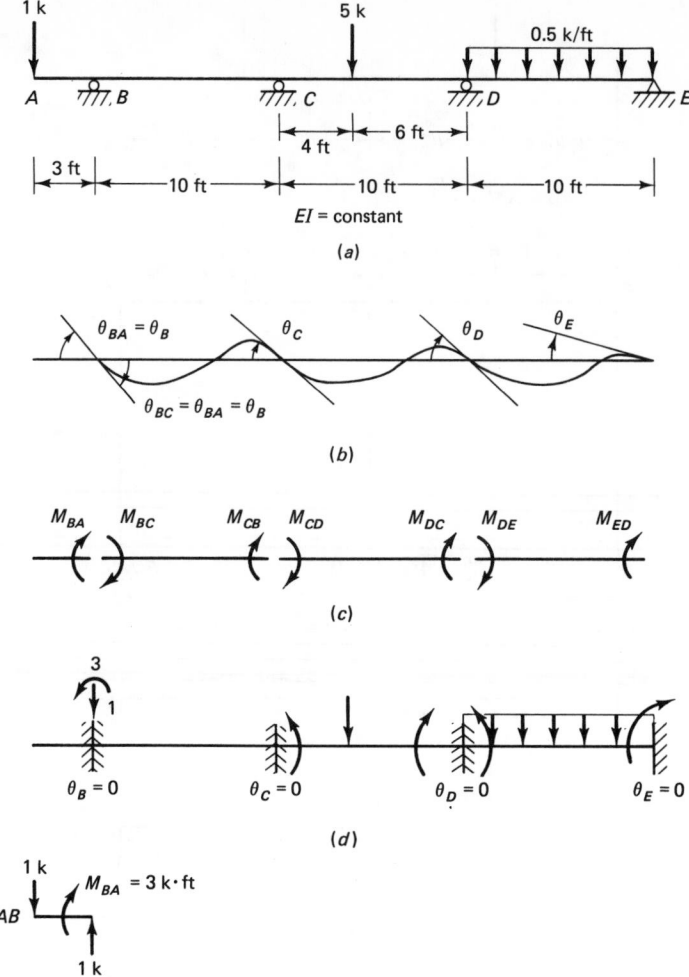

Figure 10.6 Structure for Example 10.1. (*b*) Assumed deformed shape. (*c*) Assumed end moments on members (not showing shears.) (*d*) Fixed end moments. (*e*) Free-body diagrams of joints. (*f*) Final end moments. (*g*) BMD. (*h*) Total shears. (*i*) Reactions. (*j*) SFD.

drawn to conform with the positive sense used in the slope-deflection method. Here, the overhanging load of 1 k is transferred to act on joint *B*, as a downward load of 1 k and a counterclockwise moment of 3 k · ft (see Fig. 10.6*d*).

Step 1. Fixed-End Moments. Here we fix all joints as shown in Fig. 10.6*d*, apply the external loads, and then calculate the fixed-end moments at ends of members. Thus from the Appendix:

$$\text{FEM}_{BC} = \text{FEM}_{CB} = 0$$

since there is no load on *BC*.

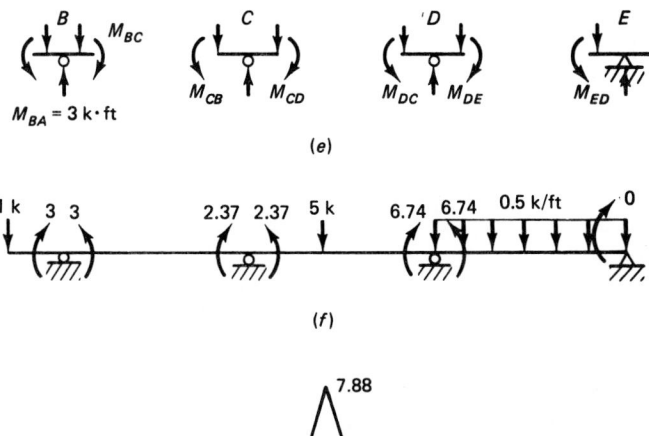

(e)

(f)

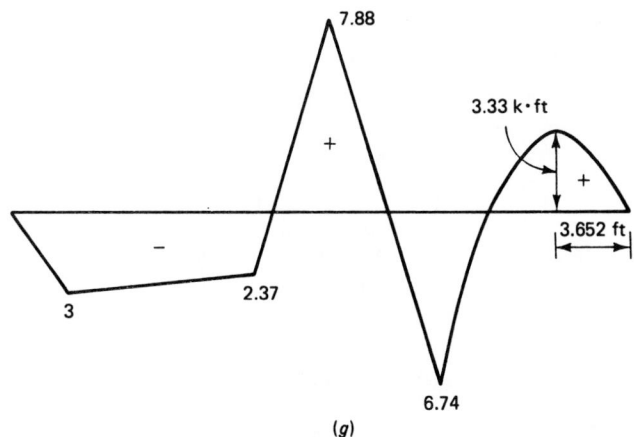

(g)

For span DE:

$+M_{max}$ is at $x = \dfrac{1}{w}\left(\dfrac{wl}{2} - \dfrac{M}{l}\right)$

$\qquad x = \dfrac{1}{0.5}\left[\dfrac{(0.5)(10)}{2} - \dfrac{6.74}{10}\right]$

$\qquad = 3.652$ ft

and

$\qquad M_{max} = \dfrac{1}{2w}\left(\dfrac{wl}{2} - \dfrac{M}{l}\right)^2$

$\qquad = \dfrac{1}{(2)(0.5)}\left[\dfrac{(0.5)(10)}{2} - \dfrac{6.74}{10}\right]^2$

$\qquad = 3.33$ k·ft

Figure 10.6 (*Continued*)

$$\text{FEM}_{CD} = -\frac{Wab^2}{l^2} = -\frac{(5)(4)(6)^2}{(10)^2} = -7.2 \text{ k·ft}$$

(negative sign is because the sense of the FEM is counterclockwise)

$$\text{FEM}_{DC} = +\frac{Wa^2b}{l^2} = +\frac{(5)(4)^2(6)}{(10)^2} = +4.8 \text{ k·ft}$$

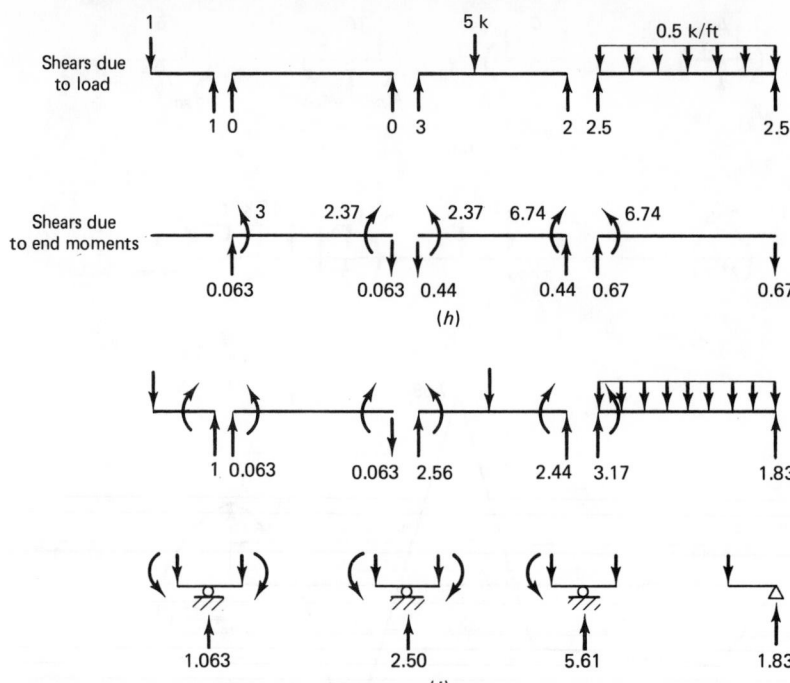

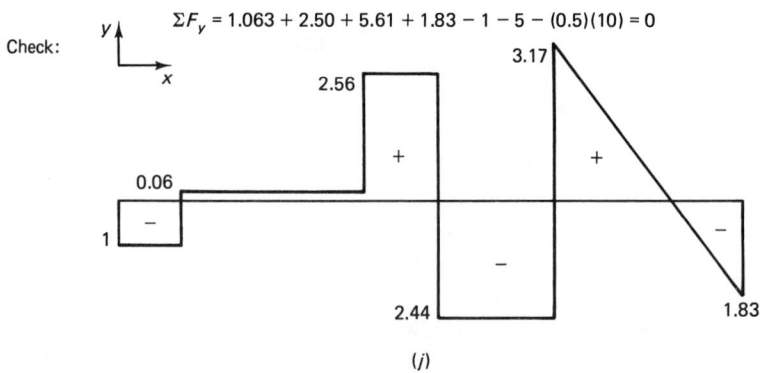

Figure 10.6 (*Continued*)

$$\text{FEM}_{DE} = -\frac{wl^2}{12} = -\frac{(0.5)(10)^2}{12} = -4.17 \text{ k} \cdot \text{ft}$$

$$\text{FEM}_{ED} = +\frac{wl^2}{12} = +\frac{(0.5)(10)^2}{12} = +4.17 \text{ k} \cdot \text{ft}$$

At end B: Owing to the overhanging load of 1 k,

$$M_{BA} = (1)(3) = 3 \text{ k} \cdot \text{ft}$$

Therefore, from equilibrium of moments at joint B,

$$M_{BC} + M_{BA} = 0$$

or

$$M_{BC} = -M_{BA} = -3 \text{ k} \cdot \text{ft}$$

Step 2. Compatibility Equations

At B:

$$\theta_{BC} = \theta_{BA} = \theta_B$$

At C:

$$\theta_{CB} = \theta_{CD} = \theta_C$$

At D:

$$\theta_{DC} = \theta_{DE} = \theta_D$$

Step 3. Slope-Deflection Equations for End Moments of Members. Apply Eqs. (10.4) to each member:

$$M_{BC} = -3 = \frac{2EI}{10}(2\theta_B + \theta_C)$$

$$M_{CB} = \frac{2EI}{10}(2\theta_C + \theta_B)$$

$$M_{CD} = \frac{2EI}{10}(2\theta_C + \theta_D) - 7.2$$

$$M_{DC} = \frac{2EI}{10}(2\theta_D + \theta_C) + 4.8 \tag{a}$$

$$M_{DE} = \frac{2EI}{10}(2\theta_D + \theta_E) - 4.17$$

$$M_{ED} = \frac{2EI}{10}(2\theta_E + \theta_D) + 4.17$$

We should note that the Δ term is zero in the above equations since there is no relative deflection at the ends of the members.

Step 4. Equilibrium Equations at the Joints. The continuous beam $ABCDE$ has four joints where rotations are possible with zero deflections at those joints. Therefore, this beam has four degrees of freedom. Referring to Fig. 10.6e showing the free-body diagrams of the four joints, we can write

At B:

$$\sum M_B = 0: \qquad M_{BC} + M_{BA} = 0 \quad \text{that is} \quad M_{BC} = -M_{BA} = -3$$

Therefore,

$$-3 = \frac{2EI}{10}(2\theta_B + \theta_C)$$

At C:

$$\sum M_C = 0: \qquad M_{CB} + M_{CD} = 0 \qquad 0 = -7.2 + \frac{2EI}{10}(4\theta_C + \theta_B + \theta_D)$$

At D:

$$\sum M_D = 0: \qquad M_{DC} + M_{DE} = 0 \qquad 0 = +0.63 + \frac{2EI}{10}(4\theta_D + \theta_C + \theta_E)$$

At E:

$$\sum M_E = 0: \qquad M_{ED} = 0 \qquad 0 = 4.17 + \frac{2EI}{10}(2\theta_E + \theta_D)$$

Solving for θ_B, θ_C, θ_D, and θ_E gives

$$EI\theta_B = -13.943 \qquad EI\theta_C = +12.887 \qquad EI\theta_D = -1.603 \quad \text{and} \quad EI\theta_E = -9.623$$

Step 5. Substitution of the joint displacements (θ's) into the member equations (a):

$$-3 = \frac{(4)(-13.943) + (2)(12.887)}{10} = -3 \text{ k} \cdot \text{ft} \qquad \text{(checks)}$$

$$M_{CB} = \frac{(4)(+12.887) + (2)(-13.943)}{10} = +2.37 \text{ k} \cdot \text{ft}$$

$$M_{CD} = \frac{(4)(12.887) + (2)(-1.603)}{10} - 7.2 = -2.37 \text{ k} \cdot \text{ft}$$

$$M_{DC} = \frac{(4)(-1.603) + (2)(12.887)}{10} + 4.8 = +6.74 \text{ k} \cdot \text{ft}$$

$$M_{DE} = \frac{(4)(-1.603) + (2)(-9.623)}{10} - 4.17 = -6.74 \text{ k} \cdot \text{ft}$$

$$M_{ED} = \frac{(4)(-9.623) + (2)(-1.603)}{10} + 4.17 = 0 \text{ k} \cdot \text{ft}$$

(as it should be).

These final moments are shown at the ends of each member, (Fig. 10.6f). Note that the positive sign for M_{CB} means it acts on CB at end C in the clockwise direction; while the negative sign for M_{CD}, acting on CD at end C, means it is counterclockwise. Also it should be noted that the action of these moments causes tension at top and compression at the bottom of the beam at C. The same applies for the other support moments in this example. Combining these support moments with those due to the external loads on simple spans will result in the bending-moment diagram shown in

Fig. 10.6g. The shears and reactions and the shear-force diagram are shown in Fig. 10.6h to j. ∎

■ **EXAMPLE 10.2**

Analyze the previous example, assuming support C settles down $\Delta = 1$ in. Given $EI = 10 \text{ k} \cdot \text{ft}^2$.

Solution. Steps 1 to 5 are the same as before except that, owing to the settlement of support C, member equations for BC and CD will change:

$$M_{BC} = -3 = \frac{2EI}{10}\left[2\theta_B + \theta_C - \frac{(3)(+\frac{1}{12})}{10}\right]$$

$$M_{CB} = \frac{2EI}{10}\left[2\theta_C + \theta_B - \frac{(3)(+\frac{1}{12})}{10}\right]$$

$$M_{CD} = \frac{2EI}{10}\left[2\theta_C + \theta_D - \frac{(3)(-\frac{1}{12})}{10}\right] - 7.2$$

$$M_{DC} = \frac{2EI}{10}\left[2\theta_D + \theta_C - \frac{(3)(-\frac{1}{12})}{10}\right] + 4.8$$

Note that the Δ term for BC and CB is positive since such settlement imparts a clockwise rotation; in contrast, the Δ term for CD and DC is negative since it imparts counterclockwise rotation.

The expressions for M_{DE} and M_{ED} remain the same as before. ∎

■ **EXAMPLE 10.3**

Calculate the end moments for each member of the rigid frame shown loaded in Fig. 10.7a.

Solution. Here $\theta_A = \theta_D = 0$. Therefore, the number of degrees of freedom = 3, namely, θ_B, θ_C, and Δ, that is, three unknowns. This single degree of freedom of joint translation is explained as follows:

It should be noted that since we neglect any small change in the length of a member due to axial loads, and since the rotations (θ) of members are small, the joint B moves horizontally and perpendicular to member AB an amount Δ. Furthermore, C moves also horizontally and perpendicular to member DC; and since the change in the axial length of BC is neglected, the horizontal movement of C must also be Δ. That is, $BB' = CC' = \Delta$ (Fig. 10.7b).

Here none of the members are loaded and therefore

$$\text{FEM}_{AB} = \text{FEM}_{BA} = \text{FEM}_{BC} = \text{FEM}_{CB} = \text{FEM}_{CD} = \text{FEM}_{DC} = 0$$

Using Eqs. (10.4), we write:

For AB

$$M_{AB} = \frac{2EI}{l}\left(2\theta_A + \theta_B - \frac{3\Delta}{l}\right) = \frac{2EI}{10}\left(\theta_B - \frac{3\Delta}{10}\right)$$

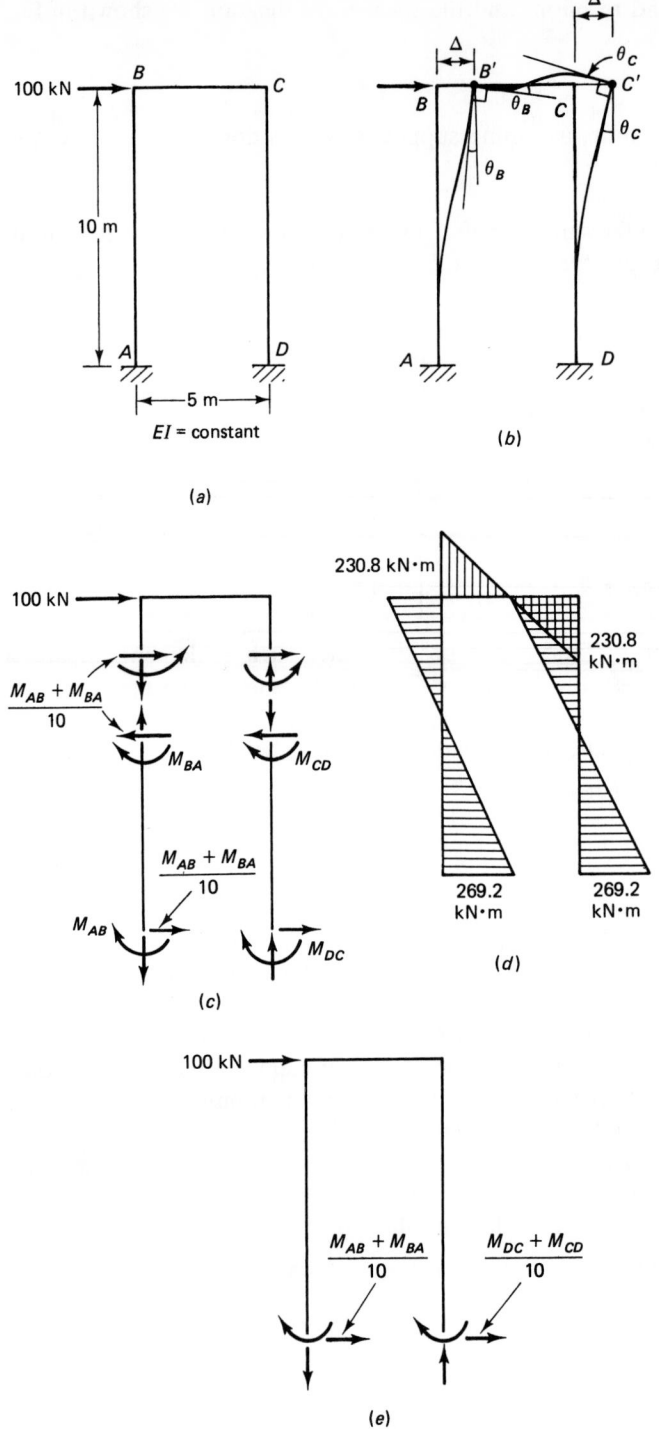

Figure 10.7 Structure for Example 10.3.

$$M_{BA} = \frac{2EI}{10}\left(2\theta_B - \frac{3\Delta}{10}\right)$$

For BC

$$M_{BC} = \frac{2EI}{5}(2\theta_B + \theta_C)$$

$$M_{CB} = \frac{2EI}{5}(2\theta_C + \theta_B)$$

(Note: There is no Δ term in both M_{BC} and M_{CB} since there is no relative displacement between the ends B and C.)

For CD

$$M_{CD} = \frac{2EI}{l}\left(2\theta_C + \theta_D - \frac{3\Delta}{l}\right) = \frac{2EI}{10}\left(2\theta_C - \frac{3\Delta}{10}\right)$$

$$M_{DC} = \frac{2EI}{10}\left(\theta_C - \frac{3\Delta}{10}\right)$$

Joint-Equilibrium Equations

$$\Sigma M_{\text{Joint }B} = M_{BA} + M_{BC} = 0$$

$$\frac{2EI}{10}\left(2\theta_B - \frac{3\Delta}{10}\right) + \frac{2EI}{5}(2\theta_B + \theta_C) = 0$$

Therefore,

$$6\theta_B + 2\theta_C - \frac{3\Delta}{10} = 0 \tag{b}$$

$$\Sigma M_{\text{Joint }C} = M_{CB} + M_{CD} = 0$$

$$\frac{2EI}{5}(2\theta_C + \theta_B) + \frac{2EI}{10}\left(2\theta_C - \frac{3\Delta}{10}\right) = 0$$

Therefore,

$$6\theta_C + 2\theta_B - \frac{3\Delta}{10} = 0 \tag{c}$$

The third equation is derived from the equilibrium of the frame as a *whole*. Consider the free-body diagram of the frame shown in Fig. 10.7c. From $\Sigma F_x = 0$:

$$100 + \frac{M_{AB} + M_{BA}}{10} + \frac{M_{DC} + M_{CD}}{10} = 0$$

Therefore,

$$100 + \frac{(2EI/10)(3\theta_B - 3\Delta/5)}{10} + \frac{(2EI/10)(3\theta_C - 3\Delta/5)}{10} = 0$$

$$10,000 + 2EI\left(3\theta_B + 3\theta_C - \frac{6\Delta}{5}\right) = 0$$

or

$$-\frac{3}{10}\theta_B - \frac{3}{10}\theta_C + \frac{3}{25}\Delta = +\frac{500}{EI} \tag{d}$$

Solving Eqs. (b), (c), and (d) yields

$$\theta_B = \frac{192.31}{EI} \qquad \theta_C = \frac{192.31}{EI} \qquad \Delta = \frac{5128.21}{EI}$$

Substituting these values into the equations for moments at the ends of members gives

$$M_{AB} = -269.2 \text{ kN} \cdot \text{m} \qquad M_{BA} = -230.8 \text{ kN} \cdot \text{m} \qquad M_{BC} = +230.8 \text{ kN} \cdot \text{m}$$

$$M_{CB} = +230.8 \text{ kN} \cdot \text{m} \qquad M_{CD} = -230.8 \text{ kN} \cdot \text{m} \qquad M_{DC} = -269.2 \text{ kN} \cdot \text{m}$$

The bending-moment diagram, drawn on the compression side, is shown in Fig. 10.7 d. The reactions at A and D can be found from statics by using Fig. 10.7 e. ■

Equation (d) is sometimes referred to as the "sway" equation; for frames with *vertical columns* and *horizontal beams,* these sway equations are of the type:

$$\Sigma \frac{\text{sum of column end moments}}{\text{column height}} = \Sigma \text{ lateral external loads}$$

This equation does not contain any vertical external loads, since in this case the columns rotate, but not the horizontal members. However, if the frame has sloping members, as shown in Fig. 10.8a, all members of the frame will rotate and they will have a Δ term. Assume an analogous frame with pinned joints and distorted as shown in Fig. 10.8b. Thus if joint B moves a horizontal distance δ, joint C will also move horizontally a distance of δ as shown in Fig. 10.8b, since the change in axial length of BC is neglected. Thus the Δ term for AB is BB', or $+\delta/\cos \alpha_1$; for DC it is CC', or $+\delta/\cos \alpha_2$; and for BC it is $(Bb + C'c)$, or $-(\delta \tan \alpha_1 + \delta \tan \alpha_2)$. Δ term for BC is negative since relative movement of the two ends of BC is counterclockwise. With $\text{FEM}_{BC} = -Wab^2/l^2 = -80 \text{ k} \cdot \text{ft}$ and $\text{FEM}_{CB} = +Wa^2b/l^2 = +320 \text{ k} \cdot \text{ft}$, the member equations can now be written in terms of the three unknown displacements, namely, θ_B, θ_C, and δ:

$$M_{AB} = 2k_{AB}\left[2\overset{0}{\theta_A} + \theta_B - \frac{3\Delta}{l}\right] = 4\left[\theta_B - \frac{(3)(\delta)}{(13)(\frac{12}{13})}\right]$$

$$M_{BA} = 4\left[2\theta_B - \frac{(3)(\delta)}{(13)(\frac{12}{13})}\right]$$

$$M_{BC} = 2k_{BC}\left[2\theta_B + \theta_C - \frac{3\Delta}{l}\right] + \text{FEM}_{BC}$$

$$= 8\left[2\theta_B + \theta_C - \frac{3}{10}\left(-\frac{5}{12}\delta - \frac{3.5}{12}\delta\right)\right] - 80$$

$$M_{CB} = 8\left[2\theta_C + \theta_B - \frac{3}{10}\left(-\frac{5}{12}\delta - \frac{3.5}{12}\delta\right)\right] + 320$$

$$M_{CD} = 2k_{CD}\left[2\theta_C + \overset{0}{\theta_D} - \frac{3\Delta}{l}\right] = 4\left[2\theta_C - \frac{3}{12.5}\left(\frac{\delta}{12/12.5}\right)\right]$$

$$M_{DC} = 4\left[\theta_C - \frac{3}{12.5}\left(\frac{\delta}{12/12.5}\right)\right]$$

The joint moment equilibrium equations are $\Sigma M = 0$ for joints B and C:

$$M_{BA} + M_{BC} = 0 \tag{e}$$

$$M_{CB} + M_{CD} = 0 \tag{f}$$

The third equation (due to sway deflection) is derived as follows:
From the FBD of the entire frame (Fig. 10.8c),

$$\Sigma F_x = 0: \quad H_A + H_D + 15 = 0 \tag{1}$$

$$\Sigma M \text{ about } D = 0: \quad (V_A)(18.5) + (15)(12)$$
$$- (250)(5.5) + M_{AB} + M_{DC} = 0 \tag{2}$$

$$\Sigma M \text{ about } A = 0: \quad -(V_D)(18.5) + (15)(12)$$
$$+ (250)(13) + M_{AB} + M_{DC} = 0 \tag{3}$$

Using the free-body diagrams of the two columns in Fig. 10.8d:

$$\Sigma M \text{ about } A = 0: \quad (V_A)(5) - (H_A)(12) + M_{AB} + M_{BA} = 0 \tag{4}$$

$$\Sigma M \text{ about } D = 0: \quad -(V_D)(3.5) - (H_D)(12) + M_{DC} + M_{CD} = 0 \tag{5}$$

From expressions (1) to (5), one can deduce the sway equation in terms of the column moments as

$$M_{BA} + M_{CD} + \frac{10}{18.5}(M_{AB} + M_{DC}) = 145.95 \tag{g}$$

The three unknown displacements θ_B, θ_C, and δ can be found from solving the three simultaneous equations (e) to (g). Thus,

$$\theta_B = +10.128 \qquad \theta_C = -14.872 \qquad \delta = -63.014.$$

Note that these are relative deformations and not absolute, since relative k values are used in this problem. Back substitution into member equations gives

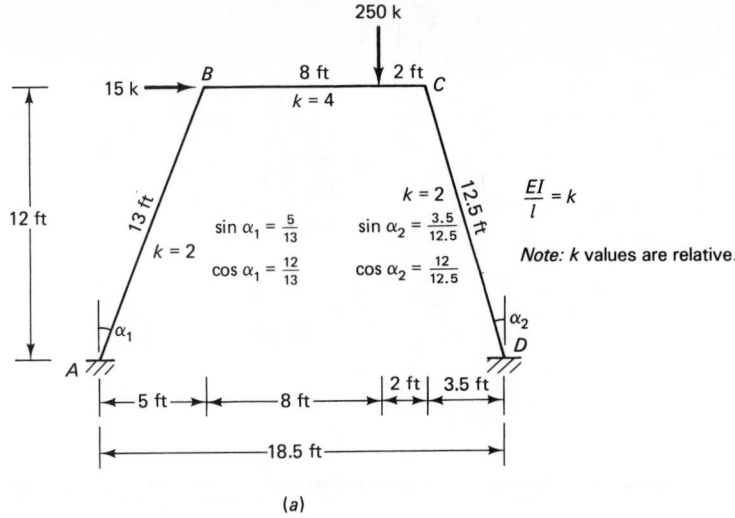

(a)

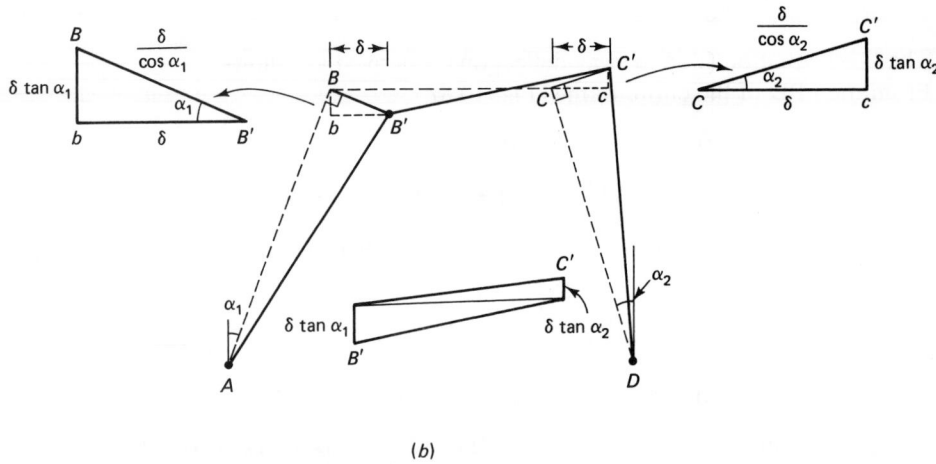

(b)

Figure 10.8 Rigid frame with sloping columns.

$$M_{AB} = +103.53 \text{ k} \cdot \text{ft} \qquad M_{BA} = +144.04 \text{ k} \cdot \text{ft} \qquad M_{BC} = -144.05 \text{ k} \cdot \text{ft}$$

$$M_{CB} = +55.95 \text{ k} \cdot \text{ft} \qquad M_{CD} = -55.96 \text{ k} \cdot \text{ft} \qquad M_{DC} = +3.53 \text{ k} \cdot \text{ft}$$

The bending-moment diagram is shown drawn on the compression side in Fig. 10.8 e.

It is interesting to note that the sway equation (g) can also be derived from the virtual-work principle. Figure 10.8 f shows the frame in a swayed position after the rigid joints have been replaced by pin joints; the moments at the ends of each member, the applied external loads, as well as the resulting rotations of each member are also

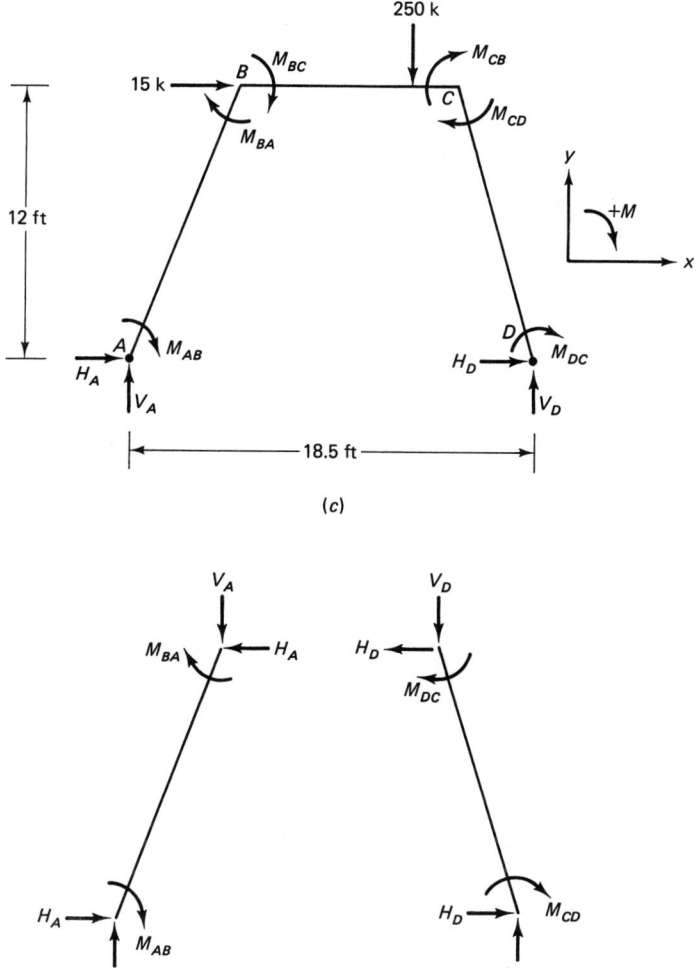

Figure 10.8 (*Continued*)

shown. By expressing the condition of equilibrium through the principle of virtual work, it is possible to relate the end moments to the loads. Thus

$$(M_{AB} + M_{BA})\phi_A + (M_{BC} + M_{CB})(-\phi_B) + (M_{CD} + M_{DC})(\phi_D)$$
$$+ 15\delta + 250[\tfrac{2}{10}(\delta \tan \alpha_1) - \tfrac{8}{10}(\delta \tan \alpha_2)] = 0$$

From geometry,

$$\phi_A = \frac{\delta/\cos \alpha_1}{13} \qquad \phi_B = \frac{\delta \tan \alpha_1 + \delta \tan \alpha_2}{10} \quad \text{and} \quad \delta_D = \frac{\delta/\cos \alpha_2}{12.5}$$

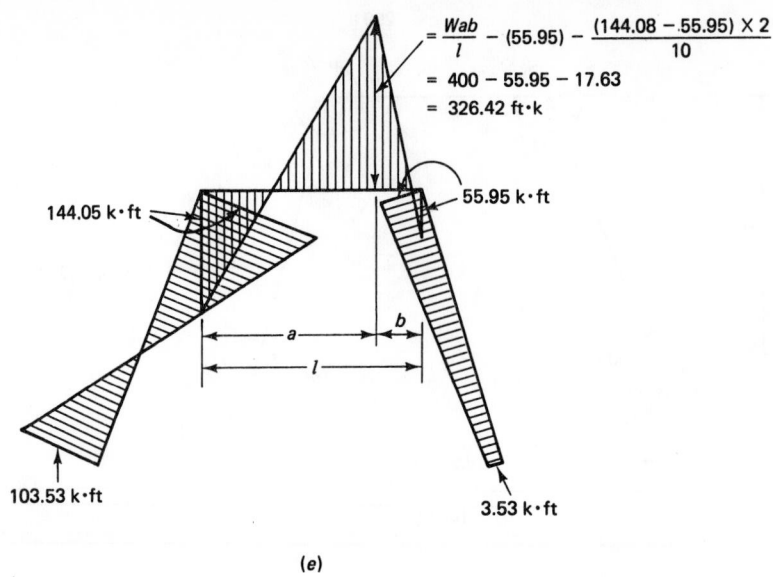

$$= \frac{Wab}{l} - (55.95) - \frac{(144.08 - .55.95) \times 2}{10}$$
$$= 400 - 55.95 - 17.63$$
$$= 326.42 \text{ ft·k}$$

-55.95 k·ft

144.05 k·ft

103.53 k·ft

3.53 k·ft

(e)

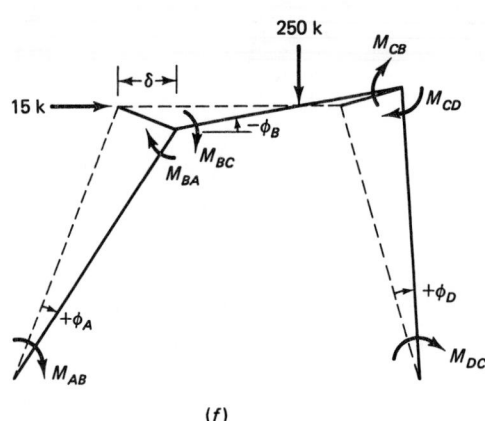

250 k

M_{CB}

M_{CD}

15 k

$-\delta-$

$-\phi_B$

M_{BC}

M_{BA}

$+\phi_A$

$+\phi_D$

M_{AB}

M_{DC}

(f)

Figure 10.8 (*Continued*)

Also from the moment equilibrium at joints B and C, we have

$$M_{BC} = -M_{BA} \quad \text{and} \quad M_{CB} = -M_{CD}$$

Thus substituting in the above virtual-work equation yields

$$M_{BA} + M_{CD} + \frac{10}{18.5} (M_{AB} + M_{DC}) = 145.95 \qquad \text{as before!}$$

Another simpler method of deriving the sway equation is by applying $\sum M = 0$ about the center of moment O as shown in Fig. 10.8g. After locating O from the

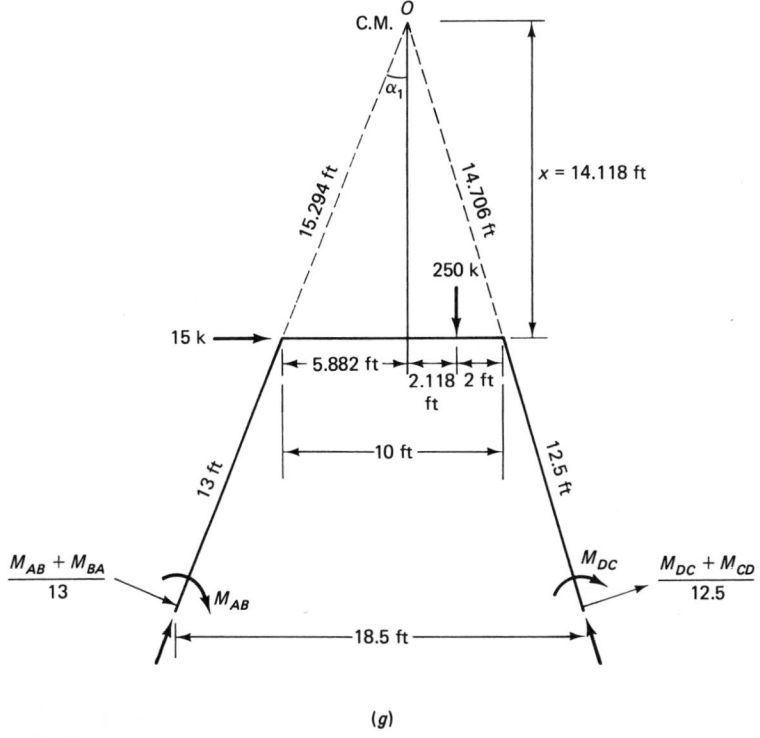

(g)

Figure 10.8 (*Continued*)

geometry of the frame, the moment-equilibrium equation about O of the FBD shown in Fig. 10.8 g gives

$$-\left(\frac{M_{AB} + M_{BA}}{13}\right)(28.294) - (15)(14.118) + (250)(2.118)$$

$$M_{AB} + M_{DC} - \left(\frac{M_{DC} + M_{CD}}{12.5}\right)(27.206) = 0$$

This yields

$$(M_{BA} + M_{CD}) + (0.541)(M_{AB} + M_{DC}) = 145.93 \qquad \text{as before!} \qquad \blacksquare$$

10.1.4 Rigid Frames with Two Degrees of Freedom of Joint Translation

For rigid frames with two degrees of freedom of joint translation, the geometrical relationship between the Δ terms of the individual members and those independent joint translations can be found as demonstrated by the following examples.

Two-Story Frame   Figure 10.9a shows the joint displacements in a two-story frame. Here the displacement of the first story is independent of the second story. As shown

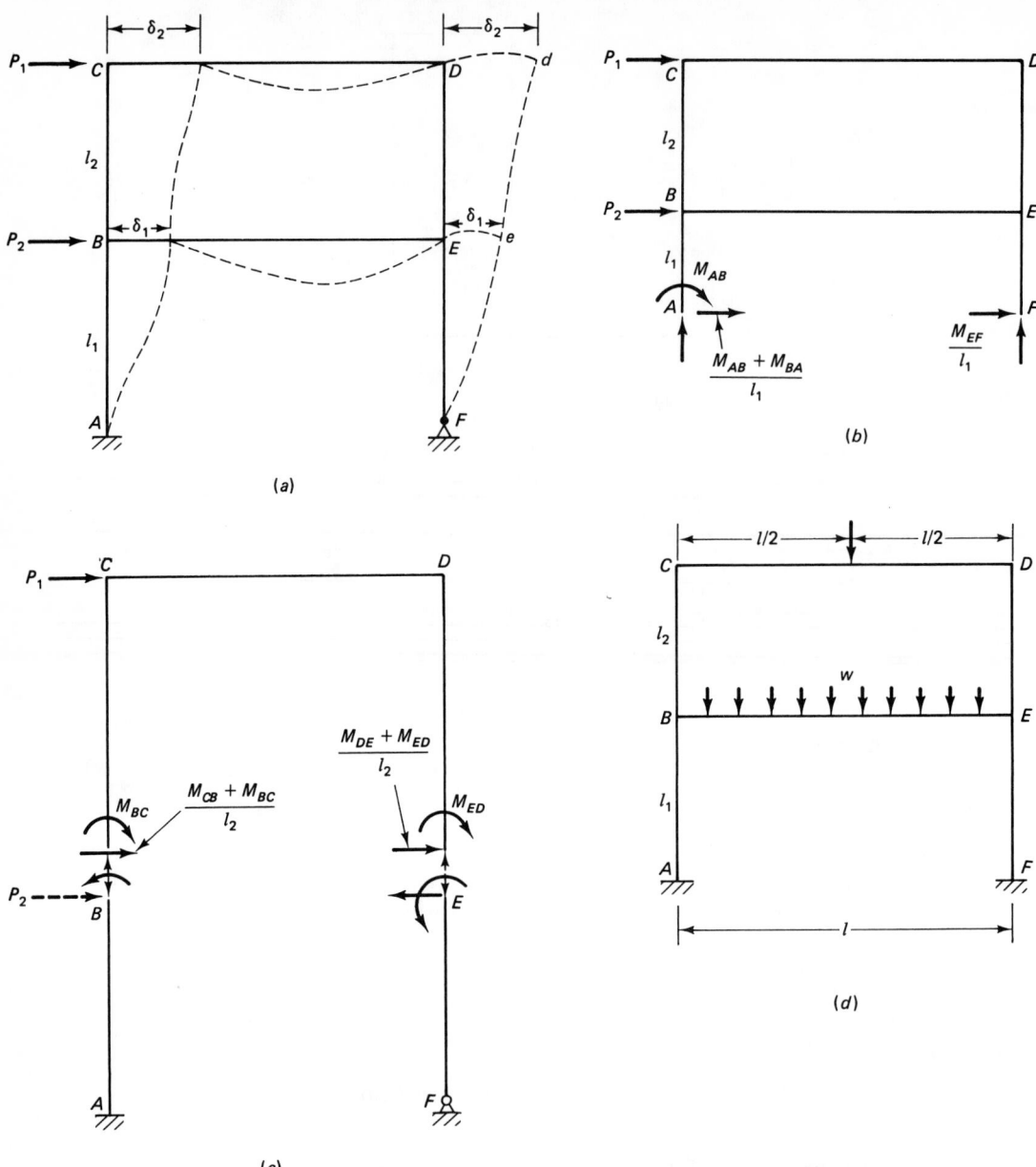

Figure 10.9 Two-story frame (with two degrees of freedom in translation).

in Fig. 10.9a, the horizontal movement δ_1 of joint B is the same for joint E and that of joint $C(\delta_2)$ is the same for joint D. Thus, the Δ term for members AB and EF is δ_1, for members BC and ED is $(\delta_2 - \delta_1)$, and for members BE and CD is zero.

Besides the four static equations of $\Sigma M = 0$ at the four joints B, C, D, and E, two sway equations are also needed; these equations are derived from the shear balance of the structure. Referring to Fig. 10.9b, the shear balance ($\Sigma F_x = 0$) of the two-story structure gives

$$P_1 + P_2 + \frac{M_{AB} + M_{BA}}{l_1} + \frac{M_{EF}}{l_1} = 0$$

The other equation is obtained from the shear balance of the FBD of the top story when cut out by a horizontal section just above BE (Fig. 10.9c). Thus from $\Sigma F_x = 0$,

$$P_1 + \frac{M_{CB} + M_{BC}}{l_2} + \frac{M_{DE} + M_{ED}}{l_2} = 0$$

It should be noted that for a multistory frame loaded only by horizontal loads at the joints, as in Fig. 10.9a, antisymmetry exists. Thus, for the structure in Fig. 10.9a we can write that

$$\theta_B = \theta_E \quad \text{and} \quad \theta_C = \theta_D$$

whereas for frames loaded as shown in Fig. 10.9d, symmetry exists, and in this case we can write

$$\theta_B = -\theta_E \quad \text{and} \quad \theta_C = -\theta_D$$

Gable Frames An unsymmetrical gable frame is shown in Fig. 10.10a in its displaced configuration when loaded. Let us assume that joints B and D move horizontally distances δ_1 and δ_2, respectively, with $\delta_2 > \delta_1$. To locate the position of joint C, let us imagine the two members CB and CD are disconnected at joint C. Thus when joint B moves to b, C will move to c'; similarly, when joint D moves to d, C will move to c''; however, the final position of joint C is c where the perpendicular to BC through c' and the perpendicular to CD through c'' intersect, as shown in Fig. 10.10b. From geometry of Fig. 10.10b, using the sine law we can write

$$\frac{cc'}{\sin(90° - \phi_2)} = \frac{cc''}{\sin(90° - \phi_1)} = \frac{\delta_2 - \delta_1}{\sin(\phi_1 + \phi_2)}$$

Hence cc' and cc'' can be deduced from the assumed displacements δ_1 and δ_2.

Thus the Δ term for member AB is δ_1, for member BC is cc', for member CD is cc'', and for member ED is δ_2.

There are five unknown displacements for the loaded gable bent as in Fig. 10.10a. There are three joint moment-equilibrium equations, $\Sigma M = 0$, at joints B, C, and D. The two "sway" equations are provided by the shear balance for the entire frame (Fig. 10.10c); thus from $\Sigma F_x = 0$

$$\frac{M_{AB} + M_{BA}}{l_3} + \frac{M_{ED} + M_{DE}}{l_3} + P_1 = 0$$

The other equation is derived from $\Sigma M = 0$ about point O; thus using Fig. 10.10d we can write

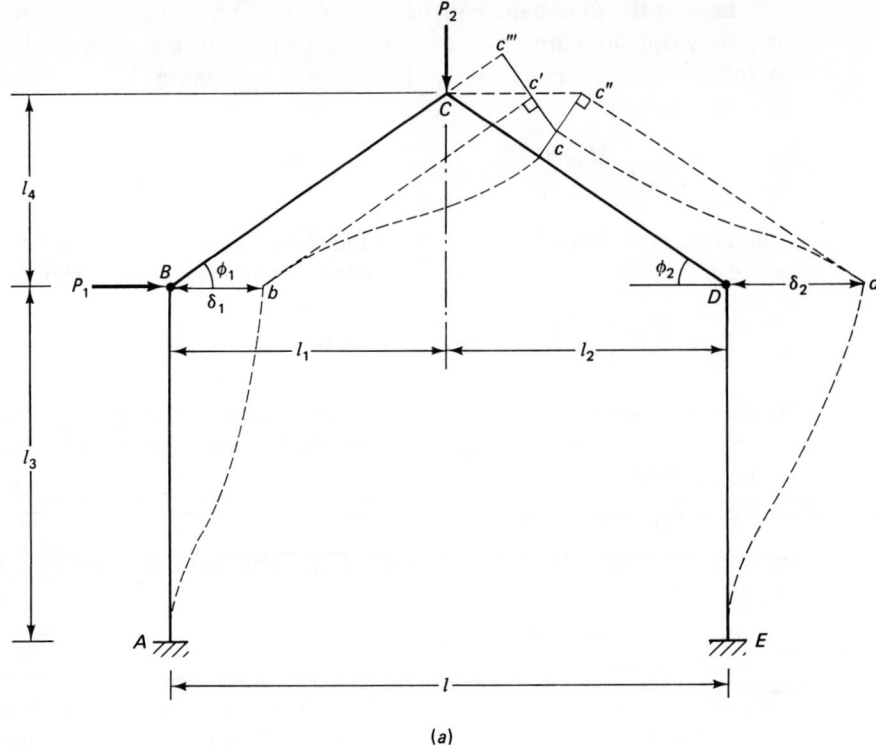

(a)

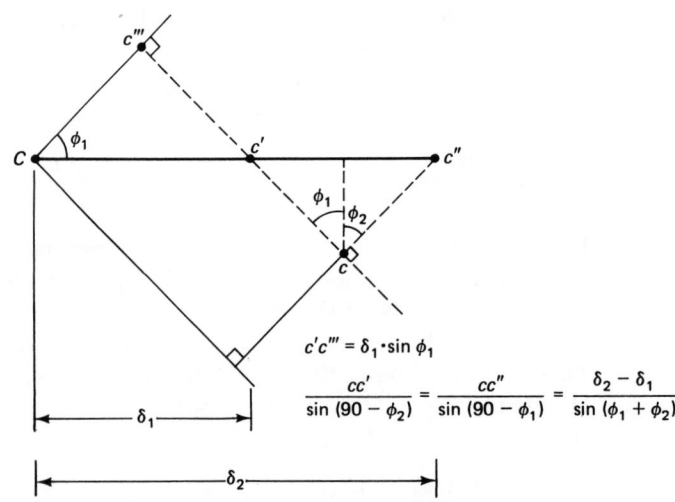

$$c'c''' = \delta_1 \cdot \sin \phi_1$$

$$\frac{cc'}{\sin(90 - \phi_2)} = \frac{cc''}{\sin(90 - \phi_1)} = \frac{\delta_2 - \delta_1}{\sin(\phi_1 + \phi_2)}$$

Figure 10.10 Gable frame. (*b*) Enlarged Δ's $Cc'c'''$ and $cc'c'''$ of (*a*).

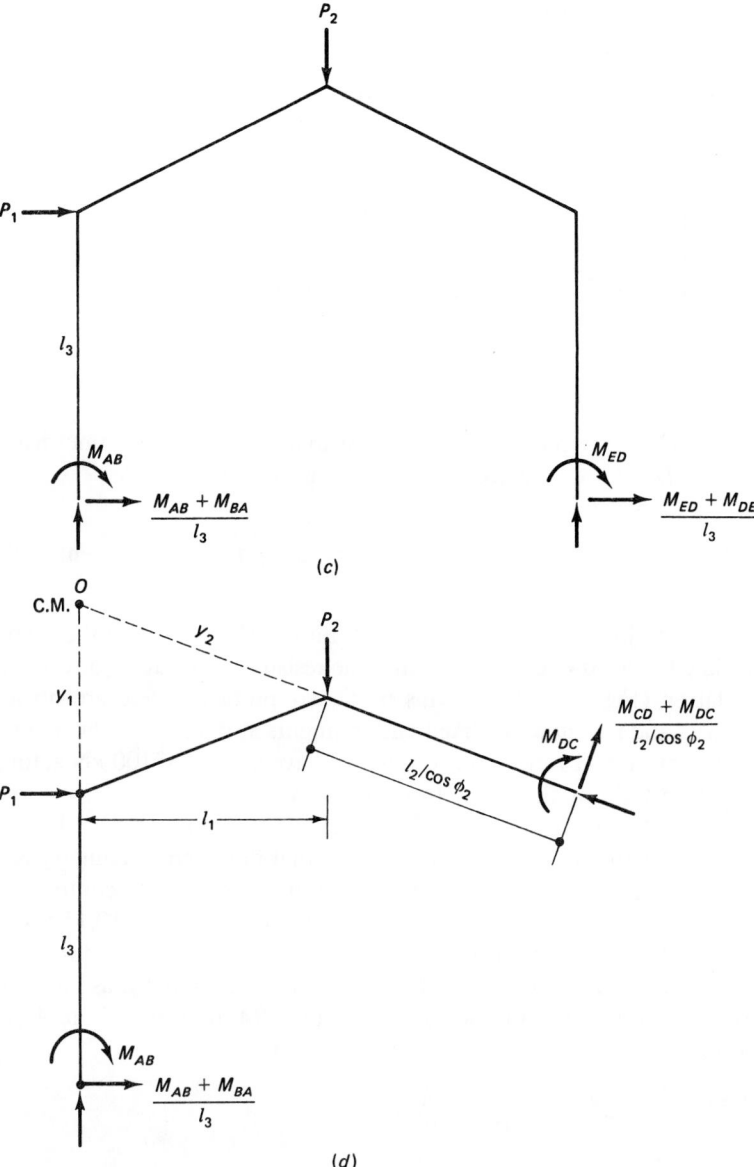

(c)

(d)

Figure 10.10 (*Continued*)

$$P_2(l_1) + M_{AB} + M_{DC} - \frac{M_{AB} + M_{BA}}{l_3}(l_3 + y_1)$$

$$- \frac{M_{CD} + M_{DC}}{l_2/\cos \phi_2}\left(\frac{l_2}{\cos \phi_2} + y_2\right) - P_1 y_1 = 0$$

The distances y_1 and y_2 are readily derived from the geometry of the gable bent.

10.1.5 Matrix Structural Analysis and the Slope-Deflection Method

Let us rewrite in matrix form the three simultaneous equations relating the θ_B, θ_C, and Δ for the rigid frame shown in Fig. 10.7a. Thus from Eqs. (b), (c), and (d) of Example 10.3

$$EI \begin{bmatrix} \frac{6}{5} & \frac{2}{5} & -\frac{3}{50} \\ \frac{2}{5} & \frac{6}{5} & -\frac{3}{50} \\ -\frac{3}{50} & -\frac{3}{50} & +\frac{3}{125} \end{bmatrix} \begin{bmatrix} \theta_B \\ \theta_C \\ \Delta \end{bmatrix} = \begin{bmatrix} 0 \\ 0 \\ +100 \end{bmatrix}$$ (h)

or

$$S^*D = P$$ (10.5)

where S^* = external or global stiffness matrix, which is symmetrical
D = joint displacements (rotations and translations)
P = external joint load matrix

We can derive the above equation by using the displacement (stiffness) method of structural analysis as follows:

1. With joints B, C fixed and with no translation, apply the member loads and calculate the fixed-end moments and the resisting force at B (or C) to prevent lateral translation (Fig. 10.11b). In this particular problem there are no loads along the members and therefore the fixed-end moments at the ends of the members and hence at the joints are all zero. However, we do have a force of 100 kN acting from right to left at C restraining the frame from sidesway.

2. Apply joint rotations and translations to each joint that has been artificially restrained (joints B and C in this case) so that the altered geometry is restored to the actual geometry of the loaded structure, thus satisfying the compatibility condition. The rigid frame is then subjected to unit rotation at B (Fig. 10.11c) while preventing rotation at the joints C and A. As was shown earlier, the resulting end moments for member BC at B and C are $4EI/l_1$ and $2EI/l_1$, respectively, as shown in Fig. 10.11c; and the resulting end moments for member BA at B and A are $4EI/l$ and $2EI/l$, respectively. The induced lateral force at joint B is

$$\frac{4EI/l + 2EI/l}{l} = \frac{6EI}{l^2}$$

acting from right to left. The resulting moments and lateral forces due to the application of a unit rotation at joint C (Fig. 10.11d) and due to the application of a unit lateral displacement (Fig. 10.11e) can be easily deduced.

Applying the equations of equilibrium for moments ($\Sigma M = 0$) at the joints and for forces ($\Sigma F_x = 0$) at joint B (or C) and making use of the principle of superposition, we can write:

At joint B:

$$\Sigma M = 0: \quad \left(\frac{4EI}{l_1} + \frac{4EI}{l}\right)\theta_B + \left(\frac{2EI}{l_1}\right)\theta_C - \left(\frac{6EI}{l^2}\right)\Delta = 0$$

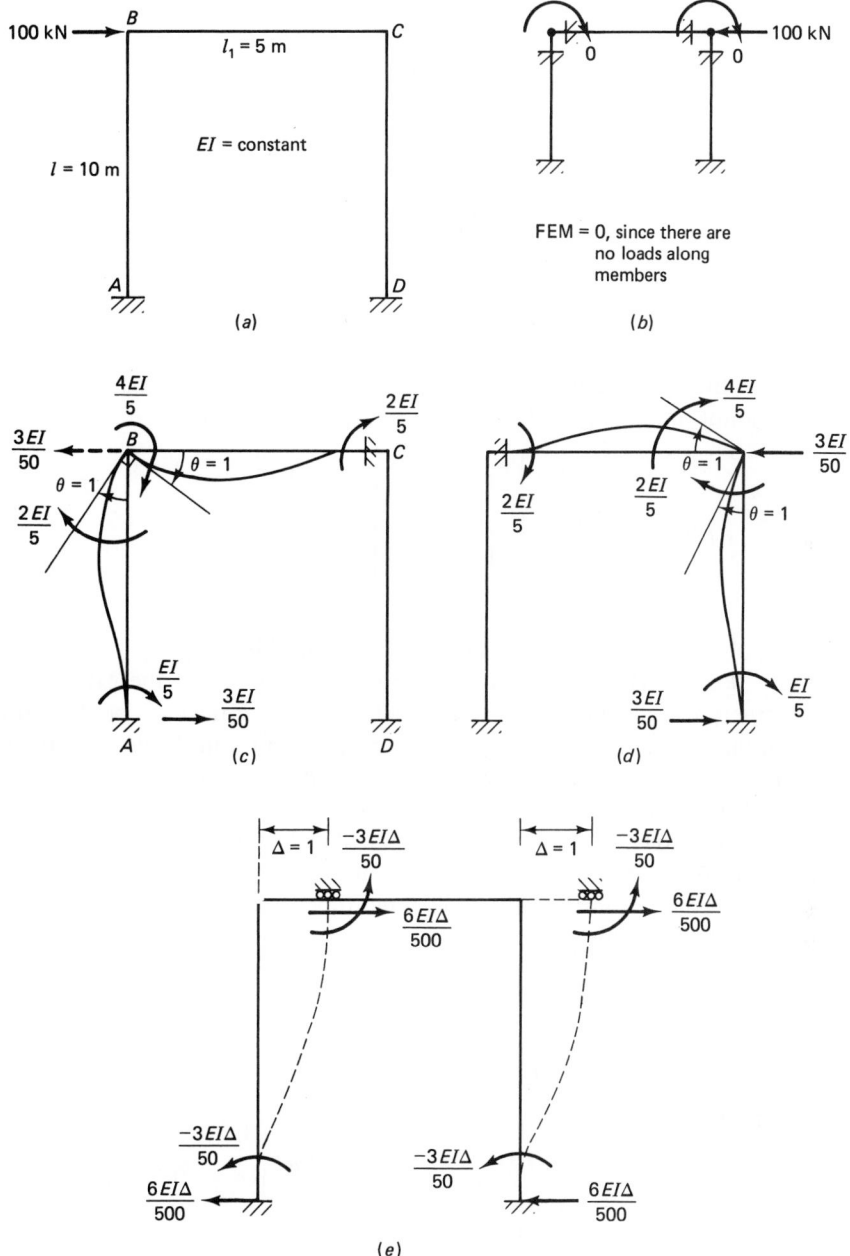

Figure 10.11 Development of stiffness coefficients for the structure in Example 10.3.

At joint C:

$$\sum M = 0: \qquad \left(\frac{2EI}{l_1}\right)\theta_B + \left(\frac{4EI}{l_1} + \frac{4EI}{l}\right)\theta_C - \left(\frac{6EI}{l^2}\right)\Delta = 0$$

At joint B:

$$\sum F_x = 0: \qquad -\left(\frac{6EI}{l^2}\right)\theta_B - \left(\frac{6EI}{l^2}\right)\theta_C + \left(\frac{12EI}{l^3} + \frac{12EI}{l^3}\right)\Delta = 100$$

Putting $l_1 = 5$ m, $l = 10$ m in the above equations yields in matrix form

$$EI \begin{bmatrix} \frac{6}{5} & \frac{2}{5} & -\frac{3}{50} \\ \frac{2}{5} & \frac{6}{5} & -\frac{3}{50} \\ -\frac{3}{50} & -\frac{3}{50} & +\frac{3}{125} \end{bmatrix} \begin{bmatrix} \theta_B \\ \theta_C \\ \Delta \end{bmatrix} = \begin{bmatrix} 0 \\ 0 \\ 100 \end{bmatrix}$$

which is the same as Eq. (h) above.

For the Example in Fig. 10.6 We can rewrite in matrix form the moment-equilibrium equations for the continuous beam in Fig. 10.6. Thus

$$EI \begin{bmatrix} \frac{4}{10} & \frac{2}{10} & 0 & 0 \\ \frac{2}{10} & \frac{8}{10} & \frac{2}{10} & 0 \\ 0 & \frac{2}{10} & \frac{8}{10} & \frac{2}{10} \\ 0 & 0 & \frac{2}{10} & \frac{4}{10} \end{bmatrix} \begin{bmatrix} \theta_B \\ \theta_C \\ \theta_D \\ \theta_E \end{bmatrix} = \begin{bmatrix} -3 \\ 7.2 \\ -0.63 \\ -4.17 \end{bmatrix} \qquad \text{(i)}$$

Apply the same procedure as above to the problem:

Fix joints B, C, D, and E against rotation, apply the external load along the structure, and calculate the fixed-end moments at the ends of all members as shown in Fig. 10.12a. We then calculate the equivalent joint loads from the moment equilibrium at the joints as shown in Fig. 10.12b and c. To determine the moments at the ends of members due to these joint moments we proceed by first applying unit joint rotations to each joint that has been artificially restrained as shown in Fig. 10.12d to g. Since a unit rotation results in the moments shown in Fig. 10.12d to g, rotations θ will produce moments equal to θ times those values since the structure is linear; thus, using the principle of superposition, we can write:

At joint B:

$$\sum M = 0: \qquad \frac{4EI}{10}(\theta_B) + \frac{2EI}{10}(\theta_C) = -3$$

At joint C:

$$\sum M = 0: \qquad \left(\frac{2EI}{10}\right)(\theta_B) + \left(\frac{4EI}{10} + \frac{4EI}{10}\right)(\theta_C) + \left(\frac{2EI}{10}\right)(\theta_D) = +7.2$$

At joint D:

$$\sum M = 0: \qquad \left(\frac{2EI}{10}\right)(\theta_C) + \left(\frac{4EI}{10} + \frac{4EI}{10}\right)(\theta_D) + \left(\frac{2EI}{10}\right)(\theta_E) = -0.63$$

At joint E:

$$\sum M = 0: \qquad \left(\frac{2EI}{10}\right)(\theta_D) + \left(\frac{4EI}{10}\right)(\theta_E) = -4.17$$

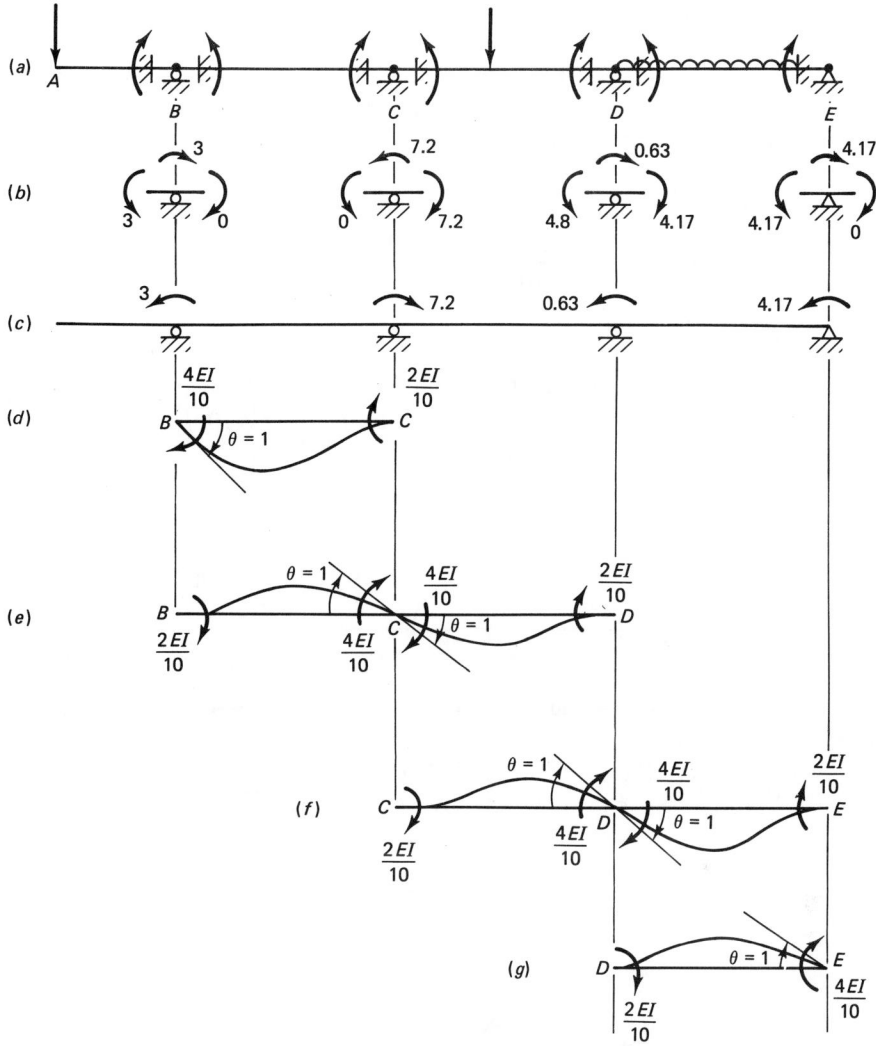

Figure 10.12 Development of stiffness coefficients for the structure in Example 10.1. (*a*) All joints are fixed, showing loads and fixed-end member forces. (*b*) Joint equilibrium. (*c*) Equivalent joint loads. (*d*) Unit rotation at *B*. (*e*) Unit rotation at *C*. (*f*) Unit rotation at *D*. (*g*) Unit rotation at *E*.

In matrix form:

$$EI \begin{bmatrix} \frac{4}{10} & \frac{2}{10} & 0 & 0 \\ \frac{2}{10} & \frac{8}{10} & \frac{2}{10} & 0 \\ 0 & \frac{2}{10} & \frac{8}{10} & \frac{2}{10} \\ 0 & 0 & \frac{2}{10} & \frac{4}{10} \end{bmatrix} \begin{bmatrix} \theta_B \\ \theta_C \\ \theta_D \\ \theta_E \end{bmatrix} = \begin{bmatrix} -3 \\ +7.2 \\ -0.63 \\ -4.17 \end{bmatrix}$$

which is identical to equation (i) obtained earlier.

Since there is no moment at the simple support E, we need apply unit rotations at joints B, C, and D only as shown in Fig. 10.13d, e, and f. Of course the fixed-end moment for DE will change at D since E is considered to be simply supported. This is shown in Fig. 10.13a. Following the same procedure we can write the moment-equilibrium equations at joints B, C, and D as follows:

At joint B:

$$\sum M = 0: \qquad \left(\frac{4EI}{10}\right)(\theta_B) + \left(\frac{2EI}{10}\right)(\theta_C) = -3$$

At joint C:

$$\sum M = 0: \qquad \left(\frac{2EI}{10}\right)(\theta_B) + \left(\frac{8EI}{10}\right)(\theta_C) + \left(\frac{2EI}{10}\right)(\theta_D) = +7.2$$

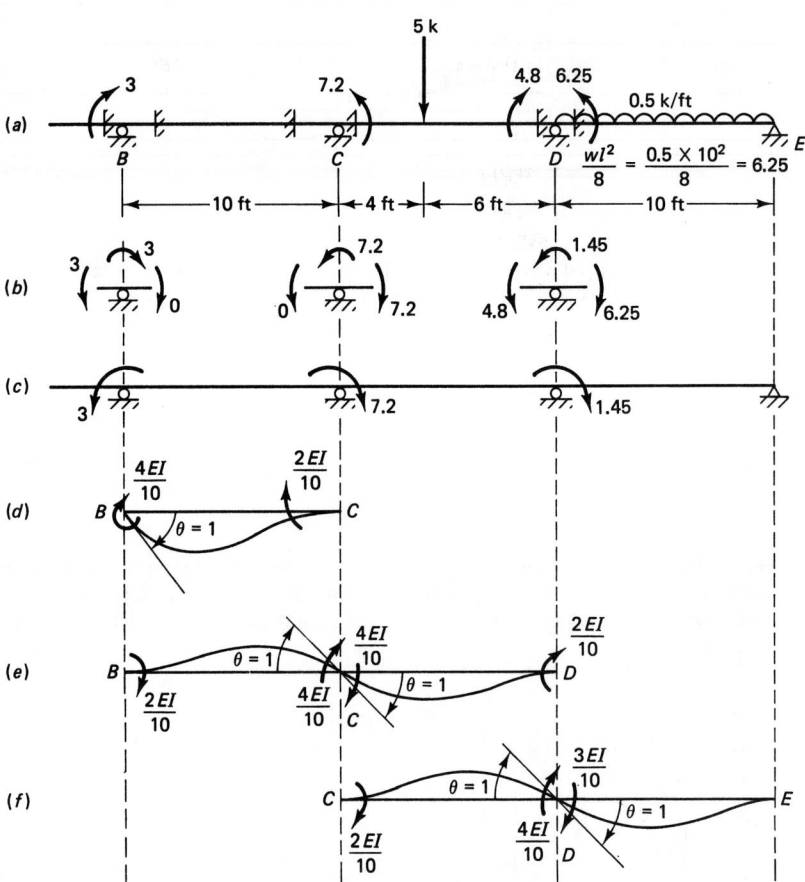

Figure 10.13 Treatment of simple end support in the development of stiffness coefficients for the structure in Example 10.1.

At joint D:

$$\sum M = 0: \quad \left(\frac{2EI}{10}\right)(\theta_C) + \left(\frac{7EI}{10}\right)(\theta_D) = +1.45$$

In matrix form:

$$EI \begin{bmatrix} \frac{4}{10} & \frac{2}{10} & 0 \\ \frac{2}{10} & \frac{8}{10} & \frac{2}{10} \\ 0 & \frac{2}{10} & \frac{7}{10} \end{bmatrix} \begin{bmatrix} \theta_B \\ \theta_C \\ \theta_D \end{bmatrix} = \begin{bmatrix} -3 \\ +7.2 \\ +1.45 \end{bmatrix}$$

Solving the above equations will yield the same values for θ_B, θ_C, and θ_D as before. If θ_E is required it can be deduced from any one of the classical methods for deflections, that is, conjugate-beam, moment-area, etc. (see Chapter 7) after determining the final moment at D from the member equations.

10.2 THE MOMENT-DISTRIBUTION METHOD

10.2.1 General

It was in 1930 that Professor Hardy Cross introduced the method of moment distribution to the profession. In this method, just as in the slope-deflection method, the joint rotations and displacements are used as unknowns in the analysis of a structure; however, unlike the slope-deflection method, the method does not require the solution of simultaneous equations. Instead, the method involves a procedure of successive approximations and iterations leading to the results obtained to any desired precision. By its use, the student is given an opportunity to "feel" how loads are transferred from one part of the structure to another and how the structure seeks its own equilibrium, to visualize the deformation of the structure and the importance of the factors that affect this deformation.

10.2.2 Introduction

Consider the rigid frame in Fig. 10.14a. Let us first remove the external loads on the structure (assumed weightless) and then clamp each joint that is capable of rotation (such as joints B and C) in order to prevent rotation; upon reapplying the load as shown in Fig. 10.14b, moments will be developed at each end of the members that are loaded, for example, FEM_{AB}, FEM_{BA}, FEM_{BC}. These fixed-end moments can be either calculated from first principles or obtained from the Appendix.

If the fixed-end moments at a clamped joint, say B, are not balanced (i.e., $\sum M \neq 0$) (see Fig. 10.14c), then upon the release of a clamp at one joint, such as B, the presence of an unbalanced moment M at joint B will cause joint B to rotate until equilibrium ($\sum M = 0$) at joint B is reached.

Such rotation will also cause moments to be developed at the far ends of all members meeting at joint B, whose clamp has been released. Joint B is then reclamped

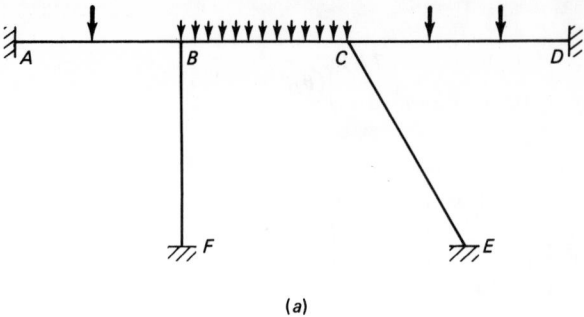

(a)

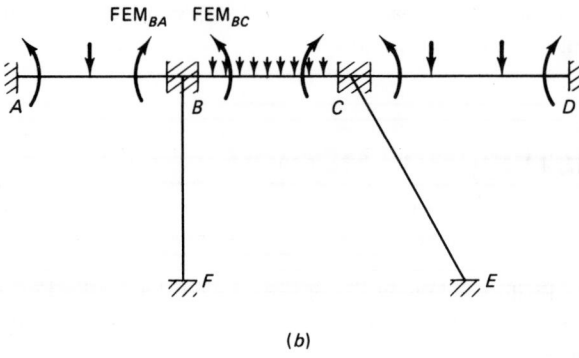

(b)

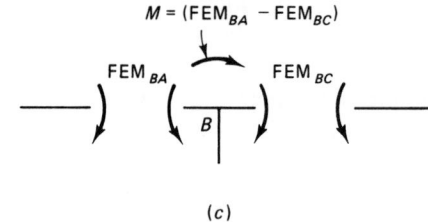

(c)

Figure 10.14 A loaded rigid frame showing fixed-end moments.

and joint C is now released. The process is repeated for several cycles until the moments at joint B and at joint C are very nearly in equilibrium, joints B and C have rotated to their actual position, and the unbalanced moment at a joint is considered to be negligibly small. It should be noted that after each cycle, the clamping moment at a joint is less than that at the previous cycle since the joint is clamped into a position closer to its own true position when no clamping moment will be acting.

Each time we balance the moment at a joint, say B, we need to know the proportion of that moment that will be carried by the members (say, BA, BC, and BF) meeting at that joint. If this joint rotates an angle θ to reach its equilibrium position, then the unclamping moment will be distributed to BA, BC, and BF according to their stiffnesses;

therefore, we need to know the distribution factors which are based on the stiffness of each member at a joint. Similarly, the amount of moment developed at the other ends of members *BA, BC,* and *BF* due to the rotation of the near ends of these members is some proportion of the moment at the near end of these members. The ratio of the far-end moment to the near-end moment is called the *carryover factor.*

In order to develop the method of moment distribution we need to adopt a sign convention as well as develop expressions for the stiffness, carryover, and distribution factors.

10.2.3 Sign Convention

We will adopt the same sign convention as the one in the slope-deflection method; that is, all moments at the *ends of members* (not at the joint) and displacements (rotations and translations) of one end of a member relative to the other will be considered positive if they are clockwise. In Fig. 10.14*b* with clamps at *B* and *C,* the fixed-end moment at *A,* FEM_{AB}, is negative while FEM_{BA} is positive; similarly, slopes θ_A, θ_B, and translation Δ, shown in Fig. 10.1 (slope-deflection method) are all positive.

10.2.4 Stiffness and Carryover Factors

Consider the unit rotation of the beam in Fig. 10.15*a*. In the moment-distribution method the stiffness *S* is defined as the moment necessary to rotate a supported end of a member through a unit angle $\theta_A = 1$ when the far end of the member *B* is fixed, $\theta_B = 0$. The carryover factor c_{ab} is defined as the ratio of the moment induced at the far end of a member, that is, at *B,* to the moment producing the rotation at end *A.* From the BMD in Fig. 10.15*b* and using the slope-deflection equations (10.4) we can write

$$M_{AB} = \mathrm{FEM}_{AB} + \frac{2EI}{l}\left(2\theta_A + \theta_B - \frac{3\Delta}{l}\right)$$

or

$$M_{AB} = \frac{4EI}{l}\cdot\theta_A$$

Putting $\theta_A = 1$ and $M_{AB} = S_{AB}$ yields the *absolute stiffness*

$$S_{AB} = \frac{4EI}{l} = 4EK_{AB} \tag{10.6}$$

which is the stiffness of *AB* at joint *A,* and $K = I/l$ is referred to as the *relative stiffness.* Also from the slope-deflection equations (10.4)

$$M_{BA} = \mathrm{FEM}_{BA} + \frac{2EI}{l}\left(2\theta_B + \theta_A - \frac{3\Delta}{l}\right)$$

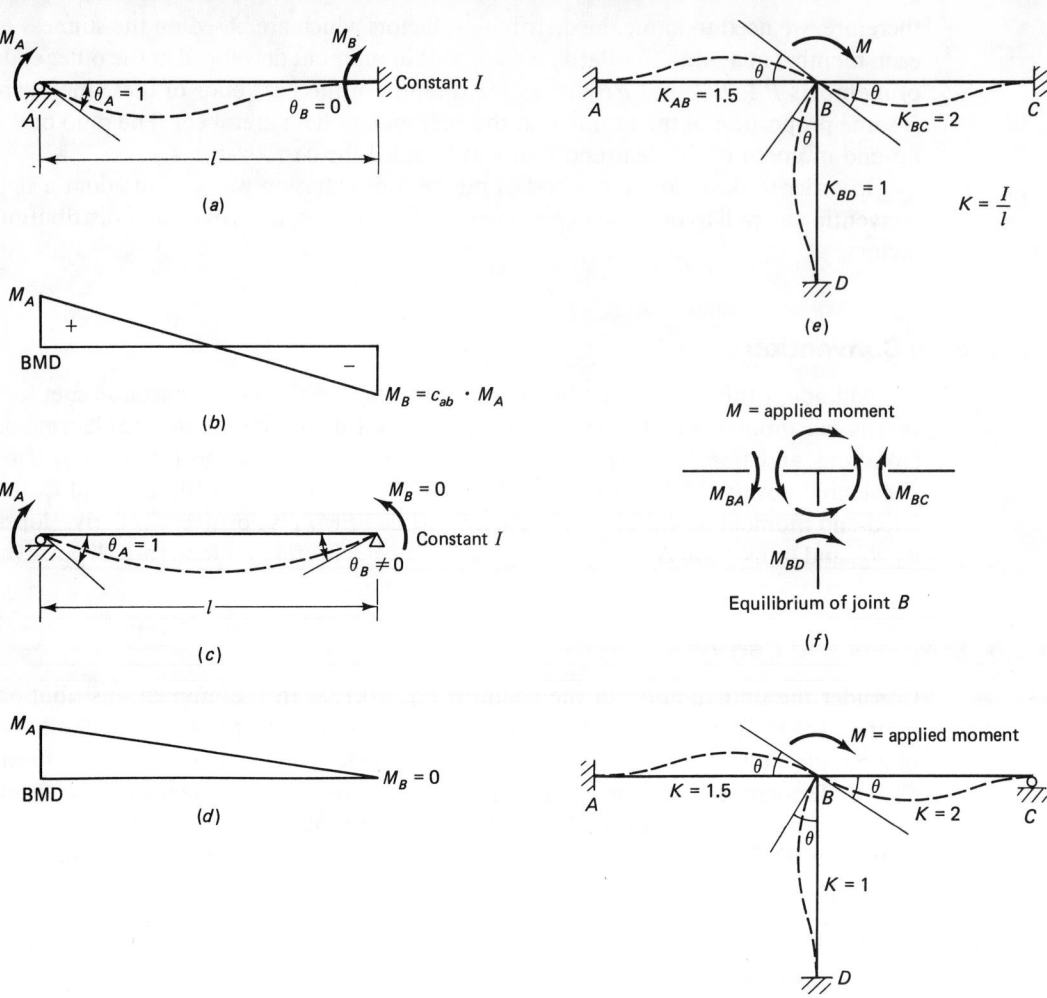

Figure 10.15 Development of stiffness and carryover factors for a beam.

or

$$M_{BA} = \frac{2EI}{l} \cdot \theta_A$$

Putting $\theta_A = 1$,

$$M_{BA} = \frac{2EI}{l} \tag{10.7}$$

Hence the carryover factor from A to B,

$$c_{ab} = \frac{M_{BA}}{M_{AB}} = \frac{1}{2} \tag{10.8}$$

If the far end of member AB is simply supported (Fig. 10.15c and d), then $M_B = 0$; thus from the slope-deflection equations:

$$M_{AB} = \overset{0}{\cancel{FEM_{AB}}} + \frac{2EI}{l}\left(2\theta_A + \theta_B - \overset{0}{\cancel{\frac{3\Delta}{l}}}\right)$$

or

$$M_{AB} = \frac{2EI}{l}(2\theta_A + \theta_B) \tag{10.9}$$

and

$$M_{BA} = 0 = \overset{0}{\cancel{FEM_{BA}}} + \frac{2EI}{l}\left(2\theta_B + \theta_A - \overset{0}{\cancel{\frac{3\Delta}{l}}}\right)$$

or

$$0 = \frac{2EI}{l}(2\theta_B + \theta_A) \tag{10.10}$$

From Eq. (10.10),

$$\theta_B = -\tfrac{1}{2}\theta_A \tag{10.11}$$

Substituting Eq. (10.11) in Eq. (10.9),

$$M_{AB} = \frac{2EI}{l}\left(2\theta_A - \frac{1}{2}\theta_A\right) = \frac{3EI}{l}\cdot\theta_A$$

Putting $\theta_A = 1$, $M_{AB} = S_{AB}$. Then

$$S_{AB} = \frac{3EI}{l}$$

or

$$S_{AB} = 3EK_{AB} \tag{10.12}$$

In this particular case, it is evident that $c_{ab} = 0$ since end B is simply supported. Comparing Eqs. (10.6) and (10.12) shows that the stiffness of the beam AB is reduced by one-fourth when the far end is simply supported and not fixed. The stiffness given by Eq. (10.12) is said to be *modified stiffness*.

10.2.5 Distribution Factor

Figure 10.15e shows three members connected rigidly at B while their far ends are rigidly fixed and with no possible translation. If a positive moment M is applied at joint B, then joint B will rotate an angle θ as shown in Fig. 10.15e. The moment M will be resisted by the three members meeting at joint B. Furthermore, since joint B

is rigid, each of the connected ends of the three members will rotate the same angle θ. From moment equilibrium of joint B (Fig. 10.15f), we have

$$M_{BA} + M_{BC} + M_{BD} = M \tag{10.13}$$

Applying the slope-deflection equation to each member meeting at B, we have

$$M_{BA} = \left(\frac{2EI}{l}\right)_{BA} (2\theta) = 4EK_{BA}\theta = S_{BA} \cdot \theta$$

$$M_{BC} = \left(\frac{2EI}{l}\right)_{BC} (2\theta) = 4EK_{BC}\theta = S_{BC} \cdot \theta \tag{10.14}$$

$$M_{BD} = \left(\frac{2EI}{l}\right)_{BD} (2\theta) = 4EK_{BD}\theta = S_{BD} \cdot \theta$$

Substituting Eq. (10.14) into Eq. (10.13) yields

$$(S_{BA} + S_{BC} + S_{BD})\theta = M$$

or

$$4E(K_{BA} + K_{BC} + K_{BD})\theta = M$$

or

$$\theta = \frac{M}{4E \sum K} \tag{10.15}$$

where $\sum K = K_{BA} + K_{BC} + K_{BD}$. Substituting Eq. (10.15) in Eq. (10.14), we have

$$M_{BA} = \frac{K_{BA}}{\sum K} \cdot M$$

$$M_{BC} = \frac{K_{BC}}{\sum K} \cdot M \tag{10.16}$$

$$M_{BD} = \frac{K_{BD}}{\sum K} \cdot M$$

The ratios $K_{BA}/\sum K$, $K_{BC}/\sum K$, and $K_{BD}/\sum K$ are defined as the *distribution factors*.

For example, in Fig. 10.15e, if $M = 100$ k $\cdot$ ft and the value for the relative stiffness K of each member is as shown, then

$$M_{BA} = \frac{K_{BA}}{\sum K} \cdot M = \frac{1.5}{(1.5 + 2.0 + 1.0)} (100) = \left(\frac{1.5}{4.5}\right)(100) = +33.3 \text{ k} \cdot \text{ft}$$

$$M_{BC} = \left(\frac{2.0}{4.5}\right)(100) = +44.4 \text{ k} \cdot \text{ft}$$

$$M_{BD} = \left(\frac{1.0}{4.5}\right)(100) = +22.2 \text{ k} \cdot \text{ft}$$

To obtain the carryover moments at A, C, and D, then according to Eq. (10.8) for prismatic members, $c = \frac{1}{2}$. Thus

$$M_{AB} = (\tfrac{1}{2})(33.3) = +16.7 \text{ k} \cdot \text{ft}$$

$$M_{CB} = (\tfrac{1}{2})(44.4) = +22.2 \text{ k} \cdot \text{ft}$$

$$M_{DB} = (\tfrac{1}{2})(22.2) = +11.1 \text{ k} \cdot \text{ft}$$

If one end of a member was simply supported, such as support C in Fig. 10.15g, then the relative stiffness of BC, $K = 2.0$, must be reduced to $\frac{3}{4}K$ or 1.5. Thus, in this case, for $M = 100$ k $\cdot$ ft,

$$M_{BA} = \frac{K_{BA}}{\Sigma K} \cdot M = \frac{1.5}{1.5 + (\frac{3}{4})(2.0) + 1.0}(100) = \left(\frac{1.5}{4}\right)(100) = 37.5 \text{ k} \cdot \text{ft}$$

$$M_{BC} = \frac{(\frac{3}{4})(2.0)}{4}(100) = +37.5 \text{ k} \cdot \text{ft}$$

$$M_{BD} = \left(\frac{1.0}{4}\right)(100) = +25.0 \text{ k} \cdot \text{ft}$$

The corresponding moments at the other ends are

$$M_{AB} = (\tfrac{1}{2})(37.5) = +18.75 \text{ k} \cdot \text{ft}$$

$$M_{CB} = 0$$

since the carryover factor is zero here with C being simply supported, and

$$M_{DB} = (\tfrac{1}{2})(25.0) = +12.5 \text{ k} \cdot \text{ft}$$

It is to be noted that the moment carried over to the far end is always of the same sign as that on the near end of the member, for example, if M_{BA} is positive, so is M_{AB}.

We demonstrate the moment-distribution method by solving several examples; in so doing we use the following abbreviations: D.F. = distribution factor; D.M. = distributed moment; C.O.M. = carryover moment. The following procedure will be used to analyze the various structures in the examples:

1. Calculate the distribution factors at all joints capable of rotation.
2. All joints capable of rotation are first locked.
3. The loads are applied, fixed-end moments are calculated, and the joint restraining moments equal to the algebraic sum of the fixed-end moments are determined.
4. Joints are unlocked one by one; that is, at each joint a moment equal and opposite to the restraining moment is applied and distributed between the members meeting at the joint according to the member stiffnesses.
5. Moment to the far end of each member thus released is carried over.
6. The procedure of locking and unlocking is continued, joint by joint, until the carryover moments are small enough to be neglected. The final moment at the end of each member is obtained by summing up the fixed-end moments, balancing moments, and carryover moments.

■ EXAMPLE 10.4

Determine the end moments in the continuous beam loaded as shown in Fig. 10.16.

Solution. First, we must calculate the distribution factors for all members at all joints.

$$\text{D.F.} = \frac{K}{\sum K}$$

where $K = I/l$ for the member. Since joint A is fixed, $\sum K = 17 + \infty$ and hence D.F. for member AB at joint $A = 17/(17 + \infty) = 0$. For joint C, which is simply supported, $\sum K = 0 + 13 = 13$; hence D.F. for member CB is $K/\sum K = \frac{13}{13} = 1$. For joint B, since C is simply supported, D.F. for BC is

$$\frac{\frac{3}{4}K_{BC}}{K_{BA} + \frac{3}{4}K_{BC}} = \frac{(\frac{3}{4})(13)}{17 + (\frac{3}{4})(13)} = 0.364$$

D.F. for BA is

$$\frac{K_{BA}}{K_{BA} + \frac{3}{4}K_{BC}} = \frac{17}{17 + (\frac{3}{4})(13)} = 0.636$$

The second step is to assume that all the joints are locked, the external loads are applied, and the fixed-end moments at the ends of the members are calculated, using, for example, the information in the Appendix.

$$\text{FEM}_{AB} = -\frac{PL}{8} = -\frac{(48)(13)}{8} = \quad -78.0 \text{ k} \cdot \text{ft}$$

$$\text{FEM}_{BA} = \quad\quad\quad\quad\quad\quad\quad\quad\quad +78.0 \text{ k} \cdot \text{ft}$$

$$\text{FEM}_{BC} = -\frac{wL^2}{12} = -\frac{(4.8)(17)^2}{12} = -115.6 \text{ k} \cdot \text{ft}$$

$$\text{FEM}_{CB} = \quad\quad\quad\quad\quad\quad\quad\quad\quad +115.6 \text{ k} \cdot \text{ft}$$

The above quantities are listed in Table 10.1. To start the moment distribution, we must begin by unlocking joint C (since it is at a simple support) and balancing by applying a moment of -115.6 k · ft to eliminate the artificial moment of $+115.6$ k · ft; in so doing we carry over a moment of $\frac{1}{2}(-115.6) = -57.8$ k · ft to joint B, which is still locked. We should note that after the distribution is carried out, a horizontal line is drawn below to signify that the joint is balanced. Next, we consider joint B, which is subjected to an artificial moment of $(-115.6 - 57.8 + 78.0)$ or (-95.4) k · ft. We eliminate this moment by unlocking joint B and allowing it to rotate; this requires a

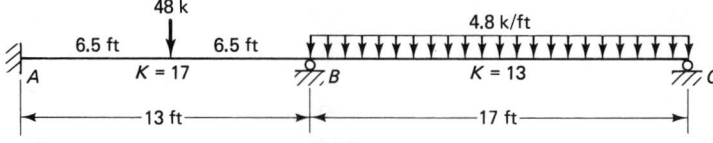

Figure 10.16 Structure for Example 10.4.

TABLE 10.1 MOMENT DISTRIBUTION FOR EXAMPLE 10.4

Joint	A	B		C
End of member	AB	BA	BC	CB
K	17	17	13	13
D.F.	$\dfrac{17}{17+\infty}=0$	$\dfrac{17}{17+(\frac{3}{4})(13)}=0.636$	$\dfrac{(\frac{3}{4})(13)}{17+(\frac{3}{4})(13)}=0.364$	$\dfrac{13}{13}=1$
FEM (lock all joints)	-78.0	$+78.0$	-115.6	$+115.6$
B locked, C unlocked: D.M. (i.e., balance moment at C) Carryover moment to B: C.O.M. B unlocked, C unlocked: D.M. i.e., balance moment at B and carryover to A C.O.M. (no C.O.M. to C since it is simply supported)	$+30.4$	$+60.7$	-57.8 $+34.7$	-115.6 0
Total moment = final end moment	-47.6	$+138.7$	-138.7	0
Change from FEM	$+30.4$	$+60.7$	-23.1	-115.6
$(-\frac{1}{2})$(change at far end)	-30.35	-15.2	$+57.8$	$+11.55$
Sum	≈ 0	$+45.5$	$+34.7$	-104.05
$3EK$	$51E$	$51E$	$39E$	$39E$
$\theta=\dfrac{\text{sum}}{3EK}$	≈ 0	$\dfrac{+0.892}{E}$	$\dfrac{+0.890}{E}$	$\dfrac{-2.668}{E}$

balancing moment of $+95.4$ k·ft applied to the members at B as shown in Table 10.1. Member BA will carry (D.F. for BA) times $(+95.4)=+60.7$ k·ft (half of which is carried to end A, that is, $+30.4$ k·ft) and member BC will carry (D.F. for BC) times $(+95.4)=+34.7$ k·ft, and no moment is carried to end C, since it is simply supported. Now B is balanced and C is balanced. Since joint A is naturally fixed, it is never released and therefore there will not be any unbalanced moment there, nor can there be a carryover moment from the fixed joint A.

The procedure is completed by adding the moments at the end of each member to obtain the final moments at the *ends* of the members as shown in Table 10.1. ∎

■ EXAMPLE 10.5

Determine the member end moments for the rigid frame loaded as shown in Fig. 10.17a.

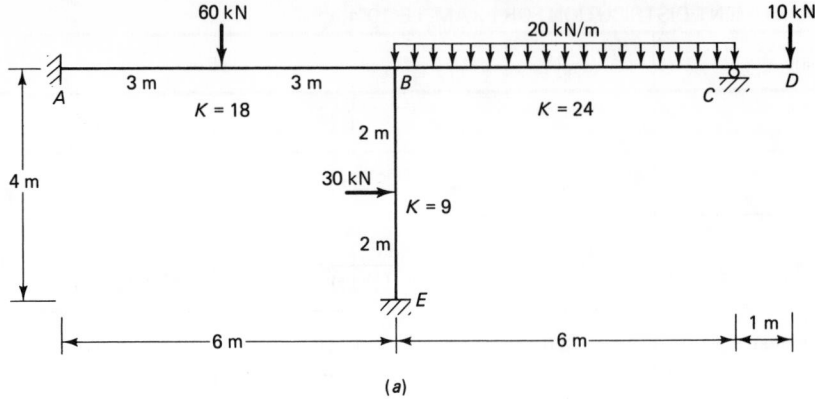

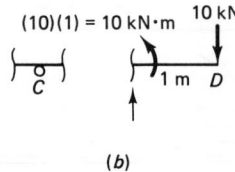

(b)

Figure 10.17 Structure for Example 10.5.

Solution. We first note that the cantilever CD offers no resistance to rotation; that is, no moment is required to rotate end C of member CD when no support or restraint is present at D. Therefore, the effective stiffness of member CD at end C is zero, and thus no unbalanced moment at C can be distributed to the cantilever. From Fig. 10.17b, the moment at end C of CD must be -10 kN $\cdot$ m (negative since it is counterclockwise). Next, we lock all joints and calculate the fixed-end moments. Using the Appendix,

$$\text{FEM}_{AB} = -\frac{PL}{8} = -\frac{(60)(6)}{8} = -45 \text{ kN} \cdot \text{m} \qquad \text{FEM}_{BA} = +45 \text{ kN} \cdot \text{m}$$

$$\text{FEM}_{BC} = -\frac{wL^2}{12} = -\frac{(20)(6)^2}{12} = -60 \text{ kN} \cdot \text{m} \qquad \text{FEM}_{CB} = +60 \text{ kN} \cdot \text{m}$$

$$\text{FEM}_{EB} = -\frac{PL}{8} = -\frac{(30)(4)}{8} = -15 \text{ kN} \cdot \text{m} \qquad \text{FEM}_{BE} = +15 \text{ kN} \cdot \text{m}$$

These are entered in Table 10.2; the moment of -10 kN $\cdot$ m at C of member CD is also entered in the same row. The distribution factors (D.F.) are calculated; since support C is considered as simply supported with a known moment, the stiffness of span BC or CB is taken as $(\frac{3}{4})(24)$.

TABLE 10.2 MOMENT DISTRIBUTION FOR EXAMPLE 10.5

Joint	A	B			C		E
End of member	AB	BA	BE	BC	CB	CD	EB
K	18	18	9	24	24		9
D.F.	$\dfrac{18}{18+\infty}=0$	$\dfrac{18}{18+(\frac{3}{4})(24)+9}$ $=\dfrac{18}{45}=0.4$	$\dfrac{9}{45}=0.2$	$\dfrac{(\frac{3}{4})(24)}{45}$ $=0.4$	1	0	$\dfrac{9}{9+\infty}=0$
FEM (lock all joints) Balance C (D.M.)	-45	$+45$	$+15$	-60	$+60$ -50	-10 0	-15
C.O.M. Balance B (D.M.) (B unlocked and C kept unlocked)		$+10$	$+5$	-25 $+10$			
C.O.M.	$+5$		$\frac{1}{2}(5)$ to EB		0		$+2.5$ (from BE)
Total moment = final end moment	-40.0	$+55.0$	$+20.0$	-75.0	$+10.0$	-10.0	-12.5
Change in moment = $(M-\text{FEM})$ $(-\frac{1}{2})$(change to far end) Sum $3EK$ $\theta=\dfrac{\text{sum}}{3EK}$	$+5$ -5 0 $54E$ 0	$+10$ -2.5 $+7.5$ $54E$ $\dfrac{+0.139}{E}$	$+5$ -1.25^* $+3.75$ $27E$ $\dfrac{+0.139}{E}$	-15 $+25$ $+10$ $72E$ $\dfrac{+0.139}{E}$	-50 $+7.5$ -42.5 $72E$ $\dfrac{-0.590}{E}$	— —	$+2.5$ $-2.5\dagger$ 0 $27E$ 0

* From end E of member EB.

† From end B of member BE.

The first step in the moment-distribution method is to balance joint C and then carry over one-half of the balancing moment of -50 to end B of member BC. Joint B is now unlocked, balanced by moments of $(0.4)(25)=10$ at end B of member BA, a moment of $(0.4)(25)$ at end B of member BC, and a moment of $(0.2)(25)$ at end B of member BE. Half of these distributed moments are carried over to ends A and E, with no carryover to end C. Joint C is kept unlocked during the remainder of the procedure since it is simply supported.

This completes the analysis since all the joints are unlocked with balanced moments; therefore, the final moments are obtained by summing up the moments in each column entry as shown in Table 10.2.

Sometimes a joint in an indeterminate continuous beam rotates or translates in a direction transverse to the axis of connected members, as in the case of settlement or rotation of supporting foundation. Such displacements can be taken into account in the form of fixed-end moments at the start of the moment-distribution procedure.   ∎

10.2.6 Check on Results from Moment-Distribution Method

Any error during the moment-distribution procedure can be checked by calculating the slopes at each side of a joint. From the slope-deflection equation (10.4), one can write

$$M_{AB} = \text{FEM}_{AB} + \frac{4EI}{l}\theta_A + \frac{2EI}{l}\theta_B$$

and   (10.17)

$$M_{BA} = \text{FEM}_{BA} + \frac{2EI}{l}\theta_A + \frac{4EI}{l}\theta_B$$

where M_{AB} and M_{BA} are the final moments, FEM_{AB} and FEM_{BA} are the fixed-end moments, and θ_A and θ_B are the end slopes of member AB, these being positive when assumed to be clockwise. Solving for θ_A and θ_B from Eqs. (10.17) yields

$$\theta_A = \frac{(M_{AB} - \text{FEM}_{AB}) - \frac{1}{2}(M_{BA} - \text{FEM}_{BA})}{3EI/l}$$

                                                                                             (10.18)

$$\theta_B = \frac{(M_{BA} - \text{FEM}_{BA}) - \frac{1}{2}(M_{AB} - \text{FEM}_{AB})}{3EI/l}$$

Thus one can see that the slope at one end of a member equals the *change* in the moment at that end minus one-half the *change* in the moment at the far end of the member, divided by $3EI/l$ or $3EK$. This check has been applied to Examples 10.4 and 10.5 as shown in Tables 10.1 and 10.2. For instance, the calculation of the end slope given at the bottom of Table 10.2 shows that $\theta_A = 0$ (as it should be since end A is fixed) and $\theta_{BA} = \theta_{BC} = \theta_{BE} = 0.139/E$, which proves the compatibility of slopes at joint B. One should be cautioned that the satisfaction of this check is no guarantee that the initial fixed-end moments are correct!

10.2.7 Symmetry and Antisymmetry

For geometrically symmetrical structures subjected to symmetrical loading as in Fig. 10.18a and b, symmetrical joints rotate the same amount but in the opposite direction. In Fig. 10.18a and b, $\theta_B = -\theta_C$ and $\theta_A = -\theta_D$. Writing the slope-deflection equation for member BC in Fig. 10.18a,

$$M_{BC} = \frac{2EI}{L}(2\theta_B + \theta_C)$$

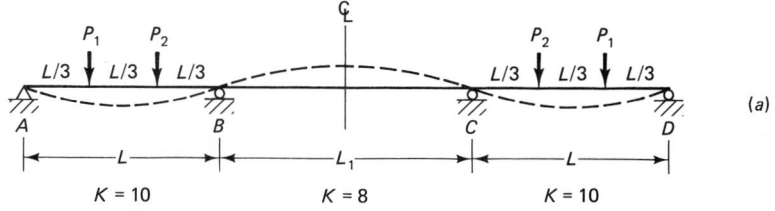

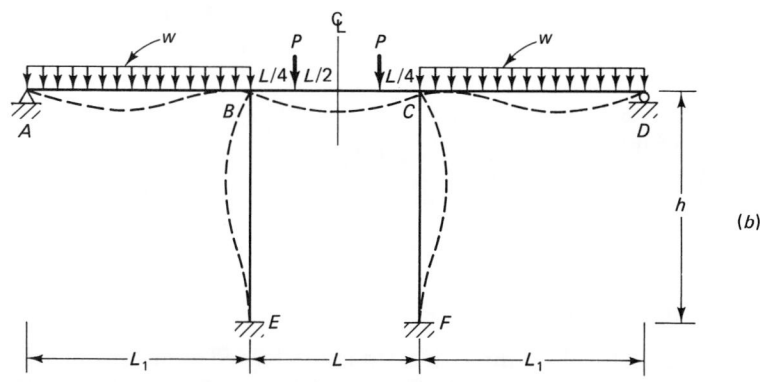

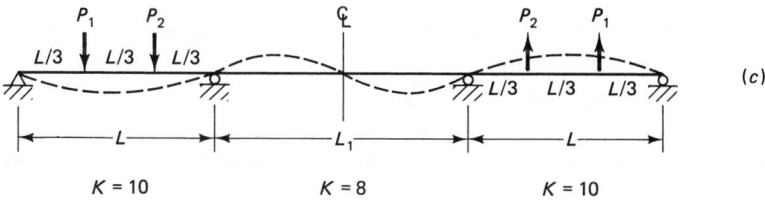

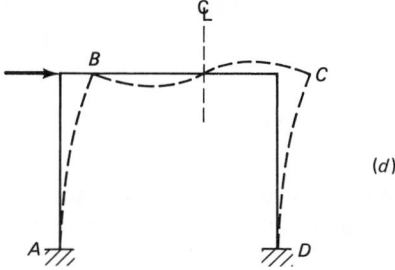

Figure 10.18 Demonstration of symmetry and antisymmetry.

Putting $\theta_B = -\theta_C$ yields

$$M_{BC} = \frac{2EI}{L}\theta_B$$

If $\theta_B = 1$, then $M_{BC} =$ stiffness of $BC = 2EI/L$ which is one-half of $4EI/L$, which is the normal value of stiffness used in calculating the distribution factors at B and C.

Thus when such symmetry exists, one can use one-half of the usual stiffness of the center span when calculating the distribution factors for the joints at its end, and balance the moments for half the structure with no carryover moments across the axis of symmetry. Thus, for example, the distribution factors (D.F.) for the structure in Fig. 10.18*a* become, after the above modification,

For *BA*:

$$\frac{(\frac{3}{4})(10)}{(\frac{3}{4})(10) + (\frac{1}{2})(8)} = \frac{7.5}{7.5 + 4} = 0.652$$

For *BC*:

$$\frac{(\frac{1}{2})(8)}{(\frac{3}{4})(10) + (\frac{1}{2})(8)} = \frac{4}{7.5 + 4} = 0.348$$

For geometrically symmetrical structures loaded by antisymmetrical loading (Fig. 10.18*c* and *d*), a point of contraflexure exists at the center of span *BC*, with $\theta_B = \theta_C$. Thus, using the slope-deflection equation,

$$M_{BC} = \frac{2EI}{L}(2\theta_B + \theta_C) = \frac{6EI}{L}\theta_B$$

If $\theta_B = 1$, $M_{BC} =$ stiffness of $BC = 6EI/L$, which is 1.5 times $4EI/L$, the usual stiffness. In such structures, therefore, we can adjust the stiffness of the center span to one and one-half times the usual value and proceed with the moment distribution for only half the structure, with no carryover moment across the axis of antisymmetry. Thus, for the continuous beam in Fig. 10.18*c*, the adjusted distribution factors become

For *BA*:

$$\frac{(\frac{3}{4})(10)}{(\frac{3}{4})(10) + (\frac{3}{2})(8)} = \frac{7.5}{7.5 + 12} = \frac{7.5}{19.5} = 0.385$$

and for *BC*:

$$\frac{(\frac{3}{2})(8)}{(\frac{3}{4})(10) + (\frac{3}{2})(8)} = \frac{12}{19.5} = 0.615$$

For geometrically symmetrical structures, much computational work is eliminated if one makes use of the symmetry or antisymmetry of the loading, if it exists.

■ **EXAMPLE 10.6**

Determine the moments at the ends of the members in the rigid frame loaded as shown in Fig. 10.19.

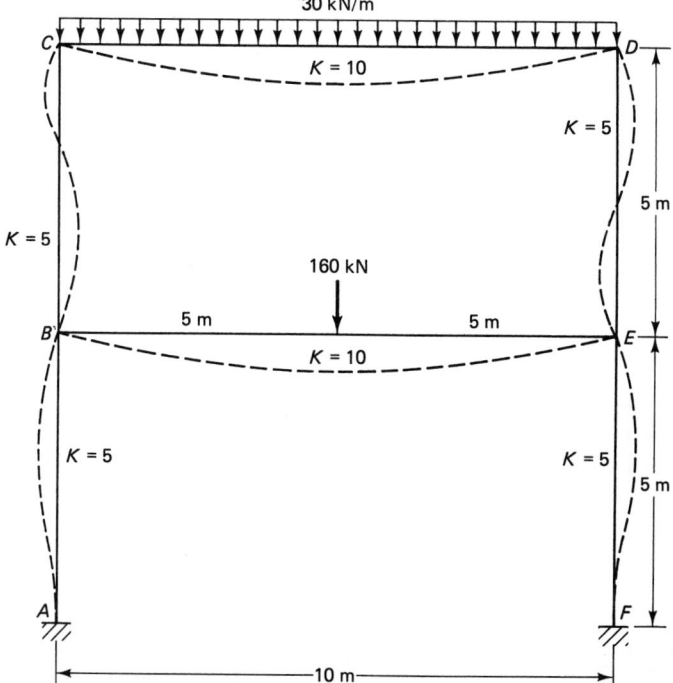

Figure 10.19 Structure for Example 10.6.

Solution. We first lock all joints and calculate the fixed-end moments.

$$\text{FEM}_{BE} = -\frac{PL}{8} = -\frac{(160)(10)}{8} = -200 \text{ kN} \cdot \text{m} \qquad \text{FEM}_{EB} = +200 \text{ kN} \cdot \text{m}$$

$$\text{FEM}_{CD} = -\frac{wL^2}{12} = -\frac{(30)(10)^2}{12} = -250 \text{ kN} \cdot \text{m} \qquad \text{FEM}_{DC} = +250 \text{ kN} \cdot \text{m}$$

We also observe that the frame is geometrically symmetrical and symmetrically loaded about the vertical centerline, with no translation of joints. Therefore, we can reduce the stiffness of *CD* and *BE* by one-half and deal only with joints *C, B,* and *A,* that is, half the structure. Thus, using the *K* values given, the distribution factors become, after the above modifications,

For *CD*:

$$\frac{(\frac{1}{2})(10)}{(\frac{1}{2})(10) + 5} = \frac{5}{10} = 0.5 \qquad \text{For } CB: \frac{5}{(\frac{1}{2})(10) + 5} = 0.5$$

For *BE*:

$$\frac{(\frac{1}{2})(10)}{(\frac{1}{2})(10) + 5 + 5} = \frac{5}{15} = 0.333 \qquad \text{For } BA: \frac{5}{(\frac{1}{2})(10) + 5 + 5} = 0.333$$

TABLE 10.3 MOMENT DISTRIBUTION FOR EXAMPLE 10.6

Joint	A	B			C	
End of member	AB	BA	BE	BC	CB	CD
K	5	5	10	5	5	10
D.F.	$\dfrac{5}{5+\infty}=0$	$\dfrac{5}{5+(\frac{1}{2})(10)+5}=0.333$	$\dfrac{(\frac{1}{2})(10)}{5+(\frac{1}{2})(10)+5}=0.333$	$\dfrac{5}{5+(\frac{1}{2})(10)+5}=0.333$	$\dfrac{5}{5+(\frac{1}{2})(10)}=0.5$	$\dfrac{(\frac{1}{2})(10)}{5+(\frac{1}{2})(10)}=0.5$
FEM			−200			−250
Balance C					+125	+125
C.O.M.				+62.5		
Balance B		+45.8	+45.8	+45.9		
C.O.M.	+22.9				+22.9	
Balance C					−11.5	−11.4
C.O.M.				−5.7		
Balance B		+1.9	+1.9	+1.9		
C.O.M.	+1.0				+0.9	
Balance C					−0.4	−0.5
C.O.M.				−0.2*		
Total moment = final end moment	+23.9	+47.7	−152.3	+104.6	+136.9	−136.9
Change from FEM	+23.9	+47.7	+47.7	+104.6	+136.9	+113.1
$(-\frac{1}{2})$(change at far end)	−23.9	−11.9	+23.9†	−68.5	−52.3	+56.6‡
Sum	0	+35.8	+71.6	+36.1	+84.6	+169.7
$\theta=\dfrac{\text{sum}}{3EK}$	0	$\dfrac{+2.39}{E}$	$\dfrac{+2.39}{E}$	$\dfrac{+2.41}{E}$	$\dfrac{+5.64}{E}$	$\dfrac{+5.66}{E}$

* Considered not large enough to distribute.

† From end E of member EB.

‡ From end D of member DC.

For BC:

$$\frac{5}{(\frac{1}{2})(10) + 5 + 5} = 0.333$$

The procedural steps for the moment distribution are shown in Table 10.3. The entries are self-explanatory. A check on the slopes, θ, shows the correctness of the calculations. ∎

10.2.8 Structures with Joint Translations

So far we have considered structures in which no joint translations occurred. We now apply the moment-distribution method to structures with joints that can rotate as well as translate. If a structure is unsymmetrical owing either to the layout of its members or to the imposed loading, then joint rotations will be accompanied by translation or sidesway, as shown in Fig. 10.20a. We can analyze such structures in two stages: First we subject the structure to a number of artificial restraints so that no joint translations can occur; for the structure in Fig. 10.20a, a horizontal restraint at C has been added as shown in Fig. 10.20b; the moment distribution can now be carried out in the usual manner yielding the member moments in the artificially restrained structure as well as the magnitude of the restraining force R. In the second stage we subject the structure to an equal and opposite force to that of the artificial restraint as shown in Fig. 10.20c, and the member moments thus produced are added to those in the first stage. This is

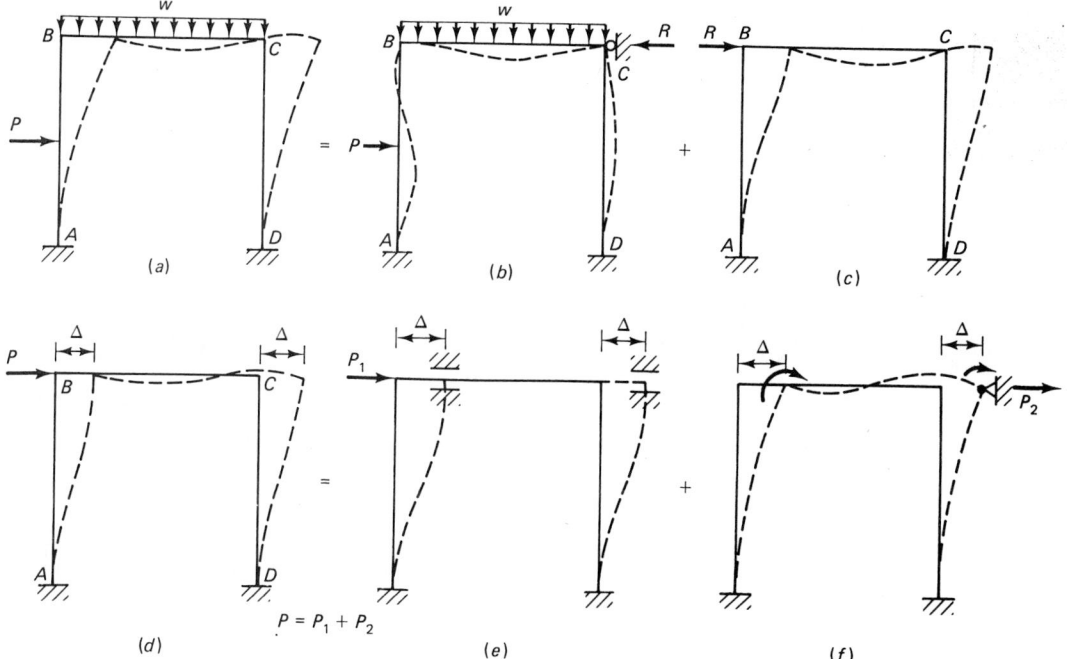

Figure 10.20 Treatment of structure with joint translation.

permitted based on the principle of superposition since we are dealing with linear structures. However, this procedure is cumbersome since in order to calculate the end moments due to the force R, deflections and rotations at the joints must be first determined. A much more convenient method is to assume that the frame is subjected to an arbitrary lateral displacement Δ due to an unbalanced horizontal force P as shown in Fig. 10.20d; in so doing joints B and C will undergo both translation and rotation; the translation of joints B and C will be the same amount Δ, since axial deformation of member BC is neglected. The deformation of joints B and C is composed of two parts, translation without rotation (Fig. 10.20e) and rotation without translation (Fig. 10.20f). The necessary horizontal forces for the deformations shown in Fig. 10.20e and f are P_1 and P_2, respectively. Thus the horizontal force $P = P_1 + P_2$ is found after determining P_1 and P_2 and hence the end moments for the frame in Fig. 10.20d. Since the structure is linear, the moments for the loading in Fig. 10.20c can be obtained by multiplying the end moments for the frame in Fig. 10.20d by the ratio R/P. The procedure will be clarified by the following solved examples.

■ **EXAMPLE 10.7**

Determine the member end moments for the loaded frame shown in Fig. 10.21a.

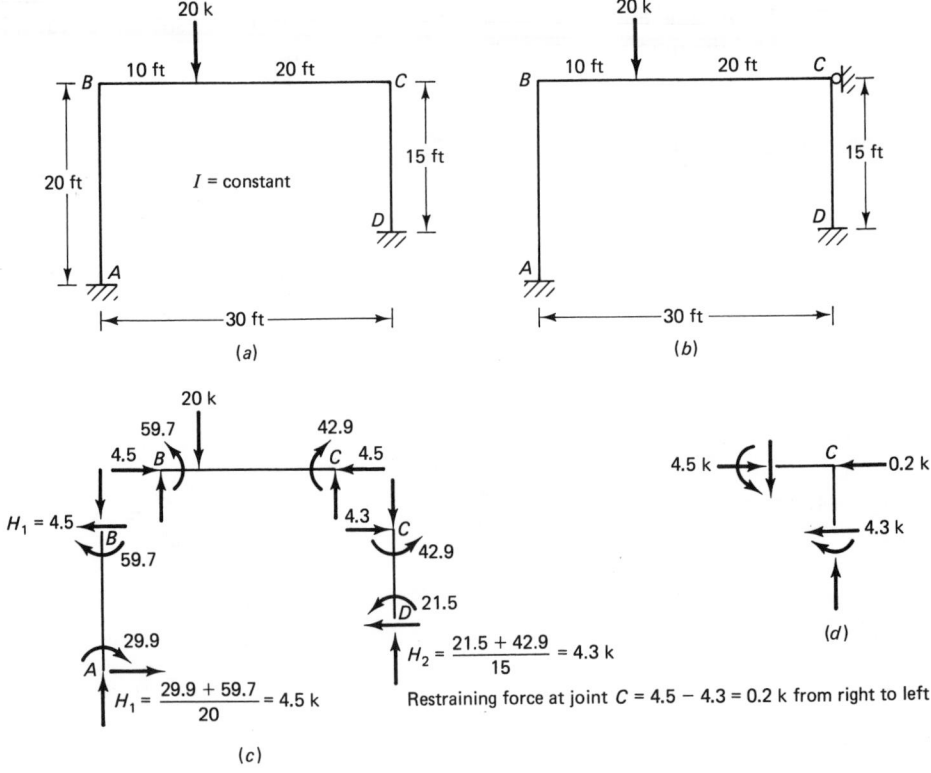

Figure 10.21 Structure for Example 10.7. (c) FBD of columns and beam. (d) FBD of joint C.

Solution. For the first stage, we artificially restrain joint C (or joint B) as shown in Fig. 10.21b and carry out the moment-distribution procedure. We first calculate the fixed-end moments. Thus, from the Appendix,

$$\text{FEM}_{BC} = -\frac{Pab^2}{L^2} = -\frac{(20)(10)(20)^2}{(30)^2} = -88.9 \text{ k} \cdot \text{ft}$$

$$\text{FEM}_{CB} = +\frac{Pa^2b}{L^2} = +\frac{(20)(10)^2(20)}{(30)^2} = +44.4 \text{ k} \cdot \text{ft}$$

The distribution factors (D.F.) are as shown in Table 10.4. Carrying out the moment-distribution procedure yields the member end moments shown in Table 10.4.

The next step is to calculate the magnitude of the artificial restraining force at C. Using the member end moments in Table 10.4, we calculate the horizontal shear at the ends of the columns as shown in Fig. 10.21c; from the free-body diagram of joint C (Fig. 10.21d), we find the restraining force at $C = 0.2$ k acting to the left. The moments determined in Table 10.4 are for the structure loaded as shown in Fig. 10.22a. We have to superimpose onto these the moments obtained from the structure loaded as shown in Fig. 10.22b to yield the moments for the given loaded structure

TABLE 10.4 FIRST MOMENT DISTRIBUTION FOR EXAMPLE 10.7

Joint	A	B		C	D
End of member	AB	BA	BC	CB	CD	DC
$K = I/l$	$I/20$	$I/20$	$I/30$	$I/30$	$I/15$	$I/15$
D.F.	$\dfrac{I/20}{I/20 + \infty} = 0$	$\dfrac{3}{5}$	$\dfrac{2}{5}$	$\dfrac{1}{3}$	$\dfrac{2}{3}$	0
FEM	0	0	-88.9	$+44.4$	0	0
Balance B and C.O.M.	$+26.7$	$+53.3$	$+35.6$			
Balance C and C.O.M.				$+17.8$ -20.7	-41.5	-20.8
Balance B and C.O.M.	$+3.1$	$+6.2$	-10.4 $+4.2$			
Balance C and C.O.M.				$+2.1$ -0.7	-1.4	-0.7
Balance B and C.O.M.	$+0.1$	$+0.2$	-0.3 $+0.1$	(negligible)		
Total moment (k·ft) from first distribution	$+29.9$	$+59.7$	-59.7	$+42.9$	-42.9	-21.5

in Fig. 10.22*c*; that is, this cancels out the effects of the artificial restraining force $R = 0.2$ k.

To Determine the Moments for the Structure Loaded as Shown in Fig. 10.22b. Let us assume that the joints *B* and *C* in Fig. 10.22*e* are initially locked against rotation but are free to translate. Then an arbitrary lateral force applied to the top level of the columns will produce a sidesway Δ. According to the slope-deflection equations (10.4), end moments are produced, given by

$$M_{AB} = \overset{0}{\cancel{FEM_{AB}}} + \frac{2EI}{l}\left(2\overset{0}{\cancel{\theta_A}} + \overset{0}{\cancel{\theta_B}} - \frac{3\Delta}{l} \right) = -\frac{6EI\Delta}{l^2}$$

$$= M_{BA}$$

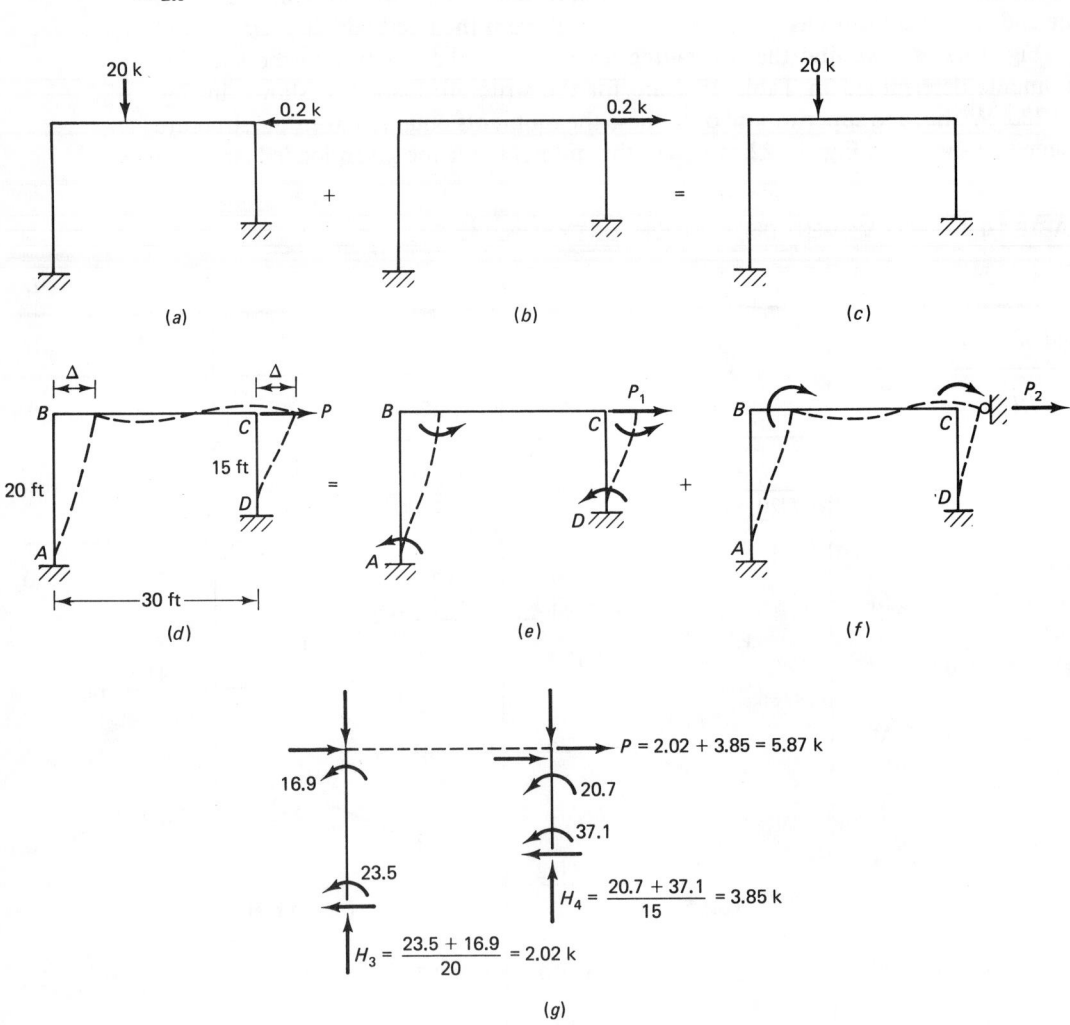

Figure 10.22 Treatment of sidesway. (*g*) FBD of columns.

Thus,

$$M_{AB} = -\frac{6EI\Delta}{(20)^2} = -0.0150\,EI\Delta = M_{BA}$$

and

$$M_{CD} = -\frac{6EI\Delta}{(15)^2} = -0.0267\,EI\Delta = M_{DC}$$

Not knowing Δ, we can arbitrarily assign a value to the term $EI\Delta = 2000$, and we have

$$M_{AB} = M_{BA} = -30.0 \text{ k} \cdot \text{ft}$$

$$M_{CD} = M_{DC} = -53.4 \text{ k} \cdot \text{ft}$$

The choice of $EI\Delta = 2000$ was made because it is desirable to assume values for the fixed-end moments that lie in the range of the moments found in Table 10.4. To unlock the joints at B and C we have to apply joint moments shown in Fig. 10.22f, which is tantamount to the moment-distribution procedure. Table 10.5 shows this procedure applied to fixed-end moments. From the end moments on the columns shown as an FBD in Fig. 10.22g:

$$P = 2.02 + 3.85 = 5.87 \text{ k}$$

TABLE 10.5 SECOND MOMENT DISTRIBUTION FOR EXAMPLE 10.7

Joint	A	B			C	D
End of member	AB	BA	BC	CB	CD	DC
D.F.	0	$\frac{3}{5}$	$\frac{2}{5}$	$\frac{1}{3}$	$\frac{2}{3}$	0
FEM	−30	−30	0	0	−53.4	−53.4
Balance B and C.O.M.	+9	+18	+12	+6		
Balance C and C.O.M.			+7.9	+15.8	+31.6	+15.8
Balance B and C.O.M.	−2.4	−4.7	−3.2	−1.6		
Balance C and C.O.M.			+0.3	+0.5	+1.1	+0.5
Balance B and C.O.M.	−0.1	−0.2	−0.1	(negligible)		
Total moments from second distribution	−23.5	−16.9	−16.9	+20.7	−20.7	−37.1

The assumed sidesway Δ was produced by a horizontal force of $P = 5.87$ k but we needed only 0.2 k. However, since member moments are proportional to the applied load, the desired moments for the loaded structure in Fig. 10.22b can be obtained by multiplying the moments in Table 10.5 by the ratio $0.2/5.87 = 0.034$. Thus the final member moments in the structure (Fig. 10.22c) are obtained by adding the corrected moments from the second moment distribution to the moments determined in the first moment distribution. Thus,

$$M_{AB} = +29.9 + 0.034(-23.5) = +29.1 \text{ k} \cdot \text{ft}$$

$$M_{BA} = +59.7 + 0.034(-16.9) = +59.1 \text{ k} \cdot \text{ft}$$

$$M_{BC} = -59.1 \text{ k} \cdot \text{ft}$$

$$M_{CD} = -42.9 + 0.034(-20.7) = -43.6 \text{ k} \cdot \text{ft}$$

$$M_{CB} = +43.6 \text{ k} \cdot \text{ft}$$

$$M_{DC} = -21.5 + 0.034(-37.1) = -22.8 \text{ k} \cdot \text{ft}$$

The FBD of the loaded frame and the corresponding bending-moment diagram are shown in Fig. 10.23a and b. ∎

■ EXAMPLE 10.8
Determine by the method of moment distribution the member end moments for the loaded rigid frame shown in Fig. 10.24a.

Solution. The frame will be initially restrained against lateral translation by providing a temporary support at the top right-hand joint C. Thus, the fixed-end moments for the restrained structure are

$$\text{FEM}_{BC} = -\frac{Pab^2}{l^2} = -\frac{(250)(8)(2)^2}{10^2} = -80 \text{ k} \cdot \text{ft}$$

$$\text{FEM}_{CB} = +\frac{Pa^2b}{l^2} = +\frac{(250)(8)^2(2)}{10^2} = +320 \text{ k} \cdot \text{ft}$$

Applying the moment-distribution procedure will lead to the first set of end moments as shown in Table 10.6. From the FBD of the left-hand column AB (Fig. 10.24b), summing up moments about joint B yields

$$\sum M_{\text{at } B} \Big\rangle^+ = 0 = H_A(12) - (44)(5) - 35 - 70$$

or

$$H_A = 27.1 \text{ k} \rightarrow$$

Similarly, summing up moments about joint C of the forces acting on the right-hand column CD, we have

$$\sum M_{\text{at } C} \Big\rangle^+ = 0 = -(H_D)(12) + (206)(3.5) + 64.9 + 130$$

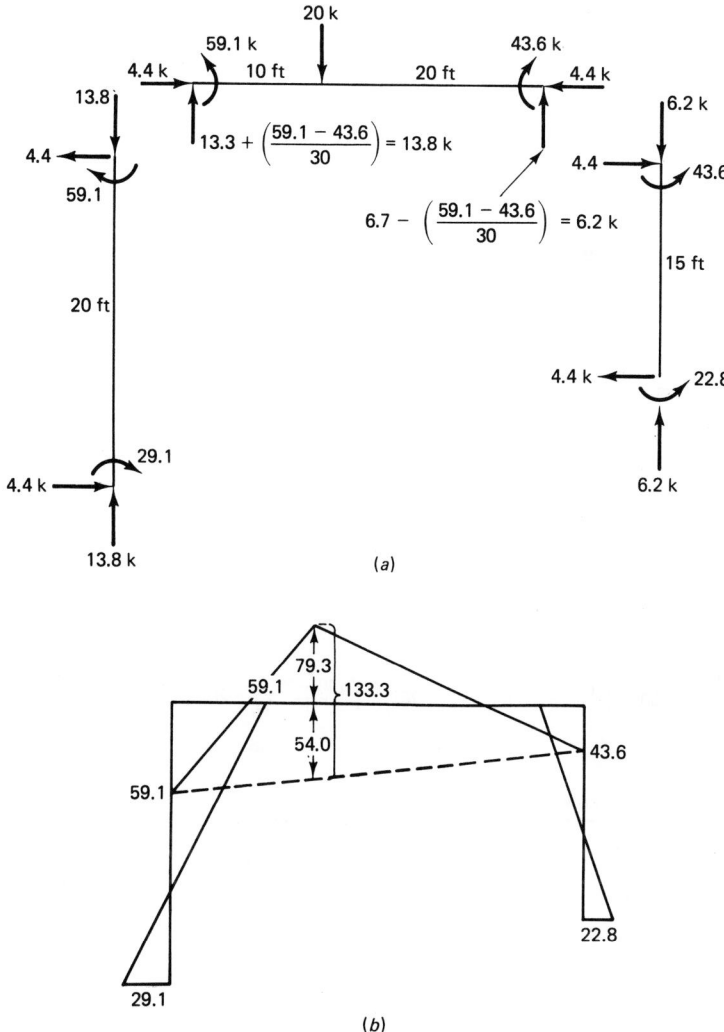

Figure 10.23 Free-body and bending-moment diagrams for the structure in Fig. 10.21. (*b*) BMD on compression side.

or

$$H_D = 76.3 \text{ k} \leftarrow$$

Total artificial restraining force $R = 76.3 - 27.1 - 15.0$
$$= 34.2 \text{ k from left to right}$$

We have to cancel the effect of this restraining force; we first assume that the frame sways to the right a distance δ with the joints prevented from rotating. To calculate the fixed-end moments due to sway, we must determine the displacement Δ terms for

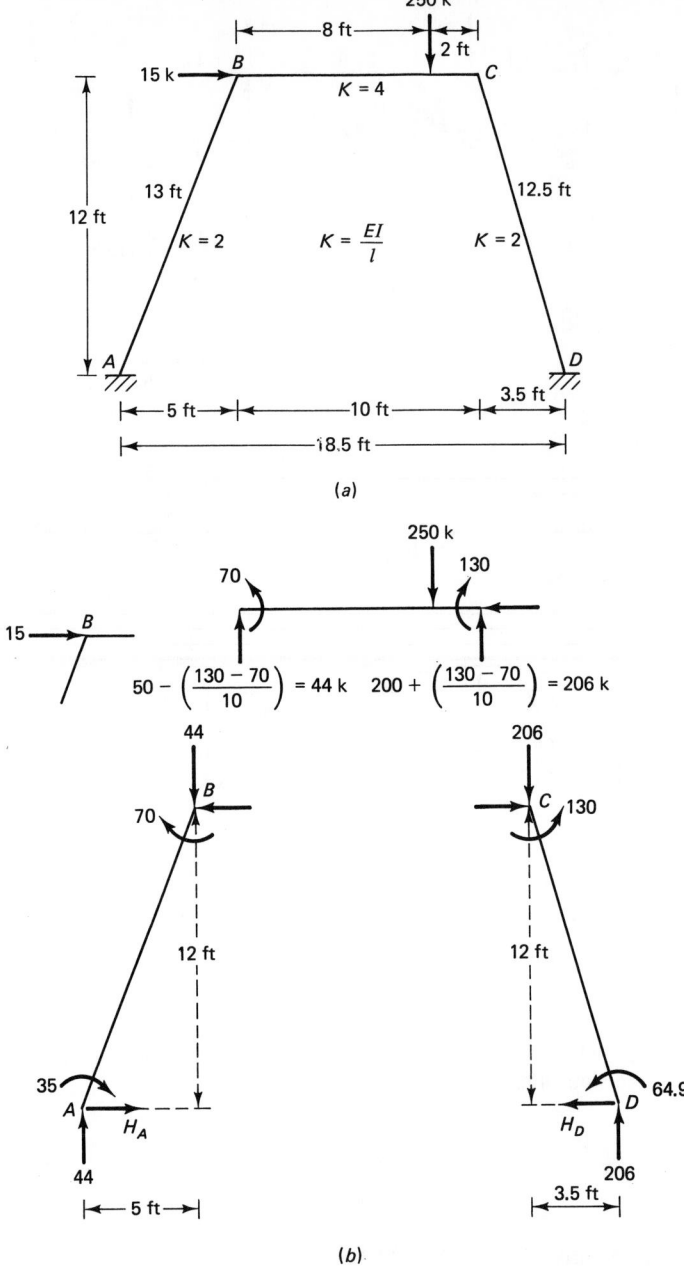

Figure 10.24 Structure for Example 10.8. (*b*) FBD (after first distribution).

each member. To simplify the derivation of the Δ terms, assume an analogous frame with pinned joints distorted as shown in Fig. 10.25*a*. Now $\Delta BB'b$ is similar to ΔABG, and $\Delta CC'g$ is similar to ΔCDH. Therefore, $BB' = bB'/\cos \angle BB'b = 13\delta/12$, and so

TABLE 10.6 FIRST MOMENT DISTRIBUTION FOR STRUCTURE IN EXAMPLE 10.8

Joint	A		B		C	D
End of member	AB	BA	BC	CB	CD	DC
K	2	2	4	4	2	2
D.F.	0	$\frac{1}{3}$	$\frac{2}{3}$	$\frac{2}{3}$	$\frac{1}{3}$	0
FEM	0	0	−80	+320	0	0
Balance C and C.O.M.				−213.3	−106.7	
			−106.7			−53.3
Balance B and C.O.M.		+62.2	+124.5			
	+31.1			+62.2		
Balance C and C.O.M.				−41.5	−20.7	
			−20.7			−10.4
Balance B and C.O.M.		+6.9	+13.8			
	+3.5			+6.9		
Balance C and C.O.M.				−4.6	−2.3	
			−2.3			−1.1
Balance B and C.O.M.		+0.8	+1.5			
	+0.4			+0.8		
Balance C and C.O.M.				−0.5	−0.3	
			−0.3			−0.1
Balance B and C.O.M.		+0.1	+0.2			
	(negligible)			+0.1		
Total moment from first distribution	+35.0	+70.0	−70.0	+130.1	−130.0	−64.9

on. Thus, for a horizontal displacement δ, the Δ term for member AB is BB', or $+13\delta/12$; for CD it is CC', or $+12.5\delta/12$; and for member BC it is $(Bb + C'g) = -(5\delta/12 + 3.5\delta/12) = -8.5\delta/12$; it is negative since the displacement is counterclockwise. Thus, because of sway, the fixed-end moments at the ends of each member are

$$\text{FEM}_{AB} = -\frac{6EI\Delta}{l^2} = -\frac{6EK\Delta}{l} = -\frac{(6)(2)}{13}\left(\frac{13\delta}{12}\right)E = -E\delta = \text{FEM}_{BA}$$

$$\text{FEM}_{BC} = -\frac{6EK\Delta}{l} = -\frac{(6)(4)}{10}\left(-\frac{8.5\delta}{12}\right)E = +1.7E\delta = \text{FEM}_{CB}$$

$$\text{FEM}_{CD} = -\frac{6EK\Delta}{l} = -\frac{(6)(2)}{12.5}\left(\frac{12.5\delta}{12}\right)E = -E\delta = \text{FEM}_{DC}$$

Assume $E\delta = 200$; then $\text{FEM}_{AB} = \text{FEM}_{BA} = -200$; $\text{FEM}_{BC} = \text{FEM}_{CB} = +340$; $\text{FEM}_{CD} = \text{FEM}_{DC} = -200$. The moment distribution is carried out in Table 10.7.

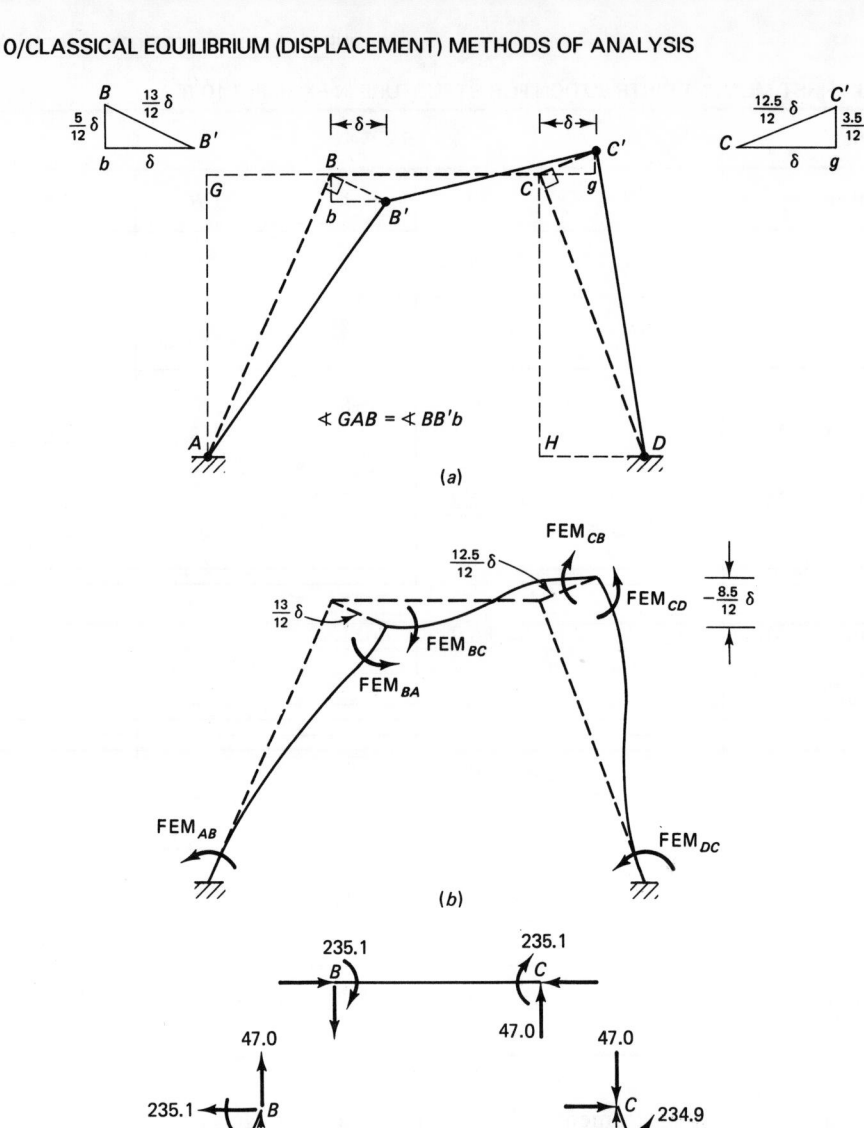

Figure 10.25 Treatment of sidesway in a frame with sloping columns.

TABLE 10.7   SECOND MOMENT DISTRIBUTION FOR THE STRUCTURE IN EXAMPLE 10.8

Joint	A	B		C	D
End of member	AB	BA	BC	CB	CD	DC
D.F.	0	$\frac{1}{3}$	$\frac{2}{3}$	$\frac{2}{3}$	$\frac{1}{3}$	0
FEM	-200	-200	$+340$	$+340$	-200	-200
Balance B and C.O.M.	-23.3	-46.7	-93.3	-46.7		
Balance C and C.O.M.			-31.1	-62.2	-31.1	-15.6
Balance B and C.O.M.	$+5.2$	$+10.4$	$+20.7$	$+10.3$		
Balance C and C.O.M.			-3.4	-6.9	-3.4	-1.7
Balance B and C.O.M.	$+0.5$	$+1.1$	$+2.3$	$+1.2$		
Balance C and C.O.M.			-0.4	-0.8	-0.4	-0.2
Balance B and C.O.M.	(negligible)	$+0.1$	$+0.3$	$+0.1$		
Total moments from second distribution	-217.6	-235.1	$+235.1$	$+235.0$	-234.9	-217.5

The FBD of the members of the frame are shown in Fig. 10.25 c. Again summing up moments of forces on member AB about joint B, we find

$$\sum M_{\text{at } B} \, \rangle^+ = 0 = (H_A)(12) + (47.0)(5) + 217.6 + 235.1$$

or

$$H_A = -57.3 \text{ k or } 57.3 \text{ k} \leftarrow$$

For the FBD of member CD, taking moments about joint C, we get

$$\sum M_{\text{at } C} \, \rangle^+ = 0 = -(H_D)(12) + (47.0)(3.5) + 217.5 + 234.9$$

or

$$H_D = 51.4 \text{ k} \leftarrow$$

Therefore, the lateral force causing sway = $57.3 + 51.4 = 108.7$ k from left to right. The moments in Table 10.7 must be multiplied by a factor of $(-34.2/108.7) = -0.315$ to obtain the true moments for the lateral force of $(-R)$ or 34.2 k acting from right to left. Therefore, by superposition, the final moments for the loaded frame in Fig. 10.24a are

$$M_{AB} = +35.0 + (-0.315)(-217.6) = +103.5 \text{ k} \cdot \text{ft}$$

$$M_{BA} = +70.0 + (-0.315)(-235.1) = +144.0 \text{ k} \cdot \text{ft}$$

$$M_{BC} = -70.0 + (-0.315)(235.1) = -144.0 \text{ k} \cdot \text{ft}$$

$$M_{CB} = +130.1 + (-0.315)(235.1) = +56.1 \text{ k} \cdot \text{ft}$$

$$M_{CD} = -130.0 + (-0.315)(-234.9) = -56.1 \text{ k} \cdot \text{ft}$$

$$M_{DC} = -64.9 + (-0.315)(-217.5) = +3.5 \text{ k} \cdot \text{ft} \qquad \blacksquare$$

10.2.9 Multistory and Gable Frames

The problem of sidesway of structures dealt with up to now was defined in terms of *one* lateral displacement, that is, one degree of freedom of joint translation. However, if a building frame consists of more than one story, say n stories, and is loaded unsymmetrically or is itself unsymmetrical and subjected to symmetrical loading, then the solution to this problem can be obtained by breaking it down to n independent cases in each of which only one degree of freedom of joint translation is allowed to occur. Consider a two-story frame shown in Fig. 10.26a; the loaded frame, with two degrees of freedom in translation, will deflect sideways at B and C distances Δ_1 and Δ_2, respectively. If Δ_1 and Δ_2 are known in advance, the frame can be deflected into this position while all joints are fixed against rotation; the unlocking of joints by means of the moment-distribution method would yield the final moments. However, Δ_1 and Δ_2 are unlikely to be known. To handle this problem, let us deal with two separate cases as shown in Fig. 10.26b and c, each involving one degree of freedom of joint translation. In Fig. 10.26b, the bottom story of the frame is pushed sideways a distance Δ_1' with all joints fixed against rotation and assuming that the second story is infinitely rigid, as shown by the solid outlines. While in this translated position, the joints are

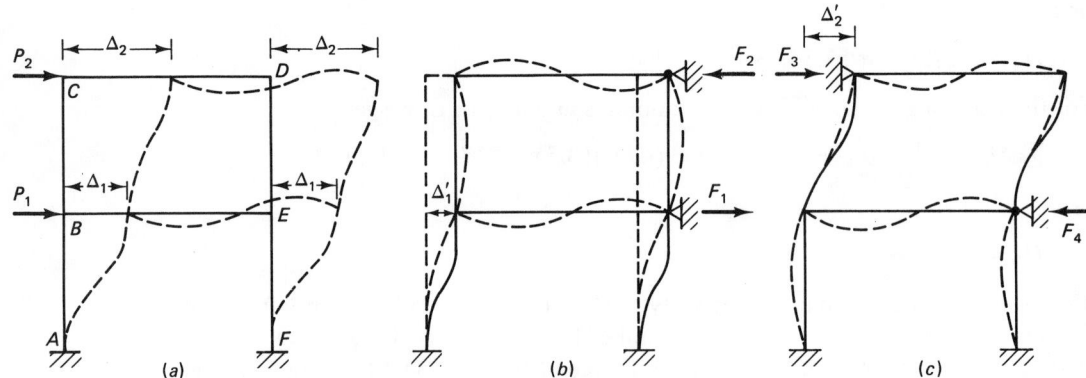

Figure 10.26 Two-story frame with sidesway.

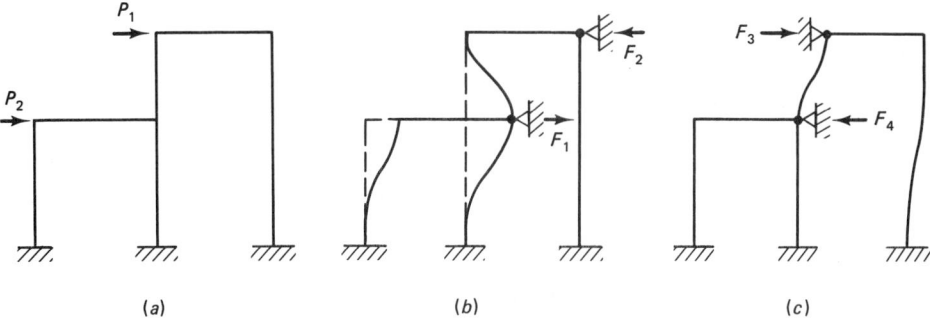

Figure 10.27 Two-story split-level frame with sidesway.

relaxed by means of the moment-distribution method; column shears and hence the forces F_1 and F_2 are determined from the moments in columns in this relaxed position shown by the dashed configuration in Fig. 10.26b. A similar solution can be carried out for the case in Fig. 10.26c where, with all the joints fixed against rotation, the second story is pushed a distance Δ_2' relative to the first floor, which is assumed to be infinitely rigid. Applying the method of moment distribution will produce the pushing force F_3 and the reaction force F_4 in terms of the column shears.

We must now determine how F_1, F_2, F_3, and F_4 should be modified so that the actual loading is obtained. If the forces in Fig. 10.26b are multiplied by a factor K_1 and the forces in Fig. 10.26c by a factor K_2, then applying the force-equilibrium equations at each story will yield

At the first-floor level:

$$K_1 F_1 - K_2 F_4 = P_1$$

At the second-floor level: (10.19)

$$-K_1 F_2 + K_2 F_3 = P_2$$

Solving these two simultaneous equations produces the values of K_1 and K_2. Therefore, the final moments for the frame in Fig. 10.26a are obtained by superimposing K_1 times the moments in Fig. 10.26b and K_2 times the moments in Fig. 10.26c. A similar procedure is used to analyze other frames such as the split-level frame shown in Fig. 10.27a and the gabled frame in Fig. 10.28a. The procedure is the same for frames having any number of stories or gables; the number of simultaneous equations in the K's will be the same as the number of degrees of freedom of joint translation for the given frame.

It should be mentioned that in the analysis of frames not loaded only at the joints, for example, the frame in Fig. 10.29a, a moment distribution must be first carried out with sidesway prevented; the lateral restraining forces will be required at each story such as R_1, R_2, and R_3 in Fig. 10.29b, and these can be found from the resulting moments. The next step is to eliminate all these artificial restraining forces by applying a set of equal and opposite forces as shown in Fig. 10.29c. As explained earlier, this can be accomplished by applying arbitrary sidesway to each story as shown in Fig.

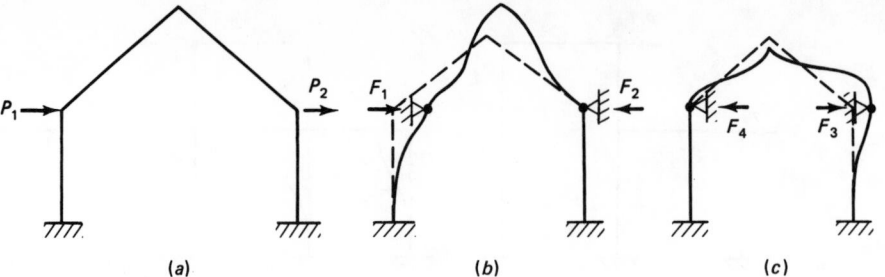

Figure 10.28 Gable frame with sidesway.

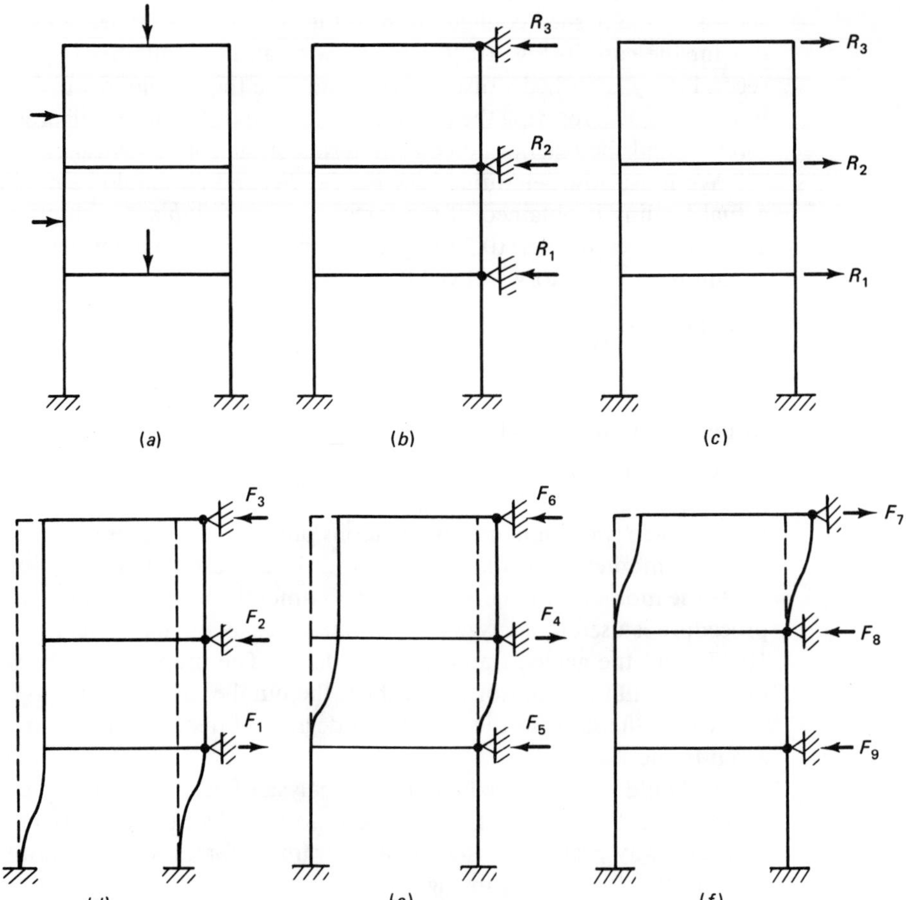

Figure 10.29 Three-story single-bay rigid frame with sidesway.

10.29 d, e, and f. The equations for the K's [see Eq. (10.19)] in this case will be as follows:

$$K_1 F_1 - K_2 F_5 - K_3 F_9 = R_1$$

$$-K_1 F_2 + K_2 F_4 - K_3 F_8 = R_2 \qquad (10.20)$$

$$-K_1 F_3 - K_2 F_6 + K_3 F_7 = R_3$$

With K_1, K_2, K_3 determined, we superpose the results as explained before to find the final moments for the frame in Fig. 10.29 a.

■ EXAMPLE 10.9

Determine the end moments in the two-story frame shown in Fig. 10.30 a.

Solution. The distribution factors (D.F.) are calculated in the usual way and are shown in Table 10.8. Since there are no lateral loads applied to the members, we will proceed with the first stage of sidesway; that is, we apply sway Δ_1 to the first story as shown in Fig. 10.30 b, with all the joints locked and the second story assumed infinitely rigid. The resulting FEMs are as follows:

$$\text{FEM}_{AB} = \frac{-6E(4I)}{(16)^2} \, \Delta_1 = \frac{-3}{32} \, EI\Delta_1 = \text{FEM}_{BA}$$

$$\text{FEM}_{FE} = \frac{-6E(2I)}{(8)^2} \, \Delta_1 = \frac{-3}{16} \, EI\Delta_1 = \text{FEM}_{EF}$$

Assuming $\frac{3}{32} EI\Delta_1 = 100$,

$$\text{FEM}_{AB} = \text{FEM}_{BA} = -100 \text{ k} \cdot \text{ft} \quad \text{and} \quad \text{FEM}_{FE} = \text{FEM}_{EF} = -200 \text{ k} \cdot \text{ft}$$

The procedure of releasing the locked joints by the method of moment distribution is shown in Table 10.8. From the final end moments given in Table 10.8, the shears and hence the lateral forces at joints D and E are determined from statics as shown in Fig. 10.30 c.

Next we apply to the second story a sway of Δ_2 (Fig. 10.30 d), while all the joints are locked and the members of the first story are considered infinitely rigid. The fixed-end moments are

$$\text{FEM}_{BC} = \text{FEM}_{CB} = \frac{-6E(2I)\Delta_2}{(16)^2} = \frac{-3}{64} \, EI\Delta_2$$

$$\text{FEM}_{ED} = \text{FEM}_{DE} = \frac{-6E(2I)\Delta_2}{(16)^2} = \frac{-3}{64} \, EI\Delta_2$$

If we let $\frac{3}{64} EI\Delta_2 = 100$, then $\text{FEM}_{BC} = \text{FEM}_{CB} = \text{FEM}_{ED} = \text{FEM}_{DE} = -100 \text{ k} \cdot \text{ft}$.

The procedure for distributing these moments by the moment-distribution method is shown in Table 10.9; from the final moments, the shears and the joint forces at E and D are calculated as shown in Fig. 10.30 e. Applying Eq. (10.19), we have

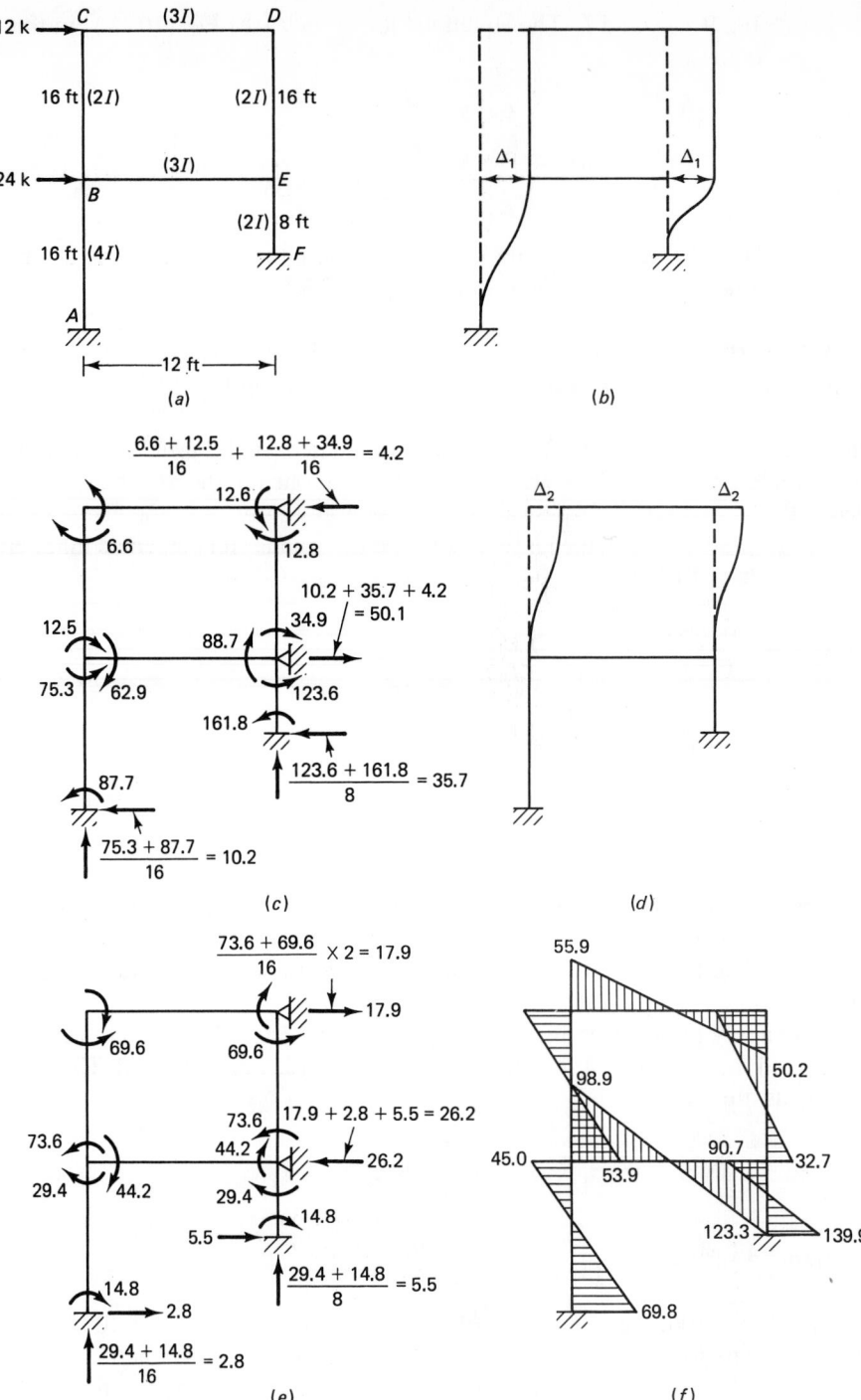

Figure 10.30 Structure for Example 10.9. (f) BMD (on compression side).

TABLE 10.8 FIRST MOMENT DISTRIBUTION FOR EXAMPLE 10.9

Joint	A	B			C		D		E			F
End of member	AB	BA	BE	BC	CB	CD	DC	DE	ED	EB	EF	FE
D.F.	0	0.4	0.4	0.2	0.33	0.67	0.67	0.33	0.2	0.4	0.4	0
FEM	−100	−100									−200	−200
Balance B and C.O.M.	+20	+40	+40 (→to EB)	+20	+10					+20		
Balance E and C.O.M.			+36					+18	+36	+72 (→to BE)	+72	+36
Balance D and C.O.M.						−6	−12	−6	−3			
Balance C and C.O.M.				−0.5	−1	−3	−1.5					
Balance B and C.O.M.	−7.1	−14.2	−14.2 (→to EB)	−7.1	−3.5					−7.1		
Balance C and C.O.M.				+0.6	+1.2	+2.3	+1.2					
Balance D and C.O.M.						+0.1	+0.2	+0.1	(negligible)			
Balance E and C.O.M.			+2.0					+1.0	+2.0	+4.0 (→to BE)	+4.1	+2.0
Balance D and C.O.M.						−0.4	−0.7	−0.3	(negligible)			
Balance C and C.O.M.				(negligible)	+0.1	+0.2	+0.1					
Balance B and C.O.M.	−0.6	−1.1	−1.0 (→to EB)	−0.5	−0.3					−0.5		
Balance E and C.O.M.			+0.1					(negligible)	+0.1 (negligible)	+0.3 (→to BE)	+0.3	+0.2
Balance C and C.O.M.				(negligible)	+0.1	+0.2	+0.1					
Total moments first case of sway	−87.7	−75.3	+62.9	+12.5	+6.6	−6.6	−12.6	+12.8	+34.9	+88.7	−123.6	−161.8

TABLE 10.9 SECOND MOMENT DISTRIBUTION FOR EXAMPLE 10.9

Joint	A	B			C		D		E			F
End of member	AB	BA	BE	BC	CB	CD	DC	DE	ED	EB	EF	FE
D.F.	0	0.4	0.4	0.2	0.33	0.67	0.67	0.33	0.2	0.4	0.4	0
FEM				-100.0	-100.0			-100.0	-100.0			
Balance B, C, D, and E		+40.0	+40.0	+20.0	+33.3	+66.7	+66.7	+33.3	+20.0	+40.0	+40.0	
C.O.M.	+20.0		+20.0*	+16.7	+10.0	+33.3	+33.3	+10.0	+16.7	+20.0†		+20.0
Balance B, C, D, and E		-14.7	-14.7	-7.3	-14.4	-28.9	-28.9	-14.4	-7.3	-14.7	-14.7	
C.O.M.	-7.3		-7.3*	-7.2	-3.7	-14.4	-14.4	-3.7	-7.2	-7.3†		-7.3
Balance B, C, D, and E		+5.8	+5.8	+2.9	+6.0	+12.1	+12.1	+6.0	+2.9	+5.8	+5.8	
C.O.M.	+2.9		+2.9*	+3.0	+1.5	+6.0	+6.0	+1.5	+3.0	+2.9†		+2.9
Balance B, C, D, and E		-2.3	-2.4	-1.2	-2.5	-5.0	-5.0	-2.5	-1.2	-2.4	-2.3	
C.O.M.	-1.1		-1.2*	-1.3	-0.6	-2.5	-2.5	-0.6	-1.3	-1.2†		-1.1
Balance B, C, D, and E		+1.0	+1.0	+0.5	+1.0	+2.1	+2.1	+1.0	+0.5	+1.0	+1.0	
C.O.M.	+0.5		+0.5*	+0.5	+0.2	+1.1	+1.1	+0.2	+0.5	+0.5†		+0.5
Balance B, C, D, and E		-0.4	-0.4	-0.2	-0.4	-0.9	-0.9	-0.4	-0.2	-0.4	-0.4	
C.O.M.	-0.2		-0.2*	-0.2	-0.1	-0.5	-0.5	-0.1	-0.2	-0.2†		-0.2
Balance B, C, D, and E		+0.1	+0.2	+0.1	+0.2	+0.4	+0.4	+0.2	+0.1	+0.2	+0.1	
C.O.M.	(negligible)		+0.1*	+0.1	(negligible)	+0.2	+0.2	(negligible)	+0.1	+0.1†		(negligible)
Balance B, C, D, and E		-0.1	-0.1	-0.1 / 0	-0.1	-0.1	-0.1	-0.1	-0.1 / 0	-0.1	-0.1	
Total moments from second case of sway	+14.8	+29.4	+44.2	-73.6	-69.6	+69.6	+69.6	-69.6	-73.6	+44.2	+29.4	+14.8

* Carryover moment from EB.

† Carryover moment from BE.

$K_1(50.1) - K_2(26.2) = 24.0$

$-K_1(4.2) + K_2(17.9) = 12.0$

Solving for K_1 and K_2: $K_1 = +0.946$, $K_2 = +0.892$.

The final moments are obtained by superimposing K_1 times moments from the first sway onto K_2 times moments from the second sway as shown in Table 10.10. The bending-moment diagram is shown in Fig. 10.30f.   ■

10.2.10 Morris's Method for Frames with Sidesway

In 1932, Professor C. T. Morris[1] of Ohio State University presented an ingenious method for analyzing frames with sidesway. The method, involving a series of successive corrections, is based on the total horizontal shear at each level of a frame with vertical columns and horizontal members. At any one level the total moment at the top and bottom of all columns must equal the product of the total horizontal shear on that level and the column height (assumed constant at any one level). The procedure is as follows: (1) The aforementioned total moment at any one level is calculated and then distributed between the columns at that level in proportion to their I/l^2 values. (2) The moment allocated to each column is then divided equally between the top and bottom ends of that column. (3) The fixed-end moments due to load (if any) on the members are then calculated. (4) All joints in the frame are balanced in the usual manner. (5) Then the carryovers for the entire frame are made. (6) As a result, the total of the column moment at each level is changed. This total is now corrected by adding (or subtracting) an amount to bring it to the initial value. The value of the added (or subtracted) correction moment at each level is distributed again to the columns at that level in proportion to their I/l^2 values. Steps 4 to 6 are repeated until the amounts of the correction moments are negligibly small (say, not exceeding 1 percent of the initial moments).

■ **EXAMPLE 10.10**
Determine the joint moments for the frame in Fig. 10.31a by Morris's method.

Solution.   The solution is presented in Table 10.11. Since there are no loads along the members, the fixed-end moments are zero. At level 2, the total column moment is 10 k (shear) × 10 ft (height of column) = −100; it is negative since the shear causes a sway to the right, and thus counterclockwise moments. At level 1, the total column moment is (10 − 5) k × 20 ft = −100. Each of these total column moments is distributed to the columns in proportion to their I/l^2; for level 2, it is (−50) to column BC and (−50) to column $B'C'$, and each column moment is then divided equally to the top and bottom of the column, that is, −25 to the top (C) and −25 to the bottom (B) as shown in Table 10.11. All joints are then balanced in the usual manner followed by the carryovers. We must now correct the column moment totals. For level 2 we now have −25 + 25 + 8.3 − 25 + 16.7 + 12.2 − 25 + 25 + 8.3 − 25 + 16.7

[1] C. T. Morris, "Morris on Analysis of Continuous Frames," *Transactions of the ASCE,* vol. 96, pp. 66–69, 1932.

TABLE 10.10 FINAL MOMENTS FOR EXAMPLE 10.9

Joint	A		B		C		D		E		F	
End of member	AB	BA	BE	BC	CB	CD	DC	DE	ED	EB	EF	FE
$k_1 \times$ moments from first sway	−83.0	−71.2	+59.5	+11.8	+6.2	−6.2	−11.9	+12.1	+33.0	+83.9	−116.9	−153.1
$k_2 \times$ moments from second sway	+13.2	+26.2	+39.4	−65.7	−62.1	+62.1	+62.1	−62.1	−65.7	+39.4	+26.2	+13.2
Final moments	−69.8	−45.0	+98.9	−53.9	−55.9	+55.9	+50.2	−50.0	−32.7	+123.3	−90.7	−139.9

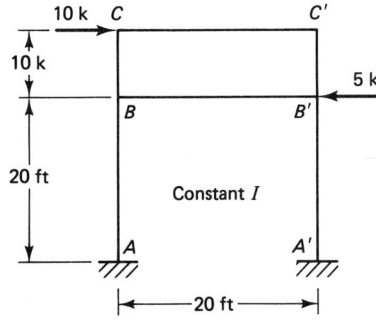

Figure 10.31 Structure for Example 10.10.

+ 12.2 = +24.40; but we require −100 at that level; therefore, apply a correction moment = −124.40 and distribute to columns BC and $B'C'$ according to their I/l^2 as shown in Table 10.11. The same procedure is applied to level 1. All the joints of the frame are then balanced, and the procedure is repeated. ∎

10.2.11 Secondary Stresses in Pin-Jointed Structures

The members in a pin-jointed structure subjected to joint loads carry purely axial load inducing tensile or compressive *primary stresses;* as a result of the elongation or shortening of such members, the geometry of the structure changes and the angles between the members also change. If, however, the joints are not pinned but riveted to gusset plates or welded, restraining moments are generated at the ends of the members and thus give rise to *secondary bending stresses.* In relatively short but deep trusses, secondary bending stresses can be quite significant and must be considered in design. A convenient method of determining the secondary stresses is by the method of moment distribution. The procedure is as follows:

1. Calculate the primary stresses by assuming that the structure is pin-connected.
2. Determine the deformed position of the joints by the Williot diagram[2] and hence the relative end deflection Δ for each member.
3. Assuming that all joints are locked against rotation, calculate the fixed-end moments $(-6EI\Delta/l^2)$.
4. Carry out the moment-distribution procedure by unlocking all joints until joint balance is obtained. The final moments give the first approximation of the secondary end moments and hence the secondary stresses.
5. These new stresses incur additional joint deflections, causing additional end moments and hence additional secondary stresses.

One can repeat the procedure of moment distribution based on step 5, but in most practical cases only the first approximation is significant.

---

[2] C. H. Norris, J. B. Wilbur, and S. Utku: "Elementary Structural Analysis," 3d ed., McGraw-Hill, New York, 1976.

TABLE 10.11 SOLUTION FOR EXAMPLE 10.10

Joint	A	B			C		C'		B'			A'
Member	AB	BA	BB'	BC	CB	CC'	C'C	C'B'	B'C	B'B	B'A'	A'B'
D.F.	0	$\frac{1}{4}$	$\frac{1}{4}$	$\frac{1}{2}$	$\frac{2}{3}$	$\frac{1}{3}$	$\frac{1}{3}$	$\frac{2}{3}$	$\frac{1}{2}$	$\frac{1}{4}$	$\frac{1}{4}$	0
Carryover	↓	←	to B'B	→	↓	→	←	↓	←	to BB'	→	↓
FEM	0	0	0	0	0	0	0	0	0	0	0	0
Column moments: level 2: $-10 \times 10 = (-100)$; Level 1: $-(10-5)(20) = (-100)$	−25	−25		−25	−25			−25	−25		−25	−25
Balance all joints		+12.5	+12.5	+25	+16.7	+8.3	+8.3	+16.7	+25	+12.5	+12.5	
Carryover	+6.2		+6.2	+8.3	+12.5	+4.1	+4.1	+12.5	+8.3	+6.2		+6.2
Correction moments: level 2: apply −124.4; Level 1: apply −37.4	−9.4	−9.4		−31.1	−31.1			−31.1	−31.1		−9.4	−9.4
Balance all joints		+6.5	+6.5	+13.0	+9.8	+4.9	+4.9	+9.8	+13.0	+6.5	+6.5	
Carryover	+3.2		+3.2	+4.9	+6.5	+2.4	+2.4	+6.5	+4.9	+3.2		+3.2
Correction moments: level 2: apply −68.4; Level 1: apply −19.2	−4.8	−4.8		−17.1	−17.1			−17.1	−17.1		−4.8	−4.8
Balance all joints		+3.4	+3.4	+6.9	+5.4	+2.7	+2.7	+5.4	+6.9	+3.4	+3.4	
Carryover	+1.7		+1.7	+2.7	+3.4	+1.3	+1.3	+3.4	+2.7	+1.7		+1.7

Moment distribution table (k·ft). Values reproduced by row; columns are member-end positions 1–12 (left to right).

Operation	Col 1	Col 2	Col 3	Col 4	Col 5	Col 6	Col 7	Col 8	Col 9	Col 10	Col 11	Col 12
Correction moments: level 2: apply −36.8			−9.2	−9.2					−9.2	−9.2		
Level 1: apply −10.2	−2.5	−2.5									−2.5	−2.5
Balance all joints		+1.8	+3.7	+3.0	+1.5			+1.5	+3.0	+3.7	+1.8	
Carryover	+0.9	+0.9	+1.5	+1.8	+0.7	+0.7	+0.7	+0.7	+1.8	+1.5	+0.9	+0.9
Correction moments: level 2: apply −20.0			−5.0	−5.0					−5.0	−5.0		
Level 1: apply −5.6	−1.4	−1.4									−1.4	−1.4
Balance all joints		+1.0	+2.0	+1.7	+0.8			+0.8	+1.7	+2.0	+1.0	
Carryover	+0.5	+0.5	+0.8	+1.0	+0.4	+0.4	+0.4	+0.4	+1.0	+0.8	+0.5	+0.5
Correction moments: level 2: apply −11.0			−2.8	−2.7					−2.7	−2.8		
Level 1: apply −3.0	−0.8	−0.7									−0.7	−0.8
Balance all joints		+0.5	+1.1	+0.9	+0.4			+0.4	+0.9	+1.1	+0.5	
Carryover	+0.3	+0.3	+0.4	+0.5	+0.2	+0.2	+0.2	+0.2	+0.5	+0.4	+0.3	+0.3
Correction moments: level 2: apply −5.8			−1.4	−1.5					−1.5	−1.4		
Level 1: apply −1.6	−0.4	−0.4									−0.4	−0.4
Balance all joints		+0.3	+0.6	+0.5	+0.3			+0.3	+0.5	+0.6	+0.3	
Carryover	+0.1	+0.1	+0.2	+0.3	+0.1	+0.1	+0.1	+0.1	+0.3	+0.2	+0.1	+0.1
Correction moments: level 2: apply −3.20			−0.8	−0.8					−0.8	−0.8		
Level 1: apply −0.8	−0.2	−0.2									−0.2	−0.2
Balance all joints		+0.2	+0.4	+0.3	+0.1			+0.1	+0.3	+0.4	+0.2	
Σ = final moments in k·ft	−31.6	−18.2	+39.1	−20.9	−28.4	+28.2	+28.2	−28.4	−20.9	+39.1	−18.2	−31.6

The effect of self-weight of members or any lateral load carried directly by the members can be also treated by the moment-distribution procedure; the end moment due to such lateral load is simply added to the first end moments due to the relative displacement Δ of the joints, followed by step 4.

10.3 SOLVED PROBLEMS

■ SOLVED PROBLEM 10.1

Using the slope-deflection method, determine the bending moments in the structure shown in Fig. 10.32a; find H_A, V_A, and V_D; draw the BMD and sketch the elastic-deflection curve.

Solution. From symmetry:

$$\Delta = 0$$

$$\theta_A = -\theta_F$$

$$\theta_B = -\theta_E$$

$$\theta_C = 0$$

$$\theta_A = \theta_D = \theta_F = 0$$

Fixed-end moments:

Span AB: Refer to Fig. 10.32b; from the Appendix,

$$\text{FEM}_{AB} = \frac{M}{4} = \frac{90 \times 0.6}{4} = 13.50 \text{ kN} \cdot \text{m}$$

$$\text{FEM}_{BA} = 13.50 \text{ kN} \cdot \text{m} \qquad (\text{clockwise})$$

Span BC: Refer to Fig. 10.32c; from the Appendix,

$$\text{FEM}_{BC} = -\frac{wl^2}{12} = -\frac{60 \times (3)^2}{12} = -45 \text{ kN} \cdot \text{m} \qquad (\text{counterclockwise})$$

$$\text{FEM}_{CB} = \frac{wl^2}{12} = \frac{6 \times (3)^2}{12} = 45 \text{ kN} \cdot \text{m} \qquad (\text{clockwise})$$

Member equations:

$$M_{AB} = 13.5 + \frac{2E(2I)}{6}(2\theta_A + \theta_B)$$

$$M_{BA} = 13.5 + \frac{2E(2I)}{6}(2\theta_B + \theta_A)$$

$$M_{BC} = -45 + \frac{2E(3I)}{3}(2\theta_B + \theta_C)$$

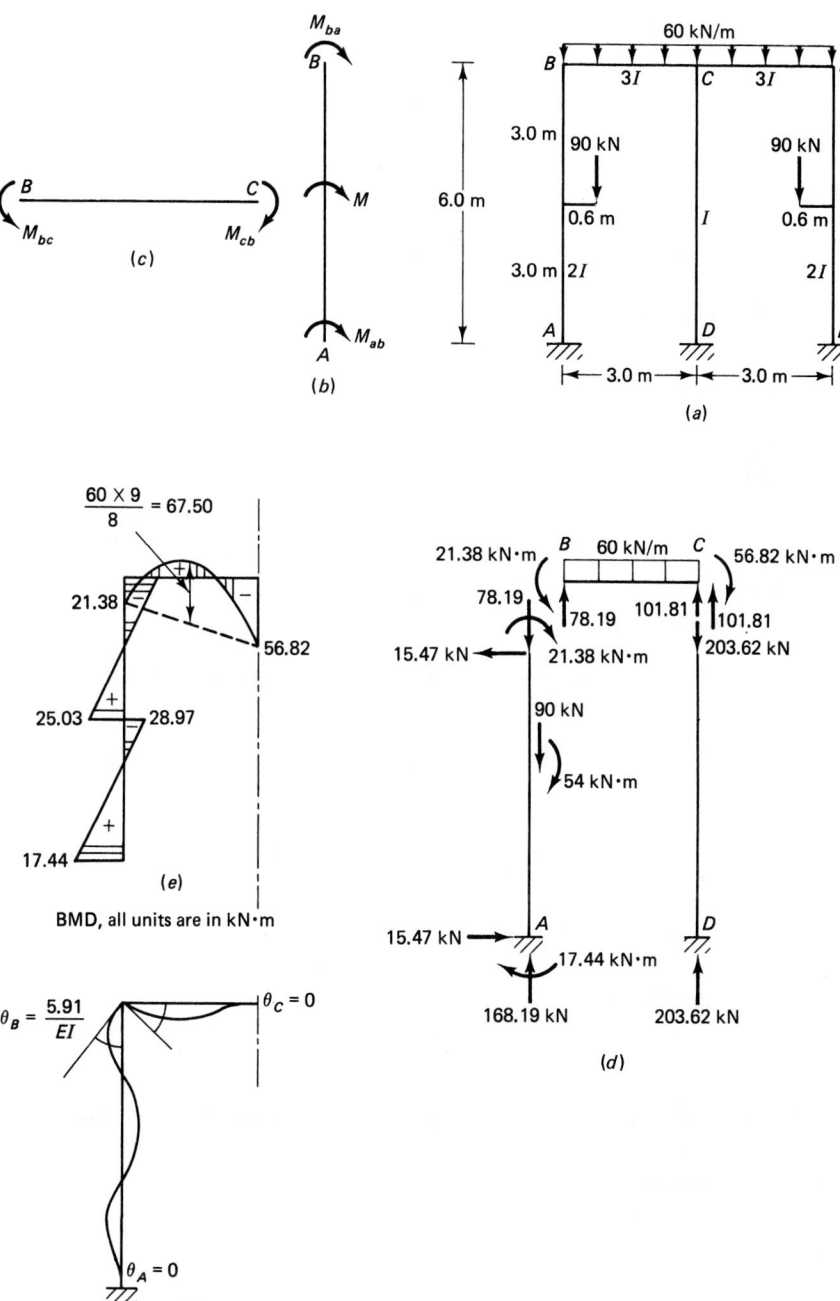

Figure 10.32 Structure for Solved Problem 10.1. (f) Elastic curve.

$$M_{CB} = 45 + \frac{2E(3I)}{3}(2\theta_C + \theta_B)$$

$$M_{CD} = 0 \quad \text{from symmetry}$$

Joint equation:

At joint B

$$\sum M_B = M_{BA} + M_{BC} = 0$$

$$13.5 + \frac{4EI}{6}(2\theta_B) - 45 + \frac{6EI}{3}(2\theta_B) = 0$$

$$31.50 = 5.33EI\theta_B$$

$$\theta_B = \frac{5.91}{EI}$$

Therefore,

$$M_{AB} = 13.50 + \frac{4EI}{6} \times \frac{5.91}{EI} = 17.44 \text{ kN} \cdot \text{m} \qquad \text{clockwise}$$

$$M_{BA} = 13.5 + \frac{4EI}{6} \times 2 \times \frac{5.91}{EI} = 21.38 \text{ kN} \cdot \text{m} \qquad \text{clockwise}$$

$$M_{BC} = -45 + \frac{6EI}{3} \times 2 \times \frac{5.91}{EI} = -21.38 \text{ kN} \cdot \text{m} \qquad \text{counterclockwise}$$

$$M_{CB} = 45 + \frac{6EI}{3} \times \frac{5.91}{EI} = 56.82 \text{ kN} \cdot \text{m} \qquad \text{clockwise}$$

BMD is shown in Fig. 10.32e while the elastic curve is shown in Fig. 10.32f.   ∎

■ SOLVED PROBLEM 10.2

Figure 10.33a shows rigid frame $ABCD$ with a steel tie EB pinned at B. By moment distribution:

(a) Calculate the magnitude of the force P to cause at C a horizontal displacement $\Delta = 1$ in. Neglect axial deformation in the members of the frame.
(b) Calculate the force in the steel tie.
(c) Draw the bending-moment diagram for the structure.

Given: Cross-sectional area of steel tie $EB = 1$ in^2, E of tie $= 30 \times 10^3$ ksi. EI of frame $=$ constant $= 3 \times 10^6$ k $\cdot$ in^2.

Solution.   Referring to Fig. 10.33b, c, and d,

$$B'B'' = \text{extension in tie} = \Delta(\tfrac{16}{20}) = 0.8\Delta = 0.8 \text{ in}$$

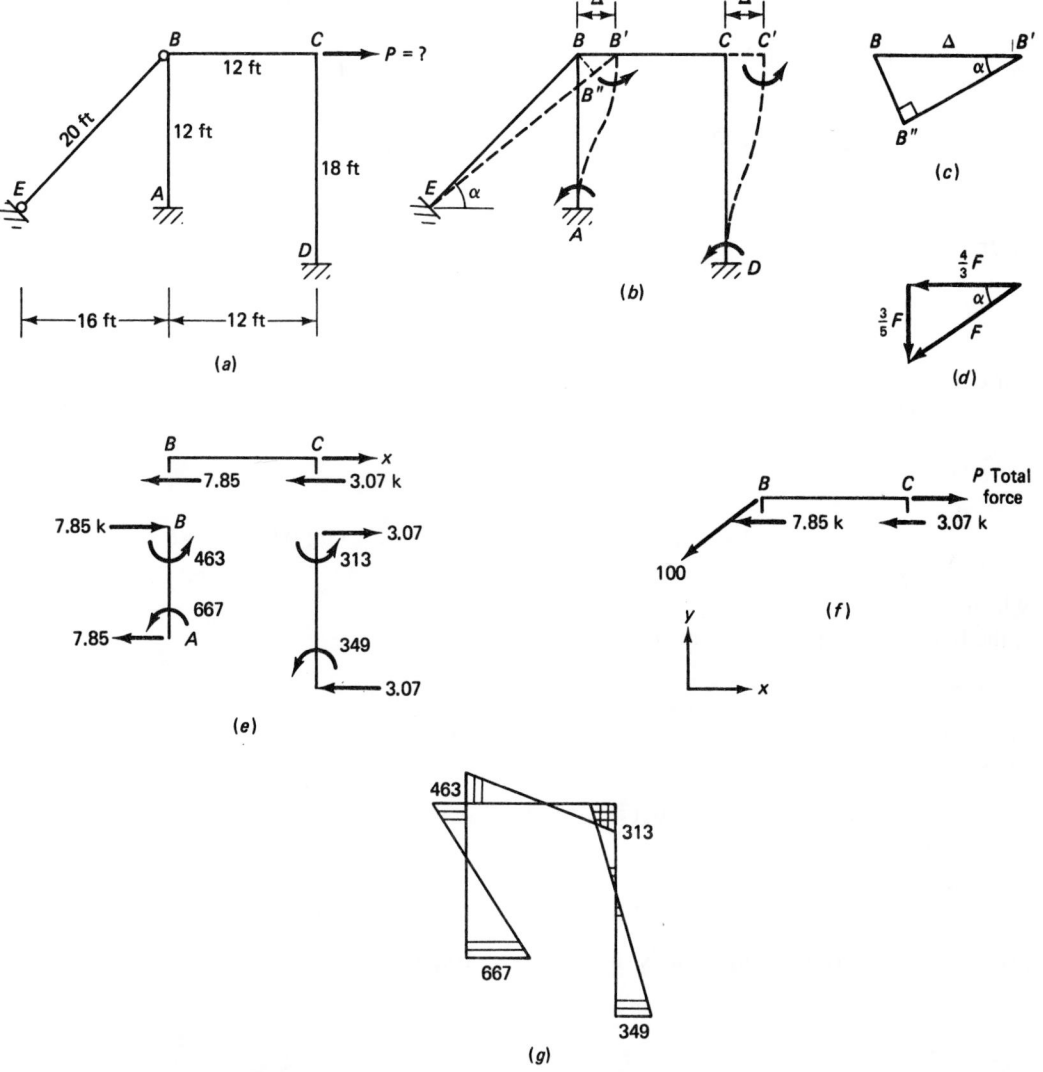

Figure 10.33 Structure for Solved Problem 10.2.

$$\text{Extension in tie} = \frac{(\text{force in tie})(\text{length of tie})}{(\text{area of tie})(\text{Young's modulus})}$$

$$\text{Force in tie} = \frac{0.8 \times 1 \times 30 \times 10^3}{20 \times 12} = 100 \text{ k}$$

Therefore,

Horizontal component of force in tie $= 100 \times \frac{4}{5} = 80$ k

Now calculate force required to displace frame horizontally by $\Delta = 1.0$ in.

$$\text{FEM}_{AB} = -\frac{6EI\Delta}{L^2} = \frac{-6 \times 3 \times 10^6 \times 1}{12 \times 12 \times 12 \times 12} = -868 \text{ k} \cdot \text{in} = \text{FEM}_{BA}$$

$$\text{FEM}_{DC} = -\frac{6EI\Delta}{L^2} = \frac{-6 \times 3 \times 10^6 \times 1}{18 \times 18 \times 12 \times 12} = -386 \text{ k} \cdot \text{in} = \text{FEM}_{CD}$$

Distribution factor for $BA = \dfrac{I/L}{\Sigma\, I/L} = \dfrac{\frac{1}{12}}{\frac{1}{12} + \frac{1}{12}} = 0.5$

For $BC = 0.5$

For $CB = \dfrac{\frac{1}{12}}{\frac{1}{12} + \frac{1}{18}} = 0.6$

For $CD = \dfrac{\frac{1}{18}}{\frac{1}{12} + \frac{1}{18}} = 0.4$

Carrying out the moment distribution we obtain $M_{AB} = -667$ k $\cdot$ in; $M_{BA} = -463$ k $\cdot$ in; $M_{BC} = +463$ k $\cdot$ in; $M_{CB} = +313$ k $\cdot$ in, $M_{CD} = -313$ k $\cdot$ in; and $M_{DC} = -349$ k $\cdot$ in.

From the free-body diagram of beam BC (Fig. 10.33e)

$$x = 7.85 + 3.07 = 10.92 \text{ k} \rightarrow$$

which is the force required to displace frame horizontally $\Delta = 1$ in, or, referring to Fig. 10.33f,

$$\Sigma\, F_x = 0 \qquad 7.85 + 3.07 + (100)(\tfrac{4}{5}) = 90.92 \text{ k} \qquad \text{(total force)}$$

Therefore,

Total force $P = 10.92 + 80 = 90.92$ k

BMD drawn on the compression side is shown in Fig. 10.33g.   ∎

PROBLEMS

Use the slope-deflection method for the solution of Problems 10.1 to 10.27.

10.1 to 10.4 Draw shear-force and bending-moment diagrams for the beams shown.

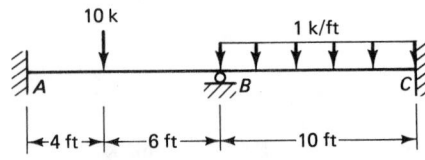

Figure P10.1

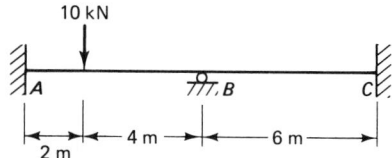

Figure P10.2

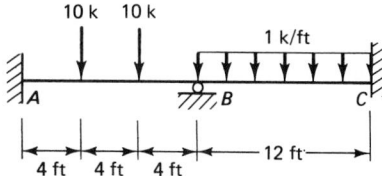

Figure P10.3

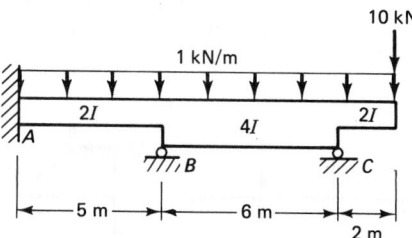

Figure P10.4

10.5 Support B of the two-span continuous beam loaded as shown sinks 0.5 in below A and support C sinks a further 1 in below B. Determine the support moments at A, B, and C and draw shear-force and bending-moment diagrams. Assume $EI = 11,000 \text{ k} \cdot \text{ft}^2$.

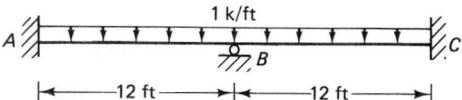

Figure P10.5

10.6 Support B of the two-span continuous beam loaded as shown sinks 13 mm below A and support C settles a further 26 mm below B. Determine the support moments at A, B, and C and draw shear-force and bending-moment diagrams. Assume $EI = 9000 \text{ kN} \cdot \text{m}^2$.

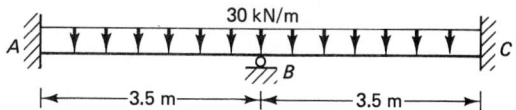

Figure P10.6

10.7 Calculate the support moments for the structure shown, assuming support B sinks 0.1 in. Assume $EI = 10,000 \text{ k} \cdot \text{ft}^2$.

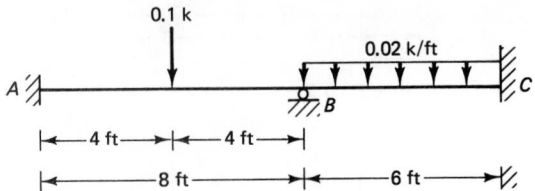

Figure P10.7

10.8 Calculate the support moments for the two-span continous beam shown, assuming support B settles by 2 in below supports A and C. Take $EI = 20,000 \text{ k} \cdot \text{ft}^2$.

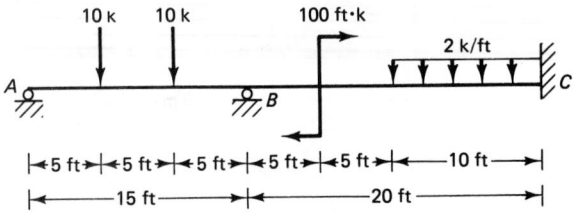

Figure P10.8

10.9 Draw the bending-moment diagram for the rigid frame shown.

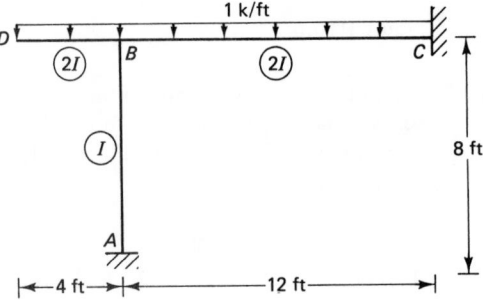

Figure P10.9

10.10 Draw the bending-moment diagram for the rigid frame shown.

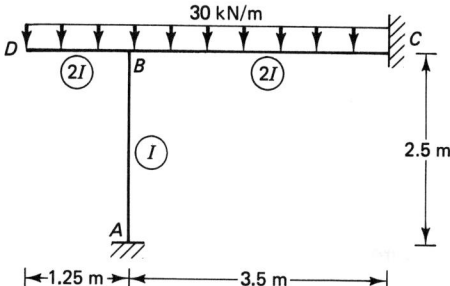

Figure P10.10

10.11 to 10.16 Draw bending-moment diagrams for the rigid frames shown.

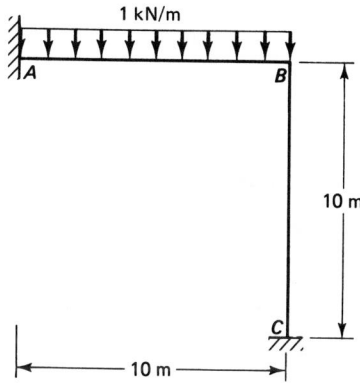

Figure P10.11

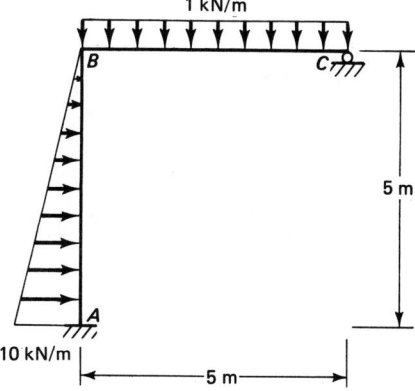

Figure P10.12

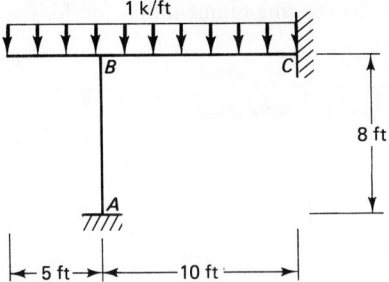

Figure P10.13

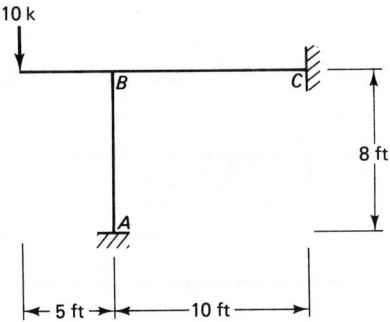

Figure P10.14

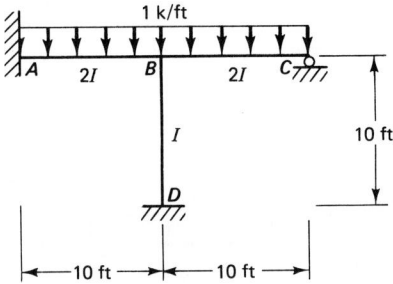

Figure P10.15

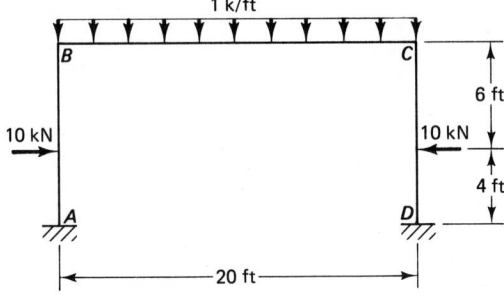

Figure P10.16

10.17 Solve for the deflection of B and the rotation of B and C of the frame shown. Draw the bending-moment diagram and sketch the elastic curve.

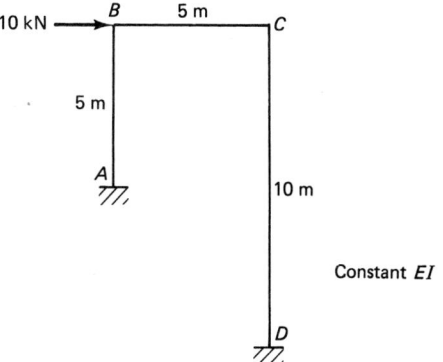

Figure P10.17

10.18 (a) Determine the joint moments in the frame shown with support D settling 1 in downward. (b) Sketch the bending-moment diagram and determine the shears and axial forces at the ends of members AC and CD. (c) Calculate the slope of member AC at B.

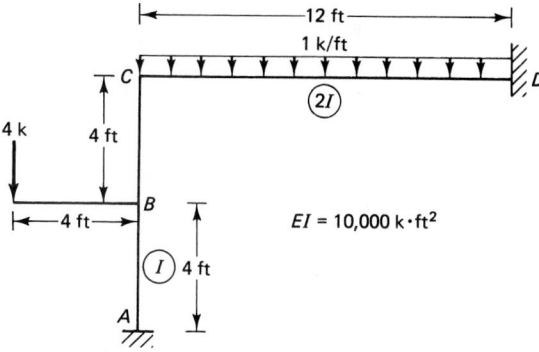

Figure P10.18

10.19 Determine the support moments and draw the bending-moment diagram for the rigid frame shown if support D settles down by 1 in and rotates clockwise 0.002 radian.

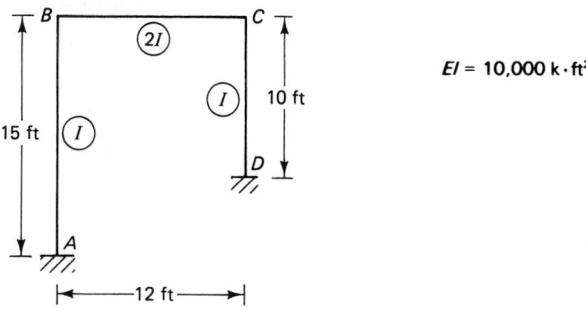

Figure P10.19

10.20 The rigid frame ABC is subjected to an external moment on joint B of 20 k·ft, as well as the other loadings shown. It is also assumed that the rigid foundation at C sinks 0.1 in and rotates clockwise, $\theta = 0.002$ radian. (a) Determine the degree of indeterminacy of the structure. (b) Calculate the joint moments. (c) Draw the bending-moment diagram. (d) Find the reactions at support A. *Note:* Neglect axial deformation.

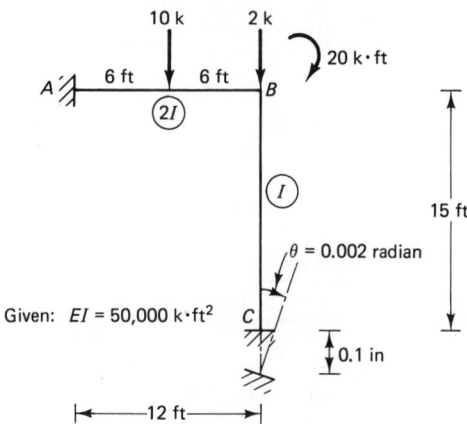

Given: $EI = 50,000$ k·ft²

Figure P10.20

10.21 (a) Determine the joint moments in the frame. (b) Sketch the bending-moment diagram and determine the shears and axial forces at the end of members AB and BC. (c) Calculate the deflection of member BC at the inflection point along BC. (d) State the degree of indeterminacy.

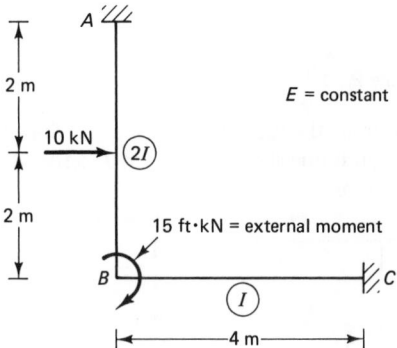

Figure P10.21

10.22 The rigid frame structure is loaded as shown. The foundation at A is expected to yield *1 in* downward. (**a**) State the degree of indeterminacy of the structure. (**b**) Determine the moments at the ends of the three members. Hence, calculate the vertical and horizontal reactions at A, draw the bending-moment diagram for member AD, and calculate the slope and horizontal deflection at G. *Note:* Neglect the effect of change in length of CD as well as axial and shear deformations.

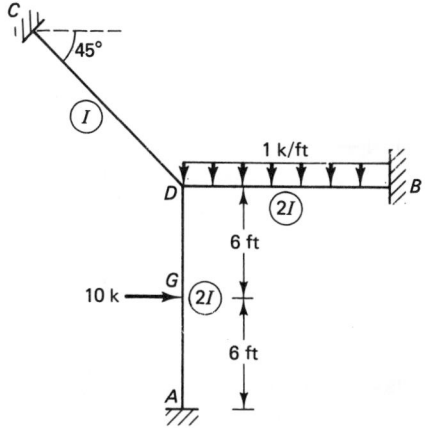

Given: $EI = 10,000$ k·ft². All members are 12 ft long.

Figure P10.22

10.23 Draw the bending-moment diagram for the symmetrically loaded box culvert shown.

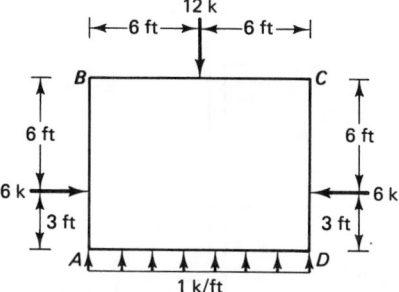

Figure P10.23

10.24 Find the joint moments and reactions in the rigid frame structure shown. Draw the bending-moment diagram.

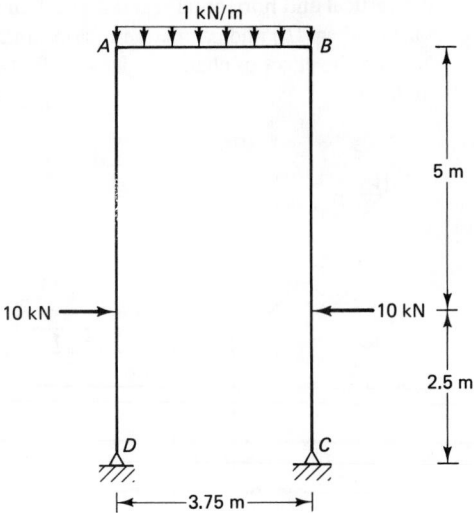

Figure P10.24

10.25 Calculate the bending moments in the square 12×12 ft concrete box culvert shown.

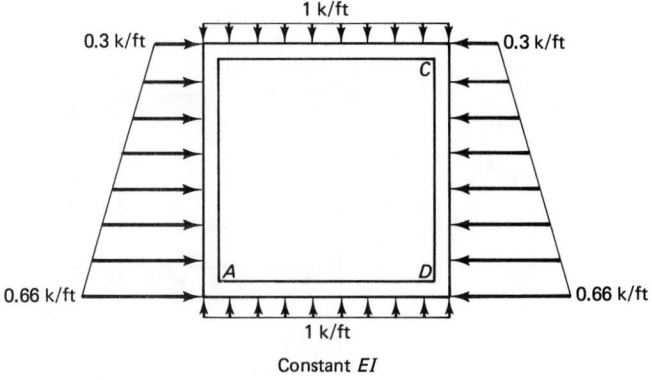

Figure P10.25

10.26 For the rigid frame structure shown, the foundations at A and E are assumed to settle 1 in downward. Determine the moments at the ends of the four members. Hence, calculate the vertical and horizontal reactions at A and draw the bending-moment diagram.

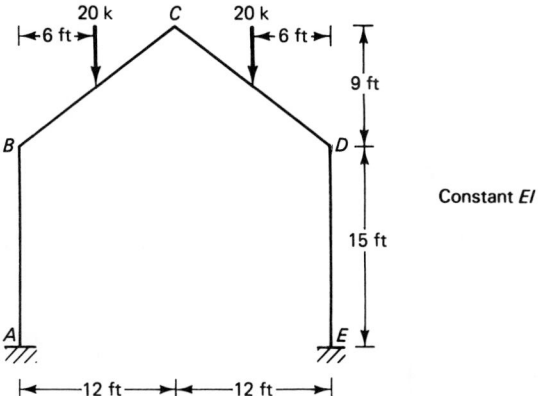

Figure P10.26

10.27 Determine the bending moments at the joints of the two-span rigid frame shown. Determine H_A, V_A, and V_D. Draw the bending-moment diagram and sketch the deflected shape.

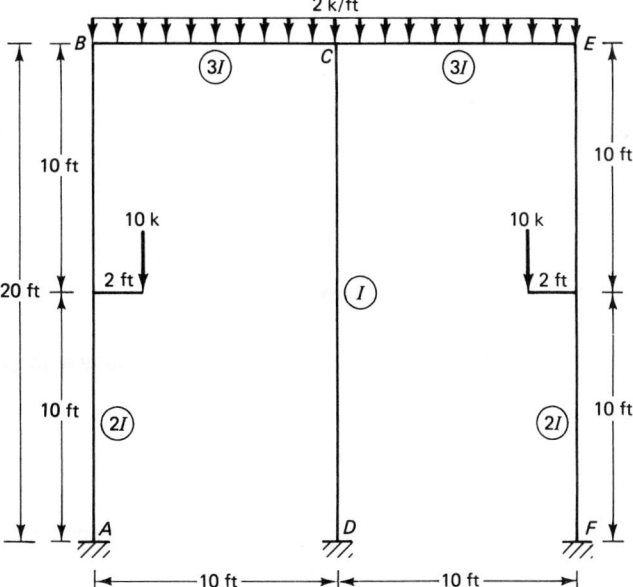

Figure P10.27

Use the moment-distribution method for the solution of Problems 10.28 to 10.48.

10.28 Determine the support moments for the continuous beam shown.

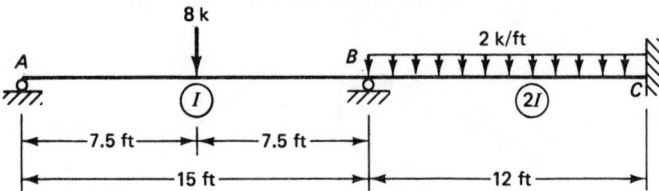

Figure P10.28

10.29 Draw the shear-force and bending-moment diagrams for the continuous beam shown.

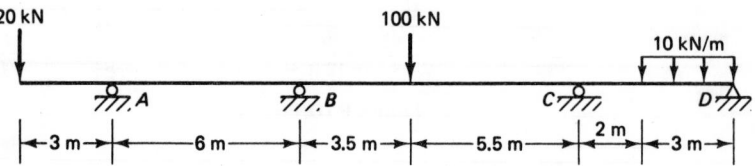

Figure P10.29

10.30 Determine the support moments of the two-span continuous beam shown.

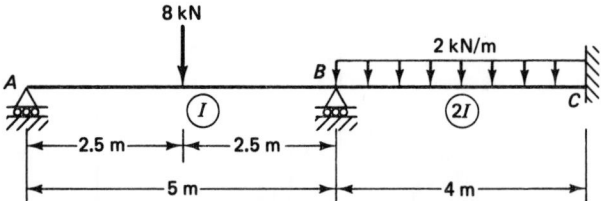

Figure P10.30

10.31 Draw shear-force and bending-moment diagrams for the three-span continuous beam shown.

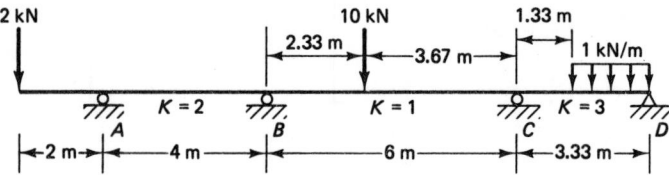

Figure P10.31

10.32 Determine the support moments for the three-span continuous beam shown.

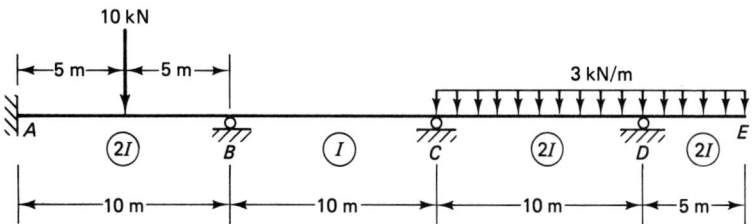

Figure P10.32

10.33 Determine moments at B and C for the frame shown.

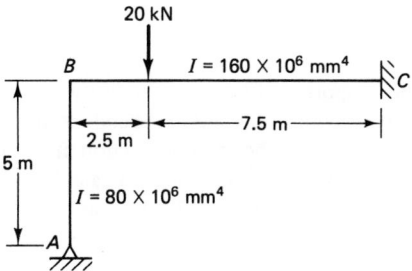

Figure P10.33

10.34 Draw the bending-moment diagram for the rigid frame shown.

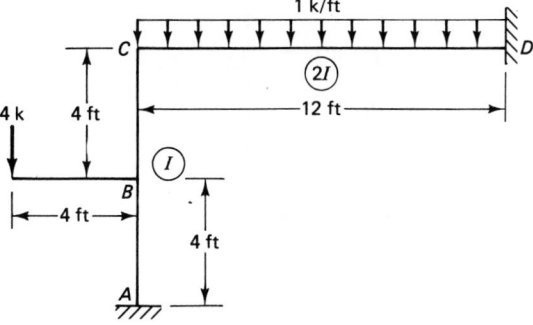

Figure P10.34

10.35 Determine the joint moments for the rigid frame shown if support A rotates 0.01 radian clockwise and support D settles vertically by 1 in. $E = 30,000$ ksi.

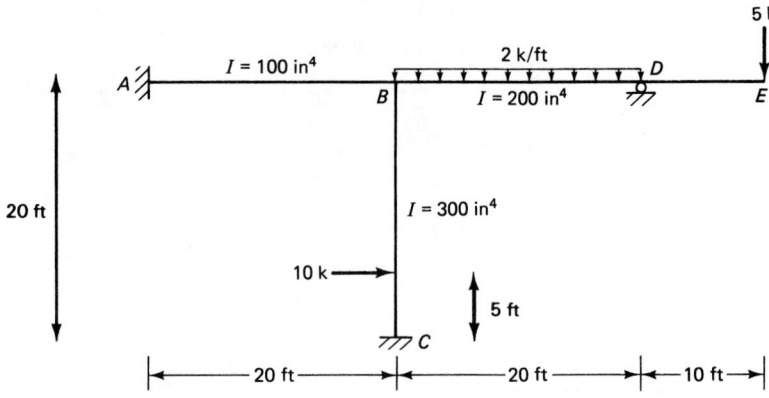

Figure P10.35

10.36 The support A of the rigid frame shown is expected to settle 1 in downward. Determine the moments at the ends of the three members. Hence, **(a)** calculate the vertical and horizontal reaction at A, **(b)** draw the bending-moment diagram for member AD, and **(c)** calculate the slope and horizontal deflection at G. All members are 12 ft long and $EI = 10,000 \text{ k} \cdot \text{ft}^2$.

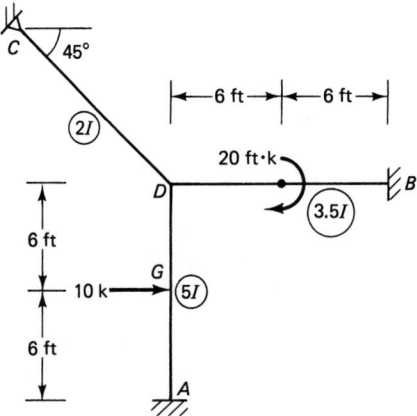

Figure P10.36

10.37 A rigid frame is loaded as shown; the *foundation at C is expected to settle 0.5 in downward.* Determine the moments at the ends of the three members. Hence, **(a)** calculate the vertical and horizontal reactions at C, **(b)** draw the bending-moment diagram, **(c)** calculate the slope at G. Assume $EI = 10,000$ k·ft^2.

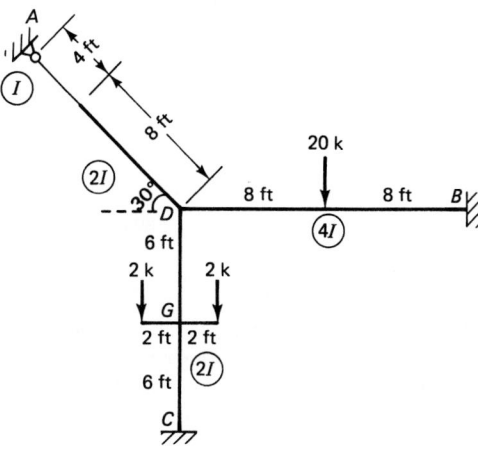

Figure P10.37

10.38 Determine the bending moments in the loaded structure and draw the bending-moment diagram.

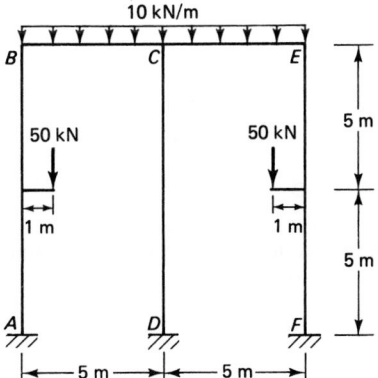

Figure P10.38

10.39 Draw the bending-moment diagram for the portal frame shown, if support D sinks 1 in downward. Assume $EI = 4000$ k·ft².

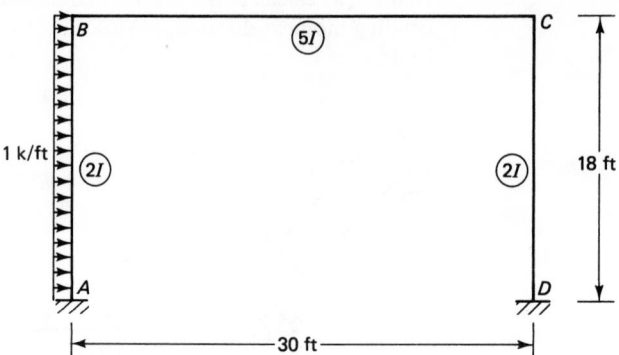

Figure P10.39

10.40 The rigidly jointed frame $ABCD$ is flexibly restrained by a spring at B attached to a rigid wall. This spring compresses or extends 1 in under a force $EI/1200$, where EI is the flexural rigidity of the members of the frame. Find the moments in the frame due to a horizontal load of 100 lb.

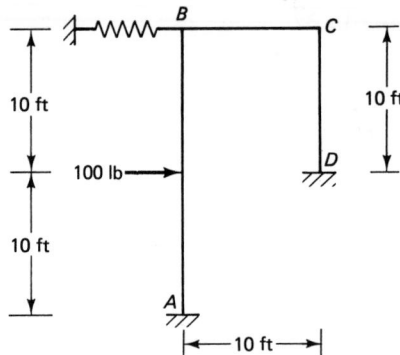

Figure P10.40

10.41 Determine the moments in the rigid frame shown. Also draw the bending-moment diagram.

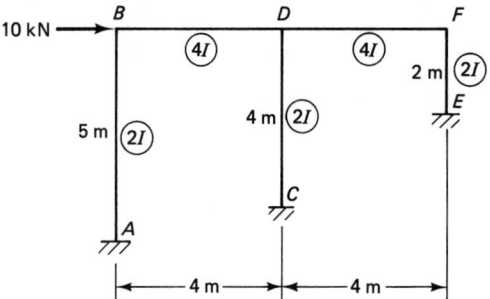

Figure P10.41

10.42 Determine the joint moments for the rigid frame shown.

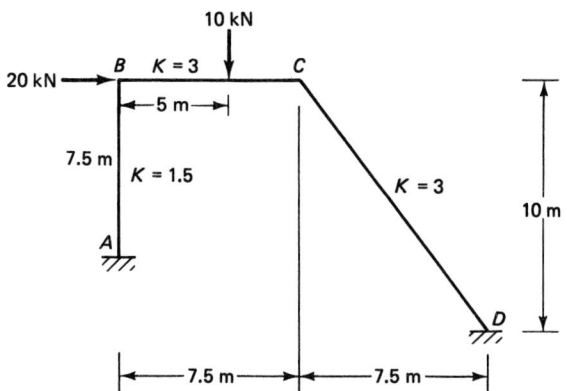

Figure P10.42

10.43 Determine the moments in the rigid frame shown.

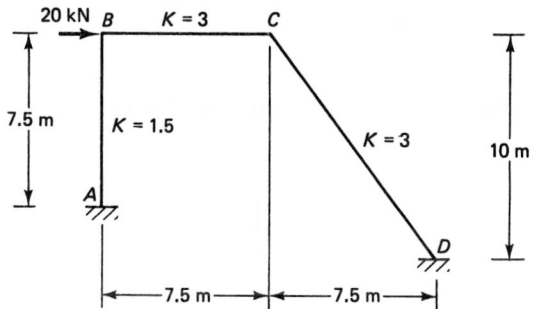

Figure P10.43

10.44 Draw the bending-moment diagram and sketch the deflected shape for the rigid frame shown.

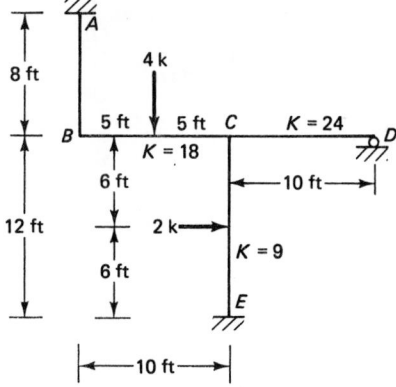

Figure P10.44

10.45 Determine the moments at joints B, C, and D for the hinged gable frame shown. Assume EI is constant.

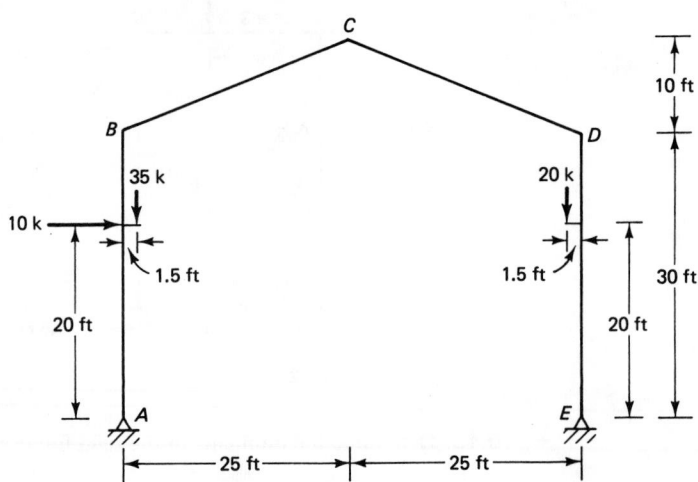

Figure P10.45

10.46 Determine the joint moments of the frame shown.

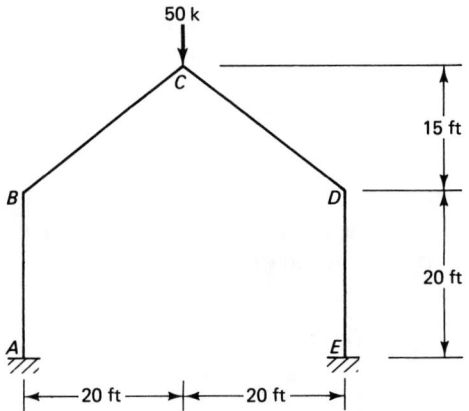

Figure P10.46

10.47 Draw the bending-moment diagram for the two-story rigid frame shown, using Morris's method.

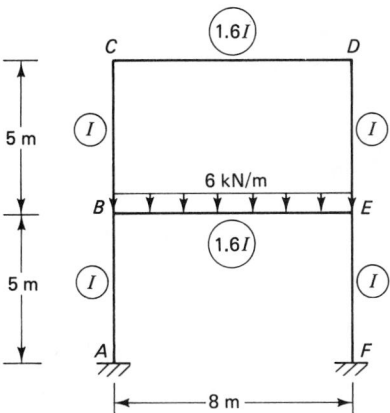

Figure P10.47

10.48 Determine the moments at joints B, C, D, and F for the structure shown.

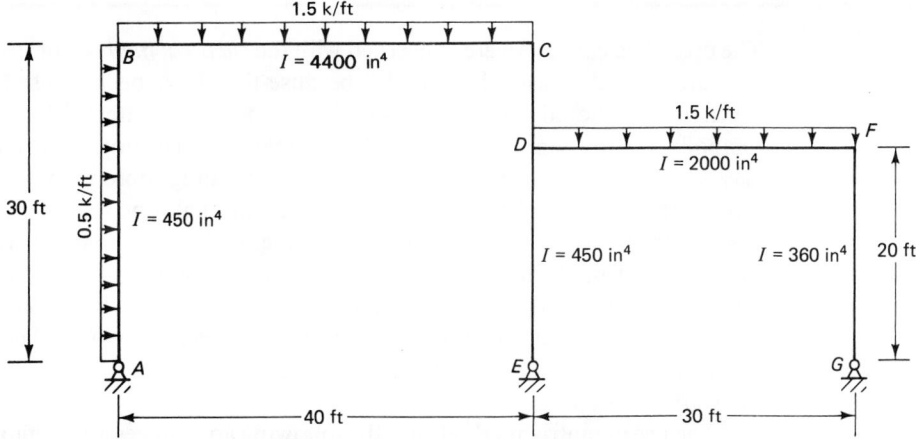

Figure P10.48

Chapter **11**

Matrix-Displacement Method

A new language is a riddle before it is conquered, a power in the hand afterwards.

Pollock

11.1 INTRODUCTION

The preceding chapters were concerned with the *classical methods* of structural analysis; they are termed "classical methods" because they have been in use for a long time, much before the advent of high-speed electronic computers. The so-called matrix methods use the same fundamental principles in analyzing a structure as those in classical methods; matrix methods are in fact nothing more than *mathematical procedures* that are intended to reduce the computational time required to analyze structures by the classical methods. The development of matrix methods was prompted by improvement in electronic computers; this freed engineers from tedious hand computations and allowed them to use more exact methods of analysis instead of timesaving approximate methods of analyzing large and complex structures. The efficient use of electronic computers required the introduction of *matrix algebra* and hence *matrix methods.*

The use of matrix methods has the following advantages: (1) It allows the engineer to analyze structures not only in a general way but also in a compact and *consistent* manner; furthermore, the fundamental principles of equilibrium, compatibility, and stress-strain relations are not obscured by computational devices or by physical differences between structures. (2) It provides a *systematic* approach to the analysis of structures culminating in the development of computer solution programs. Notwithstanding the above advantages, solutions obtained by hand calculations are still extremely valuable for preliminary results and for critical checking of results from the computer by an engineer with a sound knowledge of structural behavior.

As was hinted earlier, there are three conditions that loads and displacements must satisfy: (1) The loads acting on the ends of each member and the displacements of these ends must satisfy equations based on the stress-strain relationship of the material of the member. (2) The displacements of the ends of each member must be compatible with the displacements of the points to which that member is attached. These are called the *conditions of compatibility*. (3) The loads acting on the ends of each member must be such as to maintain the equilibrium of the member; also, the sum of the forces on the ends of the members meeting at any point must be equal to the external loads on that point. These are called the *conditions of equilibrium*.

Generally, we can classify the matrix methods of structural analysis (as well as the classical methods) according to the *order* in which the conditions of compatibility and equilibrium are applied. Matrix methods in which *compatibility conditions* are satisfied *first* give rise to equations of *joint equilibrium* and are called *matrix-equilibrium* or *-displacement methods*. In such methods, the forces are put in terms of the unknown joint displacements by means of the stiffness matrix (see Section 10.1.5) and the number of algebraic equations to be solved is equal to the number of degrees of freedom; this is why the matrix-displacement method is referred to as the stiffness method. On the other hand, matrix methods in which equilibrium conditions are used first lead to equations of *displacement compatibility* and are called *matrix-compatibility* or *-force methods*. In such methods the displacements are put in terms of forces by means of the flexibility matrix (see Section 9.2) and the number of algebraic equations to be solved is equal to the number of redundants; this is why the matrix-force method is referred to as the *flexibility method*.

As was mentioned earlier, the use of matrix methods in structural analysis entails the application of (1) the principle of equilibrium, (2) the principle of compatibility, and (3) load-deformation relations. Such applications must be performed in a consistent and systematic manner using matrix algebra. In applying the principle of equilibrium, the *external loads P* at a joint are related to the *internal resisting forces F* at the ends of members meeting at that joint. This relationship is established by the *static-equilibrium matrix* **E** when all the joints with possible displacements have been considered, or

$$\mathbf{P} = \mathbf{EF} \tag{11.1}$$

Furthermore, when applying the principle of compatibility, the *joint displacements D* are related to the end displacements δ of members meeting at that joint. When all the joints with possible displacements are considered, the relationship between the **D** and δ matrices is governed by the deformation compatibility matrix **C**, or

$$\delta = \mathbf{CD} \tag{11.2}$$

Now the relationship between the internal forces F and the end displacements δ of a member is a function of Young's modulus (E), length (l), cross-sectional area (A), and/or the second moment of area (I); thus we can write

$$\mathbf{F} = \mathbf{S}\delta \tag{11.3}$$

in which **S** is the member *stiffness matrix* defined in Section 10.1.5, or we can express the relationship as

$$\delta = \mathbf{fF} \tag{11.4}$$

in which **f** is the member *flexibility matrix* defined in Section 9.2.

Solution of any structural problem by matrix methods involves the generation of matrices **P, E, C, S,** or **f.** Once these are known, the solution to the problem can be readily determined by matrix algebra, as will be shown later. But first we must demonstrate how to form the **E, C, S,** and **P** matrices for use in the displacement method. The formation of the **f** matrix will be discussed in Chapter 13.

11.2 TRUSSES

11.2.1 The Static-Equilibrium Matrix E

Let us consider the truss in Fig. 11.1a. To generate matrix **E,** we include all possible external joint forces as shown in Fig. 11.1b; all the internal forces are assumed to be in tension and are denoted in Fig. 11.1c as $F_1, F_2, \ldots, F_9$. The P values are now

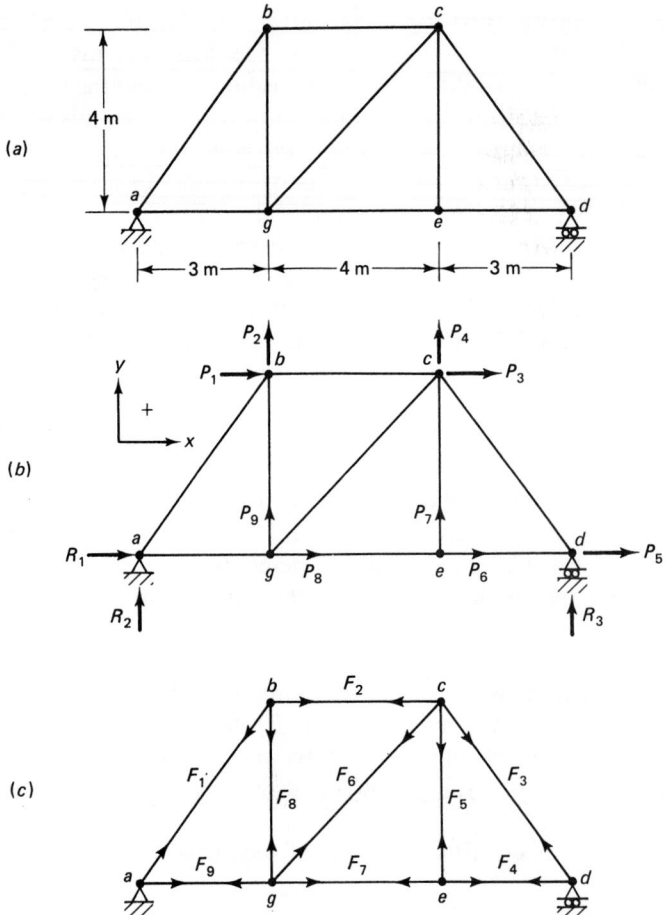

Figure 11.1 A statically determinate truss showing P and F numbering System. (b) P numbers. (c) F numbers.

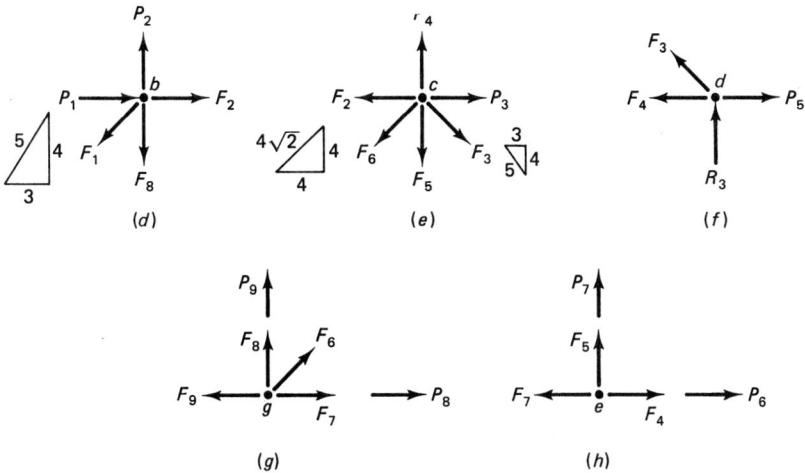

Figure 11.1 (*Continued*)

expressed in terms of the F values by applying the two equilibrium equations $\Sigma F_x = 0$ and $\Sigma F_y = 0$ for all joints with two degrees of freedom and applying only one equilibrium equation for any joint with one degree of freedom; for example, each of joints b, c, e, and g has two degrees of freedom, while joint d has only one; thus using the free-body diagrams of these joints as shown in Fig. 11.1d to h, we can write

For Joint b

$$P_1 = 0.6F_1 - F_2$$
$$P_2 = 0.8F_1 + F_8$$

For Joint c

$$P_3 = F_2 - 0.6F_3 + \frac{1}{\sqrt{2}}F_6$$

$$P_4 = 0.8F_3 + F_5 + \frac{1}{\sqrt{2}}F_6$$

For Joint d

$$P_5 = 0.6F_3 + F_4$$

For Joint e

$$P_6 = -F_4 + F_7$$
$$P_7 = -F_5$$

For Joint g

$$P_8 = -F_7 + F_9$$

$$P_9 = -\frac{1}{\sqrt{2}} F_6 - F_8$$

The above equations can be written in matrix form as

$$\{P\}_{9\times1} = [E]_{9\times9}\{F\}_{9\times1} \qquad (11.5)$$

$$
\begin{Bmatrix} P_1 \\ P_2 \\ P_3 \\ P_4 \\ P_5 \\ P_6 \\ P_7 \\ P_8 \\ P_9 \end{Bmatrix}
=
\begin{bmatrix}
0.6 & -1 & 0 & 0 & 0 & 0 & 0 & 0 & 0 \\
0.8 & 0 & 0 & 0 & 0 & 0 & 0 & +1 & 0 \\
0 & +1 & -0.6 & 0 & 0 & +\frac{1}{\sqrt{2}} & 0 & 0 & 0 \\
0 & 0 & +0.8 & 0 & +1 & +\frac{1}{\sqrt{2}} & 0 & 0 & 0 \\
0 & 0 & +0.6 & +1 & 0 & 0 & 0 & 0 & 0 \\
0 & 0 & 0 & -1 & 0 & 0 & +1 & 0 & 0 \\
0 & 0 & 0 & 0 & -1 & 0 & 0 & 0 & 0 \\
0 & 0 & 0 & 0 & 0 & 0 & -1 & 0 & +1 \\
0 & 0 & 0 & 0 & 0 & -\frac{1}{\sqrt{2}} & 0 & -1 & 0
\end{bmatrix}
\begin{Bmatrix} F_1 \\ F_2 \\ F_3 \\ F_4 \\ F_5 \\ F_6 \\ F_7 \\ F_8 \\ F_9 \end{Bmatrix}
\qquad (11.6)
$$

The truss in Fig. 11.1a is a statically determinate structure, and therefore the size of matrices **P** and **F** will be the same and hence matrix **E** is square. However, in a statically indeterminate structure, the size of matrix **F** will be greater than that of **P** and therefore matrix **E** will be rectangular. One can also observe that matrix **E** depends only on the structural form and not on the values of the external load matrix **P**.

■ **EXAMPLE 11.1**
Establish the static-equilibrium matrix $\mathbb{E}$ for the statically determinate structure in Fig. 11.2a.

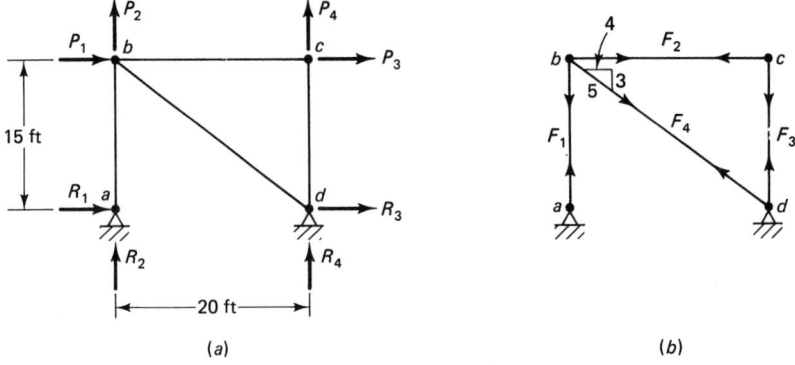

Figure 11.2 Truss for Example 11.1. (a) P numbers. (b) F numbers.

Solution. Each of joints b and c has two degrees of freedom, and therefore each joint has two possible external forces, (P_1, P_2) and (P_3, P_4). The internal bar forces are denoted as F_1, F_2, F_3, and F_4. Again, by applying the two equilibrium equations at joints b and c, the P forces can be related to the F forces. Thus,

At Joint b

$$P_1 = -F_2 - 0.8F_4$$

$$P_2 = F_1 + 0.6F_4$$

At Joint c

$$P_3 = +F_2$$

$$P_4 = +F_3$$

In matrix form

$$\begin{Bmatrix} P_1 \\ P_2 \\ P_3 \\ P_4 \end{Bmatrix} = \begin{bmatrix} 0 & -1 & 0 & -0.8 \\ 1 & 0 & 0 & +0.6 \\ 0 & 1 & 0 & 0 \\ 0 & 0 & 1 & 0 \end{bmatrix} \begin{Bmatrix} F_1 \\ F_2 \\ F_3 \\ F_4 \end{Bmatrix} \tag{11.7}$$

or

$$\{P\}_{4\times1} = [E]_{4\times4}\{F\}_{4\times1} \tag{11.8}$$

hence matrix **E**. ∎

■ EXAMPLE 11.2

Derive the static-equilibrium matrix **E** for the truss in Fig. 11.3a.

Solution. Applying the two equilibrium equations at joints b and c yields

At Joint b

$$P_1 = -F_2 - 0.8F_4$$

$$P_2 = F_1 + 0.6F_4$$

At Joint c

$$P_3 = F_2 + 0.8F_5$$

$$P_4 = F_3 + 0.6F_5$$

In matrix form

$$\begin{Bmatrix} P_1 \\ P_2 \\ P_3 \\ P_4 \end{Bmatrix} = \begin{bmatrix} 0 & -1 & 0 & -0.8 & 0 \\ 1 & 0 & 0 & +0.6 & 0 \\ 0 & 1 & 0 & 0 & +0.8 \\ 0 & 0 & 1 & 0 & +0.6 \end{bmatrix} \begin{Bmatrix} F_1 \\ F_2 \\ F_3 \\ F_4 \\ F_5 \end{Bmatrix} \tag{11.9}$$

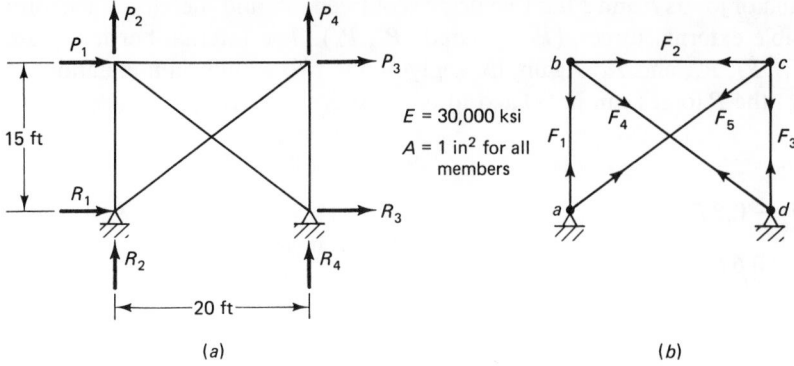

Figure 11.3 Truss for Example 11.2. (*a*) *P* numbers. (*b*) *F* numbers.

or

$$\{\mathbf{P}\}_{4\times1} = [\mathbf{E}]_{4\times5}\{\mathbf{F}\}_{5\times1} \tag{11.10}$$

hence matrix **E**; this is a rectangular matrix since the truss is statically indeterminate to the first degree. ∎

11.2.2 The Deformation Compatibility Matrix C

This matrix relates the external joint displacements *D* to the internal changes in lengths of members δ. Assuming that the displacements are small relative to the original dimensions, only the component of the joint displacement along the member's length changes its length. Consider now the truss in Fig. 11.4*a* showing the possible displacements of joints *b* and *c*. Figure 11.4*b* shows the resulting extensions δ in the five members of the truss. From the displacement diagram for member *ab* in Fig. 11.4*c*, it is clear that joint displacement D_1 does not affect the member's elongation δ_1, while D_2 does; in fact, $\delta_1 = D_2$. The effect of D_1 and D_3 on δ_2 is $\delta_2 = D_3 - D_1$; while the effect of D_1 and D_2 on δ_4 of member *bd* is $\delta_4 = -0.8D_1 + 0.6D_2$. The effects of the other *D*'s on the δ's may be similarly found. Thus, from Fig. 11.4*c* to *g*, we can write

$$\delta_1 = D_2$$

$$\delta_2 = D_3 - D_1$$

$$\delta_3 = D_4$$

$$\delta_4 = -0.8D_1 + 0.6D_2$$

$$\delta_5 = 0.8D_3 + 0.6D_4$$

In matrix form we can write

$$\begin{Bmatrix} \delta_1 \\ \delta_2 \\ \delta_3 \\ \delta_4 \\ \delta_5 \end{Bmatrix} = \begin{bmatrix} 0 & 1 & 0 & 0 \\ -1 & 0 & 1 & 0 \\ 0 & 0 & 0 & 1 \\ -0.8 & +0.6 & 0 & 0 \\ 0 & 0 & +0.8 & +0.6 \end{bmatrix} \begin{Bmatrix} D_1 \\ D_2 \\ D_3 \\ D_4 \end{Bmatrix} \tag{11.11}$$

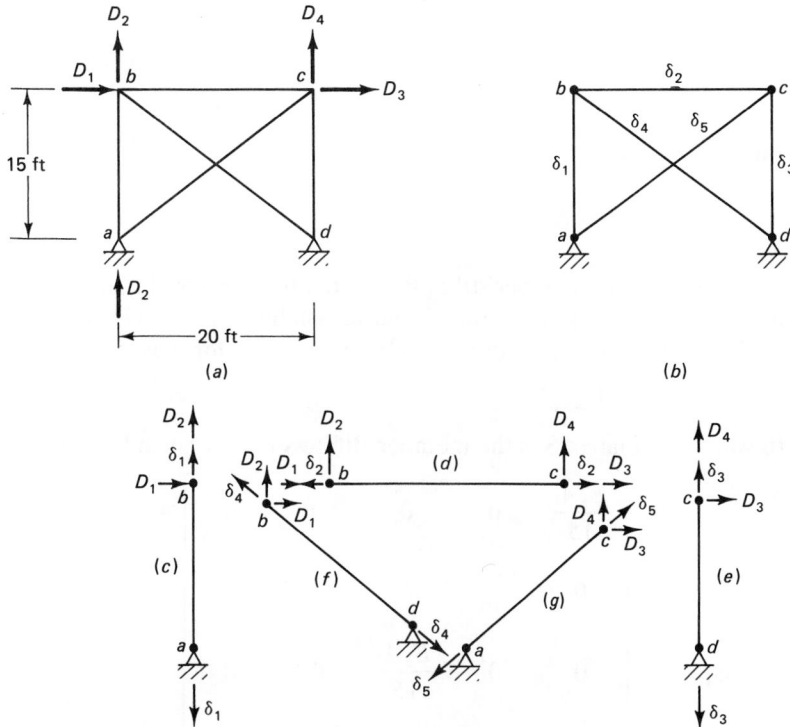

Figure 11.4 A statically indeterminate truss showing D and δ numbering system. (a) D numbers. (b) δ numbers.

or

$$\{\delta\}_{5\times1} = [\mathbf{C}]_{5\times4}\{\mathbf{D}\}_{4\times1} \qquad (11.12)$$

in which the 5×4 deformation compatibility matrix is

$$[\mathbf{C}]_{5\times4} = \begin{bmatrix} 0 & 1 & 0 & 0 \\ -1 & 0 & 1 & 0 \\ 0 & 0 & 0 & 1 \\ -0.8 & +0.6 & 0 & 0 \\ 0 & 0 & +0.8 & +0.6 \end{bmatrix} \qquad (11.13)$$

Comparison of matrix $\mathbf{E}$ in Eq. (11.9) with matrix $\mathbf{C}$ in Eq. (11.13) reveals that these two matrices are the transposes of one another; in general terms, $E_{ij} = C_{ji}$. In fact it can be shown by the principle of virtual work that

$$[\mathbf{E}] = [\mathbf{C}]^T \quad \text{or} \quad [\mathbf{C}] = [\mathbf{E}]^T \qquad (11.14)$$

11.2.3 The Member Stiffness Matrix S

In a linear elastic structure the internal stress σ_i in a member is related to its elongation δ_i by means of Hooke's law, or

$$\sigma_i = \frac{E_i \, \delta_i}{l_i}$$

But

$$\sigma_i = \frac{F_i}{A_i}$$

thus

$$F_i = \frac{E_i A_i}{l_i}\, \delta_i = S_i\, \delta_i \tag{11.15}$$

in which S_i is the member stiffness. For the truss in Fig. 11.3a, the member forces F are designated in Fig. 11.3b and the corresponding member elongations δ are presented in Fig. 11.4b. Thus, we can relate F to δ in matrix form as

$$\{F\}_{5\times1} = [S]_{5\times5}\{\delta\}_{5\times1} \tag{11.16}$$

in which the matrix **S** is the member stiffness matrix given by

$$[S]_{5\times5} = \begin{bmatrix} \dfrac{E_1 A_1}{15} & 0 & 0 & 0 & 0 \\[2mm] 0 & \dfrac{E_2 A_2}{20} & 0 & 0 & 0 \\[2mm] 0 & 0 & \dfrac{E_3 A_3}{15} & 0 & 0 \\[2mm] 0 & 0 & 0 & \dfrac{E_4 A_4}{25} & 0 \\[2mm] 0 & 0 & 0 & 0 & \dfrac{E_5 A_5}{25} \end{bmatrix} \tag{11.17}$$

This is a square matrix whose only nonzero elements lie on the principal diagonal when dealing with trusses.

11.2.4 The Global Stiffness Matrix S*

The possible external forces P are related to the joint displacements D by means of the *global or external stiffness matrix* **S***; it is denoted as global because it is a stiffness matrix for the entire structure, and not just for one member of the structure. Thus we can write

$$\{P\} = [S^*]\{D\} \tag{11.18}$$

The matrix **S*** can be derived either by matrix operations or by a direct method.

Using matrix operations, we first recall Eqs. (11.1) to (11.3):

$$P = EF \tag{11.1}$$

$$\delta = CD \tag{11.2}$$

and

$$F = S\,\delta \tag{11.3}$$

Substituting Eq. (11.2) into Eq. (11.3) yields

$$\mathbf{F} = \mathbf{S}\,\delta = \mathbf{SCD} \tag{11.19}$$

Using the relation

$$\mathbf{C} = \mathbf{E}^T \tag{11.14}$$

in Eq. (11.19), we have

$$\mathbf{F} = \mathbf{SE}^T\mathbf{D} \tag{11.20}$$

Substituting Eq. (11.20) into Eq. (11.1) gives

$$\mathbf{P} = \mathbf{ESE}^T\mathbf{D} \tag{11.21}$$

Comparing Eq. (11.21) and Eq. (11.18), we find that

$$\mathbf{S}^* = \mathbf{ESE}^T \tag{11.22}$$

It can be shown that the matrix $\mathbf{S}^*$ or matrix $[\mathbf{ESE}^T]$ is a square matrix.

To find the internal forces in a statically determinate truss, matrix $\mathbf{E}$ is square and therefore can be inverted; thus by premultiplying both sides of Eq. (11.1) by $\mathbf{E}^{-1}$ we find

$$\mathbf{F} = \mathbf{E}^{-1}\mathbf{P} \tag{11.23}$$

However, for a statically indeterminate truss, matrix $\mathbf{E}$ is not square and therefore cannot be inverted; in this case, the internal forces $\mathbf{F}$ can be obtained by first finding the joint displacements $\mathbf{D}$ from Eq. (11.21). Thus, premultiplying both sides of Eq. (11.21) by $[\mathbf{S}^*]^{-1}$ or $[\mathbf{ESE}^T]^{-1}$, we find

$$\mathbf{D} = [\mathbf{S}^*]^{-1}\mathbf{P} = [\mathbf{ESE}^T]^{-1}\mathbf{P} \tag{11.24}$$

Hence from Eq. (11.20),

$$\mathbf{F} = \mathbf{SE}^T[\mathbf{S}^*]^{-1}\mathbf{P} = \mathbf{SE}^T[\mathbf{ESE}^T]^{-1}\mathbf{P} \tag{11.25}$$

This method of solution is termed the *matrix-displacement method* or *stiffness method.*

The global matrix $\mathbf{S}^*$ can also be derived directly as follows: Let us consider the truss in Fig. 11.5a which shows the P,D numbering system, while the F,δ numbering system is given in Fig. 11.5b. Using Eq. (11.18) for P_1, say, we have

$$P_1 = S_{11}^*D_1 + S_{12}^*D_2 + S_{13}^*D_3 + S_{14}^*D_4$$

and similarly for P_2, P_3, and P_4. It becomes evident that the first column of the matrix $\mathbf{S}^*$ can be derived if $D_1 = 1$ and all the other D's are set to zero. The external joint forces for such a displacement are indeed the first column of matrix $\mathbf{S}^*$, namely, S_{11}^*, S_{21}^*, S_{31}^*, and S_{41}^*. We can find these external joint forces from force equilibrium at each joint. For example, the displacement $D_1 = 1$ (with $D_2 = D_3 = D_4 = 0$) shown in Fig. 11.5c causes internal forces F given in terms of the member stiffness ($S = EA/L$), that is, $F_1 = 0$, $F_2 = -S_2$, $F_3 = 0$, $F_4 = -(S_4)(1 \times \cos\theta) = -0.8S_4$; and $F_5 = 0$. Now from force equilibrium at joint b we have

In the x direction:

$$S_{11}^* - S_2 - (0.8S_4)\cos\theta = 0$$

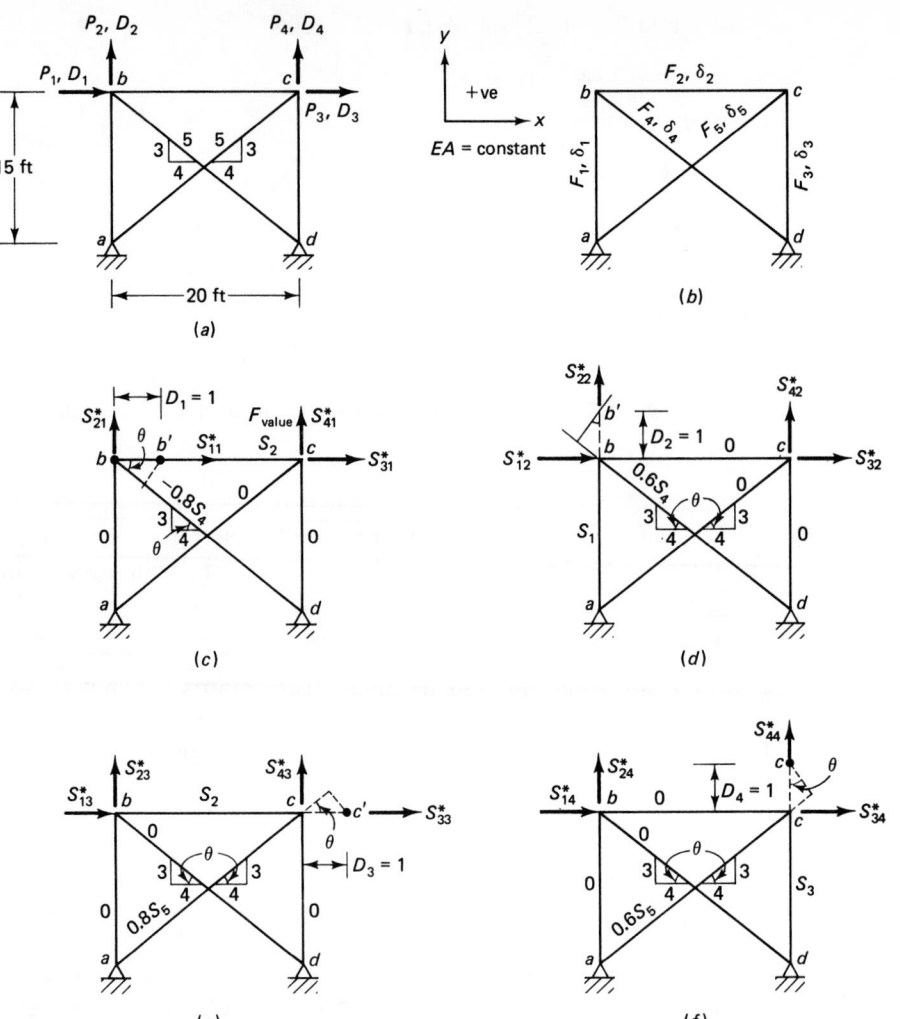

Figure 11.5 Development of the global stiffness matrix for a statically indeterminate truss. (a) P, D numbers. (b) F, δ numbers. (c) $D_1 = 1$, $D_2 = D_3 = D_4 = 0$. (d) $D_2 = 1$, $D_1 = D_3 = D_4 = 0$. (e) $D_3 = 1$, $D_1 = D_2 = D_4 = 0$. (f) $D_y = 1$, $D_1 = D_2 = D_3 = 0$.

or

$$S_{11}^* = S_2 + 0.64\,S_4$$

In the y direction:

$$S_{21}^* + (0.8\,S_4)\sin\theta = 0$$

or

$$S_{21}^* = -0.48\,S_4$$

At Joint c

In the x direction:

$$S_{31}^* + S_2 = 0 \quad \text{or} \quad S_{31}^* = -S_2$$

In the y direction:

$$S_{41}^* = 0$$

Thus, the first column of the matrix S^* is derived. If this procedure is continued, as shown in Fig. 11.5d to f, the matrix S^* can be shown to be

$$S^* = \begin{bmatrix} (S_2 + 0.64\,S_4) & -0.48\,S_4 & -S_2 & 0 \\ -0.48\,S_4 & (S_1 + 0.36\,S_4) & 0 & 0 \\ -S_2 & 0 & (S_2 + 0.64\,S_5) & 0.48\,S_5 \\ 0 & 0 & 0.48\,S_5 & (S_3 + 0.36\,S_5) \end{bmatrix} \quad (11.26)$$

which is a symmetric matrix with respect to the principal diagonal.

Substituting the values of S_1 to S_5 in Eq. (11.26) yields

$$S^* = \frac{EA}{100} \begin{bmatrix} 7.56 & -1.92 & -5.00 & 0 \\ -1.92 & 8.11 & 0 & 0 \\ -5.00 & 0 & 7.56 & 1.92 \\ 0 & 0 & 1.92 & 8.11 \end{bmatrix} \quad (11.27)$$

To prove that $S^* = ESE^T$ [see Eq. (11.22)], we substitute for E from Eq. (11.9) and for S from Eq. (11.17) and obtain

$$ESE^T = \begin{bmatrix} 0 & -1 & 0 & -0.8 & 0 \\ 1 & 0 & 0 & 0.6 & 0 \\ 0 & 1 & 0 & 0 & 0.8 \\ 0 & 0 & 1 & 0 & 0.6 \end{bmatrix} \begin{bmatrix} \dfrac{EA}{15} & 0 & 0 & 0 & 0 \\ 0 & \dfrac{EA}{20} & 0 & 0 & 0 \\ 0 & 0 & \dfrac{EA}{15} & 0 & 0 \\ 0 & 0 & 0 & \dfrac{EA}{25} & 0 \\ 0 & 0 & 0 & 0 & \dfrac{EA}{25} \end{bmatrix} \begin{bmatrix} 0 & 1 & 0 & 0 \\ -1 & 0 & 1 & 0 \\ 0 & 0 & 0 & 1 \\ -0.8 & 0.6 & 0 & 0 \\ 0 & 0 & 0.8 & 0.6 \end{bmatrix}$$

Carrying out the matrix multiplication results in the same values for S^* given by Eq. (11.27).

11.2.5 The External Load Matrix P

The elements in the external load matrix P are derived by comparing the actual loads on the truss with the possible external load P's at joints corresponding to the degrees

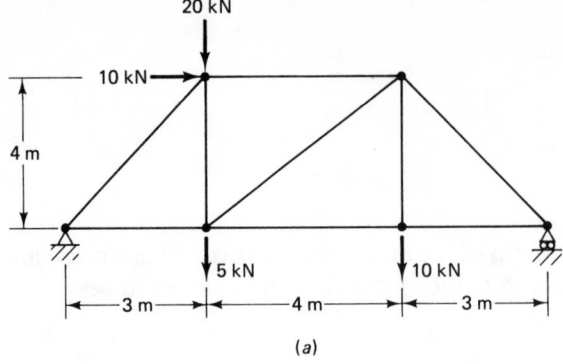

(a)

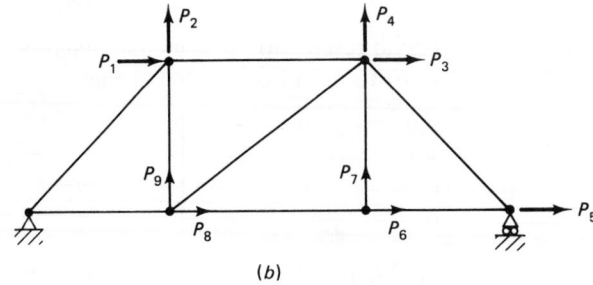

(b)

Figure 11.6 A statically determinate truss showing *P* numbering system. (*b*) *P* numbers.

of freedom of the truss. For the loaded truss in Fig. 11.6*a*, the possible *P*'s are shown in Fig. 11.6*b*. Thus comparing the loads in Fig. 11.6*a* with those in Fig. 11.6*b*, and taking proper account of direction, we can write the matrix **P** as

$$\begin{Bmatrix} P_1 \\ P_2 \\ P_3 \\ P_4 \\ P_5 \\ P_6 \\ P_7 \\ P_8 \\ P_9 \end{Bmatrix} = \begin{Bmatrix} 10 \\ -20 \\ 0 \\ 0 \\ 0 \\ 0 \\ -10 \\ 0 \\ -5 \end{Bmatrix} \quad \text{k}$$

■ EXAMPLE 11.3
(a) Derive the **P** matrix for the truss in Fig. 11.7*a*: and (b) solve for the internal forces.

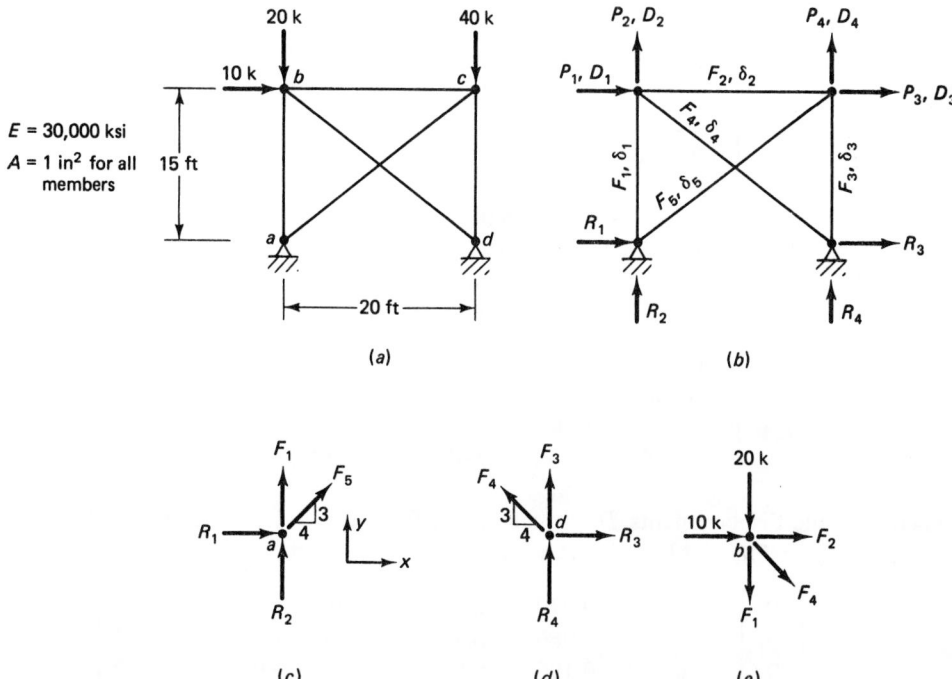

Figure 11.7 Truss for Examples 11.3 and 11.4. (*b*) *P, D* and *F, δ* numbers.

Solution

(a) All possible *P*'s corresponding to the degrees of freedom of the truss are shown numbered in Fig. 11.7*b*. Comparing the actual loads with those in Fig. 11.7*b* gives

$$\begin{bmatrix} P_1 \\ P_2 \\ P_3 \\ P_4 \end{bmatrix} = \begin{bmatrix} 10 \\ -20 \\ 0 \\ -40 \end{bmatrix} \text{ k}$$

(b) The internal forces *F* can be obtained by Eq. (11.25):

$$\mathbf{F} = \mathbf{SE}^T[\mathbf{S^*}]^{-1}\mathbf{P}$$

Now **S** is given by Eq. (11.17), $\mathbf{E}^T = \mathbf{C}$ by Eq. (11.13), and **S*** by Eq. (11.27). Thus

$$\mathbf{SE}^T = EA \begin{bmatrix} \frac{1}{15} & 0 & 0 & 0 & 0 \\ 0 & \frac{1}{20} & 0 & 0 & 0 \\ 0 & 0 & \frac{1}{15} & 0 & 0 \\ 0 & 0 & 0 & \frac{1}{25} & 0 \\ 0 & 0 & 0 & 0 & \frac{1}{25} \end{bmatrix} \begin{bmatrix} 0 & 1 & 0 & 0 \\ -1 & 0 & 1 & 0 \\ 0 & 0 & 0 & 1 \\ -0.8 & 0.6 & 0 & 0 \\ 0 & 0 & 0.8 & 0.6 \end{bmatrix}$$

$$= \frac{EA}{100} \begin{bmatrix} 0 & 6.67 & 0 & 0 \\ -5 & 0 & 5 & 0 \\ 0 & 0 & 0 & 6.67 \\ -3.2 & 2.4 & 0 & 0 \\ 0 & 0 & 3.2 & 2.4 \end{bmatrix}$$

$$[S^*]^{-1} = \frac{100}{EA} \begin{bmatrix} 7.56 & -1.92 & -5.00 & 0 \\ -1.92 & 8.11 & 0 & 0 \\ -5.00 & 0 & 7.56 & 1.92 \\ 0 & 0 & 1.92 & 8.11 \end{bmatrix}^{-1}$$

$$= \frac{100}{EA} \begin{bmatrix} 0.279 & 0.066 & 0.196 & -0.046 \\ 0.066 & 0.139 & 0.046 & -0.011 \\ 0.196 & 0.046 & 0.279 & -0.066 \\ -0.046 & -0.011 & -0.066 & 0.139 \end{bmatrix}$$

Therefore, the displacements D are calculated from Eq. (11.24) as $\mathbf{D} = [S^*]^{-1}\mathbf{P}$. Substituting for EA, $[S^*]^{-1}$, and $\mathbf{P}$, we have

$$[D] = \begin{bmatrix} D_1 \\ D_2 \\ D_3 \\ D_4 \end{bmatrix} = \frac{1}{300} \begin{bmatrix} 0.279 & 0.066 & 0.196 & -0.046 \\ 0.066 & 0.139 & 0.046 & -0.011 \\ 0.196 & 0.046 & 0.279 & -0.066 \\ -0.046 & -0.011 & -0.066 & 0.139 \end{bmatrix} \begin{bmatrix} 10 \\ -20 \\ 0 \\ -40 \end{bmatrix}$$

$$= \frac{1}{300} \begin{bmatrix} 3.31 \\ -1.68 \\ 3.67 \\ -5.81 \end{bmatrix} \quad \text{ft}$$

Hence from

$$\mathbf{F} = \begin{bmatrix} F_1 \\ F_2 \\ F_3 \\ F_4 \\ F_5 \end{bmatrix} = \mathbf{SE}^T\mathbf{D} = 300 \begin{bmatrix} 0 & 6.67 & 0 & 0 \\ -5 & 0 & 5 & 0 \\ 0 & 0 & 0 & 6.67 \\ -3.2 & 2.4 & 0 & 0 \\ 0 & 0 & 3.2 & 2.4 \end{bmatrix} \frac{1}{300} \begin{bmatrix} 3.31 \\ -1.68 \\ 3.67 \\ -5.81 \end{bmatrix}$$

$$= \begin{bmatrix} -11.21 \\ +1.80 \\ -38.75 \\ -14.62 \\ -2.20 \end{bmatrix} \quad \text{k}$$

Once the $\mathbf{F}$ matrix is obtained, the external reactions can be readily determined from the joint-equilibrium requirement in the direction of each reaction. For example, for the reactions at a, we obtain from the FBD in Fig. 11.7c

$$\sum F_x = 0: \quad R_1 + \tfrac{4}{5}F_5 = 0 \quad \text{or} \quad R_1 = -\tfrac{4}{5}F_5 = (-\tfrac{4}{5})(-2.20) = 1.76 \text{ k}$$

$$\Sigma\ F_y = 0: \qquad R_2 + F_1 + \tfrac{3}{5}F_5 = 0 \quad \text{or}$$

$$R_2 = -F_1 - \tfrac{3}{5}F_5 = -(-11.21) - (\tfrac{3}{5})(-2.20) = 12.53\ \text{k}$$

For the reaction at d, using the FBD in Fig. 11.7d:

$$\Sigma\ F_x = 0: \qquad R_3 - \tfrac{4}{5}F_4 = 0 \quad \text{or} \quad R_3 = (\tfrac{4}{5})(-14.62) = -11.70\ \text{k or } 11.70\ \text{k} \leftarrow$$

$$\Sigma\ F_y = 0: \qquad R_4 + F_3 + (\tfrac{3}{5})(F_4) = 0 \quad \text{or}$$

$$R_4 = -F_3 - (0.6)(F_4) = -(-38.75) - (0.6)(-14.62) = 47.52\ \text{k}$$

Check force equilibrium of the whole truss:

$$\Sigma\ F_x = 0: \qquad R_1 + R_3 + 10 = 0 \quad \text{or} \quad 1.76 + (-11.70) + 10 \approx 0 \quad \text{(checks)}$$

$$\Sigma\ F_y = 0: \qquad R_2 + R_4 - 20 - 40 = 0 \quad \text{or}$$

$$12.53 + 47.52 - 20 - 40 \approx 0 \qquad \text{(checks)} \quad \blacksquare$$

11.2.6 Static and Deformation Checks

It is important that the results for the joint displacements D and the internal forces F be checked to ensure that static equilibrium as well as deformation compatibility (in the case of indeterminate structures) are satisfied. The static-equilibrium check requires that $\Sigma\ F_x = 0$ and $\Sigma\ F_y = 0$ are satisfied at each joint, and that the equilibrium conditions $\Sigma\ F_x = 0$, $\Sigma\ F_y = 0$, and $\Sigma\ M = 0$ for the entire truss are proved.

The deformation compatibility check is made to ensure that the elongation of a member computed from the internal force in that member agrees with the elongation calculated from the joint displacements for each end of that member. An expression for the elongation of a member δ in terms of the joint displacements D of its ends can be derived as follows: Consider a member ab in Fig. 11.8a; after joint displacements $D_1, \ldots, D_4$ had taken place, a moves to a' and b to b', as shown in Fig. 11.8b. The

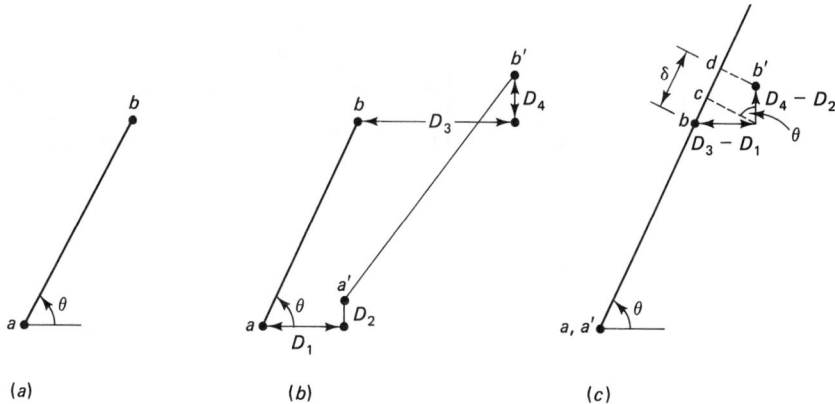

Figure 11.8 Relationship between end-member elongations and joint displacements.

relative displacement of b with respect to a is presented in Fig. 11.8c, which shows that the effect of $(D_3 - D_1)$ on δ is $(D_3 - D_1)\cos\theta$ or bc; and of $(D_4 - D_2)$ on δ is $(D_4 - D_2)\sin\theta$ or cd. Thus we can write

$$\delta = (D_3 - D_1)\cos\theta + (D_4 - D_2)\sin\theta \qquad (11.28)$$

■ EXAMPLE 11.4

Perform the static-equilibrium and deformation compatibility checks at joint b for the truss in Fig. 11.7a.

Solution. The internal forces obtained from the solution are shown in Fig. 11.9a. For the *static-equilibrium* check at joint b, apply the equilibrium equations for the FBD in Fig. 11.7e.

$$\sum F_x = 0: \quad 10 + F_2 + \tfrac{4}{5}F_4 = 0 \quad 10 + 1.80 + (0.8)(-14.62) \approx 0$$

$$\sum F_y = 0: \quad -20 - F_1 - \tfrac{3}{5}F_4 = 0 \quad -20 - (-11.21) - (0.6)(-14.62) \approx 0$$

The elongations of members and the joint displacements obtained previously are shown in Fig. 11.9b.

For *deformation compatibility* check at joint b, apply Eq. (11.28) and using Fig. 11.9b for member 2:
Here $\theta = 0$,

$$\delta_2 = (D_3 - D_1)\cos 0 + (D_4 - D_2)\sin 0 = 1.223 - 1.103 = 0.120 \qquad \text{(checks)}$$

For member 4, θ is in the fourth quadrant,

$$\delta_4 = (0 - D_1)(\tfrac{4}{5}) + (0 - D_2)(-\tfrac{3}{5})$$

$$= (-1.103)(0.8) - (-0.560)(-0.6) = -1.2184 \qquad \text{(checks)}$$

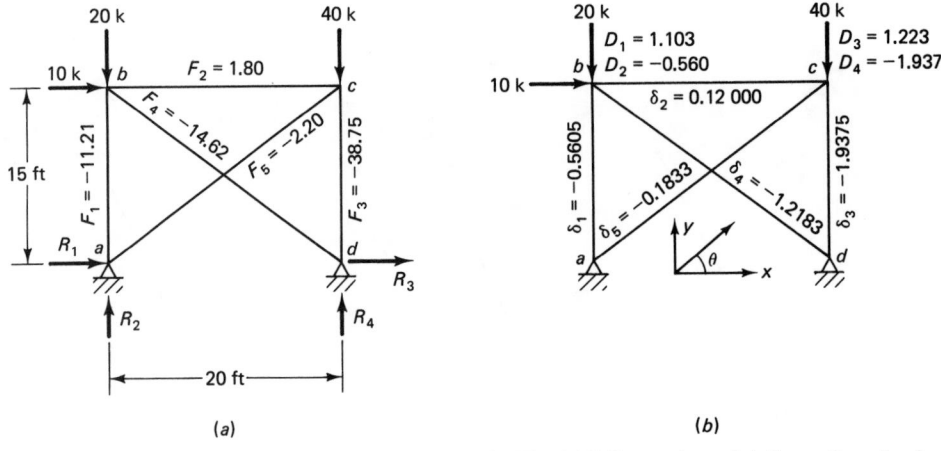

(a)   (b)

Figure 11.9 P, F, D, and δ values for the truss in Fig. 11.7 for static and deformation checks. (a) P and F values. (b) D and δ values $\times\ 10^{-2}$ ft.

For member 1, $\delta_1 = -0.5605$, which equals the joint vertical displacement of -0.560. ∎

11.2.7 Effects of Support Settlements and Lack of Fit

Support settlement and lack of fit (when any member in a structure is not fabricated to its exact length) can induce internal forces in members and/or joint deformations depending on whether the structure is statically (1) determinate, (2) internally indeterminate, and (3) externally indeterminate. This is explained succinctly in Table 11.1.

Let us assume that the roller support at d of the truss in Fig. 11.10 settles vertically a distance of 10 mm; member ed is not affected by this settlement, while member cd is lengthened by an amount $\delta_0 = 10 \sin \theta = 8$ mm; such an elongation could be provided by a pair of initial tensile forces

$$F_0 = \frac{AE\,\delta_0}{l} = 200 \text{ kN}$$

applied to cd as shown in Fig. 11.10c. After the members are fitted together so that all the joints are at their correct locations, except the settled joint d, the pair of holding forces F_0 could be released, which is equivalent to applying pulling forces on joints c and d as shown in Fig. 11.10a, b, and d. As a consequence, the external joint forces in the **P** matrix are found from the pulling forces on joints c and d, as shown in Fig. 11.10e. For example, for P_3 and P_4 at joint c, $P_3 = (F_0)(\frac{3}{5}) = 120$ kN and $P_4 = (-F_0)(\frac{4}{5}) = -160$ kN; for P_5 at joint d, $P_5 = (-F_0)(\frac{3}{5}) = -120$ kN as shown in Fig. 11.10f. Based on this **P** matrix, the displacement matrix **D** and hence the **F** matrix are determined. The final internal forces **F′** in the members are equal to the sum of the initial forces F_0 and the $\mathbf{F} = \mathbf{SE}^T\mathbf{D}$ from the computer output, or $\mathbf{F}' = \mathbf{F}_0 + \mathbf{F}$.

The same procedure can be applied to calculate member forces due to lack of fit of members in a truss. If a member is δ_0 too short, for example, a pair of initial tensile forces, $F_0 = AE\,\delta_0/l$, are applied to the member to elongate it; this leads to the determination of the **P** matrix as before. Similarly, if a member is δ_0 too long, a pair of

TABLE 11.1

Effect of	Causing	Determinate structure	Indeterminate structure	
			Internal	External
Support settlement	Joint displacement	Yes	Yes	Yes
	Internal stresses	No	No	Yes
Lack of fit	Joint displacement	Yes	Yes	Yes
	Internal stresses	No	Yes	Yes

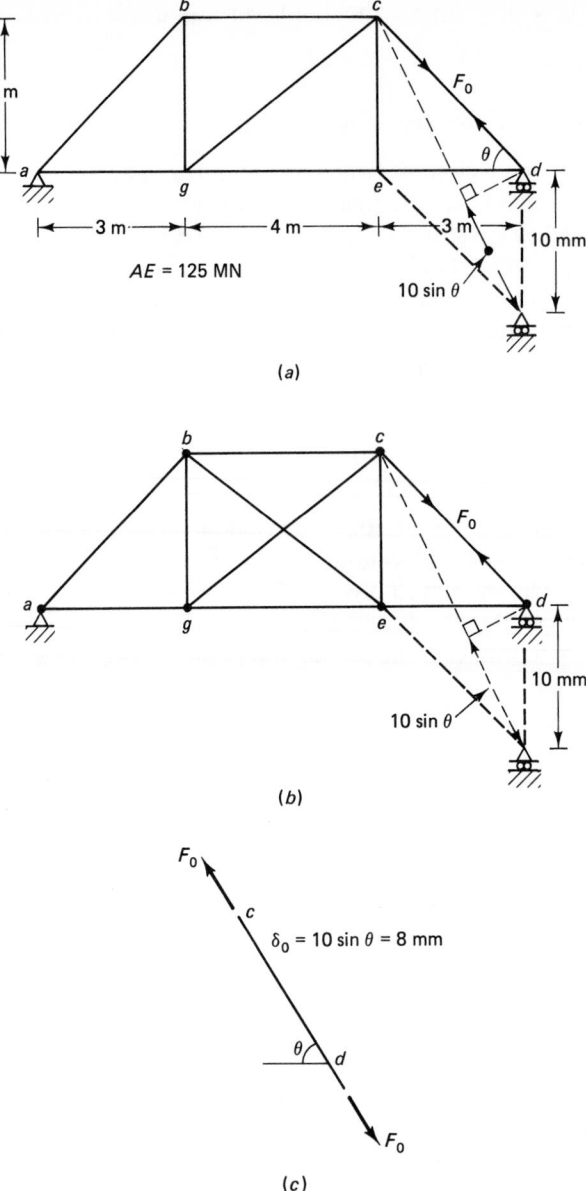

(a)

(b)

(c)

Figure 11.10 Treatment of settlement of support.

initial compressive forces, $F_0 = AE\,\delta_0/l$, are applied to the member to shorten it, and so on.

For trusses with several members that are either too long and/or too short the above procedure can be accelerated by the following matrix method: Let $\{\mathbf{F}'\}$ be the total internal forces in all members including those that do not fit, $\{\delta\}$ be the total

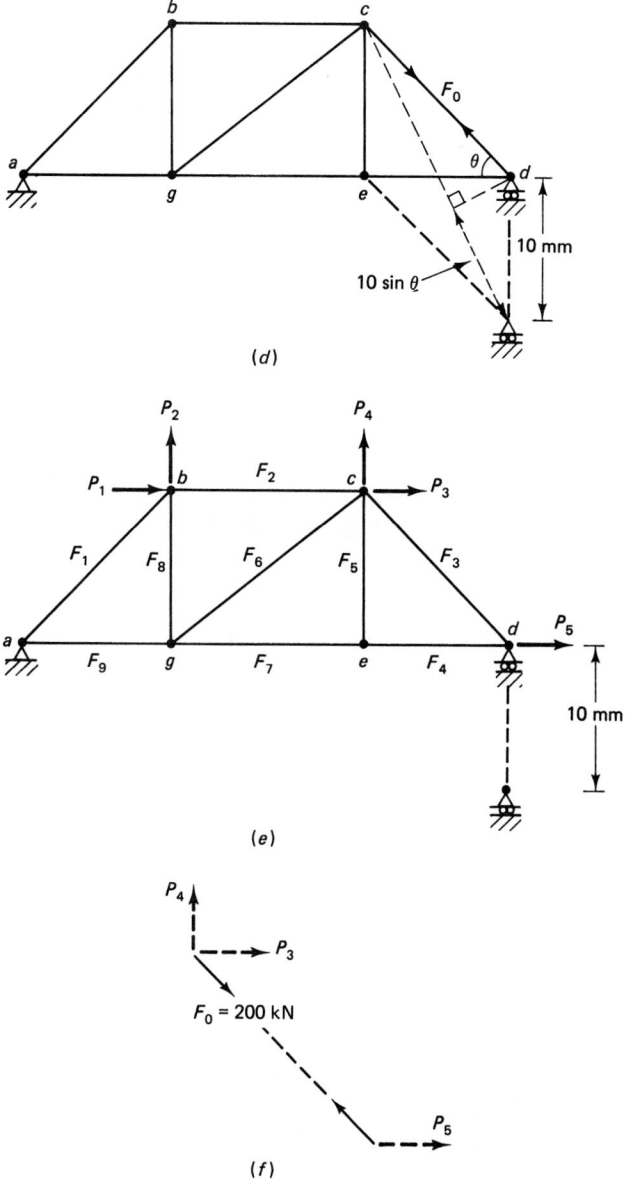

(d)

(e)

(f)

Figure 11.10 (*Continued*)

elongations in all members, and $\{\delta_0\}$ be initial elongations in all members. Then, in the absence of external loads, one can write

$$\mathbf{P} = \mathbf{EF'} = 0 \qquad (11.29)$$

$$\mathbf{F'} = \mathbf{S}(\delta - \delta_0) \qquad (11.30)$$

with

$$\delta = E^T D \tag{11.31}$$

Substituting Eq. (11.31) in Eq. (11.30),

$$F' = SE^T D - S \delta_0 \tag{11.32}$$

Substituting Eq. (11.32) in Eq. (11.29) yields

$$P = ESE^T D - ES \delta_0 = 0$$

Solving for **D** by premultiplying both sides by $(ESE^T)^{-1}$:

$$D = (ESE^T)^{-1} ES \delta_0 \tag{11.33}$$

Hence the member forces **F'** can now be found from Eq. (11.32).

■ EXAMPLE 11.5

Find the member forces in the truss of Fig. 11.3a when member 1 is made $\frac{1}{8}$ in too long, member 3 is $\frac{1}{16}$ in too short, and member 4 is $\frac{1}{12}$ in too short.

Solution.   The matrices **E**, **S**, SE^T, and $(ESE^T)^{-1}$ for this truss were established in Examples 11.2 and 11.3.

The $\{\delta_0\}$ matrix is

$$\delta_0 = \begin{bmatrix} \frac{1}{96} \\ 0 \\ -\frac{1}{192} \\ -\frac{1}{144} \\ 0 \end{bmatrix} \text{ ft } \quad \text{then} \quad S \delta_0 = \begin{bmatrix} +20.833 \\ 0 \\ -10.417 \\ -8.333 \\ 0 \end{bmatrix} \text{ k}$$

$$ES \delta_0 = \begin{bmatrix} +6.667 \\ +15.833 \\ 0 \\ -10.417 \end{bmatrix} \text{ k}$$

then

$$D = [ESE^T]^{-1}[ES \delta_0] = \begin{bmatrix} 11.294 \\ 9.185 \\ 9.105 \\ -6.440 \end{bmatrix} \times 10^{-3} \text{ ft}$$

Hence

$$SE^T D = \begin{bmatrix} 18.371 \\ -3.284 \\ -12.879 \\ -4.229 \\ 4.104 \end{bmatrix} \quad \text{Thus} \quad F' = SE^T D - S \delta_0 = \begin{bmatrix} -2.462 \\ -3.284 \\ -2.462 \\ 4.104 \\ 4.104 \end{bmatrix} \text{ k}$$

the static-equilibrium check at joint b:

$$\Sigma\, F_x = 0: \qquad -3.284 + (0.8)(4.104) \approx 0$$

and

$$\Sigma\, F_y = 0: \qquad -2.462 + (0.6)(4.104) \approx 0$$

The student can complete the static check at joint c as well as the deformation check for each member. ∎

■ EXAMPLE 11.6

For the truss in Fig. 11.3a determine the internal forces F' in the members due to a vertical settlement of $\frac{1}{2}$ in at joint d.

Solution. This displacement will incur an extension $\delta_{03} = 0.5$ in in member cd and an extension $\delta_{04} = (0.5)(\frac{3}{5}) = 0.3$ in in member bd. Such displacements can be accommodated by initial tension forces of

$$F_{03} = \frac{AE\,\delta_{03}}{l} = \frac{(30{,}000)(0.5)/12}{15} = +83.333 \text{ k}$$

and

$$F_{04} = \frac{(30{,}000)(0.3)/12}{25} = +30.00 \text{ k}$$

Such forces are equivalent to the following external loads: At joint b, $P_1 = (30) \times (0.8) = 24.0$ k; $P_2 = (-30)(0.6) = -18.0$ k. At joint c, $P_3 = 0$; $P_4 = -83.333$ k. Thus the joint external force matrix $\mathbf{P}$ can be written as

$$\mathbf{P} = \begin{bmatrix} 24.000 \\ -18.000 \\ 0 \\ -83.333 \end{bmatrix} \text{ k}$$

$[\mathbf{S^*}]^{-1}$ was derived earlier. Thus

$$\mathbf{D} = [\mathbf{S^*}]^{-1}\mathbf{P} = \begin{bmatrix} 31.250 \\ 0 \\ 31.250 \\ -41.667 \end{bmatrix} \times 10^{-3} \text{ ft}$$

Hence, knowing $\mathbf{SE}^T$ previously, one finds

$$\mathbf{F} = \mathbf{SE}^T\mathbf{D} = \begin{bmatrix} 0 \\ 0 \\ -83.333 \\ -30.000 \\ 0 \end{bmatrix} \text{ k}$$

Thus

$$
\mathbf{F}' = \mathbf{F} + \mathbf{F}_0 = \begin{bmatrix} 0 \\ 0 \\ -83.333 \\ -30.000 \\ 0 \end{bmatrix} + \begin{bmatrix} 0 \\ 0 \\ +83.333 \\ +30.00 \\ 0 \end{bmatrix} = \begin{bmatrix} 0 \\ 0 \\ 0 \\ 0 \\ 0 \end{bmatrix} \text{ k} \qquad \blacksquare
$$

This result should not be surprising since the vertical settlement of support d involves only rigid body rotation of the indeterminate truss.

11.3 CONTINUOUS BEAMS AND RIGID FRAMES

The principles of the matrix-displacement method, used in analyzing trusses, will now be applied to continuous beams and rigid frames. The degrees of freedom in such structures are the rotations at joints as well as the joint translations, wherever applicable. For example, for the continuous beam in Fig. 11.11a, there are four degrees of freedom, D_1, D_2, D_3, and D_4, whereas the rigid frame in Fig. 11.12a has three degrees of freedom—two rotations (D_1 and D_2) and one translation D_3. In this case, it is assumed that the primary action is bending and axial deformation is relatively small compared with the transverse deflection due to bending. (In other words, it is assumed that members in the frame have infinite axial stiffness.) However, in tall building frames, axial forces can be large enough to warrant consideration, especially in the lower stories, for example, in the tall multistory rigid frame shown in Fig. 11.13a. The (P, D) and (F, δ) diagrams for one story $abcdef$ are shown in Fig. 11.13b and c, respectively. Note the three degrees of freedom at each joint, one in rotation and two in translation. Another example is the case when a structure contains truss elements and beam elements with finite or infinite axial stiffness as shown in Fig. 11.14a. The (P, D) and (F, δ) diagrams are given in Fig. 11.14b and c.

11.3.1 The Treatment of Loads Applied between Nodal Points or Joints

Matrix analysis can be applied to structures loaded only at nodal points or joints. Because of this restriction, a distributed load, for example, cannot be dealt with directly (see Fig. 11.15a). In such cases one can divide the distributed load into a series of closely spaced concentrated loads and consider each concentrated load point as a node. This procedure will of course increase the number of analyzed elements in the structure, thus augmenting the size of the matrices used in the analysis. To circumvent this, one can think of the actual loads as composed of two systems: System 1 consists of the actual loads along the member, together with fixed end moments and forces preventing any translation and/or rotation of the joints (or nodes) as shown in Fig. 11.15b. These fixed-end moments, denoted by F_0, can be readily determined from the formulas given in the Appendix. System 2 is composed of the same restraining fixed-end moments and forces but with the reverse sign, applied at the joints as shown in Fig. 11.15c. It should be noted that this process of replacing a system of loads

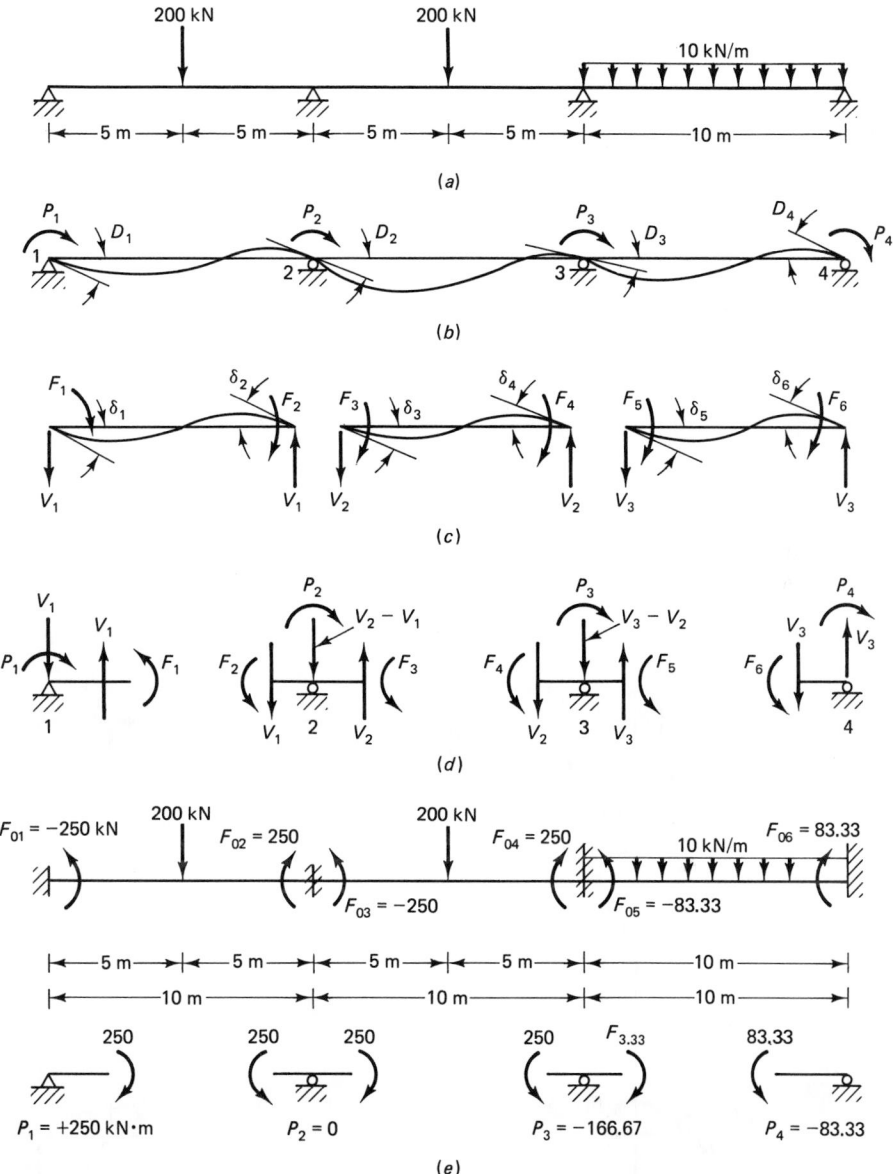

Figure 11.11 Analysis of a continuous beam by matrix stiffness method. (*b*) *P-D* diagram. (*c*) *F-δ* diagram. (*d*) Joint equilibrium diagrams. (*e*) Fixed-end moments and external joint moments.

applied between nodes or joints is dependent on the validity of the principle of superposition for linear elastic structures. Thus the final member end forces F' in the loaded structure of Fig. 11.15*a* are obtained by adding the member end forces F_0 on the restrained structure of Fig. 11.15*b* to those forces F resulting from the matrix-displacement analysis of the structure of Fig. 11.15*c* or $F' = F_0 + F$. This is the same

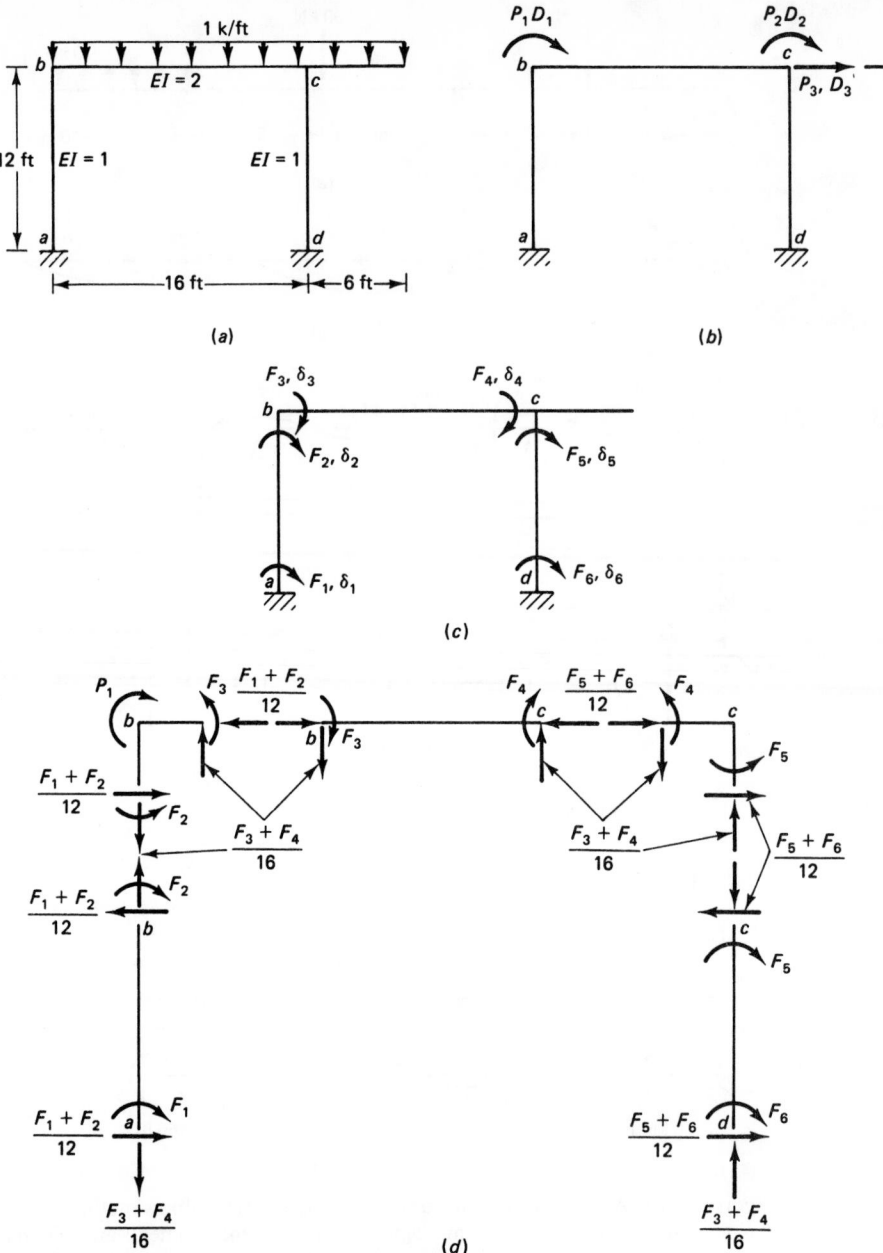

Figure 11.12 Analysis of a rigid frame by matrix stiffness method. (b) P, D diagram. (c) F, δ diagram. (d) Free-body diagram of joints and members.

concept that was used earlier in the slope-deflection method (Section 10.1). It should also be observed that some of the equivalent joint forces in a structure are absorbed directly by the supports. For example, for the continuous beam in Fig. 11.15d, the forces R_1, $(R_2 + R_3)$, and R_4 are carried directly by the supports at a, b, and c and

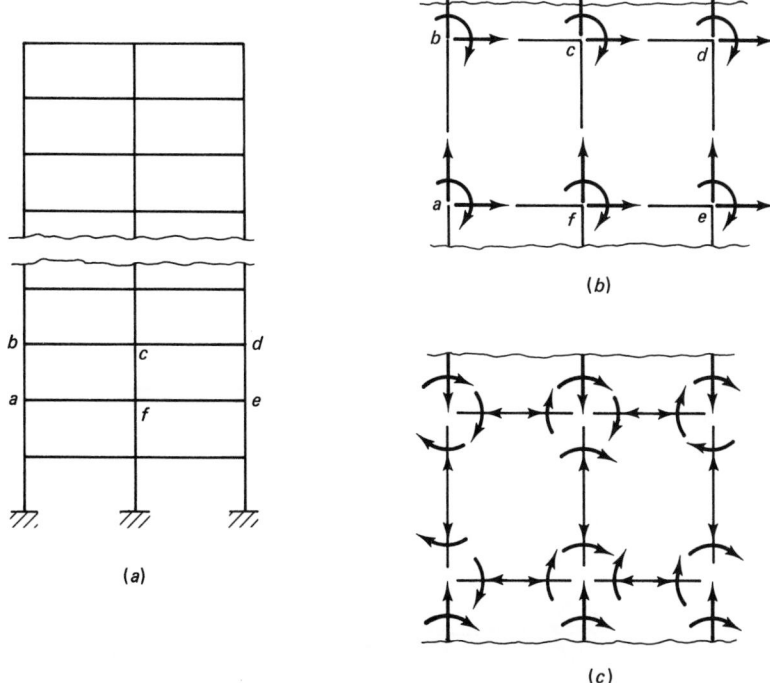

Figure 11.13 Analysis of a multistory frame by matrix stiffness method. (*b*) *P, D* diagram. (*c*) *F, δ* diagram.

therefore cause no displacement in the structure; while the joint moments F_{01}, (F_{03} $- F_{02}$), and F_{04} do incur displacements resulting in moments F at the end of members *ab, bc,* and *cd*. Such end moments give rise to shear forces V at ends of the members as shown, for example, in Fig. 11.11*c*. These shear forces are not independent since they equal the sum of the end moments divided by the member length. Thus, in a structure, a beam element with an infinite axial stiffness has two unknown end moments F; once these are known, the internal forces along the member can be completely determined from statics.

11.3.2 The Static-Equilibrium Matrix E

The matrix **E** expresses the joint forces P in terms of the end moments of members meeting at that joint. Consider the beam in Fig. 11.11*a*. From the joint equilibrium diagrams of Fig. 11.11*d* one can write the following equilibrium equations:

$$P_1 = F_1$$
$$P_2 = F_2 + F_3$$
$$P_3 = F_4 + F_5 \qquad\qquad (11.34)$$
$$P_4 = F_6$$

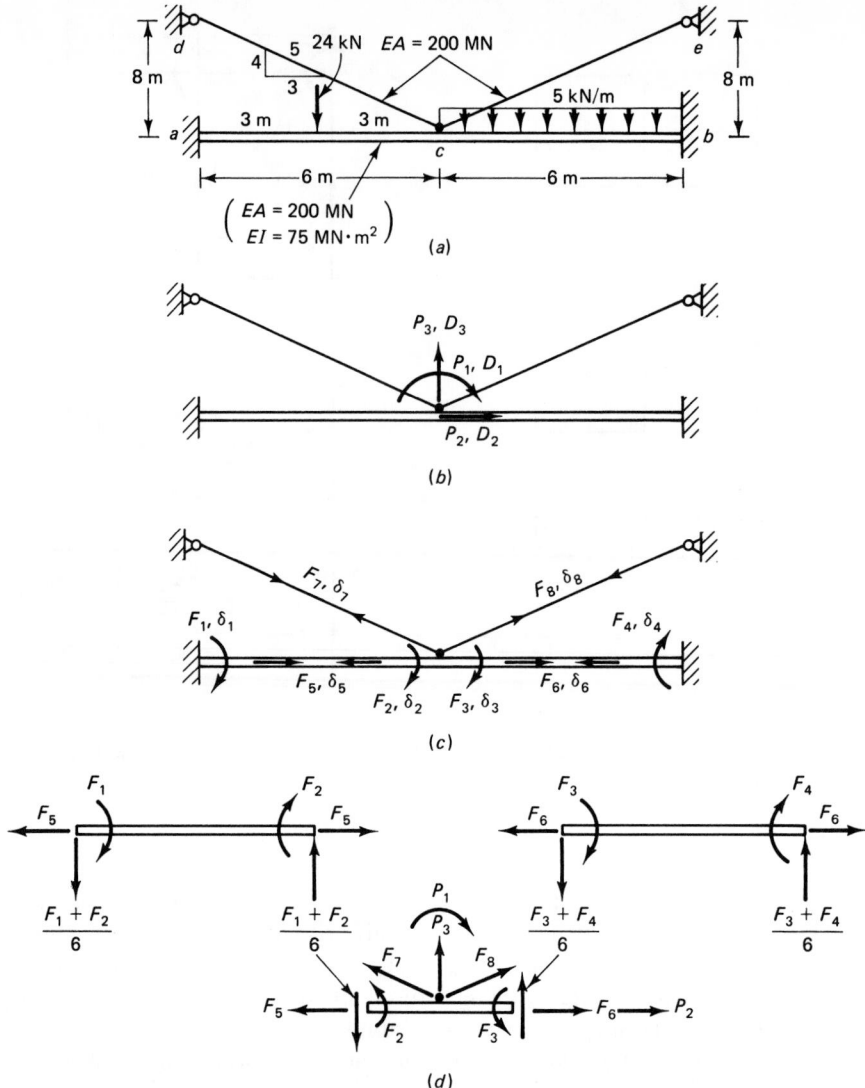

Figure 11.14 Analysis of a composite structure by matrix stiffness method. (b) P, D diagram. (c) F, δ diagram. (d) Free-body diagrams.

In matrix form $\mathbf{P} = \mathbf{EF}$ where $\mathbf{E}$ is the static-equilibrium matrix. Written in full,

$$
\begin{Bmatrix} P_1 \\ P_2 \\ P_3 \\ P_4 \end{Bmatrix} = \begin{bmatrix} 1 & 0 & 0 & 0 & 0 & 0 \\ 0 & 1 & 1 & 0 & 0 & 0 \\ 0 & 0 & 0 & 1 & 1 & 0 \\ 0 & 0 & 0 & 0 & 0 & 1 \end{bmatrix} \begin{bmatrix} F_1 \\ F_2 \\ F_3 \\ F_4 \\ F_5 \\ F_6 \end{bmatrix}
\tag{11.35}
$$

$\underset{\text{Matrix } \mathbf{E}}{\nearrow}$

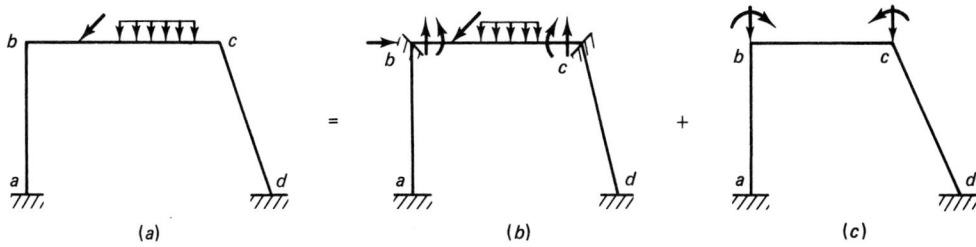

(a)   (b)   (c)

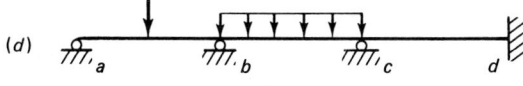

(d)

=

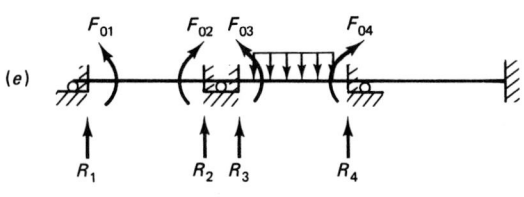

(e)

+

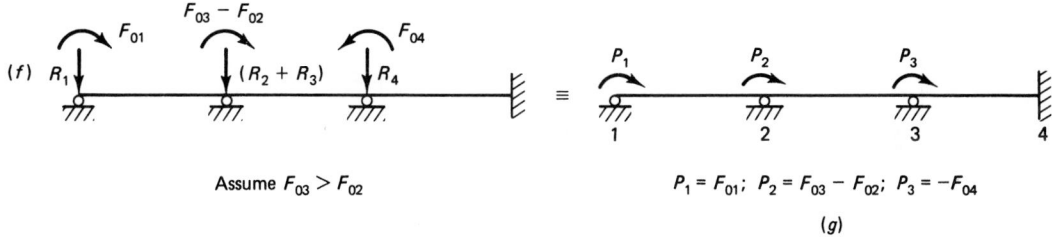

Assume $F_{03} > F_{02}$

$P_1 = F_{01};\ P_2 = F_{03} - F_{02};\ P_3 = -F_{04}$

(g)

Figure 11.15 Treatment of loads applied between joints.

■ **EXAMPLE 11.7**

Determine the **E** matrix for the structure in Fig. 11.12a.

Solution. The free-body diagrams for joints and members are shown in Fig. 11.12d. Thus, applying the equations of equilibrium at joints b and c yields

$$P_1 = F_2 + F_3$$

$$P_2 = F_4 + F_5$$

$$P_3 = -\left(\frac{F_1 + F_2}{12}\right) - \left(\frac{F_5 + F_6}{12}\right)$$

Hence for **P** = **EF**,

$$
\begin{Bmatrix} P_1 \\ P_2 \\ P_3 \end{Bmatrix} = \begin{bmatrix} 0 & 1 & 1 & 0 & 0 & 0 \\ 0 & 0 & 0 & 1 & 1 & 0 \\ -\frac{1}{12} & -\frac{1}{12} & 0 & 0 & -\frac{1}{12} & -\frac{1}{12} \end{bmatrix} \begin{Bmatrix} F_1 \\ F_2 \\ F_3 \\ F_4 \\ F_5 \\ F_6 \end{Bmatrix} \qquad (11.36)
$$

Matrix **E**   ∎

■ EXAMPLE 11.8

Deduce the **E** matrix for the structure in Fig. 11.14*a*.

Solution. Using the free-body diagrams in Fig. 11.14*d* and applying the equilibrium equations results in

$$P_1 = F_2 + F_3$$

$$P_2 = F_5 - F_6 + \tfrac{3}{5}F_7 - \tfrac{3}{5}F_8$$

$$P_3 = \left(\frac{F_1 + F_2}{6}\right) - \left(\frac{F_3 + F_4}{6}\right) - \frac{4}{5}F_7 - \frac{4}{5}F_8$$

In matrix form,

$$
\begin{Bmatrix} P_1 \\ P_2 \\ P_3 \end{Bmatrix} = \begin{bmatrix} 0 & 1 & 1 & 0 & 0 & 0 & 0 & 0 \\ 0 & 0 & 0 & 0 & 1 & -1 & \frac{3}{5} & -\frac{3}{5} \\ \frac{1}{6} & \frac{1}{6} & -\frac{1}{6} & -\frac{1}{6} & 0 & 0 & -\frac{4}{5} & -\frac{4}{5} \end{bmatrix} \begin{Bmatrix} F_1 \\ F_2 \\ F_3 \\ F_4 \\ F_5 \\ F_6 \\ F_7 \\ F_8 \end{Bmatrix} \qquad (11.37)
$$

Matrix **E**   ∎

11.3.3 The Deformation Compatibility Matrix C

This matrix expresses the member end rotations δ in terms of the joint displacements D, that is, $\delta = \mathbf{CD}$, from Eq. (11.2). For the continuous beam in Fig. 11.11*a*, and using the joint displacements shown in Fig. 11.16*a* to *d*, one can write

$$\delta_1 = D_1$$

$$\delta_2 = D_2$$

$$\delta_3 = D_2$$

$$\delta_4 = D_3$$

$$\delta_5 = D_3$$

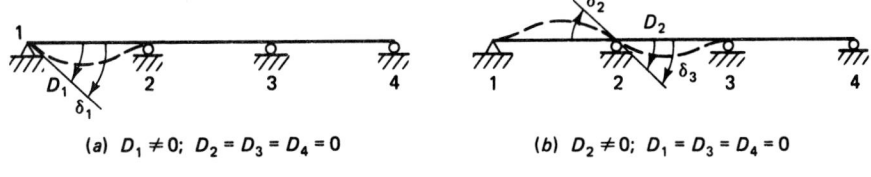

(a) $D_1 \neq 0$; $D_2 = D_3 = D_4 = 0$ (b) $D_2 \neq 0$; $D_1 = D_3 = D_4 = 0$

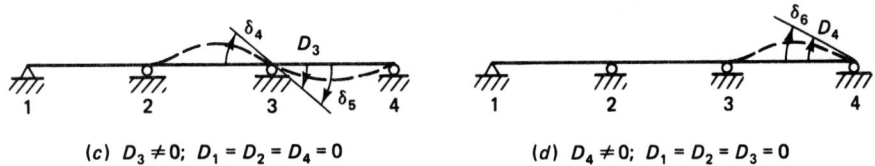

(c) $D_3 \neq 0$; $D_1 = D_2 = D_4 = 0$ (d) $D_4 \neq 0$; $D_1 = D_2 = D_3 = 0$

Figure 11.16 Development of deformation compatibility matrix **C** for a continuous beam.

and

$$\delta_6 = D_4$$

In matrix form,

$$
\begin{Bmatrix} \delta_1 \\ \delta_2 \\ \delta_3 \\ \delta_4 \\ \delta_5 \\ \delta_6 \end{Bmatrix} =
\begin{bmatrix}
1 & 0 & 0 & 0 \\
0 & 1 & 0 & 0 \\
0 & 1 & 0 & 0 \\
0 & 0 & 1 & 0 \\
0 & 0 & 1 & 0 \\
0 & 0 & 0 & 1
\end{bmatrix}
\begin{Bmatrix} D_1 \\ D_2 \\ D_3 \\ D_4 \end{Bmatrix}
\qquad (11.38)
$$

Matrix **C**

Comparing matrix **C** with matrix **E** in Eq. (11.35) again shows that

$$\mathbf{C} = \mathbf{E}^T \qquad\qquad (11.14)$$

■ EXAMPLE 11.9

Deduce the matrix **C** for the rigid frame in Fig. 11.12a.

Solution. As a result of the three joint displacements D shown in Fig. 11.17a to c, the following relations can be written between the end rotations δ and D:
From Fig. 11.17a:

$$\delta_2 = D_1 \qquad \delta_3 = D_1$$

From Fig. 11.17b:

$$\delta_4 = D_2 \qquad \delta_5 = D_2$$

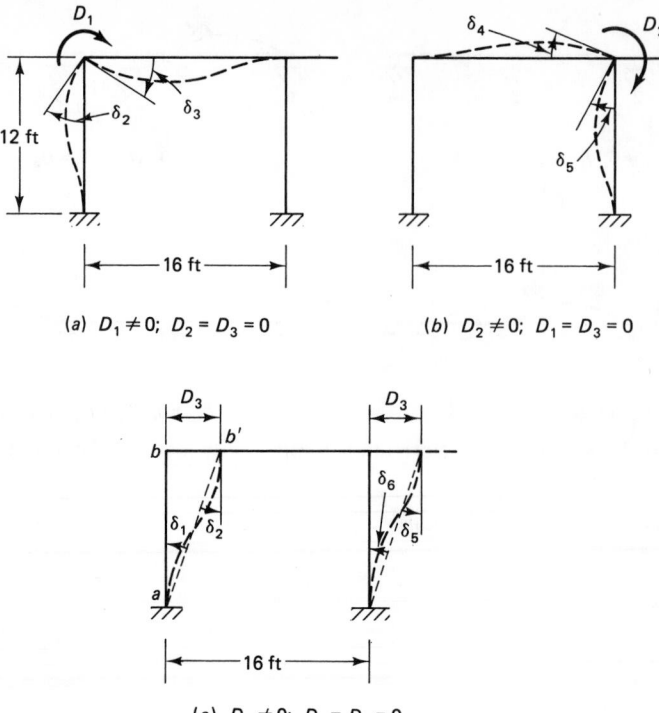

(a) $D_1 \neq 0; \ D_2 = D_3 = 0$ (b) $D_2 \neq 0; \ D_1 = D_3 = 0$

(c) $D_3 \neq 0; \ D_1 = D_2 = 0$

Figure 11.17 Structure for Example 11.9.

and from Fig. 11.17c: Member axis ab has rotated through a clockwise angle of $D_3/12$ to position ab', and member-end rotation δ is therefore measured from the sloped straight line ab' to the vertical tangent to the elastic curve; that is, δ is counterclockwise. Thus $\delta_1 = \delta_2 = -D_3/12$; similarly $\delta_5 = \delta_6 = -D_3/12$.
In matrix form,

$$
\begin{Bmatrix} \delta_1 \\ \delta_2 \\ \delta_3 \\ \delta_4 \\ \delta_5 \\ \delta_6 \end{Bmatrix} =
\begin{bmatrix}
0 & 0 & -\frac{1}{12} \\
1 & 0 & -\frac{1}{12} \\
1 & 0 & 0 \\
0 & 1 & 0 \\
0 & 1 & -\frac{1}{12} \\
0 & 0 & -\frac{1}{12}
\end{bmatrix}
\begin{Bmatrix} D_1 \\ D_2 \\ D_3 \end{Bmatrix}
\tag{11.39}
$$

Matrix **C**

Comparing this matrix **C** with matrix **E** in Eq. (11.36) again shows that $\mathbf{C} = \mathbf{E}^T$.   ■

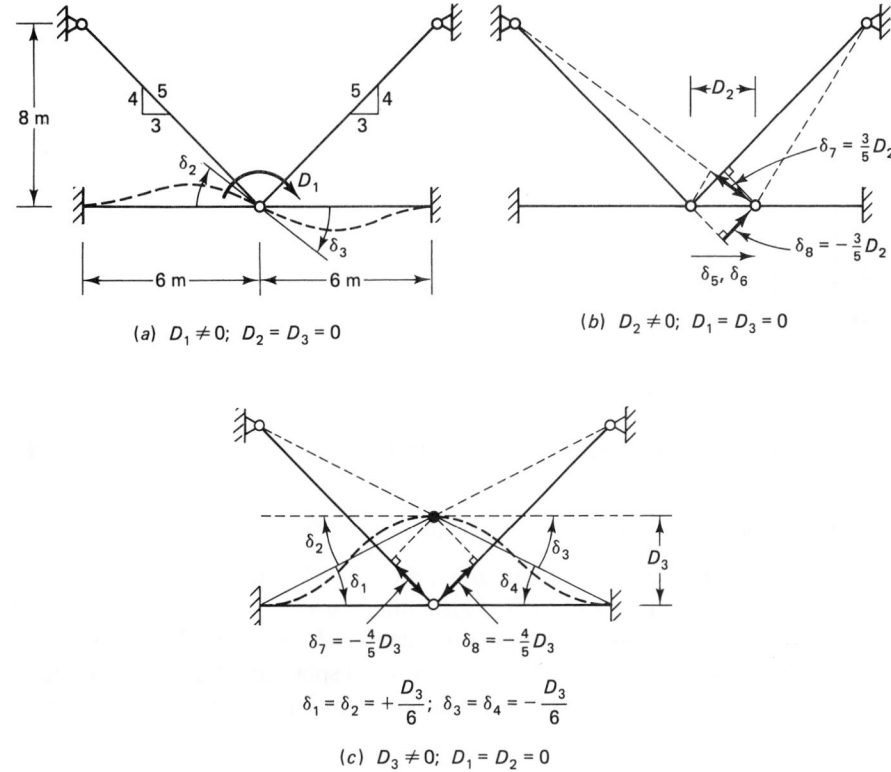

Figure 11.18 Structure for Example 11.10.

■ EXAMPLE 11.10
Determine the matrix $\mathbf{C}$ for the structure in Fig. 11.14a.

Solution. This structure has three degrees of freedom as shown in Fig. 11.14b. From the joint displacement diagrams in Fig. 11.18, one can deduce the following relations between the member end displacement δ and the joint displacement D. Thus, from Fig. 11.18a, due to joint displacement D_1:

$$\delta_2 = \delta_3 = D_1$$

From Fig. 11.18b, due to joint displacement D_2:

$$\delta_5 = D_2 \qquad \delta_6 = -D_2 \qquad \delta_7 = \tfrac{3}{5}D_2 \qquad \delta_8 = -\tfrac{3}{5}D_2$$

and from Fig. 11.18c, due to joint displacement D_3:

$$\delta_1 = \delta_2 = +\frac{D_3}{6} \qquad \delta_3 = \delta_4 = -\frac{D_3}{6}$$

$$\delta_7 = -\tfrac{4}{5}D_3 \qquad \delta_8 = -\tfrac{4}{5}D_3$$

In matrix form,

$$
\begin{Bmatrix} \delta_1 \\ \delta_2 \\ \delta_3 \\ \delta_4 \\ \delta_5 \\ \delta_6 \\ \delta_7 \\ \delta_8 \end{Bmatrix} = \begin{bmatrix} 0 & 0 & +\frac{1}{6} \\ 1 & 0 & +\frac{1}{6} \\ 1 & 0 & -\frac{1}{6} \\ 0 & 0 & -\frac{1}{6} \\ 0 & 1 & 0 \\ 0 & -1 & 0 \\ 0 & \frac{3}{5} & -\frac{4}{5} \\ 0 & -\frac{3}{5} & -\frac{4}{5} \end{bmatrix} \begin{Bmatrix} D_1 \\ D_2 \\ D_3 \end{Bmatrix} \tag{11.40}
$$

Matrix C

This matrix **C** is the transpose of matrix **E** in Eq. (11.37). ∎

11.3.4 The Member Stiffness Matrix S

The end rotations δ of a member are related to the end moments F by means of the member stiffness matrix **S**. This stiffness was introduced earlier when the method of slope deflection was discussed in Chapter 10, in which the end moments were related to the end rotations and the transverse displacement of one end relative to the other end [see Eq. (10.4)]. Applying these relations to an unloaded deformed beam element in Fig. 11.19a, one can write

$$
\begin{aligned}
M_{ab} &= \frac{2EI}{l}\left(2\theta_A + \theta_B - \frac{3\Delta}{l}\right) = \frac{2EI}{l}\left[2\left(\theta_A - \frac{\Delta}{l}\right) + \left(\theta_B - \frac{\Delta}{l}\right)\right] \\
M_{ba} &= \frac{2EI}{l}\left(2\theta_B + \theta_A - \frac{3\Delta}{l}\right) = \frac{2EI}{l}\left[2\left(\theta_B - \frac{\Delta}{l}\right) + \left(\theta_A - \frac{\Delta}{l}\right)\right]
\end{aligned} \tag{11.41}
$$

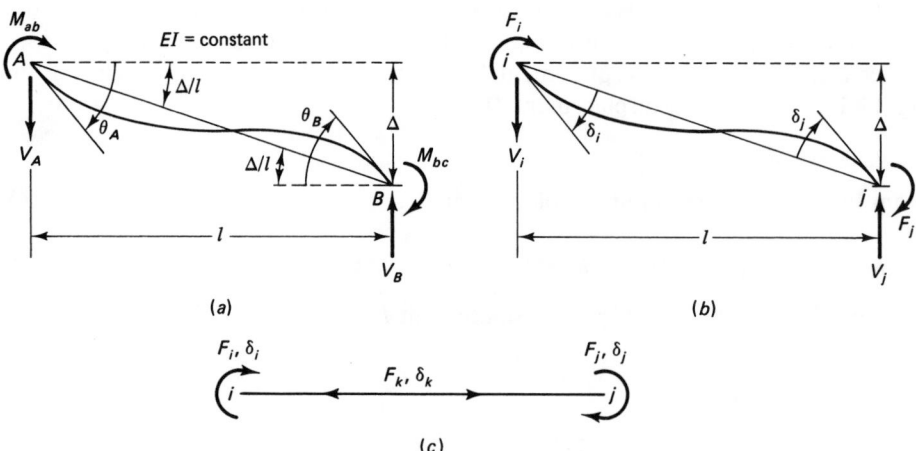

(a) (b)

(c)

Figure 11.19 Development of member stiffness matrix **S**.

In Fig. 11.19b, the symbols F_i and F_j are used for M_{ab} and M_{ba}, respectively, and $\delta_i = (\theta_A - \Delta/l)$, $\delta_j = (\theta_B - \Delta/l)$ in which δ_i and δ_j are the clockwise end rotations measured from the displaced straight segment ij to the elastic curve at i and j, respectively. Thus Eq. (11.41) becomes

$$F_i = \frac{4EI}{l}\delta_i + \frac{2EI}{l}\cdot\delta_j$$

$$F_j = \frac{2EI}{l}\delta_i + \frac{4EI}{l}\cdot\delta_j$$

(11.42)

In matrix form, $\mathbf{F} = \mathbf{S}\boldsymbol{\delta}$:

$$\begin{Bmatrix} F_i \\ F_j \end{Bmatrix} = \begin{bmatrix} \dfrac{4EI}{l} & \dfrac{2EI}{l} \\ \dfrac{2EI}{l} & \dfrac{4EI}{l} \end{bmatrix} \begin{Bmatrix} \delta_i \\ \delta_j \end{Bmatrix}$$

(11.43)

Hence

$$\mathbf{S} = \begin{bmatrix} \dfrac{4EI}{l} & \dfrac{2EI}{l} \\ \dfrac{2EI}{l} & \dfrac{4EI}{l} \end{bmatrix}$$

(11.44)

If member ij is subjected to an axial force F_k producing elongation δ_k, as shown in Fig. 11.19c, then matrix $\mathbf{S}$ can be established from

$$\begin{Bmatrix} F_i \\ F_j \\ F_k \end{Bmatrix} = \begin{bmatrix} \dfrac{4EI}{l} & \dfrac{2EI}{l} & 0 \\ \dfrac{2EI}{l} & \dfrac{4EI}{l} & 0 \\ 0 & 0 & \dfrac{EA}{l} \end{bmatrix} \begin{Bmatrix} \delta_i \\ \delta_j \\ \delta_k \end{Bmatrix}$$

(11.45)

Member stiffness matrix $\mathbf{S}$

in which EA/l is the axial stiffness of member ij. Now Eq. (11.43) can be solved to yield δ_i and δ_j in the matrix form $\boldsymbol{\delta} = \mathbf{f}\mathbf{F}$, as follows:

$$\begin{Bmatrix} \delta_i \\ \delta_j \end{Bmatrix} = \begin{bmatrix} \dfrac{l}{3EI} & -\dfrac{l}{6EI} \\ \dfrac{-l}{6EI} & \dfrac{l}{3EI} \end{bmatrix} \begin{Bmatrix} F_i \\ F_j \end{Bmatrix}$$

(11.46)

in which

$$
\mathbf{f} = \begin{bmatrix} \dfrac{l}{3EI} & -\dfrac{l}{6EI} \\[2ex] \dfrac{-l}{6EI} & \dfrac{l}{3EI} \end{bmatrix}
\tag{11.47}
$$

is the member flexibility matrix.

■ EXAMPLE 11.11
Determine the member stiffness matrix **S** of the continuous beam in Fig. 11.11a.

Solution.   Referring to Fig. 11.11c and using Eq. (11.43) (i.e., $\mathbf{F} = \mathbf{S}\boldsymbol{\delta}$) for each member yields

$$
\begin{Bmatrix} F_1 \\ F_2 \\ F_3 \\ F_4 \\ F_5 \\ F_6 \end{Bmatrix} = 2EI
\begin{bmatrix}
\frac{2}{10} & \frac{1}{10} & 0 & 0 & 0 & 0 \\
\frac{1}{10} & \frac{2}{10} & 0 & 0 & 0 & 0 \\
0 & 0 & \frac{2}{10} & \frac{1}{10} & 0 & 0 \\
0 & 0 & \frac{1}{10} & \frac{2}{10} & 0 & 0 \\
0 & 0 & 0 & 0 & \frac{2}{10} & \frac{1}{10} \\
0 & 0 & 0 & 0 & \frac{1}{10} & \frac{2}{10}
\end{bmatrix}
\begin{Bmatrix} \delta_1 \\ \delta_2 \\ \delta_3 \\ \delta_4 \\ \delta_5 \\ \delta_6 \end{Bmatrix}
\tag{11.48}
$$

Hence the required member stiffness matrix **S**.   ■

■ EXAMPLE 11.12
Deduce the matrix **S** of the rigid frame in Fig. 11.12a.

Solution.   Using Eq. (11.43) (i.e., $\mathbf{F} = \mathbf{S}\boldsymbol{\delta}$) and Fig. 11.12c, one can deduce **S** as follows:

$$
\begin{Bmatrix} F_1 \\ F_2 \\ F_3 \\ F_4 \\ F_5 \\ F_6 \end{Bmatrix} = 2
\begin{bmatrix}
\frac{2}{12} & \frac{1}{12} & 0 & 0 & 0 & 0 \\
\frac{1}{12} & \frac{2}{12} & 0 & 0 & 0 & 0 \\
0 & 0 & \frac{4}{16} & \frac{2}{16} & 0 & 0 \\
0 & 0 & \frac{2}{16} & \frac{4}{16} & 0 & 0 \\
0 & 0 & 0 & 0 & \frac{2}{12} & \frac{1}{12} \\
0 & 0 & 0 & 0 & \frac{1}{12} & \frac{2}{12}
\end{bmatrix}
\begin{Bmatrix} \delta_1 \\ \delta_2 \\ \delta_3 \\ \delta_4 \\ \delta_5 \\ \delta_6 \end{Bmatrix}
\tag{11.49}
$$

Hence matrix **S**.   ■

■ EXAMPLE 11.13
Establish the member stiffness matrix **S** for the "composite" structure in Fig. 11.14a.

Solution.   Here the members have both flexural and/or axial stiffnesses. Based on Eqs. (11.15) and (11.45), $\mathbf{F} = \mathbf{S}\boldsymbol{\delta}$, and using Fig. 11.14c one can show

$$\begin{Bmatrix} F_1 \\ F_2 \\ F_3 \\ F_4 \\ F_5 \\ F_6 \\ F_7 \\ F_8 \end{Bmatrix} = \begin{bmatrix} 50 & 25 & 0 & 0 & 0 & 0 & 0 & 0 \\ 25 & 50 & 0 & 0 & 0 & 0 & 0 & 0 \\ 0 & 0 & 50 & 25 & 0 & 0 & 0 & 0 \\ 0 & 0 & 25 & 50 & 0 & 0 & 0 & 0 \\ 0 & 0 & 0 & 0 & \frac{100}{3} & 0 & 0 & 0 \\ 0 & 0 & 0 & 0 & 0 & \frac{100}{3} & 0 & 0 \\ 0 & 0 & 0 & 0 & 0 & 0 & 20 & 0 \\ 0 & 0 & 0 & 0 & 0 & 0 & 0 & 20 \end{bmatrix} \begin{Bmatrix} \delta_1 \\ \delta_2 \\ \delta_3 \\ \delta_4 \\ \delta_5 \\ \delta_6 \\ \delta_7 \\ \delta_8 \end{Bmatrix} \qquad (11.50)$$

Hence matrix S.————————→   ∎

11.3.5 The Global Stiffness Matrix S*

The global or external stiffness matrix $\mathbf{S}^*$ relates the external joint forces $\mathbf{P}$ and the joint displacements $\mathbf{D}$ as

$$\mathbf{P} = \mathbf{S}^*\mathbf{D} \qquad (11.18)$$

In relating the matrix-displacement method to the classical slope-deflection method in Chapter 10, it was demonstrated how to generate the matrix $\mathbf{S}^*$ for a loaded continuous beam; see Section 10.1.5 and Fig. 10.6. It is suggested that the reader review Section 10.1.5. In this section, the matrix $\mathbf{S}^*$ will be established using the notation of this chapter. As with trusses, the first column of matrix $\mathbf{S}^*$ can be generated by displacing only joint 1 by an amount $D_1 = 1$, with all other D's set to zero; the resulting external joint forces will be the elements of the first column of $\mathbf{S}^*$. Applying this procedure to the beam in Fig. 11.20a, making $D_1 = 1$ and $D_2 = D_3 = D_4 = 0$, causes end moments F as shown in Fig. 11.20b. Moment equilibrium at joints 1 to 4 will result in the external joint forces P_1, P_2, P_3, and P_4, which are the elements S_{11}^*, S_{21}^*, S_{31}^*, and S_{41}^*, respectively. Following the same procedure for joints 2 to 4 will yield the other elements of $\mathbf{S}^*$ as shown in Fig. 11.20c to e. Thus the matrix $\mathbf{S}^*$ is

$$\mathbf{S}^* = EI \begin{bmatrix} \frac{4}{10} & \frac{2}{10} & 0 & 0 \\ \frac{2}{10} & \frac{8}{10} & \frac{2}{10} & 0 \\ 0 & \frac{2}{10} & \frac{8}{10} & \frac{2}{10} \\ 0 & 0 & \frac{2}{10} & \frac{4}{10} \end{bmatrix}$$

which is a symmetric matrix with respect to the principal diagonal and is identical to the first matrix on the left-hand side of Eq. (i) in Section 10.1.5.

With access to a computer it is simpler to derive $\mathbf{S}^*$ from the relation

$$\mathbf{S}^* = \mathbf{ESE}^T \qquad (11.22)$$

11.3.6 The External Force Matrix P

It was mentioned earlier that the external joint forces P are functions of the restraining forces at the end of members, as shown in Fig. 11.15g. For example, Fig. 11.15g

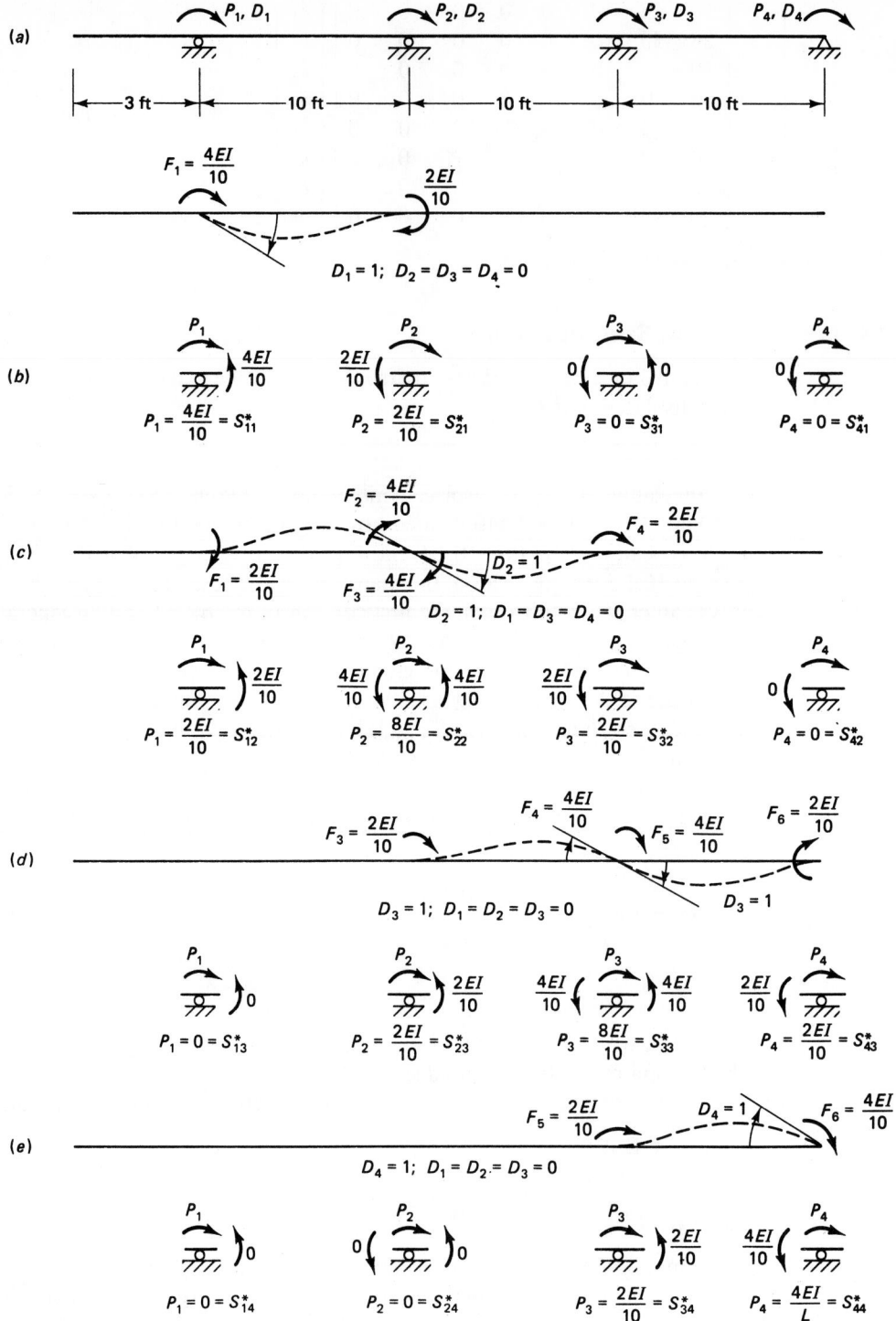

Figure 11.20 Development of global stiffness matrix **S*** for a continuous beam.

shows that the external joint force P_2 at any joint 2 that is likely to rotate is determined by combining the fixed-end moments F_0 acting at joint 2. Taking into account the reversal of sign, in general, one can write

$$P_j = -\sum \text{ fixed-end moments at ends of members meeting at } j \text{th joint} \quad (11.51)$$

In the case of rigid frames with sidesway (Fig. 11.12a), both the external joint moments and the external joint sidesway forces must be determined. First, lock every joint of the structure against rotation and translation by applying fixed-end moments and forces at ends of members that are loaded; then the external joint moments are found from Eq. (11.51). To determine the external joint sidesway forces: (1) Apply the two force equilibrium equations at all joints without the unknown sidesway, e.g., at joint b in Fig. 11.12d. (2) At the joint with the unknown sidesway, for example, joint c in Fig. 11.12d, apply the one force equilibrium equation in the direction perpendicular to that of the joint sidesway force (or unknown sidesway) as, for example, perpendicular to P_3 (or D_3) in Fig. 11.12d; in this case the sidesway force P_3 is given by

$$P_3 = \sum \text{ restraining forces in the direction of } P_3$$

$$\text{acting at joint } c \quad (\text{Fig. } 11.12d) \quad (11.52)$$

For several reasons, the supports of a continuous beam or rigid frame may undergo some movements in the form of rotation and/or linear displacement; such movements must be properly estimated and allowed for by the design engineer. In this case, the analysis proceeds with allowing the particular support(s) to displace an estimated amount and then the resulting external joint movements and sidesway forces are determined to hold all unknown joint displacements to zero. In other words, movement of support only affects the external force matrix **P,** and the rest of the procedure in the matrix-displacement method of determining the internal end moments remains the same. Figure 11.21 presents the fixed-end moments for the more common type of loadings; the fixed-end moments for other loadings can be determined from the Appendix.

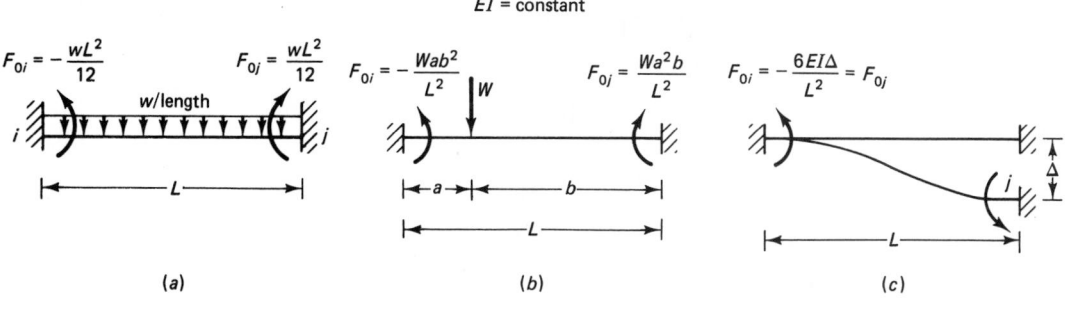

Figure 11.21 Fixed-end moments for beams.

■ **EXAMPLE 11.14**

Deduce the external joint force matrix **P** for the continuous beam in Fig. 11.11*a*.

Solution. The fixed-end moments F_0 are calculated, using the results shown in Fig. 11.21. Considering the moment equilibrium at joints 1 to 4 yields the following **P** matrix:

$$\begin{Bmatrix} P_1 \\ P_2 \\ P_3 \\ P_4 \end{Bmatrix} = \begin{Bmatrix} +250 \\ 0 \\ -166.67 \\ -83.33 \end{Bmatrix} \tag{11.53}$$

■

■ **EXAMPLE 11.15**

Determine the **P** matrix for the continuous beam in Fig. 11.22*a* due to a vertical settlement $\Delta = 5$ mm at support 2. Take $EI = 150 \times 10^3$ kN·m^2.

Solution. The fixed-end moments F_0 are calculated from Fig. 11.21*c*. See also the Appendix. Thus,

$$F_{01} = -\frac{6EI\,\Delta}{l^2} = -\frac{(6)(150{,}000)(5 \times 10^{-3})}{10^2} = -45 \text{ kN·m} = F_{02}$$

similarly

$$F_{03} = F_{04} = +45 \text{ kN·m}$$

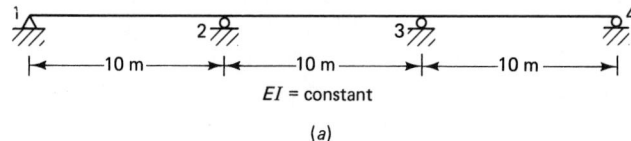

(*a*)

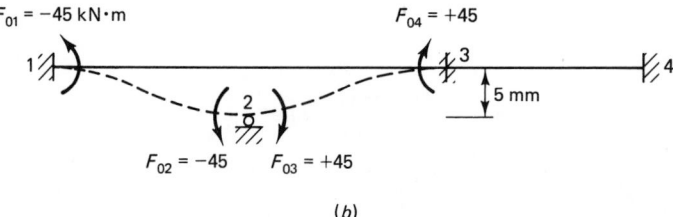

(*b*)

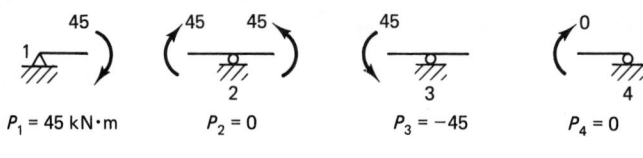

(*c*)

Figure 11.22 Beam for Example 11.15.

These moments are shown in Fig. 11.22b. The values of the **P** matrix are derived from the moment equilibrium at joints 1 to 4 (Fig. 11.22c). Thus, in matrix form

$$\begin{Bmatrix} P_1 \\ P_2 \\ P_3 \\ P_4 \end{Bmatrix} = \begin{Bmatrix} 45 \\ 0 \\ -45 \\ 0 \end{Bmatrix}$$

∎

■ EXAMPLE 11.16
Establish the **P** matrix for the rigid-frame structure in Fig. 11.23a.

Solution. Locking all the joints and applying the load yields the fixed-end moments as well as shears as shown in Fig. 11.23b. Reversing the signs of these forces and applying them to joints b and c (Fig. 11.23c) determines the **P** matrix. With the P numbering given in Fig. 11.12b, one can deduce

$P_1 = +21.33 \text{ k} \cdot \text{ft}$

$P_2 = -3.33 \text{ k} \cdot \text{ft}$

and

$P_3 = 0$ ∎

■ EXAMPLE 11.17
Deduce the **P** matrix for the structure in Fig. 11.24a.

Solution. The P, D diagram is shown in Fig. 11.24b. Locking the joints and loading the structure will produce fixed-end forces as shown in Fig. 11.24c. Reversing the signs of these fixed-end forces and using the free-body diagram of Fig. 11.24e yield the following values for the **P** matrix:

$P_1 = +30 \text{ k} \cdot \text{ft}$

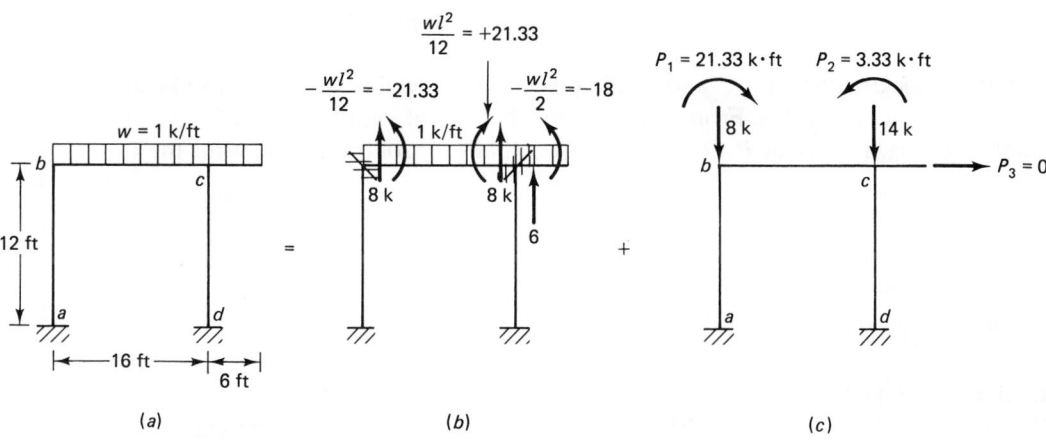

Figure 11.23 Frame for Example 11.16.

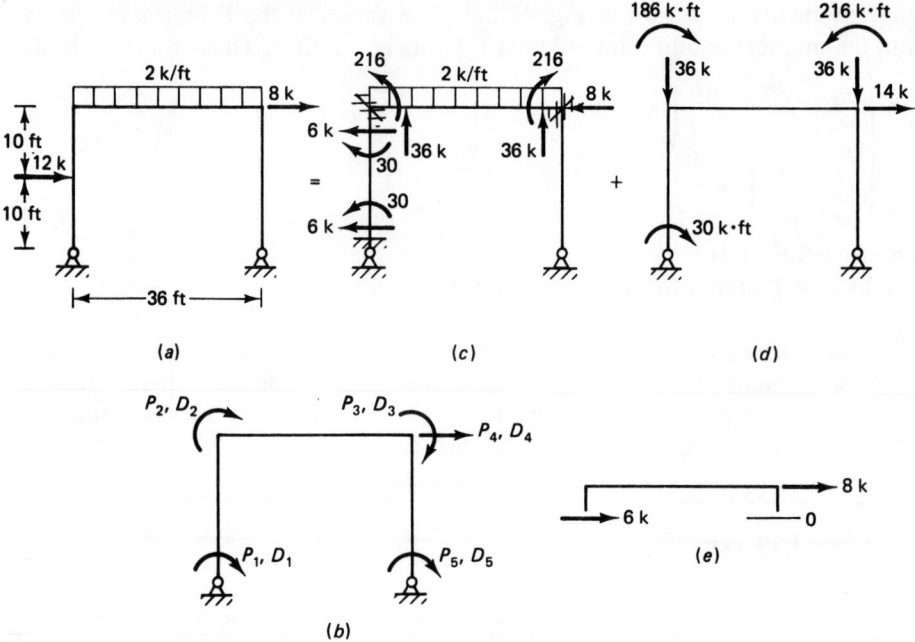

Figure 11.24 Frame for Example 11.17. (*b*) *P-D* diagram.

$P_2 = +186 \text{ k} \cdot \text{ft}$

$P_3 = -216 \text{ k} \cdot \text{ft}$

$P_4 = +14 \text{ k}$

$P_5 = 0$ ∎

■ EXAMPLE 11.18
Determine the **P** matrix for the "composite" structure in Fig. 11.14*a*.

Solution. Locking the joint *C* and loading the structure will generate the fixed-end forces shown in Fig. 11.25*a*. From the free-body diagram of joint *C* (Fig. 11.25*b*) and referring to Fig. 11.14*b* for *P, D* numbering, one finds the following values for the **P** matrix:

$P_1 = -3 \text{ kN} \cdot \text{m}$

$P_2 = 0$

$P_3 = -27 \text{ kN}$ ∎

■ EXAMPLE 11.19
Determine (a) the **P** matrix and (b) the static-equilibrium matrix **E** for the rigid frame in Fig. 11.26*a*.

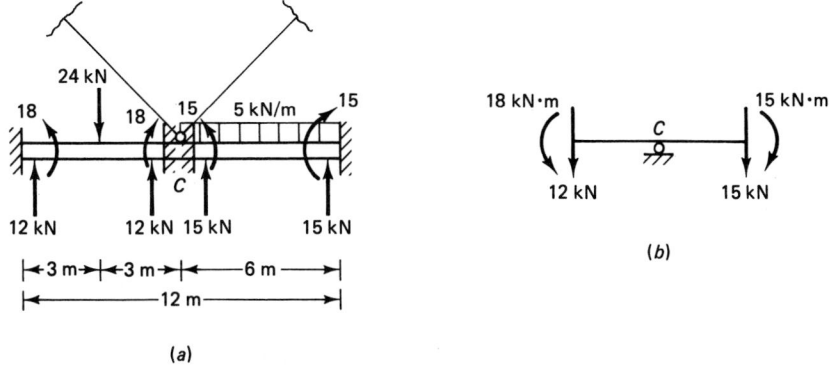

Figure 11.25 Composite structure for Example 11.18.

Solution

(a) This structure was analyzed in Section 10.1.3 (Fig. 10.8). Here we apply the procedure stated earlier in the chapter with regard to establishing the **P** matrix. The *P, D* numbering is shown in Fig. 11.26*b*. First the fixed-end moments acting on member *BC* are determined as shown in Fig. 11.26*d* (see also the Appendix). Thus,

$P_1 = -\sum$ fixed-end moments at ends of members *BA* and *BC*

$= -(-80 + 0) = 80 \text{ k} \cdot \text{ft}$

$P_2 = -(320 + 0) = -320 \text{ k} \cdot \text{ft}$

Since no sidesway is permitted at joint *B*, both equilibrium equations $\sum F_x = 0$ and $\sum F_y = 0$ must be satisfied at joint *B*, which is shown in Fig. 11.26*d*. Also, since one degree of freedom in sidesway (D_3) is allowed at joint *C* in the horizontal direction, only one equilibrium equation ($\sum F_y = 0$) should be satisfied at joint *C*; this is also shown in Fig. 11.26*d*. (Note $H_1 \neq H_2$.) Thus the sidesway force P_3 is

$P_3 = \sum$ horizontal components of all forces in the *x* direction acting at joint *C*

$= 15 + H_1 - H_2$

From the free-body diagram of *AB* in Fig. 11.26*d*:

$H_1 = \dfrac{(26)(5)}{12} = 10.833 \text{ k}$

From the free-body diagram of *CD* in Fig. 11.26*d*:

$H_2 = \dfrac{(224)(3.5)}{12} = 65.333 \text{ k}$

Hence $P_3 = 15 + 10.833 - 65.333 = -39.50$ k. Thus the **P** matrix is

$P_1 = 80 \text{ k} \cdot \text{ft}$

$P_2 = -320 \text{ k} \cdot \text{ft}$

$P_3 = -39.50 \text{ k}$

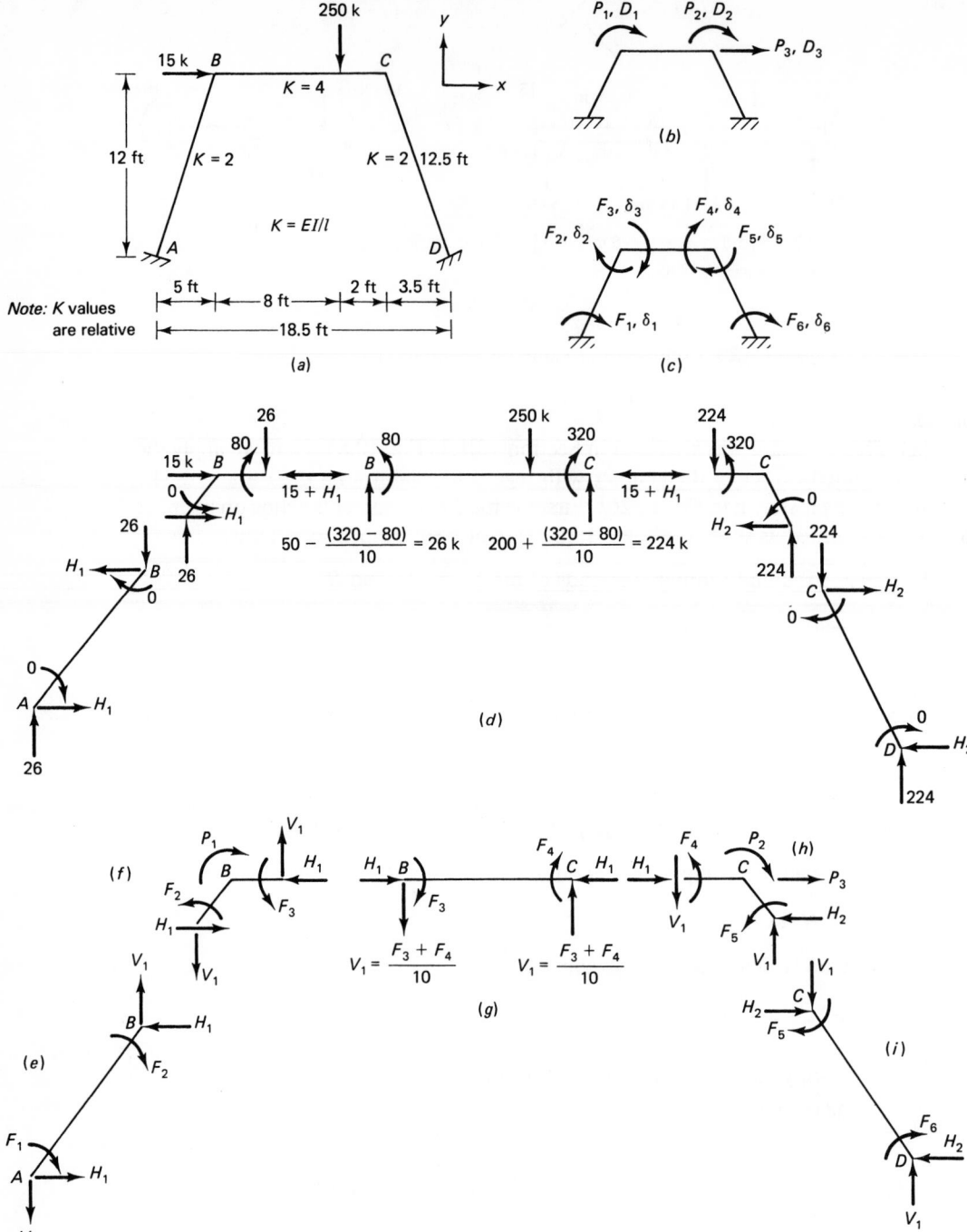

Figure 11.26 Frame for Example 11.19. (*b*) *P, D* numbers. (*c*) *F, δ* numbers.

(b) The static-equilibrium matrix $\mathbf{E}$ is deduced by expressing the joint forces P in terms of the internal forces F, or $\mathbf{P} = \mathbf{EF}$. Referring to Fig. 11.26f, moment equilibrium of joint B leads to

$$P_1 = F_2 + F_3 \tag{a}$$

Similarly, from Fig. 11.26h,

$$P_2 = F_4 + F_5 \tag{b}$$

The joint force P_3 is determined from the force-equilibrium equation in the direction of P_3 at joint C. Here again, both $\Sigma F_x = 0$ and $\Sigma F_y = 0$ must be satisfied at joint B; this is done by the pair of equal and opposite forces H_1 and V_1 acting on joint B and on ends of members BA and BC, as shown in Fig. 11.26e, f, and g. From the free-body diagram of AB in Fig. 11.26e,

$$-5V_1 - 12H_1 + F_1 + F_2 = 0$$

or

$$H_1 = \frac{-5V_1 + F_1 + F_2}{12}$$

From the free-body diagram of BC in Fig. 11.26g

$$V_1 = \frac{F_3 + F_4}{10}$$

Therefore,

$$H_1 = \tfrac{1}{12}(F_1 + F_2) - \tfrac{1}{24}(F_3 + F_4)$$

At joint C only $\Sigma F_y = 0$ should be satisfied; this is accomplished by the pair of equal and opposite forces V_1 at joint C. Taking member CD as a free body (Fig. 11.26i),

$$-3.5V_1 + 12H_2 + F_5 + F_6 = 0$$

Substituting for V_1 yields

$$H_2 = \frac{3.5}{120}(F_3 + F_4) - \frac{1}{12}(F_5 + F_6)$$

Taking joint C as a free body (Fig. 11.26h)

$$P_3 + H_1 - H_2 = 0$$

or

$$P_3 = -H_1 + H_2$$

Substituting for H_1 and H_2

$$P_3 = -\tfrac{1}{12}(F_1 + F_2) + \tfrac{17}{240}(F_3 + F_4) - \tfrac{1}{12}(F_5 + F_6) \tag{c}$$

Putting expressions (a) to (c) in matrix form:

$$
\begin{Bmatrix} P_1 \\ P_2 \\ P_3 \end{Bmatrix} = \begin{bmatrix} 0 & 1 & 1 & 0 & 0 & 0 \\ 0 & 0 & 0 & 1 & 1 & 0 \\ -\frac{1}{12} & -\frac{1}{12} & +\frac{17}{240} & +\frac{17}{240} & -\frac{1}{12} & -\frac{1}{12} \end{bmatrix} \begin{Bmatrix} F_1 \\ F_2 \\ F_3 \\ F_4 \\ F_5 \\ F_6 \end{Bmatrix}
\tag{11.54}
$$

Hence matrix **E**

The complete analysis of this rigid frame is performed as Solved Problem 11.6. ∎

■ EXAMPLE 11.20
Determine the support moments for the continuous beam in Fig. 11.11a.

Solution. The matrices **E**, **S**, and **P** for this structure are given by Eqs. (11.35), (11.48), and (11.53), respectively. Thus

$$
\mathbf{SE}^T = \frac{2EI}{10} \begin{bmatrix} 2 & 1 & 0 & 0 & 0 & 0 \\ 1 & 2 & 0 & 0 & 0 & 0 \\ 0 & 0 & 2 & 1 & 0 & 0 \\ 0 & 0 & 1 & 2 & 0 & 0 \\ 0 & 0 & 0 & 0 & 2 & 1 \\ 0 & 0 & 0 & 0 & 1 & 2 \end{bmatrix} \begin{bmatrix} 1 & 0 & 0 & 0 \\ 0 & 1 & 0 & 0 \\ 0 & 1 & 0 & 0 \\ 0 & 0 & 1 & 0 \\ 0 & 0 & 1 & 0 \\ 0 & 0 & 0 & 1 \end{bmatrix} = \frac{2EI}{10} \begin{bmatrix} 2 & 1 & 0 & 0 \\ 1 & 2 & 0 & 0 \\ 0 & 2 & 1 & 0 \\ 0 & 1 & 2 & 0 \\ 0 & 0 & 2 & 1 \\ 0 & 0 & 1 & 2 \end{bmatrix}
$$

and

$$
\mathbf{ESE}^T = \mathbf{S}^* = \frac{2EI}{10} \begin{bmatrix} 2 & 1 & 0 & 0 \\ 1 & 4 & 1 & 0 \\ 0 & 1 & 4 & 1 \\ 0 & 0 & 1 & 2 \end{bmatrix}
$$

thus

$$
[\mathbf{S}^*]^{-1} = \frac{5}{EI} \begin{bmatrix} 0.578 & -0.156 & 0.044 & -0.022 \\ -0.156 & 0.311 & -0.089 & 0.044 \\ 0.044 & -0.089 & 0.311 & -0.156 \\ -0.022 & 0.044 & -0.156 & 0.578 \end{bmatrix}
$$

From Eq. (11.24)

$$
\mathbf{D} = [\mathbf{S}^*]^{-1} \cdot \mathbf{P} = \frac{5}{EI} \begin{bmatrix} 0.578 & -0.156 & 0.044 & -0.022 \\ -0.156 & 0.311 & -0.089 & 0.044 \\ 0.044 & -0.089 & 0.311 & -0.156 \\ -0.022 & 0.044 & -0.156 & 0.578 \end{bmatrix} \begin{Bmatrix} 250.0 \\ 0 \\ -166.67 \\ -83.33 \end{Bmatrix}
$$

$$
= \frac{5}{EI} \begin{bmatrix} 138.88 \\ -27.78 \\ -27.78 \\ -27.78 \end{bmatrix} \begin{matrix} D_1 \\ D_2 \\ D_3 \\ D_4 \end{matrix} \quad \text{radians}
$$

Hence, from $\mathbf{F} = \mathbf{SE}^T\mathbf{D}$

$$\begin{Bmatrix} F_1 \\ F_2 \\ F_3 \\ F_4 \\ F_5 \\ F_6 \end{Bmatrix} = \begin{Bmatrix} 250.00 \\ 83.33 \\ -83.33 \\ -83.33 \\ -83.33 \\ -83.33 \end{Bmatrix} \quad \text{kN} \cdot \text{m}$$

Final moments $\mathbf{F}' = \mathbf{F}_0 + \mathbf{F}$, in which the fixed-end moments are shown in Fig. 11.11e. Thus,

$\mathbf{F}_0$ (from the Appendix)

$$\mathbf{F}' = \begin{Bmatrix} -250.00 \\ +250.00 \\ -250.00 \\ +250.00 \\ -83.33 \\ +83.33 \end{Bmatrix} + \begin{Bmatrix} 250.00 \\ 83.33 \\ -83.33 \\ -83.33 \\ -83.33 \\ -83.33 \end{Bmatrix} = \begin{Bmatrix} 0 \\ 333.33 \\ -333.33 \\ 166.67 \\ -166.67 \\ 0 \end{Bmatrix} \quad \text{kN} \cdot \text{m} \qquad \blacksquare$$

■ **EXAMPLE 11.21**

Deduce the support moments for the continuous beam in Fig. 11.22a when support 2 settles down a distance of 5 mm.

Solution. This beam has the same geometry as the one in Fig. 11.11a. Since the matrices $\mathbf{E}$, $\mathbf{S}$, and $\mathbf{S}^*$ are functions of the geometry of the structure only, these are identical for the two structures. Therefore, from Example 11.20,

$$[\mathbf{S}^*]^{-1} = \frac{5}{EI} \begin{bmatrix} 0.578 & -0.156 & 0.044 & -0.022 \\ -0.156 & 0.311 & -0.089 & 0.044 \\ 0.044 & -0.089 & 0.311 & -0.156 \\ -0.022 & 0.044 & -0.156 & 0.578 \end{bmatrix}$$

From Fig. 11.22c, it was deduced earlier (Example 11.15) that the matrix

$$P = \begin{Bmatrix} 45 \\ 0 \\ -45 \\ 0 \end{Bmatrix}$$

Thus, from Eq. (11.24),

$$\mathbf{D} = [\mathbf{S}^*]^{-1} \cdot \mathbf{P} = \frac{5}{EI} \begin{Bmatrix} 24.0 \\ -3.0 \\ -12.0 \\ 6.0 \end{Bmatrix} \begin{matrix} D_1 \\ D_2 \\ D_3 \\ D_4 \end{matrix} \quad \text{radians}$$

From $\mathbf{F} = \mathbf{SE}^T\mathbf{D}$,

$$\begin{Bmatrix} F_1 \\ F_2 \\ F_3 \\ F_4 \\ F_5 \\ F_6 \end{Bmatrix} = \begin{Bmatrix} 45.0 \\ 18.0 \\ -18.0 \\ -27.0 \\ -18.0 \\ 0 \end{Bmatrix} \quad \text{kN} \cdot \text{m}$$

The values of the fixed-end moments F_0 are given in Fig. 11.22 b. Thus the final moments $\mathbf{F}' = \mathbf{F}_0 + \mathbf{F}$ are

$$\mathbf{F}' = \overset{\mathbf{F}_0}{\begin{Bmatrix} -45.0 \\ -45.0 \\ +45.0 \\ +45.0 \\ 0 \\ 0 \end{Bmatrix}} + \overset{\mathbf{F}}{\begin{Bmatrix} +45.0 \\ 18.0 \\ -18.0 \\ -27.0 \\ -18.0 \\ 0 \end{Bmatrix}} = \begin{Bmatrix} 0 \\ -27.0 \\ +27.0 \\ +18.0 \\ -18.0 \\ 0 \end{Bmatrix} \quad \text{kN} \cdot \text{m} \quad \blacksquare$$

■ EXAMPLE 11.22

Determine the support moments for the rigid frame in Fig. 11.12 a.

Solution. Matrices $\mathbf{E}$ and $\mathbf{S}$ for this structure are given by Eqs. (11.36) and (11.49). Thus,

$$\mathbf{SE}^T = 2 \overset{\mathbf{S}}{\begin{bmatrix} \frac{2}{12} & \frac{1}{12} & 0 & 0 & 0 & 0 \\ \frac{1}{12} & \frac{2}{12} & 0 & 0 & 0 & 0 \\ 0 & 0 & \frac{4}{16} & \frac{2}{16} & 0 & 0 \\ 0 & 0 & \frac{2}{16} & \frac{4}{16} & 0 & 0 \\ 0 & 0 & 0 & 0 & \frac{2}{12} & \frac{1}{12} \\ 0 & 0 & 0 & 0 & \frac{1}{12} & \frac{2}{12} \end{bmatrix}} \overset{\mathbf{E}^T}{\begin{bmatrix} 0 & 0 & -\frac{1}{12} \\ 1 & 0 & -\frac{1}{12} \\ 1 & 0 & 0 \\ 0 & 1 & 0 \\ 0 & 1 & -\frac{1}{12} \\ 0 & 0 & -\frac{1}{12} \end{bmatrix}} = 2 \begin{bmatrix} \frac{1}{12} & 0 & -\frac{1}{48} \\ \frac{2}{12} & 0 & -\frac{1}{48} \\ \frac{4}{16} & \frac{2}{16} & 0 \\ \frac{2}{16} & \frac{4}{16} & 0 \\ 0 & \frac{2}{12} & -\frac{1}{48} \\ 0 & \frac{1}{12} & -\frac{1}{48} \end{bmatrix}$$

Also,

$$\mathbf{ESE}^T = \mathbf{S}^* = 2 \begin{bmatrix} \frac{5}{12} & \frac{2}{16} & -\frac{1}{48} \\ \frac{2}{16} & \frac{5}{12} & -\frac{1}{48} \\ -\frac{1}{48} & -\frac{1}{48} & +\frac{1}{144} \end{bmatrix}$$

and

$$[\mathbf{S}^*]^{-1} = \begin{bmatrix} 1.457 & -0.257 & 3.60 \\ -0.257 & 1.457 & 3.60 \\ 3.6 & 3.6 & 93.60 \end{bmatrix}$$

From Fig. 11.23 c:

$$\mathbf{P} = \begin{bmatrix} +21.33 \\ -3.33 \\ 0 \end{bmatrix} \quad \text{thus} \quad \mathbf{D} = [\mathbf{S}^*]^{-1}\mathbf{P} = \begin{bmatrix} 31.936 \\ -10.336 \\ 64.800 \end{bmatrix}$$

Note that these are relative joint displacements and not absolute values, since relative EI values are used in the example.

$$\mathbf{F} = \mathbf{SE}^T\mathbf{D}; \quad \begin{Bmatrix} F_1 \\ F_2 \\ F_3 \\ F_4 \\ F_5 \\ F_6 \end{Bmatrix} = \begin{bmatrix} 2.623 \\ 7.945 \\ 13.384 \\ 2.816 \\ -6.145 \\ -4.423 \end{bmatrix}$$

Final moments,

$$\mathbf{F}' = \mathbf{F}_0 + \mathbf{F} = \overset{\mathbf{F}_0}{\begin{Bmatrix} 0 \\ 0 \\ -21.33 \\ +21.33 \\ 0 \\ 0 \end{Bmatrix}} + \overset{\mathbf{F}}{\begin{Bmatrix} 2.623 \\ 7.945 \\ 13.384 \\ 2.816 \\ -6.145 \\ -4.423 \end{Bmatrix}} = \begin{Bmatrix} 2.623 \\ 7.945 \\ -7.946 \\ 24.146 \\ -6.145 \\ -4.423 \end{Bmatrix} \quad \text{k·ft} \quad ■$$

■ EXAMPLE 11.23

Deduce the forces in the members of the composite structure of Fig. 11.14a.

Solution. The matrices **E** and **S** have been deduced earlier and are given by Eqs. (11.37) and (11.50). Thus

$$\mathbf{SE}^T = \begin{bmatrix} 50 & 25 & 0 & 0 & 0 & 0 & 0 & 0 \\ 25 & 50 & 0 & 0 & 0 & 0 & 0 & 0 \\ 0 & 0 & 50 & 25 & 0 & 0 & 0 & 0 \\ 0 & 0 & 25 & 50 & 0 & 0 & 0 & 0 \\ 0 & 0 & 0 & 0 & \frac{100}{3} & 0 & 0 & 0 \\ 0 & 0 & 0 & 0 & 0 & \frac{100}{3} & 0 & 0 \\ 0 & 0 & 0 & 0 & 0 & 0 & 20 & 0 \\ 0 & 0 & 0 & 0 & 0 & 0 & 0 & 20 \end{bmatrix} \begin{bmatrix} 0 & 0 & \frac{1}{6} \\ 1 & 0 & \frac{1}{6} \\ 1 & 0 & -\frac{1}{6} \\ 0 & 0 & -\frac{1}{6} \\ 0 & 1 & 0 \\ 0 & -1 & 0 \\ 0 & \frac{3}{5} & -\frac{4}{5} \\ 0 & -\frac{3}{5} & -\frac{4}{5} \end{bmatrix}$$

$$= \begin{bmatrix} 25 & 0 & \frac{25}{2} \\ 50 & 0 & \frac{25}{2} \\ 50 & 0 & -\frac{25}{2} \\ 25 & 0 & -\frac{25}{2} \\ 0 & \frac{100}{3} & 0 \\ 0 & -\frac{100}{3} & 0 \\ 0 & 12 & -16 \\ 0 & -12 & -16 \end{bmatrix}$$

$$\mathbf{ESE}^T = \mathbf{S}^* = \begin{bmatrix} 100 & 0 & 0 \\ 0 & \frac{1216}{15} & 0 \\ 0 & 0 & \frac{509}{15} \end{bmatrix}$$

and

$$[S*]^{-1} = \begin{bmatrix} \frac{1}{100} & 0 & 0 \\ 0 & \frac{15}{1216} & 0 \\ 0 & 0 & \frac{15}{509} \end{bmatrix}$$

Based on Fig. 11.25b, it was shown that the matrix P is given by

$$P = \begin{bmatrix} -3 \\ 0 \\ -27 \end{bmatrix} \quad \text{thus} \quad D = [S*]^{-1}P = \begin{bmatrix} -0.03 \\ 0 \\ -0.796 \end{bmatrix} \begin{matrix} D_1 \\ D_2 \\ D_3 \end{matrix} \begin{matrix} \text{radians} \\ \text{m} \\ \text{m} \end{matrix}$$

Note that the vertical deflection (D_3) is unacceptably large for a practical structure. From

$$F = SE^T D = \begin{Bmatrix} F_1 \\ F_2 \\ F_3 \\ F_4 \\ F_5 \\ F_6 \\ F_7 \\ F_8 \end{Bmatrix} = \begin{Bmatrix} -10.70 \\ -11.45 \\ +8.45 \\ 9.20 \\ 0 \\ 0 \\ +12.74 \\ +12.74 \end{Bmatrix}$$

the final force $F' = F_0 + F$ where the fixed-end moments are shown in Fig. 11.25a. Thus,

$$F' = \underset{F_0}{\begin{Bmatrix} -18.0 \\ +18.0 \\ -15.0 \\ +15.0 \\ 0 \\ 0 \\ 0 \\ 0 \end{Bmatrix}} + \underset{F}{\begin{Bmatrix} -10.70 \\ -11.45 \\ +8.45 \\ +9.20 \\ 0 \\ 0 \\ +12.74 \\ +12.74 \end{Bmatrix}} = \begin{Bmatrix} -28.70 \\ +6.55 \\ -6.55 \\ +24.20 \\ 0 \\ 0 \\ +12.74 \\ +12.74 \end{Bmatrix} \begin{matrix} \\ \\ \text{kN} \cdot \text{m} \\ \\ \\ \\ \text{kN} \\ \end{matrix} \quad \blacksquare$$

11.3.7 Static-Equilibrium and Deformation-Compatibility Checks

These checks are necessary to verify the correctness of results. The static checks require that moments at any joint must balance and that the static equilibrium of any member and of the entire structure is maintained; such checks ensure that static-equilibrium matrix E is correct. Although the static check is necessary, it is not sufficient to ensure the correctness of the solution; in the analysis of statically indeterminate structures, checks on the compatibility of deformation are also necessary; these compatibility checks require that the rotations and displacements of ends of members are equal to the corresponding rotations and displacements of the joints obtained from the analysis.

This ensures that the matrices **P** and **S** have been formulated correctly. These checks are illustrated by the following example.

■ **EXAMPLE 11.24**

Carry out the static and deformation checks for the rigid frame in Fig. 11.27a.

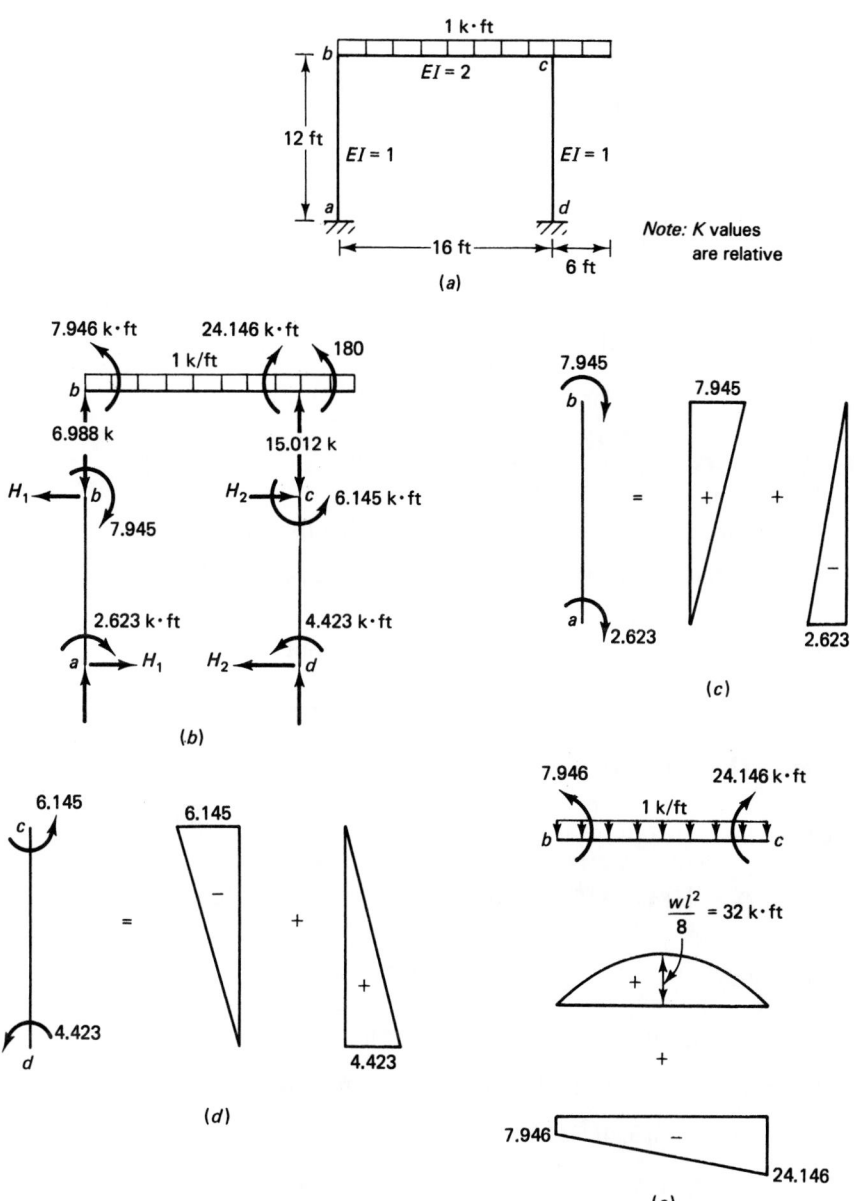

Figure 11.27 Frame for Example 11.24 for static and deformation checks.

Solution. This structure was analyzed earlier with the following results:

$D_1 = \theta_b = 31.936$ $D_2 = \theta_c = -10.336$ $D_3 = \Delta$ of joint b or $c = 64.80$

$M_{ab} = +2.623 \text{ k} \cdot \text{ft}$ $M_{ba} = +7.945 \text{ k} \cdot \text{ft}$ $M_{bc} = -7.946 \text{ k} \cdot \text{ft}$

$M_{cb} = +24.146 \text{ k} \cdot \text{ft}$ $M_{cd} = -6.145 \text{ k} \cdot \text{ft}$ $M_{dc} = -4.423 \text{ k} \cdot \text{ft}$

Static Checks. By inspection of Fig. 11.27*b* the moments at joints b and c are balanced. From the free body of *ab*,

$$H_1 = \frac{7.945 + 2.623}{12} = 0.881 \text{ k}$$

From the free body of *cd*

$$H_2 = \frac{6.145 + 4.423}{12} = 0.881 \text{ k}$$

Therefore

$H_1 - H_2 = 0.881 - 0.881 = 0$ (checks)

Compatibility Checks. For member *ab* (Fig. 11.27*c*):

Slope $\theta_b = D_1 = $ area of $\dfrac{M}{EI}$ between *a* and *b*

(based on the moment-area theorem with joint *a* being fixed).

$\theta_b = D_1 = \frac{1}{1}(7.945 - 2.623)(\frac{12}{2}) = 31.932$ (checks)

$\Delta_b = D_3 = $ moment of $\dfrac{M}{EI}$ area between *a* and *b* about *b*

$= \frac{1}{1}[(7.945)(\frac{12}{2})(\frac{12}{3}) - (2.623)(\frac{12}{2})(\frac{2}{3})(12)] = 64.78$ (checks)

For member *cd* (Fig. 11.27*d*):

Slope $\theta_c = D_2 = $ area of $\dfrac{M}{EI}$ between *c* and *d* (since joint *d* is fixed)

$\theta_c = D_2 = \frac{1}{1}[(4.423)(\frac{12}{2}) - (6.145)(\frac{12}{2})] = -10.33$ (checks)

$\Delta_c = D_3 = $ moment of $\dfrac{M}{EI}$ area between *c* and *d* about *c*

$= \frac{1}{1}[(4.423)(\frac{12}{2})(\frac{2}{3})(12) - (6.145)(\frac{12}{2})(\frac{12}{3})] = 64.82$ (checks)

For member *bc* (Fig. 11.27*e*): By the conjugate-beam method,

Slope $\theta_b = D_1 = \frac{1}{2}\{\frac{1}{2}(32)(\frac{2}{3})(16)$

$- [(7.946)(\frac{16}{2})(\frac{2}{3})(\frac{16}{16}) + (24.146)(\frac{16}{2})(\frac{1}{3})(\frac{1}{16})]\}$

$= 31.95$ (checks)

Slope $\theta_c = D_2 = \frac{1}{2}\{\frac{1}{2}(32)(\frac{2}{3})(16)$

$\qquad - [(7.946)(\frac{16}{2})(\frac{16}{3})(\frac{1}{16}) + (24.146)(\frac{16}{2})(\frac{2}{3})(\frac{16}{16})]\}$

$\qquad = -10.34 \quad$ (checks)   ∎

11.4 PARTITIONING THE GLOBAL STIFFNESS MATRIX S*

In the formulation of the matrix $\mathbf{S}^*$ we considered only the unknown joint or nodal displacements (degrees of freedom) in analyzing a structure by solving the equation

$$\mathbf{P} = \mathbf{S}^*\mathbf{D} \tag{11.18}$$

After the unknown joint displacements were determined, the internal member forces were calculated and hence the external reactions. However, if the known or specified displacements $\mathbf{D}_R$ corresponding to the unknown reactions $\mathbf{R}$ are included in the formulation of Eq. (11.18), then it is possible to determine the unknown reactions directly. Thus, we can express Eq. (11.18) in the form

Known forces $\;\rightarrow$

Unknown forces $\;\rightarrow$
(support
reactions)

$$\begin{Bmatrix} \mathbf{P} \\ \hline \mathbf{R} \end{Bmatrix} = \begin{bmatrix} \mathbf{S}^*_{11} & \mathbf{S}^*_{12} \\ \hline \mathbf{S}^*_{21} & \mathbf{S}^*_{22} \end{bmatrix} \begin{Bmatrix} \mathbf{D}_p \\ \hline \mathbf{D}_R \end{Bmatrix} \begin{matrix} \leftarrow \text{Unknown displacements} \\ \\ \leftarrow \text{Known displacements} \\ \text{(at supports)} \end{matrix} \tag{11.55}$$

The procedure is as follows: (1) Partition the joint or nodal displacements into two categories—unknown displacements $\mathbf{D}_p$ and known displacements $\mathbf{D}_R$. Thus we can rewrite Eq. (11.2), relating member end displacements to nodal displacements, as

$$\delta = [\mathbf{C}_p | \mathbf{C}_R] \begin{Bmatrix} \mathbf{D}_p \\ \hline \mathbf{D}_R \end{Bmatrix} \tag{11.56}$$

(2) Establish the global stiffness matrix $\mathbf{S}^*$ from

$$\mathbf{S}^* = \mathbf{ESE}^T \tag{11.22}$$

or

$$\mathbf{S}^* = \mathbf{C}^T\mathbf{SC} \tag{11.57}$$

$$= \begin{Bmatrix} \mathbf{C}_p^T \\ \hline \mathbf{C}_R^T \end{Bmatrix} [\mathbf{S}][\mathbf{C}_p | \mathbf{C}_R]$$

$$= \begin{bmatrix} \mathbf{C}_p^T\mathbf{SC}_p & \mathbf{C}_p^T\mathbf{SC}_R \\ \hline \mathbf{C}_R^T\mathbf{SC}_p & \mathbf{C}_R^T\mathbf{SC}_R \end{bmatrix}$$

Comparing the above components of $\mathbf{S}^*$ with those in Eq. (11.55), we find

$$\mathbf{S}_{11}^* = \mathbf{C}_p^T \mathbf{S} \mathbf{C}_p \qquad \mathbf{S}_{12}^* = \mathbf{C}_p^T \mathbf{S} \mathbf{C}_R$$
$$\mathbf{S}_{21}^* = \mathbf{C}_R^T \mathbf{S} \mathbf{C}_p \qquad \mathbf{S}_{22}^* = \mathbf{C}_R^T \mathbf{S} \mathbf{C}_R \tag{11.58}$$

Expanding Eq. (11.55), we can write

$$\mathbf{P} = \mathbf{S}_{11}^* \mathbf{D}_p + \mathbf{S}_{12}^* \mathbf{D}_R \tag{11.59}$$

and

$$\mathbf{R} = \mathbf{S}_{21}^* \mathbf{D}_p + \mathbf{S}_{22}^* \mathbf{D}_R \tag{11.60}$$

Rearranging Eq. (11.59) and premultiplying by $(\mathbf{S}_{11}^*)^{-1}$ yield the unknown displacements

$$\mathbf{D}_p = (\mathbf{S}_{11}^*)^{-1}(\mathbf{P} - \mathbf{S}_{12}^* \mathbf{D}_R) \tag{11.61}$$

Once $\mathbf{D}_p$ is determined, the unknown reactions can be found from Eq. (11.60), since the displacements $\mathbf{D}_R$ are either zero or specified. The application of Eq. (11.60) is demonstrated in Solved Problem 11.7.

Quite often some of the nodal loads in a structure are known to be zero; for example, in Fig. 11.7a there is no horizontal load at joint c. Such a condition makes it possible to reduce the global stiffness matrix $\mathbf{S}^*$ ($=\mathbf{ESE}^T$) for the restrained structure and as such it is referred to as a condensed matrix. Thus we can relate nodal loads to nodal displacements as follows:

$$\begin{array}{l} \text{Known nodal} \\ \text{forces} \\ \\ \text{Known zero} \\ \text{nodal forces} \end{array} \left\{ \begin{array}{c} \mathbf{P} \\ -- \\ \mathbf{0} \end{array} \right\} = \left[\begin{array}{c|c} \mathbf{S}_{11}^* & \mathbf{S}_{12}^* \\ \hline \mathbf{S}_{21}^* & \mathbf{S}_{22}^* \end{array} \right] \left\{ \begin{array}{c} \mathbf{D}_p \\ --- \\ \mathbf{D}_0 \end{array} \right\} \begin{array}{l} \leftarrow \text{Unknown displacements} \\ \\ \leftarrow \text{Unknown displacements} \end{array} \tag{11.62}$$

Thus,

$$\mathbf{P} = \mathbf{S}_{11}^* \mathbf{D}_p + \mathbf{S}_{12}^* \mathbf{D}_0 \tag{11.63}$$

and,

$$0 = \mathbf{S}_{21}^* \mathbf{D}_p + \mathbf{S}_{22}^* \mathbf{D}_0 \tag{11.64}$$

Premultiplying Eq. (11.64) by $(\mathbf{S}_{22}^*)^{-1}$ and rearranging:

$$\mathbf{D}_0 = -(\mathbf{S}_{22}^*)^{-1} \mathbf{S}_{21}^* \mathbf{D}_p \tag{11.65}$$

Substituting Eq. (11.65) into Eq. (11.63):

$$\mathbf{P} = (\mathbf{S}_{11}^* - \mathbf{S}_{12}^*(\mathbf{S}_{22}^*)^{-1}\mathbf{S}_{21}^*)\mathbf{D}_p \tag{11.66}$$

or

$$\mathbf{D}_p = [\mathbf{S}_{11}^* - \mathbf{S}_{12}^*(\mathbf{S}_{22}^*)^{-1}\mathbf{S}_{21}^*]^{-1}\mathbf{P} \tag{11.67}$$

where the matrix within the square brackets is known as the reduced or condensed stiffness matrix.

11.5 SOLVED PROBLEMS

■ SOLVED PROBLEM 11.1

Use the matrix-displacement method to find the forces in the members of the truss shown in Fig. 11.28a. Assume the value AE/l to be the same for all members.

Solution.   From equilibrium of joint A (Fig. 11.28b)

$$\Sigma H = 0 \qquad P_1 + F_1 \cos 30 + F_3 \cos 60 + F_4 \cos 30 = 0$$

Therefore,

$$P_1 = F_1 \frac{\sqrt{3}}{2} - F_3 \frac{1}{2} - F_4 \frac{\sqrt{3}}{2}$$

$$\Sigma V = 0 \qquad P_2 + F_1 \cos 60 + F_2 + F_3 \cos 30 + F_4 \cos 60 = 0$$

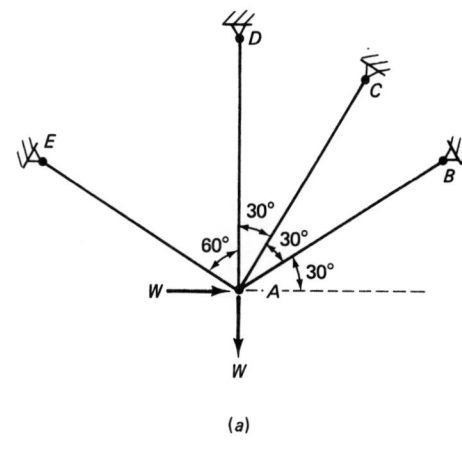

(a)

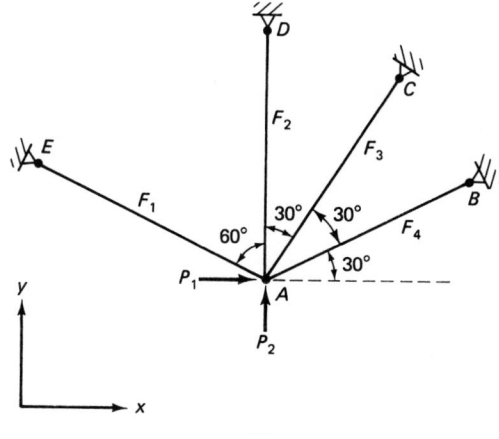

(b)

Figure 11.28 Structure for Solved Problem 11.1.

Therefore,

$$P_2 = -F_1 \frac{1}{2} - F_2 - F_3 \frac{\sqrt{3}}{2} - F_4 \frac{1}{2}$$

P = EF

Equilibrium matrix $\mathbf{E} = \begin{bmatrix} \dfrac{\sqrt{3}}{2} & 0 & -\dfrac{1}{2} & -\dfrac{\sqrt{3}}{2} \\ -\dfrac{1}{2} & -1 & -\dfrac{\sqrt{3}}{2} & -\dfrac{1}{2} \end{bmatrix}$

$$= \begin{bmatrix} 0.866 & 0 & -0.5 & -0.866 \\ -0.5 & -1 & -0.866 & -0.5 \end{bmatrix}$$

Member stiffness matrix $\mathbf{S} = \dfrac{EA}{L} \begin{bmatrix} 1 & 0 & 0 & 0 \\ 0 & 1 & 0 & 0 \\ 0 & 0 & 1 & 0 \\ 0 & 0 & 0 & 1 \end{bmatrix}$

$$\mathbf{SE}^T = \frac{EA}{L} \begin{bmatrix} 0.866 & -0.5 \\ 0 & -1 \\ -0.5 & -0.866 \\ -0.866 & -0.5 \end{bmatrix}$$

Global stiffness matrix $\mathbf{S^*} = \mathbf{ESE}^T$

Therefore,

$$\mathbf{S^*} = \frac{EA}{L} \begin{bmatrix} 0.866 & 0 & -0.5 & -0.866 \\ -0.5 & -1 & -0.866 & -0.5 \end{bmatrix} \begin{bmatrix} 0.866 & -0.5 \\ 0 & -1 \\ -0.5 & -0.866 \\ -0.866 & -0.5 \end{bmatrix}$$

$$= \frac{EA}{L} \begin{bmatrix} 1.75 & 0.433 \\ 0.433 & 2.25 \end{bmatrix}$$

$$[\mathbf{S^*}]^{-1} = 0.2667 \frac{L}{EA} \begin{bmatrix} 2.25 & -0.433 \\ -0.433 & 1.75 \end{bmatrix}$$

$$= \frac{L}{EA} \begin{bmatrix} 0.6 & -0.1155 \\ -0.1155 & 0.4667 \end{bmatrix}$$

Load matrix $\mathbf{P} = W \begin{Bmatrix} 1 \\ -1 \end{Bmatrix}$

$$\mathbf{D} = [\mathbf{S^*}]^{-1} \{\mathbf{P}\} = \frac{WL}{EA} \begin{bmatrix} 0.6 & -0.1155 \\ -0.1155 & 0.4667 \end{bmatrix} \begin{Bmatrix} 1 \\ -1 \end{Bmatrix}$$

$$= \begin{Bmatrix} 0.7155 \\ -0.5822 \end{Bmatrix} \frac{WL}{EA}$$

$$\mathbf{F} = \mathbf{SE}^T\mathbf{D} = W \begin{bmatrix} 0.866 & -0.5 \\ 0 & -1 \\ -0.5 & -0.866 \\ -0.866 & -0.5 \end{bmatrix} \begin{bmatrix} 0.7155 \\ -0.5822 \end{bmatrix}$$

$$= W \begin{bmatrix} 0.9107 \\ 0.5822 \\ 0.1464 \\ -0.3285 \end{bmatrix}$$

$F_{AE} = 0.9107\,W$ (tension)

$F_{AD} = 0.5822\,W$ (tension)

$F_{AC} = 0.1464\,W$ (tension)

$F_{AB} = -0.3285\,W$ (compression)   ■

■ SOLVED PROBLEM 11.2

Use the matrix-displacement method to find the forces in the members of the truss shown in Fig. 11.29a. The numbers in parentheses are areas in square inches. $E = 30,000$ ksi.

Solution

Load vector
$$\begin{bmatrix} P_1 \\ P_2 \\ P_3 \\ P_4 \\ P_5 \\ P_6 \\ P_7 \\ P_8 \end{bmatrix} = \begin{bmatrix} 0 \\ -10 \\ 0 \\ 0 \\ 0 \\ 0 \\ 0 \\ 0 \end{bmatrix}$$

Equilibrium matrix $\mathbf{E}$ (Fig. 11.29b and c)

At Joint B

$$P_1 + F_2 + \tfrac{3}{5}F_8 - \tfrac{3}{5}F_1 = 0$$

$$P_2 - \tfrac{4}{5}F_1 - F_7 - \tfrac{4}{5}F_8 = 0$$

At Joint C

$$P_3 - F_2 + \tfrac{3}{5}F_3 - \tfrac{3}{5}F_9 = 0$$

$$P_4 - \tfrac{4}{5}F_3 - \tfrac{4}{5}F_9 - F_{10} = 0$$

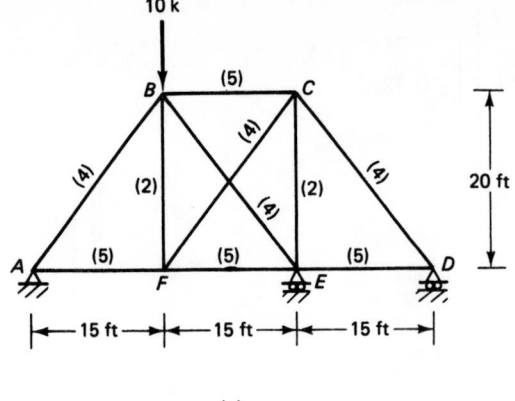

(a)

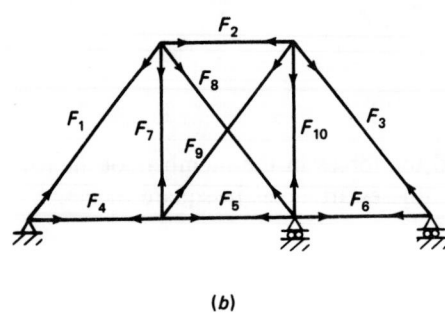

(b)

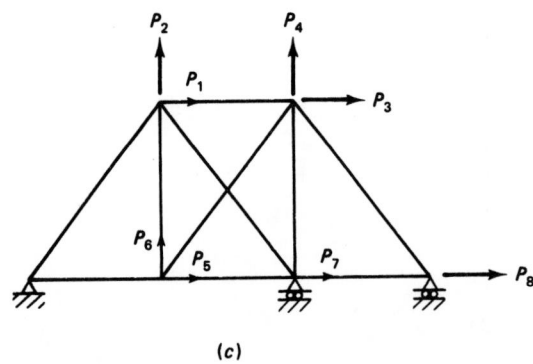

(c)

Figure 11.29 Truss for Solved Problem 11.2. (b) F, δ numbering. (c) P, D numbering.

At Joint F

$$P_5 - F_4 + F_5 + \tfrac{3}{5}F_9 = 0$$

$$P_6 + F_7 + \tfrac{4}{5}F_9 = 0$$

At Joint E

$$P_7 - F_5 + F_6 - \tfrac{3}{5}F_8 = 0$$

At Joint D

$$P_8 - F_6 - \tfrac{3}{5}F_3 = 0$$

$$
\begin{bmatrix} P_1 \\ P_2 \\ P_3 \\ P_4 \\ P_5 \\ P_6 \\ P_7 \\ P_8 \end{bmatrix}
=
\begin{bmatrix}
0.6 & -1 & 0 & 0 & 0 & 0 & 0 & -0.6 & 0 & 0 \\
0.8 & 0 & 0 & 0 & 0 & 0 & 1 & 0.8 & 0 & 0 \\
0 & 1 & -0.6 & 0 & 0 & 0 & 0 & 0 & 0.6 & 0 \\
0 & 0 & 0.8 & 0 & 0 & 0 & 0 & 0 & 0.8 & 1 \\
0 & 0 & 0 & 1 & -1 & 0 & 0 & 0 & -0.6 & 0 \\
0 & 0 & 0 & 0 & 0 & 0 & -1 & 0 & -0.8 & 0 \\
0 & 0 & 0 & 0 & 1 & -1 & 0 & 0.6 & 0 & 0 \\
0 & 0 & 0.6 & 0 & 0 & 1 & 0 & 0 & 0 & 0
\end{bmatrix}
\begin{bmatrix} F_1 \\ F_2 \\ F_3 \\ F_4 \\ F_5 \\ F_6 \\ F_7 \\ F_8 \\ F_9 \\ F_{10} \end{bmatrix}
$$

Member stiffness matrix **S**

$$
\mathbf{S} = 30000
\begin{bmatrix}
\dfrac{4}{25 \times 12} & & & & & & & & & \\
 & \dfrac{5}{15 \times 12} & & & & & & & & \\
 & & \dfrac{4}{25 \times 12} & & & & & & & \\
 & & & \dfrac{5}{15 \times 12} & & & & & & \\
 & & & & \dfrac{5}{15 \times 12} & & & & & \\
 & & & & & \dfrac{5}{15 \times 12} & & & & \\
 & & & & & & \dfrac{2}{20 \times 12} & & & \\
 & & & & & & & \dfrac{4}{25 \times 12} & & \\
 & & & & & & & & \dfrac{4}{25 \times 12} & \\
 & & & & & & & & & \dfrac{2}{20 \times 12}
\end{bmatrix}
$$

$$
[\mathbf{E}]^T =
\begin{bmatrix}
0.6 & 0.8 & 0 & 0 & 0 & 0 & 0 & 0 \\
-1 & 0 & 1 & 0 & 0 & 0 & 0 & 0 \\
0 & 0 & -0.6 & 0.8 & 0 & 0 & 0 & 0.6 \\
0 & 0 & 0 & 0 & 1 & 0 & 0 & 0 \\
0 & 0 & 0 & 0 & -1 & 0 & 1 & 0 \\
0 & 0 & 0 & 0 & 0 & 0 & -1 & 1 \\
0 & 1 & 0 & 0 & 0 & -1 & 0 & 0 \\
-0.6 & 0.8 & 0 & 0 & 0 & 0 & 0.6 & 0 \\
0 & 0 & 0.6 & 0.8 & -0.6 & -0.8 & 0 & 0 \\
0 & 0 & 0 & 1 & 0 & 0 & 0 & 0
\end{bmatrix}
$$

Global stiffness matrix $\mathbf{S^*} = \mathbf{ESE}^T$

Displacement matrix $\{\mathbf{D}\} = [\mathbf{S^*}]^{-1}\{\mathbf{P}\}$

Force matrix $\{\mathbf{F}\} = [\mathbf{SE}^T]\{\mathbf{D}\}$

$$[\mathbf{S^*}] = \begin{bmatrix} 1121.33 & 0 & -833.33 & 0 & 0 & 0 & -144 & 0 \\ & 762 & 0 & 0 & 0 & -250 & 192 & 0 \\ & & 1121.33 & 0 & -144 & -192 & 0 & -144 \\ & & & 762 & -192 & -256 & 0 & 192 \\ & & & & 1810.6 & 192 & -833.33 & 0 \\ & & & & & 506 & 0 & 0 \\ & & \text{Symmetric} & & & & 1810.67 & -833.33 \\ & & & & & & & 977.33 \end{bmatrix}$$

$$[\mathbf{S^*}]^{-1} = \begin{bmatrix} 0.003 & -0.00007 & 0.00262 & 0.00007 & 0.0007 & 0.00073 & 0.00121 & 0.00141 \\ & 0.00198 & 0.00004 & 0.00057 & -0.00045 & 0.00145 & -0.00077 & -0.00076 \\ & & 0.00332 & 0.00017 & 0.00074 & 0.00109 & 0.00124 & 0.00151 \\ & & & 0.00193 & -0.00014 & 0.00137 & -0.00047 & -0.00075 \\ & \text{Symmetric} & & & 0.00118 & -0.00046 & 0.000117 & 0.00113 \\ & & & & & 0.00398 & -0.00059 & -0.00061 \\ & & & & & & 0.00236 & 0.00223 \\ & & & & & & & 0.0033 \end{bmatrix}$$

$$\mathbf{D} = \begin{Bmatrix} 0.00066 \\ -0.01984 \\ -0.00045 \\ -0.00568 \\ 0.00446 \\ -0.01454 \\ 0.00772 \\ 0.00763 \end{Bmatrix}$$

$\blacksquare$

$$\mathbf{F} = \begin{Bmatrix} -6.18 \\ -0.92 \\ 0.12 \\ 3.71 \\ 2.72 \\ -0.07 \\ -1.33 \\ -4.65 \\ 1.66 \\ -1.42 \end{Bmatrix}$$

■ SOLVED PROBLEM 11.3

Use the displacement method to find the bending-moment diagram for the frame shown in Fig. 11.30. Given: $K = EI/l$.

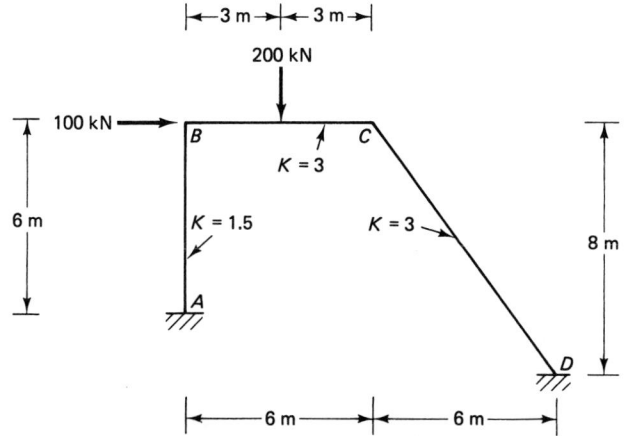

Figure 11.30 Frame for Solved Problem 11.3.

Solution. The equilibrium matrix **E**:
From Fig. 11.31 *g*:

$$8H + 6\left(\frac{F_3 + F_4}{6}\right) - F_5 - F_6 = 0$$

$$H = \frac{F_5}{8} + \frac{F_6}{8} - \frac{F_3}{8} - \frac{F_4}{8}$$

From Fig. 11.31 *f*:

$$P_3 + \frac{F_1 + F_2}{6} + H = 0$$

$$P_3 = -\frac{F_5}{8} - \frac{F_6}{8} + \frac{F_3}{8} + \frac{F_4}{8} - \frac{F_1}{6} - \frac{F_2}{6}$$

$$= \frac{F_3 + F_4}{8} - \frac{F_5 + F_6}{8} - \frac{F_1 + F_2}{6}$$

From Fig. 11.31 *c*:

$$P_1 = F_2 + F_3$$

From Fig. 11.31 *d*:

$$P_2 = F_4 + F_5$$

$$\mathbf{P} = \mathbf{EF}$$

$$\begin{bmatrix} P_1 \\ P_2 \\ P_3 \end{bmatrix} = \begin{bmatrix} 0 & 1 & 1 & 0 & 0 & 0 \\ 0 & 0 & 0 & 1 & 1 & 0 \\ -\frac{1}{6} & -\frac{1}{6} & \frac{1}{8} & \frac{1}{8} & -\frac{1}{8} & -\frac{1}{8} \end{bmatrix} \begin{bmatrix} F_1 \\ F_2 \\ F_3 \\ F_4 \\ F_5 \\ F_6 \end{bmatrix}$$

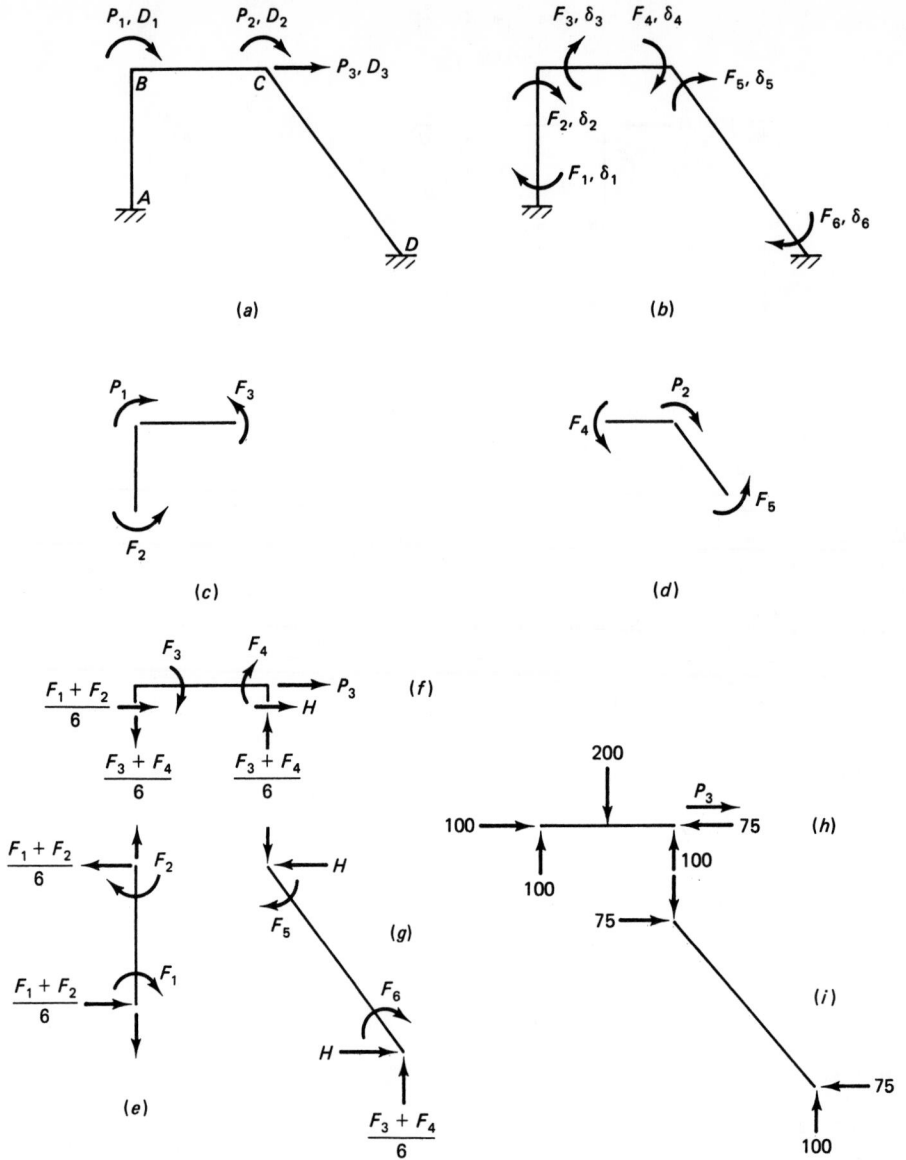

Figure 11.31 Analysis of frame shown in Fig. 11.30. (*a*) *P-D* numbering. (*b*) *F-δ* numbering.

Values of external joint forces

$$P_3 = 100 - 75 = 25 \text{ kN}$$

$$P_1 = -\left(-\frac{Wl}{8}\right) = 150 \text{ kN} \cdot \text{m}$$

$$P_2 = -\left(\frac{Wl}{8}\right) = -150 \text{ kN} \cdot \text{m}$$

Stiffness matrix **S**:

$$\mathbf{F} = \mathbf{S}\boldsymbol{\delta}$$

$$
\begin{bmatrix} F_1 \\ F_2 \\ F_3 \\ F_4 \\ F_5 \\ F_6 \end{bmatrix}
=
\begin{bmatrix}
6 & 3 & 0 & 0 & 0 & 0 \\
3 & 6 & 0 & 0 & 0 & 0 \\
0 & 0 & 12 & 6 & 0 & 0 \\
0 & 0 & 6 & 12 & 0 & 0 \\
0 & 0 & 0 & 0 & 12 & 6 \\
0 & 0 & 0 & 0 & 6 & 12
\end{bmatrix}
\begin{bmatrix} \delta_1 \\ \delta_2 \\ \delta_3 \\ \delta_4 \\ \delta_5 \\ \delta_6 \end{bmatrix}
$$

$$\underset{\mathbf{S}}{\underbrace{\qquad\qquad\qquad\qquad}}$$

$$
[\mathbf{E}][\mathbf{S}] =
\begin{bmatrix}
3 & 6 & 12 & 6 & 0 & 0 \\
0 & 0 & 6 & 12 & 12 & 6 \\
-\dfrac{3}{2} & -\dfrac{3}{2} & \dfrac{9}{4} & \dfrac{9}{4} & -\dfrac{9}{4} & -\dfrac{9}{4}
\end{bmatrix}
$$

Global stiffness matrix

$$
\mathbf{S}^* = \mathbf{E}\mathbf{S}\mathbf{E}^T =
\begin{bmatrix}
18 & 6 & 0.75 \\
6 & 24 & 0 \\
0.75 & 0 & 1.63
\end{bmatrix}
$$

$$
[\mathbf{S}^*]^{-1} =
\begin{bmatrix}
0.0619 & -0.0154 & -0.028 \\
-0.0154 & 0.0455 & 0.00712 \\
-0.0284 & 0.00712 & 0.626
\end{bmatrix}
$$

Displacement matrix

$$[\mathbf{D}] = [\mathbf{S}^*]^{-1}\{\mathbf{P}\}$$

$$
\begin{bmatrix} D_1 \\ D_2 \\ D_3 \end{bmatrix}
=
\begin{bmatrix} 10.89 \\ -8.96 \\ 10.31 \end{bmatrix}
$$

Note that these are relative joint displacements and not absolute values, since relative EI/l are used in the Solved Problem.

Force matrix

$$[\mathbf{F}] = [\mathbf{S}][\mathbf{E}]^T\{\mathbf{D}\}$$

$$
\mathbf{F} =
\begin{bmatrix} F_1 \\ F_2 \\ F_3 \\ F_4 \\ F_5 \\ F_6 \end{bmatrix}
=
\begin{bmatrix}
17.14 \\
49.82 \\
100.17 \\
-19.02 \\
-130.98 \\
-77.14
\end{bmatrix}
$$

Final moment: $\mathbf{F}' = \mathbf{F} + \mathbf{F}_0$

$$M_{AB} = 17.14 \text{ kN} \cdot \text{m}$$

$$M_{BA} = 49.82 \text{ kN} \cdot \text{m}$$

$$M_{BC} = 100.17 - 150 = -49.83 \text{ kN} \cdot \text{m}$$

$$M_{CB} = -19.02 + 150 = 130.98 \text{ kN} \cdot \text{m}$$

$$M_{CD} = -130.98 \text{ kN} \cdot \text{m}$$

$$M_{DC} = -77.14 \text{ kN} \cdot \text{m}$$

■ SOLVED PROBLEM 11.4

For the loaded rigid frame shown in Fig. 11.32a, determine (a) the static equilibrium matrix, (b) the deformation matrix, (c) the stiffness matrix, (d) the load matrix.

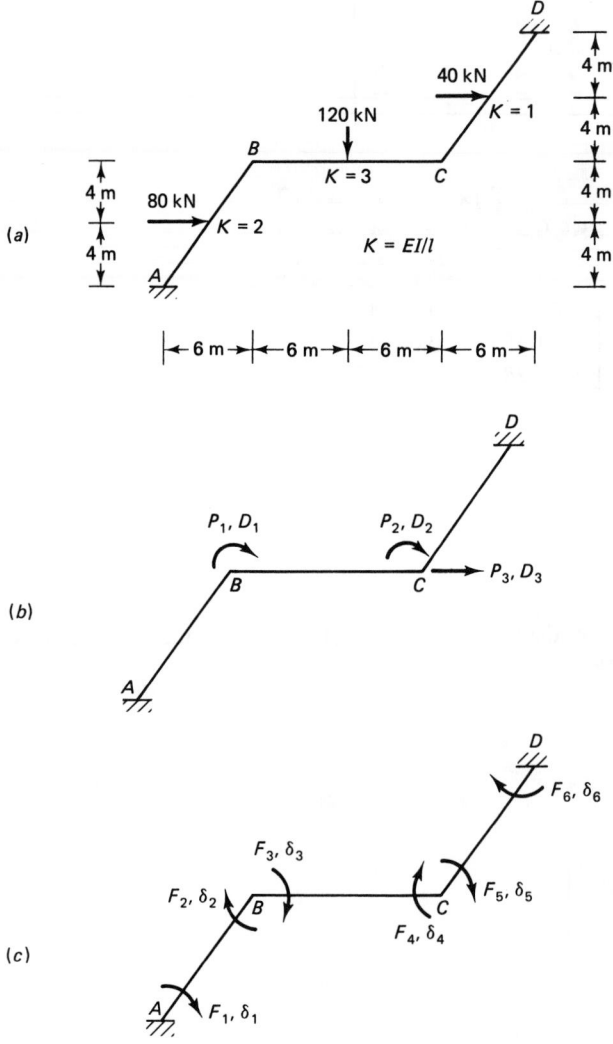

Figure 11.32 Frame for Solved Problem 11.4. (b) P-D numbering. (c) F- δ numbering.

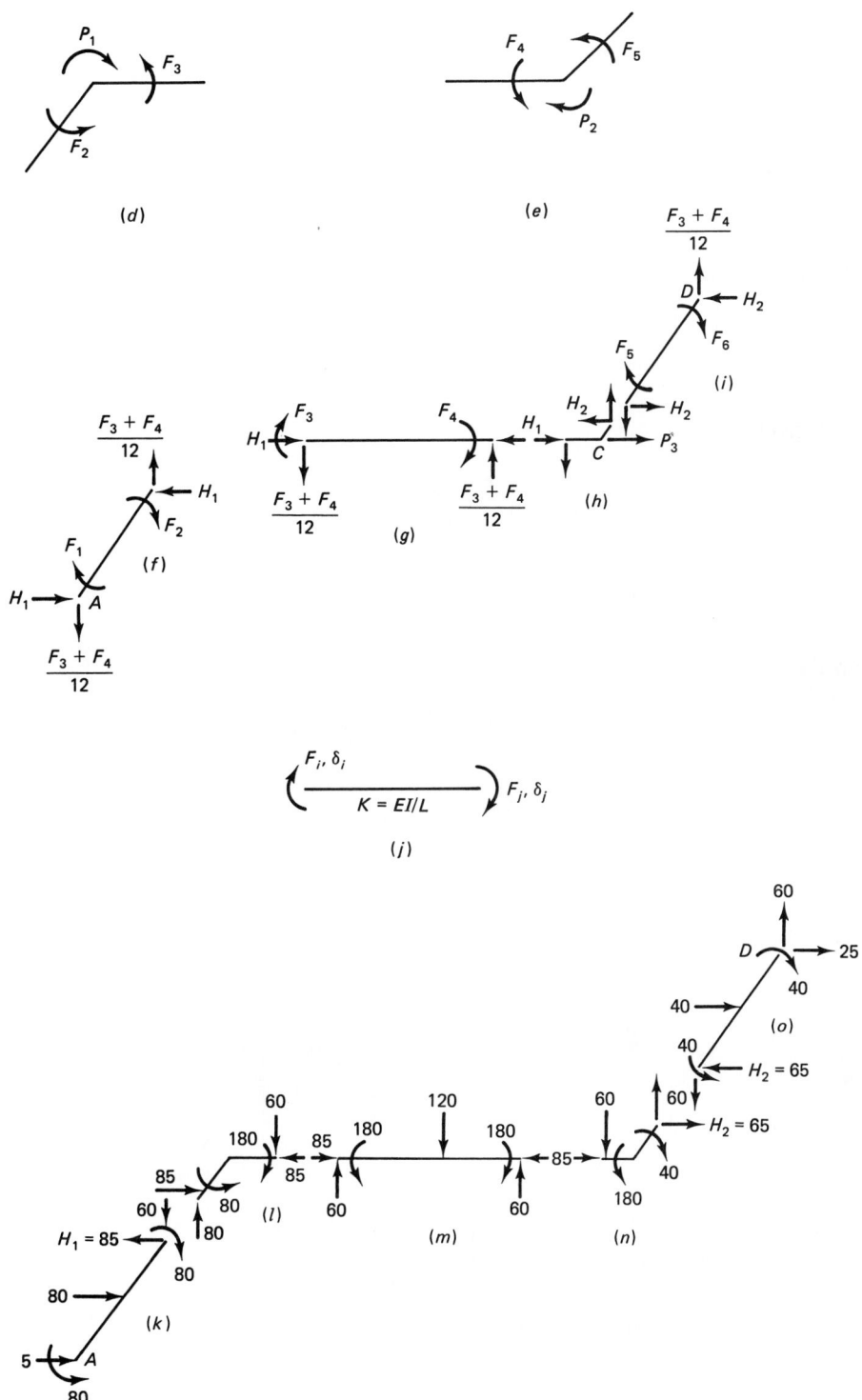

Figure 11.32 (*Continued*)

Solution

   (a) Static equilibrium matrix **E**: Apply the moment equilibrium equation at joints *B and C.*

   From Fig. 11.32*d*:

$$P_1 - F_3 - F_2 = 0$$

$$P_1 = F_2 + F_3$$

From Fig. 11.32*e*:

$$P_2 - F_4 - F_5 = 0$$

$$P_2 = F_4 + F_5$$

From Fig. 11.32*f*:

$$^+\!\!\left(\, \sum M_A = 0 \qquad F_1 + F_2 - 6\,\frac{F_3 + F_4}{12} = 8H_1\right.$$

$$H_1 = \frac{F_1 + F_2}{8} - \frac{F_3 + F_4}{16}$$

From Fig. 11.32*i*:

$$^+\!\!\left(\, \sum M_D = 0 \qquad F_6 + F_5 - 6\,\frac{F_3 + F_4}{12} = 8H_2\right.$$

$$H_2 = \frac{F_5 + F_6}{8} - \frac{F_3 + F_4}{16}$$

From Fig. 11.32*h*:

$$P_3 = H_2 - H_1$$

$$P_3 = \frac{F_5 + F_6}{8} - \frac{F_1 + F_2}{8}$$

Therefore,

   $$\mathbf{P} = \mathbf{EF}$$

$$
\begin{bmatrix} P_1 \\ P_2 \\ P_3 \end{bmatrix} =
\begin{bmatrix}
0 & 1 & 1 & 0 & 0 & 0 \\
0 & 0 & 0 & 1 & 1 & 0 \\
-\frac{1}{8} & -\frac{1}{8} & 0 & 0 & \frac{1}{8} & \frac{1}{8}
\end{bmatrix}
\begin{bmatrix} F_1 \\ F_2 \\ F_3 \\ F_4 \\ F_5 \\ F_6 \end{bmatrix}
$$

$$\underset{\text{Equilibrium matrix }\mathbf{E}}{}$$

(b) Determination matrix

$$\mathbf{C} = \mathbf{E}^T$$

$$\mathbf{C} = \begin{bmatrix} 0 & 0 & -\frac{1}{8} \\ 1 & 0 & -\frac{1}{8} \\ 1 & 0 & 0 \\ 0 & 1 & 0 \\ 0 & 1 & \frac{1}{8} \\ 0 & 0 & \frac{1}{8} \end{bmatrix}$$

(c) Member stiffness matrix **S**

$$\mathbf{F} = \mathbf{S}\,\delta$$

$$F_i = \frac{2EI}{L}(2\delta_i + \delta_j)$$

$$F_j = \frac{2EI}{L}(2\delta_j + \delta_i)$$

$$\begin{bmatrix} F_1 \\ F_2 \\ F_3 \\ F_4 \\ F_5 \\ F_6 \end{bmatrix} = \begin{bmatrix} 8 & 4 & 0 & 0 & 0 & 0 \\ 4 & 8 & 0 & 0 & 0 & 0 \\ 0 & 0 & 12 & 6 & 0 & 0 \\ 0 & 0 & 6 & 12 & 0 & 0 \\ 0 & 0 & 0 & 0 & 4 & 2 \\ 0 & 0 & 0 & 0 & 2 & 4 \end{bmatrix} \begin{bmatrix} \delta_1 \\ \delta_2 \\ \delta_3 \\ \delta_4 \\ \delta_5 \\ \delta_6 \end{bmatrix}$$

Member stiffness matrix $\big\rbrace$

(d) Load matrix

From Fig. 11.32 f:

$$\big\lbrace \sum M_A = 0 \qquad H_1(8) - (60)(6) - 80 + 80 - 80(4) = 0$$

$$H_1 = 85 \text{ kN}$$

From Fig. 11.32 i:

$$\big\lbrace \sum M_D = 0 \qquad -H_2(8) + 40(4) + 60(6) + 40 - 40 = 0$$

$$H_2 = 65 \text{ kN}$$

Load matrix: From Fig. 11.32 l: $P_1 = 180 - 80 = 100$ kN·m; from Fig. 11.32 n: $P_2 = 40 - 180 = -140$ kN·m and $P_3 = 85 + 65 = 150$ kN.

$$\begin{bmatrix} P_1 \\ P_2 \\ P_3 \end{bmatrix} = \begin{bmatrix} 100 \\ -140 \\ 150 \end{bmatrix}$$

■

■ SOLVED PROBLEM 11.5

Use the matrix-displacement method to find the bending-moment diagram for the frame shown in Fig. 11.33a. Given $K = EI/l$.

Solution. Referring to Fig. 11.21, the fixed-end moments are computed as shown in Fig. 11.33g. There are three rotations D_1, D_2, D_3, and no sidesway. Therefore,

Degrees of freedom = 3

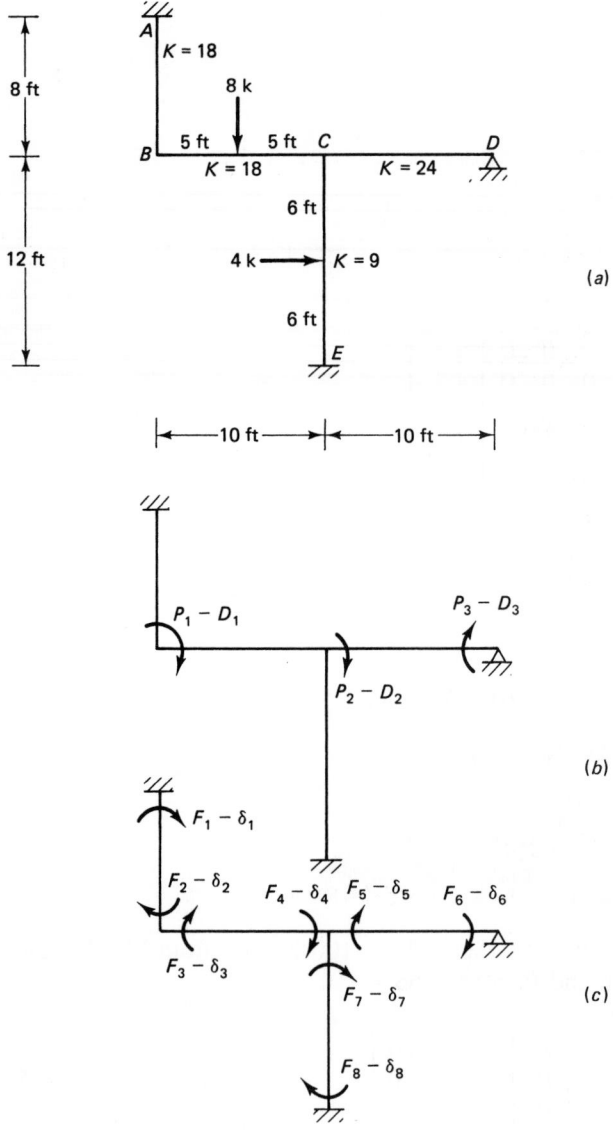

Figure 11.33 Frame for Solved Problem 11.5. (b) P-D numbering. (c) F-δ numbering.

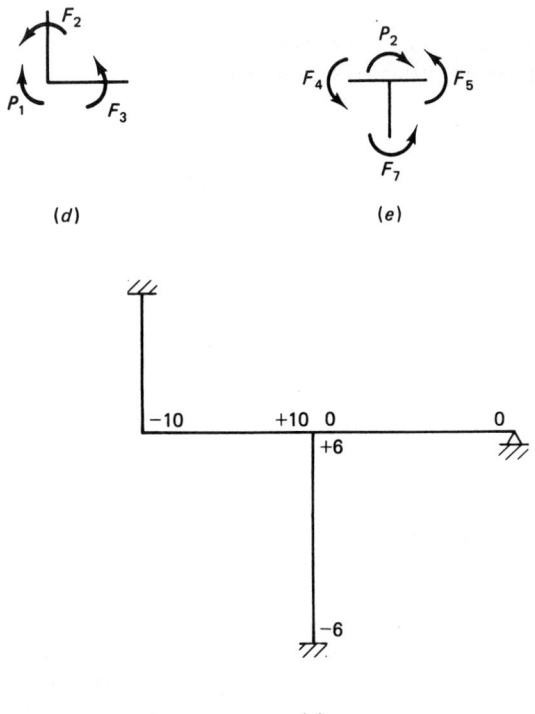

(d)   (e)   (f)

(g)

Figure 11.33 (*Continued*)

The equilibrium matrix **E**:
From Fig. 11.33 *d*:

$$P_1 - F_2 - F_3 = 0$$

From Fig. 11.33 *e*:

$$P_2 - F_4 - F_5 - F_7 = 0$$

From Fig. 11.33 *f*:

$$P_3 - F_6 = 0$$

or

$$\mathbf{P} = \mathbf{EF}$$

$$
\begin{bmatrix} P_1 \\ P_2 \\ P_3 \end{bmatrix} =
\begin{bmatrix}
0 & 1 & 1 & 0 & 0 & 0 & 0 & 0 \\
0 & 0 & 0 & 1 & 1 & 0 & 1 & 0 \\
0 & 0 & 0 & 0 & 0 & 1 & 0 & 0
\end{bmatrix}
\begin{bmatrix} F_1 \\ F_2 \\ F_3 \\ F_4 \\ F_5 \\ F_6 \\ F_7 \\ F_8 \end{bmatrix}
$$

The member stiffness matrix **S**:

$$\mathbf{F} = \mathbf{S}\,\delta$$

$$
\begin{bmatrix} F_1 \\ F_2 \\ F_3 \\ F_4 \\ F_5 \\ F_6 \\ F_7 \\ F_8 \end{bmatrix}
=
\begin{bmatrix}
72 & 36 & 0 & 0 & 0 & 0 & 0 & 0 \\
36 & 72 & 0 & 0 & 0 & 0 & 0 & 0 \\
0 & 0 & 72 & 36 & 0 & 0 & 0 & 0 \\
0 & 0 & 36 & 72 & 0 & 0 & 0 & 0 \\
0 & 0 & 0 & 0 & 96 & 48 & 0 & 0 \\
0 & 0 & 0 & 0 & 48 & 96 & 0 & 0 \\
0 & 0 & 0 & 0 & 0 & 0 & 36 & 18 \\
0 & 0 & 0 & 0 & 0 & 0 & 18 & 36
\end{bmatrix}
\begin{bmatrix} \delta_1 \\ \delta_2 \\ \delta_3 \\ \delta_4 \\ \delta_5 \\ \delta_6 \\ \delta_7 \\ \delta_8 \end{bmatrix}
$$

Member stiffness matrix S

Values of external joint forces

$$P_1 = -(0 - 10) = \quad 10 \text{ k} \cdot \text{ft}$$

$$P_2 = -(10 + 6) = -16 \text{ k} \cdot \text{ft} \qquad \text{Therefore } [P] = \begin{bmatrix} 10 \\ -16 \\ 0 \end{bmatrix}$$

$$P_3 = -(0) \qquad = \quad 0 \text{ k} \cdot \text{ft}$$

$$\mathbf{D} = [\mathbf{S}^*]^{-1}[\mathbf{P}] = [\mathbf{E}\mathbf{S}\mathbf{E}^T]^{-1}[\mathbf{P}]$$

$$
[\mathbf{E}][\mathbf{S}] =
\begin{bmatrix}
0 & 1 & 1 & 0 & 0 & 0 & 0 & 0 \\
0 & 0 & 0 & 1 & 1 & 0 & 1 & 0 \\
0 & 0 & 0 & 0 & 0 & 1 & 0 & 0
\end{bmatrix}
\begin{bmatrix}
72 & 36 & 0 & 0 & 0 & 0 & 0 & 0 \\
36 & 72 & 0 & 0 & 0 & 0 & 0 & 0 \\
0 & 0 & 72 & 36 & 0 & 0 & 0 & 0 \\
0 & 0 & 36 & 72 & 0 & 0 & 0 & 0 \\
0 & 0 & 0 & 0 & 96 & 48 & 0 & 0 \\
0 & 0 & 0 & 0 & 48 & 96 & 0 & 0 \\
0 & 0 & 0 & 0 & 0 & 0 & 36 & 18 \\
0 & 0 & 0 & 0 & 0 & 0 & 18 & 36
\end{bmatrix}
\tag{a}
$$

$$
=
\begin{bmatrix}
36 & 72 & 72 & 36 & 0 & 0 & 0 & 0 \\
0 & 0 & 36 & 72 & 96 & 48 & 36 & 18 \\
0 & 0 & 0 & 0 & 48 & 96 & 0 & 0
\end{bmatrix}
$$

$$
\mathbf{E}\mathbf{S}\mathbf{E}^T =
\begin{bmatrix}
144 & 36 & 0 \\
36 & 204 & 48 \\
0 & 48 & 96
\end{bmatrix}
\tag{b}
$$

$$
[\mathbf{E}\mathbf{S}\mathbf{E}^T]^{-1} =
\begin{bmatrix}
0.00731 & -0.001462 & 0.000731 \\
-0.001462 & 0.005848 & -0.002924 \\
0.000731 & -0.002924 & 0.011879
\end{bmatrix}
\tag{c}
$$

$$
\begin{Bmatrix} D_1 \\ D_2 \\ D_3 \end{Bmatrix}
=
\begin{bmatrix}
0.00731 & -0.001462 & 0.000731 \\
-0.00146 & 0.005848 & -0.002924 \\
0.00073 & -0.002924 & 0.011879
\end{bmatrix}
\begin{bmatrix} 10 \\ -16 \\ 0 \end{bmatrix}
$$

$$
=
\begin{bmatrix}
0.096492 \\
-0.108810 \\
0.054099
\end{bmatrix}
\text{ ft}
$$

and

$$\mathbf{F} = \mathbf{SE}^T\mathbf{D}$$

$$
\begin{bmatrix} F_1 \\ F_2 \\ F_3 \\ F_4 \\ F_5 \\ F_6 \\ F_7 \\ F_8 \end{bmatrix} =
\begin{bmatrix}
72 & 36 & 0 & 0 & 0 & 0 & 0 & 0 \\
36 & 72 & 0 & 0 & 0 & 0 & 0 & 0 \\
0 & 0 & 72 & 36 & 0 & 0 & 0 & 0 \\
0 & 0 & 36 & 72 & 0 & 0 & 0 & 0 \\
0 & 0 & 0 & 0 & 96 & 48 & 0 & 0 \\
0 & 0 & 0 & 0 & 48 & 96 & 0 & 0 \\
0 & 0 & 0 & 0 & 0 & 0 & 36 & 18 \\
0 & 0 & 0 & 0 & 0 & 0 & 18 & 36
\end{bmatrix}
\begin{bmatrix}
0 & 0 & 0 \\
1 & 0 & 0 \\
1 & 0 & 0 \\
0 & 1 & 0 \\
0 & 1 & 0 \\
0 & 0 & 1 \\
0 & 1 & 0 \\
0 & 0 & 0
\end{bmatrix}
\begin{bmatrix} 0.09649 \\ -0.108188 \\ 0.054099 \end{bmatrix}
$$

$$
= \begin{bmatrix} 3.47 \\ 6.95 \\ 3.05 \\ -4.32 \\ -7.79 \\ 0 \\ -3.89 \\ -1.95 \end{bmatrix}
$$

Final bending-moment values

$$\mathbf{F}' = \mathbf{F} + \mathbf{F}_0$$

$$F_1 = M_{AB} = 3.47 + 0 = 3.47 \ \text{k} \cdot \text{ft}$$

$$F_2 = M_{BA} = 6.95 + 0 = 6.95 \ \text{k} \cdot \text{ft}$$

$$F_3 = M_{BC} = 3.05 - 10 = -6.95 \ \text{k} \cdot \text{ft}$$

$$F_4 = M_{CB} = -4.32 + 10 = 5.68 \ \text{k} \cdot \text{ft}$$

$$F_5 = M_{CD} = -7.79 + 0 = -7.79 \ \text{k} \cdot \text{ft}$$

$$F_6 = M_{DC} = 0 + 0 = 0 \ \text{k} \cdot \text{ft}$$

$$F_7 = M_{CE} = -3.89 + 6 = 2.11 \ \text{k} \cdot \text{ft}$$

$$F_8 = M_{EC} = -1.95 - 6 = -7.95 \ \text{k} \cdot \text{ft}$$

∎

■ SOLVED PROBLEM 11.6

Determine the bending moment at the end of members of rigid frame in Fig. 11.26a. Neglect axial deformation.

Solution. The matrices **P** and **E** are determined earlier in Example 11-19. Assuming the elastic modulus = 1, the member stiffness matrix **S** is found from

$$\begin{Bmatrix} F_1 \\ F_2 \\ F_3 \\ F_4 \\ F_5 \\ F_6 \end{Bmatrix} = \underbrace{\begin{bmatrix} 8 & 4 & 0 & 0 & 0 & 0 \\ 4 & 8 & 0 & 0 & 0 & 0 \\ 0 & 0 & 16 & 8 & 0 & 0 \\ 0 & 0 & 8 & 16 & 0 & 0 \\ 0 & 0 & 0 & 0 & 8 & 4 \\ 0 & 0 & 0 & 0 & 4 & 8 \end{bmatrix}}_{[S]} \begin{Bmatrix} \delta_1 \\ \delta_2 \\ \delta_3 \\ \delta_4 \\ \delta_5 \\ \delta_6 \end{Bmatrix}$$

Thus

$$\mathbf{SE}^T = \begin{bmatrix} 8 & 4 & 0 & 0 & 0 & 0 \\ 4 & 8 & 0 & 0 & 0 & 0 \\ 0 & 0 & 16 & 8 & 0 & 0 \\ 0 & 0 & 8 & 16 & 0 & 0 \\ 0 & 0 & 0 & 0 & 8 & 4 \\ 0 & 0 & 0 & 0 & 4 & 8 \end{bmatrix} \begin{bmatrix} 0 & 0 & -\frac{1}{12} \\ 1 & 0 & -\frac{1}{12} \\ 1 & 0 & \frac{17}{240} \\ 0 & 1 & \frac{17}{240} \\ 0 & 1 & -\frac{1}{12} \\ 0 & 0 & -\frac{1}{12} \end{bmatrix}$$

$$\mathbf{SE}^T = \begin{bmatrix} 4 & 0 & -1 \\ 8 & 0 & -1 \\ 16 & 8 & 1.7 \\ 8 & 16 & 1.7 \\ 0 & 8 & -1 \\ 0 & 4 & -1 \end{bmatrix}$$

and

$$\mathbf{ESE}^T = \begin{bmatrix} 24 & 8 & 0.7 \\ 8 & 24 & 0.7 \\ 0.7 & 0.7 & 0.574 \end{bmatrix} = \mathbf{S}^*$$

where $\mathbf{S}^*$ is the global stiffness matrix. Thus,

$$\mathbf{D} = [\mathbf{ESE}^T]^{-1}\mathbf{P} = [\mathbf{S}^*]^{-1}\mathbf{P} = \begin{bmatrix} 0.04776 & -0.01474 & -0.04024 \\ -0.01474 & 0.04776 & -0.04024 \\ -0.04024 & -0.04024 & 1.83967 \end{bmatrix} \begin{bmatrix} 80 \\ -320 \\ -39.5 \end{bmatrix}$$

$$= \begin{bmatrix} 10.128 \\ -14.872 \\ -63.009 \end{bmatrix} \begin{matrix} \text{radians} \\ \text{radians} \\ \text{ft} \end{matrix}$$

Note that these are relative joint displacements and not absolute values, since relative k values are used in the problem.

Hence from $\mathbf{F} = \mathbf{SE}^T\mathbf{D}$

$$\begin{Bmatrix} F_1 \\ F_2 \\ F_3 \\ F_4 \\ F_5 \\ F_6 \end{Bmatrix} = \begin{bmatrix} 4 & 0 & -1 \\ 8 & 0 & -1 \\ 16 & 8 & 1.7 \\ 8 & 16 & 1.7 \\ 0 & 8 & -1 \\ 0 & 4 & -1 \end{bmatrix} \begin{bmatrix} 10.128 \\ -14.872 \\ -63.009 \end{bmatrix} = \begin{bmatrix} 103.52 \\ 144.04 \\ -64.04 \\ -264.04 \\ -55.96 \\ 3.52 \end{bmatrix}$$

Thus final moments $\mathbf{F}' = \mathbf{F}_0 + \mathbf{F}$, or

$$
\begin{bmatrix} F'_1 \\ F'_2 \\ F'_3 \\ F'_4 \\ F'_5 \\ F'_6 \end{bmatrix} = \begin{bmatrix} 0 \\ 0 \\ -80 \\ 320 \\ 0 \\ 0 \end{bmatrix} + \begin{bmatrix} 103.52 \\ 144.04 \\ -64.04 \\ -264.04 \\ -55.96 \\ 3.52 \end{bmatrix} = \begin{bmatrix} 103.52 \\ 144.04 \\ -144.04 \\ 55.96 \\ -55.96 \\ 3.52 \end{bmatrix} \qquad \blacksquare
$$

■ SOLVED PROBLEM 11.7

Use the method of partitioning the matrix $\mathbf{S}^*$ to determine the support reactions for the structure in Fig. 11.7.

Solution. The relationship between the member end displacements (δ) and the nodal displacements (D) has been determined previously (Eq. 11.11); thus, owing to nodal displacements at joints b and c (denoted now as D_p), the member end displacements are

$$
\begin{bmatrix} \delta_1 \\ \delta_2 \\ \delta_3 \\ \delta_4 \\ \delta_5 \end{bmatrix}_P = \underbrace{\begin{bmatrix} 0 & 1 & 0 & 0 \\ -1 & 0 & 1 & 0 \\ 0 & 0 & 0 & 1 \\ -0.8 & 0.6 & 0 & 0 \\ 0 & 0 & 0.8 & 0.6 \end{bmatrix}}_{\mathbf{C}_p \uparrow} \begin{Bmatrix} D_{p1} \\ D_{p2} \\ D_{p3} \\ D_{p4} \end{Bmatrix}
$$

The matrix $\mathbf{C}_p$ is now defined as shown above. The matrix $\mathbf{C}_R$ is developed by referring to the various imposed displacements in the direction of the reactions at joints a and d, shown in Fig. 11.34. Thus,

$$
\begin{bmatrix} \delta_1 \\ \delta_2 \\ \delta_3 \\ \delta_4 \\ \delta_5 \end{bmatrix}_R = \underbrace{\begin{bmatrix} 0 & -1 & 0 & 0 \\ 0 & 0 & 0 & 0 \\ 0 & 0 & 0 & -1 \\ 0 & 0 & 0.8 & -0.6 \\ -0.8 & -0.6 & 0 & 0 \end{bmatrix}}_{\mathbf{C}_R \uparrow} \begin{bmatrix} D_{R1} \\ D_{R2} \\ D_{R3} \\ D_{R4} \end{bmatrix}
$$

Hence

$$[\delta] = [\delta]_p + [\delta]_R$$

Having determined the matrices $\mathbf{C}_p$ and $\mathbf{C}_R$, the submatrix $\mathbf{S}^*_{21}$ can be calculated from Eq. (11.58). Thus

$$\mathbf{S}^*_{21} = \mathbf{C}^T_R \mathbf{S} \mathbf{C}_p = \mathbf{C}^T_R \mathbf{S} \mathbf{E}^T$$

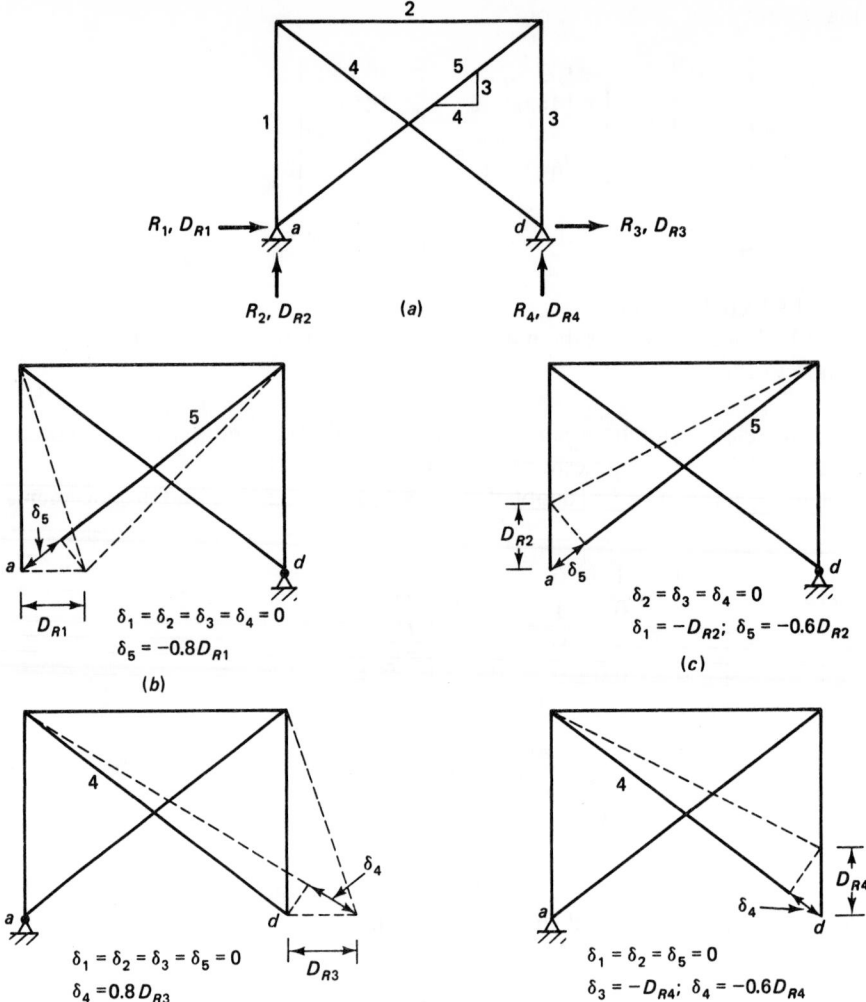

Figure 11.34 Truss for Solved Problem 11.7.

Knowing $\mathbf{C}_R$ we get $\mathbf{C}_R^T$ and $\mathbf{SC}_p = \mathbf{SE}^T$, which was determined before on page 508. Thus,

$$
\mathbf{S}_{21}^* = \frac{EA}{100}
\begin{bmatrix}
0 & 0 & 0 & 0 & -0.8 \\
-1 & 0 & 0 & 0 & -0.6 \\
0 & 0 & 0 & 0.8 & 0 \\
0 & 0 & -1 & -0.6 & 0
\end{bmatrix}
\begin{bmatrix}
0 & 6.67 & 0 & 0 \\
-5 & 0 & 5 & 0 \\
0 & 0 & 0 & 6.67 \\
-3.2 & 2.4 & 0 & 0 \\
0 & 0 & 3.2 & 2.4
\end{bmatrix}
$$

or

$$
\mathbf{S}_{21}^* = \frac{EA}{100}
\begin{bmatrix}
0 & 0 & -2.56 & -1.92 \\
0 & -6.67 & -1.92 & -1.44 \\
-2.56 & 1.92 & 0 & 0 \\
1.92 & -1.44 & 0 & -6.67
\end{bmatrix}
$$

From Eq. (11.60):

$$R = S_{21}^* D_p + S_{22}^* \overset{0}{\cancel{D}_R} = S_{21}^* D_p$$

($D_R = 0$, since points a and d are restrained.) D_p has been determined earlier (see page 508). Thus, with $EA = 30,000$ k,

$$R = \begin{bmatrix} R_1 \\ R_2 \\ R_3 \\ R_4 \end{bmatrix} = \frac{EA}{100} \begin{bmatrix} 0 & 0 & -2.56 & -1.92 \\ 0 & -6.67 & -1.92 & -1.44 \\ -2.56 & 1.92 & 0 & 0 \\ 1.92 & -1.44 & 0 & -6.67 \end{bmatrix} \begin{bmatrix} 3.31 \\ -1.68 \\ 3.67 \\ -5.81 \end{bmatrix} \times \frac{1}{300}$$

$$= \begin{bmatrix} 1.76 \\ 12.53 \\ -11.70 \\ 47.52 \end{bmatrix} \quad k$$

The reaction values are identical to those obtained previously (see Pages 508 and 509). ∎

PROBLEMS

11.1 to 11.12 Determine the static equilibrium matrix, stiffness matrix, and load matrix for each of the structures in Figs. P11.1 to P11.12.

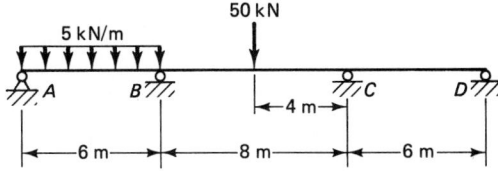

Figure P11.1

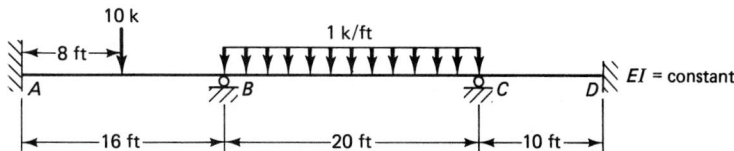

Figure P11.2

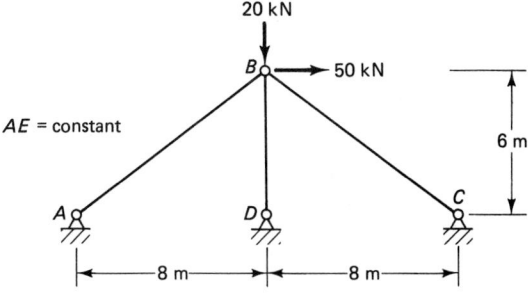

Figure P11.3

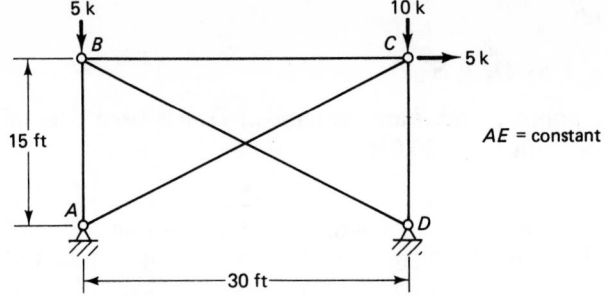

Figure P11.4

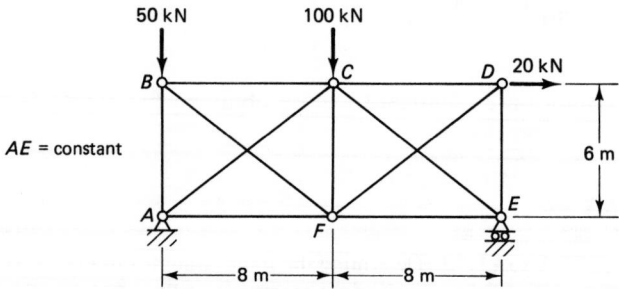

Figure P11.5

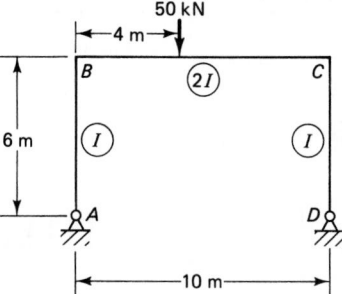

Figure P11.6

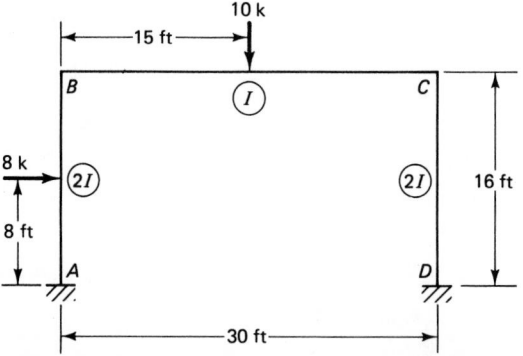

Figure P11.7

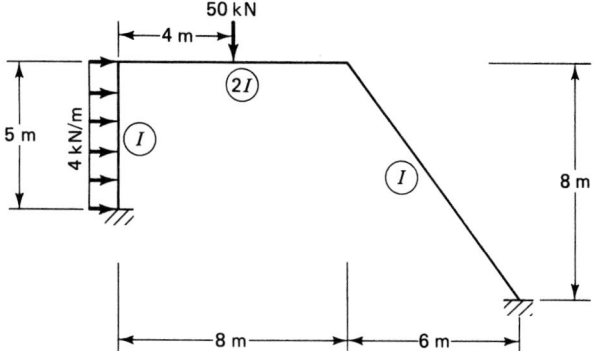

Figure P11.8

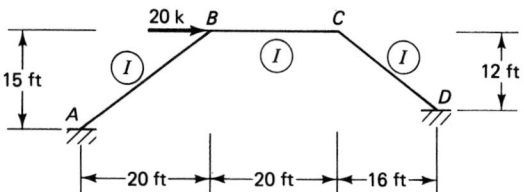

Figure P11.9

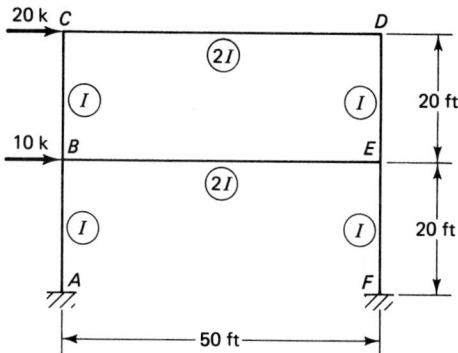

Figure P11.10

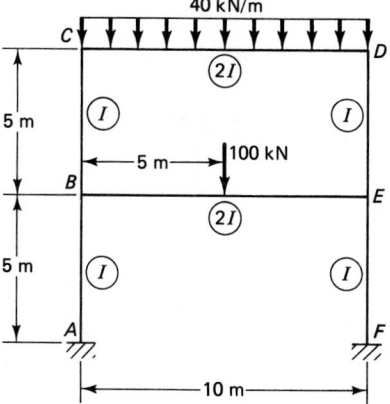

Figure P11.11

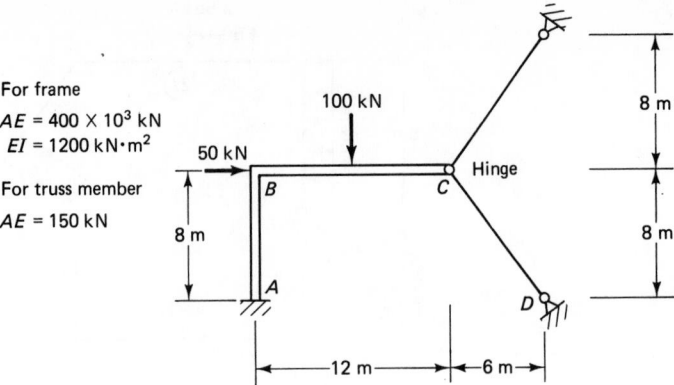

For frame

$AE = 400 \times 10^3$ kN
$EI = 1200$ kN·m²

For truss member

$AE = 150$ kN

Figure P11.12

11.13 to 11.19 Determine the global stiffness matrix **S*** for each of the structures in Figs. P11.1, P11.3, P11.5, P11.6, P11.9, P11.11, and P11.12.

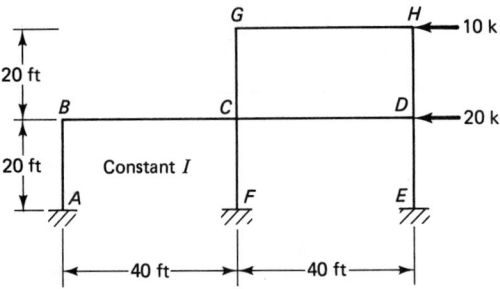

Figure P11.13

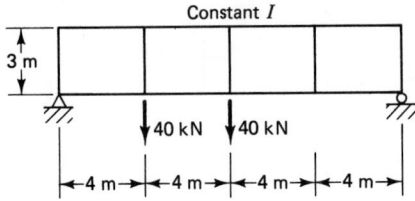

Figure P11.14

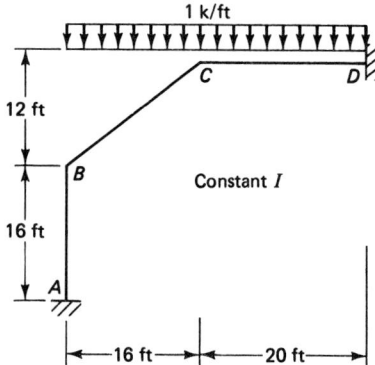

Figure P11.15

11.20 to 11.23 Calculate the final forces in each of the structures in Figs. P11.1, P11.4, P11.8, and P11.12.

11.24 to 11.25 Check equilibrium and compatibility in each of the structures in Figs. P11.1 and P11.12.

11.26 to 11.29 Use the method of partitioning the global stiffness matrix **S*** to determine the support reactions in each of the structures in Figs. P11.1, P11.5, P11.9, and P11.12.

11.30 Determine the forces in the structure in Fig. P11.4 caused by the loads shown and a vertical settlement of joint D equal to 0.2 in. Assume $AE = 10^5$ k.

11.31 Determine the forces in the structure in Fig. P11.6 caused by the loads shown and a vertical settlement of joint A of 0.004 m. Assume $EI = 1200$ kN $\cdot$ m^2.

11.32 Assume a vertical support at joint F in the structure of Fig. P11.5. Determine the final forces in the members.

11.33 to 11.35 For the structures in Figs. P11.13 to P11.15 determine (a) the static equilibrium matrix, (b) the stiffness matrix, (c) the load matrix, (d) the final forces.

Chapter 12

The Direct Stiffness Method

Imagination is more important than knowledge.

Albert Einstein

The stiffness or displacement method discussed in Chapter 11 entailed the formation of the global stiffness matrix $\mathbf{S}^*$ of the structure from the relationship

$$\mathbf{S}^* = \mathbf{ESE}^T \qquad\qquad (11.22)$$

in which matrices $\mathbf{E}$ and $\mathbf{S}$ were determined for the entire structure; equilibrium of forces at each node was expressed in the form

$$\mathbf{P} = \mathbf{S}^*\mathbf{D} \qquad\qquad (11.18)$$

relating the nodal forces $\mathbf{P}$ to the nodal displacements $\mathbf{D}$. The numerical calculations involved in the triple matrix product in Eq. (11.22) can become very laborious; furthermore, the member stiffness matrix $\mathbf{S}$ for the entire structure is relatively large and contains a large number of zero terms, requiring large computer storage space. The transformation process, implied in Eq. (11.22), can be avoided by simply calculating first the stiffness matrix of each individual member in the structure; these individual member stiffnesses are then combined in some fashion (discussed later) to arrive directly at the final global stiffness matrix $\mathbf{S}^*$, and hence the name *direct stiffness method*. This method, mentioned briefly in Chapter 10 (Pages 428–433), is suitable for the computer analysis of large structures. In this chapter we also consider the numerical aspect of the method in a manner suitable for programming a computer to analyze structures. In order to acquire a better understanding of the method, an in-depth discussion on the nature of the global stiffness matrix $\mathbf{S}^*$ is in order.

12.1 DERIVATION OF MATRIX $S*$ BY ENERGY CONSIDERATIONS

The stiffness coefficients of matrix $S*$ can be generated from energy considerations as follows: Consider the elastic body in Fig. 12.1, subjected to forces $P_1, \ldots, P_n$ and causing displacements $D_1, \ldots, D_n$. The external work W performed by these forces is

$$W = \tfrac{1}{2}(P_1 D_1 + \cdots + P_n D_n) \tag{12.1}$$

As a consequence of the external work, internal strain energy U is stored in the elastic body. From the principle of conservation of energy, we have

$$W = U \tag{12.2}$$

Now, let us assume that one displacement such as D_1 is changed by a small amount dD_1, while all the other displacements are held constant; then the change in the strain energy of the body is

$$dU = \frac{\partial U}{\partial D_1} dD_1 \tag{12.3}$$

and the corresponding change in the force P_1 is

$$dP_1 = \frac{\partial P_1}{\partial D_1} dD_1 \tag{12.4}$$

The change in the external work W, to a first order of approximation, is

$$dW = P_1 \, dD_1 \tag{12.5}$$

The conservation of energy requires that dW is equal to the change in the strain energy dU; that is,

$$dU = dW \tag{12.6}$$

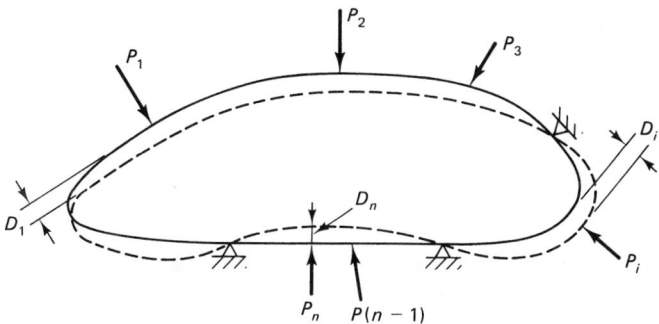

Figure 12.1 Deformation of elastic body subjected to forces $P_1, P_2, \cdots P_x$.

or

$$\frac{\partial U}{\partial D_1}\, dD_1 = P_1\, dD_1$$

hence,

$$P_1 = \frac{\partial U}{\partial D_1} \tag{12.7}$$

a result that coincides with Castigliano's theorem—part I (see Chapter 13, page 656). Differentiating the terms in Eq. (12.1) with respect to D_1 yields

$$\frac{\partial W}{\partial D_1} = \frac{1}{2}\left(P_1 + \frac{\partial P_1}{\partial D_1} D_1 + \frac{\partial P_2}{\partial D_1} D_2 + \cdots + \frac{\partial P_n}{\partial D_1} D_n \right)$$

$$= \frac{\partial U}{\partial D_1} \qquad (\text{since } U = W) \tag{12.8}$$

Using Eq. (12.7) in Eq. (12.8), we find

$$P_1 = \left(\frac{\partial P_1}{\partial D_1} D_1 + \frac{\partial P_2}{\partial D_1} D_2 + \cdots + \frac{\partial P_n}{\partial D_1} D_n \right)$$

Now, if we change D_2 by a small amount dD_2, this will yield a similar equation:

$$P_2 = \left(\frac{\partial P_1}{\partial D_2} D_1 + \frac{\partial P_2}{\partial D_2} D_2 + \cdots + \frac{\partial P_n}{\partial D_2} D_n \right)$$

Generally, one can write

$$P_n = \left(\frac{\partial P_1}{\partial D_n} D_1 + \frac{\partial P_2}{\partial D_n} D_2 + \cdots + \frac{\partial P_n}{\partial D_n} D_n \right)$$

Writing these n simultaneous equations in matrix form:

$$
\begin{bmatrix} P_1 \\ P_2 \\ \vdots \\ P_n \end{bmatrix}
=
\begin{bmatrix}
\dfrac{\partial P_1}{\partial D_1} & \dfrac{\partial P_2}{\partial D_1} & \cdots & \dfrac{\partial P_n}{\partial D_1} \\[2mm]
\dfrac{\partial P_1}{\partial D_2} & \dfrac{\partial P_2}{\partial D_2} & \cdots & \dfrac{\partial P_n}{\partial D_2} \\[2mm]
\cdots\cdots\cdots\cdots\cdots \\[2mm]
\dfrac{\partial P_1}{\partial D_n} & \dfrac{\partial P_2}{\partial D_n} & \cdots & \dfrac{\partial P_n}{\partial D_n}
\end{bmatrix}
\begin{bmatrix} D_1 \\ D_2 \\ \vdots \\ D_n \end{bmatrix}
\tag{12.9}
$$

$$\mathbf{P} = \mathbf{S}*\mathbf{D} \tag{12.10}$$

The coefficients of the stiffness matrix $\mathbf{S}*$ in Eq. (12.9) represent the rate of change of force with displacement; for a linearly elastic structure, such coefficients remain constant. Thus the partial derivative $\partial P_i/\partial D_j$ is the rate of change of force at node i (P_i) with variation of displacements at node j, with all other displacements being held constant during the change of D_j. If D_j is varied by unity, then $\partial P_i/\partial D_j$ becomes the force at node i required to maintain unit displacement at node j; whence column j of the stiffness matrix in Eq. (12.9) comprises a force system required to maintain unit displacement at node j, with all other displacements being held constant during the displacement at node j. Thus, the coefficients of the stiffness matrix S^* can be generated by calculating the force system required to maintain unit displacements at each freedom node in turn. Such a procedure was followed in Chapter 10, page 428.

To demonstrate the physical meaning of the stiffness coefficients, consider the frame shown in Fig. 12.2; the frame has four degrees of freedom and is subjected to nodal forces $P_1, \ldots, P_4$, causing displacements $D_1, \ldots, D_4$. The nodal force systems required to maintain unit displacements in the direction of each nodal freedom in turn are shown in Fig. 12.3. Assuming linear elastic behavior, the force system associated with displacement D_i will be equal to D_i times the force system associated with unit displacement in the direction of the nodal freedom i. Thus, if the force systems associated with each displacement are summed up, then, using the principle of superposition (page 34), the total restraining nodal force must be equal to the corresponding applied nodal forces. For example, for the structure in Fig. 12.2, using the information in Fig. 12.3, we have

Ontario Hydro head office, Toronto, Ontario, Canada. (Photo courtesy of Canadian Institute of Steel Construction.)

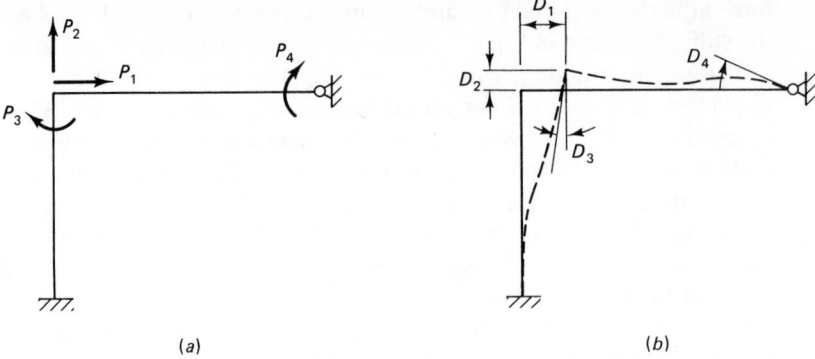

Figure 12.2 Nodal forces and displacements.

$$D_1 S_{11}^* + D_2 S_{12}^* + D_3 S_{13}^* + D_4 S_{14}^* = P_1$$
$$D_1 S_{21}^* + D_2 S_{22}^* + D_3 S_{23}^* + D_4 S_{24}^* = P_2$$
$$D_1 S_{31}^* + D_2 S_{32}^* + D_3 S_{33}^* + D_4 S_{34}^* = P_3$$
$$D_1 S_{41}^* + D_2 S_{42}^* + D_3 S_{43}^* + D_4 S_{44}^* = P_4$$

$$(12.11)$$

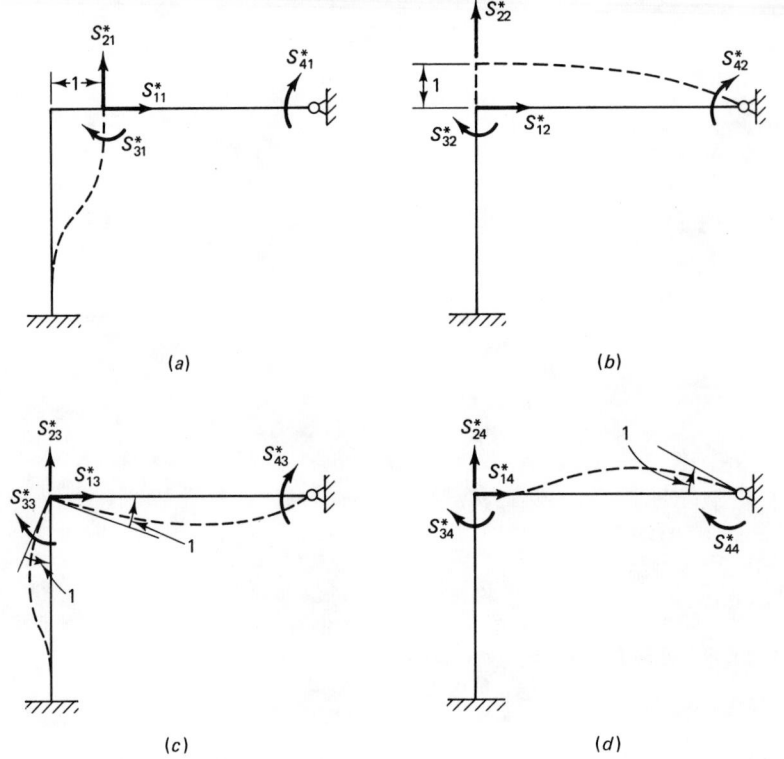

Figure 12.3 Nodal forces resulting from unit displacement at each nodal freedom.

or

$$\mathbf{P} = \mathbf{S}^*\mathbf{D} \qquad\qquad (12.10)$$

To summarize, the direct stiffness method requires (1) identification of the nodal freedoms; (2) derivation of the stiffness coefficients by applying unit displacements at each nodal freedom in turn; (3) formation of the external force vector P from the applied loads; (4) solution of the resulting simultaneous equations, for example, Eqs. (12.11), to obtain the unknown nodal displacements; and (5) evaluation of the member forces from the nodal displacements. Before embarking on demonstrating the application of the method, it is instructive to enumerate the following properties of the global stiffness matrix $\mathbf{S}^*$:

1. For a properly restrained structure the global stiffness matrix is a square matrix of dimension n which is equal to the degrees of freedom of the structure; it has an inverse. For an unrestrained structure, the global stiffness matrix is singular; that is, it has no inverse.

2. It is a symmetric matrix. From Castigliano's theorem—part I (see Chapter 13) we have

$$P_i = \frac{\partial U}{\partial D_i} \quad \text{thus} \quad \frac{\partial P_i}{\partial D_j} = \frac{\partial^2 U}{\partial D_j\,\partial D_i} \quad \text{also} \quad \frac{\partial P_j}{\partial D_i} = \frac{\partial^2 U}{\partial D_i\,\partial D_j}$$

Since the order of differentiation is of no consequence, we can put

$$\frac{\partial^2 U}{\partial D_i\,\partial D_j} = \frac{\partial^2 U}{\partial D_j\,\partial D_i} \quad \text{or} \quad \frac{\partial P_i}{\partial D_j} = \frac{\partial P_j}{\partial D_i}$$

After comparing the terms in Eqs. (12.9) and (12.11), this means that

$$S^*_{ij} = S^*_{ji} \qquad\qquad (12.12)$$

implying symmetry.

3. It is a banded matrix. The global stiffness matrix $\mathbf{S}^*$ can be formed such that the nonzero values appear only in a band along the main diagonal of the matrix. A good banding can be accomplished if the nodal freedoms are well numbered. Consider the one-story multibay rigid frame in Fig. 12.4a showing the numbering of the nodal freedoms. By definition of stiffness coefficients, each column i in the matrix $\mathbf{S}^*$ for the frame in Fig. 12.4a represents the restraining force required to maintain a unit displacement in the direction of the imposed freedom i; hence the global stiffness matrix is as shown in Fig. 12.5; for example, to maintain a unit rotation in the direction of freedom 15 (Fig. 12.4b) will require a restraining force system, composed of nine forces (stiffness coefficients), as shown in Fig. 12.4b. It can be observed from Fig. 12.5 that the nonzero stiffness coefficients in the resulting matrix are clustered around the diagonal; that is, the *stiffness matrix is banded.* Such a band would not have been formed if the nodal degrees of freedom were poorly numbered as shown in Fig. 12.4c. It is left as an exercise for the student to show this. It is important to mention that a banded matrix can save considerable computer storage space and time; for the rigid frame in Fig. 12.4a, only the upper semi-band is stored as shown in Fig. 12.5.

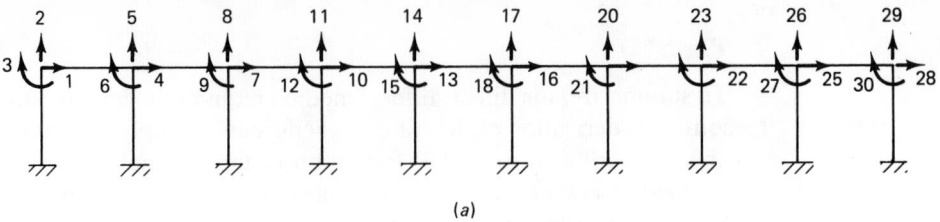

(a)

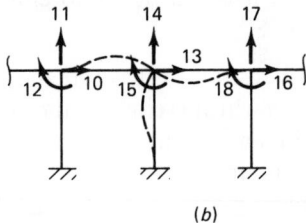

(b)

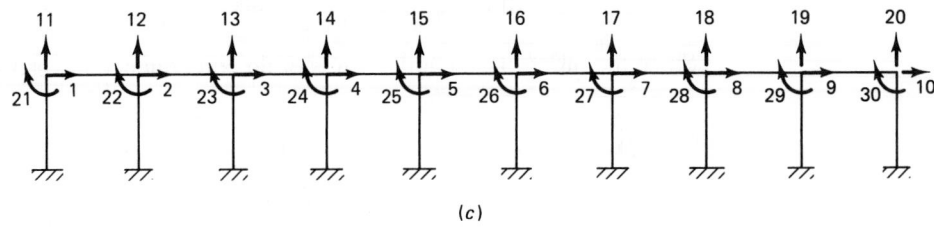

(c)

Figure 12.4 Nodal numbering schemes.

4. It is a positive definite matrix. This is a useful property in the solution of the generated simultaneous equations (in $\mathbf{P} = \mathbf{S}^*\mathbf{D}$). Consider the external work done on a stable structure by a system of external forces P during the displacement of the joints D. Such work (W) is the sum of ($\frac{1}{2}$) (load)(corresponding displacement), the sum being taken for all the external loads. To represent W in matrix form we must first transpose the load matrix $\mathbf{P}$ to make it conformable for multiplication. Thus

$$W = \tfrac{1}{2}\mathbf{P}^T\mathbf{D} \tag{12.13}$$

or [after using Eq. (12.10)],

$$W = \tfrac{1}{2}\mathbf{D}^T\mathbf{S}^{*T}\mathbf{D}$$

But from Eq. (12.12), $\mathbf{S}^{*T} = \mathbf{S}^*$. Thus

$$W = \tfrac{1}{2}\mathbf{D}^T\mathbf{S}^*\mathbf{D} \tag{12.14}$$

For a structure in a state of stable equilibrium the work W done on the structure will increase when subjected to any arbitrary small displacement ΔD. Thus

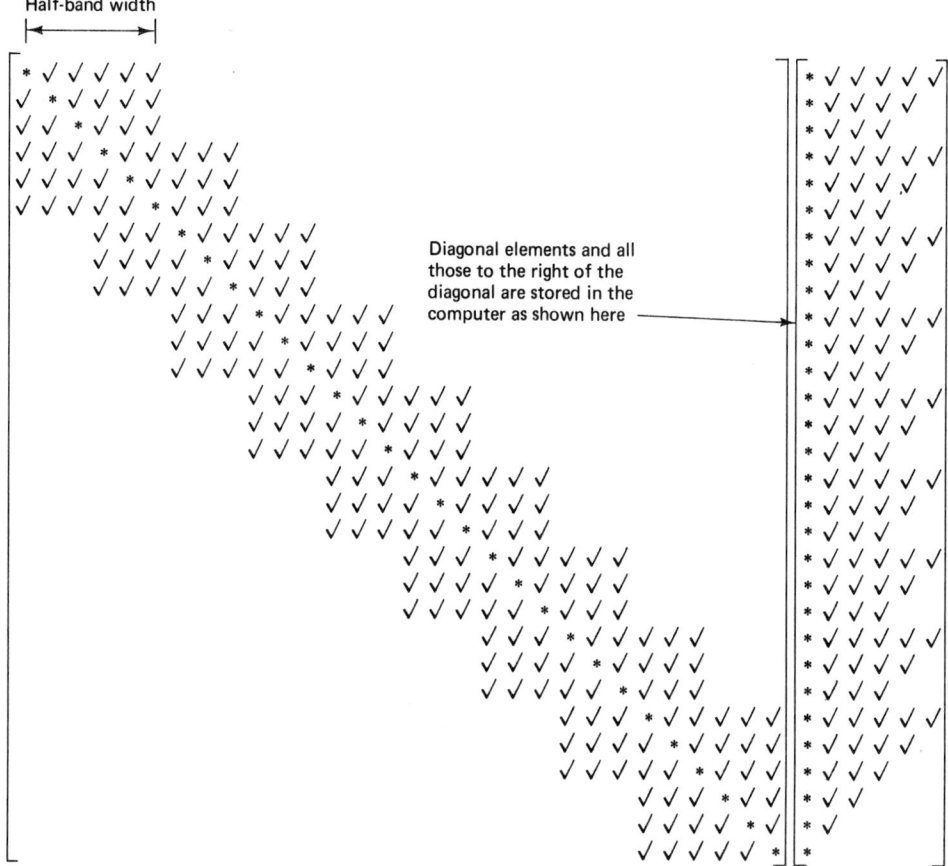

Figure 12.5 Banding of the structure stiffness matrix.

the stiffness matrix **S*** must be positive definite to meet this requirement; such a matrix will have a determinant greater than zero, that is, it is nonsingular. This property is used to advantage in the solution of simultaneous equations, **P = S*D**. The criterion of nonsingularity of a stiffness matrix is also quite often utilized to determine the buckling load of a structure by putting the determinant of the stiffness matrix to zero.

12.2 TRANSFORMATION MATRICES

To apply the direct stiffness method in analyzing a structure we require to define: (1) a local coordinate system applied to each member of the structure and (2) a global coordinate system applied to the structure as a whole. Thus, for the two-dimensional structure shown in Fig. 12.6a, the local coordinate system (x,y,z) for each member is

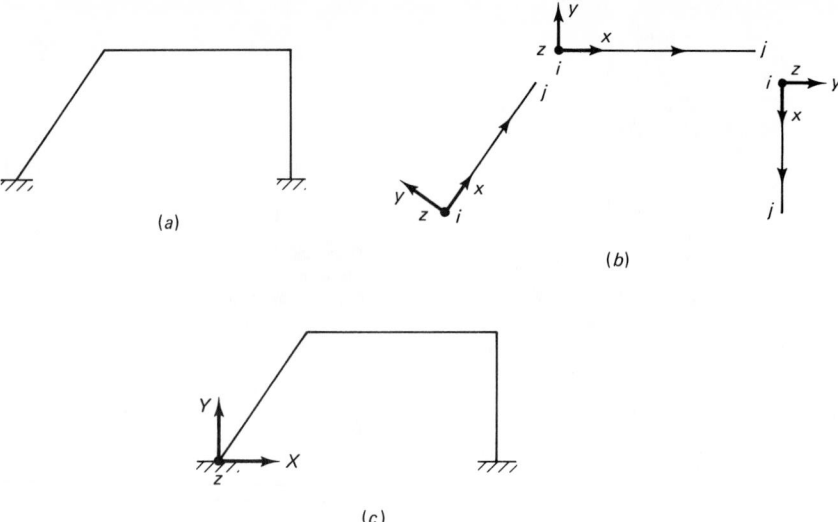

Figure 12.6 Frame with local and global coordinates.

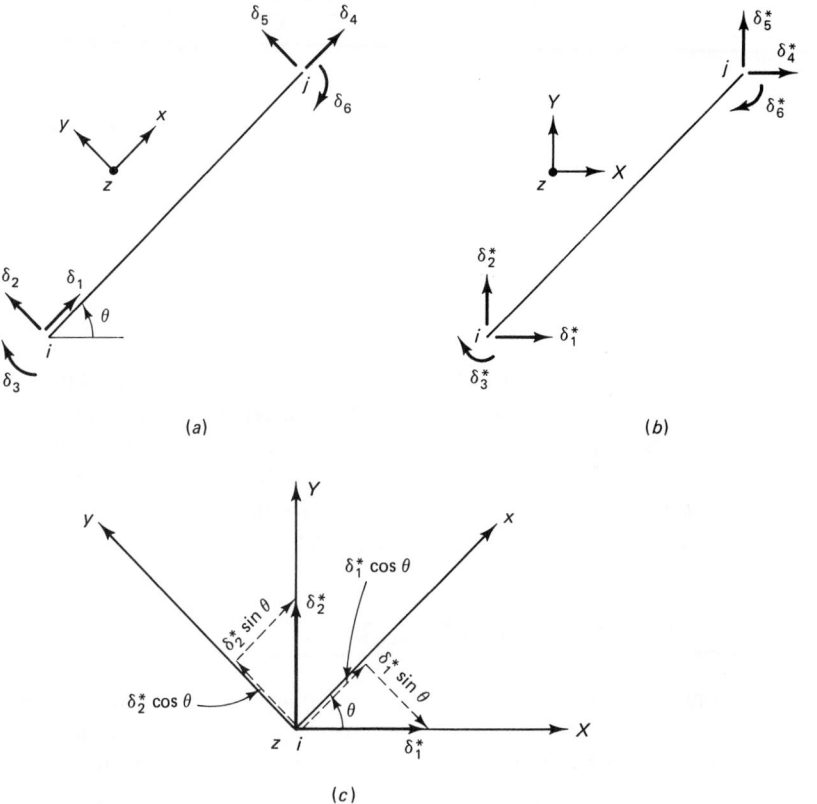

Figure 12.7 Relation between displacements in local and global coordinates.

shown in Fig. 12.6b, while the global coordinate system (X,Y,z) for the structure is given in Fig. 12.6c. Each member has a start point i and an end point j as shown. In order to be able to assemble the global stiffness matrix $\mathbf{S}^*$ we require to transform local member stiffnesses to stiffnesses in terms of global coordinates. Consider a member, in a plane, rigidly connected at both ends and inclined to the X axis by an angle θ. The member end displacement components in local coordinates and in global coordinates, that is, δ and δ^*, are shown in Fig. 12.7a and b, respectively. By superposition, it can be shown that they are related. Thus, referring to Fig. 12.7c and considering the displacements at end i, we have

$$\delta_1 = \delta_1^* \cos\theta + \delta_2^* \sin\theta$$

$$\delta_2 = -\delta_1^* \sin\theta + \delta_2^* \cos\theta \tag{12.15}$$

$$\delta_3 = \delta_3^*$$

The third equation states that rotation remains unchanged in a plane coordinate transformation. Similarly, for end j, we have

$$\delta_4 = \delta_4^* \cos\theta + \delta_5^* \sin\theta$$

$$\delta_5 = -\delta_4^* \sin\theta + \delta_5^* \cos\theta \tag{12.16}$$

$$\delta_6 = \delta_6^*$$

Combining Eqs. (12.15) and (12.16) we have, in matrix form

$$
\begin{bmatrix} \delta_1 \\ \delta_2 \\ \delta_3 \\ \delta_4 \\ \delta_5 \\ \delta_6 \end{bmatrix}
=
\underbrace{\begin{bmatrix}
\cos\theta & \sin\theta & 0 & 0 & 0 & 0 \\
-\sin\theta & \cos\theta & 0 & 0 & 0 & 0 \\
0 & 0 & 1 & 0 & 0 & 0 \\
0 & 0 & 0 & \cos\theta & \sin\theta & 0 \\
0 & 0 & 0 & -\sin\theta & \cos\theta & 0 \\
0 & 0 & 0 & 0 & 0 & 1
\end{bmatrix}}_{\mathbf{R}}
\begin{bmatrix} \delta_1^* \\ \delta_2^* \\ \delta_3^* \\ \delta_4^* \\ \delta_5^* \\ \delta_6^* \end{bmatrix}
\tag{12.17}
$$

or

$$\delta = \mathbf{R}\delta^* \tag{12.18}$$

in which $\mathbf{R}$ is the transformation matrix between the local and global coordinate systems. The square matrix $\mathbf{R}$ can be shown to be orthogonal, that is,

$$\mathbf{R}^{-1} = \mathbf{R}^T \tag{12.19}$$

Since both displacements and forces are vector quantities, the end member forces in local and global coordinates, F and F^*, respectively, for member ij, Fig. 12.8a and b, are also related by the same matrix $\mathbf{R}$; or

$$F = \mathbf{R}F^* \tag{12.20}$$

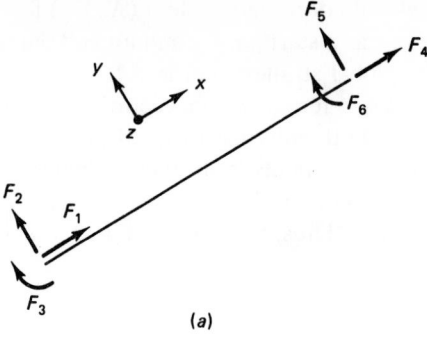

(a)

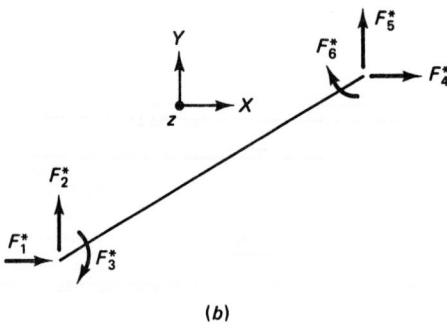

(b)

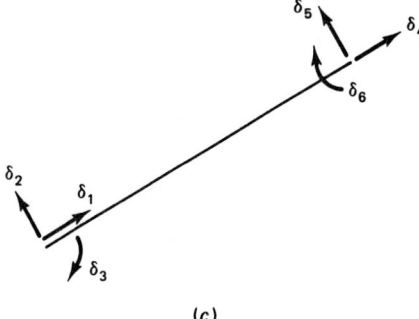

Figure 12.8 End forces in local and global coordinates.

(c)

In a plane truss, the transformation matrix for an axially loaded member can be deduced from the transformation matrix $\mathbf{R}$ in Eq. (12.17) by deleting rows and columns 3 and 6, resulting in

$$\mathbf{R} = \begin{bmatrix} \cos\theta & \sin\theta & 0 & 0 \\ -\sin\theta & \cos\theta & 0 & 0 \\ 0 & 0 & \cos\theta & \sin\theta \\ 0 & 0 & -\sin\theta & \cos\theta \end{bmatrix} \qquad (12.21)$$

12.3 MEMBER STIFFNESS MATRIX

Consider a member ij of a rigid frame subjected to bending, shear, and axial forces; the resulting member end forces (F) and displacements (δ) in local coordinates are shown in Fig. 12.8a and c, respectively. The relationship between **F** and δ is established through the member stiffness **S** in local coordinates, that is,

$$\mathbf{F} = \mathbf{S}\delta \qquad (12.22)$$

Writing Eq. (12.22) in full:

$$
\begin{bmatrix} F_1 \\ F_2 \\ F_3 \\ F_4 \\ F_5 \\ F_6 \end{bmatrix}
=
\begin{bmatrix}
S_{11} & S_{12} & \cdots & S_{16} \\
S_{21} & S_{22} & \cdots & S_{26} \\
\cdot & \cdot & \cdots & \cdot \\
\cdot & \cdot & \cdots & \cdot \\
\cdot & \cdot & \cdots & \cdot \\
S_{61} & S_{62} & \cdots & S_{66}
\end{bmatrix}
\begin{bmatrix} \delta_1 \\ \delta_2 \\ \delta_3 \\ \delta_4 \\ \delta_5 \\ \delta_6 \end{bmatrix}
\qquad (12.23)
$$

(see also Chapter 10, page 428). Using Eq. (12.18) we can write Eq. (12.22) as

$$\mathbf{F} = \mathbf{SR}\delta^* \qquad (12.24)$$

Although the meaning of the stiffness coefficients S_{ij} has been explained briefly (see pages 428–433), it is instructive to explain it more fully here. The elements S_{ij} in the j column are the restraining forces necessary to hold the member in the deformed shape caused by a displacement $\delta_j = 1$ while the other displacement components are kept at zero. For example, if the imposed displacement $\delta_1 = 1$ and $\delta_2 = \delta_3 = \delta_4 = \delta_5 = \delta_6 = 0$ (Fig. 12.9a), then from Eq. (12.23), $F_1 = S_{11}; F_2 = S_{21}; F_3 = S_{31}; F_4 = S_{41}; F_5 = S_{51}; F_6 = S_{61}$. This condition is shown in Fig. 12.9a; for such a deformation, the member strain $\varepsilon = 1/l$; member stress $\sigma = \varepsilon E = E/l$; axial force in member $= \sigma A = EA/l = S_{11}$. From force equilibrium in the x direction: $S_{11} + S_{14} = 0$; thus $S_{14} = -S_{11} = -EA/l$. Also, $S_{12} = S_{13} = S_{15} = S_{16} = 0$. For the imposed displacement $\delta_2 = 1$ (and $\delta_1 = \delta_3 = \delta_4 = \delta_5 = \delta_6 = 0$) in Fig. 12.9b, we have, from the slope-deflection equations,

$$M_{AB} = \frac{2EI}{l}\left(2\,\theta_A^{\,0} + \theta_B^{\,0} - \frac{3\Delta}{l}\right) + \mathrm{FEM}_{AB}$$

$$M_{BA} = \frac{2EI}{l}\left(2\,\theta_B^{\,0} + \theta_A^{\,0} - \frac{3\Delta}{l}\right) + \mathrm{FEM}_{BA} \qquad (10.4)$$

or

$$M_{AB} = \frac{-6EI\Delta}{l^2} \qquad M_{BA} = \frac{-6EI\Delta}{l^2} \qquad \text{since } \mathrm{FEM}_{AB} = \mathrm{FEM}_{BA} = 0.$$

Putting $\Delta = \delta_2 = 1$, then $S_{23} = -M_{AB} = 6EI/l^2$; $S_{26} = -M_{BA} = 6EI/l^2$. The end shear $S_{22} = (S_{23} + S_{26})/l = 12EI/l^3$; from force equilibrium in the vertical direction: $S_{22} + S_{25} = 0$; or $S_{25} = -S_{22} = -12EI/l^3$. Since there are no axial deformations, $S_{21} = S_{24} = 0$.

For the imposed deformation $\delta_3 = 1$ (and $\delta_1 = \delta_2 = \delta_4 = \delta_5 = \delta_6 = 0$) in Fig. 12.9$c$: From the slope-deflection equation [Eq. (10.4)], with $\text{FEM}_{AB} = \text{FEM}_{BA} = \Delta = \theta_B = 0$, and $\theta_A = \delta_3 = 1$, $M_{AB} = 4EI/l$ and $M_{BA} = 2EI/l$. Thus, $S_{33} = M_{AB} = 4EI/l$; $S_{36} = M_{BA} = 2EI/l$; the shear $S_{32} = (S_{33} + S_{36})/l = 6EI/l^2$; and, from vertical force equilibrium, $S_{32} + S_{35} = 0$; or $S_{35} = -6EI/l^2$. The other stiffness coefficients, shown in Fig. 12.9d, e, and f, can be generated in the same way. Thus, the complete member stiffness matrix in local coordinates can be written as

$$
\mathbf{S} =
\begin{bmatrix}
\dfrac{EA}{l} & 0 & 0 & \dfrac{-EA}{l} & 0 & 0 \\[2mm]
0 & \dfrac{12EI}{l^3} & \dfrac{-6EI}{l^2} & 0 & \dfrac{-12EI}{l^3} & \dfrac{-6EI}{l^2} \\[2mm]
0 & \dfrac{-6EI}{l^2} & \dfrac{4EI}{l} & 0 & \dfrac{6EI}{l^2} & \dfrac{2EI}{l} \\[2mm]
\dfrac{-EA}{l} & 0 & 0 & \dfrac{EA}{l} & 0 & 0 \\[2mm]
0 & \dfrac{-12EI}{l^3} & \dfrac{6EI}{l^2} & 0 & \dfrac{12EI}{l^3} & \dfrac{6EI}{l^2} \\[2mm]
0 & \dfrac{-6EI}{l^2} & \dfrac{2EI}{l} & 0 & \dfrac{6EI}{l^2} & \dfrac{4EI}{l}
\end{bmatrix}
\qquad (12.25)
$$

This is a 6×6 stiffness matrix in local coordinates for a member in a plane and subjected to both bending and axial load; each member has three degrees of freedom. The matrix is symmetric; that is, $\mathbf{S}_{ij} = \mathbf{S}_{ji}$, which accords with Maxwell's principle of reciprocity. For a member with an end hinged while the other end is rigidly connected, the stiffness matrix $\mathbf{S}$ reduces to 5×5 size. In the derivation of Eq. (12.25) it was tacitly assumed that shear deformation is negligible.

In an articulated structure (truss) each member's end has two degrees of freedom, horizontal and vertical displacements; in this case the rotational degree of freedom is not restrained and therefore does not activate restraining forces. Thus, if the rows and columns associated with the rotational degree of freedom δ_3 and δ_6 are deleted and if $EI = 0$ is substituted (since there is no bending in an idealized truss member) in Eq. (12.25), then

$$
\mathbf{S} =
\begin{bmatrix}
\dfrac{EA}{l} & 0 & \dfrac{-EA}{l} & 0 \\[2mm]
0 & 0 & 0 & 0 \\[2mm]
\dfrac{-EA}{l} & 0 & \dfrac{EA}{l} & 0 \\[2mm]
0 & 0 & 0 & 0
\end{bmatrix}
\qquad (12.26)
$$

which is the stiffness matrix of a truss member in local coordinates.

For a member subjected to bending and shear forces at its ends (i.e., ends that can rotate as well as translate perpendicular to the member), its stiffness matrix $\mathbf{S}$ can

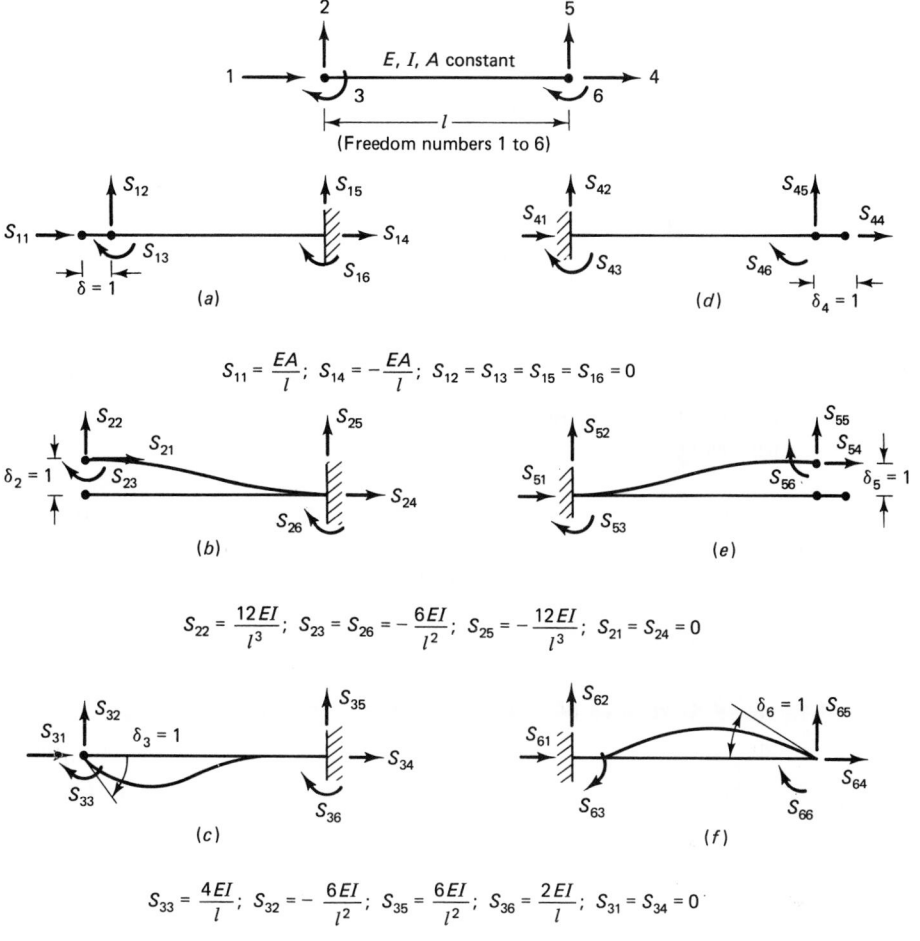

Figure 12.9 Member stiffness influence coefficients.

be deduced from Eq. (12.25) by deleting the rows and columns associated with freedoms δ_1 and δ_4; thus the stiffness matrix for such a member becomes

$$
\mathbf{S} =
\begin{bmatrix}
\dfrac{12\,EI}{l^3} & \dfrac{-6\,EI}{l^2} & \dfrac{-12\,EI}{l^3} & \dfrac{-6\,EI}{l^2} \\[2ex]
\dfrac{-6\,EI}{l^2} & \dfrac{4\,EI}{l} & \dfrac{6\,EI}{l^2} & \dfrac{2\,EI}{l} \\[2ex]
\dfrac{-12\,EI}{l^3} & \dfrac{6\,EI}{l^2} & \dfrac{12\,EI}{l^3} & \dfrac{6\,EI}{l^2} \\[2ex]
\dfrac{-6\,EI}{l^2} & \dfrac{2\,EI}{l} & \dfrac{6\,EI}{l^2} & \dfrac{4\,EI}{l}
\end{bmatrix}
\qquad (12.27)
$$

For a member subjected to bending only (ends rotate but do not translate), the member's stiffness matrix can be formed from Eq. (12.25) by deleting the rows and columns associated with freedoms δ_1, δ_2, δ_4, and δ_5 (only freedoms δ_3 and δ_6 are relevant in this case, e.g., in continuous beams); thus the member stiffness matrix in local coordinates for a member in bending becomes

$$
\mathbf{S} = \begin{bmatrix} \dfrac{4EI}{l} & \dfrac{2EI}{l} \\[2ex] \dfrac{2EI}{l} & \dfrac{4EI}{l} \end{bmatrix}
\tag{12.28}
$$

It is interesting to note that the inverse of the 6×6 member stiffness matrix $\mathbf{S}$ given by Eq. (12.25) cannot be obtained since its determinant is zero. This is due to the following identities:

1. Columns 1 and 4 are identical except for their signs being reversed; the same observation applies to columns 2 and 5.
2. The coefficients in column 6 can be determined by multiplying the coefficients of column 5 by $(-l)$ and subtracting the coefficients of column 3 from it.

12.3.1 Member Stiffness Matrix in Global Coordinates

The stiffness matrix for the whole structure can be assembled from the stiffness matrix of its members. To do this, the member stiffness matrices must be transformed from local to global coordinates. Substituting Eq. (12.18) and Eq. (12.20) into Eq. (12.22) yields

$$\mathbf{RF^*} = \mathbf{SR}\delta^*$$

Premultiplying both sides by $\mathbf{R}^{-1}$ results in

$$\mathbf{F^*} = \mathbf{R}^{-1}\mathbf{SR}\delta^*$$

From Eq. (12.19), $\mathbf{R}^{-1} = \mathbf{R}^T$; thus

$$\mathbf{F^*} = (\mathbf{R}^T\mathbf{SR})\delta^* \tag{12.29}$$

Defining

$$\mathbf{S}_m^* = \mathbf{R}^T\mathbf{SR} \tag{12.30}$$

as the member stiffness matrix in global coordinates, Eq. (12.29) can now be written as

$$\mathbf{F^*} = \mathbf{S}_m^*\delta^* \tag{12.31}$$

For a two-dimensional frame member subjected to bending, shear, and axial forces the expanded form for $\mathbf{S}_m^*$ [Eq. (12.30)] is determined from Eqs. (12.17) and (12.25).

Thus,

$$\mathbf{S}_m^* = \frac{E}{l} \times \begin{bmatrix} \left(Ac^2 + \frac{12Is^2}{l^2}\right) & \left(A - \frac{12I}{l^2}\right)sc & +\frac{6I}{l}s & \left(-Ac^2 - \frac{12I}{l^2}s^2\right) & \left(-A + \frac{12I}{l^2}\right)sc & +\frac{6I}{l}s \\ & \left(As^2 + \frac{12I}{l^2}c^2\right) & -\frac{6I}{l}c & \left(-A + \frac{12I}{l^2}\right)sc & -\left(As^2 + \frac{12I}{l^2}c^2\right) & -\frac{6I}{l}c \\ & & 4I & -\frac{6I}{l}s & +\frac{6Ic}{l} & 2I \\ & & & \left(Ac^2 + \frac{12I}{l^2}s^2\right) & \left(A - \frac{12I}{l^2}\right)sc & -\frac{6I}{l}s \\ & \text{Symmetric} & & & \left(As^2 + \frac{12I}{l^2}c^2\right) & +\frac{6I}{l}c \\ & & & & & 4I \end{bmatrix}$$

(12.30a)

in which $c = \cos\theta$ and $s = \sin\theta$.

For a two-dimensional frame member subjected to bending and shear force only, $\mathbf{S}_m^*$ can be deduced from Eq. (12.30a) by putting $A = 0$. Thus,

$$\mathbf{S}_m^* = \frac{12EI}{l^3} \times \begin{bmatrix} s^2 & -sc & ls/2 & -s^2 & sc & ls/2 \\ & c^2 & -lc/2 & sc & -c^2 & -lc/2 \\ & & l^2/3 & -ls/2 & lc/2 & l^2/6 \\ & & & s^2 & -sc & -ls/2 \\ & \text{Symmetric} & & & c^2 & lc/2 \\ & & & & & l^2/3 \end{bmatrix}$$

(12.30b)

For members with end rotation but no translation, $\mathbf{S}_m^*$ is obtained from Eq. (12.30b) by deleting rows as well as columns 1, 2, 4 and 5; thus,

$$\mathbf{S}_m^* = \frac{2EI}{l} \times \begin{bmatrix} 2 & 1 \\ 1 & 2 \end{bmatrix}$$

(12.30c)

For an axially loaded member, $\mathbf{S}_m^*$ is deduced from Eq. (12.30a) by deleting rows as well as columns 3 and 6 and putting the moment of inertia $I = 0$; thus,

$$\mathbf{S}_m^* = \frac{AE}{l} \times \begin{bmatrix} c^2 & sc & -c^2 & -cs \\ & s^2 & -sc & -s^2 \\ & \text{Symmetric} & c^2 & sc \\ & & & s^2 \end{bmatrix}$$

(12.30d)

Equations (12.30a) to (12.30d) give the element stiffness matrix in global or structure coordinates. In these equations the values of $c = \cos\theta$ and $s = \sin\theta$ for each member are known and the extremities of each member are designated by the freedom numbers at the nodes. Thus, these freedom numbers give the row and column designation of matrix $\mathbf{S}_m^*$ and are used to properly place the individual terms of $\mathbf{S}_m^*$ in the global stiffness matrix S^* of the whole structure. This procedure is demonstrated in Example 12.1.

12.4 GLOBAL STIFFNESS MATRIX

The individual member stiffness matrices in global coordinates [Eq. (12.30)] are now used to form the global stiffness matrix S^* which governs the relationship between external nodal forces and nodal displacements ($P = S^* D$). The successful formation of S^* by merging of the member stiffness matrices (S_m^*) can be demonstrated as follows: For a structure containing t nodes, Eq. (12.10) can be written in the form

$$\left\{ \begin{array}{c} P_1 \\ \cdot \\ P_i \\ \cdot \\ P_t \end{array} \right\} = \left[\cdots (S_m^*)_{in} \cdots (S_m^*)_{ii} \cdots (S_m^*)_{ir} \right] \left\{ \begin{array}{c} \cdot \\ \cdot \\ D_n \\ \cdot \\ D_i \\ \cdot \\ D_r \\ \cdot \end{array} \right\} \qquad (12.32)$$

in which P_i and D_i are the external nodal force and displacement vectors at node i, respectively. The meaning of the submatrices $(S_m^*)_{in}$, $(S_m^*)_{ii}$, and $(S_m^*)_{ir}$ can be explained by referring to Fig. 12.10. The ith node of a structure is shown in Fig. 12.10a with members $ni, \ldots, ji, \ldots, ri$ framing into node i. Now if only node n is given a

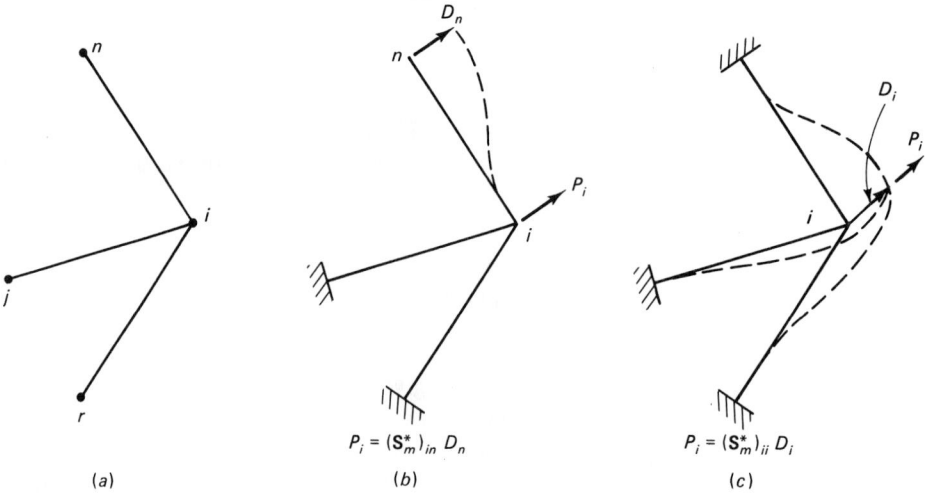

Figure 12.10 Dependency of stiffness coefficient on end displacements.

displacement D_n (Fig. 12.10b), the restraining force vector activated at node i will be $P_i = (S_m^*)_{in} D_n$; thus the submatrix $(S_m^*)_{in}$ is a stiffness matrix that relates the force vector at node i to the displacement at node n; the elements of this submatrix are determined by the stiffness properties of the deformed member ni only. Furthermore, if node i is given a displacement D_i (Fig. 12.10c), the restraining force vector activated at node i, P_i, will be given by $P_i = (S_m^*)_{ii} D_i$; the submatrix $(S_m^*)_{ii}$ relates the force vector at node i resulting from a displacement at node i; the elements of this submatrix depend on the stiffness properties of all the members framing into node i as shown in Fig. 12.10c. Thus, one can write

$$(S^*)_{ii} = (S_m^*)_{ii}^n + \cdots + (S_m^*)_{ii}^i + \cdots + (S_m^*)_{ii}^r = \sum_{j=n}^{r} (S_m^*)_{ii}^j \qquad (12.33)$$

Thus, the submatrix $(S_m^*)_{ii}^j$ connects the nodal force vector P_i induced at node i and caused by the deformation of member ij due to the displacement vector D_i at node i. Hence, we can write the following expression for the nodal force vector P at node i in terms of displacement vector D_i as well as the displacement vector D_j for all members joined at node i:

$$P_i = \sum_{j=n}^{r} [(S_m^*)_{ii}^j D_i + (S_m^*)_{ij} D_j] \qquad (12.34)$$

This means that the force vector P at node i is a function not only of the nodal stiffnesses of all the members joined at i but also of the nodal stiffnesses at the far end of each member joined at node i. Thus, the global stiffness matrix S^* is obtained by summing up the stiffness matrices of individual members, S_m^*. In so doing, both compatibility of displacements and equilibrium of forces are ensured at each node. For example, equilibrium is ensured when equating the member end forces to the applied nodal forces; furthermore, nodal compatibility is also invoked when it is implied that displacement at node i associated with member ij is the same as the displacement of node i for member ir or in.

■ **EXAMPLE 12.1**
Develop the global stiffness matrix S^* for the truss in Fig. 12.11. Use the direct stiffness method.

Solution. The degrees of freedom on all the joints, including the two restrained ones, are shown numbered. Since this is a truss structure, each joint has only two degrees of freedom. The positive direction of each member is shown by an arrow, leading from the lower joint number at the near end of the member to the higher joint number at the far end of the member; this is a useful convention that would allow the member stiffness matrices S_m^* to be added to form the global stiffness matrix S^*. Table 12.1 shows the member number, node or joint number, and direction cosines of the six members in the truss. The origin for each member is taken at the lesser joint number with the X axis pointing to the joint having a higher joint number. The member stiffness matrix in local coordinates is given by Eq. (12.26) as

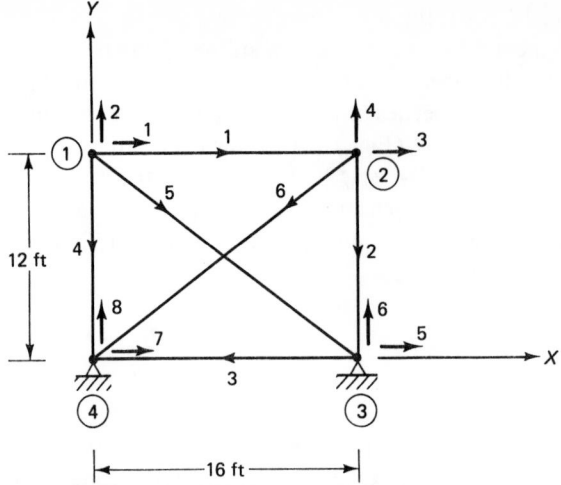

Figure 12.11 Structure for Example 12.1. Given: $AE = 144,000$ k; constant for all members.

$$
S = \begin{bmatrix}
\dfrac{EA}{l} & 0 & \dfrac{-EA}{l} & 0 \\[2mm]
0 & 0 & 0 & 0 \\[2mm]
\dfrac{-EA}{l} & 0 & \dfrac{EA}{l} & 0 \\[2mm]
0 & 0 & 0 & 0
\end{bmatrix}
$$

and the transformation (rotation) matrix for a plane truss member can be deduced from Eq. (12.21) as

$$
R = \begin{bmatrix}
\cos\theta & \sin\theta & 0 & 0 \\
-\sin\theta & \cos\theta & 0 & 0 \\
0 & 0 & \cos\theta & \sin\theta \\
0 & 0 & -\sin\theta & \cos\theta
\end{bmatrix}
$$

TABLE 12.1

Member	Joint or node number		Member inclination θ (see Fig. 12.11)	$\sin\theta$	$\cos\theta$
	Near end	Far end			
1	1	2	0	0	1
2	2	3	270°	−1.0	0
3	3	4	180°	0	−1
4	1	4	270°	−1.0	0
5	1	3	≈323°	−0.6	0.8
6	2	4	≈217°	−0.6	−0.8

Thus for members 1 and 3,

$$S_1 = S_3 = \begin{bmatrix} 1 & 0 & -1 & 0 \\ 0 & 0 & 0 & 0 \\ -1 & 0 & 1 & 0 \\ 0 & 0 & 0 & 0 \end{bmatrix} \times 750 \text{ k/in}$$

for members 2 and 4,

$$S_2 = S_4 = \begin{bmatrix} 1 & 0 & -1 & 0 \\ 0 & 0 & 0 & 0 \\ -1 & 0 & 1 & 0 \\ 0 & 0 & 0 & 0 \end{bmatrix} \times 1000 \text{ k/in}$$

and, for members 5 and 6,

$$S_5 = S_6 = \begin{bmatrix} 1 & 0 & -1 & 0 \\ 0 & 0 & 0 & 0 \\ -1 & 0 & 1 & 0 \\ 0 & 0 & 0 & 0 \end{bmatrix} \times 600 \text{ k/in}$$

The transformation matrices for the six members are

$$R_1 = \begin{bmatrix} 1 & 0 & 0 & 0 \\ 0 & 1 & 0 & 0 \\ 0 & 0 & 1 & 0 \\ 0 & 0 & 0 & 1 \end{bmatrix} \qquad R_2 = R_4 = \begin{bmatrix} 0 & -1 & 0 & 0 \\ 1 & 0 & 0 & 0 \\ 0 & 0 & 0 & -1 \\ 0 & 0 & 1 & 0 \end{bmatrix}$$

$$R_3 = \begin{bmatrix} -1 & 0 & 0 & 0 \\ 0 & -1 & 0 & 0 \\ 0 & 0 & -1 & 0 \\ 0 & 0 & 0 & -1 \end{bmatrix} \qquad R_5 = \begin{bmatrix} 0.8 & -0.6 & 0 & 0 \\ 0.6 & 0.8 & 0 & 0 \\ 0 & 0 & 0.8 & -0.6 \\ 0 & 0 & 0.6 & 0.8 \end{bmatrix}$$

$$R_6 = \begin{bmatrix} -0.8 & -0.6 & 0 & 0 \\ 0.6 & -0.8 & 0 & 0 \\ 0 & 0 & -0.8 & -0.6 \\ 0 & 0 & 0.6 & -0.8 \end{bmatrix}$$

The member stiffness matrices $S_1, \ldots, S_6$ in local coordinates are transformed to global coordinates using Eq. (12.30), that is, $S_m^* = R^T S R$. Applying Eq. (12.30) to member 1:

$$(S_m^*)_1 = \begin{bmatrix} 1 & 0 & 0 & 0 \\ 0 & 1 & 0 & 0 \\ 0 & 0 & 1 & 0 \\ 0 & 0 & 0 & 1 \end{bmatrix} \begin{bmatrix} 1 & 0 & -1 & 0 \\ 0 & 0 & 0 & 0 \\ -1 & 0 & 1 & 0 \\ 0 & 0 & 0 & 0 \end{bmatrix} \times 750 \begin{bmatrix} 1 & 0 & 0 & 0 \\ 0 & 1 & 0 & 0 \\ 0 & 0 & 1 & 0 \\ 0 & 0 & 0 & 1 \end{bmatrix}$$

Freedom
numbers ⟶ 1   2   3   4

$$(S_m^*)_1 = \begin{matrix} 1 \\ 2 \\ 3 \\ 4 \end{matrix} \begin{bmatrix} 1 & 0 & -1 & 0 \\ 0 & 0 & 0 & 0 \\ -1 & 0 & 1 & 0 \\ 0 & 0 & 0 & 0 \end{bmatrix} \times 750 \text{ k/in}$$

The freedom numbers shown next to the rows and columns of member 1 follow the convention established earlier: With the local origin at the lesser nodal number, the first two freedoms of the member correspond to the lesser joint number and the last two freedoms to the higher joint number. Similarly, for the other five members, the member stiffness matrices in global coordinates are

Freedom numbers → 3 4 5 6

$$(\mathbf{S}_m^*)_2 = \begin{array}{c} 3 \\ 4 \\ 5 \\ 6 \end{array} \begin{bmatrix} 0 & 0 & 0 & 0 \\ 0 & 1 & 0 & -1 \\ 0 & 0 & 0 & 0 \\ 0 & -1 & 0 & 1 \end{bmatrix} \times 1000 \text{ k/in}$$

$$(\mathbf{S}_m^*)_3 = \begin{array}{c} 5 \\ 6 \\ 7 \\ 8 \end{array} \begin{array}{cccc} 5 & 6 & 7 & 8 \end{array} \begin{bmatrix} 1 & 0 & -1 & 0 \\ 0 & 0 & 0 & 0 \\ -1 & 0 & 1 & 0 \\ 0 & 0 & 0 & 0 \end{bmatrix} \times 750 \text{ k/in}$$

$$(\mathbf{S}_m^*)_4 = \begin{array}{c} 1 \\ 2 \\ 7 \\ 8 \end{array} \begin{array}{cccc} 1 & 2 & 7 & 8 \end{array} \begin{bmatrix} 0 & 0 & 0 & 0 \\ 0 & 1 & 0 & -1 \\ 0 & 0 & 0 & 0 \\ 0 & -1 & 0 & 1 \end{bmatrix} \times 1000 \text{ k/in}$$

$$(\mathbf{S}_m^*)_5 = \begin{array}{c} 1 \\ 2 \\ 5 \\ 6 \end{array} \begin{array}{cccc} 1 & 2 & 5 & 6 \end{array} \begin{bmatrix} 0.64 & -0.48 & -0.64 & 0.48 \\ -0.48 & 0.36 & 0.48 & -0.36 \\ -0.64 & 0.48 & 0.64 & -0.48 \\ 0.48 & -0.36 & -0.48 & 0.36 \end{bmatrix} \times 600 \text{ k/in}$$

$$(\mathbf{S}_m^*)_6 = \begin{array}{c} 3 \\ 4 \\ 7 \\ 8 \end{array} \begin{array}{cccc} 3 & 4 & 7 & 8 \end{array} \begin{bmatrix} 0.64 & 0.48 & -0.64 & -0.48 \\ 0.48 & 0.36 & -0.48 & -0.36 \\ -0.64 & -0.48 & 0.64 & 0.48 \\ -0.48 & -0.36 & 0.48 & 0.36 \end{bmatrix} \times 600 \text{ k/in}$$

The global stiffness matrix $\mathbf{S}^*$ for the unrestrained truss is now assembled by inserting the elements from the member stiffness matrices $\mathbf{S}_m^*$ into the correct location as indicated by the freedom numbers. For example, the element $\mathbf{S}^*(1,1)$ is formed by summing element $(1,1)$ from $(\mathbf{S}_m^*)_1$, element $(1,1)$ from $(\mathbf{S}_m^*)_4$, and element $(1,1)$ from $(\mathbf{S}_m^*)_5$, that is,

$$(1 \times 750) + (0 \times 1000) + (0.64 \times 600) = 1.134 \times 10^3 \text{ k/in}$$

and so on. Since there are eight degrees of freedom in the unrestrained structure, the global stiffness matrix $\mathbf{S}^*$ will be an (8×8) matrix as shown in Eq. (12.35); the array is partitioned, as shown by the dotted lines, such that the first four rows and columns correspond to the actual unrestrained freedoms (D_1, D_2, D_3, and D_4) with the truss restrained at its supports with $D_5 = D_6 = D_7 = D_8 = 0$.

$$
\begin{array}{c}
\text{Freedom} \\
\text{numbers}
\end{array}
$$

$$
\mathbf{S^*} =
\begin{array}{c c}
 & \begin{array}{cccccccc} 1 & 2 & 3 & 4 & 5 & 6 & 7 & 8 \end{array} \\
\begin{array}{c} 1 \\ 2 \\ 3 \\ 4 \\ \\ 5 \\ 6 \\ 7 \\ 8 \end{array} &
\left[
\begin{array}{cccc|cccc}
1.134 & -0.288 & -0.750 & 0 & -0.384 & 0.288 & 0 & 0 \\
-0.288 & 1.216 & 0 & 0 & 0.288 & -0.216 & 0 & -1.000 \\
-0.750 & 0 & 1.134 & 0.288 & 0 & 0 & -0.384 & -0.288 \\
0 & 0 & 0.288 & 1.216 & 0 & -1.000 & -0.288 & -0.216 \\
\hline
-0.384 & 0.288 & 0 & 0 & 1.134 & -0.288 & -0.750 & 0 \\
0.288 & -0.216 & 0 & -1.000 & -0.288 & 1.216 & 0 & 0 \\
0 & 0 & -0.384 & -0.288 & -0.750 & 0 & 1.134 & 0.288 \\
0 & -1.000 & -0.288 & -0.216 & 0 & 0 & 0.288 & 1.216 \\
\end{array}
\right]
\end{array}
$$

$$
\times 10^3 \text{ k/in} \qquad (12.35)
$$

Following the notation in Eq. (11.55), the matrix $\mathbf{S^*}$ can be partitioned as

$$
\mathbf{S^*} = \left[
\begin{array}{c|c}
\mathbf{S^*_{11}} & \mathbf{S^*_{12}} \\
\hline
\mathbf{S^*_{21}} & \mathbf{S^*_{22}}
\end{array}
\right]
\qquad (12.36)
$$

with

$$
\mathbf{S^*_{11}} =
\begin{bmatrix}
1.134 & -0.288 & -0.750 & 0 \\
-0.288 & 1.216 & 0 & 0 \\
-0.750 & 0 & 1.134 & 0.288 \\
0 & 0 & 0.288 & 1.216
\end{bmatrix}
\times 10^3 \text{ k/in}
$$

and so on for the other submatrices. The unknown displacements $\mathbf{D}_p$ due to any external nodal load vector P can now be determined from Eq. (11.61) as

$$
\mathbf{D}_p = (\mathbf{S^*_{11}})^{-1}(\mathbf{P} - \mathbf{S^*_{12}}\mathbf{D}_R)
$$

in which $\mathbf{D}_R$ is the known or specified displacement vector. In this case, $\mathbf{D}_R = 0$ since the truss is restrained at joints 3 and 4. Thus

$$
\mathbf{D}_p = (\mathbf{S^*_{11}})^{-1}\mathbf{P}
$$

With the displacements known, the member forces F can be found from Eq. (12.24), that is, $\mathbf{F} = \mathbf{SR}\boldsymbol{\delta}^*$ in which $\boldsymbol{\delta}^*$ can be found in terms of $\mathbf{D}_p$. It is interesting to point out that the axial force in member 3 will be zero (try it as an exercise); this is due to the manner in which joints 3 and 4 are supported. However, in practice a member such as 3 may be used in a situation where the supporting footings at joints 3 and 4 are likely to move horizontally. In such a case the presence of member 3 will minimize such movement. The reactions R at joints 3 and 4 can be found from Eq. (11.59) as $\mathbf{R} = (\mathbf{S^*_{21}})\mathbf{D}_p$ since $\mathbf{D}_R = 0$; these reactions would correspond to the freedom numbers 5 and 6 at joint 3 and freedom numbers 7 and 8 at joint 4.

It is important to mention that the global stiffness matrix $\mathbf{S^*}$ of linear systems [e.g., Eq. (12.35)] will always be symmetrical following Maxwell's law of reciprocity; this property provides a check on the computed values. The inversion of matrix $\mathbf{S^*}$ by hand calculations can be a very laborious task; thus, it is best performed on a computer using a standard subroutine; the generation of an $\mathbf{S^*}$ matrix by a computer can also be done by following a simple algorithm discussed later.

To avoid any errors in the use of units, the same units should be used throughout the analysis; thus with area A being in square inches, the length l should be in inches, E in kips per square inch, and $\mathbf{P}$ in kips; the displacements $\mathbf{D}$ would then be in inches, forces in kips, and moments in kip-inches. Similarly, in using SI units, with A in square meters, l should be in meters, E in kN/m^2, and $\mathbf{P}$ in kilonewtons; the displacement $\mathbf{D}$ would be in meters, forces in kilonewtons, and moments in $kN \cdot m$. ∎

■ EXAMPLE 12.2

Apply the direct stiffness method to the continuous beam shown in Fig. 12.12a. This problem was solved in Chapter 11 (see page 538).

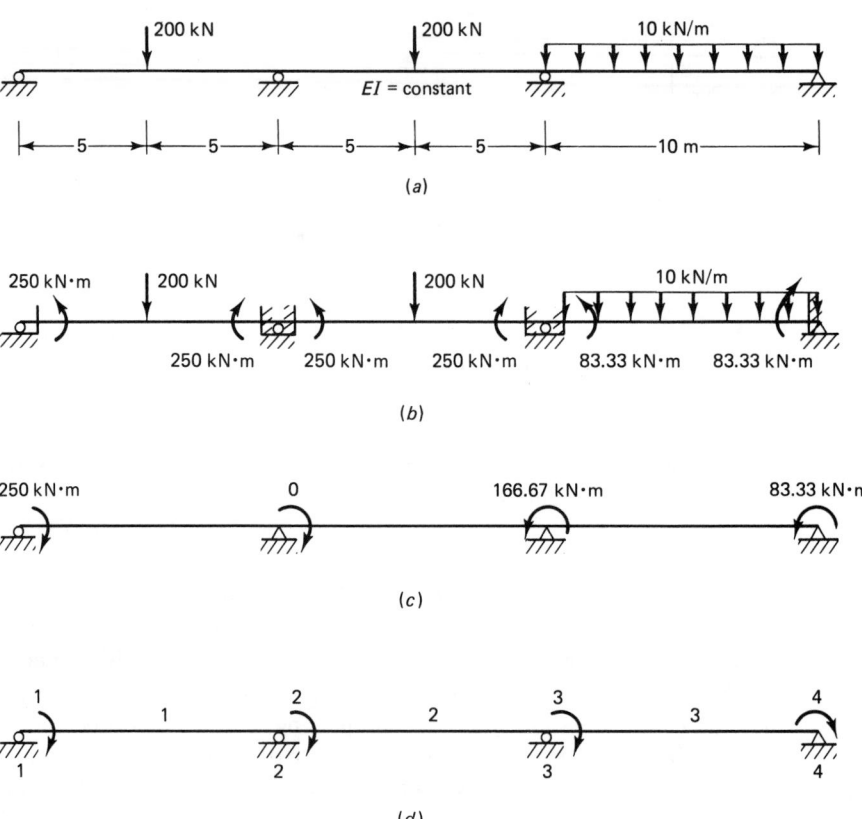

Figure 12.12 Structure for Example 12.2.

Solution. As before, the first step is to convert member loads to nodal loads as shown in Fig. 12.12*b*. The resulting nodal loads on the beam are shown in Fig. 12.12*c*; the fixed-end moments due to the member loads can be readily calculated using the Appendix. The reactions at the supports are not included in the load vector *P* since they are transmitted directly to the supporting foundation and thus do not contribute to the displacements of the structure. The degrees of freedom which include only rotations (since the joints do not translate) are shown in Fig. 12.12*d*. The restrained degrees of freedom at the joints are not included in the solution in order to minimize the size of the global matrix **S***. Using Eq. (12.28), the member stiffness matrices for the three members are calculated as

Freedom numbers ⟶ 1   2

$$(S)_1 = (S_m^*) = \begin{matrix}1\\2\end{matrix}\begin{bmatrix}2 & 1\\1 & 2\end{bmatrix} \times \frac{2EI}{10} \quad kN/m$$

2   3

$$(S)_2 = (S_m^*)_2 = \begin{matrix}2\\3\end{matrix}\begin{bmatrix}2 & 1\\1 & 2\end{bmatrix} \times \frac{2EI}{10} \quad kN/m$$

3   4

$$(S)_3 = (S_m^*)_3 = \begin{matrix}3\\4\end{matrix}\begin{bmatrix}2 & 1\\1 & 2\end{bmatrix} \times \frac{2EI}{10} \quad kN/m$$

It is observed from the above that the **S** matrices are identical to the **S*** matrices since the local and global coordinates coincide; furthermore, the corresponding freedom numbers have been indicated adjacent to the rows and columns. The global stiffness matrix is now assembled by placing each term of the member stiffness into its appropriate location in the global stiffness matrix as determined by the row and column designation of that term; thus

Freedom numbers ⟶ 1   2   3   4

$$S^* = \begin{matrix}1\\2\\3\\4\end{matrix}\begin{bmatrix}2 & 1 & 0 & 0\\1 & 2+2 & 1 & 0\\0 & 1 & 2+2 & 1\\0 & 0 & 1 & 2\end{bmatrix} \times \frac{2EI}{10} \quad kN/m$$

or

$$S^* = \begin{bmatrix}2 & 1 & 0 & 0\\1 & 4 & 1 & 0\\0 & 1 & 4 & 1\\0 & 0 & 1 & 2\end{bmatrix} \times \frac{2EI}{10} \quad kN/m$$

The **S*** matrix is of (4 × 4) size since the structure has four degrees of freedom; it is identical to that derived in Chapter 11, page 538. The displacement vector **D** is determined from **D** = (**S***)⁻¹*P*. The results for the displacements and member forces

F are shown in Chapter 11, pages 538 and 539; in this example the transformation matrix is a unit matrix and thus $D_1 = \delta_1$, $D_2 = \delta_2$, $D_3 = \delta_3$, and $D_4 = \delta_4$. The final forces **F′** at the end of the three members are obtained from

$$\mathbf{F'} = \mathbf{F} + \mathbf{F}_0$$

in which $\mathbf{F}_0$ are the fixed-end forces shown in Fig. 12.12*b*. The values of **F′** are given in Chapter 11 on page 539. ∎

■ EXAMPLE 12.3

(a) Use the direct stiffness method to determine the global stiffness matrix **S*** for the structure in Fig. 12.13*a*. (b) Hence determine the joint displacements, reactions, and element or member forces for the loading in Fig. 12.13*b*.

Solution. The degrees of freedom on all the joints are shown numbered, including the restrained joints 3 and 4; the positive direction of each member is also shown. Each rigid joint has three degrees of freedom; Table 12.2 presents the member and

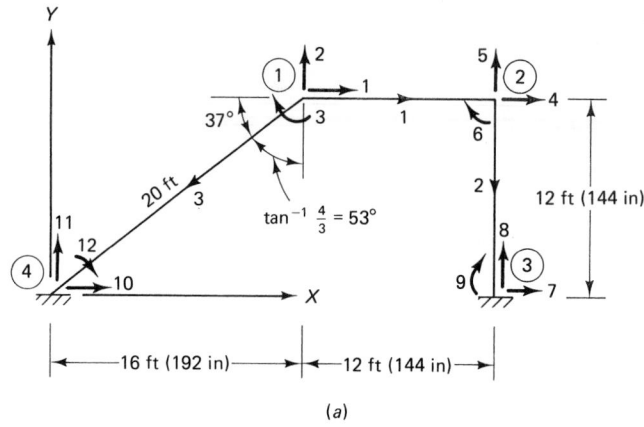

(a)

$AE = 144 \times 10^3$ k for all members

$EI = 3000 \times 10^3$ k·in² for all members

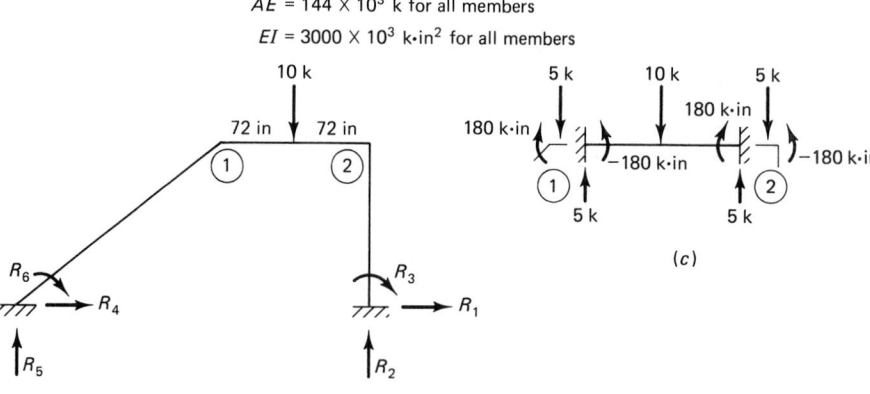

(b)

(c)

Figure 12.13 Structure for Example 12.3.

TABLE 12.2

| Member | Joint number | | Member inclination θ | $\sin\theta$ | $\cos\theta$ |
	Near end	Far end			
1	1	2	0	0	1.0
2	2	3	270°	−1.0	0
3	1	4	≈217°	−0.6	−0.8

joint numbers as well as the direction cosines of the three members. The member stiffness matrices (S) in local coordinates are determined by Eq. (12.25). Thus,

$$S_1 = S_2 = \begin{bmatrix} 1 & 0 & 0 & -1 & 0 & 0 \\ 0 & 0.0121 & -0.868 & 0 & -0.0121 & -0.868 \\ 0 & -0.868 & 83.33 & 0 & +0.868 & 41.67 \\ -1 & 0 & 0 & 1 & 0 & 0 \\ 0 & -0.0121 & +0.868 & 0 & 0.0121 & +0.868 \\ 0 & -0.868 & 41.67 & 0 & +0.868 & 83.33 \end{bmatrix} \times 10^3 \, k/in$$

and

$$S_3 = \begin{bmatrix} 0.6 & 0 & 0 & -0.6 & 0 & 0 \\ 0 & 0.0026 & -0.3125 & 0 & -0.0026 & -0.3125 \\ 0 & -0.3125 & 50 & 0 & +0.3125 & 25 \\ -0.6 & 0 & 0 & 0.6 & 0 & 0 \\ 0 & -0.0026 & +0.3125 & 0 & 0.0026 & +0.3125 \\ 0 & -0.3125 & 25 & 0 & +0.3125 & 50 \end{bmatrix} \times 10^3 \, k/in$$

The transformation matrices are calculated from Eq. (12.17) and Table 12.2; thus

$$R_1 = \begin{bmatrix} 1 & 0 & 0 & 0 & 0 & 0 \\ 0 & 1 & 0 & 0 & 0 & 0 \\ 0 & 0 & 1 & 0 & 0 & 0 \\ 0 & 0 & 0 & 1 & 0 & 0 \\ 0 & 0 & 0 & 0 & 1 & 0 \\ 0 & 0 & 0 & 0 & 0 & 1 \end{bmatrix}$$

$$R_2 = \begin{bmatrix} 0 & -1 & 0 & 0 & 0 & 0 \\ 1 & 0 & 0 & 0 & 0 & 0 \\ 0 & 0 & 1 & 0 & 0 & 0 \\ 0 & 0 & 0 & 0 & -1 & 0 \\ 0 & 0 & 0 & 1 & 0 & 0 \\ 0 & 0 & 0 & 0 & 0 & 1 \end{bmatrix}$$

$$R_3 = \begin{bmatrix} -0.8 & -0.6 & 0 & 0 & 0 & 0 \\ 0.6 & -0.8 & 0 & 0 & 0 & 0 \\ 0 & 0 & 1 & 0 & 0 & 0 \\ 0 & 0 & 0 & -0.8 & -0.6 & 0 \\ 0 & 0 & 0 & 0.6 & -0.8 & 0 \\ 0 & 0 & 0 & 0 & 0 & 1 \end{bmatrix}$$

The member stiffness matrices **S** are now transformed to $\mathbf{S}_m^*$ matrices through Eq. (12.30), that is, $\mathbf{S}_m^* = \mathbf{R}^T\mathbf{S}\mathbf{R}$.

For member 1:

$$(\mathbf{S}_m^*)_1 = \begin{bmatrix} \text{Unit} \\ \text{matrix} \end{bmatrix} \begin{bmatrix} 1 & 0 & 0 & -1 & 0 & 0 \\ 0 & 0.0121 & -0.868 & 0 & -0.0121 & -0.868 \\ 0 & -0.868 & 83.33 & 0 & +0.868 & 41.67 \\ -1 & 0 & 0 & 1 & 0 & 0 \\ 0 & -0.0121 & +0.868 & 0 & 0.0121 & +0.868 \\ 0 & -0.868 & 41.67 & 0 & +0.868 & 83.33 \end{bmatrix}$$

$$\times 10^3 \begin{bmatrix} \text{Unit} \\ \text{matrix} \end{bmatrix}$$

Freedom
numbers $\longrightarrow$

$$(\mathbf{S}_m^*)_1 = \begin{matrix} 1 \\ 2 \\ 3 \\ 4 \\ 5 \\ 6 \end{matrix}\begin{matrix} 1 & 2 & 3 & 4 & 5 & 6 \end{matrix}\begin{bmatrix} 1 & 0 & 0 & -1 & 0 & 0 \\ 0 & 0.0121 & -0.868 & 0 & -0.0121 & -0.868 \\ 0 & -0.868 & 83.33 & 0 & +0.868 & 41.67 \\ -1 & 0 & 0 & 1 & 0 & 0 \\ 0 & -0.0121 & +0.868 & 0 & 0.0121 & +0.868 \\ 0 & -0.868 & 41.67 & 0 & +0.868 & 83.33 \end{bmatrix} \times 10^3 \text{ k/in}$$

Similarly, for members 2 and 3:

$$(\mathbf{S}_m^*)_2 = \begin{matrix} 4 \\ 5 \\ 6 \\ 7 \\ 8 \\ 9 \end{matrix}\begin{matrix} 4 & 5 & 6 & 7 & 8 & 9 \end{matrix}\begin{bmatrix} 0.0121 & 0 & -0.868 & -0.0121 & 0 & -0.868 \\ 0 & 1 & 0 & 0 & -1 & 0 \\ -0.868 & 0 & 83.33 & +0.868 & 0 & 41.67 \\ -0.0121 & 0 & +0.868 & 0.0121 & 0 & +0.868 \\ 0 & -1 & 0 & 0 & +1 & 0 \\ -0.868 & 0 & 41.67 & +0.868 & 0 & 83.33 \end{bmatrix} \times 10^3 \text{ k/in}$$

$$(\mathbf{S}_m^*)_3 = \begin{matrix} 1 \\ 2 \\ 3 \\ 10 \\ 11 \\ 12 \end{matrix}\begin{matrix} 1 & 2 & 3 & 10 & 11 & 12 \end{matrix}\begin{bmatrix} 0.385 & 0.287 & -0.1875 & -0.385 & -0.287 & -0.1875 \\ 0.287 & 0.218 & +0.250 & -0.287 & -0.218 & +0.250 \\ -0.1875 & +0.250 & 50.0 & +0.1875 & -0.250 & 25.0 \\ -0.385 & -0.287 & +0.1875 & 0.385 & 0.287 & +0.1875 \\ -0.287 & -0.218 & -0.250 & 0.287 & 0.218 & -0.250 \\ -0.1875 & +0.250 & 25.0 & +0.1875 & -0.250 & 50.0 \end{bmatrix}$$

$$\times 10^3 \text{ k/in}$$

We are now ready to assemble the (12×12) global stiffness matrix $\mathbf{S}^*$ for the unrestrained frame by placing the elements from the member stiffness matrices $\mathbf{S}_m^*$ into the correct position as indicated by the freedom numbers. The $\mathbf{S}^*$ matrix is partitioned so that the first six rows and columns correspond to the actual freedoms D_1, D_2, D_3, D_4, D_5, and D_6; the frame is restrained at joints 3 and 4 and therefore D_7 to D_{12} are zero.

Freedom
numbers

	1	2	3	4	5	6	7	8	9	10	11	12
1	1.385	0.287	−0.1875	−1.0	0	0	0	0	0	−0.385	−0.287	−0.1875
2	0.287	0.230	−0.618	0	−0.0121	−0.868	0	0	0	−0.287	−0.218	+0.250
3	−0.1875	−0.618	133.33	0	+0.868	41.67	0	0	0	+0.1875	−0.250	25.0
4	−1.0	0	0	1.0121	0	−0.868	−0.0121	0	−0.868	0	0	0
5	0	−0.0121	+0.868	0	1.0121	+0.868	0	−1.0	0	0	0	0
6	0	−0.868	41.67	−0.868	+0.868	166.67	+0.868	0	41.67	0	0	0
7	0	0	0	−0.0121	0	+0.868	0.0121	0	+0.868	0	0	0
8	0	0	0	0	−1.0	0	0	+1.0	0	0	0	0
9	0	0	0	−0.868	0	41.67	+0.868	0	83.33	0	0	0
10	−0.385	−0.287	+0.1875	0	0	0	0	0	0	0.385	0.287	+0.1875
11	−0.287	−0.218	−0.250	0	0	0	0	0	0	0.287	0.218	−0.250
12	−0.1875	+0.250	25.0	0	0	0	0	0	0	+0.1875	−0.250	50.0

$\mathbf{S^*} = $ (to the left, at rows 6–7)

$\times 10^3$ k/in

(12.37)

Thus we can partition $\mathbf{S^*}$ as

$$\mathbf{S^*_*} = \begin{bmatrix} \mathbf{S^*_{11}} & \mathbf{S^*_{12}} \\ \mathbf{S^*_{21}} & \mathbf{S^*_{22}} \end{bmatrix}$$

with the values for $\mathbf{S^*_{11}}$ being

$$\mathbf{S^*_{11}} = \begin{bmatrix}
1.385 & 0.287 & -0.1875 & -1.0 & 0 & 0 \\
0.287 & 0.230 & -0.618 & 0 & -0.0121 & -0.868 \\
-0.1875 & -0.618 & 133.33 & 0 & +0.868 & 41.67 \\
-1.0 & 0 & 0 & 1.0121 & 0 & -0.868 \\
0 & -0.0121 & +0.868 & 0 & 1.0121 & +0.868 \\
0 & -0.868 & 41.67 & -0.868 & +0.868 & 166.67
\end{bmatrix}$$

$\times 10^3$ k/in

The unknown displacements $D_1, D_2, \ldots, D_6$ are then found from Eq. (11.61):

$$\mathbf{D}_p = (\mathbf{S^*_{11}})^{-1}(\mathbf{P} - \mathbf{S^*_{12}}\mathbf{D}_R)$$

in which $\mathbf{P}$ is any given nodal force vector and $\mathbf{D}_R$ is the known displacement vector (in this case, $\mathbf{D}_R = \{D_7, \ldots, D_{12}\} = 0$).

(b) From Fig. 12.13b and c, the nodal force vector P is

$$\mathbf{P} = \begin{bmatrix} P_1 \\ P_2 \\ P_3 \\ P_4 \\ P_5 \\ P_6 \end{bmatrix} = \begin{bmatrix} 0 & \\ -5 & k \\ 180 & k \cdot in \\ 0 & \\ -5 & k \\ -180 & k \cdot in \end{bmatrix}$$

Using Eq. (11.61) and the inverse of $\mathbf{S^*_{11}}$, that is, $[\mathbf{S^*_{11}}]^{-1}$ (obtained from a computer package program), and since $\mathbf{D}_R = 0$, we find

$$\mathbf{D}_P = (\mathbf{S}_{11}^*)^{-1}(P) = \begin{bmatrix} 0.156 \\ -0.221 \\ 0.112 \times 10^{-2} \\ 0.153 \\ -0.710 \times 10^{-2} \\ -0.167 \times 10^{-2} \end{bmatrix} \begin{matrix} D_1 \\ D_2 \\ D_3 \\ D_4 \\ D_5 \\ D_6 \end{matrix} \begin{matrix} \text{in} \\ \text{in} \\ \text{radian} \\ \text{in} \\ \text{in} \\ \text{radian} \end{matrix}$$

The end-member displacements δ^* are related to the point displacements D as follows:

For member 1	For member 2	For member 3

$$\begin{bmatrix} \delta_1^* \\ \delta_2^* \\ \delta_3^* \\ \delta_4^* \\ \delta_5^* \\ \delta_6^* \end{bmatrix} = \begin{bmatrix} D_1 \\ D_2 \\ D_3 \\ D_4 \\ D_5 \\ D_6 \end{bmatrix} \qquad \begin{bmatrix} \delta_1^* \\ \delta_2^* \\ \delta_3^* \\ \delta_4^* \\ \delta_5^* \\ \delta_6^* \end{bmatrix} = \begin{bmatrix} D_4 \\ D_5 \\ D_6 \\ 0 \\ 0 \\ 0 \end{bmatrix} \qquad \begin{bmatrix} \delta_1^* \\ \delta_2^* \\ \delta_3^* \\ \delta_4^* \\ \delta_5^* \\ \delta_6^* \end{bmatrix} = \begin{bmatrix} D_1 \\ D_2 \\ D_3 \\ 0 \\ 0 \\ 0 \end{bmatrix}$$

The end-member forces $\mathbf{F}$ are now calculated from Eq. (12.24):

$$\mathbf{F} = \mathbf{SR}\delta^*$$

Thus for

	Member 1		Member 2	Member 3

$$\begin{bmatrix} F_1 \\ F_2 \\ F_3 \\ F_4 \\ F_5 \\ F_6 \end{bmatrix} = \begin{bmatrix} 3.30 \\ -2.10 \\ 208.61 \\ -3.30 \\ 2.10 \\ 92.35 \end{bmatrix} \begin{matrix} \text{k} \\ \text{k} \\ \text{k-in} \\ \text{k} \\ \text{k} \\ \text{k-in} \end{matrix} \quad \begin{bmatrix} 7.10 \\ 3.30 \\ -272.33 \\ -7.10 \\ -3.30 \\ -202.61 \end{bmatrix} \begin{bmatrix} 4.34 \\ 0.35 \\ -28.61 \\ -4.34 \\ -0.35 \\ -56.53 \end{bmatrix}$$

The final end-member forces $\mathbf{F}'$ are given by $\mathbf{F}' = \mathbf{F} + \mathbf{F}_0$. The fixed-end forces F_0 for

	Member 1		Member 2	Member 3

$$(\mathbf{F}_0) = \begin{bmatrix} F_{01} \\ F_{02} \\ F_{03} \\ F_{04} \\ F_{05} \\ F_{06} \end{bmatrix} = \begin{bmatrix} 0 \\ +5 \\ -180 \\ 0 \\ +5 \\ +180 \end{bmatrix} \qquad (\mathbf{F}_0) = 0 \qquad (\mathbf{F}_0) = 0$$

See Fig. 12.13 c. Therefore, for member 1:

$$(\mathbf{F}') = \begin{bmatrix} 3.30 \\ 2.90 \\ 28.61 \\ -3.30 \\ 7.10 \\ 272.35 \end{bmatrix} \begin{matrix} \text{k} \\ \text{k} \\ \text{k} \cdot \text{in} \\ \text{k} \\ \text{k} \\ \text{k} \cdot \text{in} \end{matrix}$$

and for members 2 and 3: $\mathbf{F}' = \mathbf{F}$ since $\mathbf{F}_0 = 0$.

From Eq. (11.60) the reactions $\mathbf{R} = (\mathbf{S}_{21}^*)(\mathbf{D}_P)$. The matrix $\mathbf{S}_{21}^*$ is contained in Eq. (12.37); thus

$$
(\mathbf{R}) = \begin{bmatrix} R_1 \\ R_2 \\ R_3 \\ R_4 \\ R_5 \\ R_6 \end{bmatrix} = \begin{bmatrix} -3.30 \\ 7.10 \\ -202.61 \\ 3.30 \\ 2.92 \\ -56.53 \end{bmatrix} \begin{array}{l} \text{k} \\ \text{k} \\ \text{k} \cdot \text{in} \\ \text{k} \\ \text{k} \\ \text{k} \cdot \text{in} \end{array} \qquad \blacksquare
$$

It should be mentioned that a solution following the above procedure is very suitable to analyze structures subjected to any load as well as those with known or specified displacements such as those occurring due to anticipated settlement of foundation. This is demonstrated in the solved problem at the end of the chapter.

12.5 DISPLACEMENT BOUNDARY CONDITIONS

It should be emphasized that the global stiffness matrix $\mathbf{S}^*$ for an *unrestrained structure* [e.g., Eq. (12.37)] is singular and therefore does not possess an inverse. This reflects the fact that such a structure will accelerate under the action of any arbitrary force and therefore its degrees of freedom cannot be defined. The matrix $\mathbf{S}^*$ must therefore be modified by imposing the specified boundary conditions such as prescribed zero support movements, or anticipated support movements. It was shown in Chapter 11, Eq. (11.55), that displacements can be separated into two categories: unknown displacements $\mathbf{D}_p$ and known or specified displacements $\mathbf{D}_R$; it is implied by Eq. (11.55) that the known displacements $\mathbf{D}_R$ will occupy a position in the displacement vector $\mathbf{D}$ that always follows the unknown displacements $\mathbf{D}_p$. However, this arrangement may not be realized because of the numbering of the freedoms in a particular structure. This difficulty can be overcome by rearranging the global equilibrium equations $(\mathbf{P} = \mathbf{S}^*\mathbf{D})$; but such rearranging can lead to some complications in computer programming due to reindexing the equations. Two simpler alternatives are suggested.

Consider a structure with six degrees of freedom; writing the global equilibrium equations $(\mathbf{P} = \mathbf{S}^*\mathbf{D})$ in full we have

$$
\begin{bmatrix} S_{11}^* & S_{12}^* & S_{13}^* & S_{14}^* & S_{15}^* & S_{16}^* \\ S_{21}^* & \cdot & \cdot & \cdot & \cdot & S_{26}^* \\ \cdot & \cdot & \cdot & \cdot & \cdot & \cdot \\ \cdot & \cdot & \cdot & \cdot & \cdot & \cdot \\ \cdot & \cdot & \cdot & \cdot & \cdot & \cdot \\ S_{61}^* & S_{62}^* & S_{63}^* & S_{64}^* & S_{65}^* & S_{66}^* \end{bmatrix} \begin{bmatrix} D_1 \\ D_2 \\ D_3 \\ D_4 \\ D_5 \\ D_6 \end{bmatrix} = \begin{bmatrix} P_1 \\ P_2 \\ P_3 \\ P_4 \\ P_5 \\ P_6 \end{bmatrix} \qquad (12.38)
$$

Let us assume that freedom D_2 and D_5 are known or specified displacements, that is, $D_2 = D_{R2}$; $D_5 = D_{R5}$. We can now modify the matrix $\mathbf{S}^*$ and matrix $\mathbf{P}$ as follows:

$$
\begin{bmatrix} S_{11}^* & 0 & S_{13}^* & S_{14}^* & 0 & S_{16}^* \\ 0 & 1.0 & 0 & 0 & 0 & 0 \\ S_{31}^* & 0 & S_{23}^* & S_{34}^* & 0 & S_{36}^* \\ S_{41}^* & 0 & S_{43}^* & S_{44}^* & 0 & S_{46}^* \\ 0 & 0 & 0 & 0 & 1.0 & 0 \\ S_{61}^* & 0 & S_{63}^* & S_{64}^* & 0 & S_{66}^* \end{bmatrix} \begin{bmatrix} D_1 \\ D_2 \\ D_3 \\ D_4 \\ D_5 \\ D_6 \end{bmatrix} = \begin{bmatrix} P_1 \\ D_{R2} \\ P_3 \\ P_4 \\ D_{R5} \\ P_6 \end{bmatrix} - \begin{bmatrix} S_{12}^* & S_{15}^* \\ 0 & 0 \\ S_{32}^* & S_{35}^* \\ S_{42}^* & S_{45}^* \\ 0 & 0 \\ S_{62}^* & S_{65}^* \end{bmatrix} \begin{bmatrix} D_{R2} \\ D_{R5} \end{bmatrix} \qquad (12.39)
$$

The first, third, fourth, and sixth equations in Eqs. (12.39) are the same as in Eqs. (12.38); while the second and fifth equations are simply identities: $D_2 = D_{R2}$ and $D_5 = D_{R5}$. Thus Eq. (12.39) can now be expressed as

$$(\mathbf{S}^*)'\mathbf{D} = \mathbf{P}' \tag{12.40}$$

in which matrices $(\mathbf{S}^*)'$ and $\mathbf{P}'$ are the modified global stiffness and external force matrices of the same size as the original matrices $\mathbf{S}^*$ and $\mathbf{P}$. Thus, from Eq. (12.40) we have the displacements

$$\mathbf{D} = [(\mathbf{S}^*)']^{-1}\mathbf{P}' \tag{12.41}$$

The above $\mathbf{D}$ matrix will contain the specified displacements D_{R2} and D_{R5}; also, the reactions R_2 and R_5 [see Eq. (11.55)] will correspond to P_2 and P_5 which are determined from the matrix product of the unmodified stiffness matrix $\mathbf{S}^*$ and the specified displacements D_{R2} and D_{R5}. Therefore, in programming a solution on a computer, involving a specified displacement D_{Ri}, say, we must modify the global stiffness matrix $\mathbf{S}^*$ so that its coefficients in the ith row and ith column are zero except for S_{ii}^*, which is made equal to one; furthermore, the force vector P is modified by subtracting from P_j the term $S_{ji}^* D_{Ri}$ ($j \neq i$) and replacing P_i with D_{Ri}. See Eq. (12.39).

The second method of dealing with specified displacements is perhaps simpler than the above method as far as computer programming is concerned. The method is as follows. The equilibrium equations (12.38) are modified as shown below:

$$
\begin{bmatrix}
S_{11}^* & S_{12}^* & S_{13}^* & S_{14}^* & S_{15}^* & S_{16}^* \\
S_{21}^* & (10)^{30} & S_{23}^* & S_{24}^* & S_{25}^* & S_{26}^* \\
S_{31}^* & S_{32}^* & S_{33}^* & S_{34}^* & S_{35}^* & S_{36}^* \\
S_{41}^* & S_{42}^* & S_{43}^* & S_{44}^* & S_{45}^* & S_{46}^* \\
S_{51}^* & S_{52}^* & S_{53}^* & S_{54}^* & (10)^{30} & S_{56}^* \\
S_{61}^* & S_{62}^* & S_{63}^* & S_{64}^* & S_{65}^* & S_{66}^*
\end{bmatrix}
\begin{bmatrix}
D_1 \\ D_2 \\ D_3 \\ D_4 \\ D_5 \\ D_6
\end{bmatrix}
=
\begin{bmatrix}
P_1 \\ (10^{30})D_{R2} \\ P_3 \\ P_4 \\ (10)^{30}D_{R5} \\ P_6
\end{bmatrix}
\tag{12.42}
$$

or

$$(\mathbf{S}^*)'\mathbf{D} = \mathbf{P}' \tag{12.42a}$$

The large number inserted in the second and fifth equations in Eq. (12.42) will so dominate the other stiffness terms in these equations that the second and fifth equations effectively reduce to $(10)^{30}D_2 = (10)^{30}D_{R2}$ and $(10)^{30}D_5 = (10)^{30}D_{R5}$; or $D_2 = D_{R2}$ and $D_5 = D_{R5}$. The unknown displacement can be derived from Eq. (12.42) as

$$\mathbf{D} = [(\mathbf{S}^*)']^{-1}\mathbf{P}'$$

in which $(\mathbf{S}^*)'$ and $\mathbf{P}'$ are the matrices modified as shown in Eq. (12.42). Thus, in this method the introduction of a specified displacement D_{Ri} requires replacing S_{ii}^* in the global stiffness matrix by $(10)^{30}$ and P_i in the global external force vector by $(10)^{30}D_{Ri}$. This procedure has been found to be very successful in practice since stiffness of structural numbers is orders of magnitude less than $(10)^8$ in imperial units, and less than $(10)^{10}$ in SI units.

12.6 COMPUTER GENERATION OF MATRIX S*

In the preceding examples the global stiffness matrix S^* was generated for the *unrestrained structures.* Such a procedure can be uneconomical from the point of view of computer time and storage. It is more expeditious to account for the boundary conditions of the structure during the assembly of matrix S^* and form only the global stiffness matrix of the *restrained structure,* that is, S_{11}^* in Eq. (11.55) or in Eq. (12.36). Once the unknown displacements are determined from Eq. (11.61), the member end forces are calculated from such displacements. The reactions are then found from force equilibrium at each restrained joint acted upon by member forces, any external load, and the reactions in question. To facilitate the formation of S_{11}^* by a computer, an algorithm is necessary to place the elements of the member stiffness matrices in the proper location in matrix S_{11}^*. A *code-numbering technique,* described below, will be used for this purpose.

Consider the rigid frame in Fig. 12.14a. The nodal freedoms are shown numbered $D_1, \ldots, D_6$; the elements of the member displacement matrix in global coordinates for one member of the frame, say 1, are also numbered as shown in Fig. 12.14b. A code number is affixed to each member of the structure; the numerical value of this number depends on the nodal freedom numbers as well as on the boundary conditions at the node or joint. For example, for member 1 in Fig. 12.14b, the code numbers would be 1, 2, 3 for the near end i and 4, 5, 6 for the far end j, that is, six degrees of freedom, since the member is assumed to carry bending as well as axial and transverse forces. The ends i and j of member 1 have three degrees of freedom each, namely,

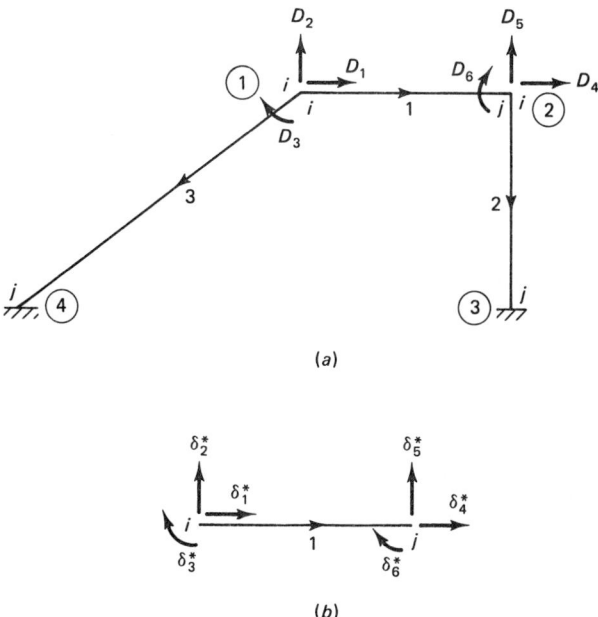

(a)

(b)

Figure 12.14 Joint and member numbering scheme.

δ_1^*, δ_2^*, and δ_3^* for end i and δ_4^*, δ_5^*, and δ_6^* for end j; these degrees of freedom of the member correspond, respectively, to the nodal freedom D_1, D_2, and D_3 at joint 1 and D_4, D_5, and D_6 at joint 2. If there is no displacement at one end of a member, then a zero code number is placed in the corresponding position. For example, for member 2 in Fig. 12.14a: For end i with three degrees of freedom δ_1^*, δ_2^*, and δ_3^* the corresponding code numbers would be 4, 5, and 6, which are the numbers designating the external displacements for joint 2; for end j, $\delta_4^* = \delta_5^* = \delta_6^* = 0$, and therefore the code numbers would be zero. In tabular form, the code numbers for the frame structure in Fig. 12.14a, are

Member code numbers

Member displacement number

	1	2	3	4	5	6
Member	δ_1^*	δ_2^*	δ_3^*	δ_4^*	δ_5^*	δ_6^*
1	1	2	3	4	5	6
2	4	5	6	0	0	0
3	1	2	3	0	0	0

Numbers corresponding to the D's

The above matrix is referred to as the member code-number matrix denoted as matrix **MC**. A two-dimensional pin-connected member, which can carry only axial load, has only four degrees of freedom, two at each end, and therefore there are only four positions in the code-number matrix **MC**. For a member in space, there will be 12 positions to fill.

The global stiffness matrix S_{11}^* is formulated from the member code numbers. Each member displacement number is linked first with itself and then with all member displacement numbers to the right, thus providing the row-and-column subscript for each element in the member stiffness matrix S_m^*. Each position value of the member code number is then linked first with itself and then with all position values to the right, thus identifying the row-and-column subscript for the proper positioning of that element in the global S^* matrix. A zero value in the code numbers is ignored since it means that this particular stiffness coefficient does not contribute to the global stiffness matrix S^*. Applying this procedure to the rigid frame in Fig. 12.14a and using the member code numbers tabulated above we have

Member 1		Member 2		Member 3	
Element in $(S_m^*)_1$	Position in S^*	Element in $(S_m^*)_2$	Position in S^*	Element in $(S_m^*)_3$	Position in S^*
11	11	11	44	11	11
12	12	12	45	12	12
13	13	13	46	13	13

(Continued)

Member 1		Member 2		Member 3	
Element in $(S_m^*)_1$	Position in S^*	Element in $(S_m^*)_2$	Position in S^*	Element in $(S_m^*)_3$	Position in S^*
14	14	22	55	22	22
15	15	23	56	23	23
16	16	33	66	33	33
22	22				
23	23				
24	24				
25	25				
26	26				
33	33				
34	34				
35	35				
36	36				
44	44				
45	45				
46	46				
55	55				
56	56				
66	66				

This means

$$\mathbf{S}^*(1,1) = (\mathbf{S}_m^*)_{1,11} + (\mathbf{S}_m^*)_{3,11} = (1.0)(10^3) + (0.385)(10^3) = 1.385 \times 10^3$$

$$\mathbf{S}^*(1,2) = (\mathbf{S}_m^*)_{1,12} + (\mathbf{S}_m^*)_{3,12} = 0 + (0.287)(10^3) = 0.287 \times 10^3$$

$$\mathbf{S}^*(4,5) = (\mathbf{S}_m^*)_{1,45} + (\mathbf{S}_m^*)_{2,12} = 0 + 0 = 0$$

and so on. The complete global stiffness matrix $\mathbf{S}^*$ is then developed by observing its property of symmetry. The numerical values of matrix $\mathbf{S}^*$ are given by $\mathbf{S}_{11}^*$ in Eq. (12.37).

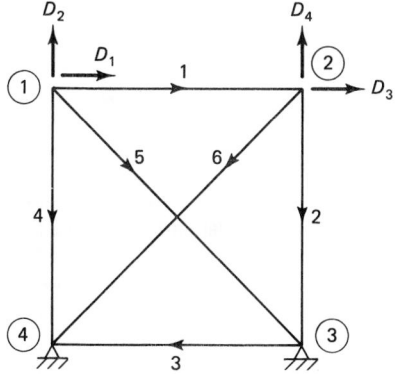

Figure 12.15 Joint displacements in structure for Example 12.1.

Consider now the truss in Fig. 12.11; since the truss is restrained at joints 3 and 4 there are only four degrees of freedom: D_1, D_2, D_3, and D_4 at joints 1 and 2 as shown in Fig. 12.15. All joints are pin-connected and therefore each member will only have four displacements δ_1^*, ..., δ_4^*. Thus, the code numbers for the truss are given below.

Member code numbers				
	Member displacement number			
	1	2	3	4
Member	δ_1^*	δ_2^*	δ_3^*	δ_4^*
1	1	2	3	4
2	3	4	0	0
3	0	0	0	0
4	1	2	0	0
5	1	2	0	0
6	3	4	0	0

Numbers corresponding to the D's

Thus, the positions in the matrix **S*** will be as follows:

Member 1		Member 2		Member 3	
Element in $(S_m^*)_1$	Position in **S***	Element in $(S_m^*)_2$	Position in **S***	Element in $(S_m^*)_3$	Position in **S***
11	11	11	33	Nil	
12	12	12	34	(no contribution)	
13	13	22	44		
14	14				
22	22				
23	23				
24	24				
33	33				
34	34				
44	44				

Member 4		Member 5		Member 6	
Element in $(S_m^*)_4$	Position in **S***	Element in $(S_m^*)_5$	Position in **S***	Element in $(S_m^*)_6$	Position in **S***
11	11	11	11	11	33
12	12	12	12	12	34
22	22	22	22	22	44

From the above we have

$$S^*(1,1) = (S^*_m)_{1,11} + (S^*_m)_{4,11} + (S^*_m)_{5,11} = 750 + 0 + (0.64)(600)$$

$$= 1.134 \times 10^3$$

$$S^*(1,2) = (S^*_m)_{1,12} + (S^*_m)_{4,12} + (S^*_m)_{5,12} = 0 + 0 + (-0.48)(600)$$

$$= -0.288 \times 10^3$$

$$S^*(1,3) = (S^*_m)_{1,13} = -750 = -0.75 \times 10^3$$

and so on. The above values are identical to those given by S^*_{11} in Eq. (12.35). The complete global stiffness matrix S^* of size (4×4) is shown assembled in Eq. (12.43).

Freedom
numbers

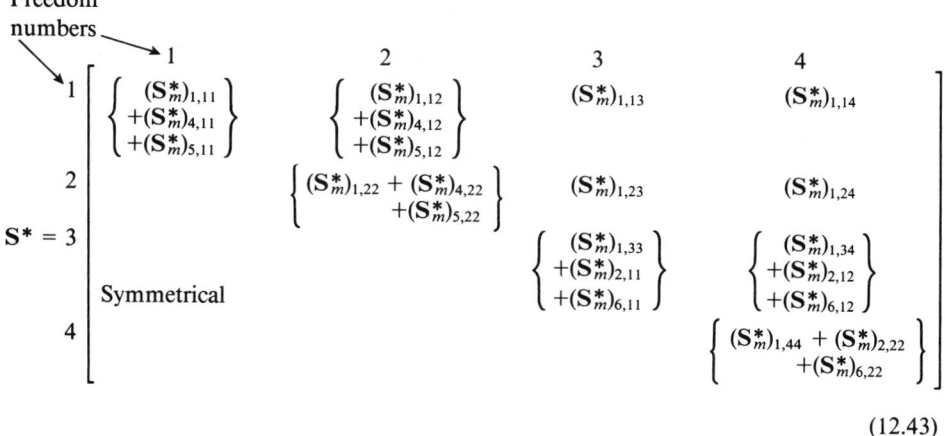

$$(12.43)$$

The unknown displacements are then determined from

$$D = (S^*)^{-1}P \tag{11.69}$$

It should be made clear that the above formulation of the global stiffness matrix S^* is for a *restrained structure* and therefore such a matrix is exactly the same as S^*_{11} developed earlier for an unrestrained structure; also the displacement vector D_p for the unrestrained structure is the same as D for the restrained structure. The member displacement vector in global coordinates δ^* can then be determined from the calculated displacements D using the MC matrix; thus, for the rigid frame in Fig. 12.14:

For member 1: $\delta^*_1 = D_1$ $\delta^*_2 = D_2$ $\delta^*_3 = D_3$ $\delta^*_4 = D_4$

$$\delta^*_5 = D_5 \qquad \delta^*_6 = D_6$$

For member 2: $\delta^*_1 = D_4$ $\delta^*_2 = D_5$ $\delta^*_3 = D_6$ $\delta^*_4 = \delta^*_5 = \delta^*_6 = 0$

For member 3: $\delta^*_1 = D_1$ $\delta^*_2 = D_2$ $\delta^*_3 = D_3$ $\delta^*_4 = \delta^*_5 = \delta^*_6 = 0$

The end-member forces (axial force, shear, and bending moment) in local coordinates can now be determined from

$$F = SR\delta^* \tag{12.24}$$

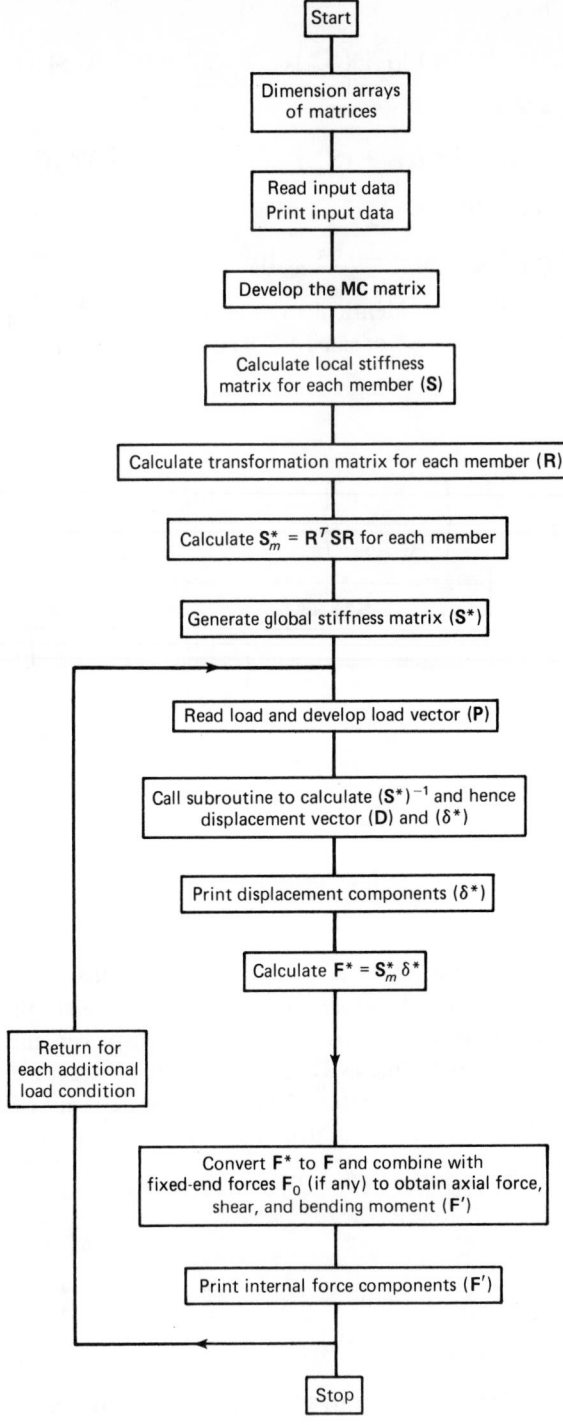

These F forces are combined with fixed-end forces $\mathbf{F}_0$ (if any) to yield the final end forces $\mathbf{F}'$. The above computations are normally performed on the computer, using one of several commercially available computer programs; a typical flow diagram for a computer solution program is shown on page 608.

12.7 TEMPERATURE AND LACK-OF-FIT PROBLEMS

The analysis of structures subjected to temperature change and/or with one or more members suffering from lack of fit can proceed in the same manner as explained above with some slight modifications. Consider a structure subjected to a change in temperature of ΔT; the change in length e of a member if free to move is

$$e = (\alpha)\Delta T(l) \tag{12.44}$$

in which α and l are the coefficient of thermal expansion of the material and the length of the member, respectively. If movement at the ends is prevented, the restraining axial force developed is $P = (AE/l)e$ at one end of the member and another equal and opposite force $P = -AEe/l$ at the other end of the member. These two member end forces can be transformed into global components of end member forces $\mathbf{F}^*$; summing up all such forces at the nodal points due to temperature changes in the members yields a column force matrix $\mathbf{P}_T$ with known values. For such a loading, Eq. (11.18) now becomes

$$\mathbf{P} = \mathbf{S}^*\mathbf{D} + \mathbf{P}_T \tag{12.45}$$

or

$$\mathbf{S}^*\mathbf{D} = \mathbf{P} - \mathbf{P}_T \tag{12.46}$$

For a lack-of-fit problem a similar procedure is followed with the amount of misfit e used to find the restraining end forces $P = AEe/l$ and $P = -(AE/l)e$. These end forces are transformed into global end-member forces and summed up, if there is more than one member with a lack-of-fit problem, to yield a column force matrix P_L; Eq. (11.18) is then adjusted to

$$\mathbf{S}^*\mathbf{D} = \mathbf{P} - \mathbf{P}_L \tag{12.47}$$

The procedure is demonstrated in the following solved problem.

■ SOLVED PROBLEM

Using the direct stiffness method generate the global stiffness matrix $\mathbf{S}^*$ for the structure in Fig. 12.16. Thus determine the nodal displacements and member forces due to: (a) given loads on the beam in Fig. 12.16; (b) an axial shortening (lack of fit) of 0.2 in in the inclined member 1. *Note:* Members 1 and 4 are axial-force members while members 2 and 3 can carry both axial and bending forces.

Solution. The freedom numbers are shown in Fig. 12.16; it should be noted that only three freedoms D_1, D_2, and D_3 are unrestrained. Therefore, we develop the (3×3) matrix $\mathbf{S}^*$ for the *restrained structure*. [A global matrix $\mathbf{S}^*$ for the *unrestrained structure*

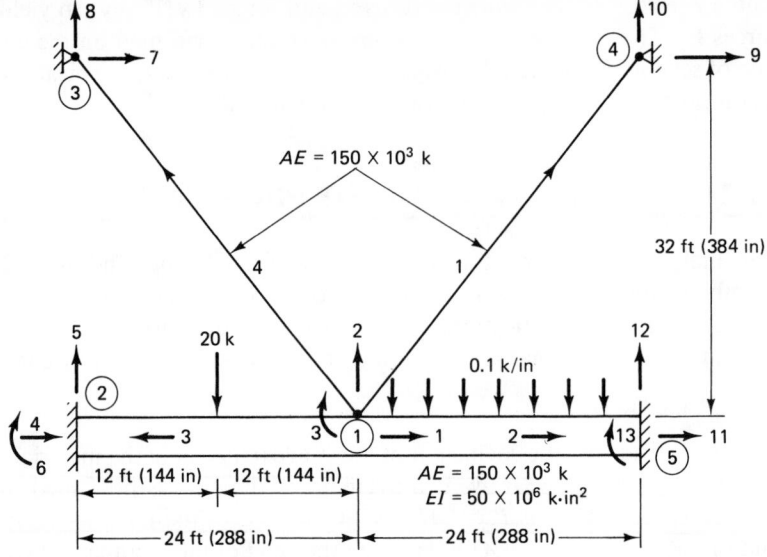

Figure 12.16 Structure for solved problem.

would have been of (13×13) size.] The member stiffness matrices for members 1, 4, 2, and 3 are given below.

$$
\mathbf{S}_1 = \mathbf{S}_4 =
\begin{bmatrix}
\dfrac{EA}{l} & 0 & \dfrac{-EA}{l} & 0 \\
0 & 0 & 0 & 0 \\
\dfrac{-EA}{l} & 0 & \dfrac{EA}{l} & 0 \\
0 & 0 & 0 & 0
\end{bmatrix}
=
\begin{bmatrix}
0.3125 & 0 & -0.3125 & 0 \\
0 & 0 & 0 & 0 \\
-0.3125 & 0 & 0.3125 & 0 \\
0 & 0 & 0 & 0
\end{bmatrix}
\times 10^3 \, \text{k/in}
$$

$$
\mathbf{S}_2 = \mathbf{S}_3 =
\begin{bmatrix}
\dfrac{EA}{l} & 0 & 0 & \dfrac{-EA}{l} & 0 & 0 \\[2mm]
0 & \dfrac{12EI}{l^3} & \dfrac{-6EI}{l^2} & 0 & \dfrac{-12EI}{l^3} & \dfrac{-6EI}{l^2} \\[2mm]
0 & \dfrac{-6EI}{l^2} & \dfrac{4EI}{l} & 0 & \dfrac{+6EI}{l^2} & \dfrac{2EI}{l} \\[2mm]
\dfrac{-EA}{l} & 0 & 0 & \dfrac{EA}{l} & 0 & 0 \\[2mm]
0 & \dfrac{-12EI}{l^3} & \dfrac{+6EI}{l^2} & 0 & \dfrac{12EI}{l^3} & \dfrac{+6EI}{l^2} \\[2mm]
0 & \dfrac{-6EI}{l^2} & \dfrac{2EI}{l} & 0 & \dfrac{+6EI}{l^2} & \dfrac{4EI}{l}
\end{bmatrix}
$$

$$= \begin{bmatrix} 0.521 & 0 & 0 & -0.521 & 0 & 0 \\ 0 & 0.0251 & -3.617 & 0 & -0.0251 & -3.617 \\ 0 & -3.617 & 694.44 & 0 & +3.617 & 347.22 \\ -0.521 & 0 & 0 & +0.521 & 0 & 0 \\ 0 & -0.0251 & +3.617 & 0 & 0.0251 & +3.617 \\ 0 & -3.617 & 347.22 & 0 & +3.617 & 694.44 \end{bmatrix} \times 10^3 \text{ k/in}$$

To generate the transformation matrices **R** we make use of the information in Table 12.3; thus from Eqs. (12.21) and (12.17):

$$\mathbf{R}_1 = \begin{bmatrix} 0.6 & 0.8 & 0 & 0 \\ -0.8 & 0.6 & 0 & 0 \\ 0 & 0 & 0.6 & 0.8 \\ 0 & 0 & -0.8 & 0.6 \end{bmatrix} \quad \mathbf{R}_4 = \begin{bmatrix} -0.6 & 0.8 & 0 & 0 \\ -0.8 & -0.6 & 0 & 0 \\ 0 & 0 & -0.6 & 0.8 \\ 0 & 0 & -0.8 & -0.6 \end{bmatrix}$$

$$\mathbf{R}_2 = \begin{bmatrix} 1 & 0 & 0 & 0 & 0 & 0 \\ 0 & 1 & 0 & 0 & 0 & 0 \\ 0 & 0 & 1 & 0 & 0 & 0 \\ 0 & 0 & 0 & 1 & 0 & 0 \\ 0 & 0 & 0 & 0 & 1 & 0 \\ 0 & 0 & 0 & 0 & 0 & 1 \end{bmatrix} \quad \mathbf{R}_3 = \begin{bmatrix} -1 & 0 & 0 & 0 & 0 & 0 \\ 0 & -1 & 0 & 0 & 0 & 0 \\ 0 & 0 & 1 & 0 & 0 & 0 \\ 0 & 0 & 0 & -1 & 0 & 0 \\ 0 & 0 & 0 & 0 & -1 & 0 \\ 0 & 0 & 0 & 0 & 0 & 1 \end{bmatrix}$$

The member stiffness matrices **S** in local coordinates are now transformed to those in global coordinates $\mathbf{S}_m^*$ using Eq. (12.30), $\mathbf{S}_m^* = \mathbf{R}^T\mathbf{S}\mathbf{R}$. Thus, for *member 1:*

$$(\mathbf{S}_m^*)_1 = \begin{bmatrix} 0.6 & -0.8 & 0 & 0 \\ 0.8 & 0.6 & 0 & 0 \\ 0 & 0 & 0.6 & -0.8 \\ 0 & 0 & 0.8 & 0.6 \end{bmatrix} \begin{bmatrix} 0.3125 & 0 & -0.3125 & 0 \\ 0 & 0 & 0 & 0 \\ -0.3125 & 0 & 0.3125 & 0 \\ 0 & 0 & 0 & 0 \end{bmatrix}$$

$$\times \begin{bmatrix} 0.6 & 0.8 & 0 & 0 \\ -0.8 & 0.6 & 0 & 0 \\ 0 & 0 & 0.6 & 0.8 \\ 0 & 0 & -0.8 & 0.6 \end{bmatrix} \times 10^3 \text{ k/in}$$

TABLE 12.3

| Member | Joint or node number | | Member inclination θ | $\sin\theta$ | $\cos\theta$ |
	Near end	Far end			
1	1	4	$\approx 53°$	0.8	0.6
2	1	5	0	0	1
3	1	2	180°	0	-1
4	1	3	$\approx 127°$	0.8	-0.6

or

$$(S_m^*)_1 = \begin{bmatrix} 0.1125 & 0.150 & -0.1125 & -0.150 \\ 0.150 & 0.20 & -0.150 & -0.20 \\ -0.1125 & -0.15 & 0.1125 & 0.15 \\ -0.15 & -0.2 & 0.15 & 0.20 \end{bmatrix} \times 10^3 \, \text{k/in}$$

$$(S_m^*)_4 = \begin{bmatrix} 0.1125 & -0.15 & -0.1125 & 0.15 \\ -0.15 & 0.20 & 0.15 & -0.20 \\ -0.1125 & 0.15 & 0.1125 & -0.15 \\ 0.15 & -0.20 & -0.15 & 0.20 \end{bmatrix} \times 10^3 \, \text{k/in}$$

$$(S_m^*)_2 = \begin{bmatrix} 0.521 & 0 & 0 & -0.521 & 0 & 0 \\ 0 & 0.0251 & -3.617 & 0 & -0.0251 & -3.617 \\ 0 & -3.617 & 694.44 & 0 & +3.617 & 347.22 \\ -0.521 & 0 & 0 & 0.521 & 0 & 0 \\ 0 & -0.0251 & +3.617 & 0 & 0.0251 & +3.617 \\ 0 & -3.617 & 347.22 & 0 & +3.617 & 694.44 \end{bmatrix}$$
$$\times 10^3 \, \text{k/in}$$

$$(S_m^*)_3 = \begin{bmatrix} 0.521 & 0 & 0 & -0.521 & 0 & 0 \\ 0 & 0.0251 & 3.617 & 0 & -0.0251 & 3.617 \\ 0 & 3.617 & 694.44 & 0 & -3.617 & 347.22 \\ -0.521 & 0 & 0 & 0.521 & 0 & 0 \\ 0 & -0.0251 & -3.617 & 0 & 0.0251 & -3.617 \\ 0 & 3.617 & 347.22 & 0 & -3.617 & 694.44 \end{bmatrix}$$
$$\times 10^3 \, \text{k/in}$$

Noting the numbering of the joints and members in Fig. 12.16, the following **MC** matrix is generated:

	Member code numbers					
	Member displacement number					
	1	2	3	4	5	6
Member	δ_1^*	δ_2^*	δ_3^*	δ_4^*	δ_5^*	δ_6^*
1	1	2	0	0	—	—
2	1	2	3	0	0	0
3	1	2	3	0	0	0
4	1	2	0	0	—	—

Numbers corresponding to the D's

Based on the above **MC** matrix, the positions of the elements in $\mathbf{S}_m^*$ matrices will be as follows:

Member 1		Member 4	
Element in $(\mathbf{S}_m^*)_1$	Position in $\mathbf{S}^*$	Element in $(\mathbf{S}_m^*)_4$	Position in $\mathbf{S}^*$
11	11	11	11
12	12	12	12
22	22	22	22

Member 2		Member 3	
Element in $(\mathbf{S}_m^*)_2$	Position in $\mathbf{S}^*$	Element in $(\mathbf{S}_m^*)_3$	Position in $\mathbf{S}^*$
11	11	11	11
12	12	12	12
13	13	13	13
22	22	22	22
23	23	23	23
33	33	33	33

Thus,

$$
\mathbf{S}^* = \begin{array}{c} \\ 1 \\ 2 \\ 3 \end{array}
\begin{array}{ccc} \quad 1 \qquad\qquad & 2 \qquad\qquad & 3 \\
\left[\begin{array}{ccc}
(0.1125 + 0.521 & (0.150 - 0.150) & 0 \\
+0.521 + 0.1125) & & \\
0 & (0.20 + 0.0251 & (-3.617 + 3.617) \\
& +0.0251 + 0.20) & \\
0 & (-3.617 + 3.617) & (694.44 + 694.44)
\end{array}\right]
\end{array}
$$

$$\times\ 10^3\ \mathrm{k/in}$$

or

$$
\mathbf{S}^* = \begin{bmatrix} 1.267 & 0 & 0 \\ 0 & 0.4502 & 0 \\ 0 & 0 & 1388.88 \end{bmatrix} \times 10^3\ \mathrm{k/in}
$$

The inverse of $\mathbf{S}^*$ can be shown to be

$$
(\mathbf{S}^*)^{-1} = \begin{bmatrix} 0.7893 & 0 & 0 \\ 0 & 2.2212 & 0 \\ 0 & 0 & 0.000720 \end{bmatrix} \times 10^{-3}\ \mathrm{in/k}
$$

For loading case (a), the P matrix can be derived based on the information shown in Fig. 12.17a and b; thus

$$\begin{bmatrix} P_1 \\ P_2 \\ P_3 \end{bmatrix} = \begin{bmatrix} 0 \\ -24.4 \\ -28.8 \end{bmatrix}$$

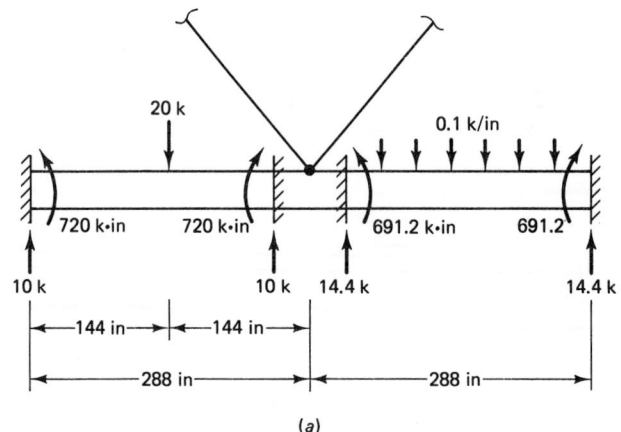

(a)

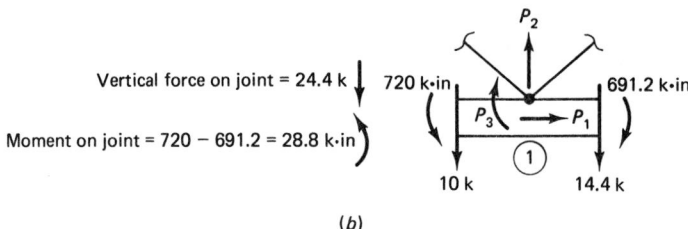

(b)

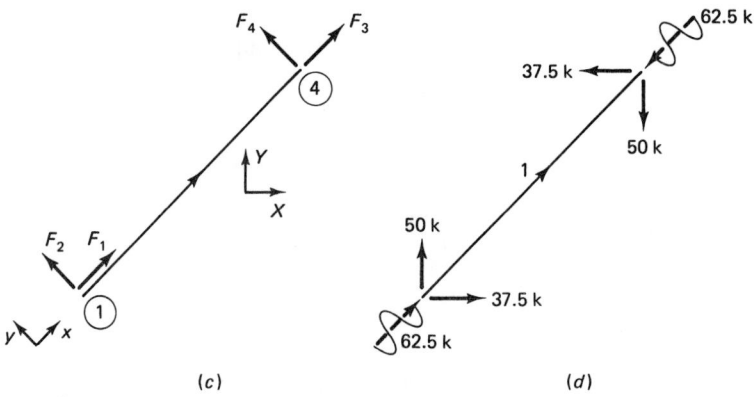

(c)   (d)

Figure 12.17 Forces on joints in structure of solved problem.

Thus,

$$\mathbf{D} = (\mathbf{S}^*)^{-1}\mathbf{P} = \begin{bmatrix} 0 \\ -0.0542 \\ -0.00002073 \end{bmatrix} \begin{matrix} D_1 \\ D_2 \\ D_3 \end{matrix} \quad \begin{matrix} \text{in} \\ \text{in} \\ \text{radian} \end{matrix}$$

Using the **MC** matrix, we can find the correspondence between the above **D** matrix and δ^* matrix for each member:

For members 1 and 4: $\qquad \delta_1^* = D_1 \qquad \delta_2^* = D_2 \qquad \delta_3^* = \delta_4^* = 0$

For members 2 and 3: $\qquad \delta_1^* = D_1 \qquad \delta_2^* = D_2 \qquad \delta_3^* = D_3$

$$\delta_4^* = \delta_5^* = \delta_6^* = 0$$

The member forces in local coordinates **F** can now be determined from Eq. (12.24): $\mathbf{F} = \mathbf{SR}\delta^*$. Thus, for *member 1:*

$$\begin{bmatrix} F_1 \\ F_2 \\ F_3 \\ F_4 \end{bmatrix} = 10^3 \times \begin{bmatrix} 0.3125 & 0 & -0.3125 & 0 \\ 0 & 0 & 0 & 0 \\ -0.3125 & 0 & 0.3125 & 0 \\ 0 & 0 & 0 & 0 \end{bmatrix}$$

$$\times \begin{bmatrix} 0.6 & 0.8 & 0 & 0 \\ -0.8 & 0.6 & 0 & 0 \\ 0 & 0 & 0.6 & 0.8 \\ 0 & 0 & -0.8 & 0.6 \end{bmatrix} \begin{bmatrix} 0 \\ -0.0542 \\ 0 \\ 0 \end{bmatrix}$$

or

$$\begin{bmatrix} F_1 \\ F_2 \\ F_3 \\ F_4 \end{bmatrix} = \begin{bmatrix} -13.55 \\ 0 \\ +13.55 \\ 0 \end{bmatrix}$$

These are shown in Fig. 12.17 c. For *member 2:*

$$\begin{bmatrix} F_1 \\ F_2 \\ F_3 \\ F_4 \\ F_5 \\ F_6 \end{bmatrix} = 10^3 \times \begin{bmatrix} 0.521 & 0 & 0 & -0.521 & 0 & 0 \\ 0 & 0.0251 & -3.617 & 0 & -0.0251 & -3.617 \\ 0 & -3.617 & 694.44 & 0 & +3.617 & 347.22 \\ -0.521 & 0 & 0 & 0.521 & 0 & 0 \\ 0 & -0.0251 & +3.617 & & 0.0251 & +3.617 \\ 0 & -3.617 & 347.22 & 0 & +3.617 & 694.44 \end{bmatrix}$$

$$\times \begin{bmatrix} 0 \\ -0.0542 \\ -0.00002073 \\ 0 \\ 0 \\ 0 \end{bmatrix} = \begin{bmatrix} 0 \\ -1.29 \\ 181.64 \\ 0 \\ 1.29 \\ 188.84 \end{bmatrix}$$

Adding the fixed-end forces, shown in Fig. 12.17, we have

$$
\begin{bmatrix} F_1' \\ F_2' \\ F_3' \\ F_4' \\ F_5' \\ F_6' \end{bmatrix} = \begin{bmatrix} 0 \\ -1.29 \\ 181.64 \\ 0 \\ 1.29 \\ 188.84 \end{bmatrix} + \begin{bmatrix} 0 \\ 14.4 \\ -691.2 \\ 0 \\ 14.4 \\ 691.2 \end{bmatrix} = \begin{bmatrix} 0 \\ 13.11 \\ -509.56 \\ 0 \\ 15.69 \\ 880.04 \end{bmatrix} \begin{matrix} k \\ k \\ k \cdot in \\ k \\ k \\ k \cdot in \end{matrix}
$$

Similarly, the final forces in members 3 and 4 can be calculated.

For loading case (b): The shortening of axial-force member 1 will lead to fixed-end axial forces, proportional to the axial stiffness of the member. Thus, the fixed-end axial forces, from Fig. 12.17d, can be calculated as (AE/l). $\Delta = [(150)(10)^3/480](0.2) = 62.5$ k. These fixed-end axial forces are now resolved into components parallel to the global axes X, Y, as shown in Fig. 12.17d. Thus, the P matrix for this loading at joint 1 is

$$
\begin{bmatrix} P_1 \\ P_2 \\ P_3 \end{bmatrix} = \begin{bmatrix} 37.5 \\ 50.0 \\ 0 \end{bmatrix}
$$

Hence

$$
\mathbf{D} = (\mathbf{S^*})^{-1}\mathbf{P} = \begin{bmatrix} 0.0296 \\ 0.1111 \\ 0 \end{bmatrix} \begin{matrix} D_1 \\ D_2 \\ D_3 \end{matrix} \begin{matrix} in \\ in \\ radians \end{matrix}
$$

The member forces in local coordinates can now be calculated following the same procedure as for loading case (a). ∎

PROBLEMS

12.1 to 12.4 Determine the transformation matrices for each of the structures in Figs. P12.1, P12.2, P12.5, P12.10.

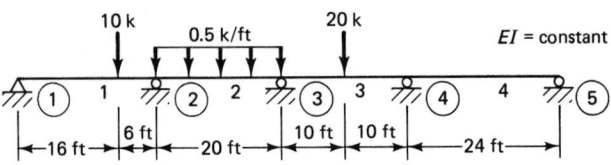

Figure P12.1

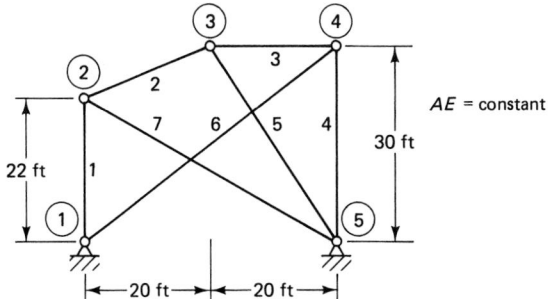

Figure P12.2

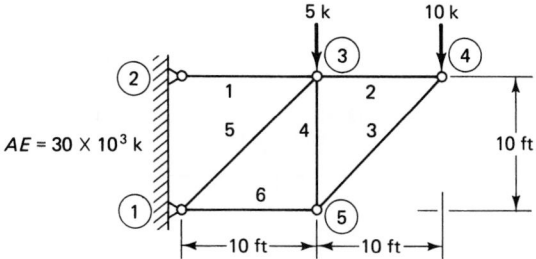

Figure P12.3

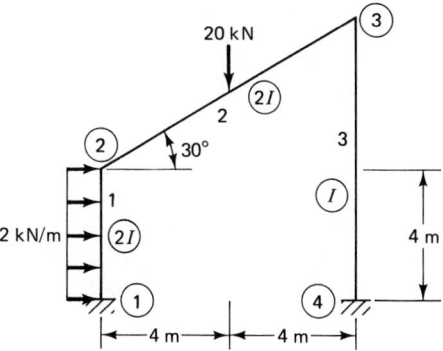

Figure P12.4

12.5 to 12.9 Determine the member stiffness matrices in **(a)** local coordinates, and in **(b)** global coordinates for each of the structures in Figs. P12.1, P12.2, P12.5, P12.7, P12.10.

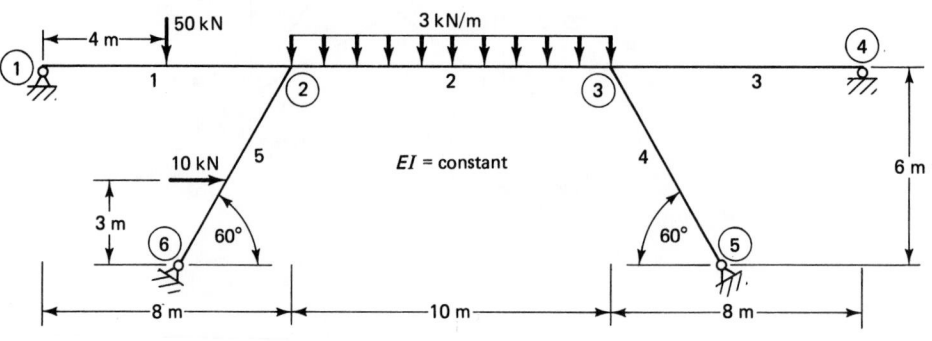

Figure P12.5

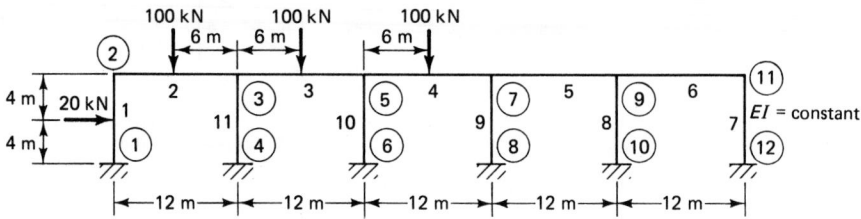

Figure P12.6

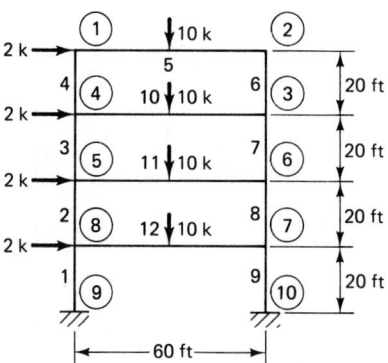

Figure P12.7

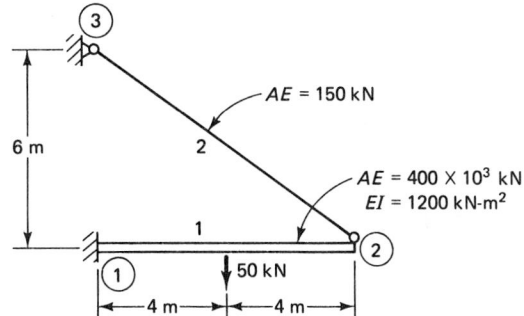

Figure P12.8

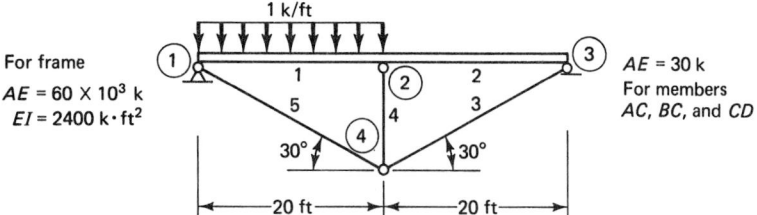

Figure P12.9

12.10 to 12.13 Apply the direct stiffness method to form the global stiffness matrix for each of the structures in Figs. P12.1, P12.2, P12.5, P12.8.

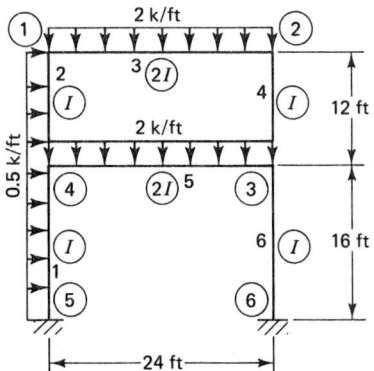

Figure P12.10

12.14 to 12.17 Apply the code-numbering technique to generate the global stiffness matrix for each of the structures in Figs. P12.3, P12.4, P12.6, P12.9.

12.18 Apply the direct stiffness method to generate the global stiffness matrix S^* for the structure in Fig. P12.10. Then determine the nodal displacements and member forces for each of the following conditions: **(a)** Given load on the structure. **(b)** Vertical settlement of support 5 equal to 0.2 in. **(c)** Temperature rise in members 3 and 5 of 20°C. Given $\alpha = 1.25 \times 10^{-5}/°C$. (*Note:* Develop a computer program to calculate S^* and then the nodal displacements and forces.) $EI = 10,000$ k·ft^2; $AE = 10^3$ k.

12.19 and 12.20 By the direct stiffness method determine the nodal displacements and member forces for each of the structures in Figs. P12.8, P12.9.

12.21 to 12.23 Apply the code-numbering technique to form the global stiffness matrix S^* for each of the structures in Figs. P11.13, P11.14 and P11.15.

12.24 to 12.26 Use the direct stiffness method to determine the nodal displacements and member forces for each of the structures in Figs. P11.13, P11.14, and P11.15.

Chapter **13**

Other Methods of Structural Analysis

A man should never be ashamed to (admit) he has been in the wrong, which is but saying, in other words, that he is wiser today than he was yesterday.

Alexander Pope

In this chapter three other methods of structural analysis are presented. These important methods may or may not be required by curricula of all engineering schools. Therefore, it was decided to include them in one chapter in three parts: Part I: Matrix-Force Method; Part II: Finite-Element Method; and Part III: Energy Methods.

PART I: MATRIX-FORCE METHOD

13.1 INTRODUCTION

The matrix-force or flexibility method of structural analysis is a generalization of the method of consistent deformation discussed in Chapter 9. In the matrix-displacement method (Chapters 11 and 12) joint displacements D were taken as the unknowns; whereas in the matrix-force method, the unknowns are the redundant forces $\mathbf{F}_x$ in the structure. The latter method possesses some advantages over the former method for certain structures, especially for those structures where the degree of freedom is high, that is, too many unknown joint displacements. In applying the matrix-force method to analyze a statically indeterminate structure, the structure is first converted to statically determinate and stable structure (primary structure) by making cuts in all the redundant internal members present and/or releasing the restraint by any redundant external reactions. Subjecting this basic structure to external loads P causes discontinuities or gaps in the displacements at all the cuts in the redundant internal members and/or at points of application of the redundant external reactions. Such gaps in the displacements must be removed by applying redundant forces and/or reactions of such magnitude as to restore continuity of the original structure. The reader is advised to review Chapter 9 where this same technique was applied in the method of consistent deformation. It is obvious that in this technique there are as many consistent deformation conditions as unknown redundant forces and reactions.

13.2 TRUSSES

13.2.1 Choice of Redundants

A primary structure is derived from the statically indeterminate structure by removing a definite number of internal members and/or external supports; such members and supports are labeled redundants. The choice of these redundants must be based on (1) stability and (2) accuracy. Stability implies that after the removal of the redundants, the resulting primary structure is kinematically stable; this also means that its static-equilibrium matrix E is a nonsingular square matrix with an inverse. If the chosen primary structure is unstable, this is revealed in the computation of the inverse of matrix E. Now the accuracy of the results is dependent on the conditioning of the flexibility matrix of the structure. If the chosen redundants cause larger displacements in their own direction than in the direction of other redundants, the resulting flexibility matrix of the structure is well conditioned with strong diagonal terms; otherwise the matrix can become strongly ill conditioned, with a nearly vanishing determinant. This makes the matrix very sensitive to inversion. Often a good choice of redundants can be made by examining the elements of the complete flexibility matrix in the computer. For structures with many redundants the selection of redundants is performed by the computer program. The indeterminate truss in Fig. 13.1a has one internal and one external redundancy; two choices for the primary truss and the redundants are shown in Fig. 13.1b–c and in Fig. 13.1d–e.

13.2.2 The Static-Equilibrium Matrix E

The external joint forces P are related to the internal forces F by means of the static-equilibrium matrix E; thus from Eq. (11.1)

$$P = EF \qquad (11.1)$$

and the member elongations δ are related to F as

$$\delta = fF \qquad (11.4)$$

in which f is the member flexibility matrix. Hence for a *statically determinate truss*, solving Eq. (11.1) for F gives

$$F = E^{-1}P \qquad (13.1)$$

and the member elongations δ become, from Eq. (11.4),

$$\delta = fE^{-1}P \qquad (13.2)$$

In a *statically indeterminate truss*, both essential and redundant internal forces are present; therefore, partitioning these forces, one can write Eq. (11.1) as

$$P = E_{Pd}F_d + E_{Px}F_x \qquad (13.3)$$

The E_{Pd} matrix is a square static-equilibrium matrix for the basic or primary truss and can be inverted. Similarly the external reactions R can be related to the internal forces F by means of the static-equilibrium matrices E_{Rd} and E_{Rx} as

$$R = E_{Rd}F_d + E_{Rx}F_x \qquad (13.4)$$

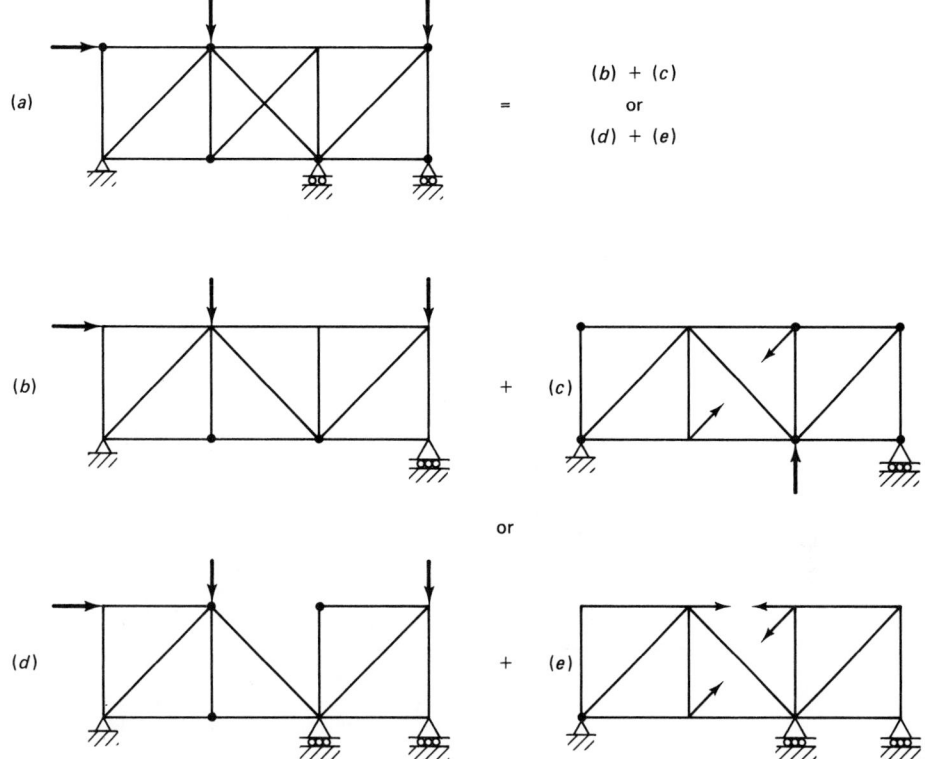

Figure 13.1 (*a*) A statically indeterminate truss. (*b*), (*d*) Two choices of primary truss. (*c*), (*e*) Two choices of redundants.

■ EXAMPLE 13.1

Deduce the static-equilibrium matrices $\mathbf{E}_{Pd}$, $\mathbf{E}_{Px}$, $\mathbf{E}_{Rd}$, and $\mathbf{E}_{Rx}$ for the loaded truss in Fig. 13.2*a*.

Solution. The two redundant internal members are chosen as shown in Fig. 13.2*c*. Using Fig. 13.2*b* and *c* and the two force-equilibrium equations at each joint, Eq. (13.3) can be written as

$$
\begin{Bmatrix} P_1 \\ P_2 \\ P_3 \\ P_4 \\ P_5 \\ P_6 \\ P_7 \\ P_8 \\ P_9 \\ P_{10} \\ P_{11} \\ P_{12} \end{Bmatrix}
=
\begin{bmatrix}
0 & -1 & 0 & 0 & 0 & 0 & 0 & 0 & 0 & 0 & 0 & 0 \\
+1 & 0 & 0 & 0 & 0 & 0 & 0 & 0 & 0 & 0 & 0 & 0 \\
0 & +1 & +0.8 & 0 & 0 & -0.8 & 0 & 0 & 0 & 0 & 0 & 0 \\
0 & 0 & 0.6 & +1 & 0 & 0.6 & 0 & 0 & 0 & 0 & 0 & 0 \\
0 & 0 & 0 & 0 & 0 & 0 & -1 & 0 & 0 & 0 & 0 & 0 \\
0 & 0 & 0 & 0 & 0 & 0 & 0 & 0 & 0 & 0 & +1 & 0 \\
0 & 0 & 0 & 0 & 0 & 0 & +1 & 0 & 0 & +0.8 & 0 & 0 \\
0 & 0 & 0 & 0 & 0 & 0 & 0 & +1 & 0 & +0.6 & 0 & 0 \\
0 & 0 & 0 & 0 & +1 & 0 & 0 & 0 & 0 & 0 & 0 & -1 \\
0 & 0 & 0 & -1 & 0 & 0 & 0 & 0 & 0 & 0 & 0 & 0 \\
0 & 0 & 0 & 0 & 0 & +0.8 & 0 & 0 & -1 & -0.8 & 0 & +1 \\
0 & 0 & 0 & 0 & 0 & 0 & 0 & 0 & +1 & 0 & 0 & 0
\end{bmatrix}
\begin{Bmatrix} F_{d1} \\ F_{d2} \\ F_{d3} \\ F_{d4} \\ F_{d5} \\ F_{d6} \\ F_{d7} \\ F_{d8} \\ F_{d9} \\ F_{d10} \\ F_{d11} \\ F_{d12} \end{Bmatrix}
+
\begin{bmatrix}
0 & 0 \\
0 & 0 \\
-1 & 0 \\
0 & 0 \\
+1 & +0.8 \\
0 & +0.6 \\
0 & 0 \\
0 & 0 \\
0 & -0.8 \\
0 & -0.6 \\
0 & 0 \\
0 & 0
\end{bmatrix}
\begin{Bmatrix} F_{x1} \\ F_{x2} \end{Bmatrix}
$$

$$\underbrace{\qquad\qquad\qquad\qquad}_{\text{Matrix } \mathbf{E}_{Pd}} \qquad\qquad\qquad\qquad \underbrace{\qquad\qquad}_{\text{Matrix } \mathbf{E}_{Px}}$$

$$
\begin{Bmatrix} R_1 \\ R_2 \\ R_3 \\ R_4 \end{Bmatrix} = \begin{bmatrix} 0 & 0 & -0.8 & 0 & -1 & 0 & 0 & 0 & 0 & 0 & 0 & 0 \\ -1 & 0 & -0.6 & 0 & 0 & 0 & 0 & 0 & 0 & 0 & 0 & 0 \\ 0 & 0 & 0 & 0 & 0 & -0.6 & 0 & 0 & 0 & -0.6 & -1 & 0 \\ 0 & 0 & 0 & 0 & 0 & 0 & 0 & -1 & 0 & 0 & 0 & 0 \end{bmatrix} \begin{Bmatrix} F_{d1} \\ F_{d2} \\ F_{d3} \\ F_{d4} \\ F_{d5} \\ F_{d6} \\ F_{d7} \\ F_{d8} \\ F_{d9} \\ F_{d10} \\ F_{d11} \\ F_{d12} \end{Bmatrix} + \begin{bmatrix} 0 & 0 \\ 0 & 0 \\ 0 & 0 \\ 0 & 0 \end{bmatrix} \begin{Bmatrix} F_{x1} \\ F_{x2} \end{Bmatrix}
$$

Matrix $\mathbf{E}_{Rd}$ ↗

Matrix $\mathbf{E}_{Rx}$ ↗ ■

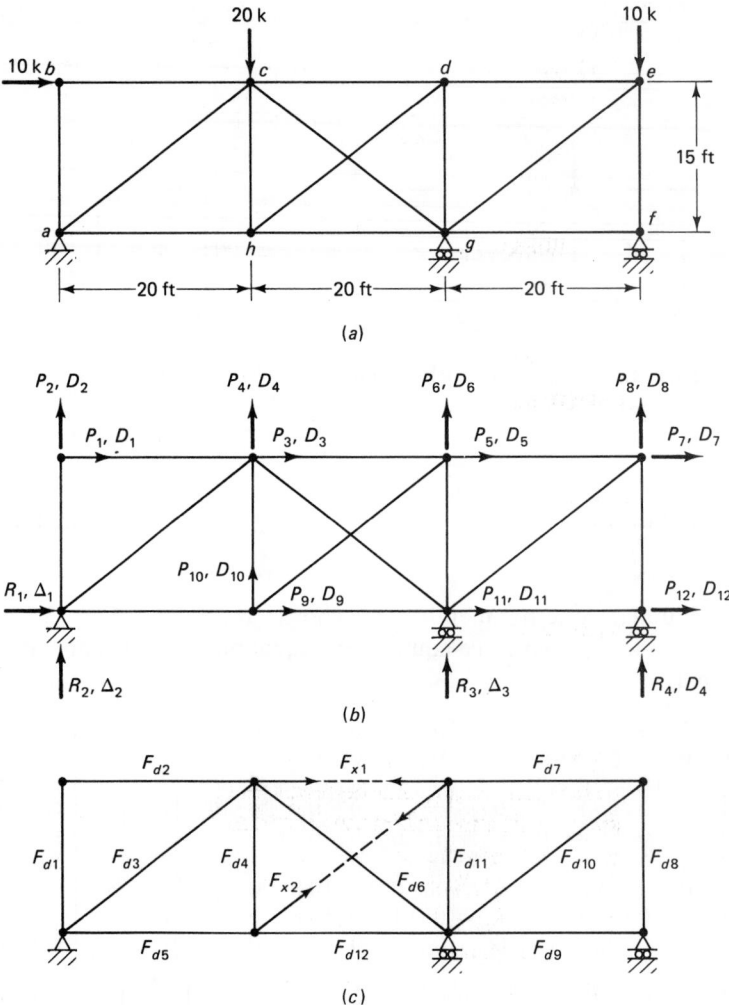

Figure 13.2 Truss for Example 13.1. (a) Indeterminate truss ($AE = 10^5$ k for all members). (b) P,D and R,Δ diagram. (c) F_d and F_x diagram.

13.2.3 The Deformation Compatibility Matrix C

From Eq. (11.2), the member elongations δ are related to the joint displacements D by means of the deformation compatibility matrix C; or

$$\delta = CD \qquad\qquad (11.2)$$

But

$$C = E^T \qquad\qquad (11.14)$$

Thus

$$\delta = E^T D \qquad\qquad (13.5)$$

However, the member elongations δ can also be affected by support settlements, and this can be accounted for in the following manner:

Matrix E^T can be partitioned into matrices E_{Pd}^T and E_{Px}^T, associated with the joint displacements D, and matrices E_{Rd}^T and E_{Rx}^T, associated with support settlements Δ if any. Thus partitioning the matrix δ into δ_d and δ_x associated with the essential and redundant internal members, respectively, one can write

$$\delta_d = E_{Pd}^T D + E_{Rd}^T \Delta \qquad\qquad (13.6)$$

$$\delta_x = E_{Px}^T D + E_{Rx}^T \Delta \qquad\qquad (13.7)$$

These matrix equations describe the geometrical compatibility in the structure.

13.2.4 The Member Flexibility Matrix f

The elongations δ in the members of a truss are functions of the axial forces F in the members; by Hooke's law: $\delta = (L/AE)F$, in which E is the modulus of elasticity and L and A are the length and cross-sectional area of the member. In Chapter 11, the matrix form of this relationship was given as

$$\delta = fF \qquad\qquad (11.4)$$

in which f is the member flexibility matrix. In terms of member flexibility matrices f_d and f_x, associated with internal forces in the essential and redundant members, respectively, Eq. (11.4) can be put in the form

$$\begin{aligned} \delta_d &= f_d F_d \\ \delta_x &= f_x F_x \end{aligned} \qquad\qquad (13.8)$$

Equations (13.8) can be adjusted to include initial elongations (member's lack of fit) in the essential and redundant members δ_{od} and δ_{ox}, and hence become

$$\delta_d = \delta_{od} + f_d F_d \qquad\qquad (13.9)$$

$$\delta_x = \delta_{ox} + f_x F_x \qquad\qquad (13.10)$$

■ EXAMPLE 13.2

Generate the f_d and f_x matrices for the truss in Fig. 13.2a. $AE = 10^5$ k for all members.

Solution

$$
\begin{Bmatrix} \delta_{d1} \\ \delta_{d2} \\ \delta_{d3} \\ \delta_{d4} \\ \delta_{d5} \\ \delta_{d6} \\ \delta_{d7} \\ \delta_{d8} \\ \delta_{d9} \\ \delta_{d10} \\ \delta_{d11} \\ \delta_{d12} \end{Bmatrix} = (10^{-3}) \underbrace{\begin{bmatrix} 1.8 & & & & & & & & & & & \\ & 2.4 & & & & & & & & & & \\ & & 3.0 & & & & 0 & & & & & \\ & & & 1.8 & & & & & & & & \\ & & & & 2.4 & & & & & & & \\ & & & & & 3.0 & & & & & & \\ & & & & & & 2.4 & & & & & \\ & & & & & & & 1.8 & & & & \\ & & & & & & & & 2.4 & & & \\ & & 0 & & & & & & & 3.0 & & \\ & & & & & & & & & & 1.8 & \\ & & & & & & & & & & & 2.4 \end{bmatrix}}_{\text{Matrix } \mathbf{f}_d \text{ (in/k)} \;\searrow} \begin{Bmatrix} F_{d1} \\ F_{d2} \\ F_{d3} \\ F_{d4} \\ F_{d5} \\ F_{d6} \\ F_{d7} \\ F_{d8} \\ F_{d9} \\ F_{d10} \\ F_{d11} \\ F_{d12} \end{Bmatrix}
$$

$$
\begin{Bmatrix} \delta_{x1} \\ \delta_{x2} \end{Bmatrix} = (10^{-3}) \underbrace{\begin{bmatrix} 2.4 & 0 \\ 0 & 3 \end{bmatrix}}_{\text{Matrix } \mathbf{f}_x \text{ (in/k)} \;\searrow} \begin{Bmatrix} F_{x1} \\ F_{x2} \end{Bmatrix}
$$

The above treatment of the matrix-force method reveals that in general there are six unknowns to be determined: F_x, F_d, δ_d, δ_x, D, and R. These unknowns are found from Eqs. (13.3) and (13.4) (static-equilibrium equations), from Eqs. (13.6) and (13.7) (deformation compatibility equations), and from Eqs. (13.9) and (13.10) (Hooke's law). Algebraic manipulation of these equations yields

$$\mathbf{F}_x = \boldsymbol{\phi}[\boldsymbol{\gamma}^T \mathbf{f}_d \mathbf{E}_{Pd}^{-1} \mathbf{P} + (\boldsymbol{\gamma}^T \boldsymbol{\delta}_{od} - \boldsymbol{\delta}_{ox}) + (\mathbf{E}_{Rx} - \mathbf{E}_{Rd}\boldsymbol{\gamma})^T \boldsymbol{\Delta}] \tag{13.11}$$

in which

$$\boldsymbol{\gamma} = \mathbf{E}_{Pd}^{-1} \mathbf{E}_{Px} \tag{13.12}$$

and

$$\boldsymbol{\phi} = (\mathbf{f}_x + \boldsymbol{\gamma}^T \mathbf{f}_d \boldsymbol{\gamma})^{-1} \tag{13.13}$$

$$\mathbf{D} = [\mathbf{E}_{Pd}^T]^{-1}[\mathbf{f}_d \mathbf{E}_{Pd}^{-1} \mathbf{P} - \mathbf{f}_d \boldsymbol{\gamma} \mathbf{F}_x + \boldsymbol{\delta}_{od} - \mathbf{E}_{Rd}^T \boldsymbol{\Delta}] \tag{13.14}$$

$$\mathbf{F}_d = \mathbf{E}_{Pd}^{-1} \mathbf{P} - \boldsymbol{\gamma} \mathbf{F}_x \tag{13.15}$$

The member elongations are then found from Eqs. (13.6) and (13.7) while the reactions R are deduced from Eq. (13.4). If the separate effects of external loads, settlements of supports, and initial lack of fit of members are required, then Eqs. (13.11), (13.14), and (13.15) reduce as follows:

For External Loads

$$\mathbf{F}_x = \boldsymbol{\phi}[\boldsymbol{\gamma}^T \mathbf{f}_d \mathbf{E}_{Pd}^{-1} \mathbf{P}]$$

$$\mathbf{F}_d = \mathbf{E}_{Pd}^{-1} \mathbf{P} - \boldsymbol{\gamma} \mathbf{F}_x \tag{13.16}$$

$$\mathbf{D} = [\mathbf{E}_{Pd}^T]^{-1}[\mathbf{f}_d \mathbf{F}_d]$$

For Settlements of Supports

$$\mathbf{F}_x = \boldsymbol{\phi}[(\mathbf{E}_{Rx} - \mathbf{E}_{Rd}\boldsymbol{\gamma})^T \boldsymbol{\Delta}]$$

$$\mathbf{F}_d = -\boldsymbol{\gamma} \mathbf{F}_x \tag{13.17}$$

$$\mathbf{D} = (\mathbf{E}_{Pd}^T)^{-1}(\mathbf{f}_d \mathbf{F}_d - \mathbf{E}_{Rd}^T \boldsymbol{\Delta})$$

For Initial Lack of Fit

$$\mathbf{F}_x = \phi(\gamma^T \delta_{od} - \delta_{ox})$$

$$\mathbf{F}_d = -\gamma \mathbf{F}_x \tag{13.18}$$

$$\mathbf{D} = (\mathbf{E}_{Pd}^T)^{-1}(\mathbf{f}_d \mathbf{F}_d + \delta_{od})$$

It should be noted that in analyzing a truss for any one of the above three conditions, matrices $\mathbf{E}_{Pd}$, $\mathbf{E}_{Px}$, $\mathbf{f}_d$, and $\mathbf{f}_x$ are required. The $\mathbf{P}$ matrix is necessary for the external loads condition; matrices $\mathbf{E}_{Rd}$, $\mathbf{E}_{Rx}$, and Δ are required for settlement of supports condition; and matrices δ_{od} and δ_{ox} are necessary for initial lack-of-fit condition. The truss in Fig. 13.2a has been analyzed for the above three conditions in Solved Problem 13.1.   ∎

∎ EXAMPLE 13.3

Use the matrix-force method to determine the member forces in the truss shown in Fig. 13.3a for the following conditions: (a) the loads at joint a; (b) member ab is "too short" by 5 mm. Assume the value AE/l to be the same for all members and equal to 10 kN/mm.

Solution. This truss is statically indeterminate to the second degree. Choose members ac and ab to be the redundants. The P,D diagram and F diagram are shown in Fig. 13.3b and c, respectively. Since there are no settlements of supports, the reactions R and displacements Δ are not shown at joints b and e in Fig. 13.3b.

Applying the two force-equilibrium equations at joint a:
From $\Sigma F_x = 0$:

$$P_1 + F_{x2} \cos 30° + F_{x1} \cos 60° - F_{d1} \cos 30° = 0$$

or

$$P_1 = \frac{\sqrt{3}}{2} F_{d1} - \frac{\sqrt{3}}{2} F_{x2} - \frac{1}{2} F_{x1}$$

From $\Sigma F_y = 0$:

$$P_2 + F_{x2} \sin 30° + F_{x1} \sin 60° + F_{d2} + F_{d1} \sin 30° = 0$$

or

$$P_2 = -\frac{1}{2} F_{d1} - F_{d2} - \frac{\sqrt{3}}{2} F_{x1} - \frac{1}{2} F_{x2}$$

Putting the above two equations in matrix form:

$$\left\{ \begin{array}{c} P_1 \\ P_2 \end{array} \right\} = \begin{bmatrix} \dfrac{\sqrt{3}}{2} & 0 \\ -\dfrac{1}{2} & -1 \end{bmatrix} \left\{ \begin{array}{c} F_{d1} \\ F_{d2} \end{array} \right\} + \begin{bmatrix} -\dfrac{1}{2} & -\dfrac{\sqrt{3}}{2} \\ -\dfrac{\sqrt{3}}{2} & -\dfrac{1}{2} \end{bmatrix} \left\{ \begin{array}{c} F_{x1} \\ F_{x2} \end{array} \right\}$$

$$\qquad\qquad\text{Matrix } \mathbf{E}_{Pd} \qquad\qquad\qquad \text{Matrix } \mathbf{E}_{Px}$$

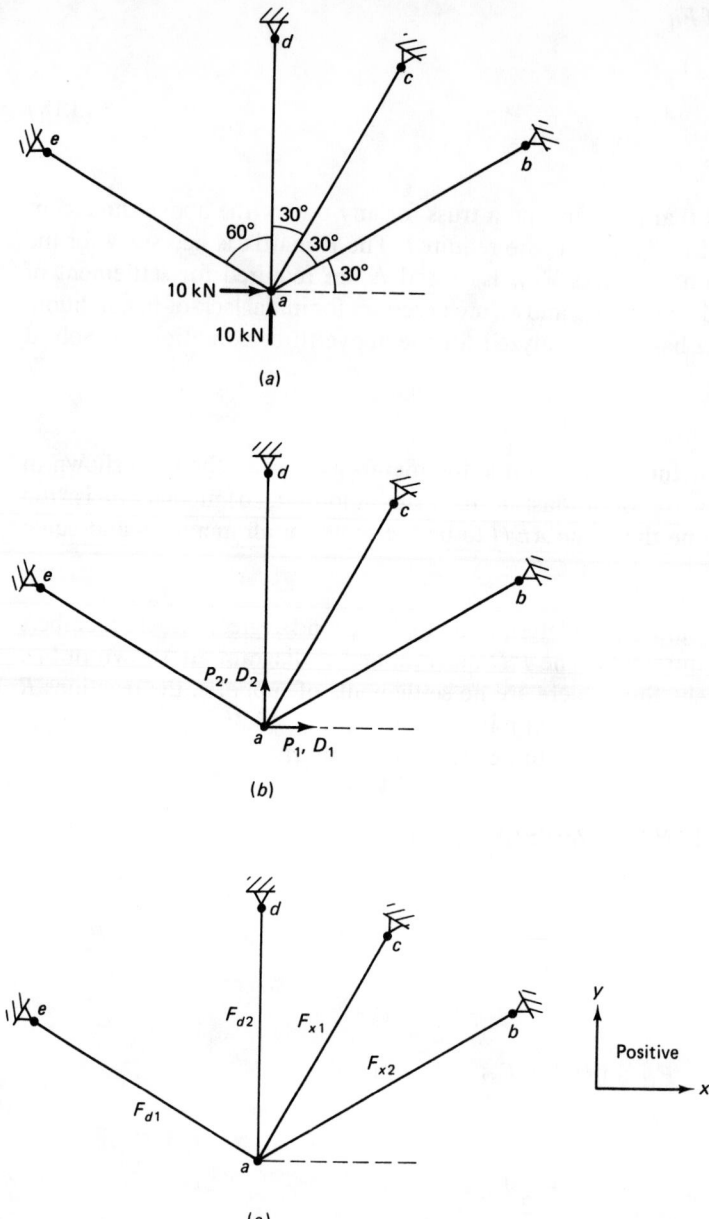

Figure 13.3 Structure for Example 13.3. (a) $\dfrac{AE}{l}$ = constant. (b) P,D numbering. (c) F numbering.

Also, from $\delta = \mathbf{f}\mathbf{F}$ and since $AE/l =$ constant,

$$\begin{Bmatrix} \delta_{d1} \\ \delta_{d2} \end{Bmatrix} = \underbrace{\frac{l}{AE}\begin{bmatrix} 1 & 0 \\ 0 & 1 \end{bmatrix}}_{\text{Matrix } \mathbf{f}_d \;\nearrow}\begin{Bmatrix} F_{d1} \\ F_{d2} \end{Bmatrix} \qquad \begin{Bmatrix} \delta_{x1} \\ \delta_{x2} \end{Bmatrix} = \underbrace{\frac{l}{AE}\begin{bmatrix} 1 & 0 \\ 0 & 1 \end{bmatrix}}_{\text{Matrix } \mathbf{f}_x \;\nearrow}\begin{Bmatrix} F_{x1} \\ F_{x2} \end{Bmatrix}$$

From Eq. (13.12):

$$\boldsymbol{\gamma} = \mathbf{E}_{Pd}^{-1}\mathbf{E}_{Px} = \begin{bmatrix} \dfrac{\sqrt{3}}{2} & 0 \\[2mm] -\dfrac{1}{2} & -1 \end{bmatrix}^{-1}\begin{bmatrix} -\dfrac{1}{2} & -\dfrac{\sqrt{3}}{2} \\[2mm] -\dfrac{\sqrt{3}}{2} & -\dfrac{1}{2} \end{bmatrix}$$

$$\mathbf{E}_{Pd}^{-1} = \begin{bmatrix} 1.1547 & 0 \\ -0.5774 & -1 \end{bmatrix}$$

Thus,

$$\boldsymbol{\gamma} = \begin{bmatrix} -0.5774 & -1 \\ 1.1547 & +1 \end{bmatrix}$$

Also, from Eq. (13.13), $\boldsymbol{\phi} = [\mathbf{f}_x + \boldsymbol{\gamma}^T\mathbf{f}_d\boldsymbol{\gamma}]^{-1}$. Now

$$(\mathbf{f}_x + \boldsymbol{\gamma}^T\mathbf{f}_d\boldsymbol{\gamma}) = \frac{l}{AE}\begin{bmatrix} 1 & 0 \\ 0 & 1 \end{bmatrix} + \frac{l}{AE}\begin{bmatrix} -0.5774 & 1.1547 \\ -1 & +1 \end{bmatrix}\begin{bmatrix} 1 & 0 \\ 0 & 1 \end{bmatrix}\begin{bmatrix} -0.5774 & -1 \\ 1.1547 & +1 \end{bmatrix}$$

$$= \frac{l}{AE}\begin{bmatrix} 2.6667 & 1.7321 \\ 1.7321 & 3 \end{bmatrix}$$

Thus

$$\boldsymbol{\phi} = \frac{AE}{l}\begin{bmatrix} 0.6 & -0.3464 \\ -0.3464 & 0.5333 \end{bmatrix}$$

(a) For the loading shown in Fig. 13.3a, the **P** matrix is

$$\begin{Bmatrix} P_1 \\ P_2 \end{Bmatrix} = \begin{Bmatrix} 10 \\ 10 \end{Bmatrix}\text{ (kN)}$$

From Eqs. (13.16):

$$\mathbf{F}_x = \boldsymbol{\phi}[\boldsymbol{\gamma}^T\mathbf{f}_d\mathbf{E}_{Pd}^{-1}\mathbf{P}] \quad \text{or} \quad \begin{Bmatrix} F_{x1} \\ F_{x2} \end{Bmatrix} = \begin{Bmatrix} -5.465 \\ -5.952 \end{Bmatrix} \quad \text{kN}$$

and

$$\mathbf{F}_d = \mathbf{E}_{Pd}^{-1}\mathbf{P} - \boldsymbol{\gamma}\mathbf{F}_x \quad \text{or} \quad \begin{Bmatrix} F_{d1} \\ F_{d2} \end{Bmatrix} = \begin{Bmatrix} 2.440 \\ -3.512 \end{Bmatrix} \quad \text{kN}$$

(b) For member ab being "too short" by 5 mm, the matrices δ_{od} and δ_{ox} are as follows:

$$\begin{Bmatrix} \delta_{od1} \\ \delta_{od2} \end{Bmatrix} = \begin{Bmatrix} 0 \\ 0 \end{Bmatrix} \qquad \begin{Bmatrix} \delta_{ox1} \\ \delta_{ox2} \end{Bmatrix} = \begin{Bmatrix} 0 \\ -5 \end{Bmatrix}$$

Therefore, from Eq. (13.18):

$$\mathbf{F}_x = \boldsymbol{\phi}(\boldsymbol{\gamma}^T \boldsymbol{\delta}_{od} - \boldsymbol{\delta}_{ox}) \quad \text{or} \quad \begin{Bmatrix} F_{x1} \\ F_{x2} \end{Bmatrix} = \frac{AE}{l} \begin{Bmatrix} -1.732 \\ +2.667 \end{Bmatrix}$$

If $AE/l = 10$ kN/mm, then $F_{x1} = -17.32$ kN (compression) and $F_{x2} = 26.67$ kN.

$$\mathbf{F}_d = -\boldsymbol{\gamma}\mathbf{F}_x \quad \text{or} \quad \begin{Bmatrix} F_{d1} \\ F_{d2} \end{Bmatrix} = \frac{AE}{l} \begin{Bmatrix} 1.667 \\ -0.667 \end{Bmatrix}$$

For $AE/l = 10$ kN/mm, then $F_{d1} = 16.67$ kN, $F_{d2} = -6.67$ kN (compression). ∎

13.2.5 The Static and Deformation Checks

To check the results, the two equilibrium equations $\Sigma F_x = 0$ and $\Sigma F_y = 0$ must be satisfied at every joint in the truss. For example, for the results from part (b) above, applying $\Sigma F_x = 0$ at joint a:

$$F_{x2} \cos 30° + F_{x1} \cos 60° - F_{d1} \cos 30°$$

$$= 23.096 - 8.660 - 14.436 = 0 \qquad \text{(checks)}$$

From $\Sigma F_y = 0$ at joint a:

$$F_{x2} \sin 30° + F_{x1} \sin 60° + F_{d2} + F_{d1} \sin 30°$$

$$= 13.335 - 15.00 - 6.67 + 8.335 \approx 0 \qquad \text{(checks)}$$

Furthermore the deformation checks must also be performed as described in Chapter 11.

13.3 CONTINUOUS BEAMS AND RIGID FRAMES

13.3.1 General

The force method has been described in Chapter 9. The basic concepts of the matrix-force method for continuous beams and rigid frames are similar to those in the matrix-displacement method described in Chapter 11. The essential differences between the two methods are that in the matrix-force method:

1. The basic unknowns are the forces at the ends of the members.
2. The flexibility matrix is used instead of its inverse, the stiffness matrix, in relating end deformations δ to end forces F.

In dealing with a statically indeterminate continuous beam or rigid frame, the structure is first transformed to a determinate one by removing the *redundants;* this transformation is effected by the insertion of i hinges at the ends of members, where i is the degree of indeterminacy of the original structure. It should be emphasized that the resulting basic structure, sometimes termed *released structure,* must be stable. Thus the primary unknowns here are the i *redundant* internal end moments F_x (exclusive of any fixed-end moments) at the i hinges. The remaining end moments are termed *essential internal end* moments F_d. For example, the continuous beam in Fig. 13.4a is indeterminate to the second degree; therefore, two hinges are introduced, as shown in Fig. 13.4d, to reduce it to a primary (basic) structure which is also stable.

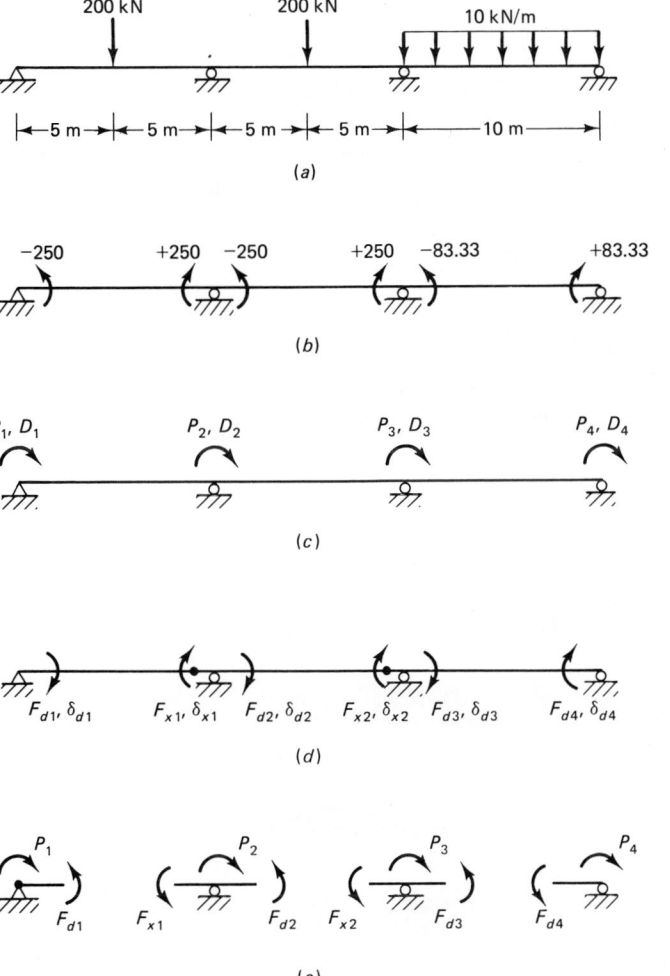

Figure 13.4 Beam for Example 13.4. (b) Fixed-end moments (kN · m). (c) The P,D diagram. (d) The F,δ diagram. (e) Joint-equilibrium diagrams (shears are not shown).

13.3.2 The Static-Equilibrium, Deformation Compatibility, and Flexibility Matrices

Based on the relation $\mathbf{P} = \mathbf{EF}$, the partitioning of the internal end-moment matrix $\mathbf{F}$ into an essential internal moment matrix $\mathbf{F}_d$ and a redundant internal moment matrix $\mathbf{F}_x$ leads to the matrices $\mathbf{E}_{Pd}$ and $\mathbf{E}_{Px}$ which are submatrices of the equilibrium matrix $\mathbf{E}$. Thus, one can write

$$\mathbf{P} = \mathbf{E}_{Pd}\mathbf{F}_d + \mathbf{E}_{Px}\mathbf{F}_x \tag{13.19}$$

Here again, $\mathbf{E}_{Pd}$ is a square matrix which has an inverse.

From the basic relation $\boldsymbol{\delta} = \mathbf{CD}$ and recognizing that $\mathbf{C} = \mathbf{E}^T$, the following matrix equations are deduced:

$$\boldsymbol{\delta}_d = \mathbf{E}_{Pd}^T\mathbf{D} \tag{13.20}$$

$$\boldsymbol{\delta}_x = \mathbf{E}_{Px}^T\mathbf{D} \tag{13.21}$$

in which $\boldsymbol{\delta}_d$ and $\boldsymbol{\delta}_x$ are the internal end rotations at the member ends where $\mathbf{F}_d$'s and $\mathbf{F}_x$'s are applied, respectively.

The relationship between end rotations δ and end moments F of internal members (Fig. 13.5) was established by Eq. (11.45), as follows:

$$\delta_i = \frac{l}{3EI}F_i - \frac{l}{6EI}F_j$$

$$\delta_j = -\frac{l}{6EI}F_i + \frac{l}{3EI}F_j \tag{11.45}$$

If we define $\mathbf{f}_d$ and $\mathbf{f}_{dx}$ = flexibility matrices relating $\boldsymbol{\delta}_d$ to $\mathbf{F}_d$ and $\mathbf{F}_x$, respectively, and $\mathbf{f}_x$ and $\mathbf{f}_{xd}$ = flexibility matrices relating $\boldsymbol{\delta}_x$ to $\mathbf{F}_x$ and $\mathbf{F}_d$, respectively, and note that

$$\mathbf{f}_{xd} = \mathbf{f}_{dx}^T \qquad \text{(by Betti's theorem)}$$

then we can write Eqs. (11.45) in the following matrix form:

$$\boldsymbol{\delta}_d = \mathbf{f}_d\mathbf{F}_d + \mathbf{f}_{dx}\mathbf{F}_x \tag{13.22}$$

$$\boldsymbol{\delta}_x = \mathbf{f}_{xd}\mathbf{F}_d + \mathbf{f}_x\mathbf{F}_x = \mathbf{f}_{dx}^T\mathbf{F}_d + \mathbf{f}_x\mathbf{F}_x \tag{13.23}$$

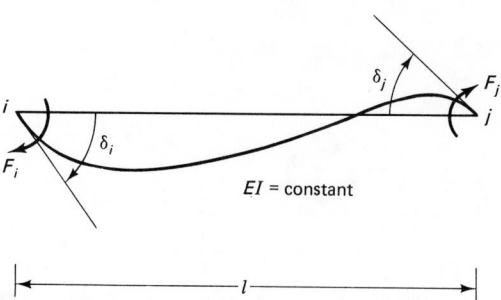

EI = constant

Figure 13.5 Relationship between end rotations δ and end moments F.

There are five unknowns: $\mathbf{F}_x$, $\mathbf{F}_d$, δ_x, δ_d, and $\mathbf{D}$; these are determined from the five matrix equations (13.19) to (13.23). Algebraic manipulation of these equations yields

$$\mathbf{F}_x = \phi(\gamma^T \mathbf{f}_d - \mathbf{f}_{dx}^T)\mathbf{E}_{Pd}^{-1}P \qquad (13.24)$$

in which

$$\gamma = \mathbf{E}_{Pd}^{-1}\mathbf{E}_{Px} \qquad (13.12)$$

and

$$\phi = [\mathbf{f}_x + (\gamma^T \mathbf{f}_d - \mathbf{f}_{dx}^T)\gamma - \gamma^T \mathbf{f}_{dx}]^{-1} \qquad (13.25)$$

$$\mathbf{F}_d = \mathbf{E}_{Pd}^{-1}P - \gamma\mathbf{F}_x \qquad (13.26)$$

$$\mathbf{D} = [\mathbf{E}_{Pd}^T]^{-1}(\mathbf{f}_d\mathbf{F}_d + \mathbf{f}_{dx}\mathbf{F}_x) \qquad (13.27)$$

To analyze a structure, six input matrices are required: $\mathbf{E}_{Pd}$, $\mathbf{E}_{Px}$, $\mathbf{f}_d$, $\mathbf{f}_x$, $\mathbf{f}_{dx}$ ($=\mathbf{f}_{xd}$), $\mathbf{P}$. We should not forget that the final end moments are $\mathbf{F}' = \mathbf{F}_0 + \mathbf{F}$, in which $\mathbf{F}_0$ are the fixed-end moments obtained from the Appendix.

■ EXAMPLE 13.4

Determine the end moments for the continuous beam in Fig. 13.4a.

Solution. The fixed-end moments F_0 are shown in Fig. 13.4b; the P,D and F,δ diagrams are given in Fig. 13.4c and d. From the joint-equilibrium diagrams in Fig. 13.4e:

$$P_1 = F_{d1} \qquad P_2 = F_{d2} + F_{x1} \qquad P_3 = F_{d3} + F_{x2} \qquad P_4 = F_{d4}$$

In matrix form

$$\begin{Bmatrix} P_1 \\ P_2 \\ P_3 \\ P_4 \end{Bmatrix} = \begin{bmatrix} 1 & 0 & 0 & 0 \\ 0 & 1 & 0 & 0 \\ 0 & 0 & 1 & 0 \\ 0 & 0 & 0 & 1 \end{bmatrix} \begin{Bmatrix} F_{d1} \\ F_{d2} \\ F_{d3} \\ F_{d4} \end{Bmatrix} + \begin{bmatrix} 0 & 0 \\ 1 & 0 \\ 0 & 1 \\ 0 & 0 \end{bmatrix} \begin{Bmatrix} F_{x1} \\ F_{x2} \end{Bmatrix}$$

$$\underbrace{\qquad\qquad}_{\text{Matrix } \mathbf{E}_{Pd}} \qquad\qquad \underbrace{\qquad\qquad}_{\text{Matrix } \mathbf{E}_{Px}}$$

$$\delta_{d1} = \frac{10}{3EI}F_{d1} - \frac{10}{6EI}F_{x1} \qquad \delta_{x1} = -\frac{10}{6EI}F_{d1} + \frac{10}{3EI}F_{x1}$$

$$\delta_{d2} = \frac{10}{3EI}F_{d2} - \frac{10}{6EI}F_{x2} \qquad \delta_{x2} = -\frac{10}{6EI}F_{d2} + \frac{10}{3EI}F_{x2}$$

$$\delta_{d3} = \frac{10}{3EI}F_{d3} - \frac{10}{6EI}F_{d4} \qquad \delta_{d4} = -\frac{10}{6EI}F_{d3} + \frac{10}{3EI}F_{d4}$$

In matrix form:

$$\begin{Bmatrix} \delta_{d1} \\ \delta_{d2} \\ \delta_{d3} \\ \delta_{d4} \end{Bmatrix} = \frac{10}{6EI}\begin{bmatrix} 2 & 0 & 0 & 0 \\ 0 & 2 & 0 & 0 \\ 0 & 0 & 2 & -1 \\ 0 & 0 & -1 & 2 \end{bmatrix}\begin{Bmatrix} F_{d1} \\ F_{d2} \\ F_{d3} \\ F_{d4} \end{Bmatrix} + \frac{10}{6EI}\begin{bmatrix} -1 & 0 \\ 0 & -1 \\ 0 & 0 \\ 0 & 0 \end{bmatrix}\begin{Bmatrix} F_{x1} \\ F_{x2} \end{Bmatrix}$$

$$\underbrace{\qquad\qquad}_{\text{Matrix } \mathbf{f}_d \ (1/\text{kN}\cdot\text{m})} \qquad\qquad \underbrace{\qquad\qquad}_{\text{Matrix } \mathbf{f}_{dx} \ (1/\text{kN}\cdot\text{m})}$$

$$\begin{Bmatrix} \delta_{x1} \\ \delta_{x2} \end{Bmatrix} = \underbrace{\frac{10}{6EI}\begin{bmatrix} -1 & 0 & 0 & 0 \\ 0 & -1 & 0 & 0 \end{bmatrix}}_{\text{Matrix } \mathbf{f}_{xd}\,(1/\mathrm{kN\cdot m})} \begin{Bmatrix} F_{d1} \\ F_{d2} \\ F_{d3} \\ F_{d4} \end{Bmatrix} + \underbrace{\frac{10}{6EI}\begin{bmatrix} 2 & 0 \\ 0 & 2 \end{bmatrix}}_{\text{Matrix } \mathbf{f}_{x}\,(1/\mathrm{kN\cdot m})}\begin{Bmatrix} F_{x1} \\ F_{x2} \end{Bmatrix}$$

By comparing matrices $\mathbf{f}_{dx}$ and $\mathbf{f}_{xd}$ it is noted that $\mathbf{f}_{xd} = \mathbf{f}_{dx}^T$ as it should be. From Fig. 13.4b, the **P** matrix is given by

$$\begin{Bmatrix} P_1 \\ P_2 \\ P_3 \\ P_4 \end{Bmatrix} = \begin{Bmatrix} +250 \\ 0 \\ -166.67 \\ -83.33 \end{Bmatrix} \quad \mathrm{kN\cdot m}$$

From the above six matrices we calculate the following:

$$\mathbf{E}_{Pd}^{-1} = \begin{bmatrix} 1 & & & \\ & 1 & & 0 \\ & & 1 & \\ 0 & & & 1 \end{bmatrix} \qquad \gamma = \begin{bmatrix} 0 & 0 \\ 1 & 0 \\ 0 & 1 \\ 0 & 0 \end{bmatrix} \qquad \gamma^T = \begin{bmatrix} 0 & 1 & 0 & 0 \\ 0 & 0 & 1 & 0 \end{bmatrix}$$

$$\phi = \frac{3EI}{5}\begin{bmatrix} 2 & 0 \\ 0 & 2 \end{bmatrix} + \left\{ \begin{bmatrix} 0 & 1 & 0 & 0 \\ 0 & 0 & 1 & 0 \end{bmatrix} \begin{bmatrix} 2 & 0 & 0 & 0 \\ 0 & 2 & 0 & 0 \\ 0 & 0 & 2 & -1 \\ 0 & 0 & -1 & 2 \end{bmatrix} - \begin{bmatrix} -1 & 0 & 0 & 0 \\ 0 & -1 & 0 & 0 \end{bmatrix} \right.$$

$$\left. \times \left[\begin{bmatrix} 0 & 0 \\ 1 & 0 \\ 0 & 1 \\ 0 & 0 \end{bmatrix} - \begin{bmatrix} 0 & 1 & 0 & 0 \\ 0 & 0 & 1 & 0 \end{bmatrix}\begin{bmatrix} -1 & 0 \\ 0 & -1 \\ 0 & 0 \\ 0 & 0 \end{bmatrix}\right]^{-1}\right.$$

$$\phi = \frac{EI}{25}\begin{bmatrix} +4 & -1 \\ -1 & +4 \end{bmatrix}$$

Therefore, from

$$\mathbf{F}_x = \phi(\gamma^T \mathbf{f}_d - \mathbf{f}_{dx}^T)\mathbf{E}_{Pd}^{-1}P \qquad \phi(\gamma^T \mathbf{f}_d - \mathbf{f}_{dx}^T)\mathbf{E}_{Pd}^{-1} = \frac{1}{15}\begin{bmatrix} 4 & 7 & -2 & 1 \\ -1 & 2 & 8 & -4 \end{bmatrix}$$

hence

$$\begin{Bmatrix} F_{x1} \\ F_{x2} \end{Bmatrix} = \frac{1}{15}\begin{bmatrix} 4 & 7 & -2 & 1 \\ -1 & 2 & 8 & -4 \end{bmatrix}\begin{bmatrix} 250 \\ 0 \\ -166.67 \\ -83.33 \end{bmatrix} = \begin{bmatrix} +83.33 \\ -83.33 \end{bmatrix} \quad \mathrm{kN\cdot m}$$

The final moments are, from $\mathbf{F}' = \mathbf{F}_0 + \mathbf{F}$,

$$\begin{Bmatrix} F'_{x1} \\ F'_{x2} \end{Bmatrix} = \begin{Bmatrix} +250 \\ +250 \end{Bmatrix} + \begin{Bmatrix} +83.33 \\ -83.33 \end{Bmatrix} = \begin{Bmatrix} 333.33 \\ 166.67 \end{Bmatrix} \quad \mathrm{kN\cdot m}$$

From Eq. (13.26):

$$
\begin{Bmatrix} F_{d1} \\ F_{d2} \\ F_{d3} \\ F_{d4} \end{Bmatrix} = \begin{bmatrix} 1 & & & \\ & 1 & 0 & \\ & 0 & 1 & \\ & & & 1 \end{bmatrix} \begin{bmatrix} 250 \\ 0 \\ -166.67 \\ -83.33 \end{bmatrix} - \begin{bmatrix} 0 & 0 \\ 1 & 0 \\ 0 & 1 \\ 0 & 0 \end{bmatrix} \begin{Bmatrix} 83.33 \\ -83.33 \end{Bmatrix}
$$

$$
= \begin{Bmatrix} 250 \\ -83.33 \\ -83.33 \\ -83.33 \end{Bmatrix} \quad \text{kN} \cdot \text{m}
$$

Hence from $\mathbf{F}' = \mathbf{F}_0 + \mathbf{F}$

$$
\begin{Bmatrix} F'_{d1} \\ F'_{d2} \\ F'_{d3} \\ F'_{d4} \end{Bmatrix} = \begin{Bmatrix} -250 \\ -250 \\ -83.33 \\ +83.33 \end{Bmatrix} + \begin{Bmatrix} 250 \\ -83.33 \\ -83.33 \\ -83.33 \end{Bmatrix} = \begin{Bmatrix} 0 \\ -333.33 \\ -166.66 \\ 0 \end{Bmatrix} \quad \text{kN} \cdot \text{m}
$$

These results check with those obtained in Example 9.4 (Fig. 9.10). ■

■ EXAMPLE 13.5

Use the matrix-force method to determine the end moments of members in the rigid frame in Fig. 13.6a.

Solution. This structure is statically indeterminate to the third degree; to convert it into a stable determinate structure, three hinges are introduced as shown in Fig. 13.6d. The fixed-end forces and the P,D diagram are shown in Fig. 13.6b and c, respectively. From the free-body diagrams of Fig. 13.6e:

$$
P_1 = F_{d1} + F_{x1} \qquad P_2 = F_{d2} + F_{x2} \qquad P_3 = -\tfrac{1}{12}(F_{d1} + F_{x3}) - \tfrac{1}{12}(F_{d2} + F_{d3})
$$

In matrix form:

$$
\begin{Bmatrix} P_1 \\ P_2 \\ P_3 \end{Bmatrix} = \underbrace{\begin{bmatrix} 1 & 0 & 0 \\ 0 & 1 & 0 \\ -\tfrac{1}{12} & -\tfrac{1}{12} & -\tfrac{1}{12} \end{bmatrix}}_{[\mathbf{E}_{Pd}]} \begin{Bmatrix} F_{d1} \\ F_{d2} \\ F_{d3} \end{Bmatrix} + \underbrace{\begin{bmatrix} 1 & 0 & 0 \\ 0 & 1 & 0 \\ 0 & 0 & -\tfrac{1}{12} \end{bmatrix}}_{[\mathbf{E}_{Px}]} \begin{Bmatrix} F_{x1} \\ F_{x2} \\ F_{x3} \end{Bmatrix}
$$

The flexibility matrix is as follows:

$$
\delta_{d1} = \frac{12}{3EI} F_{d1} - \frac{12}{6EI} F_{x3} \qquad \delta_{x3} = -\frac{12}{6EI} F_{d1} + \frac{12}{3EI} F_{x3}
$$

$$
\delta_{x1} = \frac{16}{6EI} F_{x1} - \frac{16}{12EI} F_{x2} \qquad \delta_{x2} = -\frac{16}{12EI} F_{x1} + \frac{16}{6EI} F_{x2}
$$

$$
\delta_{d2} = \frac{12}{3EI} F_{d2} - \frac{12}{6EI} F_{d3} \qquad \delta_{d3} = -\frac{12}{6EI} F_{d2} + \frac{12}{3EI} F_{d3}
$$

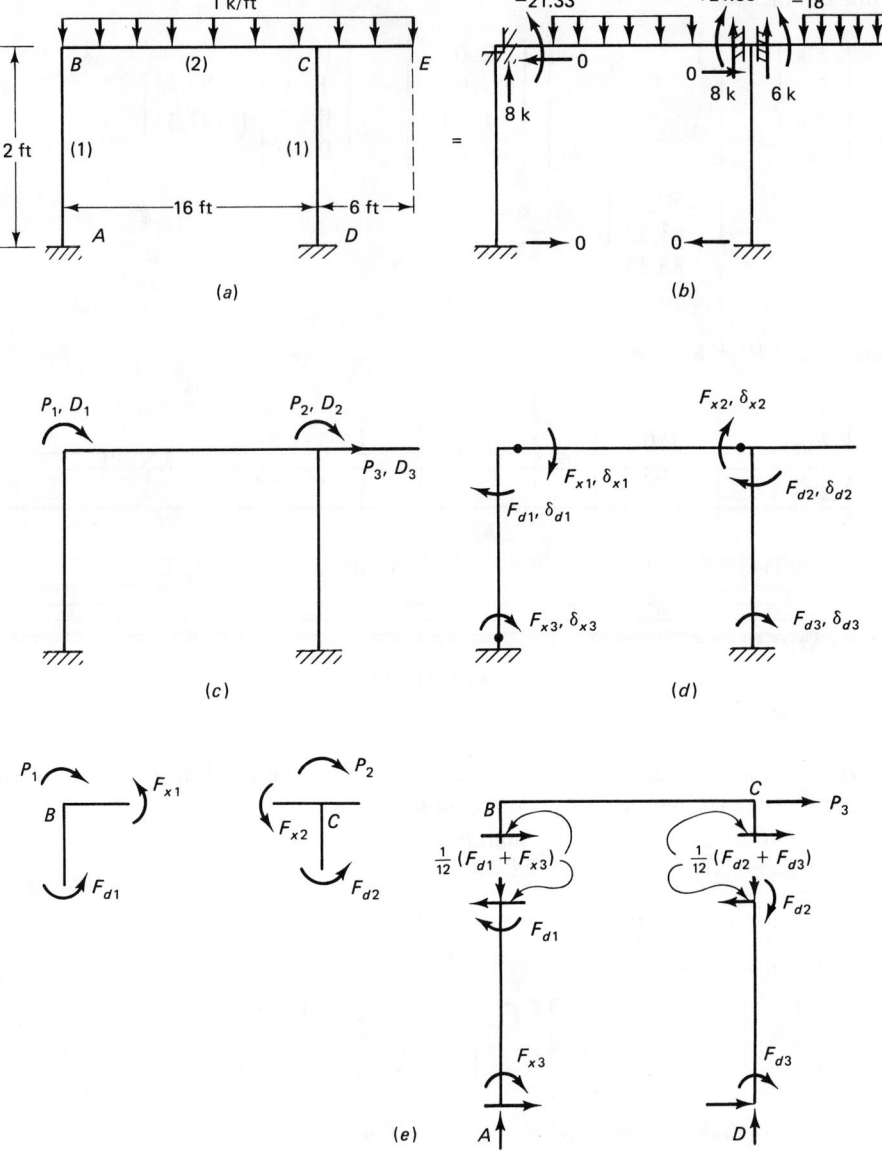

Figure 13.6 Frame for Example 13.5. (*a*) Relative value of *EI* in parentheses. (*b*) Fixed-end forces. (*c*) *P, D* diagram. (*d*) *F,δ* diagram.

In matrix form:

$$\begin{Bmatrix} \delta_{d1} \\ \delta_{d2} \\ \delta_{d3} \end{Bmatrix} = \frac{2}{EI} \begin{bmatrix} 2 & 0 & 0 \\ 0 & 2 & -1 \\ 0 & -1 & 2 \end{bmatrix} \begin{Bmatrix} F_{d1} \\ F_{d2} \\ F_{d3} \end{Bmatrix} + \frac{2}{EI} \begin{bmatrix} 0 & 0 & -1 \\ 0 & 0 & 0 \\ 0 & 0 & 0 \end{bmatrix} \begin{Bmatrix} F_{x1} \\ F_{x2} \\ F_{x3} \end{Bmatrix}$$

$$\underbrace{\qquad\qquad}_{[\mathbf{f}_d]} \qquad\qquad\qquad \underbrace{\qquad\qquad}_{[\mathbf{f}_{dx}]}$$

$$\begin{Bmatrix} \delta_{x1} \\ \delta_{x2} \\ \delta_{x3} \end{Bmatrix} = \frac{2}{EI} \begin{bmatrix} 0 & 0 & 0 \\ 0 & 0 & 0 \\ -1 & 0 & 0 \end{bmatrix} \begin{Bmatrix} F_{d1} \\ F_{d2} \\ F_{d3} \end{Bmatrix} + \frac{4}{3EI} \begin{bmatrix} 2 & -1 & 0 \\ -1 & 2 & 0 \\ 0 & 0 & 3 \end{bmatrix} \begin{Bmatrix} F_{x1} \\ F_{x2} \\ F_{x3} \end{Bmatrix}$$

$$[\mathbf{f}_{xd}] \nearrow \qquad\qquad [\mathbf{f}_x] \nearrow$$

The above results prove that $\mathbf{f}_{xd} = \mathbf{f}_{dx}^T$. From Fig. 13.6b, the **P** matrix is

$$\begin{Bmatrix} P_1 \\ P_2 \\ P_3 \end{Bmatrix} = \begin{Bmatrix} 21.33 \\ -3.33 \\ 0 \end{Bmatrix}$$

Using the above six matrices we can calculate the following:

$$\mathbf{E}_{Pd}^{-1} = \begin{bmatrix} 1 & 0 & 0 \\ 0 & 1 & 0 \\ -1 & -1 & -12 \end{bmatrix} \qquad \boldsymbol{\gamma} = \begin{bmatrix} 1 & 0 & 0 \\ 0 & 1 & 0 \\ -1 & -1 & 1 \end{bmatrix} \qquad \boldsymbol{\gamma}^T = \begin{bmatrix} 1 & 0 & -1 \\ 0 & 1 & -1 \\ 0 & 0 & 1 \end{bmatrix}$$

From Eq. (13.25), $\boldsymbol{\phi} = [\mathbf{f}_x + (\boldsymbol{\gamma}^T \mathbf{f}_d - \mathbf{f}_{dx}^T)\boldsymbol{\gamma} - \boldsymbol{\gamma}^T \mathbf{f}_{dx}]^{-1}$; hence

$$\boldsymbol{\phi} = EI \begin{bmatrix} 0.1089 & -0.0339 & 0.00178 \\ -0.0339 & 0.1089 & 0.0732 \\ 0.00178 & 0.0732 & 0.18036 \end{bmatrix}$$

From Eq. (13.24), $\mathbf{F}_x = \boldsymbol{\phi}(\boldsymbol{\gamma}^T \mathbf{f}_d - \mathbf{f}_{dx}^T)\mathbf{E}_{Pd}^{-1}P$. Thus

$$\begin{Bmatrix} F_{x1} \\ F_{x2} \\ F_{x3} \end{Bmatrix} = \begin{Bmatrix} 13.38 \\ 2.82 \\ 2.62 \end{Bmatrix}$$

Hence final moments $\mathbf{F}' = \mathbf{F}_0 + \mathbf{F}$ are

$$\begin{Bmatrix} F'_{x1} \\ F'_{x2} \\ F'_{x3} \end{Bmatrix} = \begin{Bmatrix} -21.33 \\ +21.33 \\ 0 \end{Bmatrix} + \begin{Bmatrix} 13.38 \\ 2.82 \\ 2.62 \end{Bmatrix} = \begin{Bmatrix} -7.95 \\ 24.15 \\ 2.62 \end{Bmatrix} (\text{k} \cdot \text{ft}) = \begin{Bmatrix} M_{BC} \\ M_{CB} \\ M_{AB} \end{Bmatrix}$$

From Eq. (13.26), $\mathbf{F}_d = \mathbf{E}_{Pd}^{-1}P - \boldsymbol{\gamma}\mathbf{F}_x$; or

$$\begin{Bmatrix} F_{d1} \\ F_{d2} \\ F_{d3} \end{Bmatrix} = \begin{Bmatrix} 7.95 \\ -6.15 \\ -4.42 \end{Bmatrix}$$

the final moments F' are

$$\begin{Bmatrix} F'_{d1} \\ F'_{d2} \\ F'_{d3} \end{Bmatrix} = \begin{Bmatrix} 0 \\ 0 \\ 0 \end{Bmatrix} + \begin{Bmatrix} 7.95 \\ -6.15 \\ -4.42 \end{Bmatrix} = \begin{Bmatrix} 7.95 \\ -6.15 \\ -4.42 \end{Bmatrix} (\text{k} \cdot \text{ft}) = \begin{Bmatrix} M_{BA} \\ M_{CD} \\ M_{DC} \end{Bmatrix}$$

and the moment $M_{CE} = -18$ k·ft. Static and deformation checks must always be made; the procedure has been illustrated before. ∎

■ **EXAMPLE 13.6**

Deduce the six input matrices $\mathbf{E}_{Pd}$, $\mathbf{E}_{Px}$, $\mathbf{f}_d$, $\mathbf{f}_x$, $\mathbf{f}_{dx}$, and **P** for the analysis of the rigid frame shown in Fig. 13.7a by the matrix-force method.

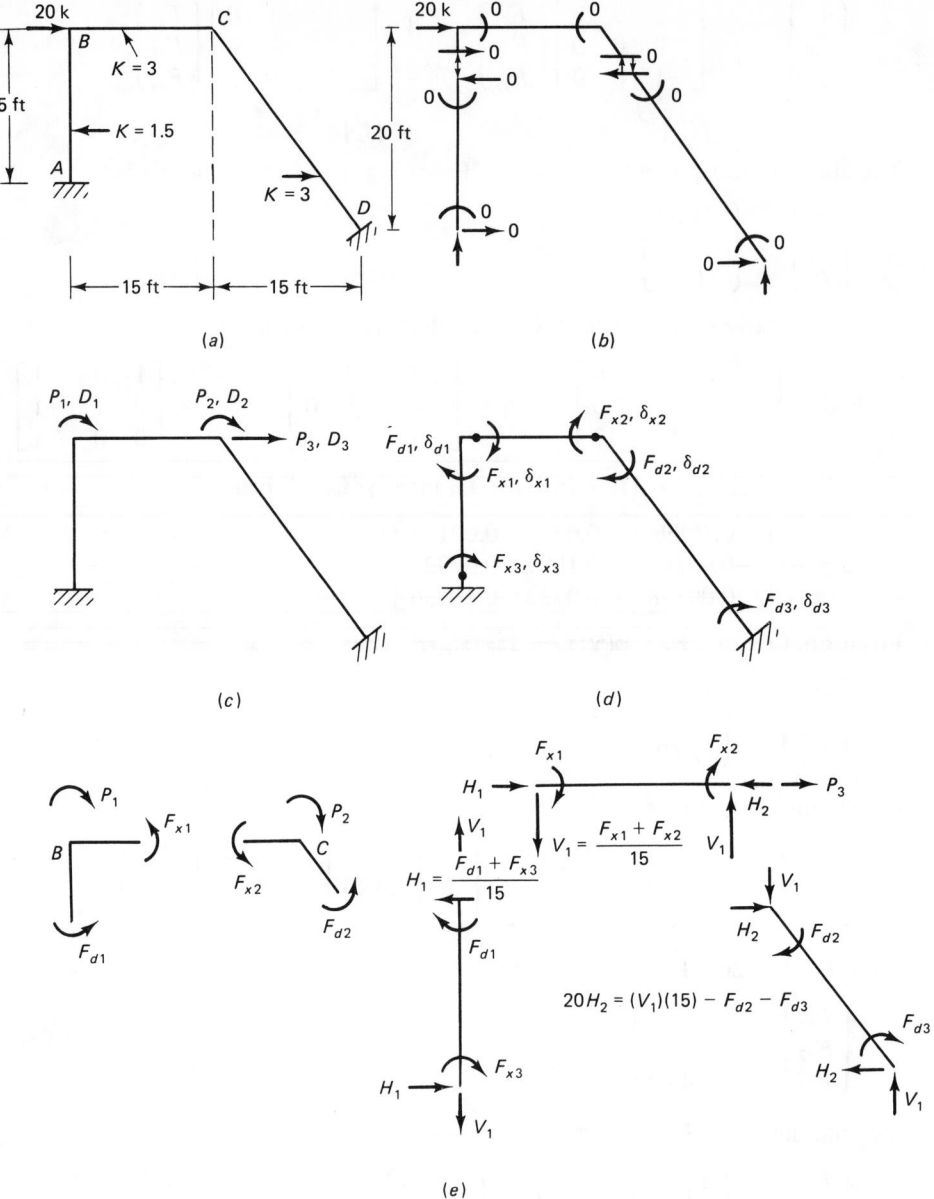

Figure 13.7 Frame for Example 13.6. (*b*) Fixed-end moments and sidesway force due to applied load.

Solution.   This rigid frame is statically indeterminate to the third degree; three hinges are inserted as shown in Fig. 13.7*d*. The fixed-end moment and sidesway force as well as the *P,D* diagrams are shown in Fig. 13.7*b* and *c*, respectively. From the free-body diagram of Fig. 13.7*e*:

$$P_1 = F_{x1} + F_{d1} \qquad P_2 = F_{x2} + F_{d2} \qquad P_3 = H_2 - H_1$$

Now

$$H_1 = \frac{F_{d1} + F_{x3}}{15} \quad \text{and} \quad H_2 = \frac{F_{x1} + F_{x2}}{20} - \frac{F_{d2}}{20} - \frac{F_{d3}}{20}$$

Hence

$$P_3 = \frac{F_{x1} + F_{x2} - F_{d2} - F_{d3}}{20} - \frac{F_{d1}}{15} - \frac{F_{x3}}{15}$$

In matrix form:

$$
\begin{Bmatrix} P_1 \\ P_2 \\ P_3 \end{Bmatrix} = \overbrace{\begin{bmatrix} 1 & 0 & 0 \\ 0 & 1 & 0 \\ -\frac{1}{15} & -\frac{1}{20} & -\frac{1}{20} \end{bmatrix}}^{[\mathbf{E}_{Pd}]} \begin{Bmatrix} F_{d1} \\ F_{d2} \\ F_{d3} \end{Bmatrix} + \overbrace{\begin{bmatrix} 1 & 0 & 0 \\ 0 & 1 & 0 \\ \frac{1}{20} & \frac{1}{20} & -\frac{1}{15} \end{bmatrix}}^{[\mathbf{E}_{Px}]} \begin{Bmatrix} F_{x1} \\ F_{x2} \\ F_{x3} \end{Bmatrix}
$$

The flexibility matrix is generated as follows:

$$\delta_{d1} = \frac{1}{4.5E} F_{d1} - \frac{1}{9E} F_{x3} \qquad \delta_{x3} = -\frac{1}{9E} F_{d1} + \frac{1}{4.5E} F_{x3}$$

$$\delta_{x1} = \frac{1}{9E} F_{x1} - \frac{1}{18E} F_{x2} \qquad \delta_{x2} = -\frac{1}{18E} F_{x1} + \frac{1}{9E} F_{x2}$$

$$\delta_{d2} = \frac{1}{9E} F_{d2} - \frac{1}{18E} F_{d3} \qquad \delta_{d3} = -\frac{1}{18E} F_{d2} + \frac{1}{9E} F_{d3}$$

In matrix form:

$$
\begin{Bmatrix} \delta_{d1} \\ \delta_{d2} \\ \delta_{d3} \end{Bmatrix} = \frac{1}{18E} \underbrace{\begin{bmatrix} 4 & 0 & 0 \\ 0 & 2 & -1 \\ 0 & -1 & 2 \end{bmatrix}}_{[\mathbf{f}_d]} \begin{Bmatrix} F_{d1} \\ F_{d2} \\ F_{d3} \end{Bmatrix} + \frac{1}{18E} \underbrace{\begin{bmatrix} 0 & 0 & -2 \\ 0 & 0 & 0 \\ 0 & 0 & 0 \end{bmatrix}}_{[\mathbf{f}_{dx}]} \begin{Bmatrix} F_{x1} \\ F_{x2} \\ F_{x3} \end{Bmatrix}
$$

$$
\begin{Bmatrix} \delta_{x1} \\ \delta_{x2} \\ \delta_{x3} \end{Bmatrix} = \frac{1}{18E} \underbrace{\begin{bmatrix} 0 & 0 & 0 \\ 0 & 0 & 0 \\ -2 & 0 & 0 \end{bmatrix}}_{[\mathbf{f}_{dx}]} \begin{Bmatrix} F_{d1} \\ F_{d2} \\ F_{d3} \end{Bmatrix} + \frac{1}{18E} \underbrace{\begin{bmatrix} 2 & -1 & 0 \\ -1 & 2 & 0 \\ 0 & 0 & 4 \end{bmatrix}}_{[\mathbf{f}_x]} \begin{Bmatrix} F_{x1} \\ F_{x2} \\ F_{x3} \end{Bmatrix}
$$

It can be observed that $[\mathbf{f}_{xd}] = [\mathbf{f}_{dx}]^T$. From Fig. 13.7b, the **P** matrix is

$$
\begin{Bmatrix} P_1 \\ P_2 \\ P_3 \end{Bmatrix} = \begin{Bmatrix} 0 \\ 0 \\ 20 \end{Bmatrix}
$$

■

13.4 COMPARISON OF THE MATRIX-FORCE METHOD AND THE MATRIX-DISPLACEMENT METHOD

The duality between these two methods becomes apparent when the basic procedures for the two methods are compared as shown in Table 13.1. Generally, the displacement method is preferred for the following reasons: (1) It uses the same procedures for

TABLE 13.1

Displacement method	Force method
Nodal displacements are the basic unknowns	End-member forces are the basic unknowns
Derive the displacement transformation matrix	Derive the force transformation matrix
Derive the member stiffness matrix	Derive the member flexibility matrix
Hence deduce the global stiffness matrix	Deduce the global flexibility matrix
Deduce the nodal forces in terms of the nodal displacements	Deduce the nodal displacements in terms of the nodal forces

analyzing statically determinate structures and statically indeterminate structures. (2) It is simpler to derive the displacement transformation matrix than the force transformation matrix. (3) The global stiffness matrix can be deduced more readily than the global flexibility matrix. (4) Generally, the method produces a well-conditioned global stiffness matrix. A well-conditioned matrix has a dominant diagonal and as such is most suitable for a computer solution; whereas a well-conditioned global flexibility matrix depends largely on the choice of redundant forces in the structure.

13.5 SOLVED PROBLEM

■ SOLVED PROBLEM 13.1

Determine the forces and joint displacements in the truss shown in Fig. 13.2a for each of the following three conditions: (a) External loads shown in Fig. 13.2a. (b) Support at g settles down vertically $\frac{1}{2}$ in. (c) Member ac is $\frac{1}{8}$ in too long and member cd is $\frac{1}{4}$ in too short.

Solution. This truss is statically indeterminate to the second degree. The choice of the two redundants is shown in Fig. 13.2c; the truss is reduced to two basic (determinate and stable) trusses connected at a common hinge g.

(a) The matrices $\mathbf{E}_{Pd}$, $\mathbf{E}_{Px}$, $\mathbf{f}_d$, and $\mathbf{f}_x$ have been derived earlier in the chapter (Examples 13.1 and 13.2). Matrix $\mathbf{P}$ is required. Therefore, by comparing Fig. 13.2b and a, we have

$$\begin{Bmatrix} P_1 \\ P_2 \\ P_3 \\ P_4 \\ P_5 \\ P_6 \\ P_7 \\ P_8 \\ P_9 \\ P_{10} \\ P_{11} \\ P_{12} \end{Bmatrix} = \begin{Bmatrix} +10 \\ 0 \\ 0 \\ -20 \\ 0 \\ 0 \\ 0 \\ -10 \\ 0 \\ 0 \\ 0 \\ 0 \end{Bmatrix} \text{ k}$$

Calculating γ and ϕ from Eq. (13.12) and Eq. (13.13), respectively, and applying Eq. (13.16) yield the following results:

$$F_{x1} = -4.34 \text{ k} \qquad F_{x2} = 11.31 \text{ k}$$

and for $\mathbf{F}_d$:

F_{d1}	F_{d2}	F_{d3}	F_{d4}	F_{d5}	F_{d6}	F_{d7}	F_{d8}	F_{d9}	F_{d10}	F_{d11}	F_{d12}	
0	-10	-7.47	-6.79	15.98	-14.55	4.71	-6.47	0	-5.89	-6.79	6.93	k

and for the joint displacements **D**:

D_1	D_2	D_3	D_4	D_5	D_6	D_7	D_8	D_9	D_{10}	D_{11}	D_{12}	
64.7	0	40.7	-91.7	30.3	-12.2	41.6	-11.6	38.3	-79.5	55.0	55.0	10^{-3} in

(b) For support g to settle downward $\Delta = \frac{1}{2}$ in the matrix Δ becomes

$$\begin{Bmatrix} \Delta_1 \\ \Delta_2 \\ \Delta_3 \\ \Delta_4 \end{Bmatrix} = \begin{Bmatrix} 0 \\ 0 \\ -500 \\ 0 \end{Bmatrix} (10^{-3} \text{ in})$$

The required matrices $\mathbf{E}_{Rd}$ and $\mathbf{E}_{Rx}$ were generated earlier in the chapter (Example 13.1). Therefore, using Eq. (13.17) yields

$$F_{x1} = -33.12 \text{ k} \qquad F_{x2} = -11.59 \text{ k}$$

and for $\mathbf{F}_d$:

F_{d1}	F_{d2}	F_{d3}	F_{d4}	F_{d5}	F_{d6}	F_{d7}	F_{d8}	F_{d9}	F_{d10}	F_{d11}	F_{d12}	
0.0	0.0	-26.50	6.96	21.20	14.90	-42.39	-31.80	0	52.99	6.96	30.47	k

and for **D**:

D_1	D_2	D_3	D_4	D_5	D_6	D_7	D_8	D_9	D_{10}	D_{11}	D_{12}	
171.9	0	171.9	-361.6	92.4	-487.5	-9.4	-57.2	50.9	-374.2	124.0	124.0	10^{-3} in

(c) For the lack of fit of members ac and cd, matrices δ_{od} and δ_{ox} are required. Considering "too long" as positive elongation and "too short" as negative elongation, one can write

$$\begin{Bmatrix} \delta_{od1} \\ \delta_{od2} \\ \delta_{od3} \\ \delta_{od4} \\ \delta_{od5} \\ \delta_{od6} \\ \delta_{od7} \\ \delta_{od8} \\ \delta_{od9} \\ \delta_{od10} \\ \delta_{od11} \\ \delta_{od12} \end{Bmatrix} = \begin{Bmatrix} 0 \\ 0 \\ 125 \\ 0 \\ 0 \\ 0 \\ 0 \\ 0 \\ 0 \\ 0 \\ 0 \\ 0 \end{Bmatrix} (10^{-3} \text{ in})$$

$$\begin{Bmatrix} \delta_{ox1} \\ \delta_{ox2} \end{Bmatrix} = \begin{Bmatrix} -250 \\ 0 \end{Bmatrix} (10^{-3} \text{ in})$$

Applying Eq. (13.18), we find

$$F_{x1} = 22.33 \text{ k} \qquad F_{x2} = -16.88 \text{ k}$$

and $\mathbf{F}_d$ as:

F_{d1}	F_{d2}	F_{d3}	F_{d4}	F_{d5}	F_{d6}	F_{d7}	F_{d8}	F_{d9}	F_{d10}	F_{d11}	F_{d12}	
0	0	5.52	10.13	−4.42	−22.40	8.83	6.62	0	−11.04	10.13	9.08	k

and $\mathbf{D}$ as:

D_1	D_2	D_3	D_4	D_5	D_6	D_7	D_8	D_9	D_{10}	D_{11}	D_{12}	
136.1	0	136.1	54.5	−60.3	18.2	−39.1	11.9	−10.6	36.3	11.2	11.2	10^{-3} in

∎

PART II: FINITE-ELEMENT METHOD

13.6 INTRODUCTION

The matrix-displacement and force methods of structural analysis discussed in the previous chapter and this chapter were based on the idealization of a structure as an assemblage of *one*-dimensional elements (see Fig. 13.8*a*) such as beams and columns. The finite-element method is an extension to the above matrix methods and applied to analyze two- or three-dimensional structures such as plates, shells, and solid bodies by means of *two*- or *three*-dimensional structural elements as shown in Fig. 13.8*b*. Therefore, the method is a technique for analyzing complicated structures by cutting

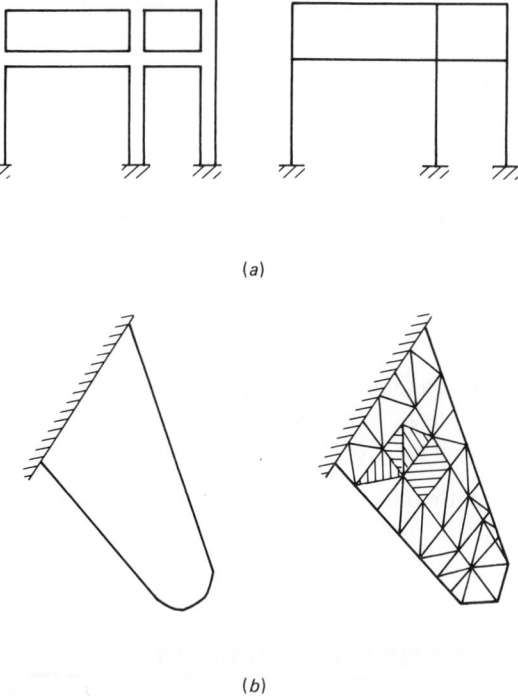

(a)

(b)

Figure 13.8 Actual and idealized structures by finite-element method. (*a*) One-dimensional elements: left, actual framed structure; right, idealized structure. (*b*) Two-dimensional elements: left, actual solid wing; right, idealized wing.

up the continuum of the actual structure into a number of small *elements* that are connected at discrete points called *nodes*. The method provides the most versatile method of analysis currently available; however, the computation can be expensive so that often the cost cannot be justified for run-of-the-mill structures. The use of the method includes three basic steps:

1. Structural idealization—where a structure is divided into elements.
2. Evaluation of the element properties, that is, stiffness or flexibility of the element which expresses the force-displacement relationship of the system.
3. Structural analysis of the element assemblage. Here one has to satisfy the usual requirements: (a) equilibrium of the internally and externally applied forces at each node of an element, (b) geometric compatibility or fit of element deformations in such a manner that they meet at the nodal points in the loaded configuration, and (3) internal force-displacement relationships established with each element as dictated by the existing geometry and material property characteristics.

The above requirements can be satisfied by two basic approaches:

1. The force method where an appropriate stress distribution is assumed to represent the actual stress distribution over the element

2. The displacement method when an assumed displacement function is considered to represent the actual deformation of the element

Unfortunately, the potentialities of the force method have not yet been explored fully. However, the displacement method seems to be the most favored, since it provides simpler formulation and computer programming labor as well as better physical relationship to the distortion of the structure under load. Therefore, only the latter method will be illustrated later. It should be emphasized that the present exposition of the finite-element method does little more than demonstrate the basic physical principles. Much more detailed and comprehensive descriptions of the method are given in books by Zienkiewicz[1] and Desai and Abel.[2] The major problems met in performing a finite-element analysis are

1. Choosing the finite-element mesh to properly model the prototype structure (wrong choice may lead, in many cases, to the wrong answer)
2. Deducing the general form of the stiffness matrix for the element shape chosen

The finite-element method is demonstrated in relation to the analysis of the plane stress (membrane) behavior of flat plates; this is one of the simplest applications of the method.

13.7 TWO-DIMENSIONAL PLANE STRESS ELEMENTS

Consider a plate containing a triangular hole shown in Fig. 13.9a, subject to uniform tension. The presence of the hole gives rise to high stress concentrations around the hole. Based on the classical theory of elasticity it can be shown that discontinuities in a continuum cause abrupt stress increases around such discontinuities; however, it is also known, both experimentally and theoretically, that the high stresses decrease rapidly away from the discontinuity. Another example is presented in Fig. 13.10, where a plate with semicircular cutouts is subjected to tension. For the analysis of the structure in Fig. 13.9a, the structure is considered to be made up of a large number of triangular elements connected together at the corners. These observations indicate that smaller-sized elements should be used around the triangular hole with the smaller elements at the corners of the hole and along a plane parallel to the applied loads, as shown in Fig. 13.9b. In the simplest of element models, it is assumed that the strains within each element are uniform during distortion with the triangular edges remaining straight. For this reason, this element is known as the *constant-strain element*. The difference between the model and the prototype structure can be reduced if the model is composed of a large number of smaller-sized elements; with an infinite number of small elements, theoretically the model effectively becomes a continuum like the prototype structure.

[1] O. C. Zienkiewicz, "The Finite Element Method: Basic Concepts and Linear Applications," 4th ed., McGraw-Hill Book Co., New York, 1988.

[2] C. S. Desai and J. F. Abel, "Introducing the Finite Element Method: A Numerical Method of Engineering Analysis," Van Nostrand and Reinhold, New York, 1972.

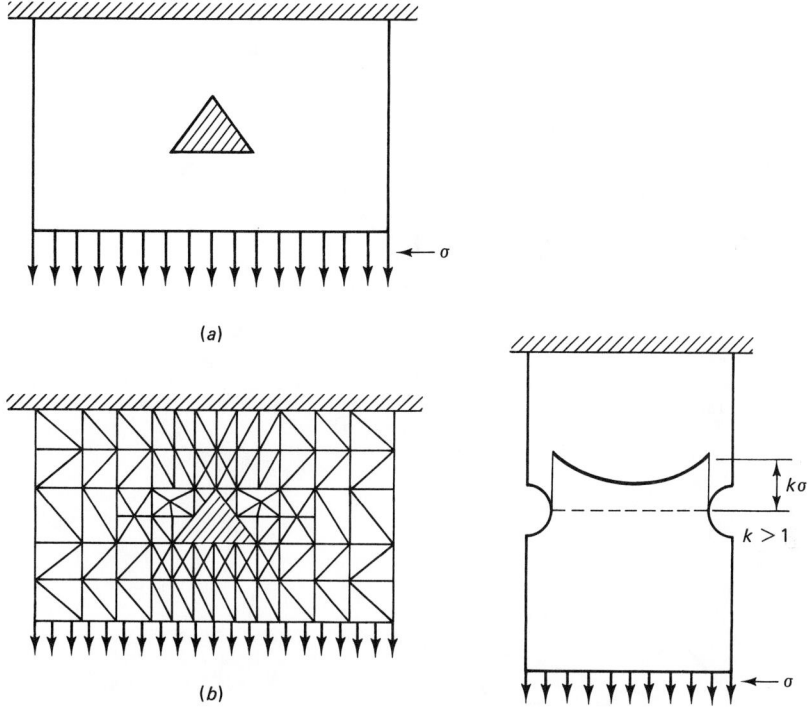

Figure 13.9 Finite-element idealization of a structure with discontinuity.

Figure 13.10 Plate with cutouts.

Having chosen an appropriate shape for the element, the next step is to develop the stiffness matrix **S** for the element. This is performed by relating:

1. The three strains ε_x, ε_y, and γ_{xy} in an element to the nodal displacements
2. The three stresses σ_x, σ_y, and τ_{xy} in an element to the nodal forces
3. The stresses to strains through Hooke's law
4. The nodal forces to the nodal displacements to yield the stiffness matrix **S** of the element

Strain-Displacement Relationship It is assumed that the triangle in Fig. 13.11*a* will distort with internal straight lines and edges remaining straight as shown in Fig. 13.11*c*. Hence strains are constant on an element (Fig. 13.11*d*). Thus the displacement field is linear and can be expressed as

$$u = A_1 + A_2 x + A_3 y$$
$$v = A_4 + A_5 x + A_6 y$$

(13.28)

where u and v are the displacements of the nodes in the x and y directions, respectively (Fig. 13.11*c*). For example, the translations of node i in Fig. 13.11*c* are u_i and v_i. From the classical theory of elasticity, it can be shown that the strains ε_x, ε_y, and γ_{xy} are related to the displacements by

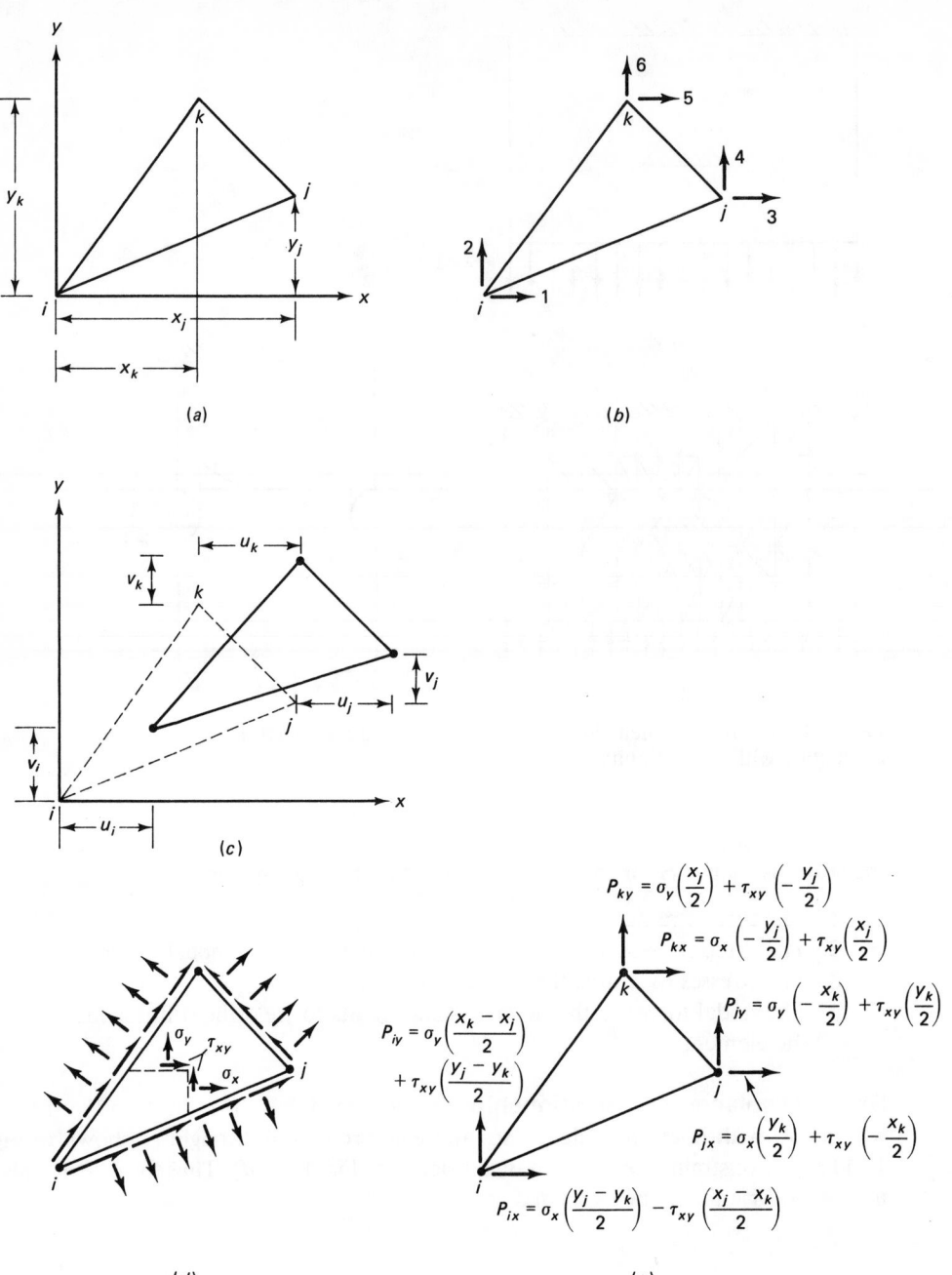

Figure 13.11 Constant-strain triangular element. (*a*) Dimension of elements. (*b*) Element references. (*c*) Nodal displacements. (*d*) Internal and external stresses. (*e*) Statically equivalent nodal forces.

$$\varepsilon_x = \frac{\partial u}{\partial x} = A_2$$

$$\varepsilon_y = \frac{\partial v}{\partial y} = A_6 \qquad (13.29)$$

$$\gamma_{xy} = \frac{\partial u}{\partial y} + \frac{\partial v}{\partial x} = A_3 + A_5$$

Now Eq. (13.28) can be written for each of the three nodes i, j, k, resulting in six equations, which when solved will yield the constants $A_1, A_2, \ldots, A_6$ as follows:

$$
\begin{Bmatrix} A_1 \\ A_2 \\ A_3 \\ A_4 \\ A_5 \\ A_6 \end{Bmatrix} = \frac{1}{x_j y_k - x_k y_j}
$$

$$
\times \begin{bmatrix} x_j y_k - x_k y_j & 0 & 0 & 0 & 0 & 0 \\ y_j - y_k & 0 & y_k & 0 & -y_j & 0 \\ x_k - x_j & 0 & -x_k & 0 & x_j & 0 \\ 0 & x_j y_k - x_k y_j & 0 & 0 & 0 & 0 \\ 0 & y_j - y_k & 0 & y_k & 0 & -y_j \\ 0 & x_k - x_j & 0 & -x_k & 0 & x_j \end{bmatrix} \begin{Bmatrix} u_i \\ v_i \\ u_j \\ v_j \\ u_k \\ v_k \end{Bmatrix} \qquad (13.30)
$$

From the results of Eq. (13.30), the strains given by Eq. (13.29) can be written as

$$
\begin{Bmatrix} \varepsilon_x \\ \varepsilon_y \\ \gamma_{xy} \end{Bmatrix} = \frac{1}{x_j y_k - x_k y_j} \begin{bmatrix} y_j - y_k & 0 & y_k & 0 & -y_j & 0 \\ 0 & x_k - x_j & 0 & -x_k & 0 & x_j \\ x_k - x_j & y_j - y_k & -x_k & y_k & x_j & -y_j \end{bmatrix} \begin{Bmatrix} u_i \\ v_i \\ u_j \\ v_j \\ u_k \\ v_k \end{Bmatrix} (13.31)
$$

which can be written as

$$\{\varepsilon\} = [\mathbf{C}]\{\mathbf{D}\} \qquad (13.32)$$

where $\{\varepsilon\} = \{\varepsilon_x, \varepsilon_y, \gamma_{xy}\}$ is related to the nodal displacements $\{\mathbf{D}\} = \{u_i, v_i, u_j, v_j, u_k, v_k\}$ by means of the compatibility matrix $[\mathbf{C}]$. It should be noted that Eqs. 13.30 and 13.31 are based on the assumption that node i coincides with the origin of the coordinate system.

Nodal Force-Stress Relationship   The internal stresses in the element σ_x, σ_y, and τ_{xy} (Fig. 13.11d) can be transferred to a statically equivalent set of nodal forces shown in Fig. 13.11e. Each nodal force is considered as the sum of the force components acting in the same direction as the force on one-half of each of the two sides connected to the node. Thus, based on the results in Fig. 13.11e, one can write

$$
\begin{Bmatrix} P_{ix} \\ P_{iy} \\ P_{jx} \\ P_{jy} \\ P_{kx} \\ P_{ky} \end{Bmatrix} = \frac{1}{2} \begin{bmatrix} y_j - y_k & 0 & -x_j + x_k \\ 0 & -x_j + x_k & y_j - y_k \\ y_k & 0 & -x_k \\ 0 & -x_k & y_k \\ -y_j & 0 & x_j \\ 0 & x_j & -y_j \end{bmatrix} \begin{Bmatrix} \sigma_x \\ \sigma_y \\ \tau_{xy} \end{Bmatrix}
$$

or

$$\{P\} = [G]\{\sigma\} \tag{13.33}$$

The above relationship is based on an element of unit thickness.

Stress-Strain Relationship   It can be shown from the theory of elasticity that the stresses $\{\sigma\}$ and strains $\{\varepsilon\}$ are related by means of Hooke's law; thus, for an element in *plane stress* (stresses in xy plane only)

$$
\begin{Bmatrix} \sigma_x \\ \sigma_y \\ \tau_{xy} \end{Bmatrix} = \frac{E}{1 - \nu^2} \begin{bmatrix} 1 & \nu & 0 \\ \nu & 1 & 0 \\ 0 & 0 & \dfrac{1-\nu}{2} \end{bmatrix} \begin{Bmatrix} \varepsilon_x \\ \varepsilon_y \\ \gamma_{xy} \end{Bmatrix} \tag{13.34}
$$

and for an element in *plane strain* (strains in xy plane only):

$$
\begin{Bmatrix} \sigma_x \\ \sigma_y \\ \tau_{xy} \end{Bmatrix} = \frac{E}{(1 + \nu)(1 - 2\nu)} \begin{Bmatrix} 1 - \nu & \nu & 0 \\ \nu & 1 - \nu & 0 \\ 0 & 0 & \dfrac{1 - 2\nu}{2} \end{Bmatrix} \begin{Bmatrix} \varepsilon_x \\ \varepsilon_y \\ \gamma_{xy} \end{Bmatrix} \tag{13.35}
$$

in which E is the elastic modulus and ν is Poisson's ratio of the element material. Both Eqs. (13.34) and (13.35) can be written as

$$\{\sigma\} = [d]\{\varepsilon\} \tag{13.36}$$

in which $[d]$ is an elasticity matrix, defined by Eq. (13.34) or (13.35) depending on whether it is a plane stress or a plane strain problem.

Nodal Force-Displacement Relationship   This can be established by substituting Eq. (13.32) into Eq. (13.36) yielding $\{\sigma\} = [d][C]\{D\}$. Substituting this into Eq. (13.33), one finds

$$\{P\} = [G][d][C]\{D\} \tag{13.37}$$

This equation relates a set of nodal forces $\{P\}$ to the corresponding set of nodal displacements $\{D\}$; whence the stiffness matrix of the element is

$$S = [G][d][C] \tag{13.38}$$

The stiffness matrix S^* of the finite-element model can be assembled by inspection from the S matrices or by means of a computer program. Details can be found in the finite-element texts cited earlier. The S^* matrix must also be corrected to account for the boundary conditions imposed on the finite-element model. External loads can be represented by equivalent concentrated loads at the nodes, yielding the **P** matrix; thus one can write

$$\{\mathbf{P}\} = [\mathbf{S^*}]\{\mathbf{D}\} \tag{13.39}$$

Solving Eq. (13.39) will yield the displacements $\{\mathbf{D}\}$ corresponding to the degrees of freedom in the finite-element model; from these results the strains and stresses in every element can be calculated using Eq. (13.31), (13.34), or (13.35).

The above approach of developing the element stiffness matrix **S** is a physical one; that is, it shows the reader the physical significance of each of the member matrices developed. For the more complex elements, it is not usually possible to relate nodal forces directly to element stresses and the assumed displacement field. In such cases the element stiffness matrix is derived from a consideration of the potential energy stored by the assumed displacement field; for each nodal force, the external work done during a virtual displacement is equated to the minimum increase in potential energy that can be stored by the displacement field. While this concept enables more complicated problems to be analyzed with computing economy, it does mean that element stresses output are *not directly* related by equilibrium to the applied loads. If the element displacement function is not appropriate for the prototype structure, the output stresses can be as low as 50 percent or less of those necessary for equilibrium with the applied loads; calculating the nodal averages does not necessarily make much difference. Therefore, it is advisable to make a hand check of the equilibrium of the output element stresses and the applied loads; a further check on the results by means of a space-frame analysis is also recommended.

The finite elements of a model do not have to be triangular; other shapes such as quadrilaterals, rectangles, parallelograms, and other polygons are used. The proper element should be selected to represent the prototype structure being analyzed. In regions of high stress or strain gradients, smaller-sized elements should be used; it is also necessary to specify proper restraints on the finite-element model to satisfy the actual boundary conditions of the structure. Figure 13.12 shows the nodal displacements and forces of a rectangular element. For this element, the displacement field can be assumed as

$$
\begin{aligned}
u &= A_1 + A_2 x + A_3 y + A_4 xy \\
v &= A_5 + A_6 x + A_7 y + A_8 xy
\end{aligned}
\tag{13.40}
$$

This displacement field has eight unknown constants which are found by solving the equations for the eight displacements $u_1, v_1, \ldots, u_4, v_4$.

In the above discussion it has been assumed that the elements used are compatible; that is, displacement functions are such that points on elements on each side of a cut always remain adjacent. On the other hand, the stresses are discontinuous at the cuts, but the stress resultants at the nodal points are in equilibrium.

Since the finite-element model is made to distort in a specified manner, it has less freedom and is therefore stiffer than the prototype structure; as a result the computed stresses are always lower than they should be, that is, lower bounds. An alternative approach is to assume a stress field for the elements so that the stresses are continuous

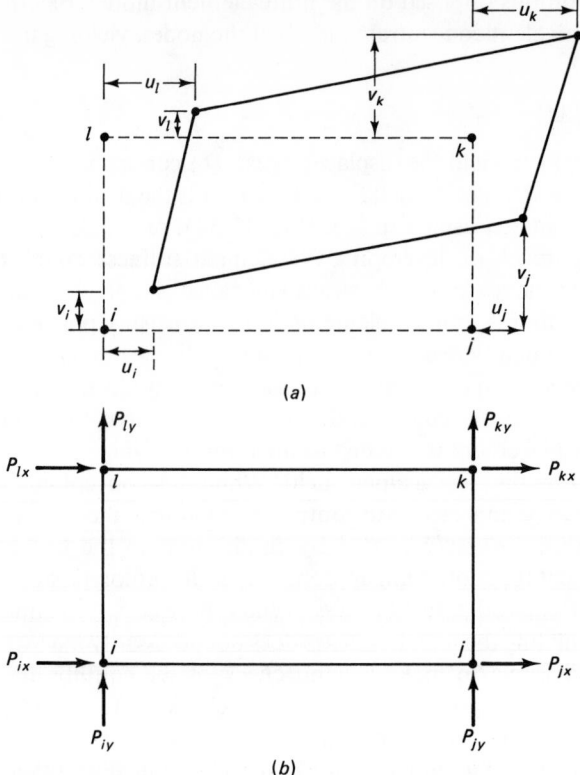

Figure 13.12 Rectangular plane stress element.

across the edge of the elements; however, the displacements become discontinuous along the edges of elements except at the nodes. The resulting model becomes more flexible than the prototype structure, leading to computed stresses greater than the actual ones, i.e., upper bounds.

13.8 PLATE BENDING ELEMENTS

In the analysis of plate bending, e.g., bridge slabs, it is assumed that each node of the finite-element model has three degrees of freedom: a vertical deflection w, rotation of node about the x axis, θ_x, and rotation of node about the y axis, θ_y. Quite often a triangular element shown in Fig. 13.13 is used for the finite-element model; a piece of the prototype structure undergoes vertical deflection (Fig. 13.13) and is subjected to bending moments, torsional moments, and vertical shears. To determine the stiffness of this triangular element in bending, the following displacement field is assumed in the form of a complete third-order polynomial:

$$w = A_1 + A_2 x + A_3 y + A_4 x^2 + A_5 xy + A_6 y^2$$

$$+ A_7 x^3 + A_8 x^2 y + A_9 xy^2 + A_{10} y^3 \qquad (13.41)$$

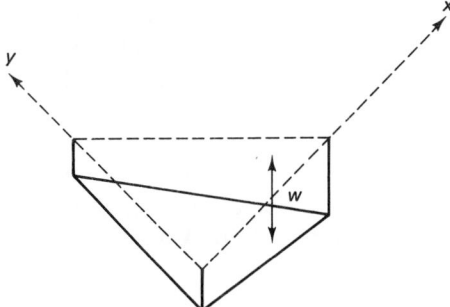

Figure 13.13 Triangular plate bending element.

Thus

$$\theta_x = \frac{\partial w}{\partial y} = A_3 + A_5 x + 2A_6 y + A_8 x^2 + 2A_9 xy + 3A_{10} y^2$$

$$\theta_y = \frac{\partial w}{\partial x} = A_2 + 2A_4 x + A_5 y + 3A_7 x^2 + 2A_8 xy + A_9 y^2$$

(13.42)

It is important to choose the function for w to satisfy the governing plate equation, since then the element will possess the same characteristics of deformability as the prototype structure.

Equation (13.41) has 10 coefficients, which cannot be found from the nine displacement variables w_i, θ_{xi}, θ_{yi}, . . . , θ_{yk} for the element. Such an element is said to be *nonconformal* since it does not completely satisfy the displacement (deflections and slopes) compatibility. Equation (13.41) must be adjusted by either removing one constant arbitrarily or making two constants equal. For example, it has been suggested that the xy term in Eq. (13.41) be omitted. However, this omission makes it impossible to represent constant twist condition within the element, resulting in an element that is far too stiff. The other suggestion is to combine the terms $x^2 y$ and xy^2. This leads to a serious lack of invariance; that is, for certain orientations of the element sides with respect to the coordinate axes, the matrix relating the nodal displacements to global displacements becomes *singular;* obviously, any assumed displacement function must form a complete polynomial in x and y in order to meet the requirements of *invariance.*

The rotations (slopes) at the nodal points and normal to the element side are given by

$$\theta_{xi} = \left(\frac{\partial w}{\partial y}\right)_i \qquad \theta_{yi} = \left(\frac{\partial w}{\partial x}\right)_i \qquad \text{etc.}$$

It can be shown that such rotations or slopes can be different for elements on two sides of an interface; thus, except at the nodal points, there will be a kink at the interface between elements even though the vertical deflections are continuous. Thus, because the complete polynomial [Eq. (13.41)] is too general for the number of defining degrees of freedom (3×3), the elements are not truly compatible and are said to be "nonconforming." Solutions obtained from such nonconforming elements are neither lower bounds nor upper bounds. A nonconforming element will give higher stresses than the lower bound obtained from a conforming element of the same shape because it has less stiffness due to its greater freedom in taking up a deflected shape. The stiffness equations for the elements are usually derived by a consideration of virtual work, using a computer program. Other bending elements have been proposed, and the results based on some of these elements have been quite accurate.

Several general-purpose finite-element computer programs are readily available. Some of the programs more commonly used in North America are

1. ANSYS developed by Swanson Analysis Systems, Inc., Elizabeth, Pennsylvania
2. GTSTRUDL developed at the Georgia Institute of Technology, Atlanta, Georgia
3. ICES STRUDL, originally developed at the Massachusetts Institute of Technology, Cambridge, Massachusetts
4. MSC-NASTRAN developed by MacNeal-Schwendler Corporation, Los Angeles, California
5. SAP IV developed at the University of California, Berkeley, California

PART III: ENERGY METHODS

In Chapter 7 some energy methods were introduced to calculate displacements in structures. In this chapter several energy methods will be used to determine redundant forces in indeterminate structures.

13.9 PRINCIPLE OF MINIMUM TOTAL POTENTIAL ENERGY

"For a stable equilibrium condition, the total potential energy of a system must be a minimum with regard to variations in the displacement." In other words, for a system to be in equilibrium, the variation in the total potential energy must vanish for any virtual displacement.

Proof We have seen earlier [see Eq. (7.51)] that for a deformable body to be in equilibrium the principle of virtual work required that

$$\delta W_e = \delta W_i \tag{13.43}$$

where δW_e and δW_i are the external and internal virtual works; δW_i is also the virtual change in the strain energy δU. Thus we can rewrite Eq. (13.43) as

$$\delta U - \delta W_e = 0 \tag{13.44}$$

or

$$\delta(U - W_e) = 0 \tag{13.45}$$

The term δW_e is the virtual work done by the applied body and surface forces; and the scalar quantity $(U - W_e)$ is called the total potential energy of the structural system and is denoted by π.[3] Thus (Eq. 13.45) can be written as

$$\delta \pi = 0 \qquad (13.46)$$

or

$$\frac{\partial}{\partial \Delta_i}(U - W_e) = 0 \qquad (13.46a)$$

for any virtual displacement Δ_i.

It should be noted that the method of minimum total potential energy is valid for linear and nonlinear structures as well as for structures experiencing gross distortion. The method is valuable when the number of degrees of freedom of the joints is less than the number of redundant bars, that is, when there are many members but few joints, for example, the spoked wheel problem shown in Fig. 13.15a. It should be noted that in order to satisfy the condition of continuity of displacements over the entire body the strain must first be expressed in terms of the displacement and the variation of strain energy U must be carried out with respect to the displacements.

■ EXAMPLE 13.7

The pin-jointed truss in Fig. 13.14a has member AD with axial stiffness equal to infinity; the elastic members AB and AC have a cross-sectional area equal to α. Calculate the force in the members.

Solution. The frame is indeterminate to the first degree. Under the load P the joint A can move to A'' distances Δx and Δy as shown. Since AD is infinitely rigid (i.e., no change in length can take place), A moves at right angles to the direction AD as shown in Fig. 13.14b. Thus, from geometry, we have $\Delta x = \Delta y = \Delta$ and the elongations in AC and AB are

$$e_{AC} = \Delta y = \Delta \qquad e_{AB} = AA'' \cos 15° = \sqrt{2}\,\Delta \cos 15°$$

The above equations relating joint movements of A to the changes in lengths of the members are the compatibility equations. Now the strain energy in terms of displacements e is [from Eq. (7.29b)]

$$U = \frac{\sum AEe^2}{2L}$$

$$= \frac{\alpha E \Delta^2}{2L} + \frac{\alpha E(\sqrt{2}\,\Delta \cos 15°)^2}{2L/\sin 60°} + 0$$

---

[3] π is also equal to $(U + V)$ where V is the potential energy of the external forces after deformation, assuming the potential energy of the external forces before deformation to be zero, that is, $W_e = -V$.

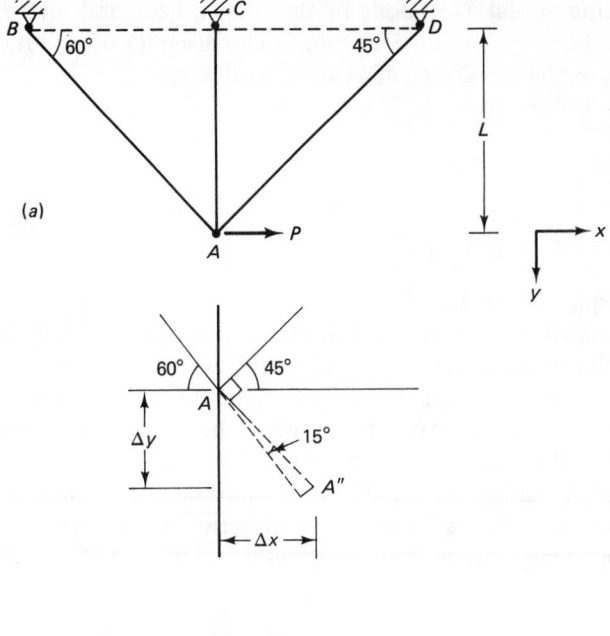

(a)

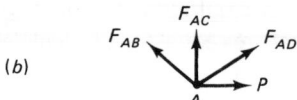

(b)

Figure 13.14 Truss for Example 13.7.

The term 0 is the strain energy for AD since its $e = 0$. The external work performed by the load P as joint A moves to A'' is $W_e = P(\Delta x) = P \cdot \Delta$. Hence the total potential energy of the system is

$$\pi = U - W_e = \frac{\alpha E \Delta^2}{2L} (1 + \sqrt{3} \cos^2 15°) - P\Delta$$

Now from the total potential energy theorem, $\partial \pi / \partial \Delta = 0$. Thus

$$\frac{\partial \pi}{\partial \Delta} = \frac{\alpha E \Delta}{L} (1 + \sqrt{3} \cos^2 15°) - P = 0$$

or

$$\Delta = \frac{PL}{2.616 \alpha E}$$

But

$$\text{Force in } AC = F_{AC} = \frac{\alpha E \Delta}{L}$$

Therefore

$$F_{AC} = 0.382P$$

Resolving forces at joint A yields:

$$\sum F_x = 0: \qquad P + F_{AD}(\cos 45°) - F_{AB}\cos 60° = 0$$

$$P + F_{AD}\frac{\sqrt{2}}{2} - F_{AB}\frac{1}{2} = 0$$

Hence

$$\sqrt{2}F_{AD} - F_{AB} = -2P \tag{a}$$

From

$$\sum F_y = 0: \qquad -F_{AC} - F_{AD}\sin 45° - F_{AB}\sin 60° = 0$$

that is,

$$-0.382P - F_{AD}\frac{\sqrt{2}}{2} - F_{AB}\frac{\sqrt{3}}{2} = 0$$

that is,

$$-\sqrt{2}F_{AD} - \sqrt{3}F_{AB} = 0.764P \tag{b}$$

From (a) and (b) we find $F_{AD} = -1.094P$ and $F_{AB} = 0.452P$. ∎

■ EXAMPLE 13.8

If in Example 13.7, the bars AB and AC followed the load-deflection relation $F = Re(1 - 2e)$, determine the force in AC where $P = R/5$ and AD is still infinitely rigid.

Solution. The compatibility relations are the same as in Example 13.7. However, the strain energy in each of bars AC and AB becomes

$$U = \int F\, de = \int_0^e Re(1 - 2e)\, de = \frac{Re^2}{6}(3 - 4e)$$

Thus the total U for the system

$$U = \frac{R\Delta^2}{6}(3 - 4\Delta) + \frac{2R\Delta^2 \cos^2 15°}{6}(3 - 4\sqrt{2}\Delta\cos 15°) + 0$$

$$= (1.433 - 2.366\Delta)R\Delta^2$$

The external work is again $W_e = P\Delta$; thus

$$\pi = (1.433 - 2.366\Delta)R\Delta^2 - P\Delta$$

From $\partial\pi/\partial\Delta = 0$ we find $(2.866\Delta - 7.098\Delta^2)R - P = 0$. Hence

$$\Delta = 0.202 \pm \sqrt{0.0408 - \frac{P}{R}(0.141)} = 0.202 \pm \sqrt{0.0126}$$

Taking the negative sign since it yields the lower value for Δ, we find

$\Delta = 0.0897$

Since $e_{AC} = \Delta$, therefore, $F_{AC} = R\Delta(1 - 2\Delta) = 0.368P$. The forces F_{AB} and F_{AD} can be found as before by the resolution of forces at A.   ∎

13.10 CASTIGLIANO'S THEOREM—PART I

The theorem states that the partial *derivative* of the *strain energy* with respect to any *displacement* is equal to the force corresponding to that displacement. The theorem is applicable to both linear and nonlinear structures, including gross distortion.

Proof   Let us assume that a body in equilibrium is subjected to a virtual change in its ith displacement Δ_i; then from Eq. (13.46)

$$\delta\pi = \delta U - \delta W_e = \frac{\partial U}{\partial \Delta_i}\, \delta\Delta_i - P_i\, \delta\Delta_i = 0 \qquad (13.47)$$

in which $\delta\Delta_i$ is the virtual change in Δ_i, P_i is the force in the direction of the displacement Δ_i, and the strain energy U is in terms of the displacements Δ.

Rewriting Eq. (13.47) yields

$$\left(\frac{\partial U}{\partial \Delta_i} - P_i\right) \delta\Delta_i = 0$$

Since $\delta\Delta_i \neq 0$, the term in parentheses must vanish. Thus

$$\frac{\partial U}{\partial \Delta_i} = P_i \qquad (13.48)$$

This is the mathematical expression for Castigliano's theorem—part I. Equation (13.48) is valid for $i = 1, 2, \ldots, n$ and thus can yield n equations of equilibrium corresponding to the n displacement degrees of freedom.

■ EXAMPLE 13.9

The spoked wheel in Fig. 13.15a has a rigid rim and n identical equally spaced spokes which are equally prestressed so that they remain in tension when the load P is applied to the hub A. Use Castigliano's theorem—part I to deduce the change in tension in the ith spoke. Determine the change in tension in one of the spokes when $n = 8$.

Solution.   Assume a downward vertical displacement Δ as shown in Fig. 13.15b. (Due to symmetry, the horizontal displacement $= 0$.) From geometry, the elongation in the ith spoke $\Delta_i = \Delta \cos \theta_i$. The strain energy in the ith spoke $= (AE/2l)\Delta_i^2 = (AE/2l)\Delta^2 \cos^2 \theta_i$. Thus the total strain energy for all the spokes is

$$U = \sum_{i=1}^{n} \frac{AE}{2l}\, \Delta^2 \cos^2 \theta_i$$

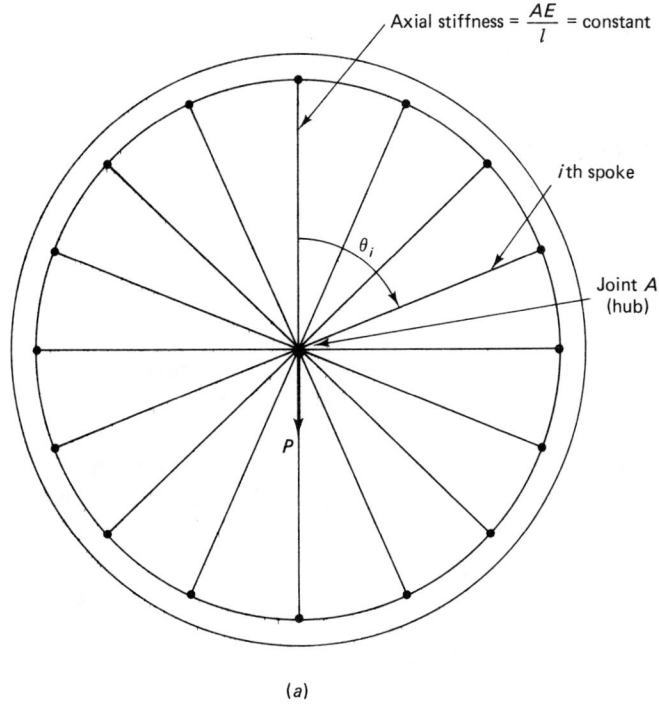

Axial stiffness = $\dfrac{AE}{l}$ = constant

*i*th spoke

θ_i

Joint *A*
(hub)

P

(a)

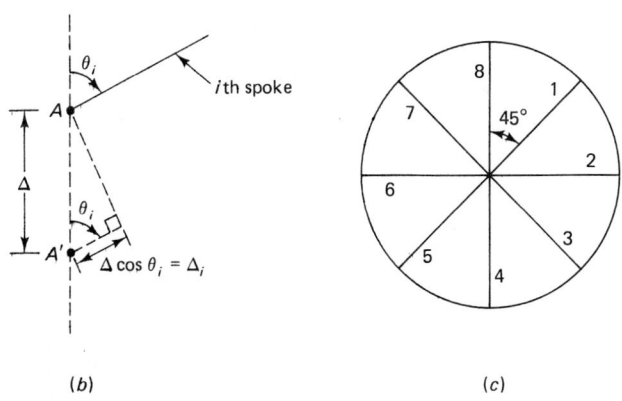

θ_i

A

Δ

*i*th spoke

θ_i

A'

$\Delta \cos \theta_i = \Delta_i$

(b)

8

7

1

45°

6

2

5

3

4

(c)

Figure 13.15 Structure for Example 13.9.

Applying Eq. (13.48) yields

$$\frac{\partial U}{\partial \Delta} = \sum_{i=1}^{n} \frac{AE}{l} \Delta \cos^2 \theta_i = P$$

or

$$\Delta = \frac{Pl}{AE \sum_{i=1}^{n} \cos^2 \theta_i}$$

But

$$F_i = \frac{AE}{l} \Delta_i = \frac{AE}{l} \Delta \cos \theta_i = \frac{P \cos \theta_i}{\sum_{i=1}^{n} \cos^2 \theta_i}$$

For $n = 8$, $\theta_1 = 45°$ (see Fig. 13.15c),

$$F_1 = \frac{P \cos 45°}{\cos^2 45° + \cos^2 90° + \cdots + \cos^2 315° + \cos^2 360°} = \frac{\sqrt{2}}{8} P = 0.177P$$

∎

■ EXAMPLE 13.10

Figure 13.16 shows three linearly elastic hinged bars subjected to concentrated loads P at joints B and C. Determine the deflection of joints B and C.

Solution. Here the structure is linearly elastic and geometrically nonlinear in its deformed state at equilibrium; if Δ is the displacement of joints B and C, and assuming that $\Delta \ll L$, the strains in bars AB and CD are

$$\varepsilon_{AB} = \frac{(\sqrt{L^2 + \Delta^2} - L)}{L} = \left[1 + \left(\frac{\Delta}{L} \right)^2 \right]^{1/2} - 1 = 1 + \frac{1}{2} \left(\frac{\Delta}{L} \right)^2 + \cdots - 1$$

or

$$\varepsilon_{AB} \approx \frac{1}{2} \left(\frac{\Delta}{L} \right)^2$$

Therefore

$$e_{AB} = \frac{1}{2} \left(\frac{\Delta}{L} \right)^2 \cdot L = \frac{\Delta^2}{2L} \quad \text{and} \quad e_{CD} = \frac{\Delta^2}{2L}$$

Thus the total strain energy of the system is

$$U = \sum \frac{AEe^2}{2L} = 2 \left[\frac{AE}{2L} \left(\frac{\Delta^2}{2L} \right)^2 \right] + 0$$

The zero term corresponds to the zero strain energy in bar BC, or

$$U = \frac{AE}{4L^3} \Delta^4$$

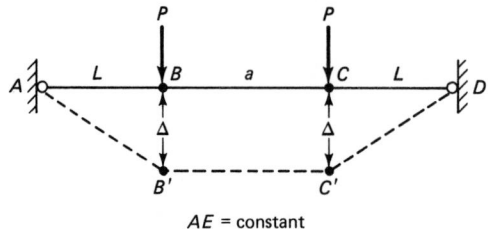

AE = constant

Figure 13.16 Beam for Example 13.10.

Using Eq. (13.48), we find

$$\frac{\partial U}{\partial \Delta} = \frac{AE}{L^3} \Delta^3 = P$$

whence $\Delta = (L)(\sqrt[3]{P/AE})$; this result shows that the displacement is not a linear function of the applied load P. ■

13.11 CASTIGLIANO'S THEOREM OF LEAST WORK

This theorem follows from Castigliano's theorem—part II discussed in Section 7.6. It is also referred to as Castigliano's *theorem of compatibility*. It states that the work done in stressing an indeterminate structure under a given system of loads is the *least possible* consistent with the maintenance of equilibrium. It is applicable only to linearly elastic structures.

Proof Let us consider the simple structure of Fig. 13.17*a*. It is indeterminate to the first degree; let member BD be the redundant member; assume also that BD is too short by an amount λ. From equilibrium of forces it can be readily shown that for an external load Q applied at joint D and letting the unknown force in BD be R, the forces in the inclined members will be $(Q - R)$. From Eq. (7.22a), the strain energy stored in an axially loaded member is

$$U = \frac{F^2 L}{2AE} \qquad\qquad (a)$$

The compatibility equation expresses the geometrical relationship between the extensions of the three members, that is, from Fig. 13.17*b*,

$$\Delta_2 - \Delta_{1,3} = \lambda \qquad\qquad (b)$$

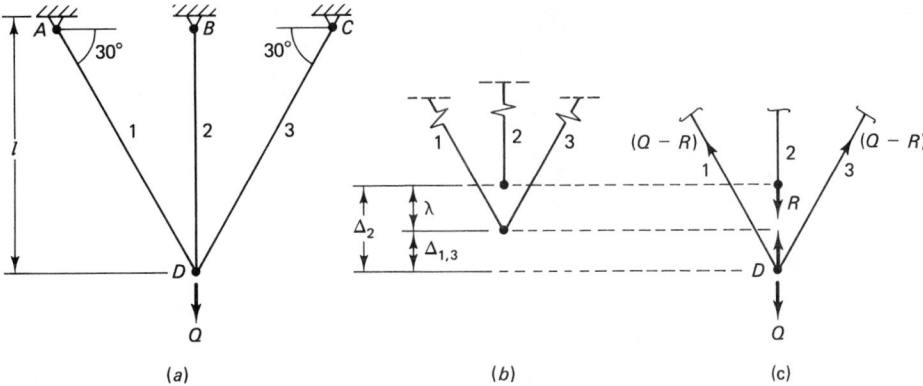

(a)   *(b)*   *(c)*

Figure 13.17 Analysis of an indeterminate truss by Castigliano's theorem of least work.

Using Castigliano's theorem—part II, Eq. (7.18), to find deflections, and referring to Fig. 13.17c, we can write

$$\Delta_2 = + \left(\frac{\partial U}{\partial R}\right)_2$$

and   (c)

$$\Delta_{1,3} = -\left(\frac{\partial U}{\partial R}\right)_{1,3}$$

The negative sign in the second equation of (c) is the result of $\Delta_{1,3}$ being opposite to R (see Fig. 13.17c). Substituting (c) in (b) yields

$$\left(\frac{\partial U}{\partial R}\right)_2 + \left(\frac{\partial U}{\partial R}\right)_{1,3} = \lambda$$

or

$$\left(\frac{\partial U}{\partial R}\right)_{1,2,3} = \lambda$$

One can generalize this result by writing

$$\frac{\partial U}{\partial R_i} = \lambda_i \tag{13.49}$$

where U is the total strain energy stored in the structure. If there is no initial lack of fit, that is, $\lambda_i = 0$, Eq. (13.49) reduces to

$$\frac{\partial U}{\partial R_i} = 0 \tag{13.50}$$

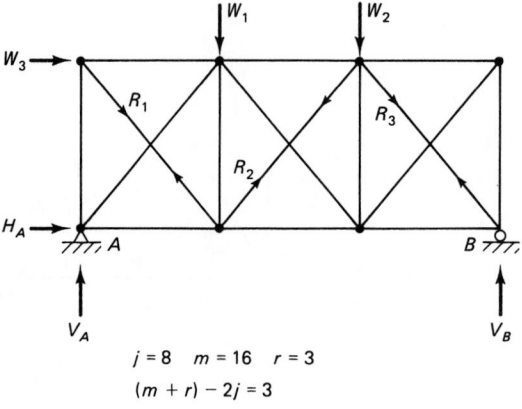

$j = 8 \quad m = 16 \quad r = 3$

$(m + r) - 2j = 3$

∴ Frame is indeterminate to third degree.

Figure 13.18 An internally indeterminate truss.

Equation (13.50) is the mathematical form of *Castigliano's theorem of least work.* For an indeterminate structure with several redundant bars with unknown forces R_1, R_2, R_3, etc., see, for example, Fig. 13.18, Eq. (13.50) becomes

$$\frac{\partial U}{\partial R_1} = 0 \qquad \frac{\partial U}{\partial R_2} = 0 \qquad \frac{\partial U}{\partial R_3} = 0 \qquad \text{etc.}$$

In this manner we can generate as many equations as there are statically indeterminate quantities.

■ **EXAMPLE 13.11**

The beam AB in Fig. 13.19a is simply supported at B, hinged at A and supported by pin-jointed members CE, DE, and EF. I for the beam is 1800 in^4; areas for members CE, DE, and EF are 4, 12, and 6 in^2, respectively. ($E = 24,000$ ksi for all members.) Calculate the force in member EF and the vertical deflection at F.

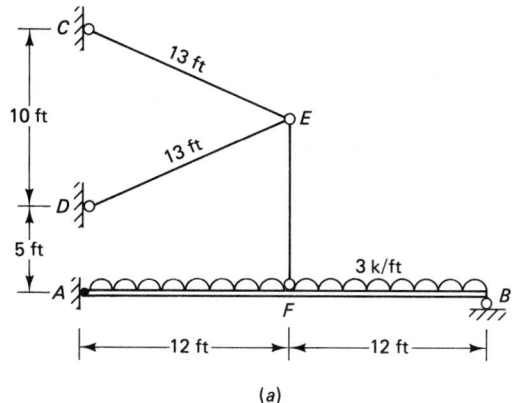

(a)

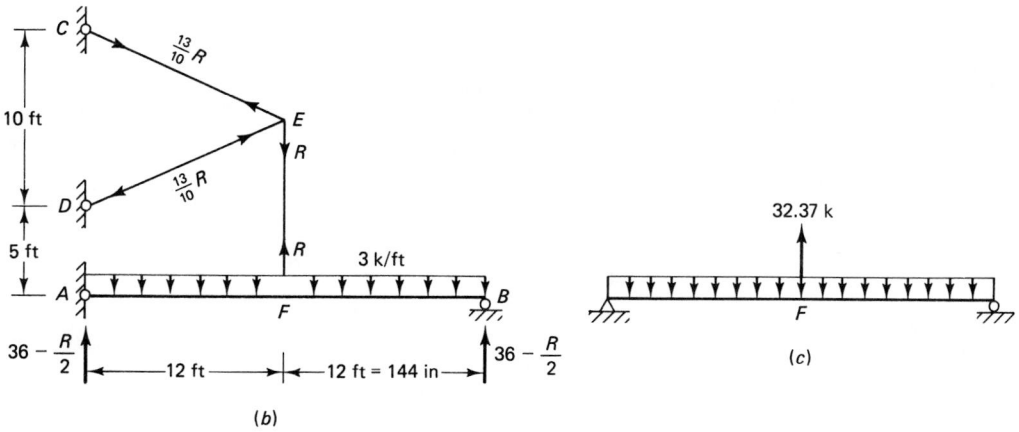

(b)

(c)

Figure 13.19 Composite structure for Example 13.11.

Solution.   Let the force in the redundant member be R. Then from Castigliano's theorem of least work,

$$\frac{\partial U}{\partial R} = 0 \qquad U = \sum \frac{F^2 L}{2AE} + \int_{AF,FB} \frac{M^2 \, dx}{2EI}$$

Therefore,

$$\sum_{CE,EF,DE} \frac{FL}{AE} \frac{\partial F}{\partial R} + \int_{AF,FB} \frac{M}{EI} \frac{\partial M}{\partial R} \, dx = 0$$

Member	F	$\dfrac{\partial F}{\partial R}$	$\dfrac{L}{A}$	$(F)\left(\dfrac{L}{A}\right)\left(\dfrac{\partial F}{\partial R}\right)$
CE	$\dfrac{13}{10}R$	$\dfrac{13}{10}$	$\dfrac{13 \times 12}{4} = 39$	$65.91\,R$
DE	$-\dfrac{13}{10}R$	$-\dfrac{13}{10}$	$\dfrac{13 \times 12}{12} = 13$	$21.97\,R$
EF	R	1	$\dfrac{10 \times 12}{6} = 20$	$20\,R$

$$\sum (F)\left(\frac{L}{A}\right)\left(\frac{\partial F}{\partial R}\right) = 107.88\,R$$

$$M_x = \left(36 - \frac{R}{2}\right)x - \frac{1}{4}\frac{x^2}{2} \qquad 0 \leqslant x \leqslant 144$$

$$\frac{\partial M}{\partial R} = -\frac{x}{2}$$

Using the symmetry of the beam about point F, we write:

$$\int M \frac{\partial M}{\partial R} \frac{dx}{I} = \frac{2}{I} \int_0^{144} \left[\left(36 - \frac{R}{2}\right)x - \frac{x^2}{8}\right]\left(-\frac{x}{2}\right) dx$$

$$= \frac{2}{1800}\left[\int_0^{144} \left(-18x^2 + \frac{Rx^2}{4} + \frac{x^3}{16}\right) dx\right]$$

$$= \frac{1}{900}\left(-6x^3 + \frac{Rx^3}{12} + \frac{x^4}{64}\right)\Big|_0^{144} = -12{,}441.6 + 276.48\,R$$

$$\frac{\partial U}{\partial R} = 107.88\,R - 12{,}441.6 + 276.48\,R = 0$$

$$R = \frac{12{,}441.6}{384.36} = 32.37 \text{ k}$$

Vertical deflection of F = (downward deflection due to 3 k/ft)

$\qquad$ − (upward deflection due to 32.37 k at F)   (see Fig. 13.19c)

$$= \frac{5wL^4}{384\,EI} - \frac{(32.37)(L^3)}{48\,EI}$$

$$= \frac{(5)(3)(24)^4(1728)}{(384)(24{,}000)(1800)} - \frac{(32.37)(24)^3(1728)}{(48)(24{,}000)(1800)}$$

$$= 0.518 \text{ in} - 0.373 \text{ in}$$

$$= 0.145 \text{ in}\downarrow \qquad\qquad\qquad\qquad \blacksquare$$

13.12 THE UNIT-LOAD METHOD

This method can be shown to be derivable from the theorem of virtual work (Chapter 7). In using this method to determine deflections, it becomes identical to the method of Castigliano's theorem—part II if instead of an imaginary force Q we use a unit force. The application of this method to determine redundant forces in an indeterminate structure is shown in Example 13.12.

■ **EXAMPLE 13.12**

Determine by (a) Castigliano's theorem of least work and (b) the unit-load method, the reaction at support B and the forces in the members of the truss ABC loaded as shown in Fig. 13.20a. AE is same for all members.

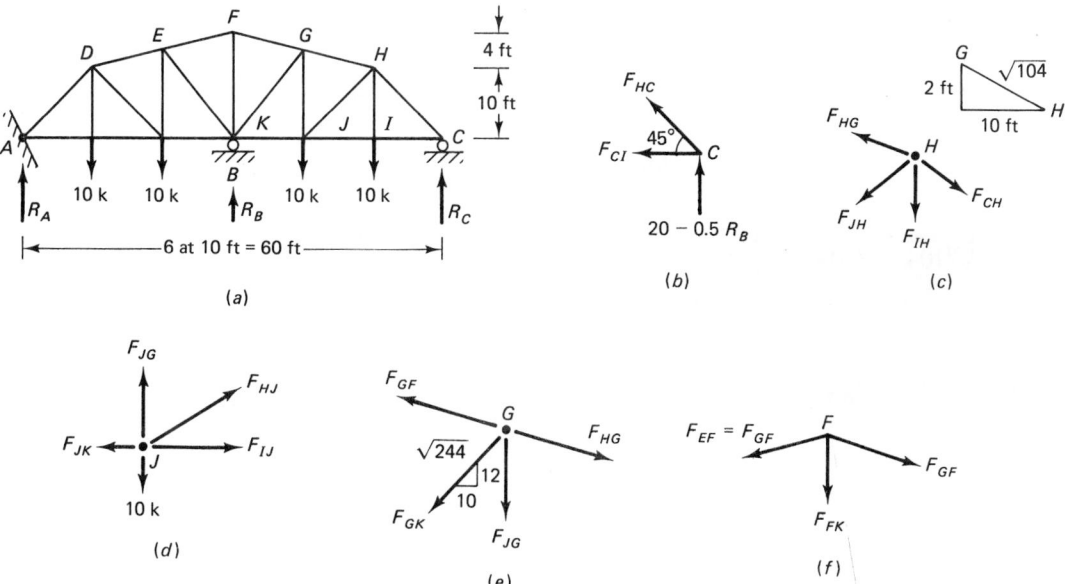

Figure 13.20 Truss for Example 13.12.

Solution

(a) *By Castigliano's theorem of least work.* Choose R_B to be redundant.

$$\Sigma \, M_{\text{about } A} = 0 \qquad (R_B)(30) + (R_C)(60) - (10)(10 + 20 + 40 + 50) = 0$$

$$R_C = 20 - 0.5 R_B$$

Joint C (see Fig. 13.20*b*)

$$F_{HC} = (\sqrt{2})(0.5 R_B - 20) \qquad F_{CI} = 20 - 0.5 R_B$$

Joint I

$$F_{JI} = F_{CI} \qquad F_{JI} = 20 - 0.5 R_B \qquad F_{HI} = 10$$

Joint H (see Fig. 13.20*c*)

From $\Sigma \, F_x = 0$:

$$(F_{CH})\left(\frac{1}{\sqrt{2}}\right) - (F_{GH})\left(\frac{10}{\sqrt{104}}\right) - F_{JH}\left(\frac{1}{\sqrt{2}}\right) = 0$$

or

$$F_{GH}\left(\frac{10}{\sqrt{104}}\right) + F_{JH}\left(\frac{1}{\sqrt{2}}\right) = 0.5 R_B - 20 \qquad \text{(a)}$$

From $\Sigma \, F_y = 0$:

$$-F_{IH} - (0.5 R_B - 20) - F_{JH}\left(\frac{1}{\sqrt{2}}\right) + F_{HG}\left(\frac{2}{\sqrt{104}}\right) = 0$$

or

$$-10 - 0.5 R_B + 20 = -F_{HG}\left(\frac{2}{\sqrt{104}}\right) + F_{JH}\left(\frac{1}{\sqrt{2}}\right) = -0.5 R_B + 10 \qquad \text{(b)}$$

Subtracting equation (b) from equation (a),

$$F_{GH}\left(\frac{10}{\sqrt{104}} + \frac{2}{\sqrt{104}}\right) = 0.5 R_B - 20 + 0.5 R_B - 10 = R_B - 30$$

$$F_{GH} = \frac{\sqrt{104}}{12} R_B - \frac{5\sqrt{104}}{2}$$

$$F_{JH} = \sqrt{2}(-\tfrac{5}{6} R_B + 25 + 0.5 R_B - 20) = \sqrt{2}(5 - \tfrac{1}{3} R_B)$$

Joint J (see Fig. 13.20*d*)

From $\Sigma \, F_x = 0$:

$$F_{IJ} + (F_{HJ})\left(\frac{1}{\sqrt{2}}\right) - F_{JK} = 0$$

$$F_{JK} = 20 - 0.5 R_B + 5 - \frac{1}{3} R_B$$

$$= 25 - \frac{5}{6} R_B$$

From $\Sigma F_y = 0$:

$$F_{JG} - 10 + F_{HJ} \frac{1}{\sqrt{2}} = 0$$

$$F_{JG} = 10 - 5 + \frac{1}{3} R_B$$

$$F_{JG} = 5 + \frac{1}{3} R_B$$

Joint G (see Fig. 13.20e)

From $\Sigma F_x = 0$:

$$F_{HG} \left(\frac{10}{\sqrt{104}} \right) - F_{GF} \left(\frac{10}{\sqrt{104}} \right) - F_{GK} \left(\frac{10}{\sqrt{244}} \right) = 0$$

or

$$F_{GF} \left(\frac{1}{\sqrt{104}} \right) + F_{GK} \left(\frac{1}{\sqrt{244}} \right) = F_{HG} \left(\frac{1}{\sqrt{104}} \right)$$

or

$$F_{GF} + F_{GK} \left(\sqrt{\frac{104}{244}} \right) = F_{HG} \tag{c}$$

From $\Sigma F_y = 0$:

$$F_{GF} \left(\frac{2}{\sqrt{104}} \right) - F_{HG} \left(\frac{2}{\sqrt{104}} \right) - F_{JG} - F_{GK} \left(\frac{12}{\sqrt{244}} \right) = 0$$

$$F_{GF} \left(\frac{2}{\sqrt{104}} \right) - F_{GK} \left(\frac{12}{\sqrt{244}} \right) = 5 + \frac{1}{3} R_B + \frac{R_B}{6} - 5 = \frac{R_B}{2}$$

$$F_{GF} - 6 F_{GK} \sqrt{\frac{104}{244}} = \frac{\sqrt{104}}{4} R_B \tag{d}$$

Subtracting equation (d) from equation (c)

$$F_{GK}(1 + 6) \left(\sqrt{\frac{104}{244}} \right) = \frac{\sqrt{104}}{12} R_B - \frac{5\sqrt{104}}{2} - \frac{\sqrt{104}}{4} R_B$$

or

$$(F_{GK})(7) \left(\sqrt{\frac{104}{244}} \right) = - \frac{\sqrt{104}}{6} R_B - \frac{5\sqrt{104}}{2}$$

$$F_{GK} = -(R_B) \left(\frac{\sqrt{244}}{42} \right) - \frac{5\sqrt{244}}{14}$$

$$F_{GF} = \frac{\sqrt{104}}{4} R_B + (6\sqrt{104})\left(-\frac{R_B}{42} - \frac{5}{14}\right)$$

$$= \left(\frac{\sqrt{104}}{4} - \frac{\sqrt{104}}{7}\right) R_B - \frac{15\sqrt{104}}{7}$$

$$F_{GF} = R_B(\tfrac{3}{28})(\sqrt{104}) - \tfrac{15}{7}\sqrt{104}$$

Joint F (see Fig. 13.20*f*)

From $\Sigma F_y = 0$:

$$F_{FK} + 2F_{GF}\left(\frac{2}{\sqrt{104}}\right) = 0$$

$$F_{FK} = -(4)\left(\frac{3R_B}{28} - \frac{15}{7}\right) = \frac{60}{7} - \frac{3R_B}{7}$$

Check Joint K

From $\Sigma F_y = 0$:

$$R_B + F_{FK} + 2F_{GK}\left(\frac{12}{\sqrt{244}}\right) = 0$$

or

$$R_B + \frac{60}{7} - \frac{3R_B}{7} + \frac{24}{\sqrt{244}}\left(-\frac{R_B\sqrt{244}}{42} - \frac{5\sqrt{244}}{14}\right) = 0$$

or

$$R_B + \frac{60}{7} - \frac{3R_B}{7} - \frac{4R_B}{7} - \frac{60}{7} = 0 \qquad \text{(check)}$$

From Eqs. (7.22b) and (13.50)

$$\frac{\partial U}{\partial R_B} = \Sigma \frac{F(\partial F/\partial R_B)l}{AE} = 0 \tag{e}$$

Values of $(Fl/AE)(\partial F/\partial R_B)$ for all members of the truss are shown in Table 13.2. Solution of equation (e) is shown on the end of the table.

(b) *By the unit-load method.* Remove the redundant reaction R_B.

Let P_i = force in the *i*th member of the determinate truss due to applied loads

p_i = force in the *i*th member of the determinate truss due to unit upward reaction at B

F_i = force in the *i*th member of the given truss due to applied load and redundant reaction

or

$$F_i = P_i + R_B \cdot p_i \quad \text{thus} \quad \frac{\partial F_i}{\partial R_B} = p_i$$

TABLE 13.2 CASTIGLIANO'S THEOREM OF LEAST WORK

Member	Force F	$\dfrac{I}{AE}$	$\dfrac{\partial F}{\partial R_B}$	$\dfrac{FI}{AE}\dfrac{\partial F}{\partial R_B}$	Multiply by	$\left(\dfrac{FI}{AE}\dfrac{\partial F}{\partial R_B}\right)$
HC	$\sqrt{2}(0.5R_B - 20)$	$10\sqrt{2}$	$\sqrt{2}/2$	$\sqrt{2}(5R_B - 200)$	2^*	$2\sqrt{2}(5R_B - 200)$
CI	$20 - 0.5R_B$	10	$-\tfrac{1}{2}$	$-100 + 2.5R_B$	2^*	$-200 + 5R_B$
IH	10	10	0	0	2^*	0
JI	$20 - 0.5R_B$	10	$-\tfrac{1}{2}$	$-100 + 2.5R_B$	2^*	$-200 + 5R_B$
JH	$\sqrt{2}(5 - \tfrac{1}{3}R_B)$	$10\sqrt{2}$	$-\dfrac{\sqrt{2}}{3}$	$\sqrt{2}\left(-\dfrac{100}{3} + \dfrac{20R_B}{9}\right)$	2^*	$(2\sqrt{2})\left(-\dfrac{100}{3} + \dfrac{20R_B}{9}\right)$
GH	$\sqrt{104}\dfrac{R_B}{12} - \dfrac{5\sqrt{104}}{2}$	$\sqrt{104}$	$\dfrac{\sqrt{104}}{12}$	$\left(\dfrac{104}{12}\right)\left(\dfrac{R_B\sqrt{104}}{12} - \dfrac{5\sqrt{104}}{2}\right)$	2^*	$\left(\dfrac{104}{6}\right)\left(\dfrac{R_B\sqrt{104}}{12} - \dfrac{5\sqrt{104}}{2}\right)$
JG	$5 + \tfrac{1}{3}R_B$	12	$\tfrac{1}{3}$	$20 + \tfrac{4}{3}R_B$	2^*	$40 + \tfrac{8}{3}R_B$
GK	$-\dfrac{R_B\sqrt{244}}{42} - \dfrac{5\sqrt{244}}{14}$	$\sqrt{244}$	$-\dfrac{\sqrt{244}}{42}$	$\left(\dfrac{244}{42}\right)\left(\dfrac{R_B\sqrt{244}}{42} + \dfrac{5\sqrt{244}}{14}\right)$	2^*	$\left(\dfrac{244}{21}\right)\left(\dfrac{R_B\sqrt{244}}{42} + \dfrac{5\sqrt{244}}{14}\right)$
GF	$\dfrac{3R_B\sqrt{104}}{28} - \dfrac{15\sqrt{104}}{7}$	$\sqrt{104}$	$\dfrac{3\sqrt{104}}{28}$	$\dfrac{(3)(104)}{28}\left(\dfrac{3R_B\sqrt{104}}{28} - \dfrac{15\sqrt{104}}{7}\right)$	2^*	$\left(\dfrac{312}{14}\right)\left(\dfrac{3R_B\sqrt{104}}{28} - \dfrac{15\sqrt{104}}{7}\right)$
JK	$25 - \tfrac{5}{6}R_B$	10	$-\tfrac{5}{6}$	$-\dfrac{1250}{6} + \dfrac{250R_B}{36}$	2^*	$-\dfrac{1250}{3} + \dfrac{250R_B}{18}$
FK	$-\tfrac{3}{7}R_B + \tfrac{60}{7}$	14	$-\tfrac{3}{7}$	$\dfrac{18R_B}{7} - \dfrac{360}{7}$	1^*	$\dfrac{18R_B}{7} - \dfrac{360}{7}$

$$\Sigma = R_B\left[10\sqrt{2} + 5 + 5 + \frac{40\sqrt{2}}{9} + \frac{104\sqrt{104}}{72} + \frac{8}{3} + \frac{244\sqrt{244}}{(21)(42)} + \frac{936\sqrt{104}}{(14)(28)} + \frac{250}{18} + \frac{18}{7}\right] - 400\sqrt{2} - 200 - 200 - \frac{200\sqrt{2}}{3}$$

$$-\frac{(15)(312)(\sqrt{104})}{98} - \frac{1250}{3} - \frac{360}{7} = 0 \quad \text{or} \quad (92.96)R_B = 2352.83 \quad \text{or} \quad R_B = 25.32 \text{ k}\uparrow$$

* Number of similar members.

667

Strain energy of the truss $= U = \sum \dfrac{F^2 l}{2AE}$ (7.28b)

From Eq. (13.50),

$$\frac{\partial U}{\partial R_B} = 0 \quad \text{that is} \quad \sum F \frac{\partial F}{\partial R_B}\frac{l}{AE} = 0$$

TABLE 13.3 UNIT-LOAD METHOD

Member	Force P_i due to applied load	Force p_i due to unit load at B	$\dfrac{l}{AE}$	$\left(\dfrac{Ppl}{AE}\right)$ (No. of similar members)	$\left(\dfrac{p^2 l}{AE}\right)$ (No. of similar members)
HC	$-20\sqrt{2}$	$0.5\sqrt{2}$	$10\sqrt{2}$	$(-200\sqrt{2}) \times 2$	$(5\sqrt{2}) \times 2$
CI	20	-0.5	10	$(-100) \times 2$	$(2.5) \times 2$
IH	10	0	10	$(0) \times 2$	$(0) \times 2$
JI	20	-0.5	10	$(-100) \times 2$	$(2.5) \times 2$
JH	$5\sqrt{2}$	$-\dfrac{\sqrt{2}}{3}$	$10\sqrt{2}$	$\left(-\dfrac{100\sqrt{2}}{3}\right) \times 2$	$\left(\dfrac{20\sqrt{2}}{9}\right) \times 2$
GH	$-\dfrac{5\sqrt{104}}{2}$	$\dfrac{\sqrt{104}}{12}$	$\sqrt{104}$	$-\left[\left(\dfrac{104}{12}\right)\left(\dfrac{5\sqrt{104}}{2}\right)\right] \times 2$	$\left[\left(\dfrac{104}{12}\right)\left(\dfrac{\sqrt{104}}{12}\right)\right] \times 2$
JG	5	$\dfrac{1}{3}$	12	$(20) \times 2$	$(\tfrac{4}{3}) \times 2$
GK	$-\dfrac{5\sqrt{244}}{14}$	$-\dfrac{\sqrt{244}}{42}$	$\sqrt{244}$	$\left[\left(\dfrac{244}{42}\right)\left(\dfrac{5\sqrt{244}}{14}\right)\right] \times 2$	$\left[\left(\dfrac{244}{42}\right)\left(\dfrac{\sqrt{244}}{42}\right)\right] \times 2$
GF	$-\dfrac{15\sqrt{104}}{7}$	$\dfrac{3\sqrt{104}}{28}$	$\sqrt{104}$	$\left[-\left(\dfrac{312}{28}\right)\left(\dfrac{15\sqrt{104}}{7}\right)\right] \times 2$	$\left[\left(\dfrac{312}{28}\right)\left(\dfrac{3\sqrt{104}}{28}\right)\right] \times 2$
JK	25	$-\dfrac{5}{6}$	10	$(-\tfrac{1250}{6}) \times 2$	$(\tfrac{250}{36}) \times 2$
FK	$\dfrac{60}{7}$	$-\dfrac{3}{7}$	14	$(-\tfrac{360}{7}) \times 1$	$(\tfrac{18}{7}) \times 1$

$\sum \dfrac{Ppl}{AE} = -400\sqrt{2} - 200 - 200 - \dfrac{200\sqrt{2}}{3} - \dfrac{520\sqrt{104}}{12} + 40 +$

$\dfrac{1220\sqrt{244}}{(21)(14)} - \dfrac{(15)(312)(\sqrt{104})}{98} - \dfrac{1250}{3} - \dfrac{360}{7} = -2352.83$

$\sum \dfrac{p^2 l}{AE} = 10\sqrt{2} + 5 + 5 + \dfrac{40\sqrt{2}}{9} + \dfrac{104\sqrt{104}}{72} + \dfrac{8}{3}$

$+ \dfrac{244\sqrt{244}}{(21)(42)} + \dfrac{936\sqrt{104}}{(14)(28)} + \dfrac{250}{18} + \dfrac{18}{7} = 92.96$

$$R_B = \frac{-\sum Ppl/AE}{\sum p^2 l/AE} = \frac{2352.83}{92.96} = 25.32 \text{ k}\uparrow$$

Substituting:

$$\sum (P + R_B \cdot p) \frac{pl}{AE} = 0$$

that is,

$$\sum \left(\frac{P \cdot p \cdot l}{AE} + \frac{R_B p^2 l}{AE} \right) = 0$$

Thus

$$R_B = - \frac{\sum P \cdot p \cdot l / AE}{\sum p^2 l / AE} \tag{f}$$

The calculations are shown in Table 13.3.   ∎

13.13 ENGESSER'S THEOREM OF COMPATIBILITY

This theorem states that for a linearly or nonlinearly elastic indeterminate stable structure, the first partial derivative of the internal complementary energy U^* with respect to any particular redundant is zero.

Proof   The derivation of this theorem follows the steps taken in the derivation of Castigliano's theorem of least work and substituting the internal complementary energy U^* for the strain energy U. Following the same procedure as in the derivation of Eq. (13.49), it can be shown that for linear and nonlinear elastic structures

$$\frac{\partial U^*}{\partial R_i} = \lambda_i \tag{13.51}$$

where λ_i is the initial lack of fit in a member i with an internal force R_i assumed positive. If there is no initial lack of fit, $\lambda_i = 0$ and Eq. (13.51) reduces to

$$\frac{\partial U^*}{\partial R_i} = 0 \tag{13.52}$$

which is the form in which Engesser originally stated the theorem.

In an indeterminate structure with several unknown redundants R_1, R_2, R_3, etc., present, Eq. (13.52) becomes

$$\frac{\partial U^*}{\partial R_1} = 0 \qquad \frac{\partial U^*}{\partial R_2} = 0 \qquad \frac{\partial U^*}{\partial R_3} = 0 \qquad \text{etc.}$$

Thus we can generate as many equations as there are redundants. It is readily observed that if the structure is linearly elastic, then $U^* = U$ and hence Eq. (13.52) reduces to Eq. (13.50) which is applicable only to linearly elastic structures. A summary of the energy theorems presented in this chapter and their limitations are shown in Table 13.4.

TABLE 13.4   SUMMARY OF ENERGY THEOREMS

Theorem	Applications
Minimum total potential energy:	Linear, nonlinear, and gross-distortion problems
$\dfrac{\partial}{\partial \Delta_i}(U - W_e) = 0$ Eq. (13.46a)	
Castigliano's theorem—part I:	Same as above
$\dfrac{\partial U}{\partial \Delta_i} = P_i$ Eq. (13.48)	
Castigliano's theorem of least work:	
$\dfrac{\partial U}{\partial R_i} = \lambda_i$ Eq. (13.49)	
for deflection;	Linear problems
$\dfrac{\partial U}{\partial R_i} = 0$ Eq. (13.50)	
for redundant.	
Engesser's theorem of compatibility (theorem of complementary energy)	
$\dfrac{\partial U^*}{\partial R_i} = \lambda_i$ Eq. (13.51)	
for deflection or lack of fit;	Linear and nonlinear problems (excluding gross distortion)
$\dfrac{\partial U^*}{\partial R_i} = 0$ Eq. (13.52)	
for redundant.	

■ **EXAMPLE 13.13**

The cross-sectional areas of the members (in circles) of a truss are indicated in Fig. 13.21a. The stress-strain relation for each member is shown in Fig. 13.21b. All members are of length L and are under stresses that exceed the "elastic limit" of $\frac{20}{3}$ k/in². Find the force in the central vertical member. Use Engesser's theorem of compatibility.

Solution. Let the horizontal reaction be H as shown in Fig. 13.21c. The bar forces are as shown.

$$F_{AB} = -100 + \frac{H}{\sqrt{3}} \qquad F_{BC} = -100 + \frac{H}{\sqrt{3}} \qquad F_{CD} = -100 + \frac{H}{\sqrt{3}}$$

$$F_{DE} = -100 + \frac{H}{\sqrt{3}} \qquad F_{AF} = -\frac{2H}{\sqrt{3}} \qquad F_{EF} = -\frac{2H}{\sqrt{3}}$$

$$F_{CF} = -100 - \frac{H}{\sqrt{3}} \qquad F_{BF} = +100 - \frac{H}{\sqrt{3}} \qquad F_{DF} = +100 - \frac{H}{\sqrt{3}}$$

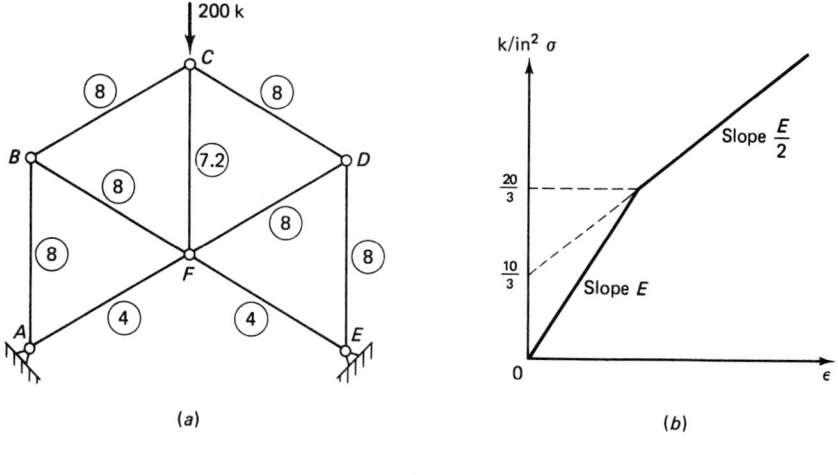

(a)   (b)

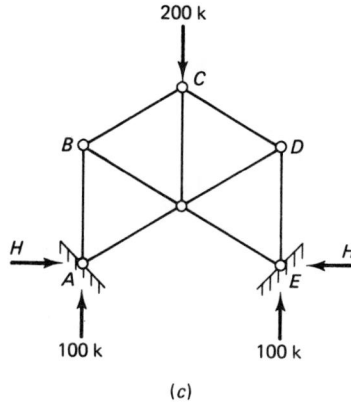

(c)

Figure 13.21 Truss for Example 13.13.

For $\sigma \leqslant \frac{20}{3}$:

$$\varepsilon = \frac{\sigma}{E}$$

$$e = \varepsilon L = \frac{FL}{AE}$$

For $\sigma > \frac{20}{3}$:

$$\sigma = \frac{10}{3} + \frac{E}{2} \cdot \varepsilon \quad \text{or} \quad \varepsilon = \frac{2}{E}\left(\sigma - \frac{10}{3}\right)$$

$$e = \varepsilon L = \frac{2L}{E}\left(\frac{F}{A} - \frac{10}{3}\right)$$

$$U^* \text{ (for one member)} = \int_0^F e \, dF = \int_0^{(20/3)A} e \, dF + \int_{(20/3)A}^F e \, dF$$

$$= \int_0^{(20/3)A} \left(\frac{FL}{AE}\right) dF + \int_{(20/3)A}^F \frac{2L}{E}\left(\frac{F}{A} - \frac{10}{3}\right) dF$$

$$= \left(\frac{F^2L}{2AE}\right)_0^{(20/3)A} + \frac{2L}{E}\left(\frac{F^2}{2A} - \frac{10F}{3}\right)_{(20/3)A}^F$$

$$= \frac{200}{9}\frac{AL}{E} + \frac{2L}{E}\left(\frac{F^2}{2A} - \frac{10F}{3}\right) - \frac{2L}{E}\left(\frac{200}{9}A - \frac{200}{9}A\right)$$

$$= \frac{200}{9}\frac{AL}{E} + \frac{F^2L}{AE} - \frac{20FL}{3E}$$

$$U^* \text{ (truss)} = \Sigma\left(\frac{200}{9}\frac{AL}{E} + \frac{F^2L}{AE} - \frac{20FL}{3E}\right)$$

From Eq. (13.52)

$$\frac{\partial U^*}{\partial H} = 0 = \Sigma\left(\frac{2FL}{AE} - \frac{20}{3}\frac{L}{E}\right)\frac{\partial F}{\partial H}$$

or

$$\Sigma\left(\frac{2F}{A}\frac{\partial F}{\partial H} - \frac{20}{3}\frac{\partial F}{\partial H}\right) = 0$$

since $L/E = $ constant for all members. In tabular form:

Member	Force F	$\dfrac{\partial F}{\partial H}$	A	$\dfrac{2F}{A}\dfrac{\partial F}{\partial H}$	$\dfrac{20}{3}\dfrac{\partial F}{\partial H}$
AB	$-100 + \dfrac{H}{\sqrt{3}}$	$\dfrac{1}{\sqrt{3}}$	8	$-\dfrac{1}{4\sqrt{3}}\left(100 - \dfrac{H}{\sqrt{3}}\right)$	$\dfrac{20}{3\sqrt{3}}$
BC	$-100 + \dfrac{H}{\sqrt{3}}$	$\dfrac{1}{\sqrt{3}}$	8	$-\dfrac{1}{4\sqrt{3}}\left(100 - \dfrac{H}{\sqrt{3}}\right)$	$\dfrac{20}{3\sqrt{3}}$
CD	$-100 + \dfrac{H}{\sqrt{3}}$	$\dfrac{1}{\sqrt{3}}$	8	$-\dfrac{1}{4\sqrt{3}}\left(100 - \dfrac{H}{\sqrt{3}}\right)$	$\dfrac{20}{3\sqrt{3}}$
DE	$-100 + \dfrac{H}{\sqrt{3}}$	$\dfrac{1}{\sqrt{3}}$	8	$-\dfrac{1}{4\sqrt{3}}\left(100 - \dfrac{H}{\sqrt{3}}\right)$	$\dfrac{20}{3\sqrt{3}}$
EF	$-\dfrac{2H}{\sqrt{3}}$	$-\dfrac{2}{\sqrt{3}}$	4	$\dfrac{2H}{3}$	$-\dfrac{40}{3\sqrt{3}}$
FA	$-\dfrac{2H}{\sqrt{3}}$	$-\dfrac{2}{\sqrt{3}}$	4	$\dfrac{2H}{3}$	$-\dfrac{40}{3\sqrt{3}}$

(Continued)

Member	Force F	$1-$	A	$\dfrac{2F}{A}\dfrac{\partial F}{\partial H}$	$\dfrac{20}{3}\dfrac{\partial F}{\partial H}$
BF	$+100 - \dfrac{H}{\sqrt{3}}$	$-\dfrac{1}{\sqrt{3}}$	8	$-\dfrac{1}{4\sqrt{3}}\left(100 - \dfrac{H}{\sqrt{3}}\right)$	$-\dfrac{20}{3\sqrt{3}}$
DF	$+100 - \dfrac{H}{\sqrt{3}}$	$-\dfrac{1}{\sqrt{3}}$	8	$-\dfrac{1}{4\sqrt{3}}\left(100 - \dfrac{H}{\sqrt{3}}\right)$	$-\dfrac{20}{3\sqrt{3}}$
CF	$-100 - \dfrac{H}{\sqrt{3}}$	$-\dfrac{1}{\sqrt{3}}$	7.2	$\dfrac{1}{(3.6)\sqrt{3}}\left(100 + \dfrac{H}{\sqrt{3}}\right)$	$-\dfrac{20}{3\sqrt{3}}$

$$\sum\left(\frac{\partial F}{\partial H}\frac{2F}{A} - \frac{20}{3}\frac{\partial F}{\partial H}\right) = 0$$

$$= -\frac{3}{2\sqrt{3}}\left(100 - \frac{H}{\sqrt{3}}\right) + \frac{4}{3}H + \frac{1}{(3.6)\sqrt{3}}\left(100 + \frac{H}{\sqrt{3}}\right) + \frac{20}{\sqrt{3}} = 0$$

that is,

$$-\frac{150}{\sqrt{3}} + \frac{H}{2} + \frac{4}{3}H + \frac{100}{(3.6)\sqrt{3}} + \frac{H}{10.8} + \frac{20}{\sqrt{3}} = 0$$

$$H\left(\frac{1}{2} + \frac{4}{3} + \frac{1}{10.8}\right) = \frac{150}{\sqrt{3}} - \frac{100}{(3.6)\sqrt{3}} - \frac{20}{\sqrt{3}}$$

$$1.93H = 59.02$$

Therefore,

$$H = 30.64$$

Force in member $CF = -100 - (30.64/\sqrt{3}) = -117.7$ k   ■

13.14 DERIVATION OF FLEXIBILITY AND STIFFNESS COEFFICIENTS BY THE ENERGY METHODS

Figure 13.22a shows the force-displacement components at the two ends of a member. In order to derive the flexibility coefficients of such a member, the member must be stable and statically determinate, allowing no rigid-body motion; such conditions are satisfied by the member in Fig. 13.22b, where $\delta_4 = 0$ in order to prevent any rigid-body motion. It should be mentioned that the member could have been constrained in several other ways.

To deduce the flexibility coefficients, we use the virtual work approach. Thus, for f_{11} in Fig. 13.22c using Fig. 13.22d and e, we can write

$$f_{11} = \int_0^l m\frac{M}{EI}dx = \frac{1}{EI}\int_0^l\left(\frac{l-x}{l}\right)F_1\left(\frac{l-x}{l}\right)dx = \frac{F_1 l}{3EI}$$

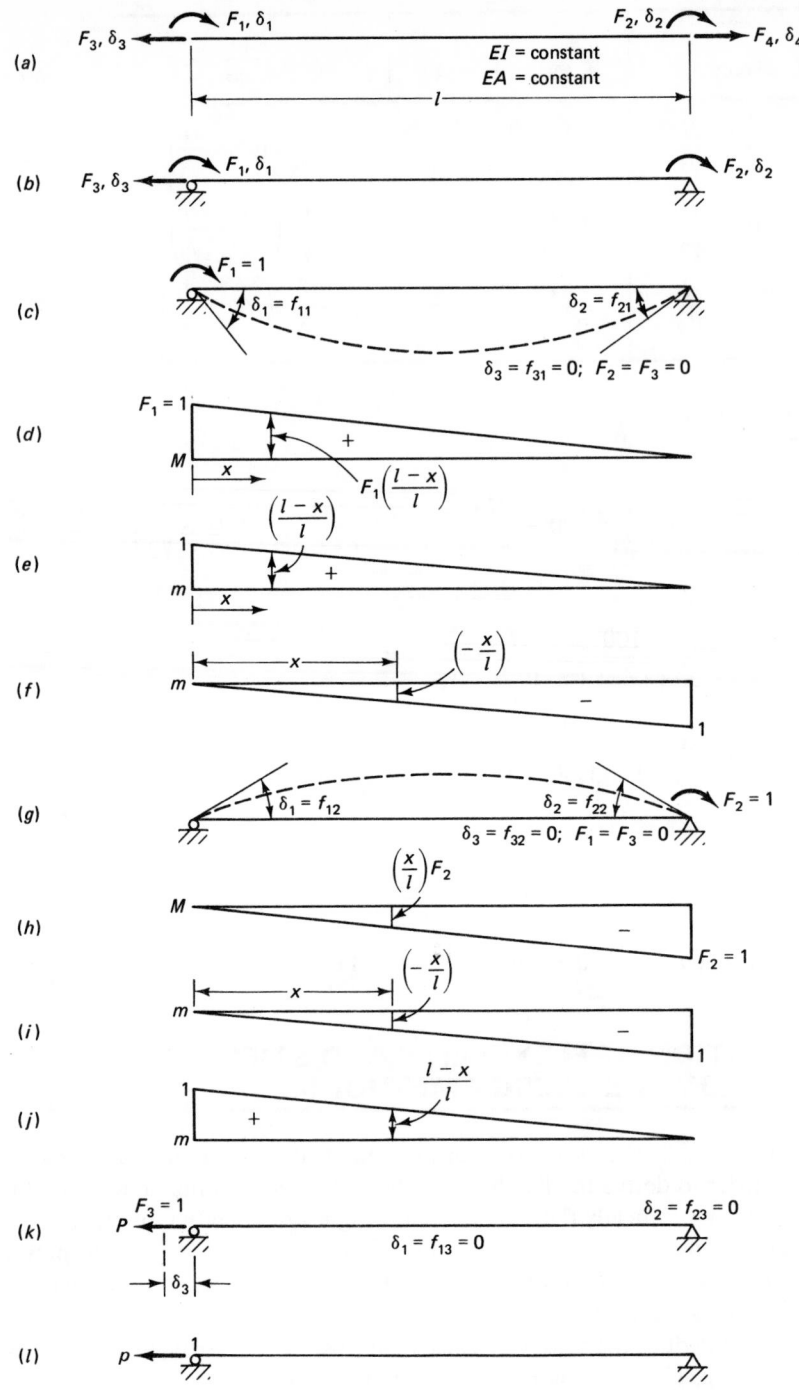

Figure 13.22 Derivation of flexibility coefficients by the method of virtual work.

With $F_1 = 1, f_{11} = l/3EI$. Using Fig. 13.22d and f,

$$f_{21} = \int_0^l m \frac{M}{EI} dx = \frac{1}{EI} \int_0^l \left(\frac{-x}{l}\right) F_1 \left(\frac{l-x}{l}\right) dx = \frac{-F_1 l}{6EI}$$

With $F_1 = 1, f_{21} = -l/6EI$. For f_{22} in Fig. 13.22g, and using Fig. 13.22h and i, we can write

$$f_{22} = \int_0^l m \frac{M}{EI} dx = \frac{1}{EI} \int_0^l \left(\frac{-x}{l}\right)\left(-F_2 \frac{x}{l}\right) dx = \frac{F_2 l}{3EI}$$

With $F_2 = 1, f_{22} = l/3EI$. From Fig. 13.22h and j,

$$f_{12} = \int_0^l m \frac{M}{EI} dx = \frac{1}{EI} \int_0^l \left(\frac{l-x}{l}\right)\left(-F_2 \frac{x}{l}\right) dx = \frac{-F_2 l}{6EI}$$

With $F_2 = 1, f_{12} = -l/6EI$. For f_{33} we find by using Fig. 13.22k and l that $f_{33} = \int_0^l p(P/AE) dx$. Since $P = F_3 = $ constant along the length of the member as well as p, then $\int_0^l dx = l$, and hence $f_{33} = [(1)(F_3)l]/AE$. With $F_3 = 1, f_{33} = l/AE$. Thus the force-displacement relation in matrix form is

$$\begin{bmatrix} f_{11} & f_{12} & f_{13} \\ f_{21} & f_{22} & f_{23} \\ f_{31} & f_{32} & f_{33} \end{bmatrix} \begin{Bmatrix} F_1 \\ F_2 \\ F_3 \end{Bmatrix} = \begin{Bmatrix} \delta_1 \\ \delta_2 \\ \delta_3 \end{Bmatrix} \tag{13.53}$$

in which the flexibility matrix is given by

$$\begin{bmatrix} f_{11} & f_{12} & f_{13} \\ f_{21} & f_{22} & f_{23} \\ f_{31} & f_{32} & f_{33} \end{bmatrix} = \begin{bmatrix} \dfrac{l}{3EI} & -\dfrac{l}{6EI} & 0 \\ -\dfrac{l}{6EI} & \dfrac{l}{3EI} & 0 \\ 0 & 0 & \dfrac{l}{AE} \end{bmatrix} \tag{13.54}$$

Equation (13.53) can be put in the form

$$\{\delta\} = [\mathbf{f}]\{\mathbf{F}\} \tag{13.55}$$

To derive the stiffness coefficients of a member with end forces and displacements as shown in Fig. 13.23a, we have to analyze an indeterminate structure. The moment and axial-force diagrams for the member with the force displacements in Fig. 13.23b are shown in Fig. 13.23c and d, respectively. The strain energy stored in the member due to bending and axial load is given by

$$U = \int_0^l \frac{M^2 dx}{2EI} + \int_0^l \frac{P^2 dx}{2AE} \tag{13.56}$$

Since $P = F_3 = F_4 = $ constant (from force-equilibrium) we can rewrite Eq. (13.56) as

$$U = \int \frac{M^2 dx}{2EI} + \frac{F_3^2 l}{2AE} \tag{13.57}$$

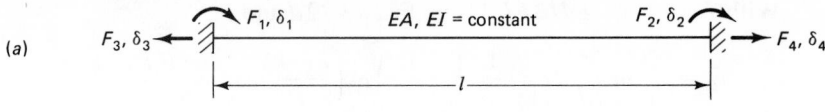

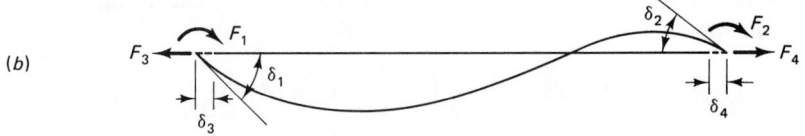

Figure 13.23 Derivation of stiffness coefficients by Castigliano's theorem—part II.

To Find $S_{11}, S_{21}, S_{31},$ and S_{41} Referring to Fig. 13.23e, by Castigliano's theorem— Part II (Chapter 7), we have

$$\frac{\partial U}{\partial F_1} = \delta_1 = 1 \quad \text{and} \quad \frac{\partial U}{\partial F_2} = \delta_2 = 0$$

as shown by the boundary conditions. From Eq. (13.57),

$$\frac{\partial U}{\partial F_1} = \int_0^l \frac{\partial M}{\partial F_1} \cdot \frac{M}{EI} \, dx + 0$$

$$\frac{\partial M}{\partial F_1} = \frac{l-x}{l}$$

Thus

$$\frac{\partial U}{\partial F_1} = \frac{1}{EI} \int_0^l \left(\frac{l-x}{l}\right) \left[F_1\left(\frac{l-x}{l}\right) - F_2 \frac{x}{l}\right] dx = 1$$

Or

$$\frac{l}{EI}\left(\frac{F_1}{3} - \frac{F_2}{6}\right) = 1 \qquad\qquad\qquad (a)$$

Also

$$\frac{\partial U}{\partial F_2} = \delta_2 = 0 = \frac{1}{EI} \int_0^l \frac{\partial M}{\partial F_2} \cdot \frac{M}{EI} \, dx$$

Substituting,

$$\frac{\partial M}{\partial F_2} = \frac{-x}{l}$$

Thus

$$\delta_2 = 0 = \frac{1}{EI} \int_0^l \left(\frac{-x}{l}\right) \left[F_1\left(\frac{l-x}{l}\right) - F_2 \frac{x}{l}\right] dx$$

or

$$\frac{l}{EI}\left(-\frac{F_1}{6} + \frac{F_2}{3}\right) = 0 \qquad\qquad\qquad (b)$$

From (a) and (b), we find

$$F_1 = S_{11} = \frac{4EI}{l} \quad \text{and} \quad F_2 = S_{21} = \frac{2EI}{l}$$

Also

$$\frac{\partial U}{\partial F_3} = \delta_3 = 0 \quad \text{or} \quad 0 = \frac{F_3 l}{AE}$$

hence

$$F_3 = S_{31} = 0 \quad \text{thus} \quad F_4 = S_{41} = 0$$

To Find $S_{12}, S_{22}, S_{32}, S_{42}$ Referring to Fig. 13.23f, we can write

$$\frac{\partial U}{\partial F_2} = \delta_2 = 1 \quad \text{and} \quad \frac{\partial U}{\partial F_1} = \delta_1 = 0$$

Following the same procedure of deducing equations (a) and (b), one can readily show that

$$F_2 = S_{22} = \frac{4EI}{l} \qquad F_1 = S_{12} = \frac{2EI}{l} \qquad F_3 = S_{32} = 0 \qquad F_4 = S_{42} = 0$$

To Find S_{13}, S_{23}, S_{33}, S_{43} Referring to Fig. 13.23g we have

$$\frac{\partial U}{\partial F_1} = \delta_1 = 0 \quad \text{or} \quad F_1 = S_{13} = 0$$

and

$$\frac{\partial U}{\partial F_2} = \delta_2 = 0 \quad \text{or} \quad F_2 = S_{23} = 0$$

$$\frac{\partial U}{\partial F_3} = \delta_3 = 1 = \frac{F_3 l}{AE} \quad \text{or} \quad F_3 = S_{33} = \frac{AE}{l}$$

Since $F_4 = F_3$,

$$F_4 = S_{43} = \frac{AE}{l}$$

To Find S_{14}, S_{24}, S_{34}, S_{44} Referring to Fig. 13.23h, we find, as in the preceding case,

$$\frac{\partial U}{\partial F_1} = \delta_1 = 0 \quad \text{or} \quad F_1 = S_{14} = 0$$

and

$$\frac{\partial U}{\partial F_2} = \delta_2 = 0 \quad \text{or} \quad F_2 = S_{24} = 0$$

Also

$$\frac{\partial U}{\partial F_4} = \delta_4 = 1 = \frac{F_4 l}{AE} \quad \text{or} \quad F_4 = S_{44} = \frac{AE}{l}$$

Since $F_3 = F_4$,

$$F_3 = F_4 = S_{34} = \frac{AE}{l} .$$

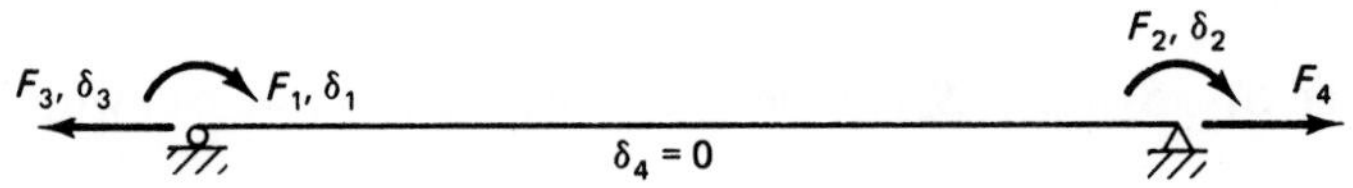

Figure 13.24 Beam made stable by support restraint (two-dimensional case).

Hence the total stiffness matrix becomes

$$
\begin{bmatrix}
S_{11} & S_{12} & S_{13} & S_{14} \\
S_{21} & S_{22} & S_{23} & S_{24} \\
S_{31} & S_{32} & S_{33} & S_{34} \\
S_{41} & S_{42} & S_{43} & S_{44}
\end{bmatrix}
=
\begin{bmatrix}
\dfrac{4EI}{l} & \dfrac{2EI}{l} & 0 & 0 \\[2mm]
\dfrac{2EI}{l} & \dfrac{4EI}{l} & 0 & 0 \\[2mm]
0 & 0 & \dfrac{AE}{l} & \dfrac{AE}{l} \\[2mm]
0 & 0 & \dfrac{AE}{l} & \dfrac{AE}{l}
\end{bmatrix}
\tag{13.58}
$$

The force-displacement relation in terms of the stiffness coefficients can be written as

$$
\begin{Bmatrix}
F_1 \\ F_2 \\ F_3 \\ F_4
\end{Bmatrix}
=
\begin{bmatrix}
S_{11} & S_{12} & S_{13} & S_{14} \\
S_{21} & S_{22} & S_{23} & S_{24} \\
S_{31} & S_{32} & S_{33} & S_{34} \\
S_{41} & S_{42} & S_{43} & S_{44}
\end{bmatrix}
\begin{Bmatrix}
\delta_1 \\ \delta_2 \\ \delta_3 \\ \delta_4
\end{Bmatrix}
\tag{13.59}
$$

The above relationship implies that the member is unstable with rigid-body motion; to render the member stable we can restrain it as shown in Fig. 13.24, where $\delta_4 = 0$. In so doing we arrive at a reduced stiffness matrix $[S]_r$ by eliminating the row and column in the $[S]$ matrix associated with the displacement δ_4. Thus we can now write Eq. (13.59) as

$$
\begin{Bmatrix}
F_1 \\ F_2 \\ F_3
\end{Bmatrix}
=
\begin{bmatrix}
S_{11} & S_{12} & S_{13} \\
S_{21} & S_{22} & S_{23} \\
S_{31} & S_{32} & S_{33}
\end{bmatrix}
\begin{Bmatrix}
\delta_1 \\ \delta_2 \\ \delta_3
\end{Bmatrix}
\tag{13.60}
$$

or

$$
\{F\} = [S]_r\{\delta\} \tag{13.61}
$$

and

$$
F_4 = S_{41}\delta_1 + S_{42}\delta_2 + S_{43}\delta_3
$$

Obviously other forms of $[S]_r$ are possible depending on the manner in which the member is made stable.

The form of matrix equations (13.58) and (13.59) tell us that the end moments F_1 and F_2 are dependent on the end rotations δ_1 and δ_2 while the axial forces F_3 and F_4 are dependent only on the axial displacements δ_3 and δ_4.

Furthermore, premultiplying both sides of Eq. (13.61) by $[S]_r^{-1}$ yields

$$
\{\delta\} = [S]_r^{-1}\{F\} \tag{13.62}
$$

Also, from Eq. (13.55),

$$
\{\delta\} = [f]\{F\}
$$

Comparing Eqs. (13.55) and (13.62), we see that

$$
[f] = [S]_r^{-1} \tag{13.63}
$$

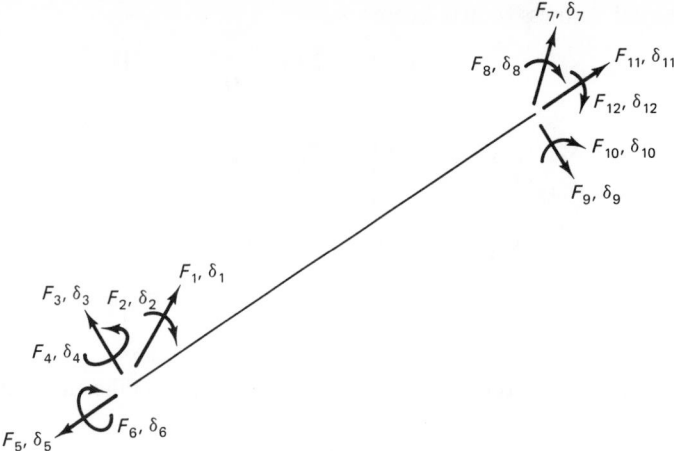

Figure 13.25 A member with six degrees of freedom (with six forces) at each end.

or

$$[S]_r[f] = [I] \tag{13.64}$$

This can be verified by substituting for $[S]_r$ and $[f]$ given by Eqs. (13.58) and (13.54); thus

$$
\begin{bmatrix}
\dfrac{4EI}{l} & \dfrac{2EI}{l} & 0 \\[2mm]
\dfrac{2EI}{l} & \dfrac{4EI}{l} & 0 \\[2mm]
0 & 0 & \dfrac{AE}{l}
\end{bmatrix}
\begin{bmatrix}
\dfrac{l}{3EI} & -\dfrac{l}{6EI} & 0 \\[2mm]
-\dfrac{l}{6EI} & \dfrac{l}{3EI} & 0 \\[2mm]
0 & 0 & \dfrac{l}{AE}
\end{bmatrix}
=
\begin{bmatrix}
1 & 0 & 0 \\
0 & 1 & 0 \\
0 & 0 & 1
\end{bmatrix}
$$

It should be noted that the above derivations for the flexibility and stiffness coefficients are for plane (two-dimensional) members subjected to axial load and moment about one axis, applied at each end with the corresponding end displacements δ. However, for the analysis of framed structures, we sometimes deal with a member that may have up to six forces at each of its two ends with the corresponding six displacements as shown in Fig. 13.25. In such cases the stiffness matrix $[S]$ can contain up to 12×12 coefficients with both $\{f\}$ and $\{\delta\}$ containing up to 12 coefficients.

PROBLEMS

For Part I

13.1 to 13.4   Use the matrix-force method to deduce the matrices $\mathbf{E}_{pd}$, $\mathbf{E}_{px}$, $\mathbf{F}_d$, $\mathbf{F}_x$, $\mathbf{F}_{dx}$, and $\mathbf{P}$ for each of the structures in Figs. P11.1, P11.5, P11.7, P11.10.

13.5 to 13.8   Determine the final forces in the structures in Figs. P11.1, P11.5, P11.7, P11.10.

13.9 Determine the forces and joint displacements in the truss in Fig. P11.4 for each of the following three conditions: (**a**) External loads shown in Fig. P11.4. (**b**) Support at *D* settles down vertically 0.5 in. (**c**) Member *BC* is 0.25 in too long and member *CA* is 0.125 in too short.

13.10 Calculate the final forces and displacements for the structure in Fig. P11.11. Draw the bending-moment diagram.

13.11 Carry out the static and deformation checks for the structure in Fig. P11.5.

13.12 to 13.14 Determine the final forces and displacements for the structures in Figs. P12.1, P12.3, P12.5.

13.15 to 13.17 Apply the matrix-force method to the structures in Figs. P11.13 to P11.15 to determine: (**a**) the matrices $\mathbf{E}_{pd}$, $\mathbf{E}_{px}$, $\mathbf{F}_d$, $\mathbf{F}_x$, $\mathbf{F}_{dx}$, and $\mathbf{P}$; (**b**) the final forces and displacements.

For Part III

Use the method of least work for Problems 13.18 to 13.30.

13.18 Find the reaction at *B* and the moment at *A* for the propped cantilever shown in Fig. P13.18.

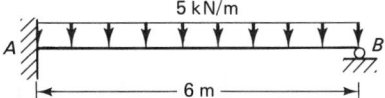

Figure P13.18

13.19 Draw bending-moment and shear-force diagrams for the beam in Fig. P13.19. Treat the reaction at *B* as redundant.

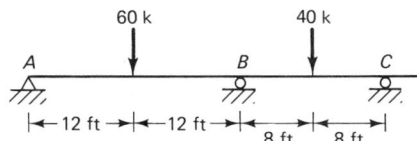

Figure P13.19

13.20 Draw bending-moment and shear-force diagrams for the beam in Fig. P13.20. Treat the reaction at *A* as redundant.

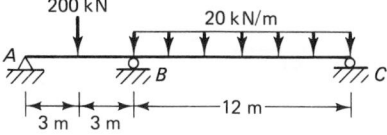

Figure P13.20

13.21 Determine **(a)** the force in the tie, and **(b)** the maximum bending moment in the beam, shown in Fig. P13.21.

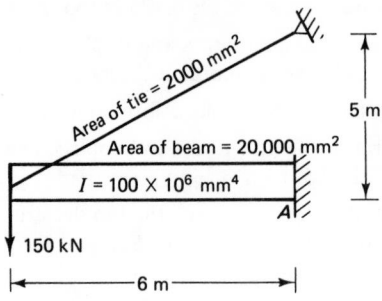

Figure P13.21

13.22 Determine, for the structure in Fig. P13.22, **(a)** the tensile force in the steel rod, **(b)** the compression force in the timber strut, **(c)** the maximum bending stress in the timber beam.

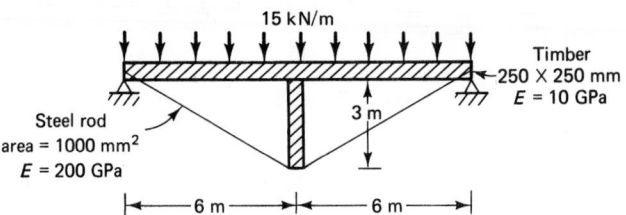

Figure P13.22

13.23 Determine the forces in all the members of the truss shown in Fig. P13.23. Choose OC as the redundant. (AE) is the same for all members.

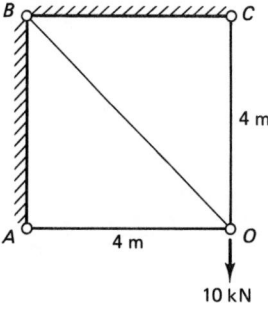

Figure P13.23

13.24 Determine the forces in the tie and all the members of the truss shown in Fig. P13.24.

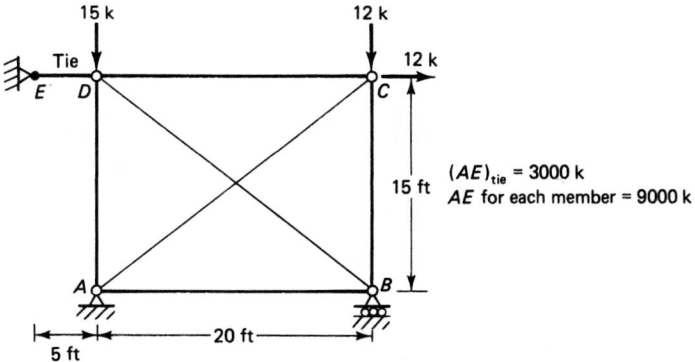

Figure P13.24

13.25 In the structure shown in Fig. P13.25, AB is a tie bar. The members in tension have a cross-sectional area of 1 in^2, and the members in compression have a cross-sectional area of 2 in^2. Find the force in the tie, and hence the resultant forces in the members.

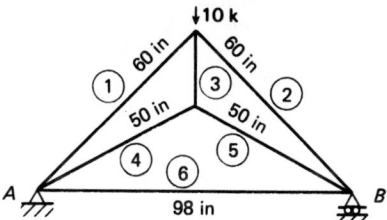

Figure P13.25

13.26 ABC is a semicircular concrete rib of uniform second moment area I (flexural rigidity $EI = 8 \times 10^4$ k · ft^2), Fig. P13.26. The ends A and B are tied together by a steel tube whose axial rigidity $AE = 8000$ k. Calculate the force in the tie AB when the rib carries the loading indicated, and the corresponding horizontal movement of end B.

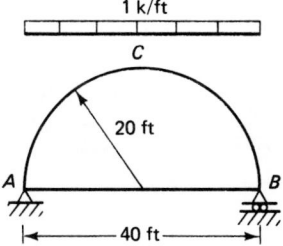

Figure P13.26

13.27 Determine the reaction for the circular arch rib shown in Fig. P13.27.

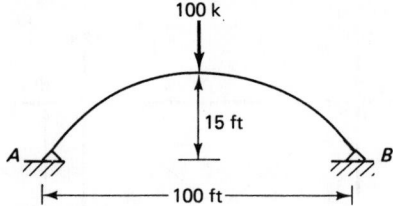

Figure P13.27

13.28 Determine the horizontal thrust for the two-hinged semicircular arch shown in Fig. P13.28.

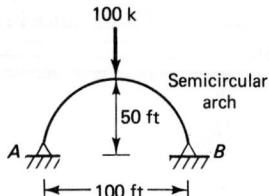

Figure P13.28

13.29 Find the horizontal thrust at A and the moment at B for the triple arches connected as shown in Fig. P13.29. A and G are hinged supports while C and E are roller supports.

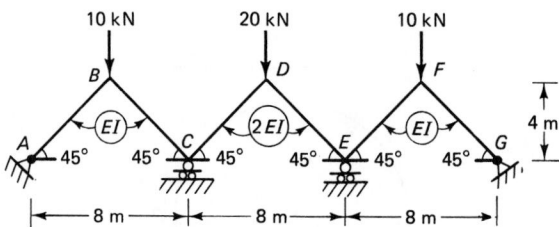

Figure P13.29

13.30 The two-pinned arch in Fig. P13.30 of constant section comprises two quarter-circular arches AB and CD connected by straight beam BC. A uniform loading of $\frac{1}{4}$ k/ft acts over the whole length of the beam. Determine the horizontal thrust at A and D. Sketch the bending-moment diagram.

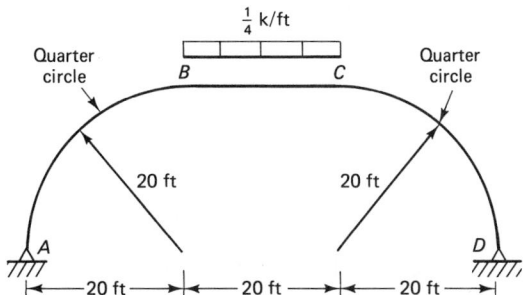

Figure P13.30

13.31 For the redundant truss shown in Fig. P13.31, determine the forces in all members by **(a)** Castigliano's theorem of least work, and **(b)** the unit-load method. Treat AD as a redundant member. Hence calculate by unit-load method the vertical and horizontal deflection of joint B. Cross-sectional area of each member $= 2500$ mm^2, $E = 200$ GPa.

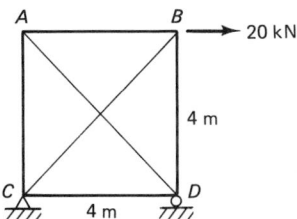

Figure P13.31

13.32 For the redundant truss shown in Fig. P13.32, determine the forces in all members by **(a)** Castigliano's theorem of least work and **(b)** the unit-load method. Treat AD as a redundant member.

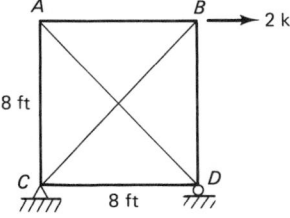

Figure P13.32

13.33 The loaded truss shown in Fig. P13.33 is hinged at A, supported on rollers at D, and subjected at C to a load of 10 k where C is constrained to move along guides. The cross-sectional areas of the members (in circles) are indicated in the figure (a). The stress-strain relation for each member is also shown in figure (b). All members are under stresses that exceed the "elastic limit" of 6 k/in^2. Find the forces in the members. Use Engesser's theorem of compatibility.

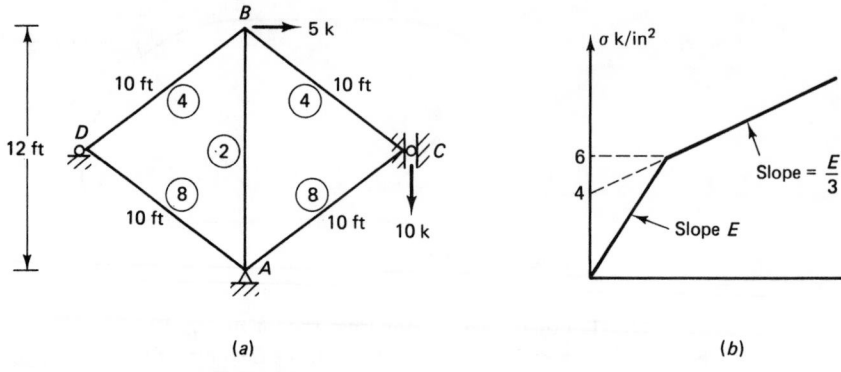

(a)   (b)

Figure P13.33

Chapter **14**

Influence Lines

My expectation is that the sky will fall. My faith is that there's another sky behind it.

Stewart Brand

14.1 INTRODUCTION

Proper design of a structure and any of its components requires the determination of the greatest stress to which they are subjected during the life of the structure. The stresses due to dead loads, which remain stationary, can be readily determined. However, maximum stresses due to live loads or to moving dead loads depend on the *position* of such loads on the structure. For example, when a truck crosses a truss bridge, the forces in the members will vary with the *position* of the truck on the bridge; often the critical position of the truck, producing the maximum stress in any specified member, cannot be readily found by inspection. The variation of force in a particular member of a structure can be determined and shown in a diagram known as an *influence line*. The concept of influence lines was first introduced by Winkler in 1868 and subsequently generalized by Müller-Breslau. An influence line can be defined as the graph of the variation of a particular function, such as a reaction, shear, moment, deflection at a particular section, or an axial force in a particular member, plotted against the position of a moving *unit load* as this load traverses the structure. It is essential to distinguish between, say, the shear-force diagram for a beam and the influence line for shear force at a particular section on the beam. Ordinates of the first diagram will correspond to the shear force at different sections on the beam for a given set of applied loads, whereas the influence line will indicate a diagram showing how the shear force varies at one particular section as a unit load moves over the beam. The same distinction applies between the bending-moment diagram and the influence-line diagram for moment, and so on.

PART I: INFLUENCE LINES FOR STATICALLY DETERMINATE STRUCTURES

14.2 INFLUENCE LINES FOR REACTIONS, SHEAR FORCE, AND BENDING MOMENT IN A SIMPLY SUPPORTED BEAM

Influence Lines for Reactions   Consider the simply supported beam AB in Fig. 14.1a subjected to a moving unit load at a distance x from A. Summing up the moments about B gives

$$(R_A)(L) - (1)(L - x) = 0$$

Hence

$$R_A = \frac{(1)(L - x)}{L} \tag{14.1}$$

White Bird Canyon bridge, White Bird, Ohio. (Photo courtesy of American Institute of Steel Construction.)

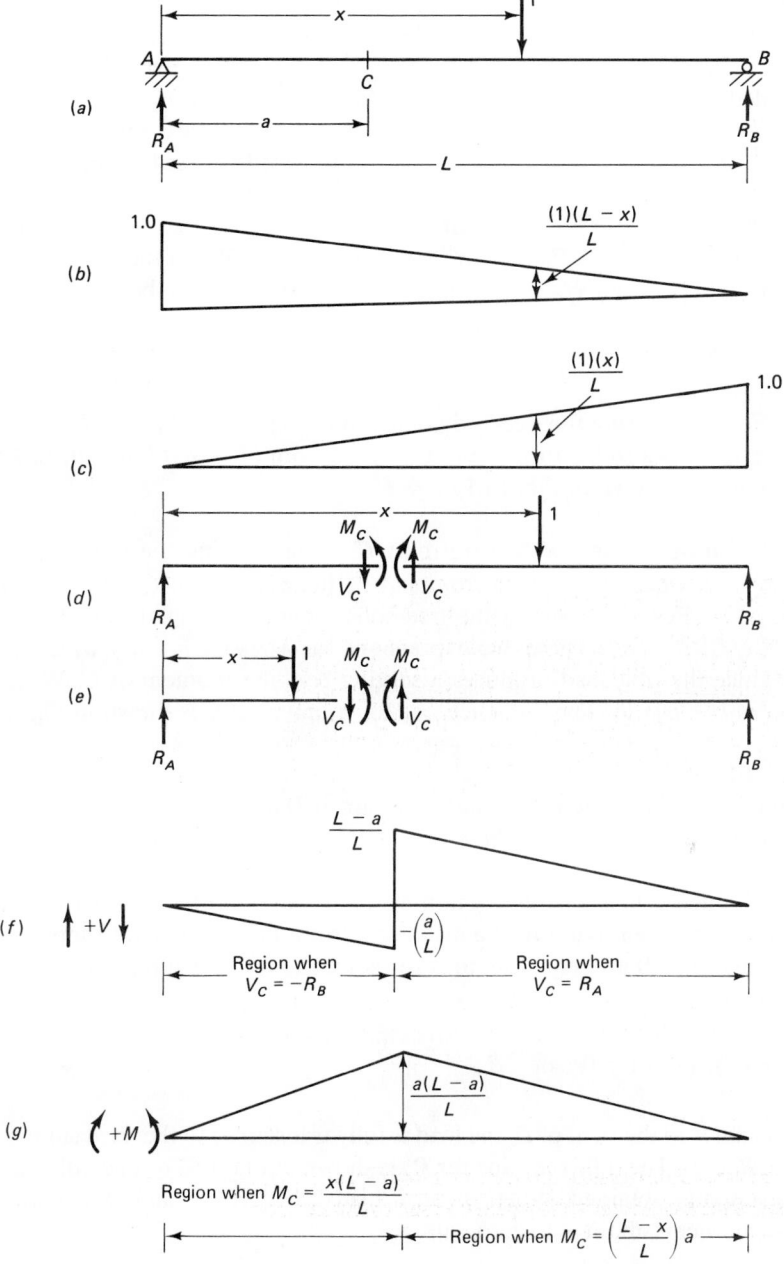

Figure 14.1 Influence lines for a simply supported beam. (b) Influence line for R_A. (c) Influence line for R_B. (d) FBD when $x > a$. (e) FBD when $x < a$.

Equation (14.1), which is a linear function of x, is the influence-line equation for the reaction A as shown in Fig. 14.1b. Thus when $x = 0$, the unit load is at A and $R_A = 1$; when $x = L$, the unit load is at B and $R_A = 0$, as it should be. Similarly, one can deduce the influence-line equation for R_B, shown in Fig. 14.1c. For equilibrium of forces in the vertical direction, the sum of the ordinates of the influence-line diagrams for R_A and R_B at any point equals the unit load (Fig. 14.1b and c).

Influence Line for Shear at C   The structure is cut at C and the shear and moment at C are shown in Fig. 14.1d in a positive sense. When the unit load is to the right of C, that is, when $x > a$, then from statics, using the free-body diagram of Fig. 14.1d,

$$V_C = R_A = \frac{L - x}{L}$$

For $x < a$, using the free-body diagram of Fig. 14.1e, $V_C = -R_B = -x/L$. When $x = a$, there is a sudden change in the value from $(L - a)/L$ to $-a/L$, a total change of unity, as shown in Fig. 14.1f.

Influence Line for Moment at C   Again, using the free-body diagram of Fig. 14.1d, when $x > a$, then from statics, the moment at C, $M_C = R_A \cdot a = [(L - x)/L] \cdot a$. For $x < a$, using the free-body diagram of Fig. 14.1e, $M_C = R_B(L - a) = (x/L)(L - a)$. These two expressions for M_C vary linearly with x and are positive since the unit load produces a sagging bending moment at C. When $x = a$, $M_C = [a(L - a)]/L$. The influence line for moment at C is shown in Fig. 14.1g.

■ **EXAMPLE 14.1**

Construct influence lines for the reaction at A, shears at D and C, and moments at B and C for the beam shown in Fig. 14.2a.

Solution.   The influence line for the reaction at A can be found by considering the effect of a unit load between A and B, at a distance x from A (Fig. 14.2a). Computing the bending moment at D and equating to zero (since the beam is hinged at D, Fig. 14.2b) yields

$$(R_A)(4) - (1)(4 - x) = 0 \quad \text{or} \quad R_A = \frac{4 - x}{4}$$

When the unit load is to the right of D, the load is fully transferred to the fixed support at B and hence $R_A = 0$. The influence line for R_A is shown in Fig. 14.2c. The influence line for V_D is similarly obtained. When the unit load is between A and D, equating the sum of the moments about A to zero will yield

$$(V_D)(4) - (1)(x) = 0 \quad \text{or} \quad V_D = \frac{x}{4} \uparrow$$

Again when the unit load is to the right of D, $V_D = 0$. The influence line for V_D is presented in Fig. 14.2d. The influence line for shear V_C is shown in Fig. 14.2e and is obtained in the same manner as for the beam shown in Fig. 14.1a. The influence line

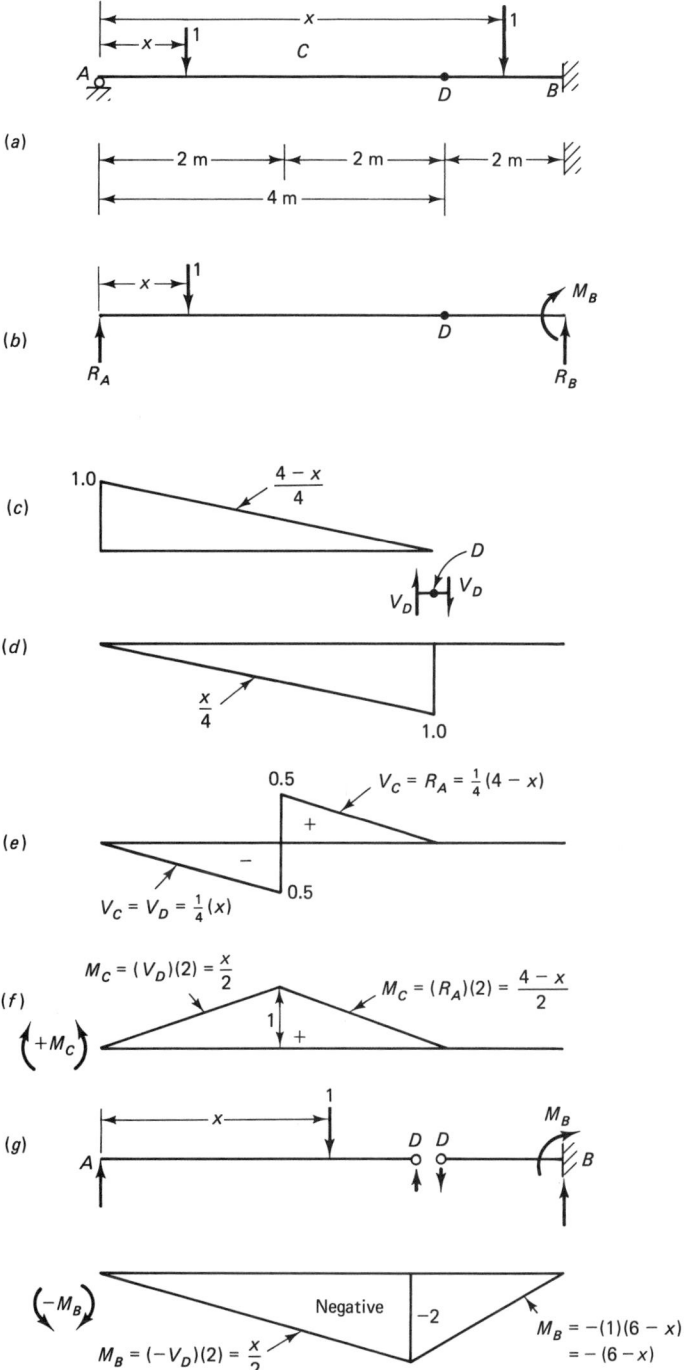

Figure 14.2 Influence lines for structure in Example 14.1. (c) Influence line for R_A. (d) Influence line for shear V_D. (e) Influence line for shear V_c. (f) Influence line for M_c. (g) Free-body diagram of parts AD and DB. (h) Influence line for M_B ($-ve$ providing hogging effect at B).

for moment at C depends on whether the unit load is to the right or left of C. When $x < 2$, $M_C = (V_D)(2) = (x/4)(2) = x/2$. When $x > 2$, $M_C = (R_A)(2) = [(4 - x)/4](2) = (4 - x)/2$. When the unit load is to the right of D, no load is transferred to span AD and therefore $M_C = 0$. The influence line for M_C is given in Fig. 14.2f. For the influence line for the support moment M_B, consider the free-body diagrams of AD and DB as shown in Fig. 14.2g; when the unit load is to the left of D,

$$M_B = -(V_D)(2) = -\left(\frac{x}{4}\right)(2) = -\frac{x}{2}$$

When the unit load is to the right of D, the reaction at the hinge $D = V_D = 0$ and $M_B = -(1)(6 - x) = -(6 - x)$. The influence line for M_B is shown in Fig. 14.2h.   ■

14.3 INFLUENCE LINES FOR GIRDERS OR BRIDGE TRUSSES WITH FLOOR BEAMS

In some cases the loads are applied directly to the beam or girder itself; in other cases the loads are applied to the girder or bridge truss by means of a stringer and floor beam system, as shown in Figs. 14.3 and 14.4. In such a system, the load P is carried by the building or bridge floor to the stringers, which are stiffened by the floor beams; the floor beams, in turn, carry the stringer reactions to the panel points, for example, 2-3, in Fig. 14.3a or b or of the main bridge truss in Fig. 14.4. The girders are supported by columns and the bridge trusses are supported by abutments or piers (Fig. 14.4a). Portions of the main girder (or truss) between the floor beams such as 1-2, 2-3, 3-4, and so on are known as *panels,* and the ends of a panel such as 1, 2, 3, etc., are known as *panel points.* To visualize the interaction of the various structural components, it

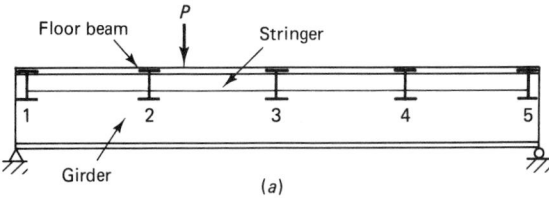

(a)

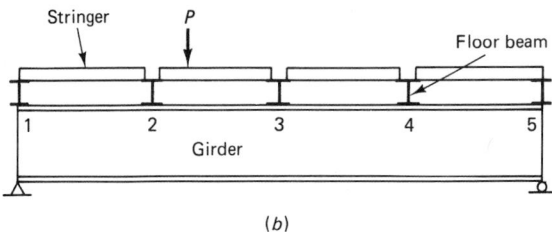

(b)

Figure 14.3 Girder-floor beam-stringer-arrangement.

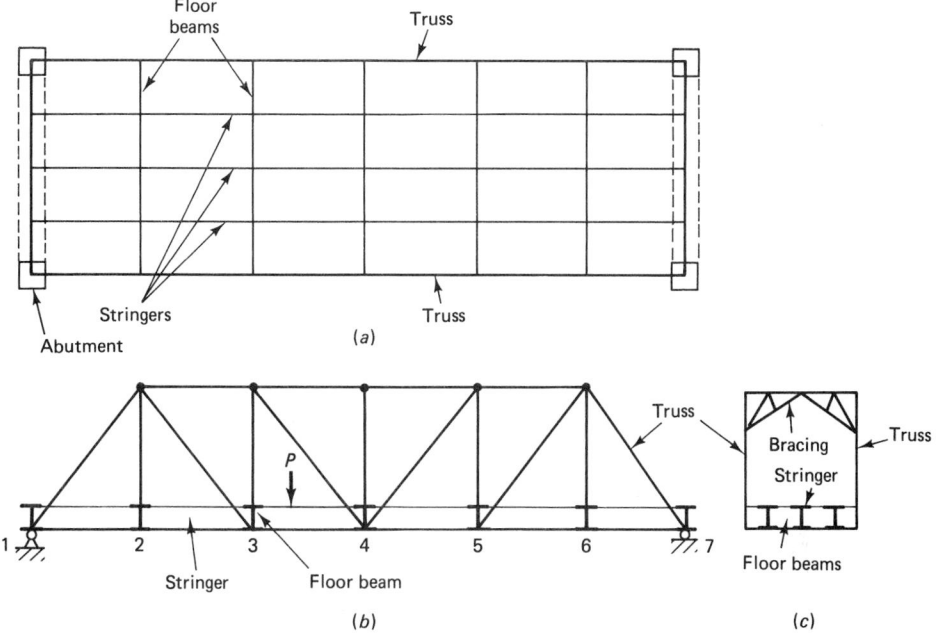

Figure 14.4 Truss-floor beam-stringer arrangement.

is assumed that the floor slab rests on top of the stringers, which rest on top of the floor beams, and these rest on top of the girder, as shown in Fig. 14.3b. To illustrate the load transfer, let us analyze the system shown in Fig. 14.5a. When the loads are transferred from the stringers to the floor beams and finally to the girder, the loads on the girder will be in the form of concentrated loads as shown in Fig. 14.5b. Taking moments about B and equating to zero will yield a gross reaction at A of 43.4 k; similarly the gross reaction at B is 36.6 k. The net reaction at A due to loads except the one applied right at the support (i.e., 7.5 k) is 35.9 k and the net reaction at B is 26.6 k. Only the net reactions are used in the calculation of shear and bending moment, whose diagrams are shown in Fig. 14.5c and d, respectively.

Let us construct the influence lines for the reactions, shears, and moments of the girder with a floor system shown in Fig. 14.6a. Since the end reactions R_A and R_B of the girder AB are unaffected by the presence of the floor system, their influence lines can be deduced as before and are shown in Fig. 14.6b and c. The influence line for shear in panel 2-3 is deduced as follows: When the unit load is in panel 1-2, the shear in panel 2-3 is equal to negative R_B, that is, line Ab in Fig. 14.6d. When the unit load is between panel points 3 and 5, the shear in panel 2-3 is equal to positive R_A, that is, line Bc in Fig. 14.6d. As the unit load traverses from point 2 to point 3, the shear in panel 2-3 has a linear variation, given by line bc shown in Fig. 14.6d, provided the stringers act as simple beams as shown in Fig. 14.6a; this can be proved as follows: When the unit load is at a distance a from panel point 2, the influence ordinate for the shear in panel 2-3 is gg' as shown in Fig. 14.6d. This unit load will be transmitted

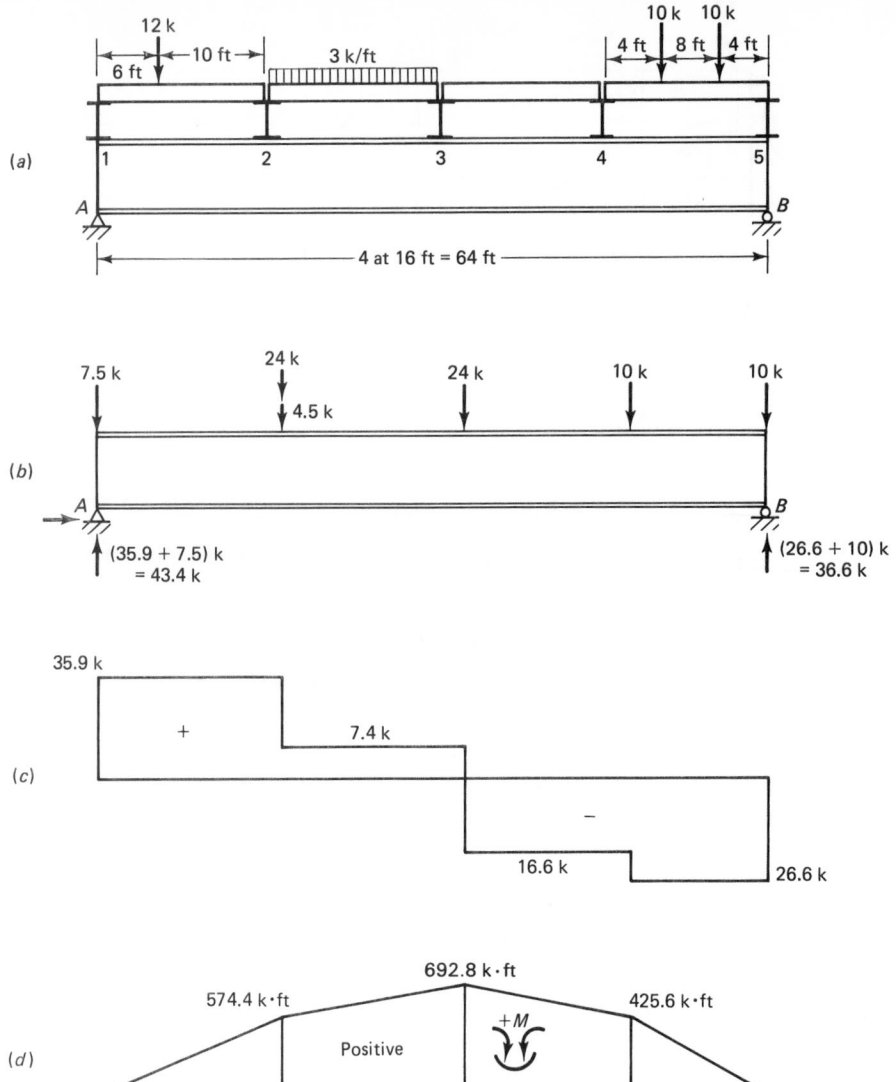

Figure 14.5 Shear-force and bending-moment diagrams for girders with floor beams.

to the girder through the floor beams at panel points 2 and 3 in inverse proportion to its distance from them, that is, $[(16 - a)/16]$ and $(a/16)$, respectively. Thus the actual effect on the girder is deduced by multiplying each of these values by the corresponding value of the influence ordinate and adding; the resulting expression is $[(16 - a)/16](-\frac{1}{4}) + (a/16)(+\frac{1}{2})$, which reduces to $(3a - 16)/64$, which is a linear expression of a, as shown in Fig. 14.6 d. This applies to beams, girders, and trusses provided the floor system consists of simply supported stringers; for example, once the influence ordinates for panel points 2 and 3 are computed, the influence-

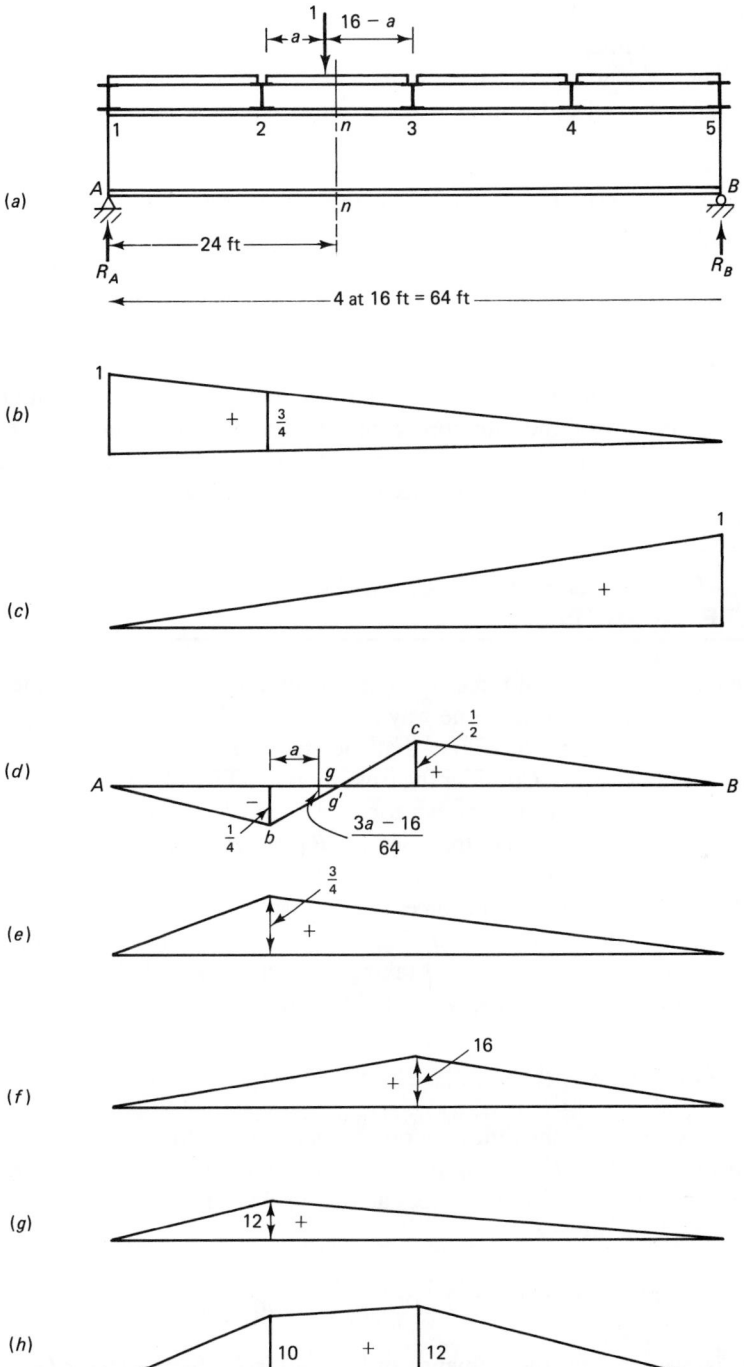

Figure 14.6 Influence-line diagrams for girders with floor beams. Influence lines for (b) R_A; (c) R_B; (d) shear in panel 2-3; (e) shear in panel 1-2; (f) moment at panel point 3; (g) moment at panel point 2; (h) moment at section n-n.

line diagram for shear in panel 2-3 can be completed by connecting these ordinates by a straight line, such as bc in Fig. 14.6d.

The influence line for shear in panel 1-2 shown in Fig. 14.6e can be deduced using the same result. For the influence line for the moment at panel point 3, we note that for the unit load between panel points 3 and 5, $M_3 = 32R_A$, which is a linear variation, with $M_3 = 16$ when the unit load is at panel point 3, that is, when $R_A = \frac{1}{2}$. The influence line for M_3 is given in Fig. 14.6f. The influence line for moment at panel point 2 shown in Fig. 14.6g can be similarly derived. To construct the influence line for the moment M_n at section n-n, we note that when the unit load is between panel points 1 and 2, $M_n = (R_B)(64 - 24) = 40R_B$, which is a straight line with zero ordinate at panel point 1 and a maximum ordinate of 10 at panel point 2. Similarly when the unit load is between panel points 3 and 5, $M_n = 24R_A$. This is a linear relation with zero ordinate at panel point 5 and maximum ordinate of 12 when unit load is at panel point 3. Following the same procedure used for the derivation of the influence line for shear in panel 2-3, we can complete the influence line for M_n by connecting the influence ordinates at panel points 2 and 3 by a straight line, as shown in Fig. 14.6h.

14.4 INFLUENCE LINES FOR STATICALLY DETERMINATE TRUSSES

Influence lines for axial forces in members of a determinate truss (Fig. 14.4b) can be constructed in much the same way as that for a girder with a floor system shown in Fig. 14.3b, assuming of course that the stringers act as simple beams between the adjacent floor beams. Consider the truss shown in Fig. 14.7a; it is required to construct the influence line for the member forces F_{DE} and F_{DF}. For equilibrium of the truss as a whole, the influence lines for reactions R_A and R_B will be the same as if A and B are connected by a simply supported beam; these influence lines are shown in Fig. 14.7b and c. The influence lines for forces in members DE and DF are derived by considering the free-body diagram shown in Fig. 14.7d. For a unit load to the right of F, cutting the truss by the x-x section and taking moments about F of all the forces acting on the left portion of the truss, we find, for $\sum M_{\text{about }F} = 0$,

$$(F_{DE})(l) + (R_A)(2a) = 0 \quad \text{or} \quad F_{DE} = R_A\left(-\frac{2a}{l}\right)$$

that is, we magnify the influence ordinates for R_A by the factor $(-2a/l)$. For the unit load to the left of F, cutting the truss by the x-x section in Fig. 14.7d and taking moments about F of all the forces acting on the right portion of the truss will yield, for $\sum M_{\text{about }F} = 0$,

$$(F_{DE})(l) + (R_B)(2a) = 0 \quad \text{or} \quad F_{DE} = R_B\left(-\frac{2a}{l}\right)$$

that is, we magnify the influence ordinates for R_B by the factor $(-2a/l)$. Thus the influence line for F_{DE} is composed of two straight portions intersecting at F, as shown in Fig. 14.7e.

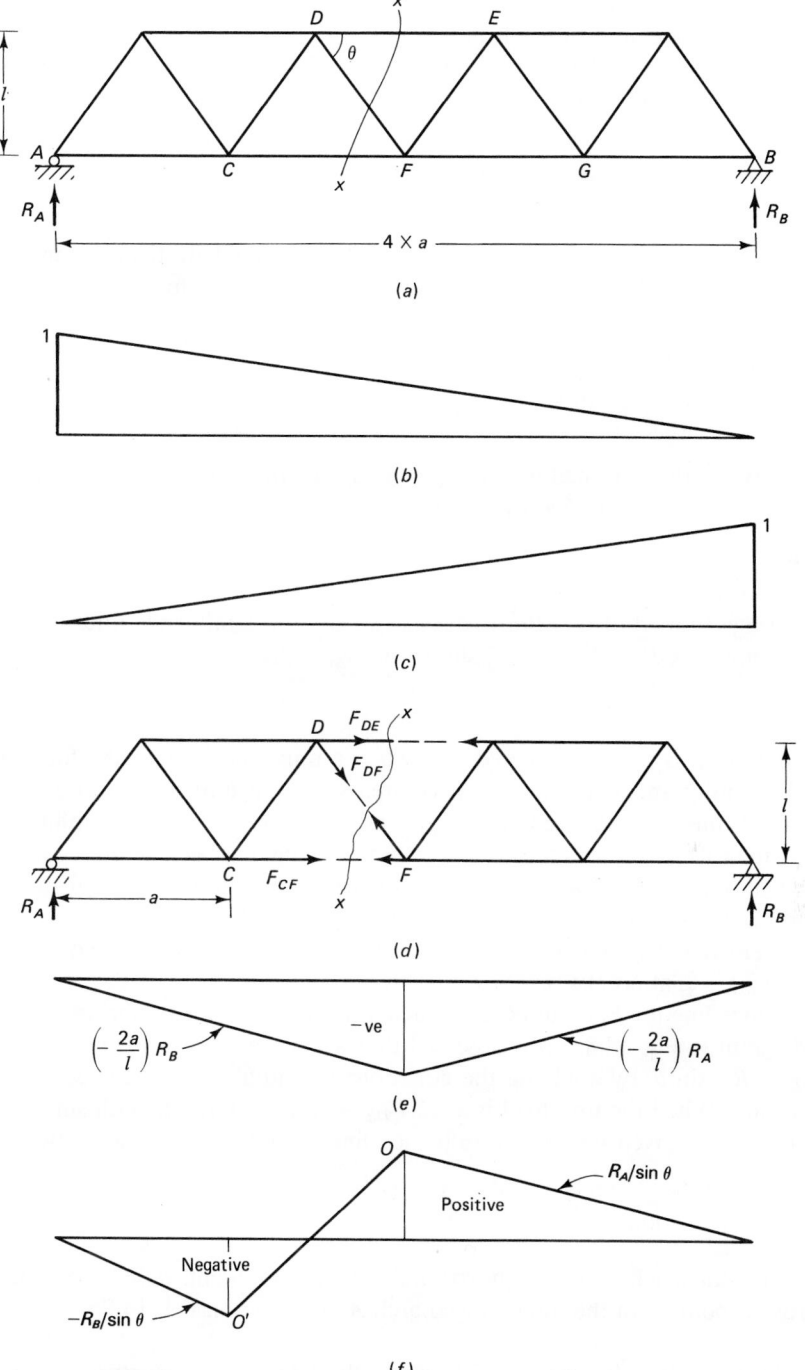

Figure 14.7 Influence-line diagrams for a statically determinate truss. (b) For R_A. (c) For R_B. (d) FBD. (e) For F_{DE}. (f) For F_{DF}.

The influence line for force in the web member DF is determined by considering the force equilibrium of the cut portions of the truss in Fig. 14.7d. Thus, when the unit load is to the left of F, applying the force-equilibrium equation $\Sigma F_y = 0$ to the right portion yields

$$R_B + F_{DF} \sin \theta = 0 \quad \text{or} \quad F_{DF} = \frac{-R_B}{\sin \theta}$$

where θ is the angle of inclination of member DF with the horizontal. When the unit load is to the right of C, we have, from $\Sigma F_y = 0$, for the forces acting on the left portion,

$$R_A - F_{DF} \sin \theta = 0 \quad \text{or} \quad F_{DF} = \frac{R_A}{\sin \theta}$$

When the unit load is in the panel CF, the influence line for F_{DF} is obtained, as was shown in Fig. 14.6d, by joining points O and O' in Fig. 14.7f.

■ EXAMPLE 14.2

Consider a bridge formed of two Pratt trusses. (a) Draw the influence lines for the forces in members CE, CD, and DE, shown in Fig. 14.8a.

Solution

By inspection, it is readily observed that when a unit load is at C, the force in member CE is unity and is tensile; furthermore, when the unit load is at D and beyond, $F_{CE} = 0$; thus the influence line for F_{CE} is as shown in Fig. 14.8b. The influence line for F_{CD} can be deduced from the portion of the truss that has been cut by a section x-x as shown in Fig. 14.8a. Taking moments of the forces acting to the left of x-x about E and equating to zero will yield $F_{CD} = R_A \cdot \left(\frac{30}{40}\right)$; as before $R_A = \frac{5}{6}, \frac{2}{3}, \frac{1}{2}, \frac{1}{3}$, and $\frac{1}{6}$ when the unit load is at C, D, F, G, and H, respectively; hence the maximum ordinate for $F_{CD} = \left(\frac{5}{6}\right)\left(\frac{30}{40}\right) = \frac{5}{8}$, at point C as shown in Fig. 14.8c.

The influence line for F_{DE} can be constructed from the force equilibrium of the free-body diagram in Fig. 14.8c; it is observed that when the unit load is between D and B, $F_{DE} = R_A/\sin \theta$, by applying the equations of equilibrium of forces in the vertical direction. When the unit load is at C, $F_{DE} = (R_A - 1)/\sin \theta$. With $\sin \theta = \frac{4}{5}$ and the values of R_A given above, the influence line for DE is drawn as shown in Fig. 14.8d. ■

■ EXAMPLE 14.3

Determine the influence lines for the horizontal reaction, moment, radial shear, and normal thrust at point O in the three-hinged arch ACB shown in Fig. 14.9a.

Solution. Since the three-hinged arch is a statically determinate structure, we can apply the same principles discussed above to construct the required influence lines.

Let the unit load act at a horizontal distance aL from A; then taking moments

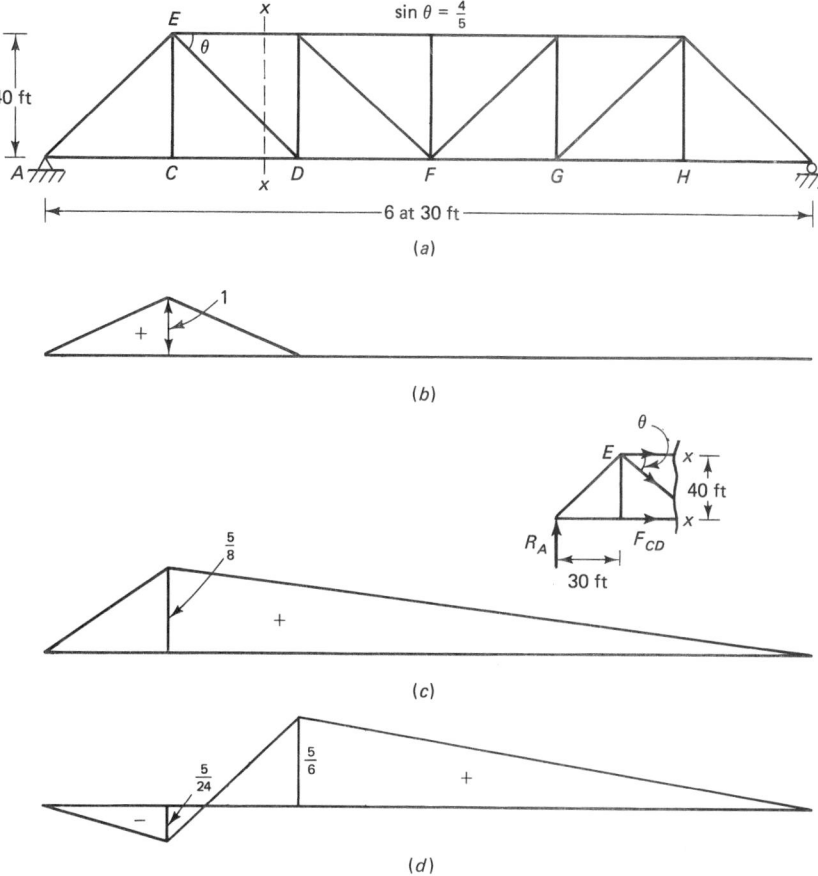

Figure 14.8 Influence lines for structure in Example 14.2. (*b*) For F_{CE}. (*c*) For F_{CD}. (*d*) For F_{DE}.

about B and equating to zero yields $R_A = 1 - a$. Taking moments about A and equating to zero, we find $R_B = a$. To determine H, we compute the bending moment about the hinge at C and equate to zero. Thus, when $0 \le aL \le L/2$ (taking moments about C of all forces acting to the right of C and equating to zero),

$$R_B \cdot \left(\frac{L}{2}\right) - H(h) = 0 \quad \text{or} \quad H = R_B \cdot \frac{L}{2h} = \frac{aL}{2h}$$

When $L/2 \le aL \le L$ (taking moments about C of all forces acting to the left of C and equating to zero),

$$(R_A)\left(\frac{L}{2}\right) - H(h) = 0 \quad \text{or} \quad H = R_A \cdot \frac{L}{2h} = \frac{(1 - a)L}{2h}$$

The above expressions for H are the influence-line equations for H and are shown plotted in Fig. 14.9b; it is observed that H is directly proportional to the distance of

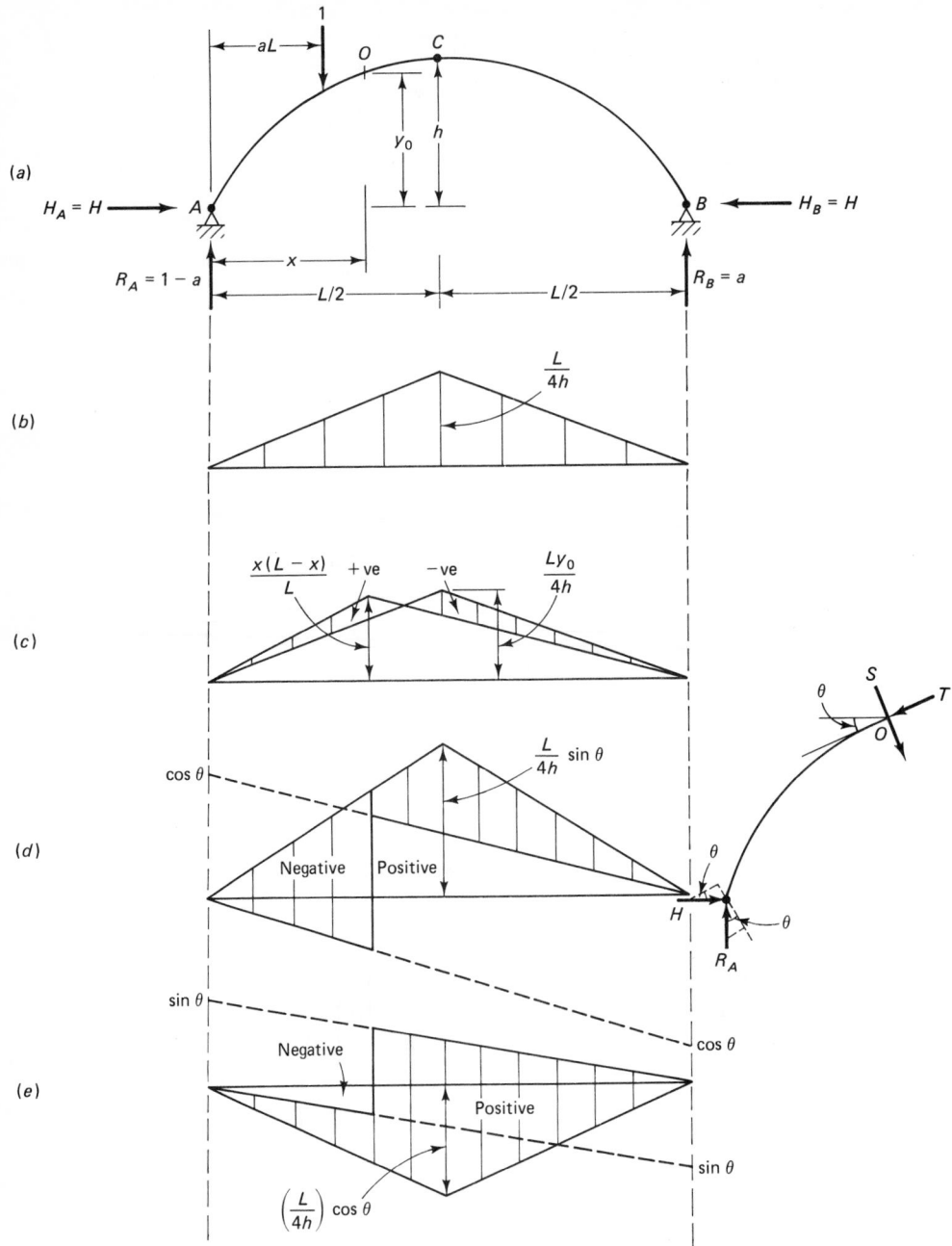

Figure 14.9 Influence lines for structure in Example 14.3. (b) For H. (c) For moment at O. (d) For radial shear S at O. (e) For normal thrust T at O.

the unit load from support A, aL; H becomes a maximum when $aL = L/2$ or $H_{max} = L/4h$. The influence line for moment at point O is obtained by superimposing the influence line for moment of a simply supported curved beam M_S and the influence line for moment M_H due to H. Now

$$M_S = R_A \cdot x - (1)(x - aL) \qquad \text{when } 0 \le aL \le x$$

$$= a(L - x)$$

and

$$M_S = R_A \cdot x \qquad \text{when } x \le aL \le L$$

$$= (1 - a)x$$

When the unit load is at O, $M_S = (x/L)(L - x)$, as shown in Fig. 14.9c.

$$M_H = -H \cdot y_0 = -\frac{aL}{2h} \cdot y_0 \qquad \text{when } 0 \le aL \le \frac{L}{2}$$

$$M_H = -H \cdot y_0 = -\frac{(1 - a)L}{2h} \cdot y_0 \qquad \text{when } \frac{L}{2} \le aL \le L$$

When the unit load is at C,

$$M_H = -\frac{L}{4h} y_0$$

Since $M_O = M_S + M_H$, the shaded area in Fig. 14.9c is the influence-line diagram for moment at O.

To construct the influence line for the radial shear S and axial thrust T in the arch at point O, we let the inclination of the arch profile to the horizontal be θ as shown by the free-body diagram in Fig. 14.9d. Resolving forces in the directions of S and T and equating to zero for equilibrium, we find for $aL > x$,

$$S = R_A \cos \theta - H \sin \theta$$

and

$$T = R_A \sin \theta + H \cos \theta$$

For $aL < x$,

$$S = (R_A - 1)\cos \theta - H \sin \theta$$

$$T = (R_A - 1)\sin \theta + H \cos \theta$$

Knowing the influence lines for R_A (reaction of a simply supported curved beam) and H (Fig. 14.9b), the influence-line graphs for S and T can be readily constructed from the above equations as shown in Fig. 14.9d and e, respectively; the positive sign for T implies compression as assumed.

Quite often the loads are not transmitted directly to the arch rib but rather through the floor beams; in this case the influence lines are modified in a manner similar to that used in girders and trusses with floor beams as discussed previously. ∎

14.5 INFLUENCE LINES BY MÜLLER-BRESLAU'S PRINCIPLE

Influence lines for statically determinate structures can be deduced quite readily by the use of the Müller-Breslau principle, which applies to statically determinate structures as well as to statically indeterminate structures (provided that the principle of superposition is valid). The principle states that if an internal-force component or a reaction component is considered to act through some small distance and thereby to deflect or displace a structure, the curve of the deflected or displaced structure will be, to some scale, the influence line for the internal force or reaction component. In case the internal-force component is a moment, the corresponding small displacement through which this moment acts will be a rotation.

To demonstrate this principle, let us consider the simply supported beam AB with a unit load at D, shown in Fig. 14.10a. To obtain the influence line for the reaction at B, we remove the support at B and make a small positive virtual displacement Δ_B of its point of application as shown by the dashed line in Fig. 14.10b. Applying the virtual-work equation we obtain

$$(R_B)(\Delta_B) = (1)(\Delta_D) \quad \text{or} \quad R_B = \frac{\Delta_D}{\Delta_B}$$

Since Δ_B is arbitrary, we can put $\Delta_B = 1$; and hence

$$R_B = \Delta_D \tag{14.2}$$

Since Δ_D is the deflection of beam AB where the unit load is applied and is also the value of the reaction at B, we can conclude that the deflected shape AG is by definition the influence line for R_B when $\Delta_B = 1$. Note that the imposed displacement is made in the positive direction of the reaction R_B.

To construct the influence line for shear at section C, we cut the section at C as shown in Fig. 14.10c; imagine a slide device (or pinned-bar connection) (Fig. 14.10d) is inserted at C, permitting a relative transverse displacement between the two beams ends at the cut, but which at the same time will always require these two ends to be parallel; that is, shearing resistance is removed but the capability to resist moment and transfer moment across the cut still exists. With the unit load at D, applying a pair of shear forces V_C produces a displacement as shown in Fig. 14.10e. Assuming that the displacements impressed are virtual, then by the virtual-work equation, we have

$$V_C(bj) + V_C(ai) + M_C \cdot j + M_C(-i) - (1)(\Delta_D) = 0$$

Since $i = j$ (parallel lines) and $bj + ai = \Delta_C$, the above reduces to

$$V_C \Delta_C = (1)(\Delta_D) \quad \text{or} \quad V_C = \frac{\Delta_D}{\Delta_C}$$

Taking $\Delta_C = 1$ yields

$$V_C = \Delta_D$$

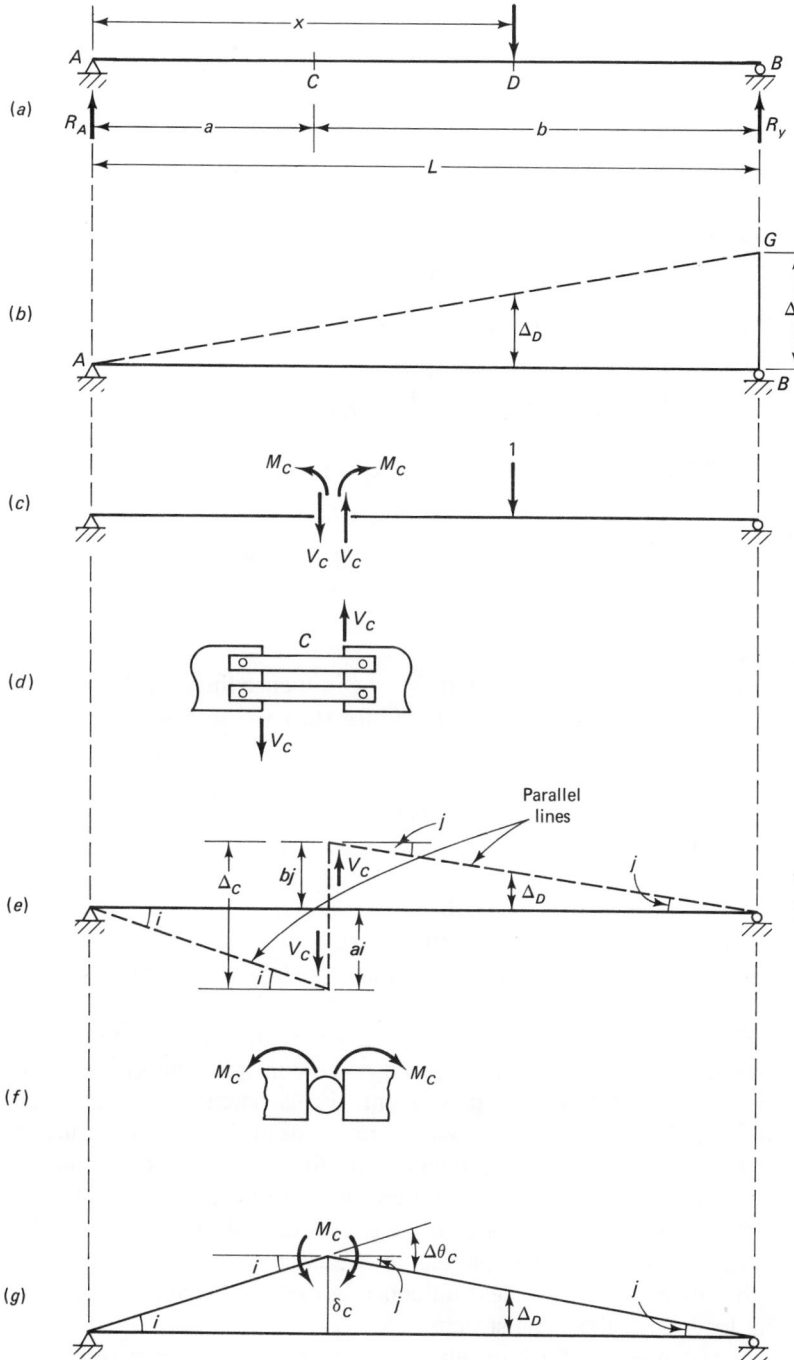

Figure 14.10 Influence lines by the Müller-Breslau principle.

which proves that the diagram in Fig. 14.10e is the influence line for the shear at C when $\Delta_C = 1$.

For the influence line for moment at C, M_C, we insert the pinned connection shown in Fig. 14.10f. With unit load at D, we then apply a pair of couples M_C, one to the right- and the other to the left-hand segments of the hinge, causing the beam to rotate an angle $\Delta\theta_C$ at C, as shown in Fig. 14.10g. We note that the rotation at C is made in the same direction as the positive moment at C. From the virtual-work equation, we can write

$$M_C \cdot (j) + M_C(i) + V_C(\delta_C) + V_C(-\delta_C) - (1)(\Delta_D) = 0$$

Putting $i + j = \Delta\theta_C$, the above reduces to

$$M_C \cdot \Delta\theta_C = (1)(\Delta_D) \quad \text{or} \quad M_C = \frac{\Delta_D}{\Delta\theta_C}$$

Putting $\Delta\theta_C = 1$ yields

$$M_C = \Delta_D$$

that is, the deflected shape shown in Fig. 14.10g is the influence line for moment at C when $\Delta\theta_C = 1$.

■ EXAMPLE 14.4

Applying the Müller-Breslau principle, determine the influence lines for the reactions at A and B, for moments at sections G, B, H, and for shears at sections H and J for the structure shown in Fig. 14.11a.

Solution. Applying the Müller-Breslau principle by removing the support reaction at A and imposing a unit displacement in the line of action of the reaction R_A will yield the influence line for R_A shown in Fig. 14.11b. It should be noted that portion FCD will remain unaffected by displacing A. In the same manner the influence line for R_B is constructed as shown in Fig. 14.11c. The influence line for M_G is obtained by cutting the beam at G and rotating the left portion GA a unit angle with respect to the right portion GBE, with restraint from support B and rotation of EF about hinge F; the influence line for M_G is given in Fig. 14.11d. Similarly the influence lines for M_B and M_H can be obtained as shown in Fig. 14.11e and f. To construct the influence line for V_H, we cut the beam at H and impose a unit displacement between the two portions HE and HF, without introducing relative rotations of these two portions; the deflected shape shown in Fig. 14.11g is the influence line for V_H. To derive the influence line for V_J, we cut the beam at J and impose an upward unit displacement of the right portion of the beam at J with respect to the left portion of the beam at J while maintaining the displaced portion $J'E'$ parallel to BJ, as shown in Fig. 14.11h.

It is left to the student to check these influence lines by the elementary method of moving a unit load along the given structure.

Inspection of influence lines for statically determinate structures reveals that they are invariably composed of straight lines; the reason for this is that when constructing such influence lines an internal or external restraint is first removed, rendering the

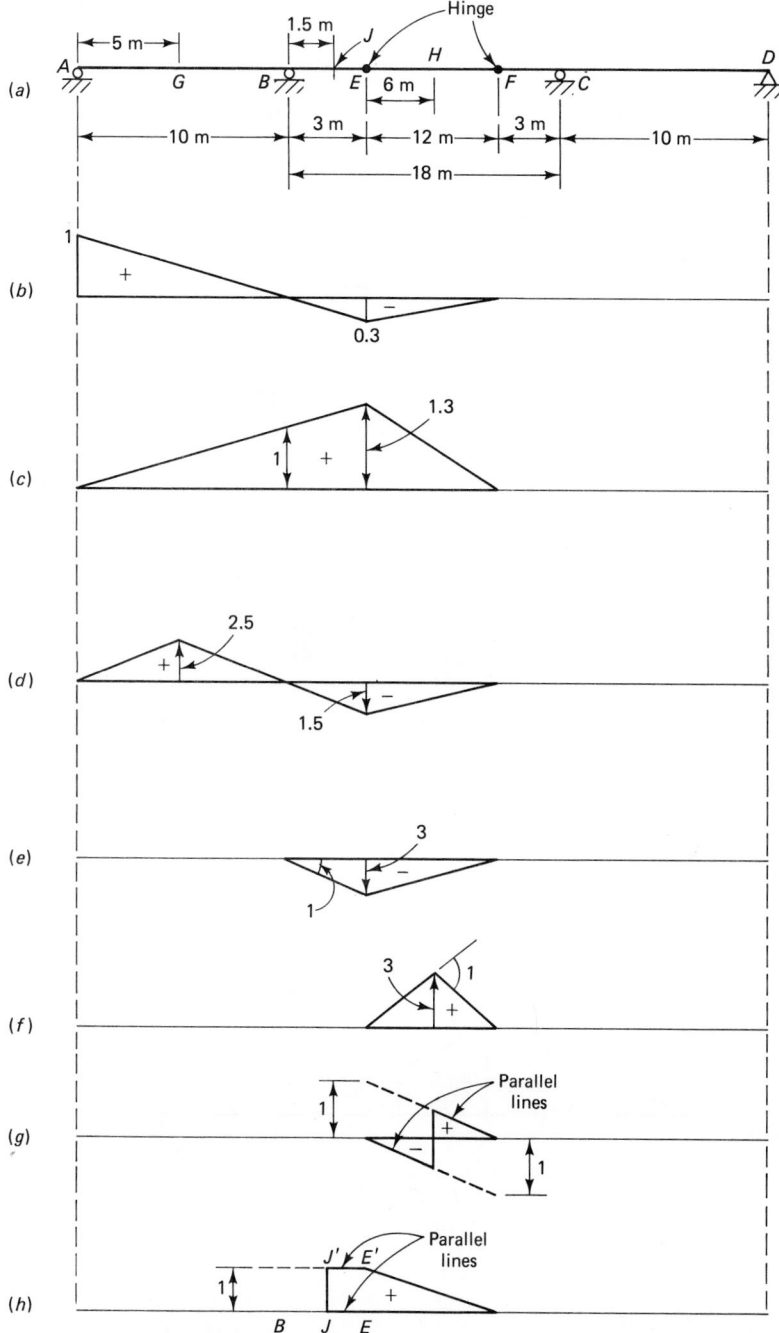

Figure 14.11 Influence lines for structure in Example 14.4. (*b*) For R_A. (*c*) For R_B. (*d*) For moment at *G*. (*e*) For moment at *B* (M_B). (*f*) For moment at *H* (M_H). (*g*) For shear at *H* (V_H). (*h*) For shear at *J* (V_J).

structure a mechanism. Thus, when a displacement is impressed at a section, the structure behaves as a rigid body rotating about support or internal hinges as demonstrated in Example 14.4 (Fig. 14.11). In contrast, influence lines for indeterminate structures consist of curved lines; this is explained further in Section 14.10. ∎

14.6 USES OF INFLUENCE LINES

Once an influence line for a function, such as reaction, shear, moment, or deflection, is determined, it can be used to find the value of the function when any loading system is applied to the structure. This is possible provided the structure is linearly elastic and the principle of superposition is valid.

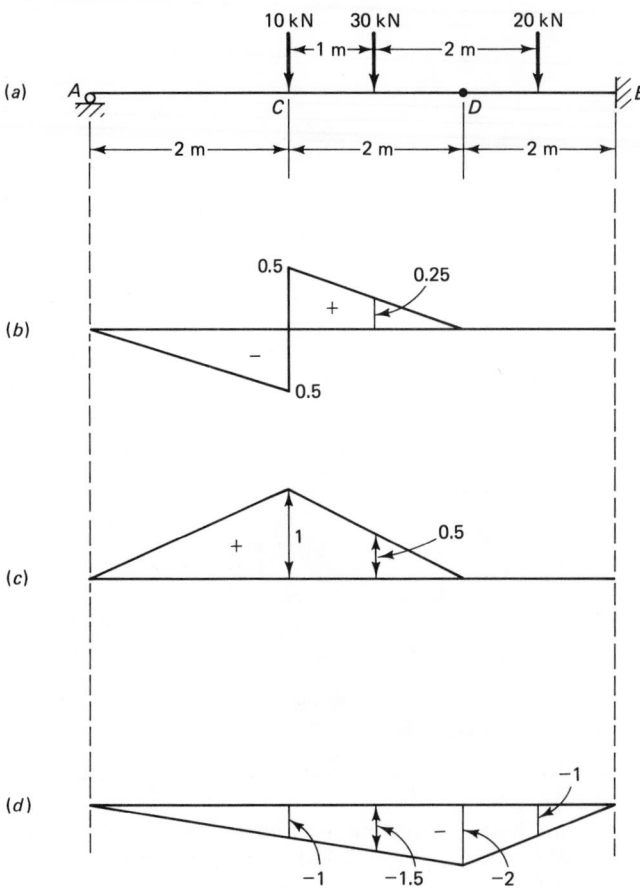

Figure 14.12 Uses of influence lines. (*b*) For V_C. (*c*) For M_C. (*d*) For M_B.

■ **EXAMPLE 14.5**

Determine the maximum shear and moment at section C for the beam in Fig. 14.2a for a concentrated load of 20 kN applied at 1 m from A.

Solution. The required shear and moment are determined by multiplying the respective influence-line ordinates under the load by the value of the load.

Referring to Fig. 14.2e, the ordinate of the influence line for V_C at a distance of 1 m from $A = -[(0.5)/2](1) = -0.25$; this is the value for a unit load; therefore, for a load of 20 kN,

Shear at $C = V_C = (-0.25)(20) = -5$ kN

For moment, referring to Fig. 14.2f, the ordinate of the influence-line diagram for M_C at a distance of 1 m from $A = [+(1)/2](1) = +0.5$. Thus the value of M_C for a load of 20 kN $= (+0.5)(20) = +10.0$ kN · m.

The same procedure is followed if the beam is subjected to a series of loads, given in Fig. 14.12a. Let us calculate the shear and moment at C as well as the moment at B due to the given loading. The influence lines for shear and moment at C and moment at B are shown in Fig. 14.12b, c, and d, respectively. Thus

$$V_C = (10)(0.5) + (30)(0.25) + (20)(0) = +12.5 \text{ kN}$$

$$M_C = (10)(1) + (30)(0.5) + (20)(0) = +25 \text{ kN} \cdot \text{m} \qquad \text{(sagging moment)}$$

$$M_B = (10)(-1) + (30)(-1.5) + (20)(-1) = -75 \text{ kN} \cdot \text{m}$$

$$\text{(hogging moment)} \qquad ■$$

14.7 DISTRIBUTED LOADS

Let us now consider distributed loads. Consider the beam AB loaded by the distributed load of varying intensity as shown in Fig. 14.13a. Let us assume that the influence-line ordinates for M_C are represented by the function $h(x)$ in Fig. 14.13b. The load

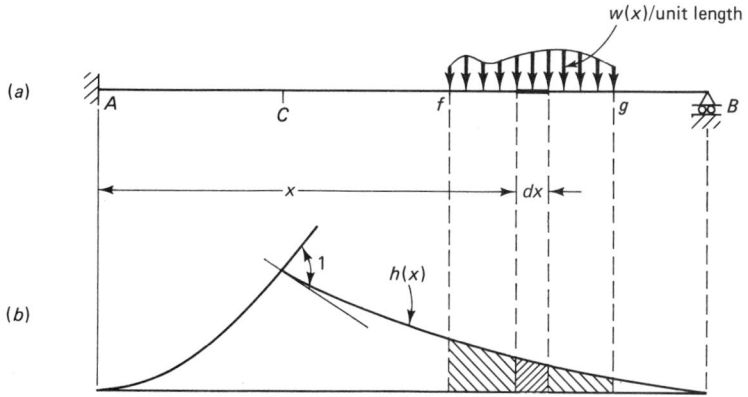

Figure 14.13 Application of influence-line diagram for the analysis of a structure subjected to distributed loads shorter than the span. (b) Influence line for M_C.

on a small element of length dx is $w(x)dx$ and can be taken as a concentrated load. The effect of this infinitesimal load on M_C is

$$dM_C = w(x) \cdot dx \cdot h(x) \tag{14.3}$$

The value of M_C for the total load acting between f and g will be

$$M_C = \int_f^g w(x)h(x)dx \tag{14.4}$$

For a uniformly distributed load $w(x) = w$, Eq. (14.4) becomes

$$M_C = w \int_f^g h(x)dx \tag{14.5}$$

which means that M_C is equal to the product of the intensity of uniform load and the area of the influence-line diagram below the distributed load, shown shaded in Fig. 14.13b.

For a uniformly distributed load w, longer than the span of a simply supported beam, shown in Fig. 14.14a, and moving from left to right, the maximum negative shear at C, V_C, will occur when the front of the load has just reached C; it is equal to

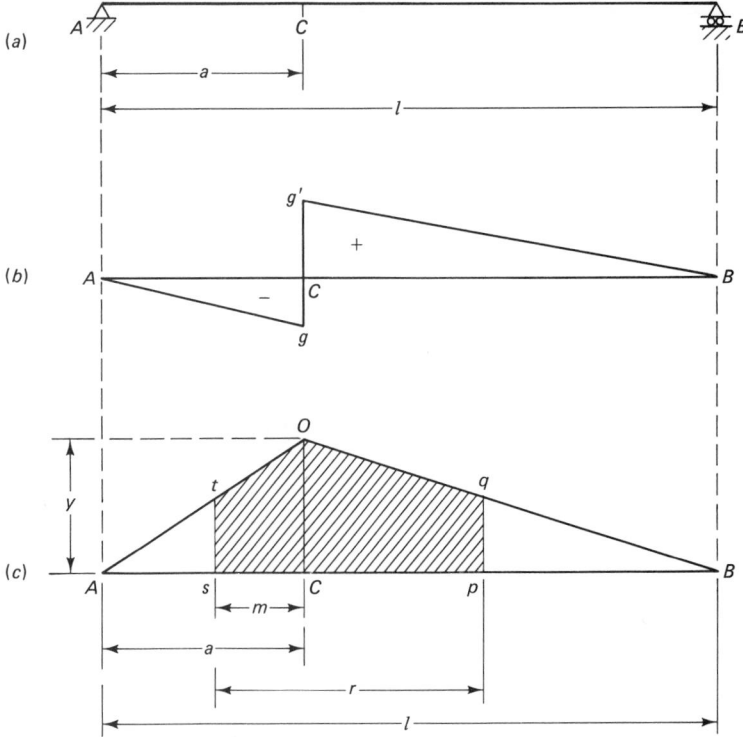

Figure 14.14 Use of influence lines to determine maximum shear and moment due to moving distributed load shorter than the span. (b) Influence line for V_C. (c) Influence line for M_C.

w times the area AgC. Similarly the maximum positive V_C will occur when the rear of the load has reached C, and it is equal to w times the area of $Cg'B$, given in Fig. 14.14b. For a uniformly distributed load that is shorter than the span, again two cases have to be considered, the front and rear of the load at C. The maximum moment M_C due to uniformly distributed load longer than the span is equal to w times the area AOB shown in Fig. 14.14c. If the length of load is r and it is shorter than the span l, maximum M_C will be found when the area $stqp$, shown in Fig. 14.14c, is maximum, that is, when the rear of the load is at a distance m from the point C. To derive this, we make the sum of the areas Ats and Bpq a minimum. Thus

$$\phi = \text{area } Ats + \text{area } Bpq$$

$$= \frac{y(a-m)^2}{2a} + \frac{y(l-a-r+m)^2}{2(l-a)}$$

Then $d\phi/dm = 0$ will yield, after simplification,

$$\frac{m}{r} = \frac{a}{l} \tag{14.6}$$

which means that the point C under consideration divides both the length of the load and the span in the same ratio; or it can be shown that ordinate st = ordinate pq.

14.8 ROLLING OR MOVING LOADS

So far we have only dealt with influence-line diagrams for single concentrated loads and uniformly distributed loads. Quite often a structure is subjected to a series of rolling or moving concentrated loads, which are of various magnitudes and spacings, such as a moving train crossing a bridge. For such complicated loads, one can only deduce the critical position of the loads by trial-and-error method using criteria based on the concept of influence lines. For each category of influence lines, there will be a corresponding criterion for a maximum. We derive such criteria for the common types of influence-line diagrams, those shown in Fig. 14.15.

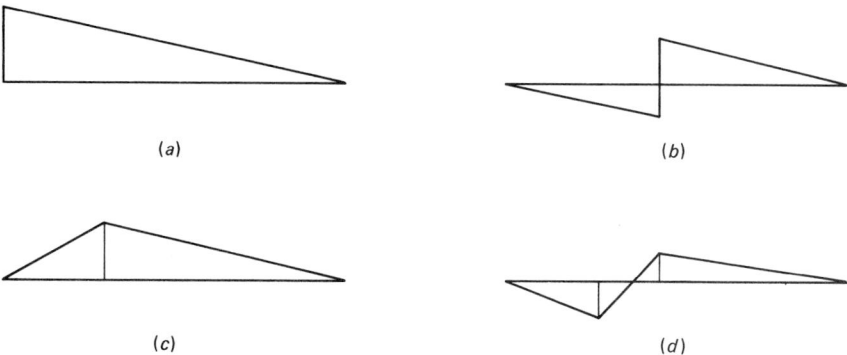

Figure 14.15 Common types of influence-line diagrams.

Maximum Value of a Function with Influence-Line Diagram Shown in Fig. 14.15a   This type of influence line corresponds to that for the end reaction of a simply supported beam, girder, or truss. Let us consider the simply supported beam AB subjected to the passage of the train loads $P_1, P_2, P_3, \ldots$, with resultant R acting at a distance x from B, as shown in Fig. 14.16a. The maximum value of the reaction R_A is found by constructing the influence-line diagram for R_A as shown in Fig. 14.16b; the influence-line ordinate corresponding to the resultant R is x/L. Thus

$$R_A = R\left(\frac{x}{L}\right)$$

The maximum value of R_A is obtained when the term x/L is the greatest and when there is no change in the value of R (i.e., there is no wheel entering or leaving the span during the train movement). In case there is a change in R, the greatest value of $R(x/L)$ will yield the maximum reaction in question. Such a maximum will always occur when one wheel is placed over the support in question.

■ **EXAMPLE 14.6**

Find the maximum reaction R_A for the beam and loading shown in Fig. 14.17a.

Solution.   First we determine the resultant of the wheel loads and its location, relative to a datum. Referring to Fig. 14.17a,

$$R = \sum W = 12 + 20 + 20 + 8 = 60 \text{ k}$$

Taking moment about wheel 4 as a datum:

$$(60)(p) = (12)(28) + (20)(20) + (20)(10) \quad \text{or} \quad p = 15.6 \text{ ft}$$

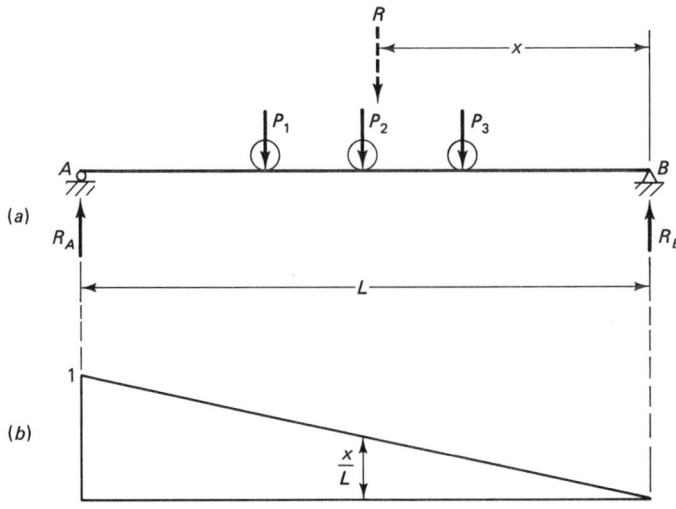

Figure 14.16 Position of the moving concentrated loads to produce maximum reaction. (*b*) Influence line for R_A.

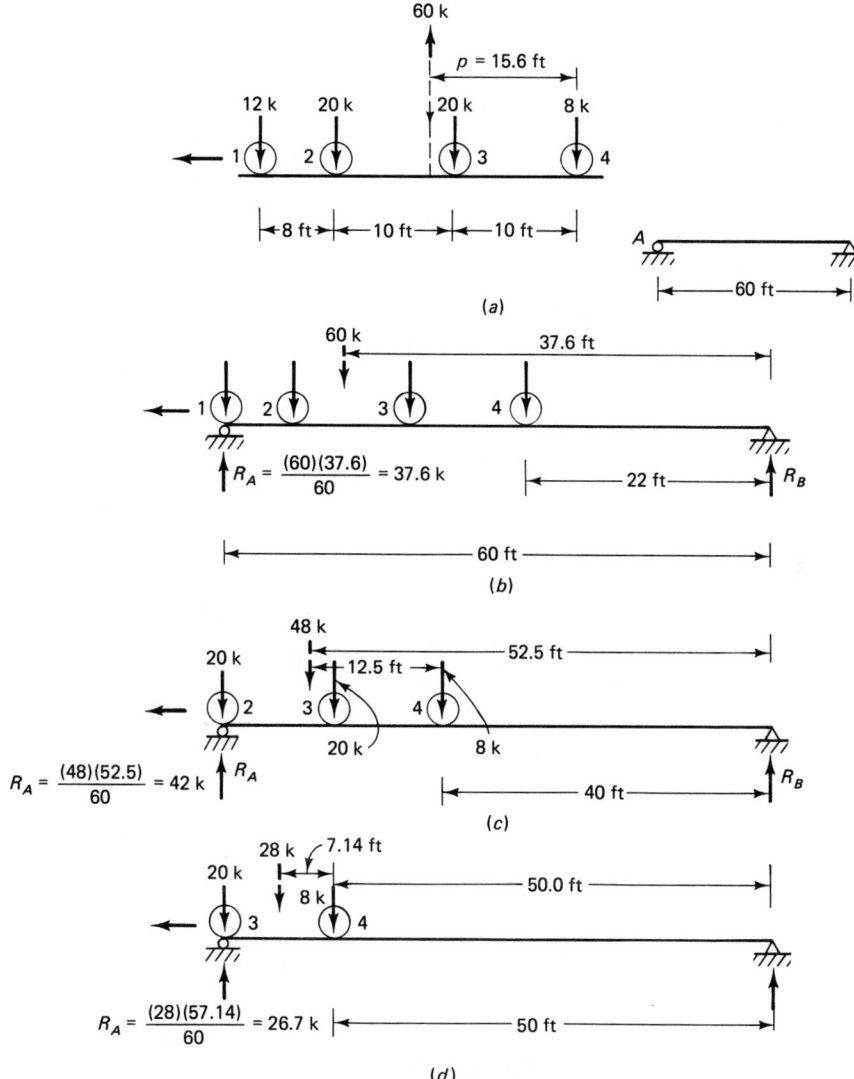

Figure 14.17 Maximum reaction for the structure in Example 14.6.

Making three trials for R_A shown in Fig. 14.17b to d, we have

For Fig. 14.17b: $R_A = \dfrac{(60)(37.6)}{60} = 37.6 \text{ k}$

For Fig. 14.17c: $R_A = \dfrac{(48)(52.5)}{60} = 42.0 \text{ k}$

For Fig. 14.17d: $R_A = \dfrac{(28)(57.14)}{60} = 26.7 \text{ k}$

Comparison of results leads to the conclusion that maximum reaction $R_A = 42.0$ k, when wheel 2 is over support A.   ■

Maximum Value of a Function with Influence-Line Diagram Shown in Fig. 14.15b   Such an influence-line diagram relates to the shear at a section in simply supported structures. Let us determine the maximum shear at a particular section C in a simply supported beam due to the passage of a train load, as shown in Fig. 14.18a. The influence line for shear force at C is shown in Fig. 14.18b. As the train loads move across the beam, the positive value of the shear force will increase until the wheel load p_1 is just to the right of C. In this position (Fig. 14.18c), the positive shear at C is

$$(V_C)_1 = R_A = \frac{R \cdot x}{L}$$

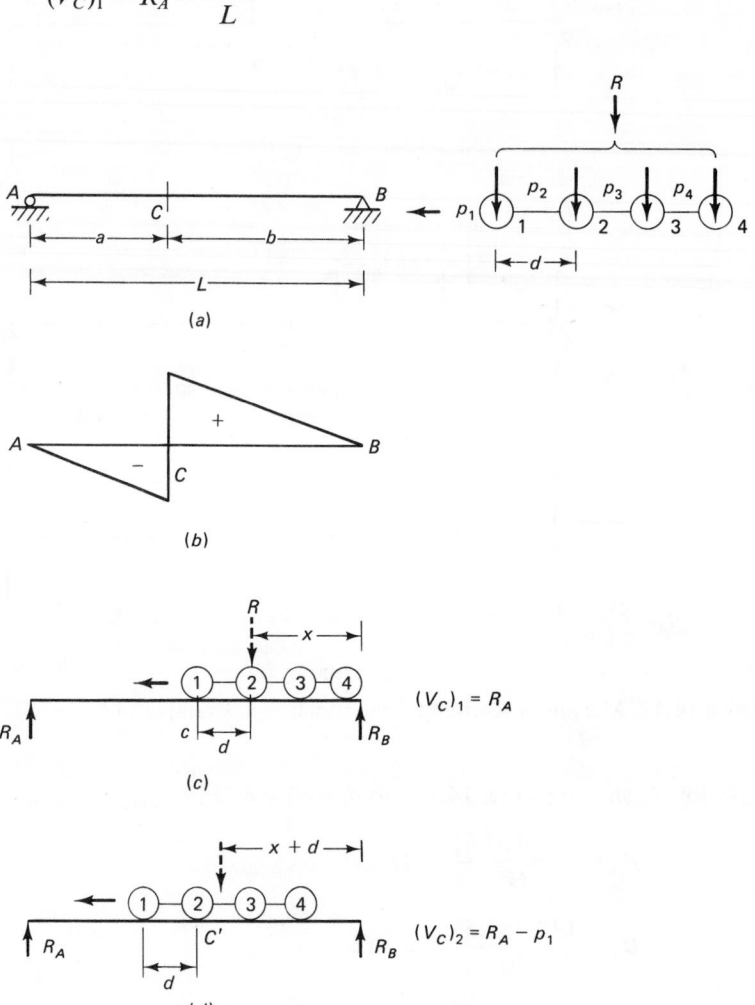

Figure 14.18 Position of the moving concentrated loads to produce maximum shear.

When load p_1 passes C, there will be a sudden decrease in shear force by an amount p_1. As the train loads move to the left, the value of the positive shear will again increase (since the influence ordinates increase toward the left) and this will continue until the wheel load p_2 is just to the right of C (Fig. 14.18d) and the shear at C is

$$(V_C)_2 = R_A - p_1 = \frac{R(x + d)}{L} - p_1$$

in which d = distance between p_1 and p_2.

The total change in shear force is given by

$$\Delta V_C = (V_C)_2 - (V_C)_1 = \frac{Rd}{L} - p_1$$

Then, if ΔV_C is positive, the shear force will have been increased by moving the train loads to the left. The loads are thus advanced until there is a change of sign in ΔV_C. Once the sign becomes negative, the maximum value of the shear force at C has been passed. From the above expression for ΔV_C, we can say that if

$$\frac{Rd}{L} > p_1 \quad \text{or} \quad \frac{R}{L} > \frac{p_1}{d} \qquad (V_C)_2 \text{ will be greater} \qquad (14.7a)$$

$$\frac{Rd}{L} < p_1 \quad \text{or} \quad \frac{R}{L} < \frac{p_1}{d} \qquad (V_C)_1 \text{ will be greater} \qquad (14.7b)$$

The above criterion is valid provided there is no wheel load entering or leaving the span.

■ EXAMPLE 14.7

For the train load shown in Fig. 14.19a, determine the maximum and minimum shears at the midspan C of the simply supported beam AB.

Solution. The influence line for shear at the midspan C is shown in Fig. 14.19b. For maximum positive shear, applying the above criterion (14.7) to the first trial position of the train load as shown in Fig. 14.9c, we find that $R \cdot d/L = (60)(8)/60 = 8 < p_1$ ($=12$). Therefore, this position produces the maximum shear. This can be calculated as $R_A = (60)(17.6)/60 = 17.6$ k. Using the product of the loads and the corresponding influence-line ordinates shown in Fig. 14.19b will yield $(12)(0.5) + (20)(0.37) + (20)(0.20) + (8)(0.03) = 17.6$ k, as before. The train loads in the position shown in Fig. 14.19d will produce a positive shear at $C = R_A - 12 = 25.6 - 12 = 13.6$ k. Maximum negative shear is produced by the train loads in the position shown in Fig. 14.19e and is equal to $R_B = -(60)(12.4 + 2)/60 = -14.4$ k.   ■

Maximum Value of a Function with Influence Line of the Form Shown in Fig. 14.15c
This type of influence line corresponds to that for bending moment in a simply supported beam and at a panel point in a simply supported girder bridge, and for an axial force in a chord member of some trusses. To establish the maximum value of such a function, let us consider the train load in Fig. 14.20a passing over the simply supported beam AB. The influence line for moment at C is shown in Fig.

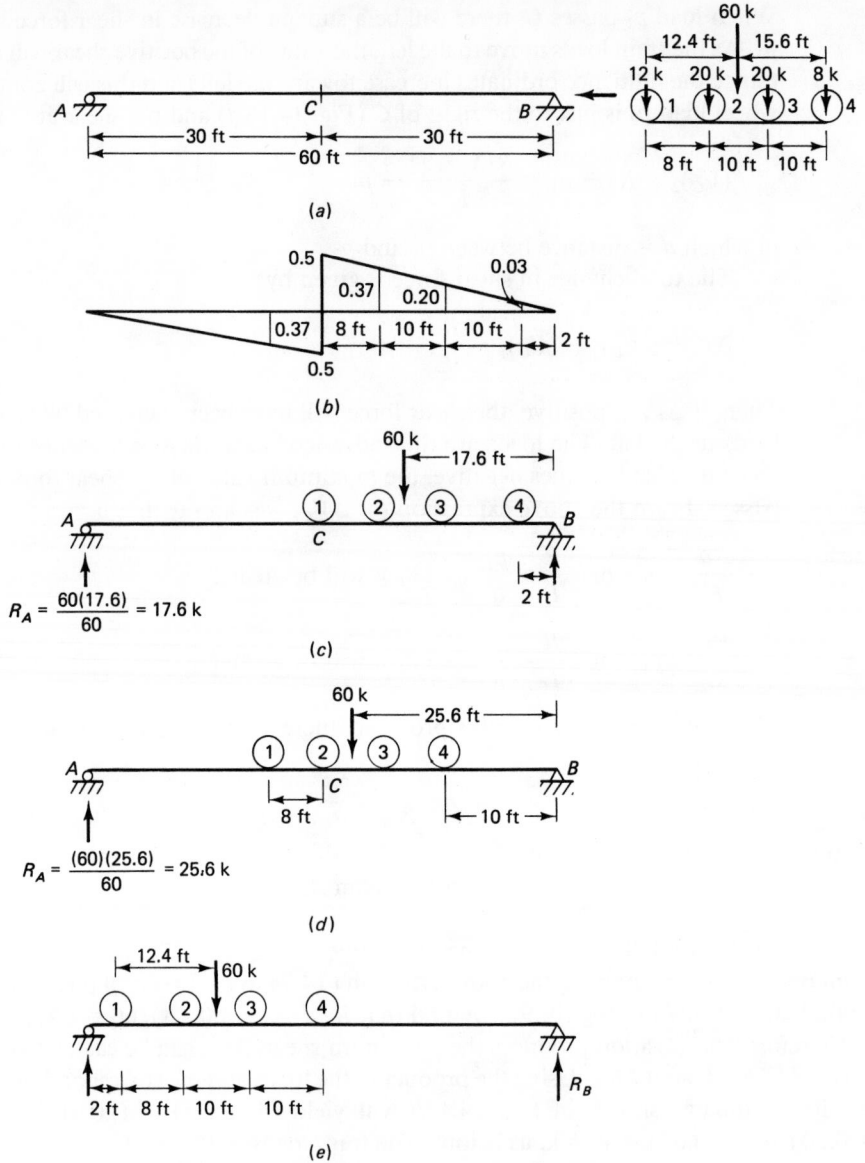

Figure 14.19 Maximum shear for the structure in Example 14.7.

14.20*b*. Where should the loads be placed such that the bending moment at *C* is a maximum? Let R_1 be the resultant of all the loads on the beam to the left of *C* and R_2 of those loads to the right of *C*. The bending moment at *C* for this load position is given by

$$M_C = R_1 y_1 + R_2 y_2 = R_1 \left(\frac{x_1}{a}\right) d + R_2 \left(\frac{x_2}{b}\right) d$$

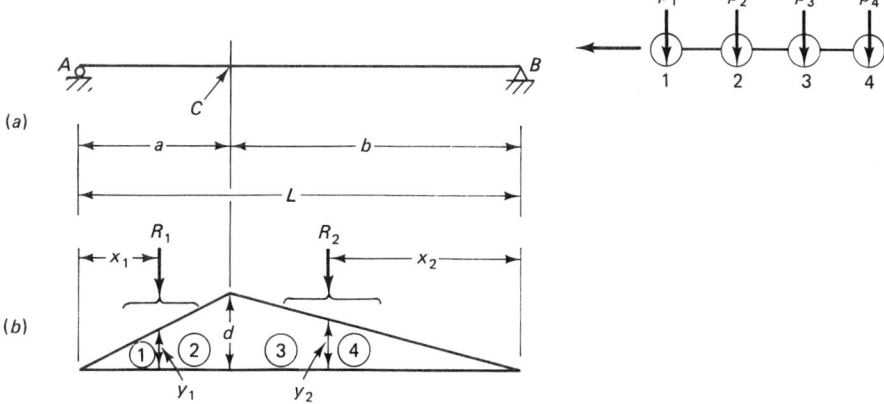

Figure 14.20 Position of moving concentrated loads to produce maximum moment.

Let the load move a small distance dx to the left such that $dx = dx_1 = dx_2$, with the value R_1 and R_2 remaining unchanged. The change in moment M_C will be

$$dM_C = R_1 \cdot \frac{d}{a}(-dx) + R_2\left(\frac{d}{b}\right)(+dx)$$

$$= \left[R_2\left(\frac{d}{b}\right) - R_1\left(\frac{d}{a}\right)\right]dx \qquad (14.8)$$

Thus as long as $R_2/b > R_1/a$, the bending moment at C will increase and the movement of the load to the left must be continued until one wheel load in group R_2 is at point C.

■ **EXAMPLE 14.8**
Determine the maximum bending moment at C for the simply supported beam AB loaded as shown in Fig. 14.21a.

Solution. The influence line for bending moment at C is shown in Fig. 14.21b. As a first trial, we place wheel load 2 at C (Fig. 14.21b) and find that

$$\frac{R_1}{a} = \frac{12}{30} < \frac{R_2}{b} = \frac{48}{30}$$

if wheel 2 is included in R_2, and

$$\frac{R_1}{a} = \frac{32}{30} > \frac{R_2}{b} = \frac{28}{30}$$

if wheel 2 is included in R_1. Therefore, the load position in Fig. 14.21b causes the maximum bending moment at C, which is calculated as

$$M_C = \frac{(60)(25.6)}{60}(30) - (12)(8) = 672.0 \text{ k} \cdot \text{ft}$$

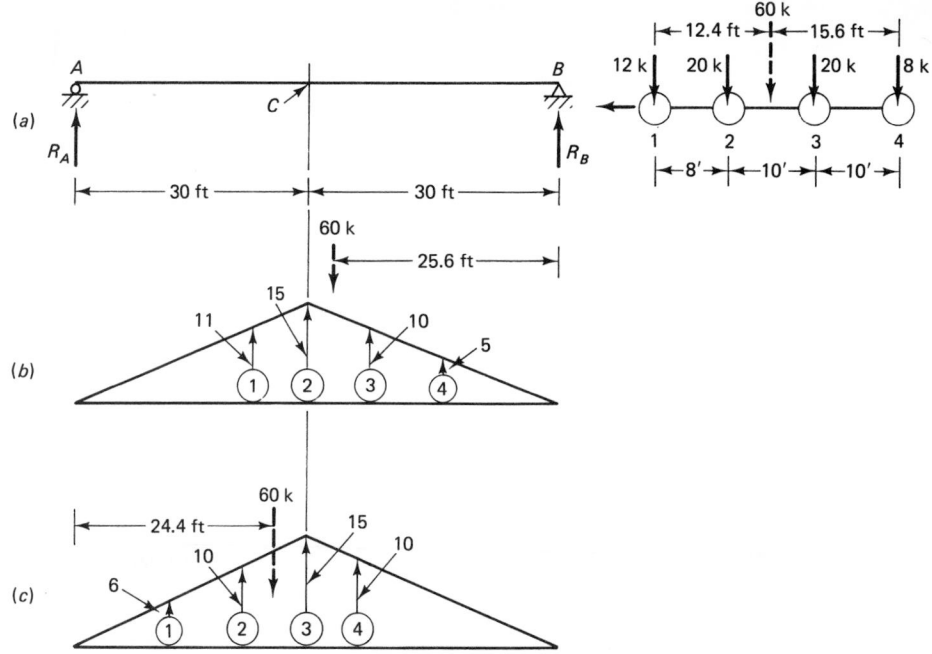

Figure 14.21 Maximum moment for the structure in Example 14.8. (b) Influence line for M_C.

or, from the influence-line ordinates shown in Fig. 14.21b, we have $M_C = (12)(11) + (20)(15) + (20)(10) + (8)(5) = 672.0$ k · ft, as before.

It is left as an exercise for the student to show that the load position given in Fig. 14.21c will not satisfy the criterion for maximum bending moment at C [Eq. (14.8)].   ∎

Maximum Value of a Function with Influence Line of the Form in Fig. 14.15d

Such an influence line corresponds to that for the shear in an interior panel of a girder bridge and axial force in the diagonal member of some bridge trusses. A typical influence line is shown in Fig. 14.22b. A criterion can be derived, based on the slopes of the influence-line segments, to compute the increase or decrease in the function. However, in this case it is much easier to try several loading positions and compare the values of the function sought. Let us consider the following example.

■ EXAMPLE 14.9

Determine the maximum shear in panel 2-3 of the girder bridge subjected to the train load shown in Fig. 14.22a.

Solution. The required influence line is given in Fig. 14.22b, where the ordinates have been determined from geometry. As wheel 1 approaches panel point 3, the influence line shows that the shear in panel 2-3 will keep increasing. Thus, considering the load position in Fig. 14.22c, with wheel 1 on panel point 3, the value of the shear

$$V_{2\text{-}3} = (12)(0.667) + (20)(0.578) + (20)(0.467) + (8)(0.356) = 31.75 \text{ k}$$

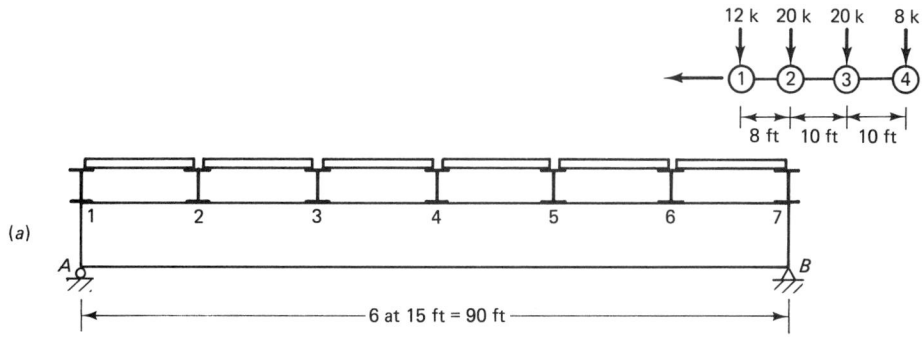

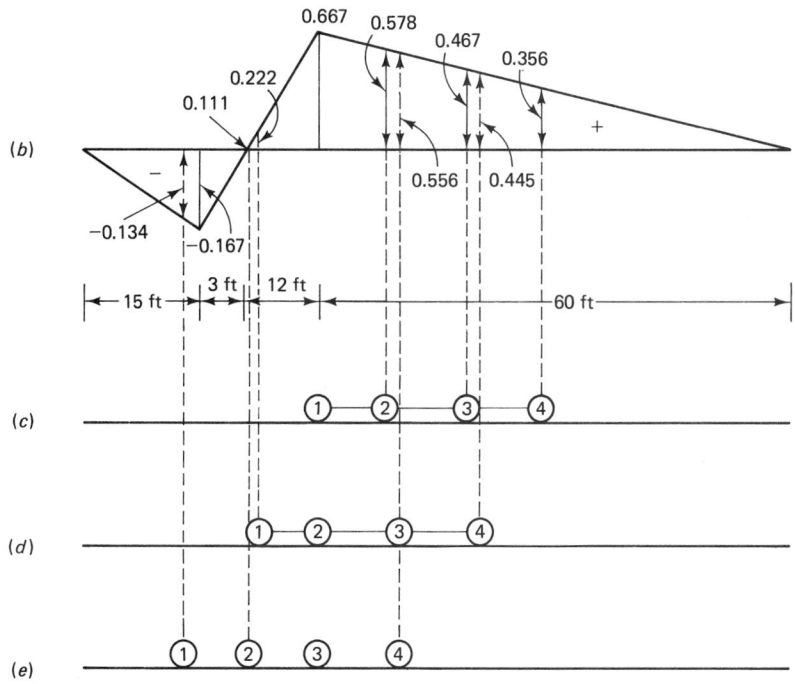

Figure 14.22 Position of moving concentrated loads to produce maximum shear in a girder with floor beams. (*b*) Influence line for shear in panel 2-3.

If the movement to the left is continued, the influence-line ordinates corresponding to successive positions of wheel 1 will decrease linearly whereas the ordinates corresponding to the positions of wheel 2, 3, and 4 will increase. This will continue until wheel 2 reaches panel point 3, as shown in Fig. 14.22 *d*. The panel shear will then be

$$V_{2\text{-}3} = (12)(0.222) + (20)(0.667) + (20)(0.556) + (8)(0.445) = 30.68 \text{ k}$$

If the movement of the train load continues to the left until wheel 3 reaches panel point 3, the panel shear will be

$$V_{2\text{-}3} = (12)(-0.134) + (20)(0.111) + (20)(0.667) + (8)(0.556) = 18.4 \text{ k}$$

By comparing the three values for $V_{2\text{-}3}$, the maximum panel shear is 31.75 k corresponding to loading in Fig. 14.22b. ■

Generally it is not necessary to try many load positions; it is obvious that the maximum shear will occur in a panel when one of the leading wheels (the first or second) is placed at the panel point adjoining the panel under study. The above method of locating a load system for the maximum effect is quite general and may be used in studies where the influence-line diagrams are more complicated. By inspection, one can start with load position close to that which produces the maximum value of the function desired.

14.9 ABSOLUTE MAXIMA

In design, we require the absolute maximum value of shear and moment that can occur in a structure. *Absolute maximum shear* will always occur at a section or panel immediately adjacent to one of the supports. In a simply supported beam subjected to a train load, the maximum bending moment does not usually occur at midspan. To obtain the absolute maximum bending moment under a certain wheel load during the passage of train loads, let us consider the simply supported beam *AB* shown in Fig. 14.23. It is required to determine the position of a particular load *W* that will cause a maximum moment under this load. Let *R* represent the resultant of all loads on the beam and R_L represent the resultant of all loads to the left of *W* but still on the beam. The left end reaction R_A obtained by taking moments about *B* is

$$R_A = \frac{R \cdot x}{L}$$

Hence the moment under *W* is

$$M = R_A(L - x - b) - R_L \cdot a = \frac{R \cdot x}{L}(L - x - b) - R_L \cdot a$$

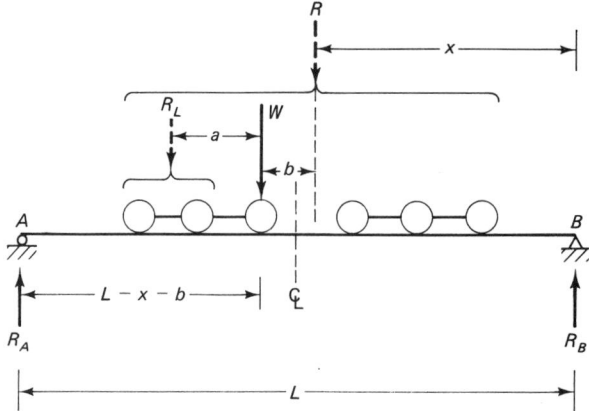

Figure 14.23 Position of the moving concentrated loads to produce absolute maximum moment.

The maximum value of M is obtained by setting $dM/dx = 0$, or

$$\frac{dM}{dx} = \frac{R}{L}(L - 2x - b) = 0$$

Thus

$$L - x - b = x \qquad\qquad (14.9)$$

This result states that in order to obtain the *maximum moment under a given wheel load, the distance from one beam support to that wheel load should be equal to the distance from the other support to the resultant of wheel loads R on the span. Or, the centerline of the span bisects the distance between W and R.*

For a single wheel load on the beam, this criterion implies locating the wheel load at midspan. With more than one wheel load on the beam, the maximum moment does not occur at midspan, as shown by the above criterion. Generally, since the section of absolute maximum moment is always near midspan, such moment will occur under the load that is closest to the resultant R.

■ **EXAMPLE 14.10**

Determine the absolute maximum bending moment for the simply supported beam loaded as shown in Fig. 14.24a.

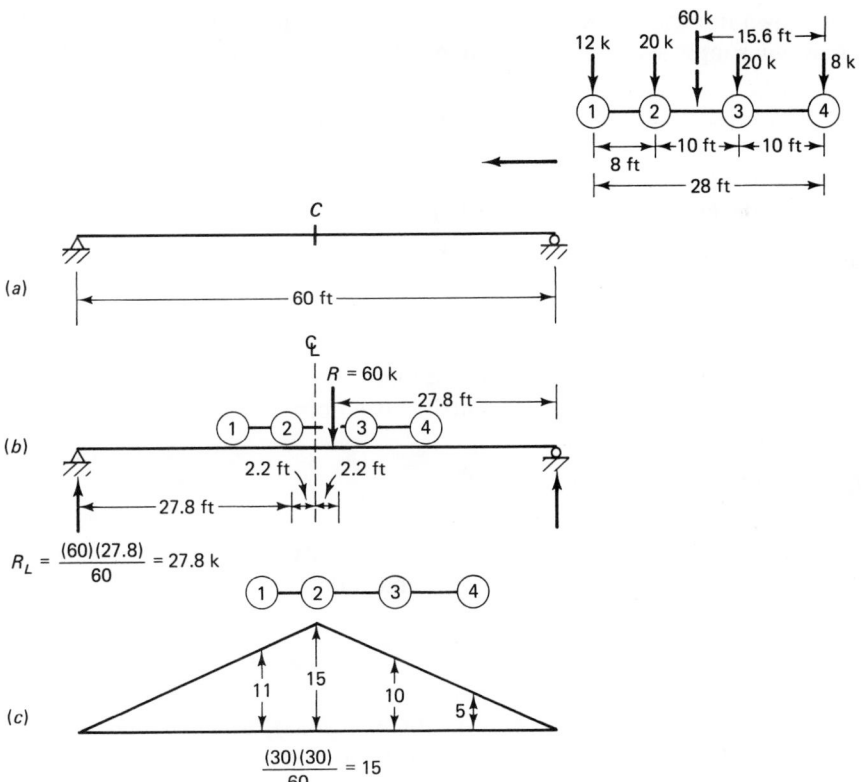

Figure 14.24 Absolute maximum bending moment for the structure in Example 14.10. (*c*) Influence line for M at C.

Solution. Since wheel 2 is closest to the resultant $R = 60$ k, we place the loading as shown in Fig. 14.24b in which the centerline of the span bisects the distance between wheel 2 and the resultant $R = 60$ k. This distance is equal to $10 + 10 - 15.6 = 4.4$ ft. Thus the absolute maximum moment under wheel 2

$$M_2 = (R_L)(27.8) - (12)(8) = 676.8 \text{ k} \cdot \text{ft}$$

It is interesting to compare this absolute maximum moment with the maximum moment at midspan. From the influence line for the moment at midspan, given in Fig. 14.24c, and with the position of the wheel load shown, the latter moment is

$$M_C = (12)(11) + (20)(15) + (20)(10) + (8)(5) = 672.0 \text{ k} \cdot \text{ft}$$

Notice the relatively small difference between the maximum moment at midspan and the absolute maximum moment; the former is, by definition, always smaller than the latter. ∎

The following two examples further illustrate the construction and use of influence-line diagrams for statically determinate structures.

■ EXAMPLE 14.11

Construct the influence-line diagram for moment at E of the simply supported girder bridge AF shown in Fig. 14.25a. The floor beams are located at A, B, E, and F. The stringers are *not* simply supported; they are overhanging as shown in Fig. 14.25a.

Solution. The bending moment at E is zero when the unit load is at A.

Unit Load at B (refer to Fig. 14.25b). The floor beam at B transfers the unit load directly to the girder. Floor beams at A, E, and F do not carry any load. Support reactions at A and F are $\frac{2}{3}$ k and $\frac{1}{3}$ k, respectively.

$$M_E = (\tfrac{1}{3})(12) = 4.0 \text{ k} \cdot \text{ft}$$

Unit Load at C (refer to Fig. 14.25c). Floor beams A and B transfer the unit load to the girder as shown. Floor beams at E and F do not carry any load. Support reactions at A and F are each $\frac{1}{2}$ k.

$$M_E = (\tfrac{1}{2})(12) = 6.0 \text{ k} \cdot \text{ft}$$

Unit Load at D (refer to Fig. 14.25d). Floor beams E and F transfer the unit load to the girder as shown. Floor beams at A and B do not carry any load. Support reactions at A and F are each $\frac{1}{2}$ k.

$$M_E = (\tfrac{1}{2} + \tfrac{1}{2})(12) = 12.0 \text{ k} \cdot \text{ft}$$

Unit Load at E (refer to Fig. 14.25e). Floor beam E transfers the unit load directly to the girder. Floor beams at A, B, and F do not carry any load. Support reactions at A and F are $\frac{1}{3}$ k and $\frac{2}{3}$ k, respectively.

$$M_E = (\tfrac{2}{3})(12) = 8.0 \text{ k} \cdot \text{ft}$$

Unit Load at F (refer to Fig. 14.25*f*).   Floor beam *F* transfers the unit load directly to the girder. Floor beams at *A, B,* and *E* do not carry any load. Support reaction at *A* is zero and that at *F* is 1.0 k.

$$M_E = (1 - 1)(12) = 0$$

Unit Load at G (refer to Fig. 14.25*g*).   Floor beams *E* and *F* transfer the unit load to the girder as shown. Floor beams at *A* and *B* do not carry any load. Support reaction at *A* is $\frac{1}{6}$ k downward and that at *F* is $\frac{7}{6}$ k upward.

$$M_E = (\tfrac{7}{6} - \tfrac{3}{2})(12) = -4.0 \text{ k} \cdot \text{ft}$$

The influence line for moment at *E* is obtained by plotting the above computed ordinates as shown in Fig. 14.25*h*.   ■

■ EXAMPLE 14.12

The cantilever bridge shown in Fig. 14.26*a* is hinged to support *F*. Supports *A, B,* and *E* are roller supports. Span *CD* is a suspended span. Draw the influence-line diagram for the axial force in the member *GH*. The loading on the bridge consists of the following: (a) uniform dead load of 2 k/ft covering the whole bridge, (b) uniformly distributed rolling load of 4 k/ft longer than the span, and (c) a concentrated rolling load of 20 k. Determine the maximum tensile and compressive forces in the member *GH* due to the dead load plus rolling load.

Solution

Influence Line for Force in Member GH.   Divide the structure into two segments by passing a vertical section *X-X* through the panel *GH* and consider the equilibrium of the left segment shown in Fig. 14.26*b*. The unit load is at a distance *x* from *A*.

(a) *When the unit load is between A and G* ($0 \leqslant x \leqslant 72$ ft). The load is transferred to the foundation through supports *A* and *B*. (Supports *E* and *F* do not carry any load.)

$$R_A = \frac{(1)(144 - x)}{144}$$

The force in member *GH* is obtained by taking moments of all forces acting on the free-body diagram shown in Fig. 14.26*b*(i) about *J*.

$$\curvearrowleft_+$$
$$\Sigma M_J = 0$$

$$-(R_A)(72) + (1)(72 - x) + (F_{GH})(24) = 0$$

$$F_{GH} = \frac{72 R_A - (72 - x)}{24}$$

$$= \frac{[72(144 - x)/144] - (72 - x)}{24}$$

$$= \frac{x}{48}$$

When $x = 0$, $F_{GH} = 0$; when $x = 72$ ft, $F_{GH} = 1.5$ k.

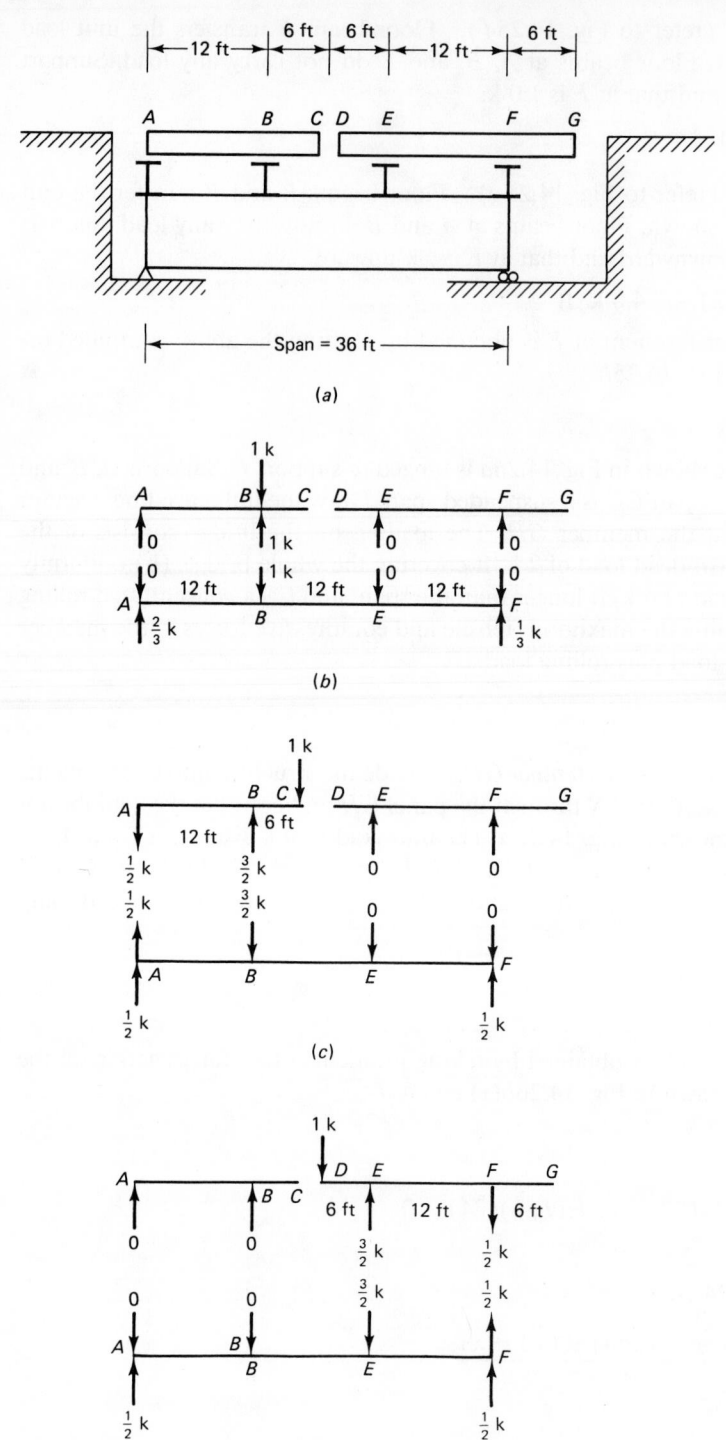

Figure 14.25 Influence line for moment at E for the structure in Example 14.11.

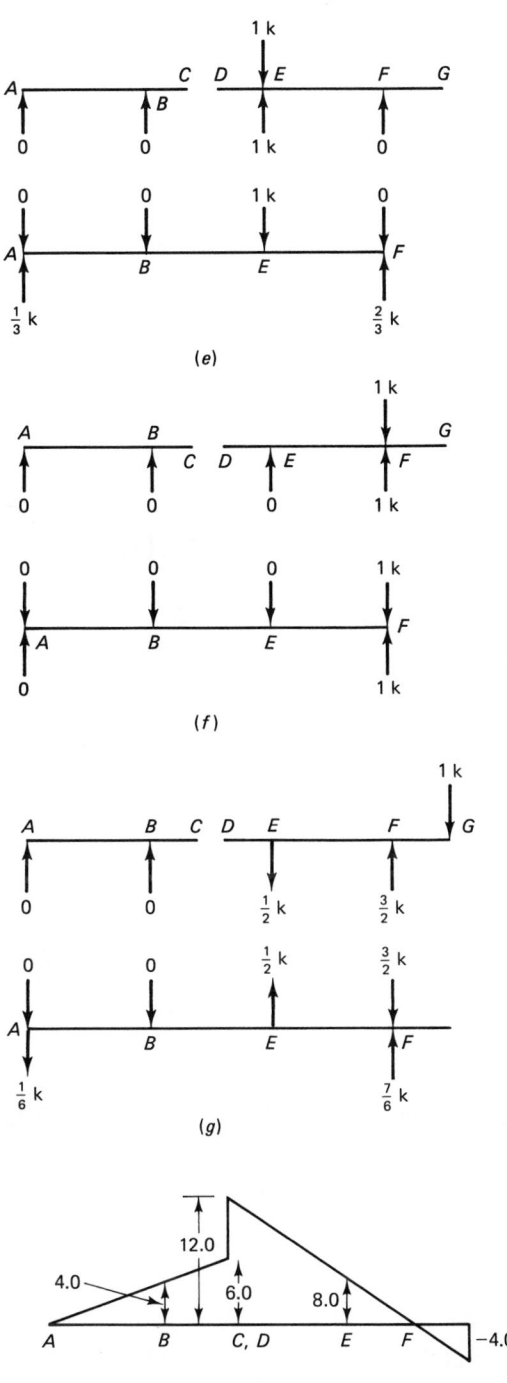

Figure 14.25 (*Continued*)

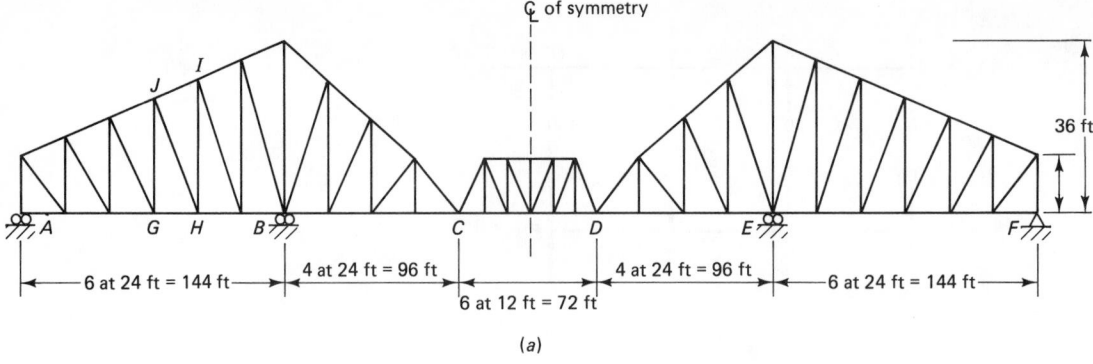

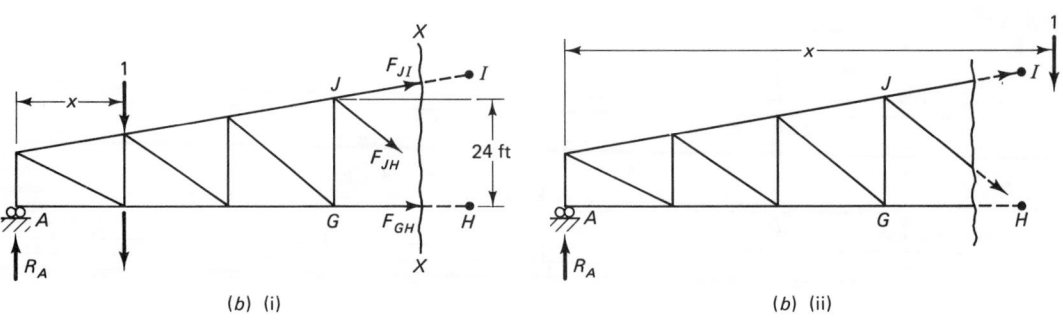

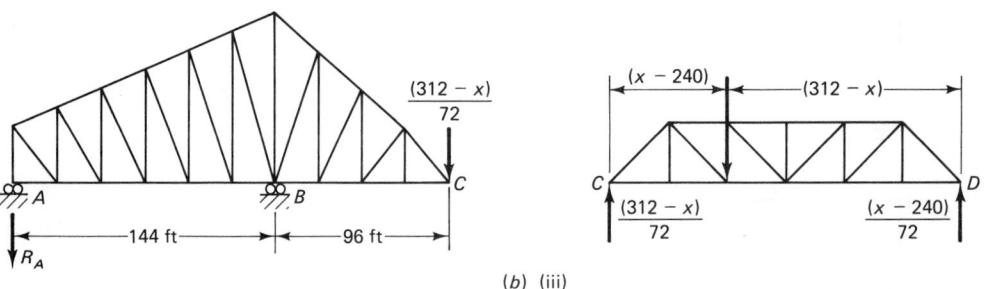

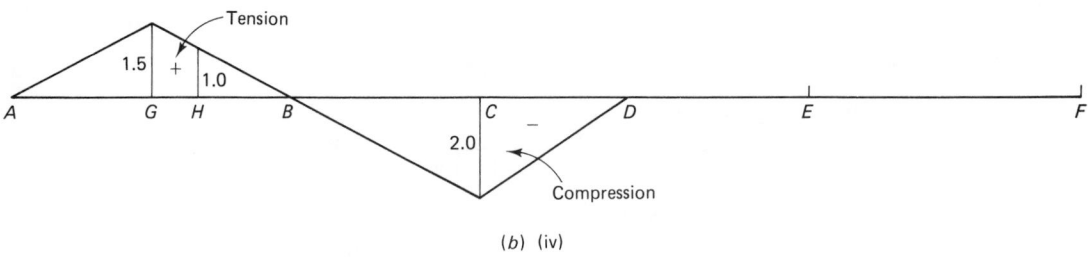

Figure 14.26 Influence lines for the structure in Example 14.12.

(b) *When the unit load is between H and C* (96 ft $\leqslant x \leqslant$ 240 ft). As before, the load is transferred to the foundation through supports A and B. (Supports E and F do not carry any load.)

$$R_A = \frac{(1)(144 - x)}{144}$$

The force in member GH is obtained by taking moments of all forces acting on the free-body diagram shown in Fig. 14.26b(ii) about J.

$$\curvearrowleft + \\ \sum M_J = 0$$

$$-(R_A)(72) + (F_{GH})(24) = 0$$

$$F_{GH} = 3R_A$$

$$= \frac{144 - x}{48}$$

When $x = 96$ ft, $F_{GH} = 1.0$ k, when $x = 144$ ft, $F_{GH} = 0$. When $x = 240$ ft, $F_{GH} = -2.0$ k.

(c) *When the unit load is between C and D* (240 ft $\leqslant x \leqslant$ 312 ft). When the unit load is at C, the load is transferred to the foundation through supports A and B. When the unit load is at D, the load is transferred to the foundation through supports E and F. When the unit load is between C and D, the load is transferred to the foundation through all the supports (A, B, E, and F).

Load transferred through hinge C to the supports A and B = $R_C = \dfrac{312 - x}{72}$

[Refer to Fig. 14.26b(iii)]

Reaction at A = $-\dfrac{312 - x}{72} \times \dfrac{96}{144} = -\dfrac{312 - x}{108}$ that is $\dfrac{312 - x}{108}\downarrow$

Referring to Fig. 14.26b(ii), taking moments about J of all the forces acting to the left of section X-X and equating to zero,

$$-(R_A)(72) + (F_{GH})(24) = 0$$

$$F_{GH} = 3R_A$$

$$= -\frac{312 - x}{36}$$

when $x = 240$ ft, $F_{GH} = -2.0$ k; when $x = 312$ ft, $F_{GH} = 0$.

(d) *When the unit load is between D and F* (312 ft $\leqslant x \leqslant$ 552 ft). When the unit load passes D, the load is transferred to the foundation through supports E and F. No load is transferred through supports A and B.

$$R_A = 0 \qquad F_{GH} = 0$$

The influence-line diagram for F_{GH} is obtained by plotting the ordinates at salient points A, G, H, B, C, D, E, and F as shown in Fig. 14.26b(iv).

Axial Force in Member GH Due to Dead Load and Rolling Loads

(a) *Axial force in member GH due to dead load.*

F_{GH} due to dead load of 2 k/ft = intensity of uniformly distributed load

$$\times \text{ area of influence-line diagram under the load}$$

$$= 2(\tfrac{1}{2} \times 144 \times 1.5 - \tfrac{1}{2} \times 168 \times 2.0)$$

$$= -120 \text{ k} \quad \text{that is, 120 k } compression$$

(b) *Axial force in member GH due to uniformly distributed rolling load of 4 k/ft.* Maximum tension occurs when the load occupies portion *AB* of the structure

$$F_{GH} = (4)(\tfrac{1}{2} \times 144 \times 1.5) = 432 \text{ k} \qquad (\text{tension})$$

Maximum compression occurs when the load occupies portion *BF* of the structure

$$F_{GH} = (4)(-\tfrac{1}{2} \times 168 \times 2.0) = -672 \text{ k} \qquad (\text{compression})$$

(c) *Axial force in member GH due to concentrated rolling load of 20 k.* Maximum tension occurs when the load is at *G*.

F_{GH} = magnitude of load $\times$ ordinate of influence-line diagram under the load

$$= (20)(1.5) = 30 \text{ k}$$

Maximum compression occurs when the load is at *C*.

$$F_{GH} = (20)(-2.0) = -40 \text{ k}$$

(d) *Maximum tension in member GH due to all loads combined.*

$$-120 + 432 + 30 = 342 \text{ k}$$

Maximum compression in member *GH* due to all loads combined

$$-120 - 672 - 40 = -832 \text{ k}$$

Note that the compressive force of 120 k due to dead load is always present and is to be considered when computing maximum compressive force as well as maximum tensile force. ∎

PART II: INFLUENCE LINES FOR STATICALLY INDETERMINATE STRUCTURES

14.10 GENERAL

Influence lines for determinate structures were discussed in Part I. They are essentially graphs showing how the various functions—shear, bending moment, axial force, etc.—vary at any section in a structure under the influence of a moving unit load. The student may recall that the Müller-Breslau principle (Section 14.5) was used to determine these influence lines, which consisted of straight-line segments; as such only one computed ordinate and the known shape were sufficient to determine such influence lines exactly. The Müller-Breslau principle will now be used to obtain the shape of the influence lines for statically indeterminate structures.

Figure 14.27 shows various influence lines for a continuous beam derived by means of the Müller-Breslau principle. For example, to construct the influence line for the vertical reaction at A, shown in Fig. 14.27a, we impose a unit vertical displacement at A in the same assumed direction as R_A, while the beam is restrained to stay

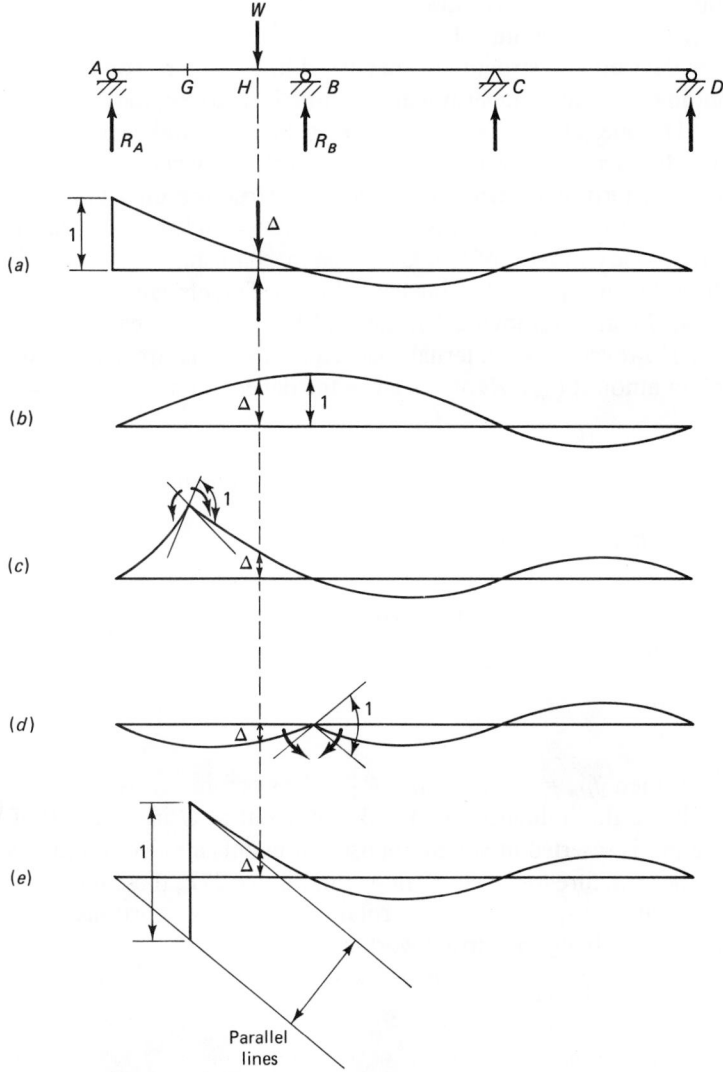

Figure 14.27 Influence lines for reactions, moments, and shear for a three-span continuous beam. (a) For R_A. (b) For R_B. (c) For moment at G. (d) For moment at B. (e) For shear at G.

on the supports at B, C, and D. The resulting displaced curve can be derived from the beam-deflection theory. From virtual work we have

$$R_A \times 1 = W \times \Delta$$

in which Δ is the deflection at point H (Fig. 14.27a). If W is set equal to unity, then

$$R_A = \Delta$$

which means that the deflected shape is the influence line for the vertical reaction R_A. The same principle can be applied to find the influence lines for the vertical reaction at B, moment at section G in span AB, moment at intermediate support B, and shear at section G in span AB, as shown in Fig. 14.27b to e, respectively. It is interesting to note that the influence lines for statically indeterminate structures, unlike those for statically determinate structures, are curved lines. The reason for this is that when a restraint in a statically indeterminate structure is removed, the degree of indeterminacy is reduced by one. The remaining structure is still stable and responds to the inducement of a unit displacement by a curved deflected shape, whereas removing a restraint in a statically determinate structure renders the structure unstable. Thus, when a unit displacement is imposed, the unstable structure which has no resistance only responds with a rigid-body motion of a linkage-type mechanism.

The following approach using flexibility coefficients can also be utilized. Referring to Fig. 14.27a, after removing R_A, the load W at H would cause a deflection equal to $(f_{AH})W$. However, if an external load equal to R_A is applied, point A will deflect upward an amount $(f_{AA})R_A$. Now since the deflection at A is zero, we can write

$$f_{AA}R_A = f_{AH}W$$

or

$$R_A = \frac{f_{AH}}{f_{AA}} W$$

The terms f_{AA} and f_{AH} are the flexibility coefficients. Now $f_{AH} = f_{HA}$, and putting $W = 1$, we have

$$R_A = \frac{f_{HA}}{f_{AA}}$$

If $f_{AA} = 1$, then $f_{HA} = \Delta$, and hence $R_A = \Delta$ as before.

To derive the influence line for the moment at G in span AB of Fig. 14.27, as before, a pin is inserted at G and opposite moments applied to either side of the pin, causing the structure to deflect as shown in Fig. 14.27c; these moments are increased until the relative displacement θ ($=$ rotation) of the two portions GA and GB is equal to 1 radian. Applying the virtual-work equation, we have

$$(M_G)(\theta) = (W)(\Delta)$$

or

$$M_G = \frac{W\Delta}{\theta}$$

If we put $W = 1$ and $\theta = 1$, then

$$M_G = \Delta$$

Thus Fig. 14.27c represents to scale the influence line for moment at G.

Before proceeding further, we should examine Betti's law of reciprocal work and Maxwell's law of reciprocal deflections (briefly discussed in Section 7.11).

14.11 BETTI'S THEOREM AND MAXWELL'S LAW OF RECIPROCAL DEFLECTIONS

The law is quite general (see Section 7.11) and it applies to a moment or couple instead of a force and to rotation instead of a linear displacement, as shown in Fig. 14.28; here, we can write

Rotation (in radians) Δ_{mn} at m due to a unit load at n

$$= \text{deflection } \Delta_{nm} \text{ at } n \text{ due to a unit couple at } m$$

Betti's law applies to any type of structure with unyielding supports and at constant temperature.

Let us demonstrate Betti's law by deducing the influence line for the moment at G in span AB of the structure in Fig. 14.29a. First we insert a hinge at G and apply unit couples on either side of the pin as shown in Fig. 14.29b, and the resulting displaced geometry of the beam is given in Fig. 14.29c. Here θ_{GA} is the slope of GA at G and θ_{GB} is the slope of GB at G, and Δ is the deflection at H, where the unit load is applied. Let us assume that owing to the application of the unit load at H (Fig. 14.29a), the moment at G is M_G and the slope at G is θ_G. Applying Betti's law, we have

$$M_G\theta_{GA} + M_G\theta_{GB} - (1)(\Delta) = (1)(\theta_G) - (1)(\theta_G)$$

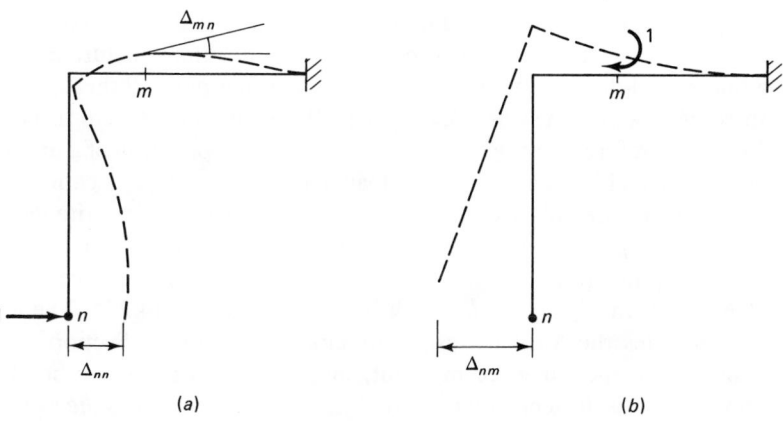

Figure 14.28 Illustration of Maxwell's law of reciprocal deflections for a unit moment. (a) Unit load at n. (b) Unit moment at m.

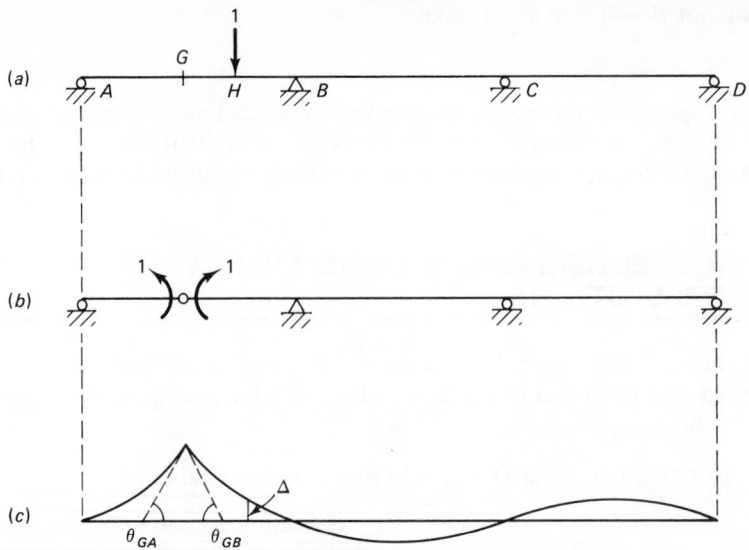

Figure 14.29 Determination of influence line by Betti's law.

or

$$M_G = \frac{\Delta}{\theta_{GA} + \theta_{GB}}$$

If we set $(\theta_{GA} + \theta_{GB})$ equal to unity, we arrive at the same expression obtained by virtual work and given earlier.

   The construction of influence lines can also be performed by the following procedure: Let us construct the influence line for the vertical reaction R_B of the three-span continuous beam $ABCD$ shown in Fig. 14.30a. The first step is to remove the vertical reaction R_B, keeping all the other restraints, and then apply a unit load (need not be a unit load) in the direction of R_B, moving point B a distance δ_{BB} and point H a distance δ_{HB} as shown in Fig. 14.30b. If a unit load is applied at H, the deflections at H and at B are δ_{HH} and δ_{BH}, respectively. Since the structure is assumed to follow a linear load-displacement law, to move the same point B through a distance δ_{BH}, a force of δ_{BH}/δ_{BB} must be applied at B. If the support B were in position as in Fig. 14.30a, therefore, the reaction upon it due to the application of a unit load at H would be δ_{BH}/δ_{BB}. This is also true for a load acting at any other location; furthermore, by Maxwell's reciprocal theorem, $\delta_{BH} = \delta_{HB}$ and hence we can write the reaction at B as $R_B = \delta_{HB}/\delta_{BB}$. Thus if the ordinates of the deflected curve shown in Fig. 14.30b are divided by the ordinate δ_{BB}, the resulting diagram is the influence line for the reaction at B shown in Fig. 14.30d, which is identical to Fig. 14.27b. The above procedure demonstrates the Müller-Breslau principle, discussed in Section 14.5. In a similar manner influence lines for moment, shear, etc., can be established by applying this principle. Thus, in general terms, the *influence-line ordinate is the ratio of the deflection ordinate to the impressed displacement, wherein the impressed displacement may be a deflection due to a force or an angle due to a moment of any arbitrary value.*

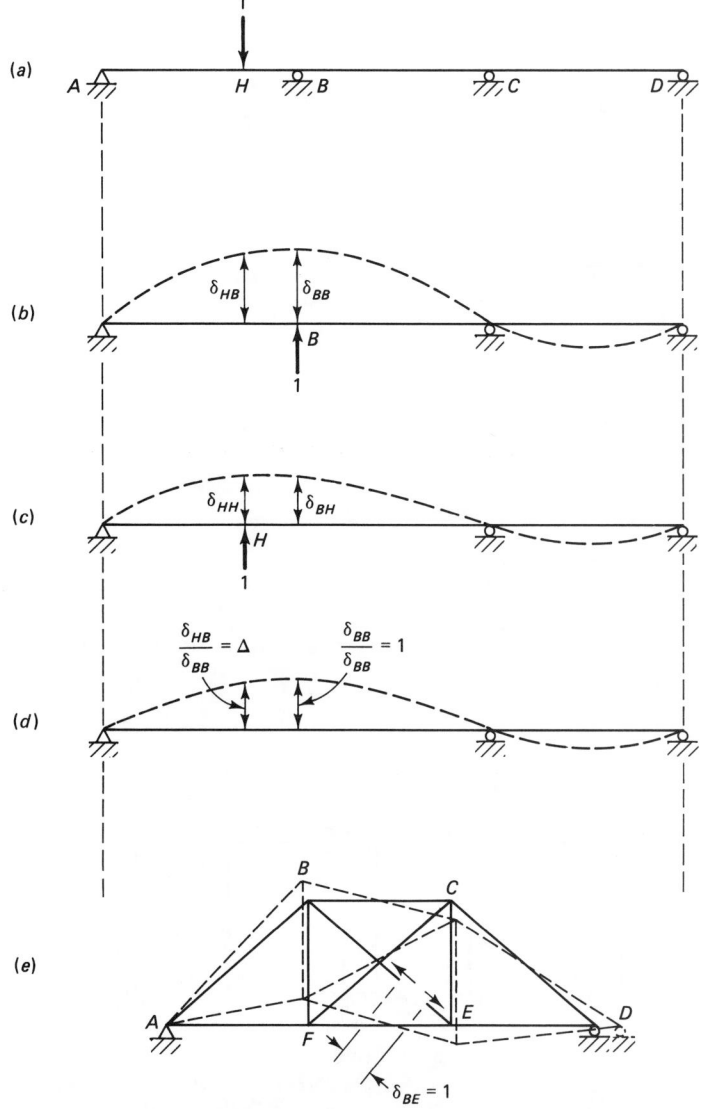

Figure 14.30 Determination of influence lines by Müller-Breslau Theorem.

Figure 14.30*e* demonstrates the construction of the influence line for the bar force *BE* in the internally indeterminate truss; the bar *BE* is first cut and the two cut ends are displaced relative to each other a distance $\delta_{BE} = 1$.

■ **EXAMPLE 14.13**

Sketch the influence lines for the reaction R_A and moment M_A for the beam *AB* in Fig. 14.31*a*.

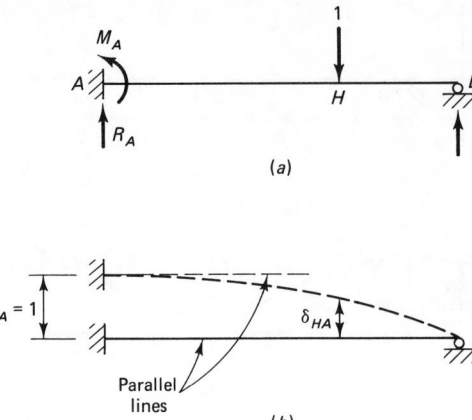

(a)

(b)

(c)

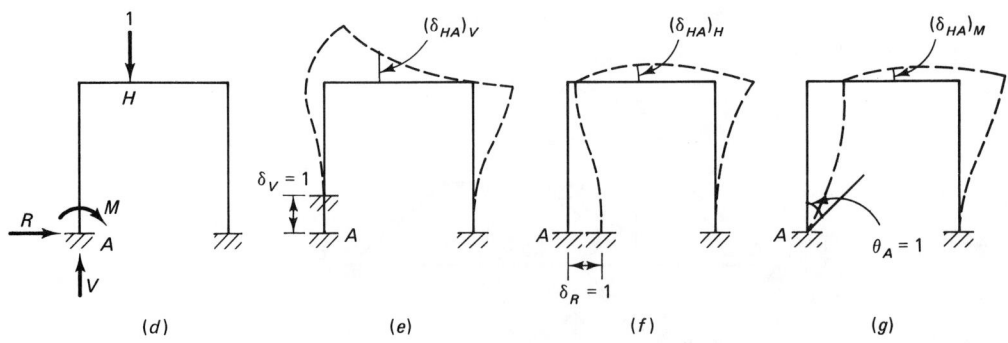

(d) (e) (f) (g)

Figure 14.31 Influence lines for the structure in Example 14.13 and for a rigid frame.

Solution. The influence line for R_A is obtained by removing the restraint R_A (keeping the moment restraint M_A) and allowing the reaction to move through a unit displacement. The deflected structure will then be the influence line for the reaction R_A as shown by the dashed lines in Fig. 14.31*b*.

The influence line for the moment M_A is derived by replacing the fixed support at A by a hinge and by introducing a unit rotation $\theta_A = 1$. The dashed line shown in Fig. 14.31*c* is the influence line for M_A.

The same principle applies to rigid frames; for example, to find the influence line for the vertical reaction V for the rigid frame in Fig. 14.31*d*, we displace A vertically one unit, keeping end A fixed against rotation and against horizontal displacement; the resulting displaced structure is the influence line for the reaction V shown by the dashed line in Fig. 14.31e. Similar procedure is followed for the influence lines for the horizontal reaction R and moment M, and these are shown in Fig. 14.31f and g, respectively. To emphasize the procedure only one displacement in the direction of only one redundant quantity can be allowed at a time, since it is essential to prevent the other redundants (in this case two redundants) from doing work. This method of drawing influence lines can be applied to intermediate sections of a rigid frame. For example, to find the influence line for the axial force at C of the rigid frame shown in Fig. 14.32a, first the beam is cut at C; then a pair of equal and opposite axial forces are applied at C, producing a horizontal displacement of $(\delta_C)_A = 1$, while keeping the rotation as well as the vertical movement of the two cut ends equal. This is necessary in order to produce zero work by the moment as well as the shear at the cut ends. The resulting displaced structure is the influence line for the axial thrust at C as shown by the dashed line in Fig. 14.32b.

Figure 14.32c shows the influence line for shear at C; here the beam is cut at C and the two cut ends are separated vertically as shown, ensuring that tangents to the two cut ends remain parallel and that there is no horizontal movement of the two cut ends.

Figure 14.32d presents the influence line for moment at C; this is obtained by inserting a hinge at C and rotating the two ends at the hinge relative to each other by an angle $(\theta_C)_M = 1$; in this case no work is performed by the shear or the axial force at C.   ∎

■ **EXAMPLE 14.14**

Determine the influence lines for the reactions R_B, R_C, and bending moment at the midspan section D of span BC for the beam in Fig. 14.33a.

Solution. First the reaction constraint at B is removed and a unit force is applied in the direction of reaction R_B as shown in Fig. 14.33b. The resulting deflection curve divided by δ_{BB} gives the influence line for R_B. To evaluate deflections, we use the conjugate-beam method (Chapter 7). The corresponding conjugate beam and support conditions, and the loading on the conjugate beam are shown in Fig. 14.33c.

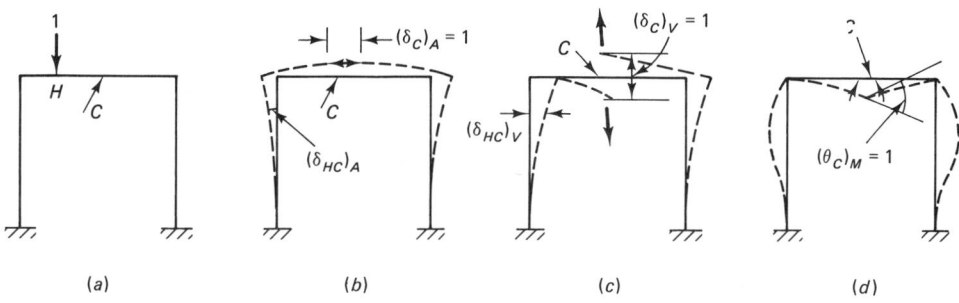

Figure 14.32 Influence lines for a rigid frame by the Müller-Breslau theorem.

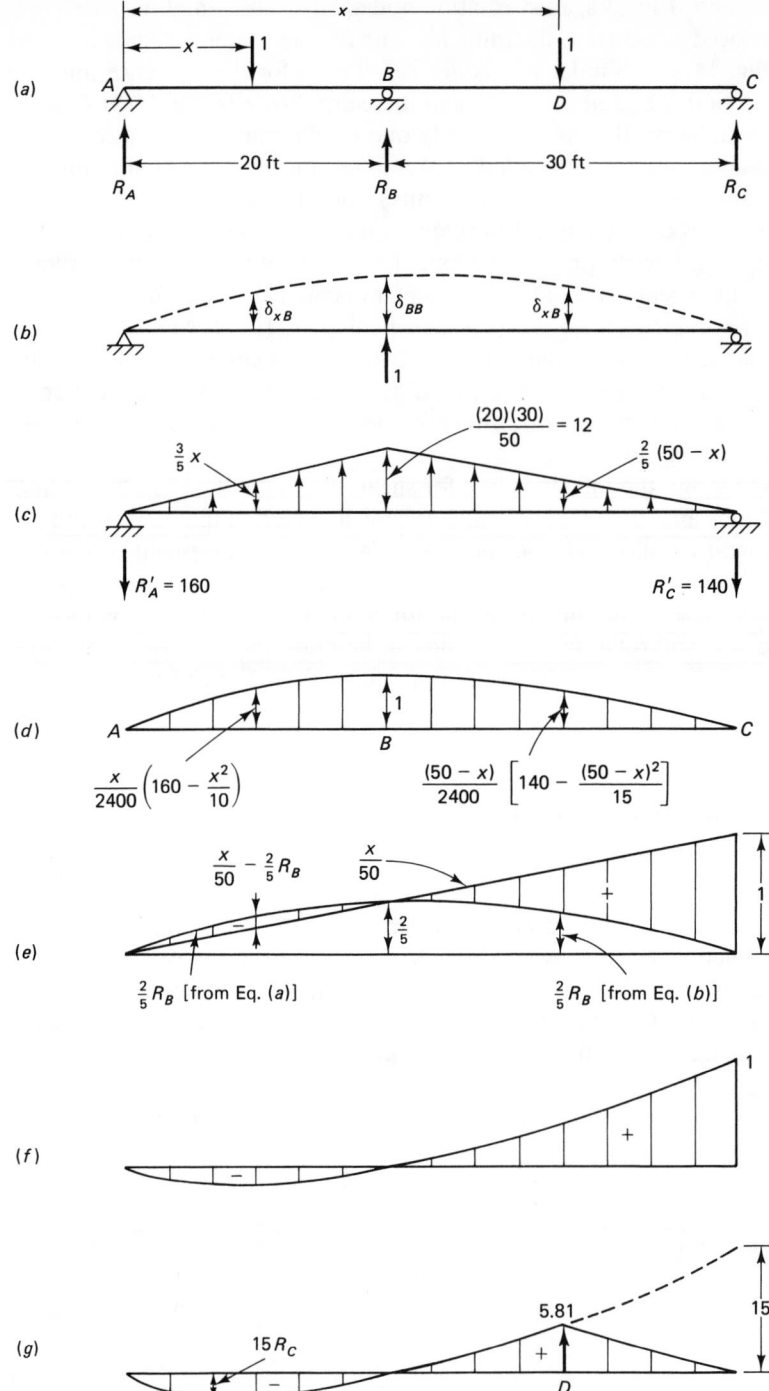

Figure 14.33 Influence lines for the structure in Example 14.14. (*d*) For R_B. (*e*) For R_C. (*f*) For R_C (drawn on a horizontal baseline). (*g*) For M_D.

Thus for $0 \leqslant x \leqslant 20$, assuming $EI = 1$,

$$\delta_{xB} = (160)(x) - \left(\frac{3}{5}x\right)\left(\frac{x}{2}\right)\left(\frac{x}{3}\right) = x\left(160 - \frac{x^2}{10}\right)$$

Substituting $x = 20$ in the above expression yields

$$\delta_{BB} = 2400$$

For $20 \leqslant x \leqslant 50$,

$$\delta_{xB} = (140)(50 - x) - \frac{2}{5}(50 - x)\frac{(50 - x)}{2}\frac{(50 - x)}{3}$$

$$= (50 - x)\left[140 - \frac{(50 - x)^2}{15}\right]$$

Therefore, influence-line ordinates for points $0 \leqslant x \leqslant 20$,

$$R_B = \frac{\delta_{xB}}{\delta_{BB}} = \frac{x}{2400}\left(160 - \frac{x^2}{10}\right) \qquad\qquad\qquad\qquad \text{(a)}$$

and for points $20 \leqslant x \leqslant 50$,

$$R_B = \frac{\delta_{xB}}{\delta_{BB}} = \frac{50 - x}{2400}\left[140 - \frac{(50 - x)^2}{15}\right] \qquad\qquad \text{(b)}$$

as shown in Fig. 14.33 d.

The continuous beam given is indeterminate externally to the first degree; once the influence line for R_B has been determined, the influence line for the reaction R_C and that for the moment at D can be found from statics. For R_C we take moments about A and equate to zero. Thus from $\Sigma M_A = 0$,

$$(R_C)(50) + (R_B)(20) - (1)(x) = 0$$

or

$$R_C = \frac{x}{50} - \frac{2}{5}R_B$$

$$= \frac{x}{50} - \frac{x}{6000}\left(160 - \frac{x^2}{10}\right) \qquad \text{for } 0 \leqslant x \leqslant 20$$

and

$$R_C = \frac{x}{50} - \frac{2}{5}R_B$$

$$= \frac{x}{50} - \frac{50 - x}{6000}\left[140 - \frac{(50 - x)^2}{15}\right] \qquad \text{for } 20 \leqslant x \leqslant 50$$

This influence-line equation for R_C is shown plotted in Fig. 14.33 e; the same equation is shown plotted with a horizontal baseline in Fig. 14.33 f.

Influence-Line Equation for M_D

(a) *When the unit load is to the left of D:*

$$M_D = (R_C)(15)$$

$$= \frac{3x}{10} - \frac{x}{400}\left(160 - \frac{x^2}{10}\right)$$

$$= -\frac{x}{200}\left(20 - \frac{x^2}{20}\right) \qquad \text{for } 0 \leqslant x \leqslant 20$$

and

$$M_D = \frac{3x}{10} - \frac{50 - x}{400}\left[140 - \frac{(50 - x)^2}{15}\right] \qquad \text{for } 20 \leqslant x \leqslant 35$$

(b) *When the unit load is to the right of D:*

$$M_D = (R_C)(15) - (1)(x - 35)$$

$$= \frac{3x}{10} - \frac{50 - x}{400}\left[140 - \frac{(50 - x)^2}{15}\right] - (x - 35) \qquad \text{for } 35 \leqslant x \leqslant 50$$

These three expressions are shown plotted in Fig. 14.33*g*. ■

■ EXAMPLE 14.15

Determine, by the Müller-Breslau principle, the influence lines for moment and shear at D, located at midspan of BC, for the beam in Fig. 14.34*a*.

Solution. To construct the influence line for M_D, we cut the beam at D, where a hinge is inserted and unit couples are applied as shown in Fig. 14.34*b*. The deflected shape is the influence line for M_D to some scale. The unit couples on each side of the hinge produce the relative rotation θ_D; the deflection at D is δ_{DD} and that at distance x from A is δ_{XD}. Such displacements are best determined by the conjugate-beam method. From the free-body diagram in Fig. 14.34*c*, the loading on the conjugate beam is shown in Fig. 14.34*d*. The reactions $R_{A'}$, $R_{D'}$, and $R_{C'}$ are derived from the equilibrium equations and the condition equation (bending moment at $B' = 0$). Thus, with $EI = 1$, since the bending moment at $B' = 0$,

$$(R_{A'})(20) + (2)(\tfrac{20}{2})(\tfrac{20}{3}) = 0 \qquad R_{A'} = -\tfrac{20}{3} \quad (\downarrow)$$

$\Sigma M_{C'} = 0$:

$$(R_{A'})(50) + (2)\left(\frac{50}{2}\right)\frac{50 + 30}{3} + (R_{D'})(15) = 0 \qquad R_{D'} = -\frac{200}{3} \quad (\downarrow)$$

$\Sigma M_{A'} = 0$:

$$(R_{C'})(50) + (2)\left(\frac{50}{2}\right)\frac{50 + 20}{3} + (R_{D'})(35) = 0 \qquad R_{C'} = \frac{70}{3} \quad (\uparrow)$$

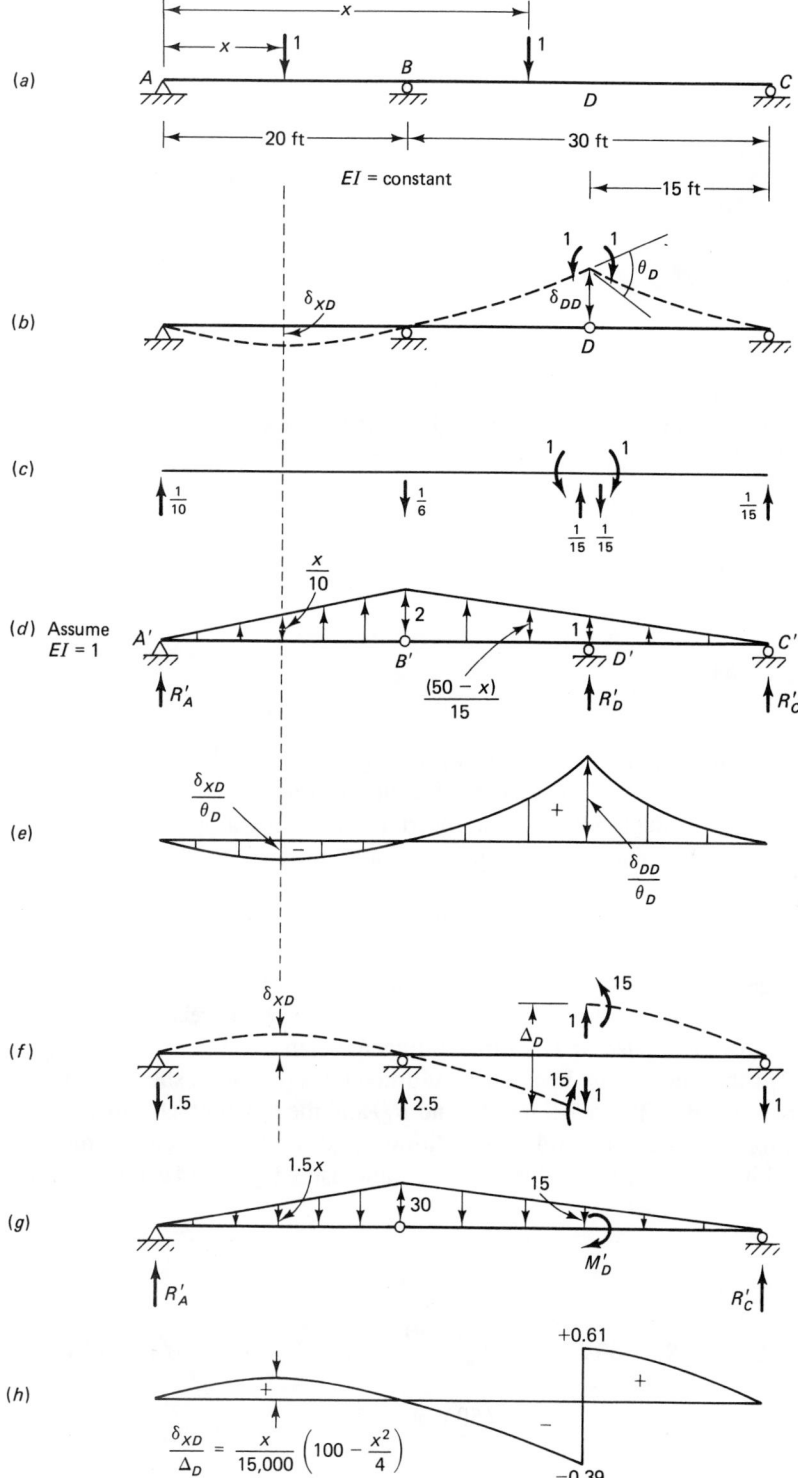

Figure 14.34 Influence lines for the structure in Example 14.15.

Now, for $0 \leqslant x \leqslant 20$,

$$\delta_{xD} = (R_{A'})(x) + \left(\frac{x}{10}\right)\left(\frac{x}{2}\right)\left(\frac{x}{3}\right) = -\frac{20}{3}x + \frac{x^3}{60} = -\frac{x}{3}\left(20 - \frac{x^2}{20}\right)$$

It should be noted that the absolute value of $R_{D'}$ is the difference in shears between the right and left sides of D' of the conjugate beam and therefore equals the relative rotation θ_D. Thus the influence-line ordinate for

$$M_D = \frac{\delta_{xD}}{\theta_D} = -\frac{x}{3}\frac{20 - x^2/20}{\frac{200}{3}} = -\frac{x}{200}\left(20 - \frac{x^2}{20}\right)$$

as before.

The other influence-line ordinates for the beam portions BD and DC can be derived in a similar manner. For example, when the unit load is at D, the influence-line ordinate for $M_D = \delta_{DD}/\theta_D$,

$$\delta_{DD} = (R_{C'})(15) + (1)(\tfrac{15}{2})(\tfrac{15}{3})$$

$$= (\tfrac{70}{3})(15) + \tfrac{75}{2} = (\tfrac{5}{2})(155)$$

Hence,

$$\frac{\delta_{DD}}{\theta_D} = \frac{(\tfrac{5}{2})(155)}{\frac{200}{3}} = +5.81$$

as before. The influence line for M_D is shown in Fig. 14.34e.

To construct the influence-line diagram for the shear at D, the midspan of BC in Fig. 14.34a, we first cut the beam at D and insert a slide device shown in Fig. 14.10d. Next we apply a pair of equal and opposite unit forces to separate the ends vertically a distance Δ_D, without causing relative rotation, as shown in Fig. 14.34f. The deflected shape and the induced reactions and moments required by equilibrium are also shown in Fig. 14.34f.

The conjugate beam and its loading are shown in Fig. 14.34g. Since we imposed a relative movement of the two cut ends equal to Δ_D without relative rotation, a moment $M_{D'}$ must be applied at D' of the conjugate beam. Such a moment satisfies the requirement that there must be a moment difference (without a shear difference) between the two cut ends. The reactions $R_{A'}$ and $R_{C'}$ and the moment $M_{D'}$ are obtained from the equilibrium conditions and the condition equation that the bending moment at the internal hinge B' is zero. Thus, with $EI = 1$, and using Fig. 14.34g, the condition equation gives

$$(R_{A'})(20) - (30)(\tfrac{20}{2})(\tfrac{20}{3}) = 0 \qquad R_{A'} = 100$$

$$\sum M_{C'} = 0: \quad (R_{A'})(50) - (30)\left(\frac{50}{2}\right)\frac{50 + 30}{3} + M_{D'} = 0 \quad M_{D'} = 15,000$$

$$\sum M_{A'} = 0: \quad (R_{C'})(50) - (30)\left(\frac{50}{2}\right)\frac{50 + 20}{3} - M_{D'} = 0 \quad R_{C'} = 650$$

We should note that $M_{D'}$ represents the relative deflection Δ_D between the two cut ends at D.

Now, referring to Fig. 14.34f and g,

$$\delta_{xD} = (R_{A'})(x) - (1.5x)\left(\frac{x}{2}\right)\left(\frac{x}{3}\right) = x\left(100 - \frac{x^2}{4}\right)$$

hence the influence-line ordinate for shear at D when the unit load is at distance x from A in span AB $(0 \leqslant x \leqslant 20)$ is

$$\frac{\delta_{xD}}{\Delta_D} = \frac{x(100 - x^2/4)}{15{,}000}$$

When the unit load is just to the right of D,

$$\delta_{xD} = (R_{C'})(15) - (15)(\tfrac{15}{2})(\tfrac{15}{3}) = 9187.5$$

Hence

$$\frac{\delta_{xD}}{\Delta_D} = \frac{9187.5}{15{,}000} = 0.61$$

When the unit load is just to the left of D,

$$\delta_{xD} = (R_{C'})(15) - (15)(\tfrac{15}{2})(\tfrac{15}{3}) - M_{D'} = -5812.5$$

Hence

$$\frac{\delta_{xD}}{\Delta_D} = \frac{-5812.5}{15{,}000} = -0.39$$

These ordinates are shown in Fig. 14.34h.   ∎

■ EXAMPLE 14.16

Determine the influence lines for the reaction R_A and R_C and for the force in member BC in the through Warren truss shown in Fig. 14.35a. All members have the same length l, cross-sectional area A, and elastic modulus E.

Solution. The truss is indeterminate externally to the first degree. Let the reaction R_C be the redundant quantity; we remove it and introduce a unit force in the direction of R_C as shown in Fig. 14.35b. To find the influence line for R_C, we require the deflections of joints B (or D) and C. These deflections are found using the virtual-work method discussed in Chapter 7, by finding first the forces in all the members of the truss when it is subjected to a unit load at B and these forces are k_1 forces. Next the forces in all the members of the truss are found when a unit load is applied at C; such forces are denoted k_2 forces. These forces are tabulated in Table 14.1.

By virtual work,

$$\delta_{BC} = \sum k_1 k_2 \left(\frac{L}{AE}\right) = \frac{23}{6}\left(\frac{l}{AE}\right)$$

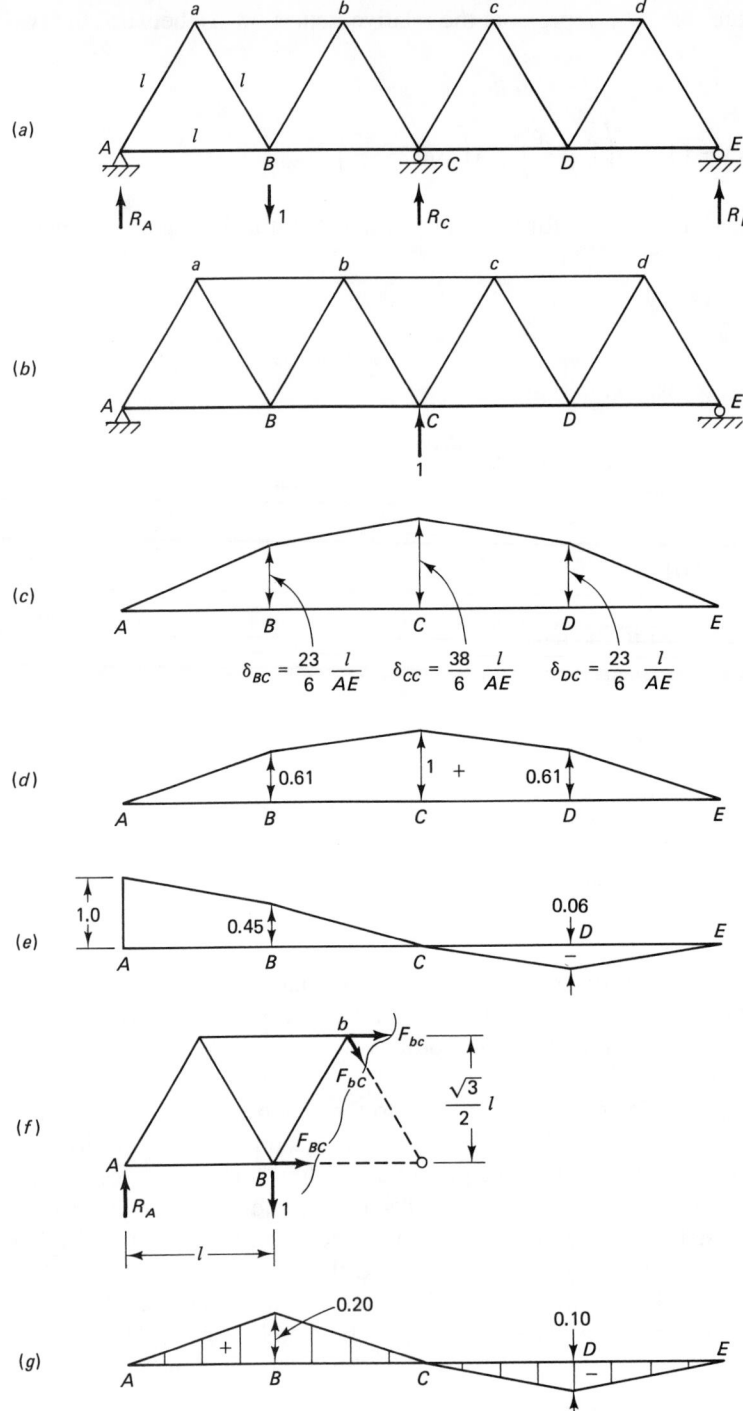

Figure 14.35 Influence lines for the structure in Example 14.16. (d) For R_C.

TABLE 14.1

Member	k_1 forces	k_2 forces	$k_1 k_2$	k_2^2
Aa	$-\sqrt{3}/2$	$-1/\sqrt{3}$	$+\frac{1}{2}$	$+\frac{1}{3}$
ab	$-\sqrt{3}/2$	$-1/\sqrt{3}$	$+\frac{1}{2}$	$+\frac{1}{3}$
bc	$-1/\sqrt{3}$	$-2/\sqrt{3}$	$+\frac{2}{3}$	$+\frac{4}{3}$
cd	$-1/2\sqrt{3}$	$-1/\sqrt{3}$	$+\frac{1}{6}$	$+\frac{1}{3}$
dE	$-1/2\sqrt{3}$	$-1/\sqrt{3}$	$+\frac{1}{6}$	$+\frac{1}{3}$
AB	$+\sqrt{3}/4$	$+1/2\sqrt{3}$	$+\frac{1}{8}$	$+\frac{1}{12}$
BC	$+5/4\sqrt{3}$	$+\sqrt{3}/2$	$+\frac{5}{8}$	$+\frac{3}{4}$
CD	$+3/4\sqrt{3}$	$+\sqrt{3}/2$	$+\frac{3}{8}$	$+\frac{3}{4}$
DE	$+1/4\sqrt{3}$	$+1/2\sqrt{3}$	$+\frac{1}{24}$	$+\frac{1}{12}$
aB	$+\sqrt{3}/2$	$+1/\sqrt{3}$	$+\frac{1}{2}$	$+\frac{1}{3}$
Bb	$+1/2\sqrt{3}$	$-1/\sqrt{3}$	$-\frac{1}{6}$	$+\frac{1}{3}$
bC	$-1/2\sqrt{3}$	$+1/\sqrt{3}$	$-\frac{1}{6}$	$+\frac{1}{3}$
cC	$+1/2\sqrt{3}$	$+1/\sqrt{3}$	$+\frac{1}{6}$	$+\frac{1}{3}$
cD	$-1/2\sqrt{3}$	$-1/\sqrt{3}$	$+\frac{1}{6}$	$+\frac{1}{3}$
Dd	$+1/2\sqrt{3}$	$+1/\sqrt{3}$	$+\frac{1}{6}$	$+\frac{1}{3}$
		Σ	$\frac{23}{6}$	$\frac{38}{6}$

From Maxwell's reciprocal theorem, $\delta_{BC} = \delta_{CB}$; furthermore, in this particular problem $\delta_{BC} = \delta_{DC}$; also

$$\delta_{CC} = \Sigma\, k_2^2 \left(\frac{L}{AE}\right) = \frac{38}{6}\left(\frac{l}{AE}\right)$$

These deflection ordinates are given in Fig. 14.35c; dividing these ordinates by δ_{CC} will yield the influence line for R_C as shown in Fig. 14.35d.

To find the influence line for R_A, we can use the results in Fig. 14.35d. Thus, when the unit load is at B, take the sum of the moments about E and we have

$$\Sigma\, M_E = 0: \qquad (R_A)(4l) + (R_C)(2l) - (1)(3l) = 0$$

or

$$R_A = \frac{3}{4} - \frac{R_C}{2} = \frac{3}{4} - \frac{0.61}{2} = 0.45$$

When the unit load is at C, $R_A = 0$. When the unit load is at D, sum the moments about E; thus

$$\Sigma\, M_E = 0: \qquad (R_A)(4l) + (R_C)(2l) - (1)(l) = 0$$

or

$$R_A = \frac{1}{4} - \frac{R_C}{2} = \frac{1}{4} - \frac{0.61}{2} = -0.06$$

Thus the influence line for R_A is given in Fig. 14.35e.

To determine the influence line for the force in member BC, we use the cut section of the truss shown in Fig. 14.35f. Thus, when the unit load is at B, taking moments about b, we have

$$\sum M_b = 0: \qquad (R_A)\left(\frac{3l}{2}\right) - (1)\left(\frac{l}{2}\right) - F_{BC}\left(\frac{l\sqrt{3}}{2}\right) = 0$$

or

$$F_{BC} = \sqrt{3}\,(0.45) - \frac{1}{\sqrt{3}} = 0.20$$

When the unit load is at D, again $\sum M_B = 0$, that is,

$$(R_A)\left(\frac{3l}{2}\right) - F_{BC}\left(\frac{l\sqrt{3}}{2}\right) = 0$$

or

$$F_{BC} = \sqrt{3}\,R_A = (\sqrt{3})(-0.06) = -0.10$$

It is obvious that when the unit load is at A, C, or E, $F_{BC} = 0$. The influence line for F_{BC} is shown in Fig. 14.35g. ∎

14.12 INFLUENCE LINES FOR TWO-HINGED ARCHES

The two-hinged arch, shown in Fig. 14.36a, is statically indeterminate to the first degree; the redundant quantity is usually taken as the horizontal thrust H at the two supports. *To determine H we consider the horizontal movement of the abutments if H were removed as shown in Fig. 14.36b.* The arch in Fig. 14.36a is separated into two systems a and b as shown in Fig. 14.36b and c. From the virtual-work equation,

$$\Delta = \int \frac{M_0 m \, ds}{EI}$$

We have, for system a,

$$\Delta_{BW} = \int \left(\frac{M_0 \, ds}{EI}\right)(1 \cdot y)$$

and for system b,

$$\Delta_{BH} = \int \left(\frac{Hy \, ds}{EI}\right)(1 \cdot y)$$

If support B in Fig. 14.36a is restrained from horizontal movement, the compatibility of deformations requires that

$$\Delta_{BW} = \Delta_{BH} \tag{14.10}$$

or

$$H = \frac{\int M_0 y \, ds / EI}{\int y^2 \, ds / EI} \tag{14.11}$$

in which M_0 = bending moment at any point along the arch in system a (Fig. 14.36b), in which the arch is considered as a simply supported curved beam subjected to the given external loading; y = height of the arch at any point; I = moment of inertia of arch section at any point; ds = elemental length along the curve of the arch rib; E = elastic modulus of the arch material. Equation (14.11) could have been derived by applying Castigliano's theorem of least work (see Chapter 13):

$$\frac{\partial U}{\partial H} = 0 \qquad \text{if the span } AB \text{ does not change}$$

Now, in terms of bending moment M in the arch (Fig. 14.36a),

$$U = \int \frac{M^2 \, ds}{2EI} \qquad \text{in which } M = M_0 - Hy$$

Thus

$$\frac{\partial U}{\partial H} = \frac{\partial}{\partial H} \int \frac{M^2 \, ds}{2EI} = \int \frac{M}{EI} \frac{\partial M}{\partial H} \, ds = \int \frac{M_0 - Hy}{EI} (-y) ds = 0$$

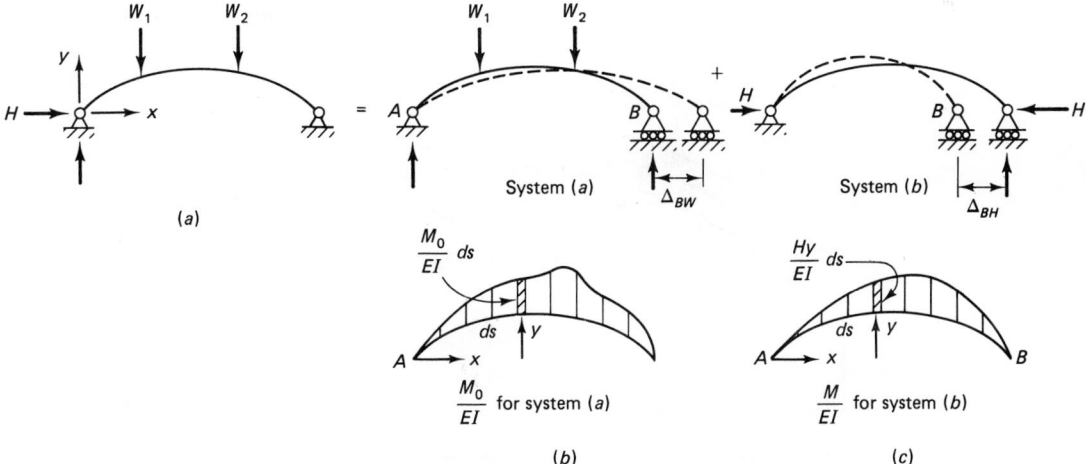

Figure 14.36 Determination of horizontal thrust H in a two-hinged arch.

or

$$H = \frac{\int (M_0/EI)y\, ds}{\int (y^2/EI)\, ds}$$

which is Eq. (14.11). This equation is good for arches of simple geometrical form such as segments of a circle or a parabola. However, when dealing with arches whose outline is constructed of portions of several curves, etc., we replace the integration sign in Eq. (14.11) by the summation sign as follows:

$$H = \frac{\Sigma (M_0/EI)y\, \Delta s}{\Sigma (y^2/EI)\Delta s} \tag{14.12}$$

in which Δs = short length of the arch rib.

In practice the moment of inertia of the arch rib varies from a maximum at each abutment to a minimum at the crown. Therefore, in order to estimate H by Eq. (14.11) or (14.12), the variation of I must be known. Since proper values of I would not be available prior to design, it is assumed, for preliminary design, that I varies directly as the secant of the angle of inclination of the arch (α) as shown in Fig. 14.37. Thus we can put

$$I_x = I_c \sec \alpha \tag{14.13}$$

in which I_c is the moment of inertia at the crown of the arch. Also $ds = dx \sec \alpha$. Therefore, Eq. (14.11) becomes

$$H = \frac{\int M_0 y \dfrac{dx \sec \alpha}{EI_c \sec \alpha}}{\int y^2 \dfrac{dx \sec \alpha}{EI_c \sec \alpha}} = \frac{\int_0^L M_0 y\, dx}{\int_0^L y^2\, dx} \tag{14.14}$$

in which L = horizontal span of the arch. Equation (14.14) is much more convenient than Eq. (14.11) since the integral is taken along the horizontal span of the arch. In

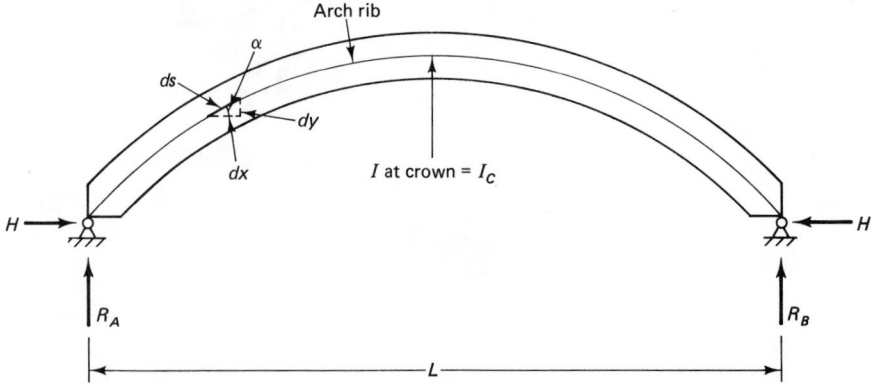

Figure 14.37 Two-hinged parabolic arch with the moment of inertia varying as the secant of the angle of inclination of the arch.

practice, the assumption made by using Eq. (14.13) has been found to lead to only a very small error in determining H and therefore is deemed acceptable.

Sometimes the foundation conditions for the abutments are not adequate to provide complete horizontal restraint against movement. In such cases, *the horizontal thrust H is provided by a tie* connecting the hinges as shown in Fig. 14.38. As a result, Eq. (14.10) is adjusted as follows to allow for the stretch in the tie which is equal to $HL/(AE)_{tie}$:

$$\Delta_{BW} - \Delta_{BH} = \frac{HL}{(AE)_{tie}} \tag{14.15}$$

Hence

$$H = \frac{\int \dfrac{M_0 y \ ds}{EI}}{\int \dfrac{y^2 \ ds}{EI} + \dfrac{L}{(AE)_{tie}}} \tag{14.16}$$

Furthermore, any *change in temperature* would have an effect on the arch by giving rise to a horizontal thrust. For example, if there is a rise t in temperature, the increase in the elemental length $ds = \alpha_t t \ ds$, in which $\alpha_t =$ coefficient of thermal expansion of the arch (Fig. 14.36a). The horizontal component of this increase of length $= \alpha_t t \ ds \cos \alpha = \alpha_t t \ dx$. Therefore, the horizontal component of increase of length of the entire arch $= \int_0^L \alpha_t t \ dx = \alpha_t tL$; this we denote as Δ_{Bt}, and therefore we can rewrite Eq. (14.10) as

$$\Delta_{Bt} = \Delta_{BH}$$

or

$$H = \frac{\alpha_t tL}{\int y^2 \ ds/EI} \tag{14.17}$$

in the absence of external loads on the arch. If external loads are present together with a temperature rise t, then Eq. (14.17) can be modified to

$$H = \frac{\int (M_0 y \ ds/EI) + \alpha_t tL}{\int y^2 \ ds/EI} \tag{14.18}$$

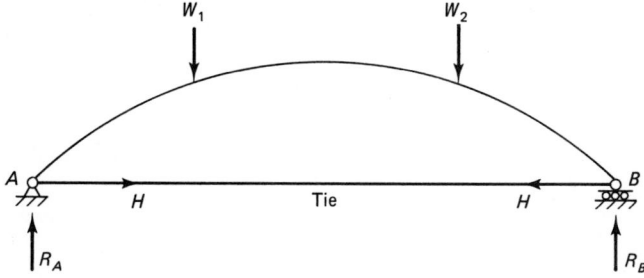

Figure 14.38 Two-hinged tied arch.

It should be noted that the expressions for the horizontal thrust H were obtained by considering the strain energy due to bending only; in so doing we have *neglected the strain energy due to shear S and axial thrust T in the arch rib*. The effect of shear can be shown to be extremely small and therefore negligible compared with that of bending; however, when the rise of the arch is rather small compared with the span, the effect of axial thrust T should not be neglected. Referring to Fig. 14.39, and considering the force equilibrium of portion AO, we have

$$S = R_A \cos \theta - (\sum W)\cos \theta - H \sin \theta = (R_A - \sum W)\cos \theta - H \sin \theta$$

and

$$T = R_A \sin \theta - (\sum W)\sin \theta + H \cos \theta = (R_A - \sum W)\sin \theta + H \cos \theta$$

in which $\sum W$ = sum of loads to the left of O. Now the *shortening of the arch rib due to thrust T* can be derived from strain energy considerations. However, the following treatment is another alternative. Now an arc ds of the arch rib is shortened by $T\, ds/AE$; the projection of this shortening on the horizontal x axis is $(T\, ds \cos \theta)/AE$. Hence the total horizontal component of shortening of rib,

$$\Delta_{BT} = \int \frac{T \cos \theta}{AE}\, ds$$

Substituting for T, we have

$$\Delta_{BT} = \int \frac{[(R_A - \sum W)\sin \theta + H \cos \theta]\cos \theta}{AE}\, ds$$

In practice, the term $(R_A - \sum W)$ can be shown to be negligible. Thus

$$\Delta_{BT} = H \int \frac{\cos^2 \theta}{AE}\, ds$$

Therefore, we can now rewrite Eq. (14.10) in the form

$$\Delta_{BW} = \Delta_{BH} + \Delta_{BT} \tag{14.19}$$

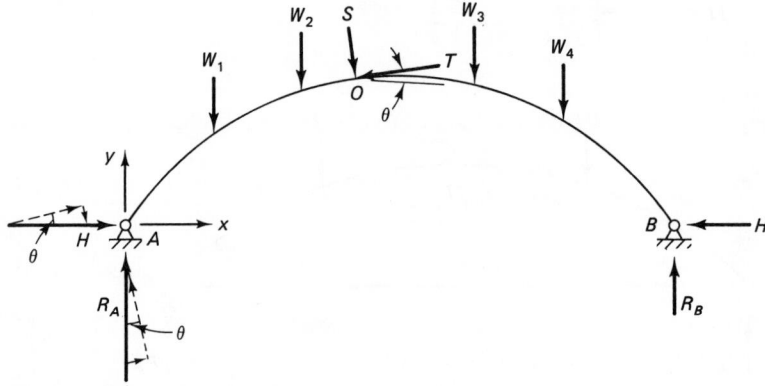

Figure 14.39 Shear and thrust in a two-hinged arch.

Substituting for Δ_{BW}, Δ_{BH}, and Δ_{BT} yields

$$H = \frac{\int M_0 y \, ds / EI}{\int y^2 \, ds / EI + \int (\cos^2 \theta \, ds) / AE} \tag{14.20}$$

It is observed that shortening of the arch rib reduces H and hence the bending moment in the arch is increased. Once H is found we can proceed with determining influence lines of the various design quantities.

■ **EXAMPLE 14.17**

Construct the influence lines for the horizontal thrust H, moment M_q, thrust T_q, and shear S_q at a section q distant one-fourth span from the left springing of the parabolic arch in Fig. 14.40a, with $I = I_c \sec \theta$.

Solution. For a parabolic arch with origin at A (Fig. 14.40a),

$$y = \frac{4 \, Dx(L - x)}{L^2} = \frac{0.4x(L - x)}{L}$$

where D = rise = $0.1 L$. Since EI varies as secant θ, we use Eq. (14.14) to determine the *influence line for H.*

For region A to K, $M_0 = R_A x = \left(\dfrac{L - a}{L} \right) x$

For region K to B (with x measured from B), $M_0 = R_B x = \left(\dfrac{a}{L} \right) x$

Thus the numerator of Eq. (14.14) becomes

$$\int_0^L M_0 y \, dx = \int_0^a \left(\frac{L - a}{L} \right) x \left[\frac{0.4x(L - x)}{L} \right] dx + \int_0^{L-a} \left(\frac{a}{L} \right) (x) \left[\frac{0.4x}{L} (L - x) \right] dx$$

$$= \frac{a}{30 L} (L^3 - 2 La^2 + a^3)$$

The denominator of Eq. (14.14) yields

$$\int_0^L y^2 \, dx = \frac{0.16}{L^2} \int_0^L x^2 (L - x)^2 \, dx = \frac{0.16}{30} L^3$$

Thus the influence-line equation for H is

$$H = \frac{\int_0^L M_0 y \, dx}{\int_0^L y^2 \, dx} = \frac{a}{0.16 L^4} (L^3 - 2 La^2 + a^3) \tag{a}$$

A sketch of the influence line for H is shown in Fig. 14.40b.

Influence Line for M_q. From the parabolic equation for the arch, when $x = 0.25 L$, $y = 0.075 L$. When the unit load is to the right of q,

$$M_q = (R_A)(0.25 L) - (H)(0.075 L) = (0.25)(L - a) - (H)(0.075 L) \tag{b}$$

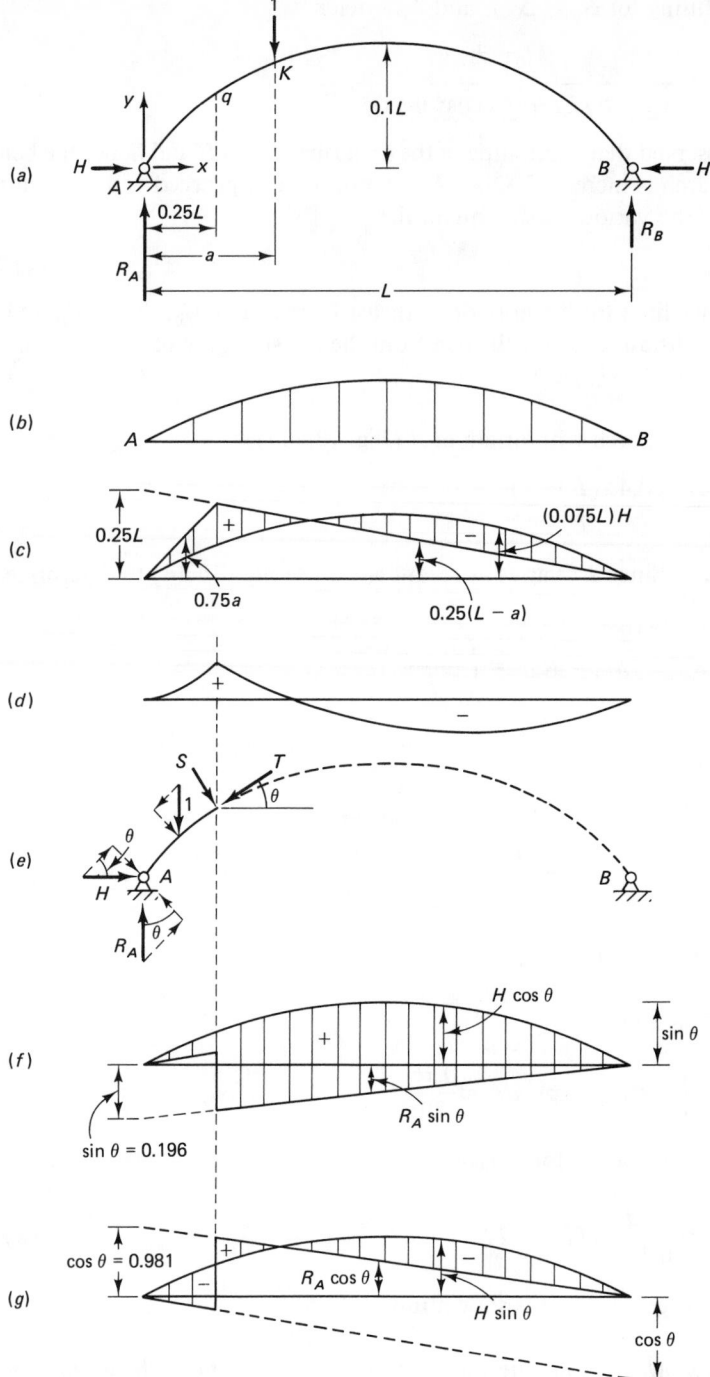

Figure 14.40 Influence lines for the structure in Example 14.17. (b) For H. (c) For M_q. (d) For $M_{\hat{q}}$. (f) For T_q. (g) For $S_{\hat{q}}$.

When the unit load is to the left of q,

$$M_q = (R_A)(0.25L) - (H)(0.075L) - (1)(0.25L - a)$$

$$= 0.25(L - a) - (H)(0.075L) - 0.25L + a$$

$$= 0.75a - (H)(0.075L) \tag{c}$$

A sketch for the influence line for M_q is given in Fig. 14.40c; this is also shown drawn on a horizontal baseline in Fig. 14.40d.

The influence lines for T_q and S_q are determined by considering the free-body diagram of the arch in Fig. 14.40e and resolving forces in the direction of T and S, respectively. Thus, when the unit load is to the right of q,

$$T_q = H \cos \theta + R_A \sin \theta \tag{d}$$

When the unit load is to the left of q,

$$T_q = H \cos \theta + R_A \sin \theta - (1)\sin \theta \tag{e}$$

Now

$$\tan \theta = \frac{dy}{dx} = \frac{4D}{L^2}(L - 2x) \qquad \text{at } q, \, x = 0.25L$$

Hence $\tan \theta = 0.2$. Therefore, at q, $\theta = 11.31°$; $\sin \theta = 0.196$ and $\cos \theta = 0.981$. The influence line for T is shown in Fig. 14.40f. The influence line for shear S_q is obtained in a similar manner:

When the unit load is to the right of q,

$$S_q = R_A \cos \theta - H \sin \theta \tag{f}$$

and when the unit load is to the left of q,

$$S_q = R_A \cos \theta - H \sin \theta - (1)\cos \theta \tag{g}$$

A sketch of the influence line for S_q is given in Fig. 14.40g. ∎

■ EXAMPLE 14.18

Determine the value of H for the parabolic arch in Fig. 14.40a when a vertical load of 100 k is applied at the crown. Hence determine M_q, T_q, and S_q; given $L = 100$ ft.

Solution. Using expression (a) of Example 14.17 for H, we have for $a = 50$ ft,

$$H = \frac{50}{(0.16)(100)^4}[100^3 - (2)(100)(50)^2 + (50)^3]$$

$$= 1.95 \text{ k}$$

From expression (b) of Example 14.17,

$$M_q = (0.25)(100 - 50) - (1.95)(0.075)(100) = -2.15 \text{ k} \cdot \text{ft}$$

From expression (d) of Example 14.17,

$$T_q = (1.95)(0.981) + \left(\frac{100 - 50}{100}\right)(0.196)$$

$$= 2.01 \text{ k} \qquad \text{(compression, as assumed)}$$

From expression (f) of Example 14.17,

$$S_q = \left(\frac{100 - 50}{100}\right)(0.981) - (1.95)(0.196)$$

$$= -0.108 \text{ k} \qquad \text{(opposite to assumed direction)} \qquad \blacksquare$$

The above values are for a unit load. For a load of 100 k: $H = (100)(1.95) = 195$ k; $M_q = (100)(-2.15) = -215$ k·ft; $T_q = (100)(2.01) = 201$ k; $S_q = (100)(0.108) = 10.8$ k.

14.13 INFLUENCE LINES FOR HINGELESS ARCHES

The hingeless arch shown in Fig. 14.41a is statically indeterminate to the third degree. To make the structure statically determinate, we require to remove three reactions; let us choose the three redundants at B, namely, R_B, H_B, and M_B, as shown in Fig. 14.41b and remove them. Now we apply a unit load along the arch and let us denote the resulting displacements at B as Δx, Δy, and θ (Fig. 14.41c). Now apply a unit vertical load at B in the direction of R_B (Fig. 14.44d), and calculate the vertical and horizontal displacements and the rotation at B, that is, Δyy, Δxy, and $\Delta\theta y$.

Now apply a unit horizontal load at B (Fig. 14.41e) and calculate Δyx, Δxx, and $\Delta\theta x$. Finally, apply a unit moment at B (Fig. 14.41f), and calculate $\Delta y\theta$, $\Delta x\theta$, and $\Delta\theta\theta$. The three compatibility conditions that must be satisfied are

1. The vertical displacement at B in Fig. 14.41a = 0; or

$$\Delta y + R_B \Delta yy + H_B \Delta yx + M_B \Delta y\theta = 0$$

Therefore,

$$\int \frac{Mm_1 \, ds}{EI} + f_{11} R_B + f_{12} H_B + f_{13} M_B = 0 \qquad (14.21)$$

2. The horizontal displacement at B in Fig. 14.41a = 0; or

$$\Delta x + R_B \Delta xy + H_B \Delta xx + M_B \Delta x\theta = 0$$

or

$$\int \frac{Mm_2 \, ds}{EI} + f_{21} R_B + f_{22} H_B + f_{23} M_B = 0 \qquad (14.22)$$

3. The rotation at B in Fig. 14.41a = 0; or

$$\theta + R_B \Delta\theta y + H_B \Delta\theta x + M_B \Delta\theta\theta = 0$$

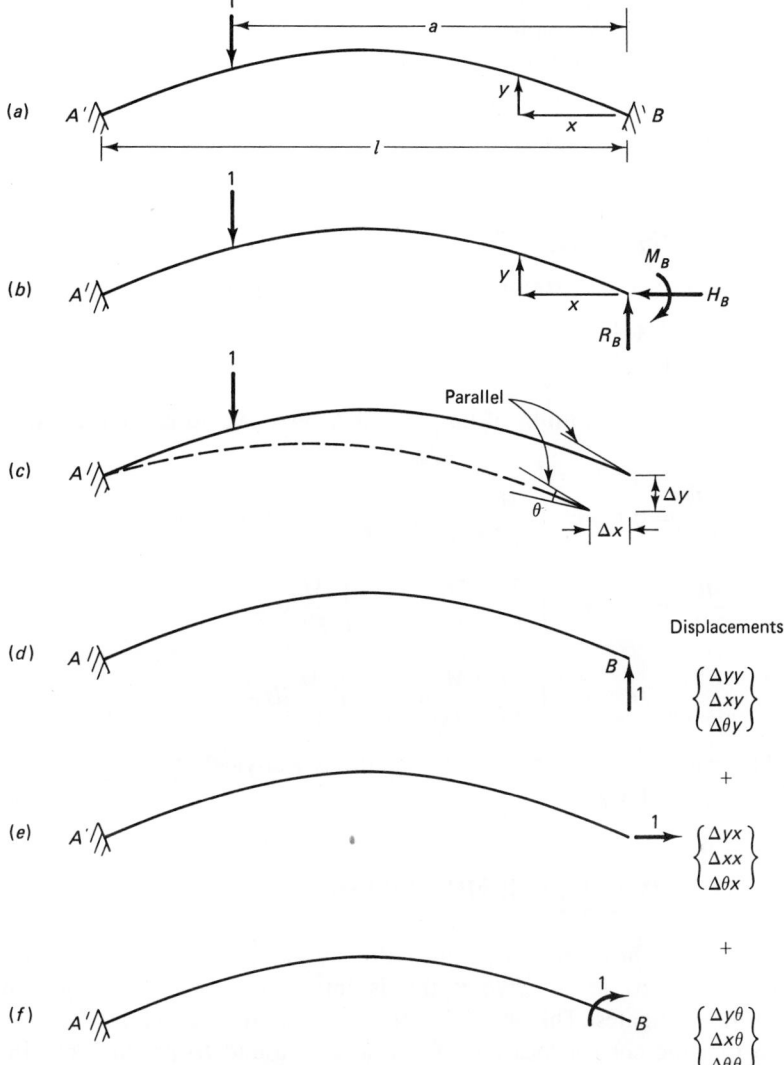

Figure 14.41 Treatment of hingeless arch.

or

$$\int \frac{Mm_3\,ds}{EI} + f_{31}R_B + f_{32}H_B + f_{33}M_B = 0 \qquad (14.23)$$

in which M is the moment at any section caused by the applied unit load on the arch in Fig. 14.41b; m_1, m_2, and m_3 are the moments caused by unit loads applied in the direction of the redundants (Fig. 14.41 d, e, and f, respectively); and $f_{11}, \ldots, f_{33}$ are the flexibility coefficients. The solution of Eqs. (14.21) to (14.23) will yield the required influence lines for the redundants. This is not a convenient method, and the analysis will be made much easier by following the column-analogy method (Section 15.4).

Equations (14.21) to (14.23) can also be derived by energy considerations as follows: For any section, distance x from B (Fig. 14.41b), the moment M is given by

$$M = R_B x - H_B y - M_B - (1)(x - a) \qquad \text{for } x > a;$$

and

$$M = R_B x - H_B y - M_B \qquad \text{for } x < a$$

Accounting for the strain energy due to bending only, we have

$$U = \int \frac{M^2 \, ds}{2EI}$$

Applying the condition of no yield at B, vertical, horizontal, or rotational, we can write

$$\frac{\partial U}{\partial R_B} = 0 \quad \text{or} \quad \int \frac{M}{EI} \frac{\partial M}{\partial R_B} \, ds = \int \frac{Mx \, ds}{EI} = 0$$

$$\frac{\partial U}{\partial H_B} = 0 \quad \text{or} \quad \int \frac{M}{EI} \frac{\partial M}{\partial H_B} \, ds = -\int \frac{M}{EI} y \, ds = 0$$

$$\frac{\partial U}{\partial M_B} = 0 \quad \text{or} \quad \int \frac{M}{EI} \frac{\partial M}{\partial M_B} \, ds = -\int \frac{M}{EI} \, ds = 0$$

The solution of the above three equations will yield the influence-line equations for R_B, H_B, and M_B.

14.14 QUALITATIVE INFLUENCE LINES

It has been shown that the Müller-Breslau principle is valuable in the determination of *quantitative influence lines,* that is, influence lines with definite numerical values for the ordinates. The principle can also be used to sketch *qualitative influence lines* so that the correct location of live load is found to produce maximum values of moment, thrust, shear, etc., in a given structure. Once the loading pattern is known, the structure so loaded can be analyzed by any of the methods studied (moment distribution, for example) to obtain the required moment, shear, etc.

Let us consider the six-span continuous beam shown in Fig. 14.42a; it is required to determine which spans should be loaded with a uniformly distributed live load to cause a maximum positive moment (compression on top side) at O, the center of span CD. We apply the Müller-Breslau principle by first removing the moment resistance at O by inserting a hinge at O and then apply equal and opposite couples at O, producing a unit relative rotation as well as compression on the top; the deflected shape is the influence line for M_0, as shown in Fig. 14.42b. Therefore, to obtain the critical live-load pattern for maximum M_0 we must *load all spans with deflections opposite to the direction of the applied load,* that is, spans *AB, CD,* and *EF* as shown

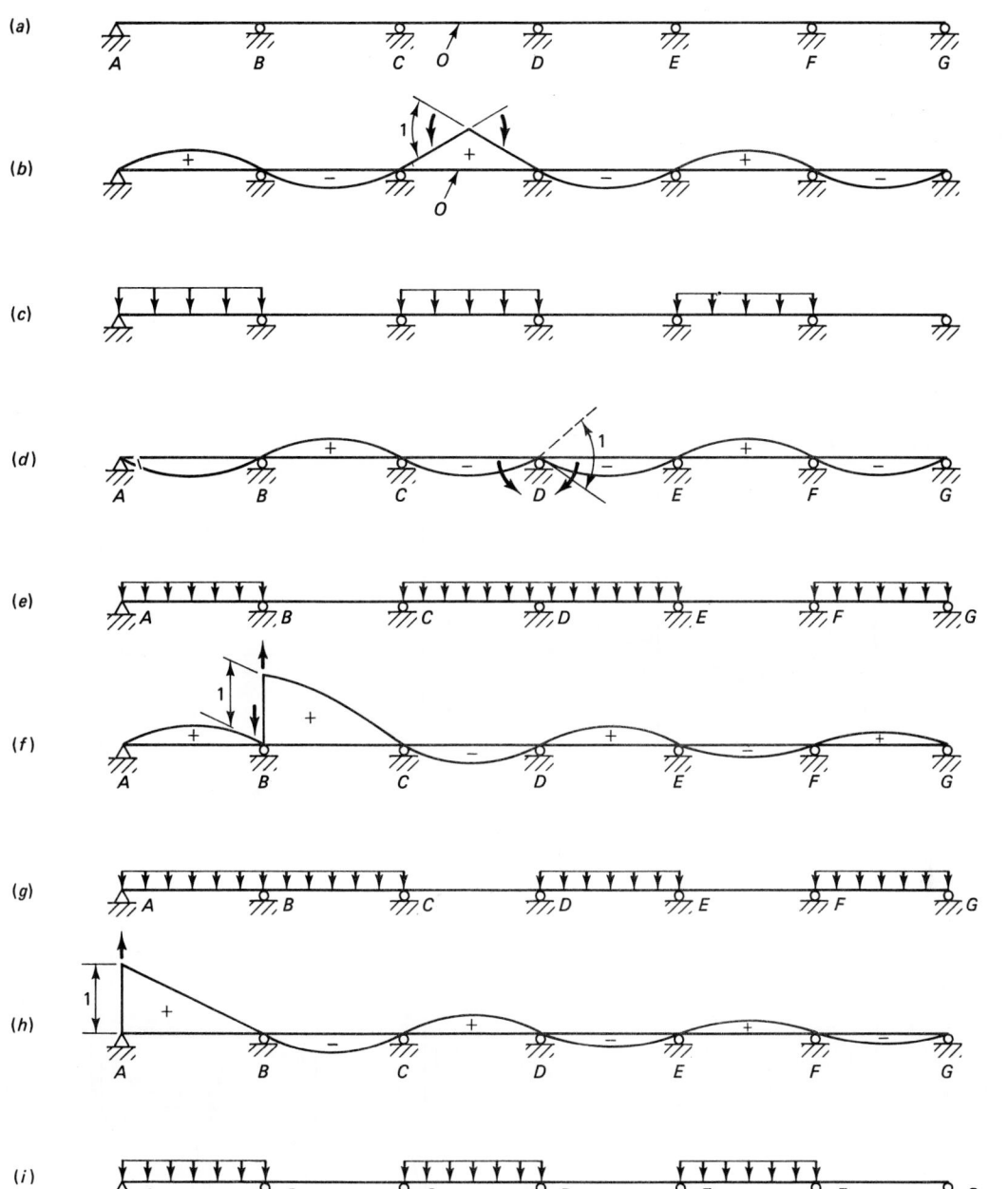

Figure 14.42 Use of qualitative influence lines for a continuous beam for the determination of critical load position. (*b*) Influence line for M_0. (*c*) Loading for maximum positive M_0. (*d*) Influence line for M_D. (*e*) Loading for maximum negative M_D. (*f*) Influence line for $(V_B)_R$. (*g*) Loading for maximum-positive $(V_B)_R$. (*h*) Influence line for R_A. (*i*) Loading for maximum positive R_A.

in Fig. 14.42c. Once the load pattern is known, we can conveniently find the maximum value of M_0 by moment distribution. It should be noted that deflections of spans remote from O are rather small and therefore the contribution of loads on such spans is normally negligible.

If the maximum negative moment at D is required, a hinge is inserted at D and equal and opposite couples (causing tension on the top side) are applied on each side of the hinge causing a relative unit rotation, deflecting the beam as shown in Fig. 14.42d; this deflected shape is the influence line for M_D. The loading pattern to produce maximum M_D is shown in Fig. 14.42e.

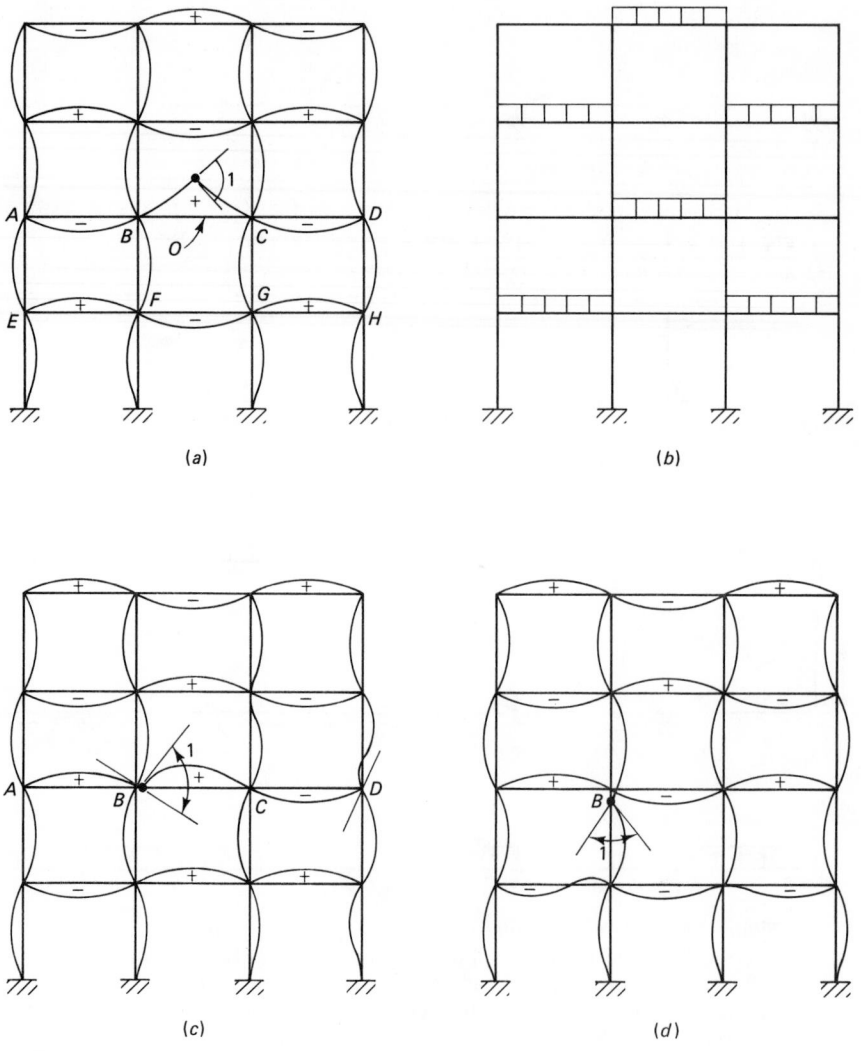

Figure 14.43 Use of qualitative influence lines for a multistory rigid frame for the determination of critical load position.

For maximum shear just to the right of *B,* the beam is first cut just to the right of *B,* a roller and slide device is inserted (see Fig. 14.10*d*), and two vertical forces of equal magnitude are applied to the beam ends, causing a unit relative transverse displacement. The resulting deflected shape shown in Fig. 14.42*f* is the influence line for $(V_B)_R$. The critical loading pattern for $(V_B)_R$ is given in Fig. 14.42*g*.

If the critical loading pattern for reaction R_A is required, the end *A* is displaced upward a unit distance and the deflected shape becomes the influence line for R_A shown in Fig. 14.42*h*; the corresponding loading pattern for maximum R_A is given in Fig. 14.42*i*.

In the analysis of multistory building frames, the qualitative influence-line method can be very useful in determining maximum live-load effects on girders and columns. Figure 14.43*a* shows a four-story rigid frame with three bays. If the loading pattern for the maximum positive moment at midspan *O* of girder *BC* is required, then by applying the Müller-Breslau principle we obtain the deflected shape, shown in Fig. 14.43*a*; the corresponding loading pattern for maximum M_O is given in Fig. 14.43*b*. The qualitative influence line for the maximum negative moment M_B at the left end of span *BC* is shown in Fig. 14.43*c*, and that for the maximum moment causing tension on the right side at the top of column *BF* is given in Fig. 14.43*d*. For maximum effects we load all spans with positive deflections.

As was mentioned earlier, the effect of the loaded span on the stress components in a span two or three bays distant is negligible. For this reason, a simplified framing of large building frames is permitted as shown in Fig. 14.44. Here it is assumed that the far ends of the column immediately above and below a given girder such as *AF* are fixed; furthermore, the ends of the girder two bays away from the section at which the stress component is being determined (e.g., at *O*) is also assumed fixed (e.g., at *D*). Thus, if the maximum positive moment at *O* is required, we load only spans *AB* and *CD*, as shown in Fig. 14.44. This simplifying procedure entails only a small loss in accuracy.

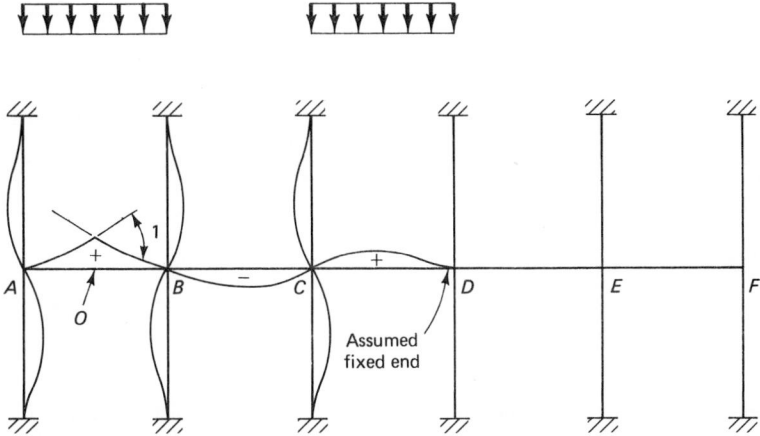

Figure 14.44 Simplified framing of large building frame for the determination of critical load position.

14.15 SOLVED PROBLEMS

■ SOLVED PROBLEM 14.1

For the continuous beam *ABCDE* shown in Fig. 14.45*a*, draw qualitative influence lines for the (a) positive moment at the middle of span *AB* (tension at bottom), (b) negative moment at support *B* (tension at top), (c) reaction at support *B* (upward direction), (d) negative moment at support *C* (tension at top). By means of moment distribution or otherwise, find the value of the maximum moment at *B* and the maximum reaction at *B*, for point loads *W*/2 at third points of span.

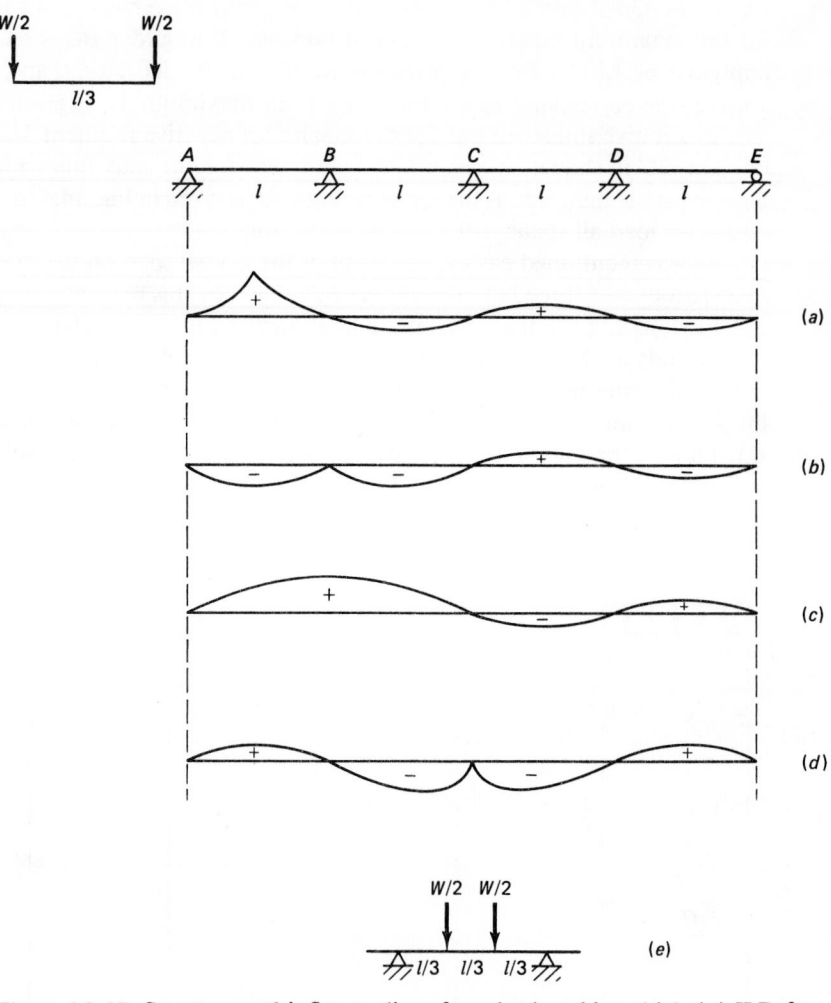

Figure 14.45 Structure and influence lines for solved problem 14.1. (*a*) ILD for maximum positive moment at midspan *AB* (load spans *AB, CD*). (*b*) ILD for maximum negative moment at *B* (load spans *AB, BC, DE*). (*c*) ILD for maximum R_B (load spans *AB, BC, DE*). (*d*) ILD for maximum negative moment at support *C* (load spans *BC, DE*).

Solution. Fixed-end moment from the Appendix,

$$FEM = \frac{2}{9}\left(\frac{W}{2}\right)l = 0.111\ Wl$$

It is left as an exercise for the student to show by the moment-distribution method that

Maximum negative moment at $B = -0.159\ Wl$

$$\text{Maximum reaction } R_B = W + \left(\frac{0.159Wl + 0}{l}\right) + \left(\frac{-0.028 + 0.159}{l}\right)Wl$$

$$= 1.29W\uparrow$$ ■

■ SOLVED PROBLEM 14.2

(a) Determine the influence-line equation for the moment at A of the structure shown in Fig. 14.46a. The arch is parabolic in form with EI varying as secant of the slope of the arch. The rigid frame ADC has the same E as that of the arch. (b) Determine the horizontal thrust at B for a concentrated vertical load of 100 kN, acting at the crown of the arch.

Solution
  (a) From Fig. 14.46b:

$$\Delta_1 = \frac{1}{EI_c}\int My\ dA$$

From Fig. 14.46c:

$$\Delta_2 = \frac{H}{EI_c}\int y^2\ dR$$

From Fig. 14.46d:

$$\Delta_3 = \frac{H}{3}\frac{(10)^3}{2EI_c}$$

Compatibility condition $\Delta_3 = \Delta_1 - \Delta_2$. Therefore,

$$\int My\ dA - H\int y^2\ dA = \frac{H(10)^3}{6}$$

M at a when $a < x$:

$$M = \frac{(L - x)(a)}{L} \qquad 0 \leqslant a \leqslant x$$

M at a when $a < L - x$, a taken from support,

$$M = \frac{x}{L}(a) \qquad 0 \leqslant a \leqslant (L - x)$$

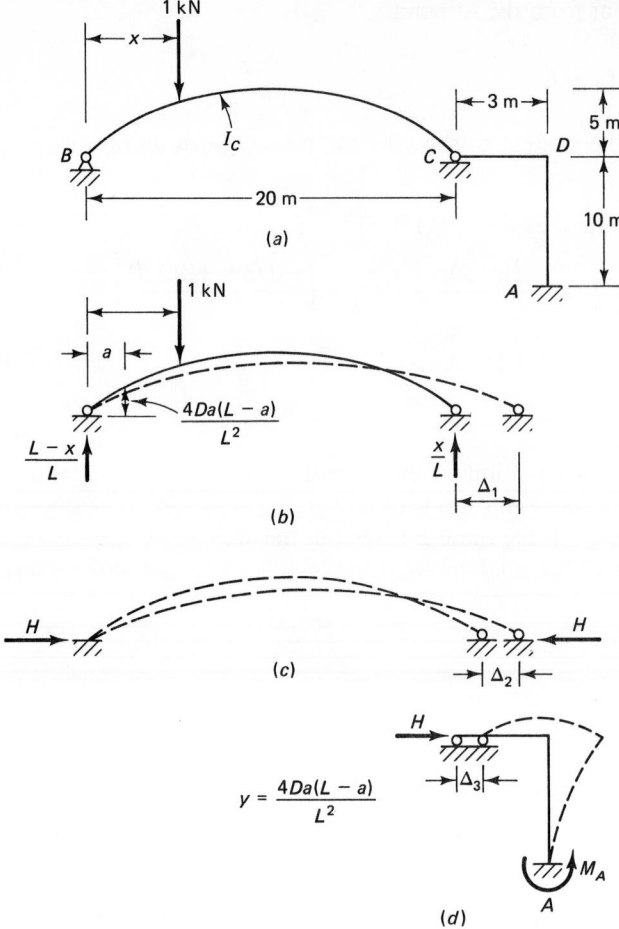

Figure 14.46 Structure for solved problem 14.2.

Therefore,

$$\int My\,da = \frac{4D}{L^2}\left[\frac{L-x}{L}\int_0^x a^2(L-a)da + \frac{x}{L}\int_0^{L-x} a^2(L-a)da\right]$$

$$= \frac{4D}{L^2}\left[\frac{L-x}{I}\left(\frac{a^3L}{3}-\frac{a^4}{4}\right)_0^x + \frac{x}{L}\left(\frac{La^3}{3}-\frac{a^4}{4}\right)_0^{L-x}\right]$$

$$= \frac{4D}{L^2}\left\{\frac{L-x}{L}\left(\frac{Lx^3}{3}-\frac{x^4}{4}\right)+\frac{x}{L}\left[\frac{L}{3}(L-x)^3-\frac{(L-x)^4}{4}\right]\right\}$$

$$= \frac{D(L-x)x}{3L^2}(L^2+Lx-x^2)$$

$$\int_0^L y^2 \, da = \frac{16D^2}{L^4} \int_0^L a^2(L-a)da = \frac{16D^2}{L^4}\left(\frac{a^3L^2}{3} - \frac{2La^4}{4} + \frac{a^5}{5}\right)_0^L$$

$$= \frac{8D^2L}{15}$$

Therefore,

$$\frac{D(L-x)x}{3L^2}(L^2 + Lx - x^2) = H\left(\frac{1000}{6} + \frac{8D^2L}{15}\right) = H\left(\frac{1000}{6} + \frac{800}{3}\right)$$

$$= H\left(\frac{1300}{3}\right)$$

I. L. for H is

$$H = \frac{1}{260(L^2)} x(L-x)(L^2 + Lx - x^2)$$

From Fig. 14.46d, $M_A = 10\,H$. I. L. for M_A is

$$M_A = \frac{1}{260(L^2)}(10x)(L-x)(L^2 + Lx - x^2)$$

(b) for a load of 100 kN at $x = 10$ m

$$H = \frac{1}{260(20)^2} 10(20-10)[20^2 + 10(20) - (10)^2] \times 100 \text{ kN}$$

$$= 48.08 \text{ kN} \qquad\qquad\qquad \blacksquare$$

■ SOLVED PROBLEM 14.3

The members of the rigidly jointed frame shown in Fig. 14.47 are of uniform cross section and material throughout. Find the influence line for bending moment in the beam at B. Use this influence line to find the approximate maximum value of the bending moment at B due to the passage from A to C of two axle loads of 2 k, 6 ft apart.

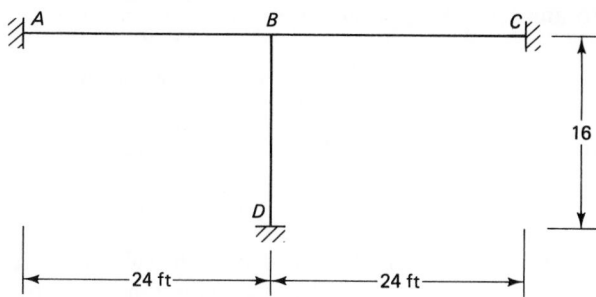

Figure 14.47 Structure for solved problem 14.3

Solution

$$\text{FEM at } A = \frac{-(1)(a)(24 - a)^2}{576}$$

$$\text{FEM at } B = \frac{+(1)(a)^2(24 - a)}{576}$$

$$\text{Distribution factor for } BA = \frac{I/L}{\sum I/L} = \frac{I/24}{I/24 + I/24 + I/16} = \frac{2}{7}$$

$$\text{Distribution factor for } BC = \tfrac{2}{7}$$

$$\text{Distribution factor for } BD = \tfrac{3}{7}$$

Joint	A	B			C
Member	*AB*	*BA*	*BD*	*BC*	*CB*
D.F.		$\frac{2}{7}$	$\frac{3}{7}$	$\frac{2}{7}$	
C.O.F.		$\leftarrow \frac{1}{7}$	$\frac{3}{14}$ to *BD*	$\frac{1}{7} \rightarrow$	
FEM	$-\dfrac{a(24-a)^2}{576}$	$\dfrac{a^2(24-a)}{576}$	0		
Balance	$-\dfrac{1}{7}\dfrac{a^2(24-a)}{576}$	$-\dfrac{2}{7}\dfrac{a^2(24-a)}{576}$	$-\dfrac{3}{7}\dfrac{a^2(24-a)}{576}$	$-\dfrac{2}{7}\dfrac{a^2(24-a)}{576}$	$-\dfrac{1}{7}\dfrac{a^2(24-a)}{576}$
		$\dfrac{5}{7}a^2\dfrac{(24-a)}{576}$		$-\dfrac{2}{7}a^2\dfrac{(24-a)}{576}$	

$$\text{I.L. ordinate for } M_{BA} \text{ due to unit load along span } AB = \frac{5}{7}a^2\frac{(24-a)}{576}$$

(*a* is measured from *A*)

$$\text{I.L. ordinate for } M_{BA} \text{ due to unit load along span } BC = -\frac{2}{7}a^2\frac{(24-a)}{576}$$

(*a* is measured from *C*)

Let the forward axle be *a* ft from *A* in the span *AB* for the maximum value of M_{BA}. The rear axle will be $(a - 6)$ ft from *A* in span *AB*. The live load is assumed to move from left to right.

$\therefore$ M_{BA} due to two axle loads of 2 k each

$$= (2)\left[\frac{5}{7}\frac{a^2(24-a)}{576}\right] + (2)\left\{\frac{5}{7}\frac{(a-6)^2[24-(a-6)]}{576}\right\}$$

$$= \frac{10}{7\times576}[a^2(24-a)+(a-6)^2(30-a)]$$

A maximum value of M_{BA} occurs when $dM_{BA}/da = 0$, or

$$\frac{10}{7\times576}[a^2(-1)+2a(24-a)+(a-6)^2(-1)+2(a-6)(30-a)] = 0$$

that is,

$$-a^2 + 48a - 2a^2 - (a-6)^2 + 2(a-6)(30-a) = 0$$

that is,

$$a^2 - 22a + 66 = 0$$

$$a = \frac{22a \pm \sqrt{(22)^2-(4)(66)}}{2} = 11 \pm \sqrt{55} = 11 \pm 7.42$$

$$= 18.42 \text{ ft} \quad \text{or} \quad 3.58 \quad \text{Therefore} \quad a = 18.42 \text{ ft}$$

The value of $a = 3.58$ is inadmissible, because the rear axle would not be in span AB.

$$\text{Maximum } M_{BA} = \frac{10}{7\times576}[a^2(24-a)+(a-6)^2(30-a)]$$

$$= \frac{10}{7\times576}[(18.42)^2(5.58)+(12.42)^2(11.58)]$$

$$= 9.126 \text{ k}\cdot\text{ft}$$

■ SOLVED PROBLEM 14.4

A unit load at B, applied to the simply supported truss AC (with reaction at B removed) (Fig. 14.48a) produces the following deflections at the lower joints: 0; 7.1; 12.4, 16.4; 12.4; 7.1; 0 (all $\times 10^{-2}$ in). Find the influence lines for R_A, R_B, and the force in DF, when the three reactions R_A, R_B, and R_C are acting.

Solution. The influence line of R_B is obtained by dividing the ordinates of the deflection curve (with unit load at B) by δ_{BB}.

I.L. for R_A

Unit load at A:

$$R_A = 1$$

Unit load at E, for value of $R_B = 0.433$:

$$R_A = \frac{1\times50-30R_B}{60} = 0.617$$

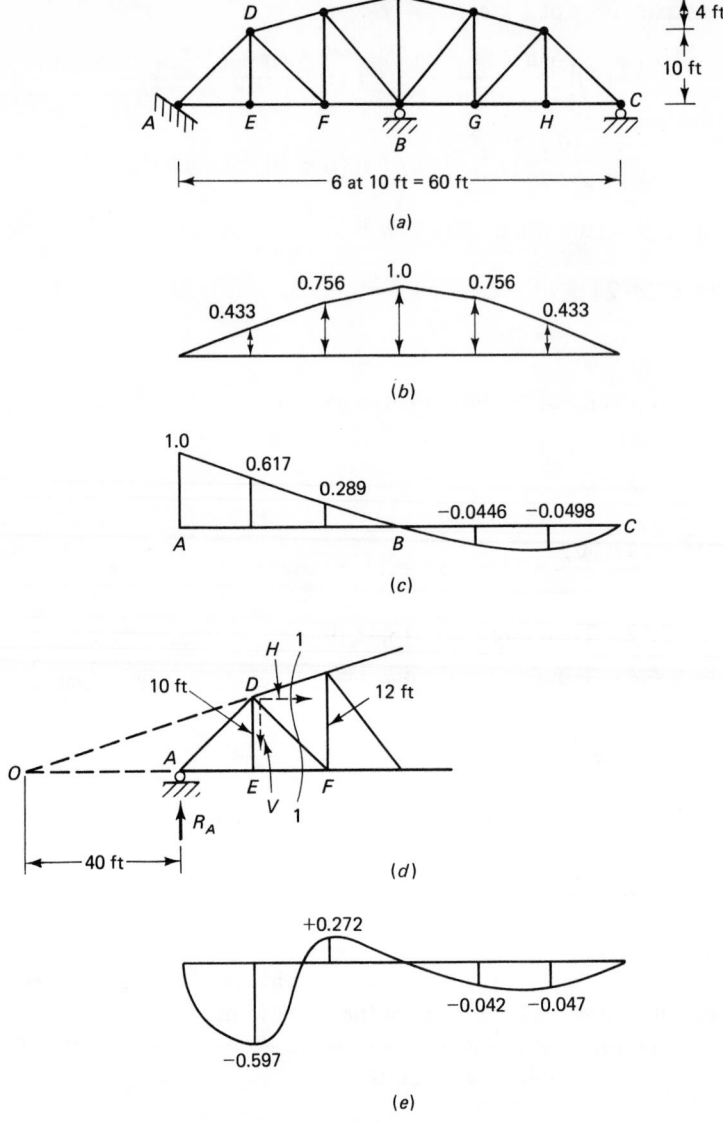

Figure 14.48 Structure and influence lines for solved problem 14.4. (*b*) Influence line for R_B. (*c*) Influence line for R_A. (*d*) Influence line for F_{DF}.

Unit load at F, for value of $R_B = 0.756$:

$$R_A = \frac{1 \times 40 - 30 R_B}{60} = 0.289$$

Unit load at B, for value of $R_B = 1.0$:

$$R_A = 0$$

Unit load at G:

$$R_A = \frac{1 \times 20 - 30R_B}{60} = -0.0446$$

Unit load at H:

$$R_A = \frac{1 \times 10 - 30R_B}{60} = -0.0498$$

Unit load at C:

$$R_A = 0$$

I.L. for F_{DF}:

Referring to Fig. 14.48*d,* from similar triangles

$$\frac{OF}{12} = \frac{OE}{10} = \frac{OF - OE}{12 - 10} = \frac{10}{2} = 5$$

$$OF = 60 \text{ ft}$$

$$OE = 50 \text{ ft}$$

To determine force in DF consider section ①-① (see Fig. 14.48*d*) and take moments about O of all the forces acting to the left of the section; let H and V be the components of F_{DF}.

Unit load at A:

$$F_{DF} = 0$$

Unit load at E, $\Sigma\, M$ about $O = 0$:

$$(H)(10) + (V)(50) + (1)(50) - R_A(40) = 0$$

But $H = V$

$$60V = -50 + 24.68$$

$$V = -0.422$$

$$F_{DF} = \sqrt{2}V = -0.597 \qquad \text{(compression)}$$

Unit load at F:

$$(H)(10) + (V)(50) - (R_A)(40) = 0$$

$$60V = 40R_A = 11.56$$

$$V = +0.193$$

$$F_{DF} = 0.272$$

Unit load at B:

$$F_{DF} = 0$$

Unit load at G:

$$(H)(10) + V(50) - R_A(40) = 0$$

$$\therefore V = -0.0297$$

$$F_{DF} = -0.042$$

Unit load at H:

$$(H)(10) + (V)(50) - R_A(40) = 0$$

$$V = -0.0332$$

$$F_{DF} = -0.047 \quad \text{(compression)}$$

PROBLEMS

14.1 to 14.4 Draw the influence-line diagrams for reactions for the beam shown.

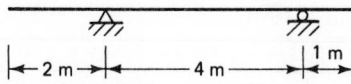

Figure P14.1

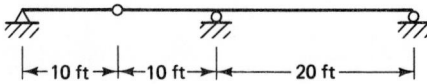

Figure P14.2

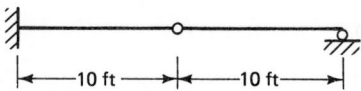

Figure P14.3

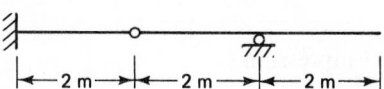

Figure P14.4

14.5 to 14.7 Draw influence lines for reactions for the trusses shown.

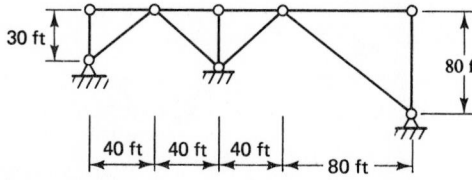

Figure P14.5

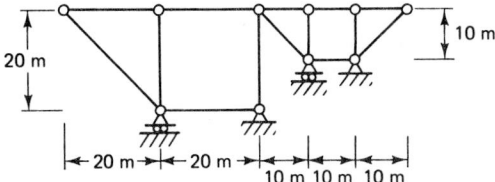

Figure P14.6

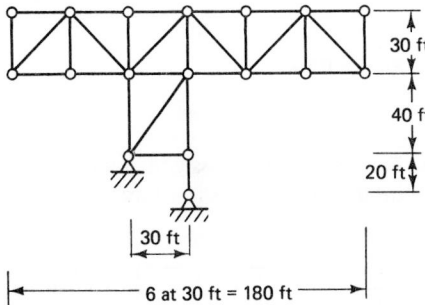

Figure P14.7

14.8 Draw influence-line diagrams for reactions at *A* and *D* as a unit load travels from *A* to *C*.

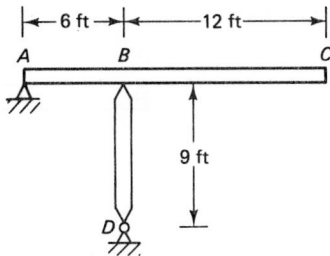

Figure P14.8

14.9 The beam *DE* supports two beams *AB* and *BC* as shown. Load can be applied to beams *AB* and *BC* only. Draw influence-line diagrams for bending moment and shear force at section *G*.

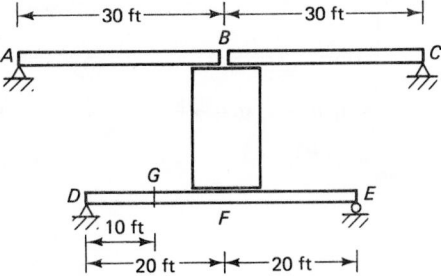

Figure P14.9

14.10 For the statically determinate beam shown, draw influence-line diagrams for **(a)** reactions at A and B, **(b)** shear force and bending moment at sections G, H, and J.

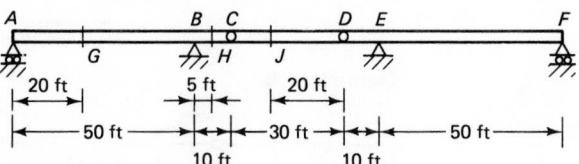

Figure P14.10

14.11 to 14.14 For the stringer–cross girder–longitudinal girder arrangements shown, draw influence lines for shear force in AB and bending moment at B.

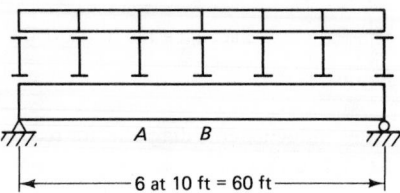

Figure P14.11

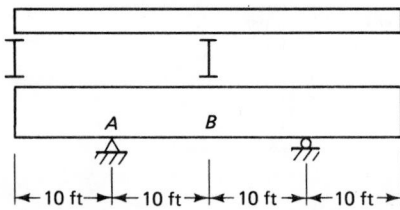

Figure P14.12

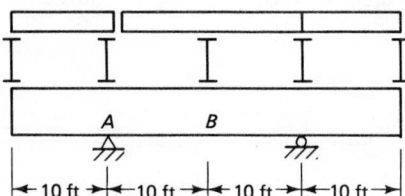

Figure P14.13

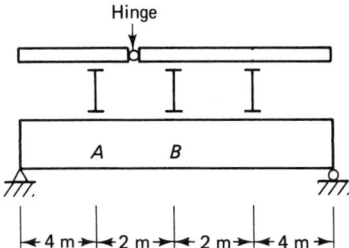

Figure P14.14

14.15 For the structure shown, draw influence-line diagrams for bending moment and shear force at sections E and F for a unit load traversing ABC.

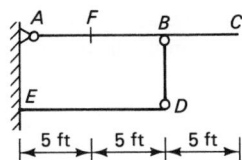

Figure P14.15

14.16 A three-hinged parabolic arch has a span of 100 m and a rise of 20 m. Draw influence-line diagrams for bending moment, shear force, and axial thrust at the left-hand quarter point.

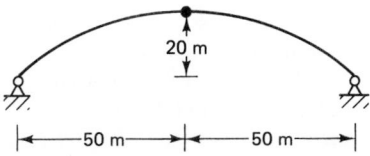

Figure P14.16

14.17 For the three-hinged portal frame shown, draw influence-line diagrams for horizontal thrust H and bending moment at section D. Determine the maximum bending moment at D due to **(a)** uniformly distributed dead load of 2 k/ft over the entire top horizontal girder, **(b)** a uniformly distributed live load of 4 k/ft covering part or all of top horizontal girder, and **(c)** a concentrated live load of 20 k which may be placed anywhere on the top horizontal girder.

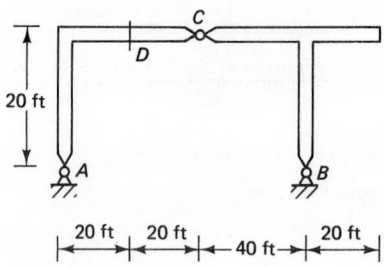

Figure P14.17

14.18 For the deck-type bridge truss shown, draw influence-line diagrams for members ①, ②, ③, and ④ as a unit load traverses from A to C. Determine the maximum tensile and compressive forces in each of the above four members for a uniformly distributed live load of 3 k/ft covering part or all of AC.

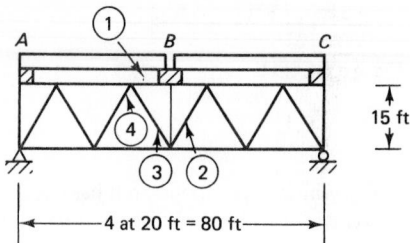

Figure P14.18

14.19 The subdivided truss shown is subjected to loading on the bottom chord. Draw the influence-line diagrams for forces in members ①, ②, ③, ④, ⑤, and ⑥. If the load consists of (a) dead load of 2 k/ft, and (b) a uniformly distributed live load of 3 k/ft covering any portion of the span, determine the maximum tensile and compressive forces in members ① to ⑥.

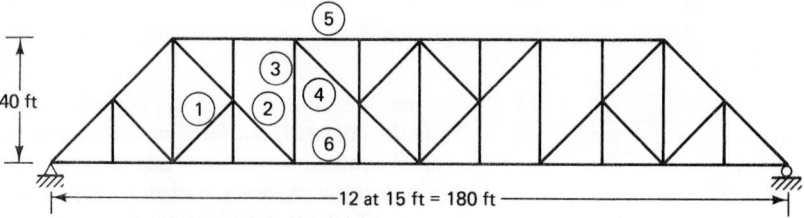

Figure P14.19

14.20 A three-hinged spandrel-braced arched bridge has a span of 100 m and a central rise of 20 m. It is hinged at the springing and crown. The load on the bridge consists of **(a)** a dead load of 10 kN/m, **(b)** a live load of 20 kN/m covering any portion of the span, and **(c)** a live load of 100 kN. Determine the maximum tensile and compressive forces in the members ①, ②, and ③.

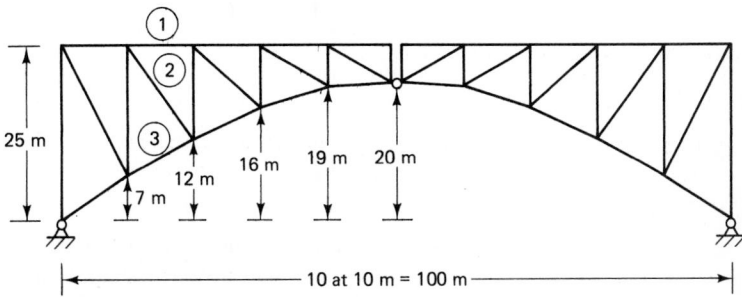

Figure P14.20

14.21 and 14.22 Draw the influence lines for the forces in bars marked ①, ②, ③, ④ of the trusses shown.

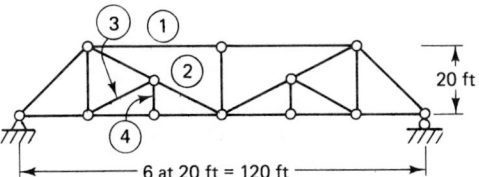

Figure P14.21

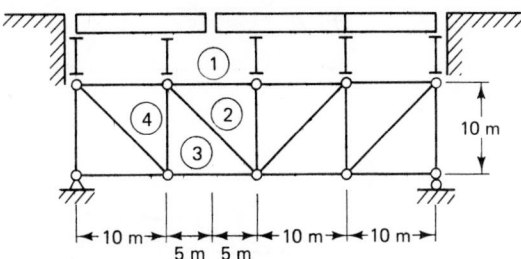

Figure P14.22

14.23 Two parabolic arches have a common hinge at B that forms part of a roller bearing there and are hinged to the springings at A and C. If EI_c is the same for both arches and $I = I_c \sec \theta$, determine the normal thrust and bending moment at the crown of the left-hand arch under the given load.

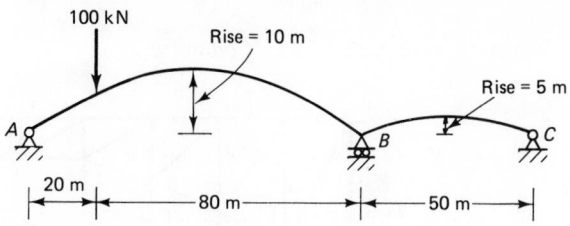

Figure P14.23

14.24 For the continuous beam $ABCDE$ shown, if a central concentrated load of 10 kN can act on any or all spans, what spans are to be loaded for producing maximum moment at B? By means of the moment-distribution method, find the value of the maximum moment at B and hence the maximum reaction at B.

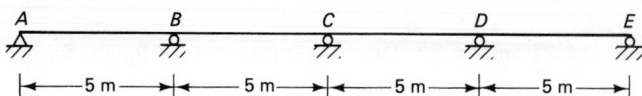

Figure P14.24

14.25 Construct the influence-line diagram for member 5-7 of the two-span continuous highway bridge truss shown in Fig. P14.25a. The influence line for the reaction at B is given in Fig. P14.25b. Also determine the maximum reaction at B and the maximum force in member 5-7 for the two-axle loading shown in Fig. P14.25c.

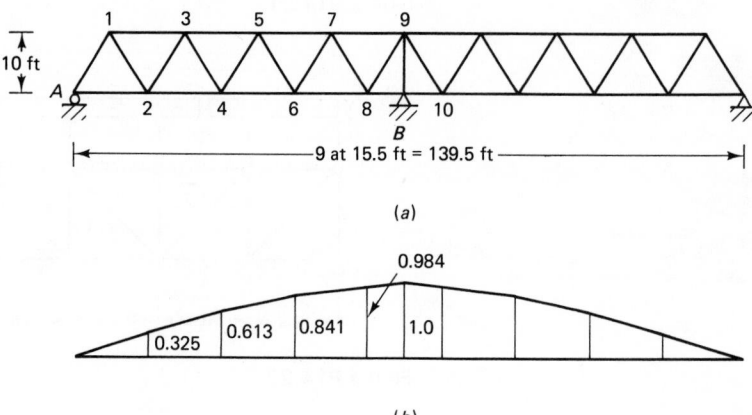

Figure P14.25

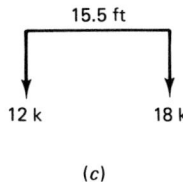

(c)

Figure 14.25 (*Continued*)

14.26 **(a)** Derive the equations for influence lines for M_{AB}, M_{BA}, V_A, and H_A of the rigid bent shown when a unit load travels along AB. The (I/L) values for members AB and BC are shown in circles. **(b)** Using the equation for the influence line for M_{BA}, compute the maximum value of M_{BA} due to the passage from A to B of two axle loads, 4 kips each, 8 ft apart.

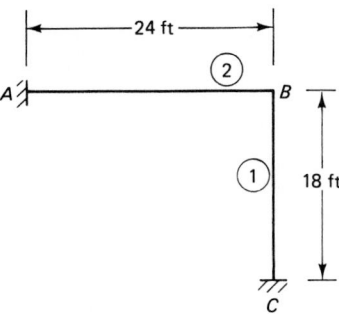

Figure P14.26

14.27 For the propped cantilever beam shown **(a)** determine the equation for the influence line for reaction at B, **(b)** draw the influence-line diagram for shear force at section C, **(c)** draw the influence-line diagram for bending moment at section C, **(d)** determine the maximum reaction at B due to the passage from A to B of a triangular load shown in Fig. P14.27b.

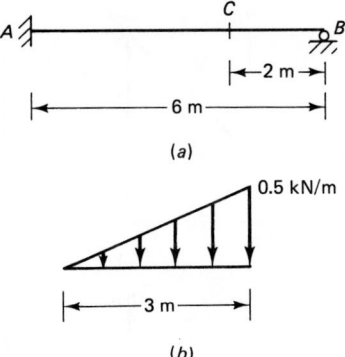

Figure P14.27

14.28 (**a**) The members of the rigidly jointed frame shown are of uniform cross section and material throughout. What is the equation for the influence line for M_{BA}? (**b**) Using the above influence line, find the maximum value of M_{BA} due to the passage from A to C of (1) two axle loads shown in Fig. P14.28b, (2) three axle loads shown in Fig. 14.28c.

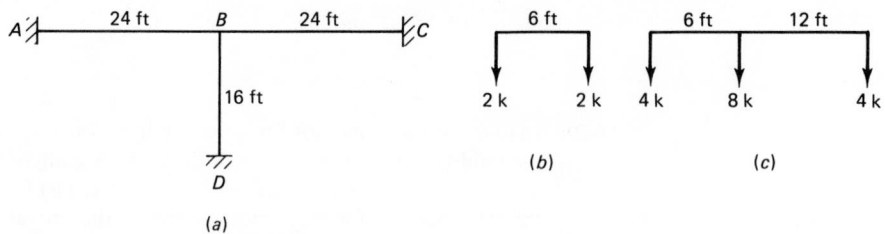

(a)

(b)   (c)

Figure P14.28

14.29 Find the equation for the influence line for the horizontal thrust H in the double parabolic arch shown. The arches are identical and their second moments of area vary according to the "secant law." Hence determine the value of H when a load of 20 kN is placed on each of the two arch crowns.

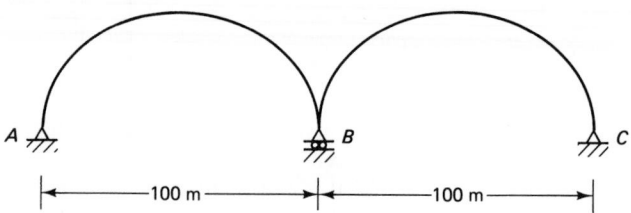

Figure P14.29

14.30 For the through-type Pratt truss shown, (**a**) draw the influence-line diagrams for R_B and R_A, (**b**) determine the maximum values of R_B and R_A for a moving trailer load shown in Fig. P14.30b.

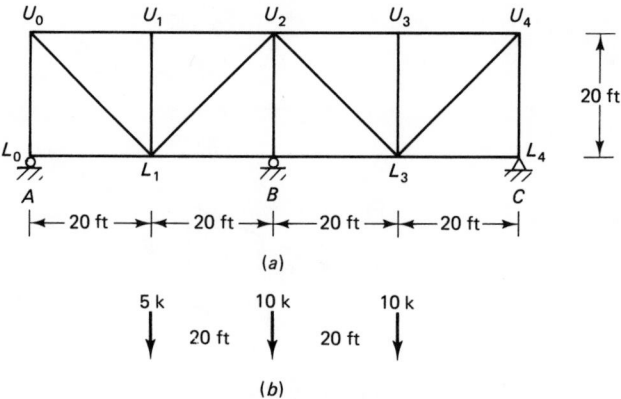

(a)

(b)

Figure P14.30

14.31 The swing bridge ABC shown is in closed position with the floor system at the lower chord level. It is known that a unit vertical load at B, applied to the simply supported bridge AC (with support B removed) produces the following vertical deflections at the lower joints: 0, 0.10, 0.16, 0.20, 0.16, 0.10, 0 in. **(a)** Sketch the influence lines for the vertical reactions R_B and R_A and for the force in member a. **(b)** What is the maximum force in member a due to the moving trailer load shown in Fig. P14.31b?

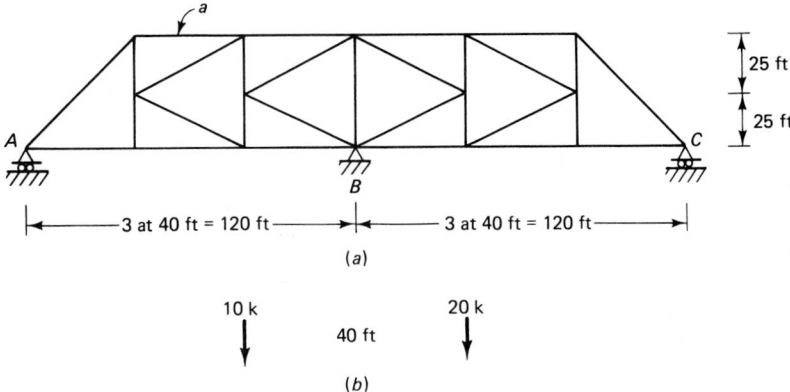

(a)

(b)

Figure P14.31

14.32 A tied parabolic arch of span 100 ft has a rise of 10 ft. Draw the influence-line diagrams for **(a)** bending moment, shear force, and thrust at a section F, 20 ft from the left-hand springing; **(b)** using these influence lines, determine the bending moment, shear force, and thrust at F due to the concentrated loads shown. Assume secant formula for the variation of moment of inertia of the arch. Given

$$\frac{E_{\text{arch}} I_{\text{crown}}}{(AE)_{\text{tie}}} = 4 \text{ ft}^2$$

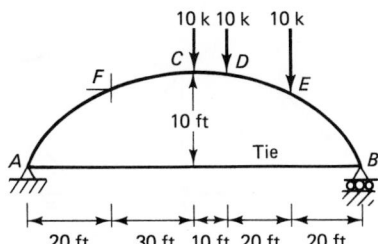

Figure P14.32

14.33 A two-hinged parabolic arch of span 100 ft has a rise of 10 ft. Draw the influence-line diagrams for bending moment, shear force, and thrust at section F, 20 ft from the left-hand springing. Using these influence-line diagrams, determine the bending moment, shearing force, and thrust at this section due to the concentrated loads shown.

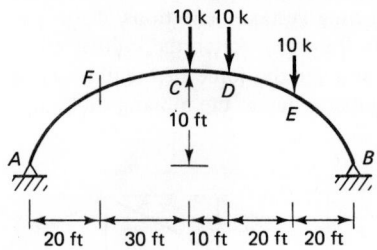

Figure P14.33

14.34 A two-hinged parabolic arch of span 30 m has a rise of 3 m. Draw the influence-line diagrams for bending moment, shear force, and thrust at a section F, 6 m from the left-hand springing. Using these influence lines, determine the bending moment, shear force, and thrust at this section due to the concentrated loads shown.

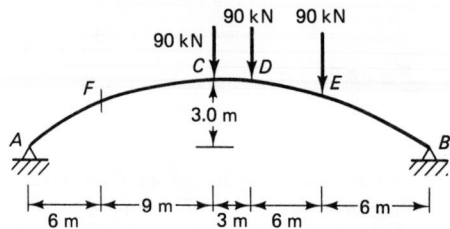

Figure P14.34

14.35 A unit load at B, applied to the simply supported truss AC (with reactions at B removed), produces the following deflections at the lower joints: 0, 1.80, 3.15, 4.15, 3.15, 1.80, 0 mm. Draw the influence-line diagrams for R_A, R_B, and the force in member DF, when the three reactions R_A, R_B, and R_C are acting.

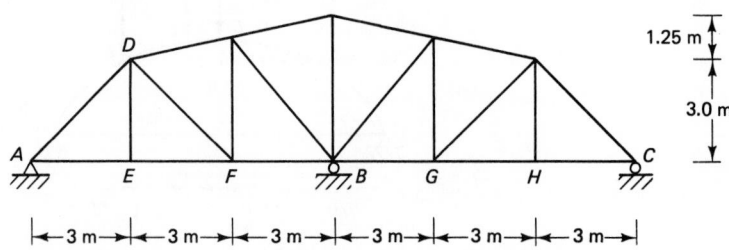

Figure P14.35

14.36 A model of the girder shown in Fig. P14.36a is constructed to a linear scale of 1:30. When point B is displaced 2.16 in, the model takes up the shape indicated in Fig. P14.36b. Determine R_A, R_B, R_C, and M_B when the prototype girder carries the loads shown in Fig. P14.36c, and sketch the bending-moment diagram.

An alternate way of carrying out the model test would have been to insert a hinge in the model girder at B and to impose an angular displacement there of BA relative to BC, keeping B at the same level. If an angular displacement of 5° had been imposed, what shape would the model have taken up?

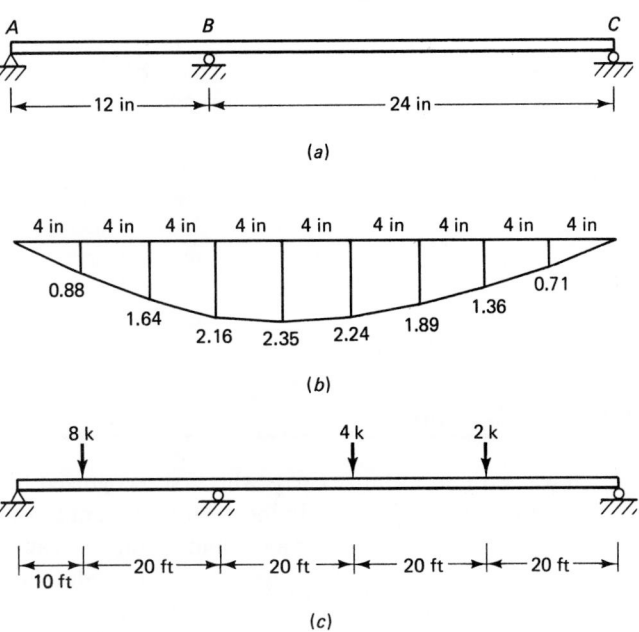

Figure P14.36 (a) Model. (b) Deflected mode. (c) Prototype girder and loading.

14.37 The figure shows a two-hinged double cantilever framed girder for a bridge. The flexural rigidity of the girder is constant between the columns and is three times that of the columns. Draw the influence-line diagrams for the horizontal thrust as a unit load passes from X to Y.

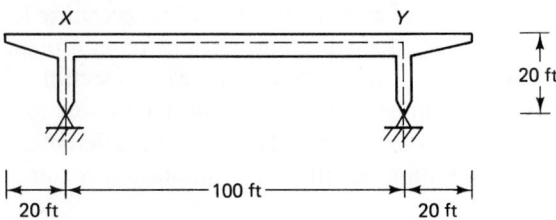

Figure P14.37

Chapter 15

The Method of Column Analogy

Where there is much desire to learn, there of necessity will be much arguing, much writing, many opinions; for opinion in good men is but knowledge in the making.

John Milton (1644)

15.1 INTRODUCTION

It was Professor Hardy Cross who first recognized the analogy between the equations for slopes and deflections in indeterminate bent beams, plane frames, arches, or closed boxes and the equations for loads and moments in an eccentrically loaded short column; the moments in these structures caused by redundant reactions (*not exceeding three*) are analogous to the fiber stresses in an eccentrically loaded short column. The method is most useful in deducing the stiffness factors and carryover factors and fixed-end moments necessary for the analysis of structures with members of varying moments of inertia.

15.2 THE COLUMN-FLEXURE EQUATION

In order to understand the analogy we restate the column-flexure equation. Consider a short column of area A subjected to an eccentric load P acting at a point $L(x_P, y_P)$ as shown in Fig. 15.1. If plane sections remain plane and if stress is proportional to strain, the stress σ at any point on the cross section of the short column shown in Fig. 15.1 will vary linearly with the x and y coordinates of that point from the centroidal axis of the cross section. If dA represents a differential area of the column cross section, then the condition equations of equilibrium require that

$$P = \int \sigma \, dA \qquad (15.1)$$

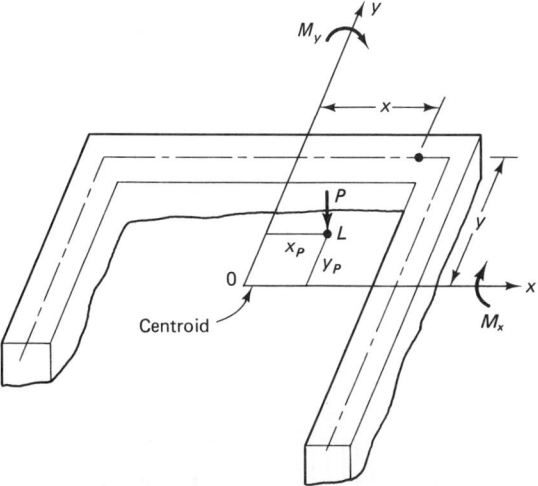

Figure 15.1 Short column subjected to an eccentric load.

$$Py_p = \int \sigma \, dA \cdot y = M_x \tag{15.2}$$

$$Px_p = \int \sigma \, dA \cdot x = M_y \tag{15.3}$$

Thus, it can be shown that the stress

$$\sigma = \frac{P}{A} + \left(\frac{M_y I_x - M_x I_{xy}}{I_x I_y - I_{xy}^2}\right)x + \left(\frac{M_x I_y - M_y I_{xy}}{I_x I_y - I_{xy}^2}\right)y \tag{15.4}$$

which is the column-flexure equation in terms of the second moments of area (I_x, I_y) and mixed product of area (I_{xy}). If the cross section is symmetrical with respect to one of the centroidal rectangular reference axes, then $I_{xy} = 0$. As a result Eq. (15.4) reduces to

$$\sigma = \frac{P}{A} + \left(\frac{M_y}{I_y}\right)x + \left(\frac{M_x}{I_x}\right)y \tag{15.5}$$

15.3 DEVELOPMENT OF THE ANALOGY

Consider the rigid frame shown in Fig. 15.2a. It is indeterminate to the third degree; the rigid frame can be resolved into two structures, a statically determinate and stable structure subjected to external loads as shown in Fig. 15.2b and an identical structure acted upon by the three unknown redundant reactions at A shown in Fig. 15.2c. The frames in Fig. 15.2b and c will displace as shown.

Let M_s be the moment at any section of the statically determinate structure in

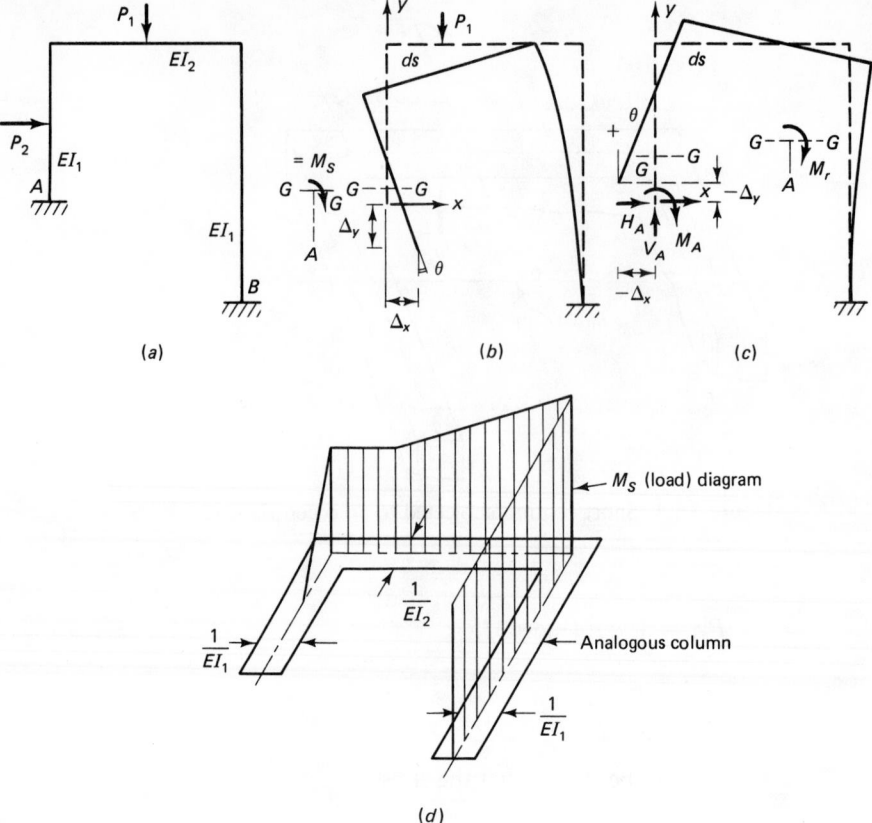

Figure 15.2 Development of the column analogy.

Fig. 15.2*b*. From moment-area considerations we can write the following expressions for the three displacement components of point A:

$$\theta = \int \frac{M_s}{EI} \, ds$$

$$\Delta_x = \int \frac{M_s}{EI} \, ds \cdot y \tag{15.6}$$

$$\Delta_y = \int \frac{M_s}{EI} \, ds \cdot x$$

in which ds is an element of length in the undistorted frame. If M_r represents the moment at any section of the structure in Fig. 15.2*c* caused by M_A, V_A, and H_A, then the three component displacements of point A must be numerically equal to those given in Eq. (15.6), since support A in the original structure (Fig. 15.2*a*) does not displace. Thus, for the structure in Fig. 15.2*c* we can write

$$-\theta = \int \frac{M_r}{EI} \, ds$$

$$-\Delta_x = \int \frac{M_r}{EI} \, ds \cdot y \tag{15.7}$$

$$-\Delta_y = \int \frac{M_r}{EI} \, ds \cdot x$$

or

$$-\int \frac{M_s}{EI} \, ds = \int \frac{M_r}{EI} \, ds$$

$$-\int \frac{M_s}{EI} \, ds \cdot y = \int \frac{M_r}{EI} \, ds \cdot y \tag{15.8}$$

$$-\int \frac{M_s}{EI} \, ds \cdot x = \int \frac{M_r}{EI} \, ds \cdot x$$

The left-hand side of the first of Eqs. (15.8) can be assumed to be an elastic load P; thus the left-hand sides of the second and third of Eqs. (15.8) are the first moments M_x and M_y of this elastic load about the x and y axes, respectively. Thus, we can rewrite Eqs. (15.8) as

$$P = \int \frac{M_r}{EI} \, ds$$

$$M_x = \int \frac{M_r}{EI} \, ds \cdot y \tag{15.9}$$

$$M_y = \int \frac{M_r}{EI} \, ds \cdot x$$

A comparison of Eq. (15.9) and Eqs. (15.1) to (15.3) shows that they are analogous, that is, M_r is analogous to the fiber stresses σ in a short column; $\int (1/EI)ds$ in a bent structure corresponds to $\int dA$ in a short column; and $\int (M_s/EI)ds$ corresponds to an elastic load P on the short column and hence the *column analogy*.

To use the method, the following conditions must be met:

1. The cross section of the *short* analogous column has a centerline similar in shape and length to the centerline of the real structure.
2. The width at any section of the column is proportional to $1/EI$ of the corresponding section of the real structure.
3. The load on any differential area of the column will be $(M_s/EI)ds$.

The loaded analogous column for the rigid frame in Fig. 15.2a is in Fig. 15.2d. Having determined M_r based on the stresses in the analogous column the final moments for the frame in Fig. 15.2a will be

$$M = M_r + M_s$$

15.4 REFERENCE AXES AND SIGN CONVENTION

For structures with both supports fixed (Fig. 15.3a and c), the rectangular reference axes will pass through the centroid of the elastic area of the corresponding analogous column as shown in Fig. 15.3b and d. If one support is hinged (Figs. 15.3e and g), the reference axes pass through the point on the analogous column corresponding to the hinge on the real structure. The reason for this is that a hinge is incapable of

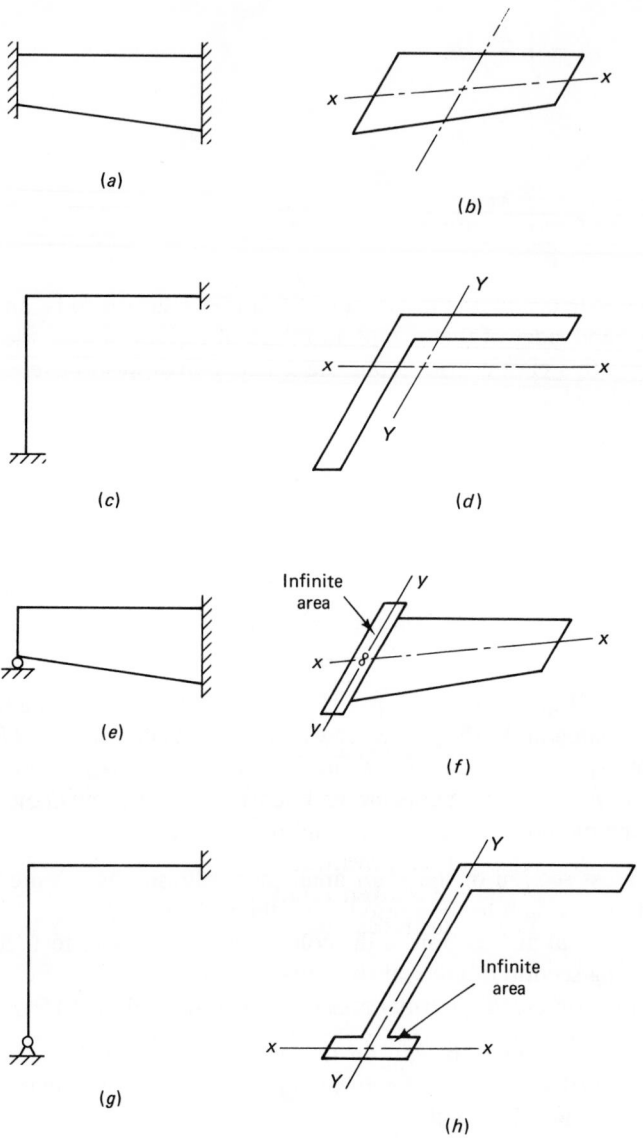

Figure 15.3 Analogous columns for different structures.

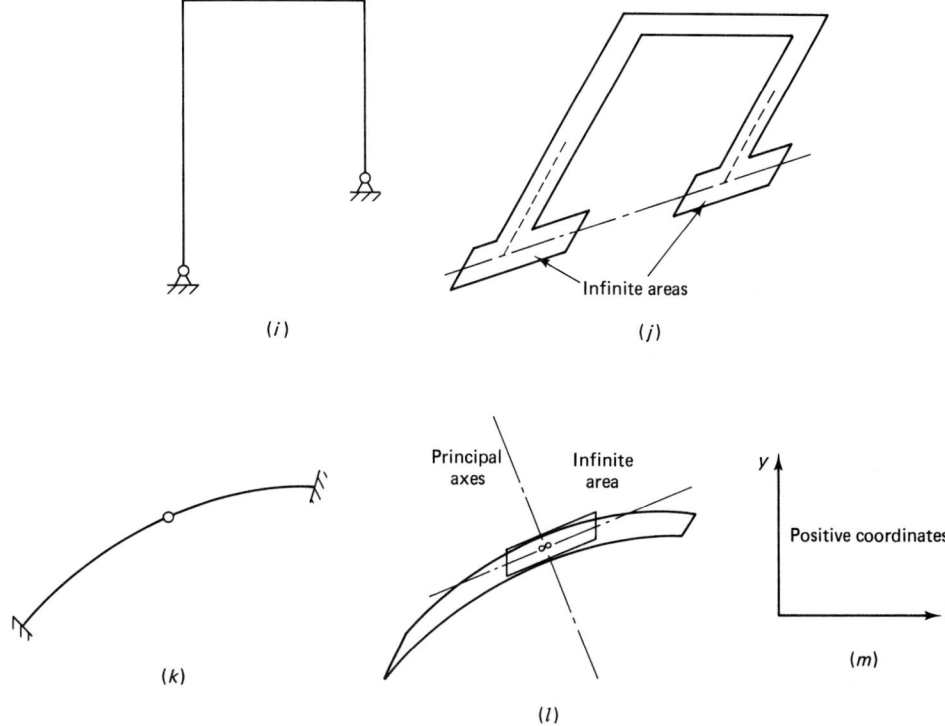

Infinite areas

(i)   (j)

Principal   Infinite
axes   area

Positive coordinates

(k)   (l)   (m)

Figure 15.3 (*Continued*)

resisting moment and therefore its $I = 0$; with $I = 0$, the width $1/EI$ of the analogous column at the hinge becomes infinite; this is shown in Fig. 15.3f, h, and l. If both supports are hinged (Fig. 15.3i), then at the points on the analogous column corresponding to the hinges in the real structure infinite elastic areas are present as shown in Fig. 15.3j.

The following sign convention is used:

1. Positive bending moments produce tension on the underside of a beam or the inside of a bent.
2. Positive (M_s/EI) load acts downward and negative (M_s/EI) load acts upward.
3. Tensile stress in column is taken as positive while compressive stress is negative.
4. Positive coordinates are measured upward and to the right of the origin, as shown in Fig. 15.3m.

The use of the column-analogy method in analyzing structures indeterminate to the third degree will be illustrated by the following example. It should be noted that the choice of a basic stable structure is purely arbitrary.

■ **EXAMPLE 15.1**
Determine the end moments for the rigid frame in Fig. 15.4a.

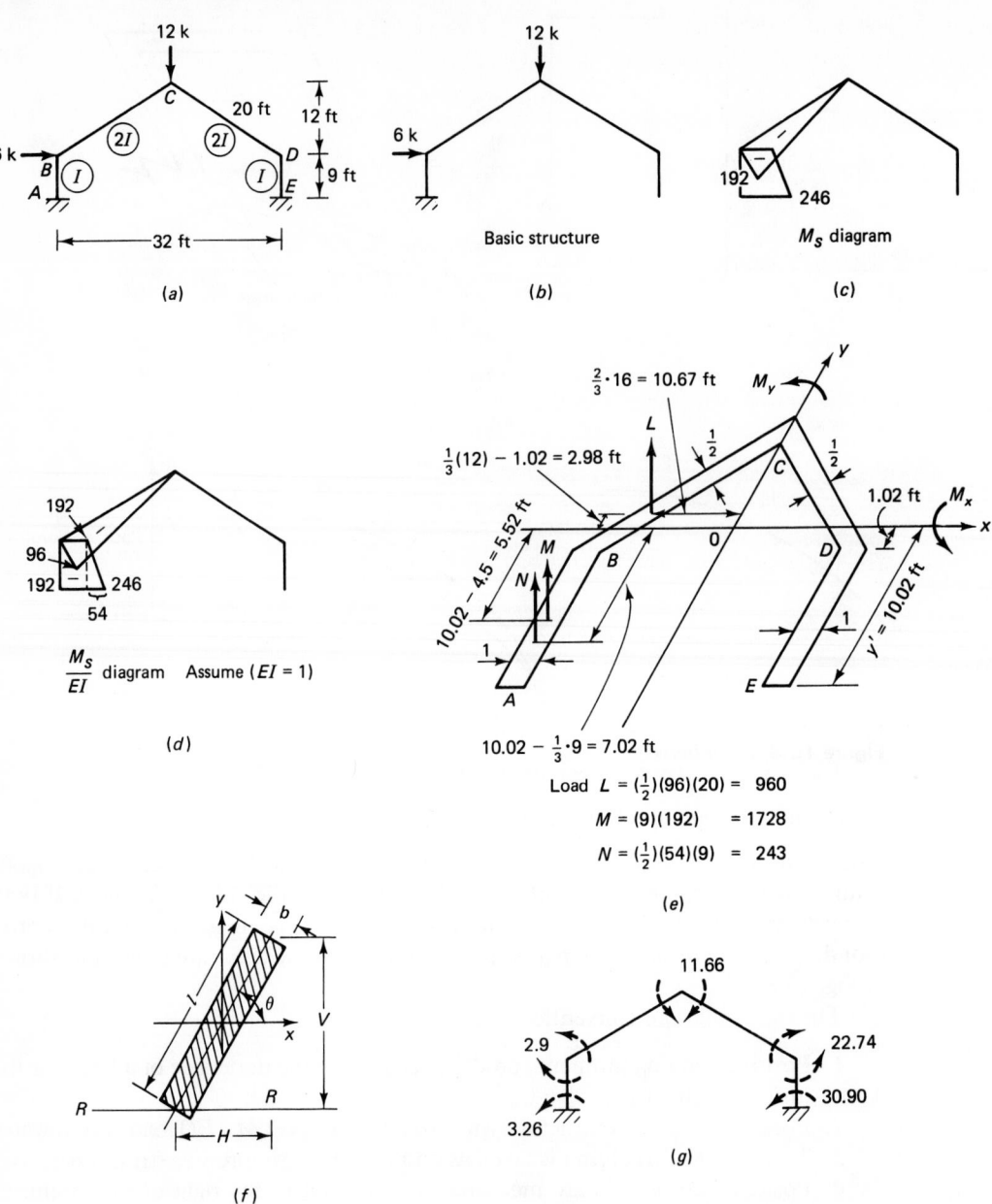

Figure 15.4 Structure for Example 15.1. (*f*) Inclined member of a frame. (*g*) End moment in k·ft.

Solution. The basic determinate structure is chosen as shown in Fig. 15.4b and its corresponding M_s and M_s/EI diagrams are presented in Figs. 15.4c and d. The analogous column is shown loaded by the M_s/EI load in Fig. 15.4e. Assuming $EI = 1$, the geometric properties of the analogous column are as follows:

$$\text{Area } A = 2[(1)(9)] + 2\left[\left(\frac{1}{2}\right)(20)\right] = 38$$

$$\bar{y} = \frac{2[(1)(9)(4.5) + (1/2)(20)(9 + 12/2)]}{38} = 10.02 \text{ ft}$$

Before calculating I_x and I_y, we must determine the moment of inertia of an inclined member such as the one shown in Fig. 15.4f; from basic mechanics, it can be shown that

$$I_x = \frac{bl}{12} V^2$$

$$I_y = \frac{bl}{12} H^2 \qquad\qquad (15.10)$$

$$I_{xy} = \frac{bl}{12} VH$$

and the moment of inertia about an arbitrary axis RR parallel to the x axis is

$$I_R = I_x + (\text{area})(\text{distance between the parallel axes})^2 \qquad (15.11)$$

Thus, for the area of the analogous column in Fig. 15.4e:

$$I_x = 2\left[\left(\frac{1}{12}\right)(1)(9^3) + (1)(9)(5.52)^2\right]$$

$$+ 2\left[\left(\frac{1}{2}\right)(20)\frac{(12)^2}{12} + (20)\left(\frac{1}{2}\right)(4.98)^2\right]$$

$$= 1406$$

$$I_y = 2\left[(1)(9)(16)^2 + \left(\frac{1}{2}\right)\left(\frac{20}{12}\right)(16)^2 + \left(\frac{1}{2}\right)(20)(8)^2\right]$$

$$= 6315$$

$$I_{xy} = 0 \qquad \text{from symmetry}$$

Loads on the analogous column are as follows:

$$L = \left(\frac{1}{2}\right)(96)(20) = 960\uparrow$$

$$M = (9)(192) = 1728\uparrow$$

TABLE 15.1

Point	$\dfrac{P}{A}$	$\dfrac{M_x}{I_x}y$	$\dfrac{M_y}{I_y}x$	M_r	M_s	$M = M_s + M_r$
A	+77.1	−5.96(−10.02) = +59.72	−6.62(−16) = +105.92	+242.74	−246	−3.26
B	+77.1	−5.96(−1.02) = +6.08	−6.62(−16) = +105.92	+189.10	−192	−2.90
C	+77.1	−5.96(+10.98) = −65.44	−6.62(0) = 0	+11.66	0	+11.66
D	+77.1	−5.96(−1.02) = +6.08	−6.62(16) = −105.92	−22.74	0	−22.74
E	+77.1	−5.96(−10.02) = +59.72	−6.62(16) = −105.92	+30.90	0	+30.90

and

$$N = \left(\frac{1}{2}\right)(54)(9) = 243\uparrow$$

Therefore,

$$P = 960 + 1728 + 243 = 2931\uparrow$$

$$\frac{P}{A} = \frac{2931}{38} = +77.1$$

Assuming that M_x and M_y act counterclockwise, producing tension in the first quadrant, we have

$$M_x = (960)(2.98) + (1728)(-5.52) + (243)(-7.02) = -8384$$

$$M_y = (960)(-10.67) + (1728)(-16) + (243)(-16) = -41,779$$

$$\frac{M_x}{I_x} = \frac{-8384}{1406} = -5.96$$

$$\frac{M_y}{I_y} = \frac{-41,779}{6315} = -6.62$$

The calculations are shown in Table 15.1
The end moments are shown in Fig. 15.4g.   ■

15.5 DETERMINATION OF STIFFNESS AND CARRYOVER FACTORS BY THE COLUMN-ANALOGY METHOD

The column-analogy method is most useful in deducing the stiffness factors and carryover factors for members of variable cross section. Let us first demonstrate the application to a beam of constant cross section.

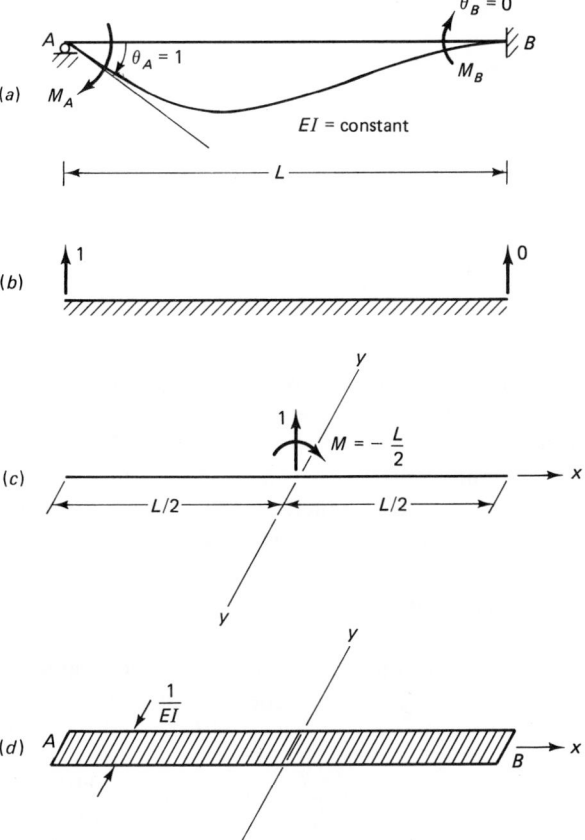

Figure 15.5 Loadings on analogous column for the determination of stiffness and carryover factors when the far end is fixed.

Consider the beam AB in Fig. 15.5a, fixed at B and subjected to moment M_A causing a rotation at A, $\theta_A = 1$ radian. Let us now recapitulate and define the stiffness factor K and the carryover factor C for the beam AB. The stiffness factor at A, K_A, is the moment applied at end A causing a unit rotation at A while end B is fixed, while the carryover factor C from A to B, C_{AB}, is the ratio of the moment induced at B due to a moment applied at A, that is, M_B/M_A. Thus from the above definitions,

$$M_A = K_A \theta_A \quad \text{and} \quad C_{AB} = \frac{M_B}{M_A}$$

For unit rotation at A, $\theta_A = 1$, $M_A = K_A$, and $M_B = C_{AB}K_A$. Since K_A and C_{AB} are *unknown*, intensity of load on the analogous column (Fig. 15.5d) is *not known*. However, we do know that:

1. $\theta_A = 1$, $\theta_B = 0$; hence from the area-moment theorem,

$$\Delta\theta = \theta_B - \theta_A = -1$$

But from the first of Eqs. (15.6),

$$\Delta\theta = \int \frac{M_s}{EI} ds = -P$$

Therefore, $P = 1$.

2. Since there is no relative vertical displacement at A, then from the third of Eqs. (15.6),

$$\Delta y = \int \frac{M_s \, ds}{EI} x = 0$$

or the first moment of the load $(M_s \, ds / EI)$ about A must be zero.

To satisfy conditions 1 and 2, the equivalent load on the analogous column must be a concentrated load of 1 radian at A, that is, at the end that is rotated (Fig. 15.5b). Therefore, it is not necessary in this case to convert the real beam into a basic determinate beam since the equivalent loading on the analogous column is now known without this conversion. It should be noted that the moment at A produces tension at the bottom of the beam and therefore it is considered positive; thus the stress in the corresponding analogous column at A must also be tensile. To accommodate this, the unit load A applied to the column must be tensile and therefore upward as shown in Fig. 15.5b.

For the constant section beam AB in Fig. 15.5a, the analogous column (Fig. 15.5d) is loaded as shown in Fig. 15.5b or the equivalent loading presented in Fig. 15.5c, where $M_y = (1)(-L/2) = -L/2$. Thus, with $I_y = (\frac{1}{12})(1/EI)L^3$,

$$\sigma_A = M_A = \frac{P}{A} + \frac{M_y}{I_y} x = \frac{1}{L/EI} + \frac{(-L/2)}{L^3/12EI}\left(-\frac{L}{2}\right) = +\frac{4EI}{L}$$

$$\sigma_B = M_B = \frac{1}{L/EI} + \frac{(-L/2)(L/2)}{L^3/12EI} = -\frac{2EI}{L}$$

The positive sign for σ_A and negative sign for σ_B indicate that the bending moments at A and B are positive and negative, respectively. Now $M_A = K_A\theta_A$ and since $\theta_A = 1$,

$$K_A = \frac{4EI}{L}$$

and

$$C_{AB} = \frac{|M_B|}{M_A} = \frac{1}{2}$$

as expected. Let us now calculate the stiffness and carryover factors for the beam AB in Fig. 15.6a, where the end B is hinged. Here, the area of the analogous column $A = \infty$ as shown in Fig. 15.6c; thus $P/A = 0$.

$$M_y = (1)(-L) = -L$$

$$I_y = \frac{L^3}{3EI}$$

(a)

(b)

(c)

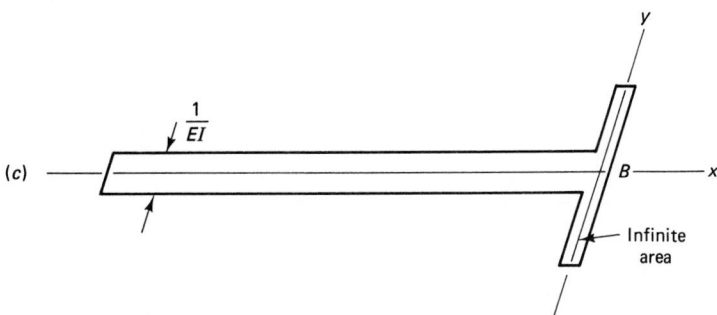

Figure 15.6 Loadings on an analogous column for the determination of stiffness and carryover factors when the far end is hinged.

Hence,

$$\sigma_A = M_A = 0 + \frac{M_y}{I_y}(-L) = \frac{-L}{L^3/3EI}(-L) = \frac{3EI}{L}$$

$$\sigma_B = M_B = 0 + \frac{M_y}{I_y}(0) = 0$$

With $\theta_A = 1$, and from $M_A = K_A\theta_A$,

$$K_A = \frac{3EI}{L}$$

and

$$C_{AB} = \frac{M_B}{M_A} = 0 \qquad \text{(there is no carryover to a hinge)}$$

We can also determine the fixed-end moments in a beam (Fig. 15.7a), when one end is displaced, without rotation, with respect to the other end as shown in Fig. 15.7b. Since there is no relative rotation between the ends A and B, the axial load on the analogous column must be zero. The relative displacement Δ must correspond to a pure moment $M = \Delta$ applied to the column cross section as shown in Fig. 15.7c; this moment is shown applied counterclockwise to produce a compressive stress in the

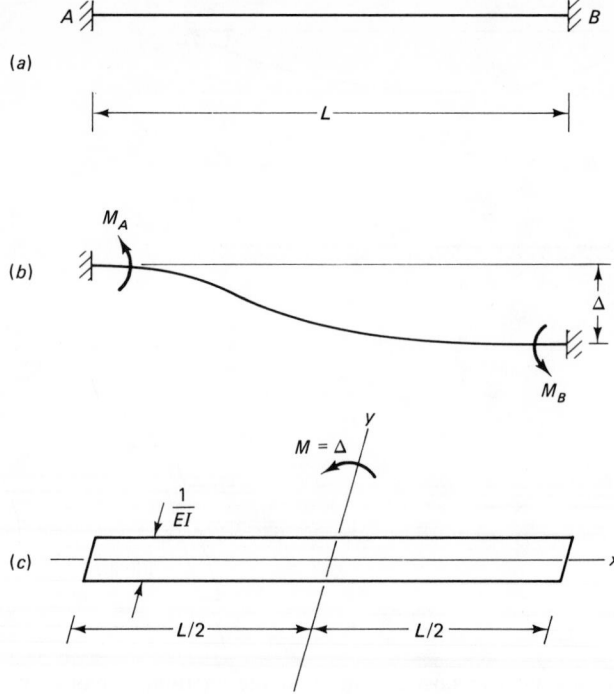

Figure 15.7 Loadings on an analogous column for the determination of fixed-end moments due to relative settlement of supports.

column at A and hence a negative moment at A in the real beam (Fig. 15.7b). Thus, with $I_y = L^3/12EI$, $M_y = \Delta$

$$\sigma_A = M_A = 0 + \frac{\Delta}{L^3/12EI}\left(-\frac{L}{2}\right) = \frac{-6EI\Delta}{L^2}$$

$$\sigma_B = M_B = 0 + \frac{\Delta}{L^3/12EI}\left(+\frac{L}{2}\right) = \frac{+6EI\Delta}{L^2}$$

The negative and positive signs agree with the sign convention of the column-analogy method.

■ **EXAMPLE 15.2**

Determine the stiffness and carryover factors for the beam AB in Fig. 15.8a.

Solution

Geometry of the Analogous Column (Fig. 15.8b).   Assuming $EI = 1$,

Area $A = (\frac{2}{3})(16) + (\frac{2}{5})(20) = 18.67$

$A\bar{x} = (\frac{2}{3})(16)(8) + (\frac{2}{5})(20)(26)$ $\bar{x} = 15.74$ ft

$$I_y = \frac{(\frac{2}{3})(16)^3}{12} + \left(\frac{2}{3}\right)(16)(7.74)^2 + \frac{(\frac{2}{5})(20)^3}{12} + \left(\frac{2}{5}\right)(20)(10.26)^2$$

$$= 1975.4$$

Loading on the Analogous Column

1. For K_A and C_{AB}: Apply a unit load at end A, as shown in Fig. 15.8c; the resultant loading applied at the center of the column is shown in Fig. 15.8d; thus, with $P = 1$ and $M = (1)(-15.74) = -15.74$,

$$\sigma_A = M_A = \frac{1}{18.67} + \frac{-15.74}{1975.4}(-15.74) = 0.179$$

Inserting EI, $M_A = 0.179\,EI$. Therefore, the stiffness at A, $K_A = 0.179\,EI$.

$$\sigma_B = M_B = \frac{1}{18.67} + \frac{-15.74}{1975.4}(20.26) = -0.108$$

Inserting EI, $M_B = -0.108\,EI$. Hence

$$C_{AB} = \frac{|-0.108\,EI|}{0.179\,EI} = 0.603$$

2. For K_B and C_{BA}: The loading on the analogous column is shown in Fig. 15.8e and the resultant loading in Fig. 15.8f; thus, with $P = 1$ and $M = (1)(+20.26) = 20.26$,

$$\sigma_B = M_B = \frac{1}{18.67} + \left(\frac{20.26}{1975.4}\right)(+20.26) = 0.261$$

Inserting EI, $M_B = 0.261\,EI$. Therefore, the stiffness at B, $K_B = 0.261\,EI$.

$$\sigma_A = M_A = \frac{1}{18.67} + \left(\frac{20.26}{1975.4}\right)(-15.74) = -0.108$$

Inserting EI, $M_A = -0.108\,EI$. Hence

$$C_{BA} = \frac{|-0.108\,EI|}{0.261\,EI} = 0.413.$$

The above calculation may be checked by applying Maxwell's reciprocal theorem (see Chapter 7) which in this case says: The rotation of end A of a member AB through an angle of 1 radian produces a moment at end B equal to the moment produced at end A if end B is rotated through an angle of 1 radian. This can be expressed as

$$K_A C_{AB} = K_B C_{BA}$$

Inserting the above derived values shows that this expression is satisfied. ■

■ EXAMPLE 15.3

Calculate the fixed-end moments for the beam in Fig. 15.8a due to a relative displacement of end B of 0.05 ft, shown in Fig. 15.8g.

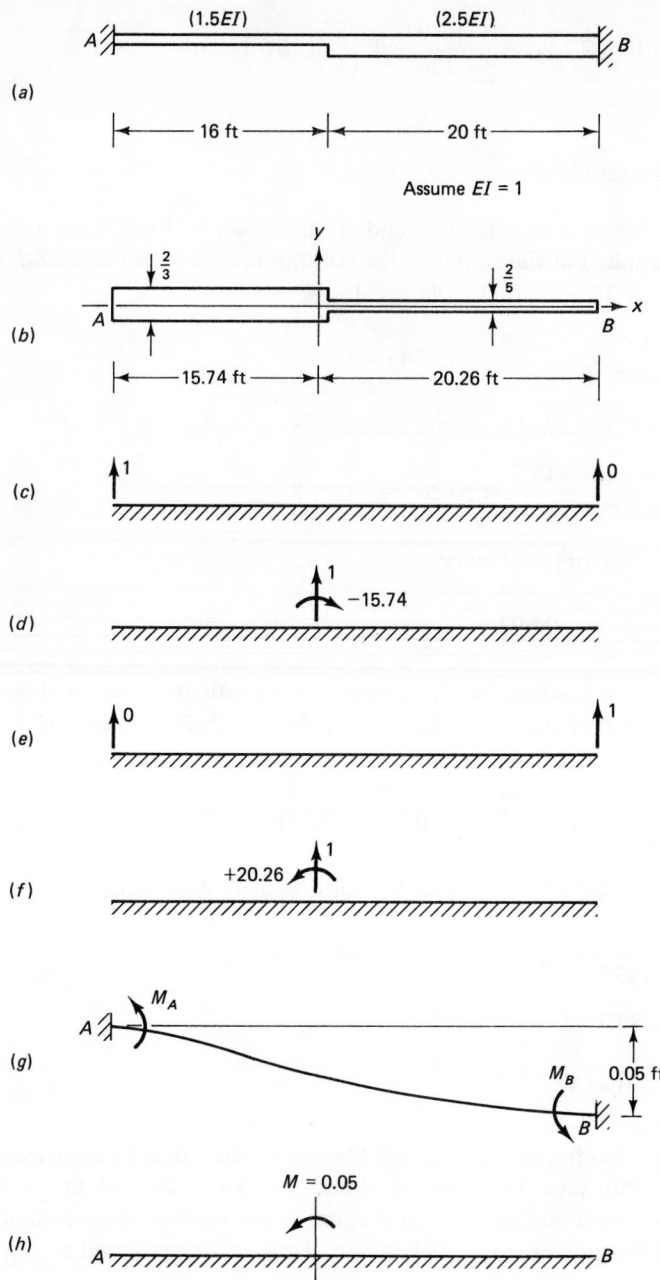

Figure 15.8 Structure for Examples 15.2, 15.3, and 15.4.

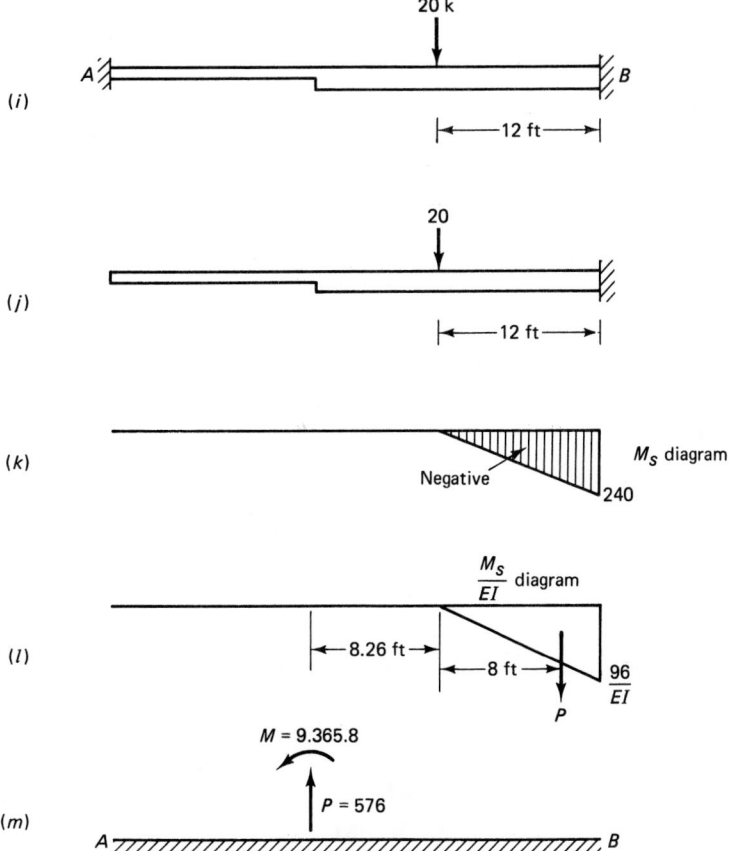

Figure 15.8 *(Continued)*

Solution.   Here the analogous column is loaded by a pure moment $M = \Delta = 0.05$ ft (Fig. 15.8h). Thus, with $P = 0$,

$$\sigma_A = M_A = \frac{M}{I_y}(-15.74) = \left(\frac{0.05}{1975.4}\right)(-15.74) = -0.0004$$

Inserting EI, $M_A = -0.0004\,EI$

$$\sigma_B = M_B = \left(\frac{0.05}{1975.4}\right)(20.26) = +0.0005$$

With EI, $M_B = 0.0005\,EI$.   ■

■ EXAMPLE 15.4
Determine the fixed-end moments for the beam loaded in Fig. 15.8i.

Solution.   The geometry of the analogous column is shown in Fig. 15.8b. The basic determinate beam is shown in Fig. 15.8j with the corresponding M_s diagram in Fig.

15.8h and the M_s/EI in Fig. 15.8l. The equivalent loading on the column is presented in Fig. 15.8m, with

$$P = \left(\frac{1}{2}\right)\left(\frac{96}{EI}\right)(12) = 576\uparrow$$

$$M = (576)(16.26) = +9365.8$$

Therefore,

$$\sigma_A = (M_r)_A = \frac{576}{18.67} + \left(\frac{9365.8}{1975.4}\right)(-15.74) = -43.7$$

$$\sigma_B = (M_r)_B = \frac{576}{18.67} + \left(\frac{9365.8}{1975.4}\right)(+20.26) = +126.9$$

$$M_A = (M_r)_A + M_s = -43.7 + 0 = -43.7 \text{ k} \cdot \text{ft}$$

$$M_B = (M_r)_B + M_s = +126.9 - 240 = -113.1 \text{ k} \cdot \text{ft}$$

Both moments M_A and M_B, being negative, produce tension at the top of the beam AB.   ∎

■ EXAMPLE 15.5

Calculate the fixed-end moments, stiffness, and carryover factors for the loaded beam in Fig. 15.9a.

Solution.   From the given geometry of the beam, one can establish the values of I for the different sections as shown in Fig. 15.9b. The basic structure is a simply supported beam loaded by 150 kN central load; the corresponding positive M_s diagram is presented in Fig. 15.9c. The values of the load on the column cross section, shown in Fig. 15.9d, and the area of the column cross section are presented in Table 15.2.

$$A = (2)\frac{4.145}{EI_0} = \frac{8.29}{EI_0} \qquad P = (2)\frac{-1248.72}{EI_0} = \frac{-2497.44}{EI_0}$$

$$I = \left(\frac{1}{12}\right)\left(\frac{0.296}{EI}\right)(13.2)^3 + \left(\frac{1}{12}\right)\left(\frac{0.126}{EI}\right)(10.8)^3 + \left(\frac{1}{12}\right)\left(\frac{0.157}{EI}\right)(8.4)^3$$

$$+ \left(\frac{1}{12}\right)\left(\frac{0.172}{EI}\right)(6)^3 + \left(\frac{1}{12}\right)\left(\frac{0.156}{EI}\right)(3.6)^3 + \left(\frac{1}{12}\right)\left(\frac{0.093}{EI}\right)(1.2)^3$$

$$= \frac{81.43}{EI_0}$$

For fixed-end moments, since $M_y = 0$ (due to symmetry of loads on the column about y axis),

$$\sigma_A = (M_r)_A = \frac{P}{A} = \frac{-2497.44/EI_0}{8.29/EI_0} = -301.3 \text{ kN} \cdot \text{m}$$

$$\sigma_B = (M_r)_B = \frac{P}{A} = -301.3 \text{ kN} \cdot \text{m}$$

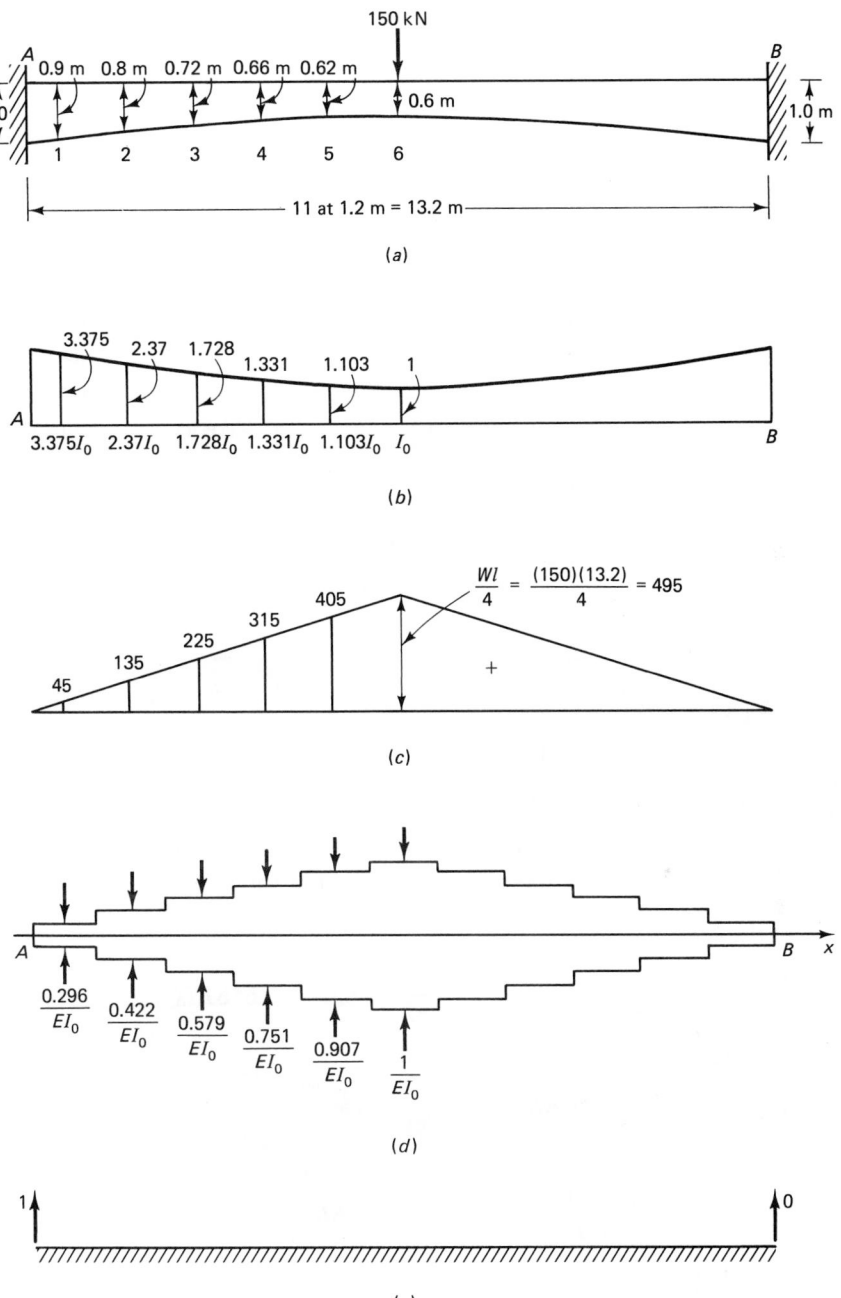

Figure 15.9 Structure for Example 15.5.

TABLE 15.2

Section	Δs	dA	dP
1	1.2	$\dfrac{(1.2)(0.296)}{EI_0} = \dfrac{0.355}{EI_0}$	$\dfrac{-(45)(1.2)}{3.375EI_0} = \dfrac{-16.0}{EI_0}$
2	1.2	$\dfrac{(1.2)(0.422)}{EI_0} = \dfrac{0.506}{EI_0}$	$\dfrac{-(135)(1.2)}{2.370EI_0} = \dfrac{-68.35}{EI_0}$
3	1.2	$\dfrac{(1.2)(0.579)}{EI_0} = \dfrac{0.695}{EI_0}$	$\dfrac{-(225)(1.2)}{1.728EI_0} = \dfrac{-156.25}{EI_0}$
4	1.2	$\dfrac{(1.2)(0.751)}{EI_0} = \dfrac{0.901}{EI_0}$	$\dfrac{-(315)(1.2)}{1.331EI_0} = \dfrac{-284.00}{EI_0}$
5	1.2	$\dfrac{(1.2)(0.907)}{EI_0} = \dfrac{1.088}{EI_0}$	$\dfrac{-(405)(1.2)}{1.103EI_0} = \dfrac{-440.62}{EI_0}$
6	0.6	$\dfrac{(0.6)(1.0)}{EI_0} = \dfrac{0.6}{EI_0}$	$\dfrac{-(472.5)(0.6)}{1EI_0} = \dfrac{-283.5}{EI_0}$
Σ		$\dfrac{4.145}{EI_0}$	$\dfrac{-1248.72}{EI_0}$

Since $(M_s)_A = (M_s)_B = 0$

$$M_A = (M_r)_A + (M_s)_A = -301.3 \text{ kN} \cdot \text{m}$$

$$M_B = -301.3 \text{ kN} \cdot \text{m}$$

The negative sign indicates tension at the top of the beam. For stiffness and carryover factors at end A, apply a unit load on the column at A as shown in Fig. 15.9 e: then

$$M_y = (1)(-6.6) = -6.6$$

$$M_A = \frac{P}{A} + \frac{M}{I}(x) = \frac{1}{8.29/EI_0} + \frac{(-6.6)}{81.43/EI_0}(-6.6) = 0.656\,EI_0$$

Since $\theta_A = 1$ by definition, from

$$M_A = K_A\theta_A \qquad K_A = 0.656\,EI_0 = K_B$$

from symmetry.

$$M_B = \frac{1}{8.29/EI_0} + \frac{(-6.6)}{81.43/EI_0}(+6.6) = -0.414\,EI_0$$

$$C_{AB} = C_{BA} = \frac{|M_B|}{M_A} = \frac{0.414\,EI_0}{0.656\,EI_0} = 0.632 \qquad \blacksquare$$

15.6 SOLVED PROBLEM

■ SOLVED PROBLEM 15.1

Determine by moment distribution the moment at the joints of the structure shown in Fig. 15.10a.

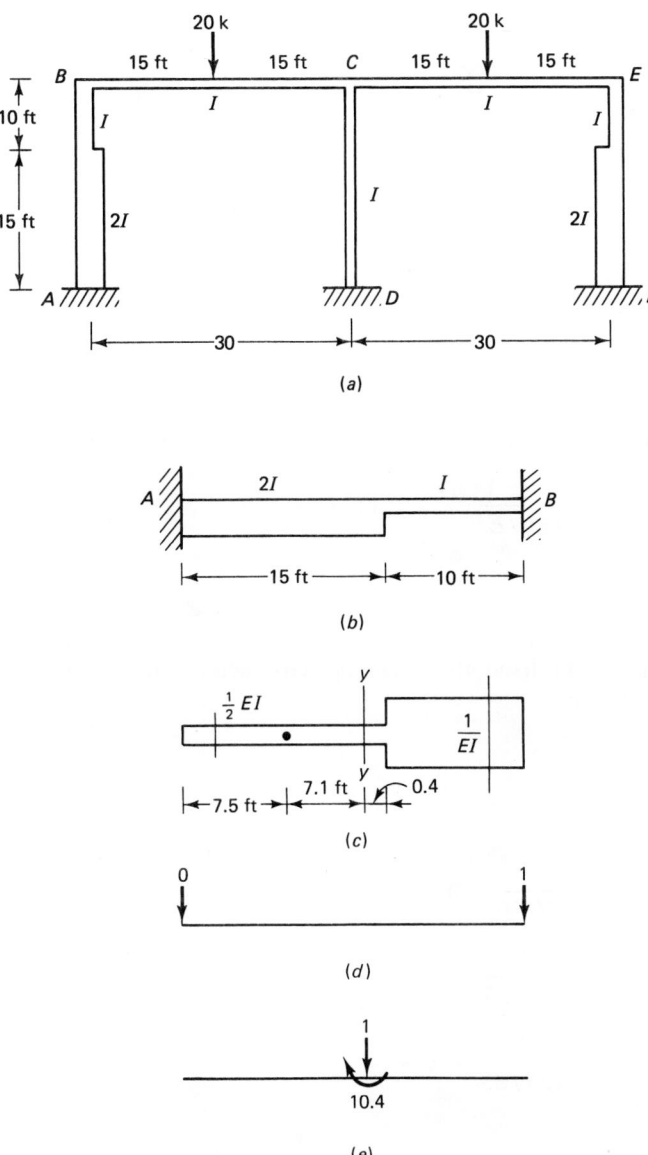

Figure 15.10 Structure for solved problem 15.1.

Solution

Step 1. Find the distribution and carryover factors for AB and EF, by calculating the stiffness and carryover moments for BA.

Analogous column (Fig. 15.10c to e)

$$A = 1 \times 10 + \tfrac{1}{2} \times 15 = 17.5$$

$$A\bar{x} = 1 \times 10 \times 20 + \tfrac{1}{2} \times 15 \times \tfrac{15}{2} = 256.25$$

$$\bar{x} = 14.6$$

$$I_{yy} = \frac{1}{2} \times \frac{15^3}{12} + \frac{1}{2} \times 15 \times 7.1^2 + \frac{1 \times 10^3}{12} + 1 \times 10 \times 5.4^2 = 893.6$$

From Fig. 15.15 d and e

$$M_y = -1 \times 10.4 = -10.4$$

$$\sigma_B = M_B = \frac{-1}{17.5/EI} + \frac{(-10.4)(10.4)}{893.6/EI} = 0.178\,EI$$

or

Stiffness for $BA = 0.18\,EI$

$$F_A = M_A = \frac{-1}{17.5/EI} + \frac{(-10.4)(-14.6)}{893.6/EI} = 0.113\,EI$$

$$C_{BA} = \frac{0.113}{0.18} = 0.63$$

Note: In this case, since support A is fixed, there is no need to calculate the distribution factor and carryover factor for AB.
Distribution factors for

$$BA = \frac{0.18\,EI}{0.18\,EI + 4\,EI/30} = 0.57$$

$$BC\,(\text{also } EC) = \frac{4\,EI/30}{0.18\,EI + 4\,EI/30} = 0.43$$

$$CB = \frac{4\,EI/30}{4\,EI/30 + 4\,EI/30 + 4\,EI/25} = 0.31$$

$$CD = \frac{4\,EI/25}{4\,EI/30 + 4\,EI/30 + 4\,EI/25} = 0.38$$

Carryover factors

$$BA = 0.63$$

$$CB = 0.5$$

$$CD = 0.5$$

Step 2. Calculate FEM.

$$FEM_{BC} = \frac{-PL}{8} = \frac{-20 \times 30}{8} = -75 \text{ k} \cdot \text{ft}$$

$$FEM_{CB} = 75 \text{ k} \cdot \text{ft}$$

$$\text{FEM}_{CE} = -75 \text{ k} \cdot \text{ft}$$

$$\text{FEM}_{EC} = 75 \text{ k} \cdot \text{ft}$$

Step 3. Moment distribution

Joint	A		B		C			E		F
Member	AB	BA	BC	CB	CD	CE	EC	EF		FE
D.F.		0.57	0.43	0.31	0.38	0.31	0.43	0.57		
C.O.F.		←0.63	0.5→	←0.5	0.5↓	0.5→	←0.5	0.63→		
FEM	0 0	+43	−75 +32	75	0	−75	75 −32	−43		
	27			→16	0	−16			→−27	
Final moments, k·ft	+27 +43		−43	+91	0	−91	+43	−43		−27

PROBLEMS

15.1 and 15.2 Using the method of column analogy, determine support moments for the fixed-ended beams shown.

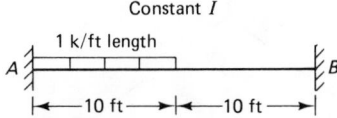

Figure P15.1

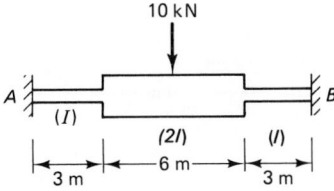

Figure P15.2

15.3 to 15.5 Using the method of column analogy, determine the moments at all the joints of the frames shown.

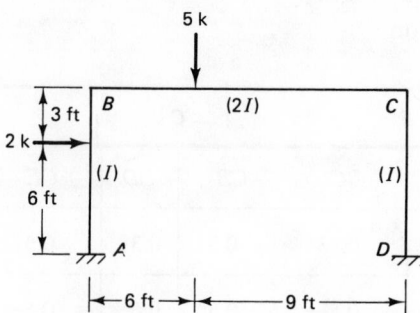

Figure P15.3

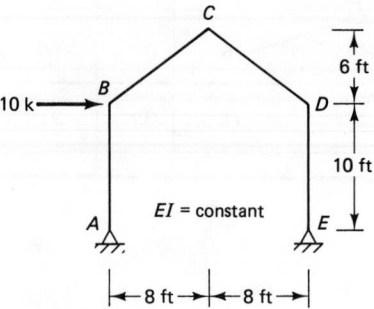

Figure P15.4

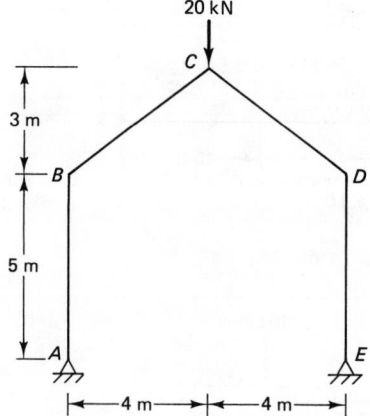

Figure P15.5

15.6 **(a)** Use the method of column analogy to determine the moments at the rigid joints A to G of the gabled structure shown. **(b)** Find the vertical and horizontal reactions at A. **(c)** Calculate the slope and horizontal deflection as well as vertical deflection at joint B. Assume EI is constant.

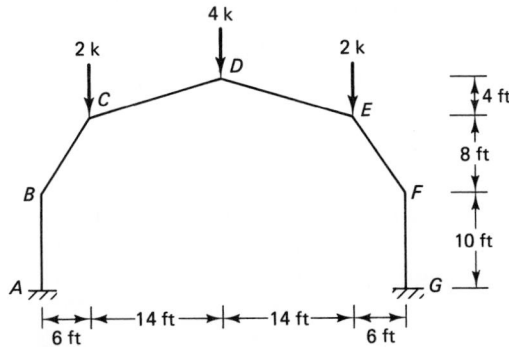

Figure P15.6

15.7 $ABCDE$ is a rigid rib of uniform cross section hinged at A and E and loaded as shown. By means of the method of column analogy: **(a)** Determine the horizontal and vertical reactions at A and E. **(b)** Calculate the bending moment at C. **(c)** Draw the bending-moment diagram.

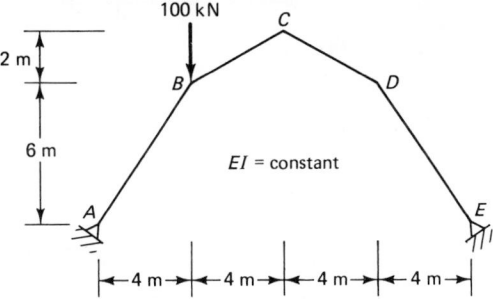

Figure P15.7

15.8 Using the column-analogy method, determine the stiffness factors at A and B, and the carryover factors from A to B and B to A for the beam shown.

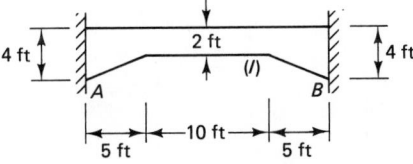

Figure P15.8

15.9 Using the column-analogy method, determine the fixed-end moments, stiffness factors, and carryover factors for the beam shown.

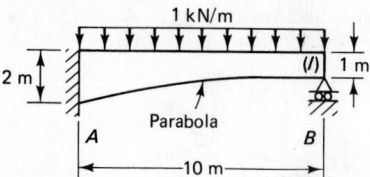

Figure P15.9

15.10 By means of the column-analogy method, calculate the moments at joints A, B, C, D, E, F, and G of the rigid frame shown. Also calculate the horizontal and vertical reactions at A.

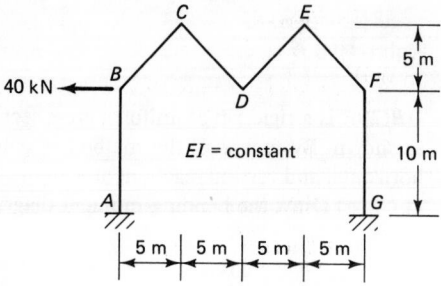

Figure P15.10

Appendix

Fixed-End Forces for Prismatic Members

α is a dimensionless constant ≤ 1

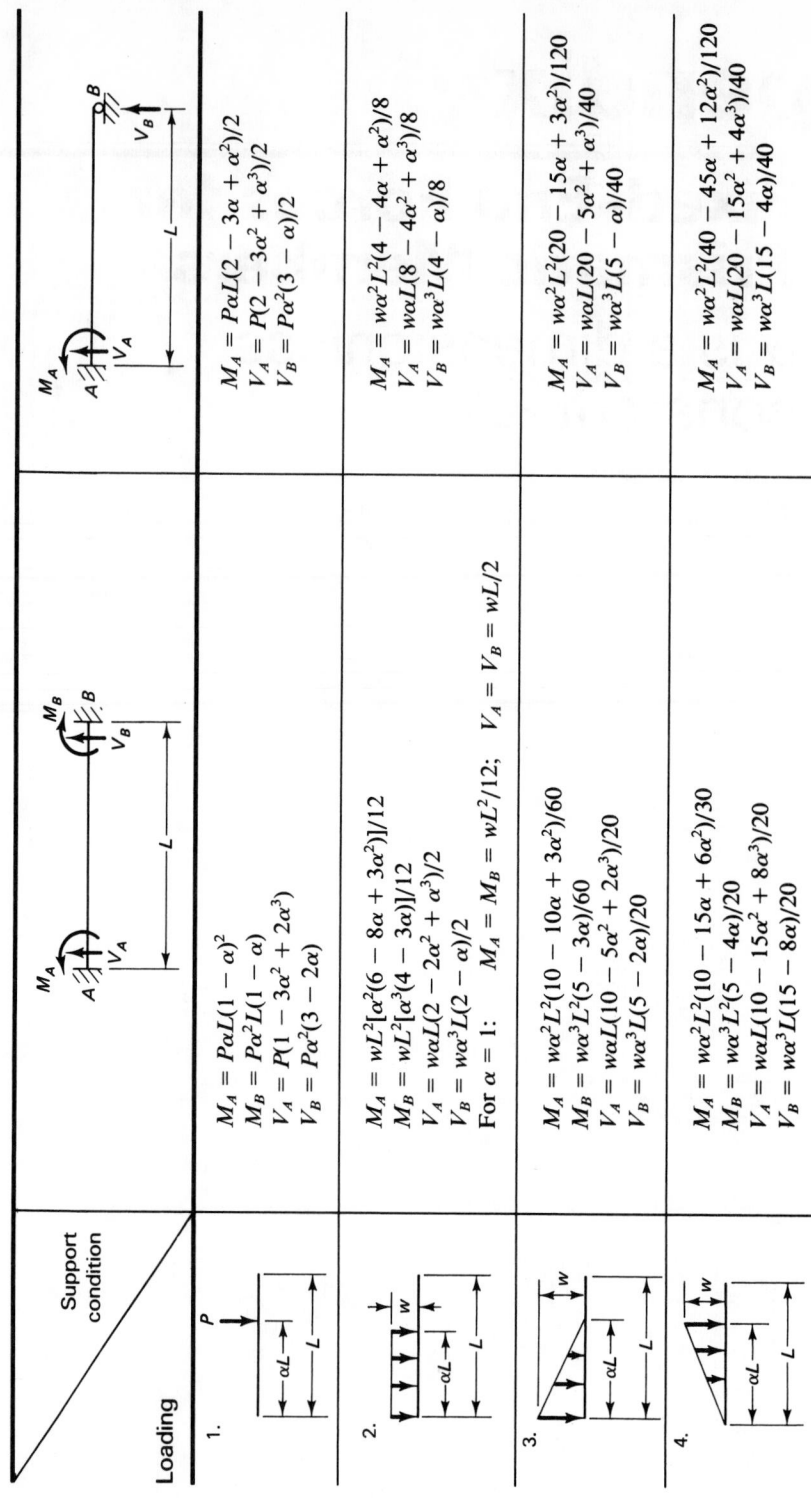

Loading	Support condition	
1. P at αL	$M_A = P\alpha L(1-\alpha)^2$ $M_B = P\alpha^2 L(1-\alpha)$ $V_A = P(1-3\alpha^2+2\alpha^3)$ $V_B = P\alpha^2(3-2\alpha)$	$M_A = P\alpha L(2-3\alpha+\alpha^2)/2$ $V_A = P(2-3\alpha^2+\alpha^3)/2$ $V_B = P\alpha^2(3-\alpha)/2$
2. w over αL	$M_A = wL^2[\alpha^2(6-8\alpha+3\alpha^2)]/12$ $M_B = wL^2[\alpha^3(4-3\alpha)]/12$ $V_A = w\alpha L(2-2\alpha^2+\alpha^3)/2$ $V_B = w\alpha^3 L(2-\alpha)/2$ For $\alpha = 1$: $M_A = M_B = wL^2/12$; $V_A = V_B = wL/2$	$M_A = w\alpha^2 L^2(4-4\alpha+\alpha^2)/8$ $V_A = w\alpha L(8-4\alpha^2+\alpha^3)/8$ $V_B = w\alpha^3 L(4-\alpha)/8$
3. w over αL	$M_A = w\alpha^2 L^2(10-10\alpha+3\alpha^2)/60$ $M_B = w\alpha^3 L^2(5-3\alpha)/60$ $V_A = w\alpha L(10-5\alpha^2+2\alpha^3)/20$ $V_B = w\alpha^3 L(5-2\alpha)/20$	$M_A = w\alpha^2 L^2(20-15\alpha+3\alpha^2)/120$ $V_A = w\alpha L(20-5\alpha^2+\alpha^3)/40$ $V_B = w\alpha^3 L(5-\alpha)/40$
4. w over αL	$M_A = w\alpha^2 L^2(10-15\alpha+6\alpha^2)/30$ $M_B = w\alpha^3 L^2(5-4\alpha)/20$ $V_A = w\alpha L(10-15\alpha^2+8\alpha^3)/20$ $V_B = w\alpha^3 L(15-8\alpha)/20$	$M_A = w\alpha^2 L^2(40-45\alpha+12\alpha^2)/120$ $V_A = w\alpha L(20-15\alpha^2+4\alpha^3)/40$ $V_B = w\alpha^3 L(15-4\alpha)/40$

5.

$M_A = w\alpha^3 L^2(5 - 4\alpha)/12$
$M_B = w\alpha^2 L^2(10 - 15\alpha + 6\alpha^2)/30$
$V_A = w\alpha^3 L(15 - 8\alpha)/20$
$V_B = w\alpha L(10 - 15\alpha^2 + 8\alpha^3)/20$

$M_A = w\alpha^3 L^2(5 - 3\alpha^2)/30$
$V_A = w\alpha^2 L(5 - \alpha^2)/10$
$V_B = w\alpha L(5 - 5\alpha + \alpha^3)/10$

6.

$M_A = w\alpha^3 L^2(5 - 3\alpha)/60$
$M_B = w\alpha^2 L^2(10 - 10\alpha + 3\alpha^2)/60$
$V_A = w\alpha^3 L(5 - 2\alpha)/20$
$V_B = w\alpha L(10 - 5\alpha^2 + 2\alpha^3)/20$

$M_A = w\alpha^2 L^2(10 - 3\alpha^2)/120$
$V_A = w\alpha^2 L(10 - \alpha^2)/40$
$V_B = w\alpha L(20 - 10\alpha + \alpha^3)/40$

7.

$M_A = M(1 - 4\alpha + 3\alpha^2)$
$M_B = M\alpha(2 - 3\alpha)$
$V_A = -V_B$
$V_B = 6\alpha M(1 - \alpha)/L$

$M_A = -M(1 - 3\alpha + 3\alpha^2/2)$
$V_A = -V_B$
$V_B = M(2 - 3\alpha + 3\alpha^2/2)/L$

8. Relative settlement Δ

$M_A = 6EI\Delta/L^2$
$M_B = -6EI\Delta/L^2$
$V_A = -V_B$
$V_B = -12EI\Delta/L^3$

$M_A = 3EI\Delta/L^2$
$V_A = -V_B$
$V_B = -3EI\Delta/L^3$

9. End rotation θ

$M_A = 4EI\theta/L$
$M_B = -2EI\theta/L$
$V_A = -V_B$
$V_B = -6EI\theta/L^2$

$M_A = 3EI\theta/L$
$V_A = -V_B$
$V_B = -3EI\theta/L^2$

Bibliography

Andersen, Paul, and Gene M. Nordby, *Introduction to Structural Mechanics.* New York: Ronald Press, 1960, 340 pp.

Azar, Jamal J., *Matrix Structural Analysis.* Elmsford, NY: Pergamon Press, 1972, 217 pp.

Bakos, Jack D., *Structural Analysis for Engineering Technology.* Columbus, OH: Merrill, 1973, 316 pp.

Bhatt, P., *Problems in Structural Analysis by Matrix Methods.* England: The Construction Press, 1981, 465 pp.

Boggs, Robert G., *Elementary Structural Analysis.* New York: Holt, Rinehart and Winston, 1984, 420 pp.

Bommer, C. M., and D. A. Symonds, *Skeletal Structures—Matrix Methods of Linear Structural Analysis Using Influence Coefficients.* England: Crosby Lockwood & Son, Ltd., 1968, 106 pp.

Carpenter, Samuel T., *Structural Mechanics.* New York: Wiley, 1960, 538 pp.

Cassie, W. Fisher, *Structural Analysis.* 2d ed. Longmans, Green & Co., 1957, 269 pp.

Chajes, Alexander, *Structural Analysis.* Englewood Cliffs, NJ: Prentice-Hall, 1983, 351 pp.

Coates, R. C., M. G. Coutie, and F. K. Kong, *Structural Analysis.* 2d ed. New York: Van Nostrand Reinhold, 1980, 579 pp.

Cook, Robert D., *Concepts and Applications of Finite Element Analysis.* 2d ed. New York: Wiley, 1981, 537 pp.

Firmage, D. Allan, *Fundamental Theory of Structures.* New York: Wiley, 1963, pp. 247–265.

Gennaro, Joseph J., *Computer Methods in Solid Mechanics.* New York: Macmillan, 1965, 292 pp.

Gerstle, Kurt H., *Basic Structural Analysis.* Englewood Cliffs, NJ: Prentice-Hall, 1974, 498 pp.

Grassie, James C., *Analysis of Indeterminate Structures.* Longmans, Green & Co., 1957, 418 pp.

Grassie, James C., *Elementary Theory of Structures.* Longmans, Green & Co., 1960, 392 pp.

Hsieh, Yuan-Yu, *Elementary Theory of Structures.* 2d ed. Englewood Cliffs, NJ: Prentice-Hall, 1982, 416 pp.

Jakkula, A. A., and Henson K. Stephenson, *Fundamentals of Structural Analysis.* New York: Van Nostrand Reinhold, 1953, 288 pp.

Kardestuncer, H., *Elementary Matrix Analysis of Structures.* New York: McGraw-Hill, 1974, 428 pp.

Kinney, J. Sterling, *Indeterminate Structural Analysis.* Reading, MA: Addison-Wesley, 1957, 655 pp.

Krishna, Prem, *Cable-Suspended Roofs.* New York: McGraw-Hill, 1978, pp. 27–57.

Laible, Jeffrey P., *Structural Analysis.* New York: Holt, Rinehart and Winston, 1985, 901 pp.

Langhaar, Henry L., *Energy Methods in Applied Mechanics.* New York: Wiley, 1962, 350 pp.

Laursen, Harold I., *Structural Analysis.* 3d ed. New York: McGraw-Hill, 1988, 475 pp.

Leonard, John W., *Tension Structures.* New York: McGraw-Hill, 1988, pp. 17–71.

Livesley, R. K., *Matrix Methods of Structural Analysis.* Elmsford, NY: Pergamon Press, 1964, 265 pp.

McCormac, Jack C., *Structural Analysis.* 4th ed. New York: Harper & Row, 1984, 641 pp.

McCormac, Jack, and Rudolf E. Elling, *Structural Analysis: A Classical and Matrix Approach.* New York: Harper & Row, 1988, 608 pp.

Martin, Harold C., *Introduction to Matrix Methods of Structural Analysis.* New York: McGraw-Hill, 1966, 331 pp.

Matheson, J. A. L., and A. J. Francis, *Hyperstatic Structures.* Stoneham, MA: Butterworth, vol. 1, 1959, 474 pp., vol. II, 1960, 282 pp.

Michalos, James, and Edward N. Wilson, *Structural Mechanics and Analysis.* New York: Macmillan, 1965, 430 pp.

Neal, B. G., *Structural Theorems and Their Applications.* Elmsford, NY: Pergamon Press, 1964, 198 pp.

Norris, C. H., J. B. Wilbur, and S. Utku, *Elementary Structural Analysis.* 3d ed. New York: McGraw-Hill, 1976, 673 pp.

Otto, Frei, *Tensile Structures.* vol. 2, *Cable Structures.* Boston: M.I.T. Press, 1969, pp. 99–107.

Podolny, W., and J. B. Scalzi, *Construction and Design of Cable-Stayed Bridges.* New York: Wiley, 1976, pp. 337–353.

Rockey, K. C., H. R. Evans, D. W. Griffiths, and D. A. Nethercot, *The Finite Element Method.* New York: Wiley (Halsted Press), 1975, 239 pp.

Sack, Ronald, *Structural Analysis.* New York: McGraw-Hill, 1984, 544 pp.

Schodek, Daniel L., *Structures.* Englewood Cliffs, NJ: Prentice-Hall, 1980, 572 pp.

Smith, J. C., *Structural Analysis.* New York: Harper & Row, 1988, 608 pp.

Stasa, Frank L., *Applied Finite Element Analysis for Engineers.* New York: Holt, Rinehart and Winston, 1985, 657 pp.

Timoshenko, S. P., and D. H. Young, *Theory of Structures.* 2d ed. New York: McGraw-Hill, 1965.

Todd, J. D., *Structural Theory and Analysis.* New York: Macmillan, 1974, 353 pp.

Tung, A., and P. Christiano, *Structural Analysis.* Englewood Cliffs, NJ: Prentice-Hall, 1987.

Vanderbilt, M. Daniel, *Matrix Structural Analysis.* Quantum Publishers, Inc., 1974, 397 pp.

Wang, Chu-Kia, and Charles G. Salmon, *Introductory Structural Analysis.* Englewood Cliffs, NJ: Prentice-Hall, 1984, 591 pp.

Weaver, William, and Paul R. Johnston, *Finite Elements for Structural Analysis.* Englewood Cliffs, NJ: Prentice-Hall, 1984, 448 pp.

West, H. H., *Analysis of Structures.* New York: Wiley, 1980.

White, R. N., P. Gergely, and R. G. Sexsmith, *Structural Engineering.* New York: Wiley, vol. 1: "Introduction to Design Concepts and Analysis," 2d ed., 1976, 316 pp., vol. 2: "Indeterminate Structures."

Willems, Nicholas, and William M. Lucas, *Structural Analysis for Engineers.* New York: McGraw-Hill, 1978, 464 pp.

Zienkiewicz, O. C., *The Finite Element Method: Basic Concepts and Linear Applications.* 4th ed. New York: McGraw-Hill, 1988, 420 pp.

Answers to Selected Problems

CHAPTER 3

3.1 Geometrically unstable, indeterminate

3.3 Stable and determinate

3.5 Unstable

3.7 Stable and determinate

3.9 Unstable

3.11 Stable and determinate

3.13 Unstable

3.15 Stable and determinate

3.17 Geometrically unstable

3.19 Geometrically unstable

3.21 Unstable

3.23 Stable and determinate

3.25 Stable and indeterminate to first degree

3.27 Critical form

3.29 Stable; does not have critical form

3.31 Critical form

3.33 $n_{ext} = 1$; $n_{int} = 0$

3.35 $n_{ext} = 3$;   internally determinate

3.37 Externally and internally determinate

3.39 $n_{ext} = 2$; $n_{int} = 0$

3.41 Unstable

3.43 $n_{ext} = 1$; $n_{int} = 4$

3.45 $n_{ext} = 1$; $n_{int} = 2$

3.47 Unstable

3.49 $n_{ext} = 2$; $n_{int} = 0$

3.51 $n_{ext} = 0$; $n_{int} = 12$

CHAPTER 4

4.1 $F_{AB} = 20$ kN(C); $F_{AC} = 22.36$ kN(T); $F_{CE} = 22.36$ kN(T); $F_{CB} = 20$ kN(C); $F_{BE} = 28.28$ kN(T); $F_{BD} = 40$ kN(C); $F_{DE} = 30$ kN(C)

4.3 $F_{U_0L_1} = 41.22$ kN(C); $F_{U_0U_1} = 40$ kN(T); $F_{U_1U_2} = 40$ kN(T); $F_{U_1L_1} = 0$; $F_{L_1U_2} = 0$; $F_{L_1L_2} = 41.22$ kN(C); $F_{U_2U_3} = 40$ kN(T); $F_{U_2L_2} = 0$; $F_{L_2U_3} = 0$; $F_{L_2L_3} = 41.22$ kN(C); $F_{L_3L_4} = 41.22$ kN(C); $F_{L_3U_3} = 0$; $F_{U_3U_4} = 28.28$ kN(T); $F_{U_3L_4} = 28.28$ kN(T)

4.5 $F_{L_0M_0} = 150$ k(C) $= F_{M_4L_4}$; $F_{L_0L_1} = 0 = F_{L_3L_4}$; $F_{U_0M_0} = 0 = F_{U_4M_4}$; $F_{U_0U_1} = 0 = F_{U_3U_4}$; $F_{M_0U_1} = 135.2$ k(C) $= F_{U_3M_4}$; $F_{M_0L_1} = 135.2$ k(T) $= F_{L_3M_4}$; $F_{U_1M_1} = 75.0$ k(T) $= F_{U_3M_3}$; $F_{U_1U_2} = 112.5$ k(C) $= F_{U_2U_3}$; $F_{L_1M_1} = 25.0$ k(T) $= F_{M_3L_3}$; $F_{L_1L_2} = 112.5$ k(T) $= F_{L_2L_3}$; $F_{M_1U_2} = 45.06$ k(C) $= F_{U_2M_3}$; $F_{M_1L_2} = 45.06$ k(T) $= F_{L_2M_3}$; $F_{L_2U_2} = 50$ k(T)

4.7 $F_{AB} = 10.0$ kN(C); $F_{AD} = 0$; $F_{CB} = 0 = F_{CF}$; $F_{BD} = 8.0$ kN(T); $F_{BF} = 8.0$ kN(C); $F_{FI} = 6.25$ kN(C); $F_{FE} = 5.0$ kN(T); $F_{DG} = 6.25$ kN(T); $F_{DE} = 25.0$ kN(C); $F_{EG} = 24.0$ kN(T); $F_{EI} = 24.0$ kN(C); $F_{IK} = 25.0$ kN(C); $F_{IH} = 15.0$ kN(T); $F_{GJ} = 25.0$ kN(T); $F_{GH} = 35.0$ kN(C); $F_{HJ} = 40.0$ kN(T); $F_{HK} = 40.0$ kN(C)

4.9 $F_{U_2U_4} = F_{U_4U_6} = 1200$ kN(C); $F_{U_2U_1} = F_{U_6U_7} = 670.8$ kN(C); $F_{U_2U_3} = F_{U_5U_6} = 670.8$ kN(T); $F_{U_3U_4} = F_{U_4U_5} = 447.2$ kN(T); $F_{L_0U_1} = F_{U_7L_8} = 782.6$ kN(C); $F_{U_1L_1} = F_{U_7L_7} = 100$ kN(T); $F_{U_1L_2} = F_{U_7L_6} = 111.8$ kN(C); $F_{L_2U_3} = F_{U_5L_6} = 335.5$ kN(T); $F_{U_3L_3} = F_{U_5L_5} = 100$ kN(T); $F_{U_3L_4} = F_{U_5L_4} = 559.0$ kN(T); $F_{U_4L_4} = 400$ kN(C); $F_{L_0L_1} = F_{L_1L_2} = F_{L_6L_7} = F_{L_7L_8} = 700$ kN(T); $F_{L_2L_3} = F_{L_3L_4} = F_{L_4L_5} = F_{L_5L_6} = 300$ kN(T)

4.11 $F_{1,2} = F_{2,3} = F_{9,10} = F_{10,11} = 750$ kN(C); $F_{3,4} = F_{7,8} = 875$ kN(C); $F_{4,5} = F_{6,7} = 1000$ kN(C); $F_{5,6} = 1041.67$ kN(C); $F_{1,12} = 1009$ kN(T) $= F_{11,32}$; $F_{2,12} = 150$ kN(C) $= F_{10,32}$; $F_{12,14} = 896.9$ kN(T) $= F_{31,32}$; $F_{3,12} = 112.1$ kN(T) $= F_{9,32}$; $F_{3,13} = 350$ kN(C); $F_{13,14} = 600$ kN(C); $F_{3,15} = F_{13,17} = F_{28,30} = 242.96$ kN(T); $F_{13,15} = F_{8,30} = F_{28,31} = 242.96$ kN(C); $F_{4,15} = F_{9,30} = 225$ kN(C); $F_{15,16} = 25$ kN(T); $F_{16,17} = 125$ kN(C); $F_{4,18} = F_{16,20} = F_{25,27} = 145.77$ kN(C); $F_{16,18} = F_{7,27} = F_{25,29} = 145.77$ kN(C); $F_{5,18} = 175$ kN(C); $F_{18,19} = 25$ kN(C); $F_{19,20} = 75$ kN(C); $F_{6,21} = 125$ kN(C); $F_{21,22} = 75$ kN(C); $F_{22,23} = 25$ kN(C); $F_{7,24} = 75$ kN(C); $F_{24,25} = 125$ kN(C); $F_{25,26} = 25$ kN(T); $F_{8,27} = 25$ kN(C); $F_{27,28} = 175$ kN(C); $F_{28,29} = 75$ kN(T); $F_{30,31} = 475$ kN(C); $F_{5,21} = F_{19,23} = F_{22,24} = 48.59$ kN(T); $F_{19,21} = F_{6,24} = F_{22,26} = 48.59$ kN(C)

4.13 $F_{AB} = F_{BC} = 5.39$ k(C); $F_{AD} = 32.0$ k(T); $F_{DB} = 4.0$ k(C); $F_{DC} = 6.39$ k(T)

4.15 $F_{AB} = 19.17$ k(T); $F_{BC} = 12.81$ k(T); $F_{CD} = F_{DE} = 8$ k(C); $F_{BD} = 10$ k(T); $F_{BE} = 14.64$ k(C)

4.17 $F_{AB} = F_{BC} = 18.76$ kN(C); $F_{AD} = F_{BE} = F_{BF} = F_{CG} = 15.63$ kN(C); $F_{AE} = F_{CF} = 15.63$ kN(T); $F_{DE} = F_{FG} = 9.38$ kN(T); $F_{EF} = 28.14$ kN(T)

4.19 $F_{AB} = 500$ kN(T); $F_{BC} = 233.33$ kN(T); $F_{CD} = 100$ kN(T); $F_{AE} = F_{DH} = 0$; $F_{AF} = 500$ kN(T); $F_{BF} = 200$ kN(C); $F_{BG} = 333.3$ kN(T); $F_{CG} = 100$ kN(C); $F_{CH} = 166.67$ kN(T); $F_{EF} = 800$ kN(C); $F_{FG} = 400$ kN(C); $F_{GH} = 133.33$ kN(C)

4.21 $F_{AB} = 5$ kN(C); $F_{AC} = 3.75$ kN(T); $F_{BE} = 3.75$ kN(C); $F_{AD} = 6.25$ kN(C); $F_{DB} = 6.25$ kN(T); $F_{CD} = F_{JK} = 0$; $F_{DE} = 10$ kN(C); $F_{CF} = 11.25$ kN(T); $F_{EH} = 11.25$ kN(C); $F_{CG} = 12.5$ kN(C); $F_{GE} = 12.5$ kN(T); $F_{FG} = 5$ kN(T); $F_{GH} = 15$ kN(C); $F_{FI} = 22.5$ kN(T); $F_{HK} = 22.5$ kN(C); $F_{FJ} = 18.75$ kN(C); $F_{JH} = 18.75$ kN(T); $F_{IJ} = 30$ kN(T)

4.23 $F_{AC} = F_{AD} = F_{CG} = F_{DG} = F_{CF} = 0$; $F_{BF} = 10.31$ kN(C); $F_{BC} = 2.5$ kN(T) = F_{DC}; $F_{EG} = 10.31$ kN(C); $F_{ED} = 2.5$ kN(T); $F_{GI} = 12.5$ kN(C); $F_{GF} = 10.0$ kN(C); $F_{FI} = 0$; $F_{FH} = 12.5$ kN(C)

4.25 $F_{AD} = 145.83$ kN(T); $F_{AB} = 16.66$ kN(C); $F_{BD} = 100$ kN(C); $F_{BC} = 16.66$ kN(C); $F_{CD} = 20.83$ kN(T)

4.27 $F_{AB} = F_{BC} = 93.66$ kN(C); $F_{AD} = 64.29$ kN(C); $F_{AE} = 102.68$ kN(T); $F_{BE} = 100$ kN(C); $F_{CE} = 27.46$ kN(T); $F_{DE} = 79.86$ kN(C)

4.29 $F_{AB} = F_{BC} = F_{AD} = F_{CE} = F_{DG} = F_{GE} = 0$; $F_{DB} = F_{BE} = 70.71$ kN(C); $F_{DF} = F_{EH} = 55.9$ kN(C); $F_{FG} = F_{GH} = 25$ kN(C); $F_{DE} = 75.0$ kN(T)

4.31 $F_{AB} = F_{BC} = F_{CD} = F_{DE} = F_{EF} = F_{FG} = F_{GH} = F_{HI} = 46.66$ k(C); $F_{AJ} = F_{IM} = 58.33$ k(T); $F_{BJ} = F_{DK} = F_{FL} = F_{HM} = 10$ k(C); $F_{CN} = F_{GP} = 20$ k(C); $F_{EO} = 0$; $F_{JC} = F_{GM} = 8.33$ k(T); $F_{CK} = F_{GL} = 8.34$ k(T); $F_{EK} = F_{EL} = 8.32$ k(C); $F_{KN} = F_{LP} = 16.66$ k(C); $F_{JN} = F_{PM} = 50$ k(T); $F_{NO} = F_{OP} = 53.33$ k(T)

4.33 $F_{AB} = 80$ kN(C); $F_{GC} = F_{FK} = 167.7$ kN(C); $F_{GH} = F_{JK} = 75$ kN(T); $F_{AD} = F_{BE} = 126.5$ kN(T); $F_{AC} = F_{BF} = 126.5$ kN(C); $F_{CH} = F_{FJ} = 33.5$ kN(T); $F_{CD} = F_{EF} = 50.0$ kN(C); $F_{HD} = F_{EJ} = 78.3$ kN(T); $F_{IJ} = F_{HI} = 55$ kN(T); $F_{DI} = F_{EI} = 55.94$ kN(T)

4.35 $F_{AB} = F_{BC} = 60$ k(C); $F_{AD} = F_{CG} = 50$ k(C); $F_{DH} = F_{GP} = 58.33$ k(C); $F_{DI} = F_{GO} = 10$ k(T); $F_{DJ} = F_{GN} = 8.33$ k(C); $F_{AJ} = F_{CN} = 15$ k(T); $F_{AE} = F_{CF} = 25$ k(T); $F_{EK} = F_{FM} = 10$ k(T); $F_{EB} = F_{BF} = 8.33$ k(T); $F_{EL} = F_{LF} = 16.67$ k(T); $F_{BL} = 10$ k(C); $F_{HI} = F_{IJ} = F_{NO} = F_{OP} = 46.67$ k(T); $F_{JK} = F_{KL} = F_{LM} = F_{MN} = 40$ k(T)

4.37 $F_1 = 40$ kN(C); $F_2 = 28.28$ kN(T); $F_3 = 22.36$ kN(T); $F_4 = 20$ kN(C)

4.39 $F_1 = 40$ kN(T); $F_2 = F_3 = F_4 = 0$

4.41 $F_1 = 112.5$ k(C); $F_2 = 75.0$ k(T); $F_3 = 45.05$ k(C); $F_4 = 25.0$ k(T)

4.43 $F_1 = 25.0$ kN(C); $F_2 = 6.25$ kN(T); $F_3 = 24.0$ kN(T); $F_4 = 5.0$ kN(T)

4.45 $F_1 = 1200$ kN(C); $F_2 = 670.8$ kN(T); $F_3 = 335.4$ kN(T); $F_4 = 300$ kN(T)

4.47 $F_1 = 1041.67$ kN(C); $F_2 = 48.59$ kN(T); $F_3 = 48.59$ kN(C); $F_4 = 48.59$ kN(T)

4.49 $F_1 = 5.39$ kN(C); $F_2 = 32.03$ kN(T); $F_3 = 4.0$ kN(C); $F_4 = 6.40$ kN(T)

4.51 $F_1 = 12.81$ kN(T); $F_2 = 14.64$ kN(C); $F_3 = 10.0$ kN(T); $F_4 = 19.17$ kN(T)

4.53 $F_1 = 18.76$ kN(C); $F_2 = 15.63$ kN(T); $F_3 = 15.63$ kN(C); $F_4 = 28.14$ kN(T)

4.55 $F_1 = 233.33$ kN(T); $F_2 = 333.33$ kN(T); $F_3 = 400.0$ kN(C); $F_4 = 100.0$ kN(C)

4.57 $F_1 = 0$; $F_2 = 12.5$ kN(C); $F_3 = 5$ kN(T); $F_4 = 10$ kN(C)

4.59 $F_1 = 2.5$ kN(T); $F_2 = 0$; $F_3 = 10.0$ kN(C); $F_4 = 0$

4.61 $F_1 = 16.66$ kN(C); $F_2 = 100$ kN(C); $F_3 = 145.83$ kN(T); $F_4 = 16.66$ kN(C)

4.63 $F_1 = 93.66$ kN(C); $F_2 = 100.0$ kN(C); $F_3 = 102.68$ kN(T); $F_4 = 79.86$ kN(C)

4.65 $F_1 = 0$; $F_2 = 70.71$ kN(C); $F_3 = 0$; $F_4 = 55.9$ kN(C)

4.67 $F_1 = 46.67$ k(C); $F_2 = 8.33$ k(C); $F_3 = 53.33$ k(T); $F_4 = 8.34$ k(T)

4.69 $F_1 = 126.5$ kN(T); $F_2 = 50.0$ kN(C); $F_3 = 78.3$ kN(T); $F_4 = 55.9$ kN(T)

4.71 $F_1 = 60.0$ k(C); $F_2 = 8.33$ k(T); $F_3 = 16.67$ k(T); $F_4 = 40.0$ k(T)

4.73 $F_1 = 123.54$ k(C); $F_2 = 120.0$ k(C)

4.75 $F_{AB} = 12.5$ kN(C); $F_{AG} = 25.0$ kN(T); $F_{GH} = 0$; $F_{BH} = 12.5$ kN(T); $F_{BC} = 127.5$ kN(C); $F_{AC} = 106.07$ kN(T); $F_{AE} = 63.75$ kN(C); $F_{EG} = 159.10$ kN(C); $F_{FG} = 63.75$ kN(T); $F_{FH} = 53.03$ kN(C); $F_{DH} = 0$; $F_{BD} = 0$; $F_{CD} = 100.0$ kN(T); $F_{CE} = 0$; $F_{EF} = 50.0$ kN(C); $F_{DF} = 0$. Reactions: $A_y = -50$ kN, $A_z = -12.5$ kN; $B_x = -87.5$ kN, $B_y = 100$ kN, $B_z = 12.5$ kN; $G_y = 100$ kN; $H_x = -12.5$ kN, $H_y = 50$ kN

4.77 $F_{CD} = 27.27$ kN(T); $F_{AD} = 0$; $F_{AB} = 45.45$ kN(T); $F_{BC} = 54.55$ kN(T); $F_{CE} = 66.80$ kN(C); $F_{DE} = 0$; $F_{AE} = 59.61$ kN(T); $F_{BE} = 115.35$ kN(C). Reactions: $A_x = -72.73$ kN, $A_y = -45.45$ kN, $A_z = 27.27$ kN; $B_y = 90.91$ kN, $C_y = 54.55$ kN, $C_z = -27.27$ kN; $D_x = -27.27$ kN

4.79 $F_{AB} = 0.494$ kN(T); $F_{BC} = 1.235$ kN(T); $F_{AC} = 1.746$ kN(T); $F_{AF} = 3.074$ kN(C); $F_{BF} = 2.590$ kN(C); $F_{CF} = 6.204$ kN(C); $F_{AD} = F_{BD} = F_{EF} = F_{DF} = F_{DE} = F_{BE} = 0$.   Reactions: $A_x = 0$, $A_y = 2.222$ kN; $B_x = 0$, $B_y = 2.222$ kN; $C_y = 5.556$ kN, $C_z = 0$

CHAPTER 5

5.1 Shear in kips: $AB = 15$, $BC = 5$, $CD = -15$.   Moment in k·ft: $M_B = 150$, $M_C = 225$

5.3 Shear in kN: $AC = 8$, $CD = -2$, $DB = -2$ to -7, $BE = 5$ to zero.   Moment in kN·m: $M_C = 8$, $M_D = 2$, $M_B = -2.5$

5.5 Shear in lb: $CA =$ zero to -14, $AD = -35.05$ to -45.05, $DE = 29.95$, $EB = -20.05$, $BF = 10$.   Moment in lb·ft: $M_A = -49$, $M_D = -249.25$, $M_E = -39.6$, $M_{B_{left}} = -200$, $M_{B_{right}} = -100$

5.7 Shear in kips: $AB = 20$, $BD = 10$, $DE = -10$.   Moment in k·ft: $M_A = -300$, $M_B = -100$, $M_D = 100$

5.9 Shear in kips: $AB = 25$, $BC = -75$, $CE = 50$, $EG = -50$, $GH = 125$ to 75, $HJ = -25$ to -75.   Moment in k·ft: $M_B = 125$, $M_C = -250$, $M_E = 250$, $M_G = -250$, $M_H = 250$

5.11 Shear in kips: $AB = -5.83$, $BC = 20$ to 10.   Moment in k·ft: $M_{B_{left}} = -46.64$, $M_{B_{right}} = -150$

5.13 Shear in kips: At B in $AB = 3.12$, left of $C = 3.74$, $CD = 6.24$ to -6.24, $DE = 3.74$ to zero.   Moment in k·ft: $M_B = -10.04$, $M_C = -21.63$, $M_G = +9.57$ (symmetric)

5.15 Axial force in kN: $AB = -13.33$, $DE = -46.67$.   Shear in kN: $AB = 15$, $BC = 5$, $CD = 13.33$ to -46.67.   Moment in kN·m: $M_B = 75$, $M_C = 100$

5.17 Axial force in kips: $CE = -1.125$ to 6.375. Shear in kips: $AB = 6.875$, $BC = 1.875$, $CE = 1.5$ to -8.5. Moment in k·ft: $M_B = 34.375$, $M_C = 43.75$

5.19 Axial force in kips: $AB = 11.67$. Shear in kips: At $A = 50$, $BC = -11.67$, $CD = -21.67$. Moment in k·ft: $M_B = 166.7$, $M_C = 108.35$

5.21 Axial force $= 0$. Shear in kips: $AB = 10$, $BC = -1$, $CD = -11$. Moment in k·ft: $M_B = 50$, $M_C = 44$

5.23 Axial force in kN: $ABC = -65$, $CD = -100$. Shear in kN: $BC = 100$, $CD = 65$ to 35. Moment in kN·m: $M_C = -300$

5.25 Axial force in kN: $AB = -5.467$, $BC = 2.533$. Shear in kN: $AB = 4.1$, $BC = -1.9$, $CD = -3.167$, $DE = 4$ to zero. Moment in kN·m: $M_B = 10.25$, $M_C = 5.5$, $M_D = -4$

5.27 Axial force in kips: $CD = -10$. Shear in kips: $AB = 10$, $BC = -10$, $DE = -10$, $EF = 10$. Moment in k·ft: $M_B = 100$, $M_E = -100$

5.29 Axial force in kips: $AB = 1$, $BC = 2$, $CD = -2.5$, $DE = -0.5$, $EF = 1.5$. Shear in kips: $AB = -\sqrt{3}$, $BC = -2\sqrt{3}$, $CD = 2.5\sqrt{3}$, $DE = 0.5\sqrt{3}$, $EF = -1.5\sqrt{3}$. Moment in k·ft: $M_B = -4$, $M_C = -12$, $M_D = 8$, $M_E = 12$

5.31 Axial force in kips: $AD = 1$, $DB = -1$. Shear in kips: $AB = -0.388$, $BC = -1.389$. Moment in k·ft: left of $D = -1.333$, right of $D = 14.67$, left of $B = 12.33$, right of $B = 8.333$

5.33 Axial force in kips: $AB = -10$, $CD = -10$. Shear in kips: $AB = 1.875$, $BC = 11.875$, $CD = 1.875$, $DE = -8.125$. Moment in k·ft: left of $B = 18.75$, right of $B = 38.75$, left of $C = 62.5$, right of $C = 82.5$, left of $D = 101.25$, right of $D = 81.25$

5.35 $x = 10.67$ ft. Shear in kips: $AB =$ zero to 69.23, $BC = -130.77$ to 215.33, $CD = -184.67$ to zero. Moment in k·ft: $M_B = 138.46$, $M_C = 984.62$

5.37 Loads in kN: $P_B = 25\downarrow$, $P_C = 95\downarrow$, $P_D = 10\downarrow$. Shear in kN: $AB = 50$, $BC = 25$, $CD = -70$, $DE = -80$

5.39 Loading in k/ft: $AB = DE = 2$, $BD = 4.8$. Shear in kips: $AB =$ zero to -20, $BD = 48$ to -48, $DE = 20$ to zero

5.41 Loading in kips: $P_A = 66.6\uparrow$, $P_B = 200\downarrow$, $P_C = 250\uparrow$, $P_D = 200\downarrow$, $P_E = 150\uparrow$, $EF = 10$ k/ft$\downarrow$, $P_F = 33.4\uparrow$. Moment in k·ft: $M_B = 333$, $M_C = -334$, $M_D = 248$, $M_E = -169$

CHAPTER 6

6.1 $H = 833.33$ k, $R_{Ay} = 666.7$ k, $R_{By} = 333.3$ k. Max. tension at A $= 1067.2$ k

6.3 80.3 k

6.5 (a) 229.5 kN, (b) 252 kN, (c) 325.05 kN

6.7 (a) 24.9 kN, (b) 18.8 m from A, (c) 6.90 m at mid-span

6.9 338.4 kN

6.11 (a) 14.03 ft, (b) 17.82 lb, (c) 20.68 lb

6.13 1,486 lb

6.15 183 ft

6.17 1.252 m

6.19 Rise at $B = 2/3$ rise at C

6.21 156.25 kN·m

6.23 16.83 in², $M_{max} = 0$

6.25 964.3 m

6.27 −38.5 kN, −631.7 kN, 14.3 kN

6.29 $R_{Ay} = 3.93$ kN, $R_{By} = 2.07$ kN, $H = 2.07$ kN, $M_D = 2.84$ kN-m.   Axial force $= -2.39$ kN,   Shear $= 0.26$ kN

6.31 $R_{Ay} = 11.67$ kN, $R_{By} = 18.33$ kN, $H = 6.67$ kN, $M_D = 33.33$ kN-m.   Axial force in kN: left of D $= -12.97$ kN, right of D $= -5.9$ kN. Shear in kN: left of $D = 3.533$ kN, right of $D = -3.54$ kN

6.33 $R_{Ay}\downarrow = R_{Ey}\uparrow = 1.25$ k, $R_{AX} = 8.333$ k←, $R_{Ex} = 1.67$ k←, $M_B = 33.33$ k-ft.   Shear at B in $AB = -1.667$ k.   Shear at B in $BC = -1.62$ k.   Axial force at B in $AB = 1.25$ k.   Axial force at B in $BC = -1.31$ k.

CHAPTER 7

7.1 $y_A = 416.67/EI$ m ($\downarrow$); $\theta_A = 125/EI$ rad $\jmath$

7.3 $y_D = 312.5/EI$ (ft)$\uparrow$; $\theta_D = 20.83/EI$ rad $\jmath$

7.5 $y_A = 324/EI$ m$\downarrow$; $\theta_A = 72/EI$ rad $\jmath$

7.7 $y_C = 55.35$ mm ($\downarrow$); $y_B = 18.9$ mm ($\downarrow$); $\theta_C = 0.0126$ rad ($\jmath$); $\theta_B = 0.0108$ rad ($\jmath$)

7.9 $y_{max} = 2.67$ mm ($\uparrow$) at a section 4 m from A

7.11 $\theta_c = 0.004446$ rad ($\jmath$); $y_C = 0.324$ in. ($\downarrow$); $\theta_B = 0.005716$ rad ($\jmath$); $y_B = 0.972$ in. ($\downarrow$)

7.13 $y_c = 380.96/EI$ ft ($\downarrow$); $\theta_{c(\text{left})} = 601.36/EI$ rad ($\jmath$); $\theta_{c(\text{right})} = 209.52$ rad$/EI$ ($\jmath$)

7.15 $y_c(\text{vertical}) = 906.7/EI$; $y_c(\text{horizontal}) = 53.33/EI \rightarrow$

7.17 $D_A^V = 19$ mm$\downarrow$

7.19 $y_C = \dfrac{10,000}{EI}\left[\dfrac{\pi}{2} + \dfrac{1}{3}\right]\downarrow$;   horizontal deflection of $c = 8000/EI(\leftarrow)$

7.21 $\theta = 8.66 \times 10^{-5}$ rad $\jmath$

7.23 Length change $= 0.5$ in.

7.25 $y_C(\text{vertical}) = 0$; $y_C(\text{horizontal}) = 0.87$ mm

7.27 $\delta_C(\text{vertical}) = 1.106$ mm$\downarrow$ $\delta_C(\text{horizontal}) = 0.36$ mm$\rightarrow$

7.29 0.04 in.$\downarrow$

CHAPTER 8

8.1 Points of inflection 0.15 L from B and C. $M_B = -67.2$ kN·m, $M_C = -67.2$ kN·m

8.3 Points of inflection 1/3 L from A and 0.2 L from C. $M_A = 5$ kN·m, $M_C = -10$ kN-m, $M_E = 15$ kN·m

8.5 Points of inflection 0.23 L and 0.3 L from E. $M_A = -17.04$ kN $\cdot$ m, $M_C = 15.06$ kN $\cdot$ m, $M_E = -12.8$ kN $\cdot$ m, $M_G = 8.58$ kN $\cdot$ m

8.7 Horizontal thrust = 200 k.   Shear: $V_A = -40$ k, $V_C = +80$ k.   Moment: $M_B = 960$ k $\cdot$ ft

8.9 Axial force: $AB = 80$ k, $BC = -20$ k, $CD = -80$ k.   Shear: $AB = 80$ k, $BC = -80$ k, $CD = 20$ k. $M_B = 800$ k $\cdot$ ft, $M_C = -400$ k $\cdot$ ft

8.11 Symmetric: Axial force: $AB = -125$ kN, $CD = EF = GH = -50$ kN.   Shear: $AB = 125$ kN, $CD = 200$ kN, $EF = 100$ kN, $GH = 0$; $AC = BD = 125$ kN, $CE = DF = 75$ kN, $EG = FH = 25$ kN.   Moments at ends of members: $AB = 250$ kN $\cdot$ m, $CD = 400$ kN $\cdot$ m, $EF = 200$ kN $\cdot$ m; $AC = BD = 250$ kN-m, $CE = DF = 150$ kN $\cdot$ m, $EG = FH = 50$ kN $\cdot$ m

8.13 Axial forces in kips: $AB = 4.167$, $BC = 0.833$, $DE = 0.833$, $EF = 0.167$, $GH = -5$, $HI = -1$, $CF = -7.5$, $FI = -2.5$, $BE = -15$, $EH = -5$.   Shear in kips: $AB = 7.5$, $BC = 2.5$, $DE = 15$, $EF = 5$, $GH = 7.5$, $HI = 2.5$, $CF = 0.8333$, $FI = 1$, $BE = 3.333$, $EH = 4$.   Moments at ends of members in k-ft: $AB = 37.5$, $BC = 12.5$, $BE = 50$, $DE = 75$, $EF = 25$, $EH = 50$, $GH = 37.5$, $HI = 12.5$, $CF = 12.5$, $FI = 12.5$

8.15 Axial force in kN (symmetric): $AB = 274.7$, $BC = 167.5$, $CD = 87.1$, $DC = 33.5$, $EF = 6.7$, $GH = 105.7$, $HI = 64.4$, $IJ = 33.5$, $JK = 12.9$, $KL = 2.6$.   Shear in kN (symmetric): $EF = 6.7$, $KL = 18.33$, $DE = 20.1$, $JK = 54.85$, $CD = 33.5$, $IJ = 91.5$, $BC = 46.9$, $HI = 128.1$, $AB = 60.3$, $GH = 164.8$, $FL = 6.7$, $LS = 9.3$, $EK = 26.8$, $KR = 37.1$, $DJ = 53.6$, $JQ = 74.2$, $CI = 80.4$, $IP = 111.3$, $BH = 107.2$, $HN = 148.5$.   Moments in kN-m at ends of members (symmetric): $FL = 13.4$, $LS = 23.3$, $EK = 53.6$, $KR = 92.8$, $DJ = 107.2$, $JQ = 185.5$, $CI = 160.8$, $IP = 278.3$, $BH = 214.4$, $HN = 371.3$, $EF = 13.4$, $KL = 36.7$, $DE = 40.2$, $JK = 109.7$, $CD = 67.0$, $IJ = 183.0$, $BC = 93.8$, $HI = 256.1$, $AB = 120.6$, $GH = 329.6$

8.17 (a) Axial force in kN: $DH = -37.5$, $HL = -12.5$, $CJ = -75$, $JK = -25$, $BF = -95.833$, $FJ = -87.5$, $JN = -41.667$, $CD = 10$, $BC = 50$, $AB = 113.333$, $KL = -10$, $JK = -50$, $IJ = -80$, $NM = -33.33$, $EF = FG = GH = 0$.   Shear in kN: $CD = KL = 12.5$, $GH = 25$, $BC = JK = 37.5$, $FG = 75$, $AB = MN = 41.667$, $EF = IJ = 83.333$, $DH = 10$, $HL = 10$, $CG = GK = 40$, $BF = FI = 63.333$, $JN = 33.333$.   Moment in kN-m at ends of members: $DH = HL = 25$, $CG = GK = 100$, $BF = FJ = 158.333$, $JN = 83.333$, $AB = 83.333$, $EF = IJ = 166.667$, $MN = 83.333$, $BC = JK = 75$, $FG = 150$, $CD = KL = 25$, $GH = 50$.   (b) Axial force in kN: $AB = -MN = 78$, $EF = -IJ = 26$, $BC = -JK = 50$, $CD = -KL = 10$, $DH = -37.5$, $HL = -12.5$, $CG = -75$, $GK = -25$, $BF = -140$, $FJ = -187.5$, $JN = -97.5$.   Shear in kN: $DH = HL = 10$, $CG = GK = 40$, $BF = 28$, $FJ = 54$, $JN = 78$, $CD = KL = 12.5$, $GH = 25$, $BC = JK = 37.5$, $FG = 75$, $AB = 2.5$, $EF = 27.5$, $IJ = 127.5$, $MN = 97.5$.   Moment in kN $\cdot$ m at ends of member: $DH = HL = 25$, $CG = GK = 100$, $BF = 70$, $FJ = 135$, $JN = 195$, $CD = KL = 25$, $GH = 50$, $BC = JK = 75$, $FG = 150$, $AB = -5$, $EF = 55$, $IJ = 255$, $MN = 195$

8.19 Axial force in kips: $ABC = 3.125$, $DEF = -3.125$, $BG = -KE = 12.5$, $CH = HJ = -11.25$, $JL = LF = 1.25$, $CG = JK = -6.988$, $GJ = KF = 6.988$.   Shear in kips: $AB = DE = 5$, $BC = FE = 7.5$.   Moments in k $\cdot$ ft: $M_A = M_D = 37.5$, $M_B = M_E = 37.5$

8.21 Axial force in kN (symmetric): $AB = -15$, $BC = -5$, $CH = 5\sqrt{5}$, $CI = -10$, $BH = 20$, $BI = -10\sqrt{5}$, $IK = -30$, $HJ = 40$, $IJ = 0$, $HK = -5\sqrt{5}$

8.23 Axial force in kN (symmetric): $A_4B_4 = -A_1B_1 = -B_4C_4 = B_1C_1 = 3.333$, $C_1D_1 = -C_4D_4 = 16.667$, $A_1A_2 = -A_3A_4 = 6.667$, $B_1B_2 = -86.667$, $B_2B_3 = -60$, $B_3B_4 = -33.333$,

$C_1C_2 = -C_3C_4 = 13.333$, $A_2A_3 = C_2C_3 = D_1D_2 = D_2D_3 = D_3D_4 = 0$.   Shear in kN
(symmetric): $A_1B_1 = A_4B_4 = 3.333$, $A_2B_2 = A_3B_3 = 6.667$, $B_1C_1 = B_4C_4 = 10$, $B_2C_2 = B_3C_3 = 20$, $C_1D_1 = C_4D_4 = 3.333$, $C_2D_2 = C_3D_3 = 6.667$, $A_1A_2 = A_2A_3 = A_3A_4 = 3.333$, $B_1B_2 = B_2B_3 = B_3B_4 = 6.667$, $C_1C_2 = C_2C_3 = C_3C_4 = 13.333$.   Moment in kN-m
(symmetric) at ends of members: $A_1B_1 = A_4B_4 = 8.333$, $A_2B_2 = A_3B_3 = 16.667$, $B_1C_1 = B_4C_4 = 25$, $B_2C_2 = B_3C_3 = 50$, $C_1D_1 = C_4D_4 = 8.333$, $C_2D_2 = C_3D_3 = 16.667$, $A_1A_2 = A_2A_3 = A_3A_4 = 8.333$, $B_1B_2 = B_2B_3 = B_3B_4 = 16.667$, $C_1C_2 = C_2C_3 = C_3C_4 = 33.333$

CHAPTER 9

9.1 $M_B = -1.90$ k·ft

9.3 $M_A = -14.28$ k·ft, $M_B = 28.57$ k·ft, $M_C = -100$ k·ft

9.5 $M_A = -10.25$ kN·m, $M_B = -4.5$ kN·m

9.7 $M_B = -3.12$ kN·m

9.9 $M_A = -30$ k·ft, $M_B = 16.25$ k·ft

9.11 $M_A = -108.57$ lb·ft; $M_B = -82.86$ lb·ft, $M_C = -48.57$ lb·ft

9.13 $M_A = -5.47$ kN·m, $M_B = -5.15$ kN·m

9.15 $M_B = 69.4$ kN·m, $M_C = -123.0$ kN·m

9.17 $M_A = -6.05$ k·ft, $M_B = -14.57$ k·ft, $M_C = -30$ k·ft; $\Delta_G = 0.010$ in.↑

9.19 See 9.7.

9.21 $M_B = -11.71$ k·ft; $M_C = -20.81$ k·ft

9.23 $M_A = -30.46$ k·ft

9.25 See 9.15.

9.27 $H_A = 5.63$ kN

9.29 $F_{AB} = 5.86$ kN; $F_{BC} = 5.86$ kN; $F_{CD} = -8.28$ kN; $F_{DB} = -8.28$ kN; $F_{AD} = 5.86$ kN

9.31 $F_{AB} = 0.79$ k, $F_{AC} = 0.79$ k, $F_{CD} = 0.79$ k, $F_{BD} = -1.21$ k, $F_{AD} = -1.12$ k, $F_{CB} = 1.71$ k

9.33 $F_{AB} = 5.73$ k, $F_1 = -9.33$ k, $F_2 = -9.33$ k, $F_3 = 0.764$ k, $F_4 = 1.92$ k, $F_5 = 1.92$ k

9.35 $V_A = -5.02$ k, $H_A = -10$ k; $V_B = 6.71$ k; $V_C = 8.31$ k

9.37 $H_A = -12.78$ kN, $V_A = -21.25$ kN, $H_B = -17.22$ kN, $V_B = 31.25$ kN

9.39 $H_B = -89.59$ kN, $H_C = 89.59$ kN, $V_C = 67.19$ kN, $V_E = 32.81$ kN

CHAPTER 10

NOTE: Clockwise moments are positive

10.1 $M_{AB} = -14.72$ k·ft, $M_{BA} = 8.97$ k·ft, $M_{CB} = 8.01$ k·ft

10.3 $M_{AB} = -30.33$ k·ft, $M_{BA} = 19.33$ k·ft, $M_{CB} = 8.33$ k·ft

10.5 $M_{AB} = -16.8$ k·ft, $M_{BA} = 21.6$ k·ft, $M_{CB} = -11.9$ k·ft

10.7 $M_{AB} = -9.22$ k·ft, $M_{BA} = -10.33$ k·ft, $M_{CB} = 12.2$ k·ft

10.9 $M_{AB} = +0.86$ k·ft, $M_{BA} = +1.71$ k·ft, $M_{CB} = 13.14$ k·ft, $M_{BC} = -9.71$ k·ft

10.11 $M_{AB} = -10.41$ kN·m, $M_{BA} = 4.17$ kN·m, $M_{CB} = -2.08$ kN·m

10.13 $M_{AB} = -1.16$ k·ft, $M_{BA} = -2.32$ k·ft, $M_{CB} = +7.41$ k·ft

10.15 $M_{AB} = -7.40$ k·ft, $M_{BA} = 10.19$ k·ft, $M_{BD} = 0.93$ k·ft, $M_{DB} = 0.46$ k·ft

10.17 $M_{AB} = -25.0$ kN·m, $M_{BA} = -16.67$ kN·m, $M_{CB} = 8.33$ kN·m, $M_{DC} = -8.33$ kN·m; $\Delta_B = \dfrac{138.89}{EI}$ m, $\theta_B = \dfrac{20.83335}{EI}$ rad, $\theta_C = 0$

10.19 $M_{AB} = -3.28$ k·ft, $M_{BA} = 5.17$ k·ft, $M_{CD} = 4.15$ k·ft, $M_{DC} = -5.15$ k·ft

10.21 (a) $M_{AB} = 11.67$ kN·m, $M_{BA} = 8.33$ kN·m, $M_{CB} = 3.33$ kN·m, $M_{BC} = 6.67$ kN·m; (c) $\Delta = 1.976/EI$, (d) third

10.23 $M_{AB} = -9.14$ k·ft, $M_{BA} = 13.14$ k·ft

10.25 $M_{AB} = -8.99$ k·ft, $M_{BA} = 8.77$ k·ft

10.27 $M_{AB} = 6.46$ k·ft, $M_{BA} = 7.92$ k·ft, $M_{CB} = 21.04$ k·ft

10.29 $M_{AB} = -60.0$ kN·m, $M_{BA} = 67.2$ kN·m, $M_{CB} = 80.5$ kN·m

10.31 $M_{AB} = -4.0$ kN·m, $M_{BA} = 5.08$ kN·m, $M_{CB} = 5.71$ kN·m

10.33 $M_{BA} = 12.10$ kN·m, $M_{CB} = 17.39$ kN·m

10.35 $M_{AB} = 48.1$ k·ft, $M_{BA} = 33.6$ k·ft, $M_{BC} = 48.4$ k·ft, $M_{CB} = -8.6$ k·ft, $M_{DB} = +50.0$ k·ft

10.37 $M_{CD} = 2.4$ k·ft, $M_{DC} = 4.8$ k·ft, $M_{GD} = 1.20$ k·ft, $M_{DB} = 6.24$ k·ft, $M_{BD} = 82.7$ k·ft, $M_{DA} = -11.0$ k·ft. (c) $\theta_G = 1.80 \times 10^{-4}$ rad counterclockwise

10.39 $M_{AB} = -75.3$ k·ft, $M_{BA} = -17.2$ k·ft, $M_{CB} = 29.1$ k·ft, $M_{DC} = -40.7$ k·ft

10.41 $M_{AB} = -1.70$ kN·m, $M_{BA} = -1.38$ kN·m, $M_{DB} = 0.44$ kN·m, $M_{DC} = -3.28$ kN·m, $M_{CD} = -3.21$ kN·m, $M_{FD} = +6.17$ kN·m, $M_{EF} = -9.35$ kN·m

10.43 $M_{AB} = -25.71$ kN·m, $M_{BA} = -27.86$ kN·m, $M_{CB} = 33.21$ kN·m, $M_{DC} = -34.29$ kN·m

10.45 $M_{BA} = -123.3$ k·ft, $M_{CB} = 31.0$ k·ft, $M_{DC} = 99.0$ k·ft

10.47 $M_{AB} = 6.74$ kN·m, $M_{BA} = 13.63$ kN·m, $M_{BE} = -25.11$ kN·m, $M_{CB} = 2.32$ kN·m

CHAPTER 11

11.1 $E = \begin{bmatrix} 1 & 0 & 0 & 0 & 0 & 0 \\ 0 & 1 & 1 & 0 & 0 & 0 \\ 0 & 0 & 0 & 1 & 1 & 0 \\ 0 & 0 & 0 & 0 & 0 & 1 \end{bmatrix}$; $S^* = EI \begin{bmatrix} 0.667 & 0.333 & 0 & 0 \\ 0.333 & 1.167 & 0.250 & 0 \\ 0 & 0.250 & 1.167 & 0.333 \\ 0 & 0 & 0.333 & 0.667 \end{bmatrix}$;

$\{P\} = \{15, \quad 35, \quad -50, \quad 0\}$

11.3 $E = \begin{bmatrix} 0.8 & 0 & -0.8 \\ 0.6 & 1 & 0.6 \end{bmatrix}$; $S^* = EA \begin{bmatrix} 0.128 & 0 \\ 0 & 0.239 \end{bmatrix}$; $\{P\} = \{50, \quad -20\}$

11.5 $E = \begin{bmatrix} 0 & -1 & 0 & 0 & -0.8 & 0 & 0 & 0 & 0 & 0 & 0 \\ 1 & 0 & 0 & 0 & 0.6 & 0 & 0 & 0 & 0 & 0 & 0 \\ 0 & 1 & 0 & 0.8 & 0 & 0 & -1 & 0 & 0 & -0.8 & 0 \\ 0 & 0 & 1 & 0.6 & 0 & 0 & 0 & 0 & 0 & 0.6 & 0 \\ 0 & 0 & 0 & 0 & 0 & 0 & 1 & 0 & 0.8 & 0 & 0 \\ 0 & 0 & 0 & 0 & 0 & 0 & 0 & 1 & 0.6 & 0 & 0 \\ 0 & 0 & 0 & 0 & 0.8 & 1 & 0 & 0 & -0.8 & 0 & -1 \\ 0 & 0 & -1 & 0 & -0.6 & 0 & 0 & 0 & -0.6 & 0 & 0 \\ 0 & 0 & 0 & 0 & 0 & 0 & 0 & 0 & 0 & 0.8 & 1 \end{bmatrix}$;

$$S^* = EA \begin{bmatrix} 0.189 & -0.048 & -0.125 & 0 & 0 & 0 & -0.064 & 0.048 & 0 \\ & 0.203 & 0 & 0 & 0 & 0 & 0.048 & -0.036 & 0 \\ & & 0.378 & 0 & -0.125 & 0 & 0 & 0 & -0.064 \\ & & & 0.239 & 0 & 0 & 0 & -0.167 & 0.048 \\ & & & & 0.189 & 0.048 & -0.064 & -0.048 & 0 \\ & \text{Symmetrical} & & & & 0.203 & -0.048 & -0.036 & 0 \\ & & & & & & 0.378 & 0 & -0.125 \\ & & & & & & & 0.239 & 0 \\ & & & & & & & & 0.189 \end{bmatrix} ;$$

$$P = \begin{bmatrix} 0 \\ -50 \\ 0 \\ -100 \\ 20 \\ 0 \\ 0 \\ 0 \\ 0 \end{bmatrix}$$

11.7 $E = \begin{bmatrix} 0 & 1 & 1 & 0 & 0 & 0 \\ 0 & 0 & 0 & 1 & 1 & 0 \\ -0.063 & -0.063 & 0 & 0 & -0.063 & -0.063 \end{bmatrix} ;$

$S^* = EI \begin{bmatrix} 0.633 & 0.067 & -0.047 \\ & 0.633 & -0.047 \\ \text{Symmetrical} & & 0.012 \end{bmatrix} ; \quad \{P\} = \{21.5, \ -37.5, \ 4\}$

11.9 $E = \begin{bmatrix} 0 & 1 & 1 & 0 & 0 & 0 \\ 0 & 0 & 0 & 1 & 1 & 0 \\ -0.067 & -0.067 & 0.133 & 0.133 & -0.083 & -0.083 \end{bmatrix} ;$

$S^* = EI \begin{bmatrix} 0.360 & 0.100 & 0.024 \\ & 0.400 & 0.015 \\ \text{Symmetrical} & & 0.017 \end{bmatrix} ; \quad \{P\} = \{0, \ 0, \ 20\}$

11.11 $E = \begin{bmatrix} 0 & 1 & 1 & 0 & 0 & 0 & 0 & 0 & 0 & 0 & 1 & 0 \\ 0 & 0 & 0 & 1 & 1 & 0 & 0 & 0 & 0 & 0 & 0 & 0 \\ 0 & 0 & 0 & 0 & 0 & 1 & 1 & 0 & 0 & 0 & 0 & 0 \\ 0 & 0 & 0 & 0 & 0 & 0 & 0 & 1 & 1 & 0 & 0 & 1 \end{bmatrix} ;$

$S^* = EI \begin{bmatrix} 2.400 & 0.400 & 0 & 0.400 \\ & 1.600 & 0.400 & 0 \\ & & 1.600 & 0.400 \\ \text{Symmetrical} & & & 2.400 \end{bmatrix} ;$

$P = \{+125, \ +333.33, \ -333.33, \ -125\}$

11.13 See 11.1

11.15 See 11.5

11.17 See 11.9

11.19 $S^* = \begin{bmatrix} 1000.00 & 200.00 & -112.50 & 49.98 \\ & 400.00 & 0 & 49.98 \\ & & 38.93 & 0 \\ \text{Symmetrical} & & & 27.53 \end{bmatrix}$

11.21 $\{F\} = \{5.478, \quad 0.955, \quad 7.978, \quad 4.524, \quad -1.069\} \, k$

11.23 $\{F'\} = \{-163.23, \quad 60.16, \quad -60.16, \quad 0, \quad -2.812, \quad -59.048\}$ kN and kN$\cdot$m

11.27 $\{R\} = \{-20.0, \quad 92.5, \quad 57.5\}$ kN

11.29 $\{R\} = \{-12.89, \quad 55.01, \quad -103.07, \quad -1.69, \quad -2.25, \quad -35.43, \quad 47.24\}$ kN and kN$\cdot$m

11.31 $\{F'\} = \{0, \quad 0, \quad 0, \quad 0, \quad 0, \quad 0\}$

11.33

$$E = \begin{bmatrix} 0 & 1 & 1 & 0 & 0 & 0 & 0 & 0 & 0 & 0 & 0 & 0 & 0 & 0 & 0 & 0 \\ 0 & 0 & 0 & 1 & 1 & 0 & 0 & 0 & 0 & 1 & 1 & 0 & 0 & 0 & 0 & 0 \\ 0 & 0 & 0 & 0 & 0 & 1 & 1 & 0 & 0 & 0 & 0 & 0 & 0 & 0 & 0 & 1 \\ -0.05 & -0.05 & 0 & 0 & 0 & 0 & -0.05 & -0.05 & -0.05 & -0.05 & 0.05 & 0.05 & 0 & 0 & 0.05 & 0.05 \\ 0 & 0 & 0 & 0 & 0 & 0 & 0 & 0 & 0 & 0 & 1 & 1 & 0 & 0 & 0 \\ 0 & 0 & 0 & 0 & 0 & 0 & 0 & 0 & 0 & 0 & 0 & 0 & 1 & 1 & 0 \\ 0 & 0 & 0 & 0 & 0 & 0 & 0 & 0 & 0 & -0.05 & -0.05 & 0 & 0 & -0.05 & -0.05 \end{bmatrix} ;$$

$$S^* = EI \begin{bmatrix} 0.30 & 0.05 & 0 & -0.015 & 0 & 0 & 0 \\ & 0.60 & 0.05 & 0 & 0.1 & 0 & -0.015 \\ & & 0.5 & 0 & 0 & 0.1 & -0.015 \\ & & & 0.008 & 0.015 & 0.015 & -0.003 \\ & \text{Symmetrical} & & & 0.30 & 0.05 & -0.015 \\ & & & & & 0.30 & -0.015 \\ & & & & & & 0.003 \end{bmatrix} ;$$

$\{P\} = \{0, \quad 0, \quad 0, \quad -20, \quad 0, \quad 0, \quad -10\} \, k;$

$\{F'\} = \{127.11, \quad 74.72, \quad -74.72, \quad -70.86, \quad -75.64, \quad -84.28, \quad 55.60, \quad 117.55, \quad 134.84, \quad 90.17, \quad 56.34, \quad 59.22, \quad -59.22, \quad -55.76, \quad 55.76, \quad 28.68\} \, k\cdot ft$

11.35 $E = \begin{bmatrix} 0 & 1 & 1 & 0 & 0 & 0 \\ -0.063 & -0.063 & 0.03 & 0.03 & 0 & 0 \\ 0 & 0 & 0 & 1 & 1 & 0 \\ 0 & 0 & -0.04 & -0.04 & 0.05 & 0.05 \end{bmatrix} ;$

$S^* = EI \begin{bmatrix} 0.45 & -0.014 & 0.10 & -0.012 \\ & 0.003 & 0.009 & -0.001 \\ & & 0.40 & 0.003 \\ \text{Symmetrical} & & & 0.002 \end{bmatrix} ;$

$\{P\} = \{21.33, \quad 0, \quad 12.0, \quad -18.0\};$

$\{F'\} = \{67.93, \quad -1.11, \quad 1.11, \quad 138.10, \quad -138.10, \quad -110.54\} \, k\cdot ft$

CHAPTER 12

12.1 $R = [I]_{6,6}$

12.3 $R_{1,2,3} = [I]_{6,6}$; $R_4 =$
$$
\begin{bmatrix}
-0.5 & 0.87 & 0 & 0 & 0 & 0 \\
-0.87 & -0.5 & 0 & 0 & 0 & 0 \\
0 & 0 & 1 & 0 & 0 & 0 \\
0 & 0 & 0 & -0.5 & 0.87 & 0 \\
0 & 0 & 0 & -0.87 & -0.5 & 0 \\
0 & 0 & 0 & 0 & 0 & 1
\end{bmatrix} ;
$$

$$
R_5 =
\begin{bmatrix}
0.5 & 0.87 & 0 & 0 & 0 & 0 \\
-0.87 & 0.5 & 0 & 0 & 0 & 0 \\
0 & 0 & 1 & 0 & 0 & 0 \\
0 & 0 & 0 & 0.5 & 0.87 & 0 \\
0 & 0 & 0 & -0.87 & 0.5 & 0 \\
0 & 0 & 0 & 0 & 0 & 1
\end{bmatrix}
$$

12.5 (a) $S_1 = EI \times 10^{-3}$
$$
\begin{bmatrix}
1.13 & 12 & -1.13 & 12 \\
 & 181.82 & -12 & 90.91 \\
 & & 1.13 & -12 \\
\text{Symmetrical} & & & 181.82
\end{bmatrix} ;
$$

$$
S_{2,3} = EI \times 10^{-3}
\begin{bmatrix}
1.5 & 15 & -1.5 & 15 \\
 & 200 & -15 & 100 \\
 & & 1.5 & -15 \\
\text{Symmetrical} & & & 200
\end{bmatrix} ;
$$

$$
S_4 = EI \times 10^{-3}
\begin{bmatrix}
0.87 & 10.42 & -0.87 & 10.42 \\
 & 166.67 & -10.42 & 83.33 \\
 & & 0.87 & -10.42 \\
\text{Symmetrical} & & & 166.67
\end{bmatrix} ;
$$

(b) Same as in (a).

12.7 (a) $S_1 = S_3 = EI$
$$
\begin{bmatrix}
0.023 & 0.094 & -0.023 & 0.094 \\
 & 0.500 & -0.094 & 0.250 \\
 & & 0.023 & -0.094 \\
\text{Symmetrical} & & & 0.500
\end{bmatrix} ;
$$

$$
S_2 = EI
\begin{bmatrix}
0.012 & 0.06 & -0.012 & 0.06 \\
 & 0.40 & -0.06 & 0.20 \\
 & & 0.012 & -0.06 \\
\text{Symmetrical} & & & 0.40
\end{bmatrix} ;
$$

$$
S_4 = S_5 = EI
\begin{bmatrix}
0.036 & 0.125 & -0.036 & 0.125 \\
 & 0.577 & -0.125 & 0.289 \\
 & & 0.036 & -0.125 \\
\text{Symmetrical} & & & 0.577
\end{bmatrix}
$$

(b) $(S_m)_1^* = (S_m)_3^* = S_1 = S_3$; $(S_m)_2^* = S_2$;

$$
(S_m)_4^* = EI
\begin{bmatrix}
0.027 & 0.016 & -0.108 & -0.027 & -0.016 & -0.108 \\
 & 0.009 & -0.063 & -0.016 & -0.009 & -0.063 \\
 & & 0.577 & 0.108 & 0.063 & 0.289 \\
 & & & 0.027 & 0.016 & 0.108 \\
 & & & & 0.009 & 0.063 \\
\text{Symmetrical} & & & & & 0.577
\end{bmatrix} ;
$$

$$(S_m)_5^* = EI \begin{bmatrix} 0.027 & -0.016 & -0.108 & -0.027 & 0.016 & -0.108 \\ & 0.009 & 0.063 & 0.016 & -0.009 & 0.063 \\ & & 0.577 & 0.108 & -0.063 & 0.289 \\ & & & 0.027 & -0.016 & 0.108 \\ & & & & 0.009 & -0.063 \\ & & \text{Symmetrical} & & & 0.577 \end{bmatrix}$$

12.9 (a) $S_1 = S_6 = EI \times 10^{-3} \begin{bmatrix} 2.93 & 23.44 & -2.93 & 23.44 \\ & 250.0 & -23.44 & 125.0 \\ & & 2.93 & -23.44 \\ \text{Symmetrical} & & & 250.0 \end{bmatrix}$;

$S_2 = S_4 = EI \times 10^{-3} \begin{bmatrix} 6.94 & 41.67 & -6.94 & 41.67 \\ & 333.33 & -41.67 & 166.67 \\ & & 6.94 & -41.67 \\ \text{Symmetrical} & & & 333.33 \end{bmatrix}$;

$S_3 = S_5 = EI \times 10^{-3} \begin{bmatrix} 1.7 & 20.8 & -1.7 & 20.8 \\ & 333.33 & -20.8 & 166.67 \\ & & 1.7 & -20.8 \\ \text{Symmetrical} & & & 333.33 \end{bmatrix}$

(b) $(S_m)_1^* = (S_m)_6^* = EI \times 10^{-3} \begin{bmatrix} 2.9 & 0 & -23.4 & -2.9 & 0 & -23.4 \\ & 0 & 0 & 0 & 0 & 0 \\ & & 250.0 & 23.4 & 0 & 125.0 \\ & & & 2.9 & 0 & 23.4 \\ & & & & 0 & 0 \\ & & \text{Symmetrical} & & & 250.0 \end{bmatrix}$;

$(S_m)_2^* = (S_m)_4^* = EI \times 10^{-3} \begin{bmatrix} 6.9 & 0 & -41.7 & -6.9 & 0 & -41.7 \\ & 0 & 0 & 0 & 0 & 0 \\ & & 333.33 & 41.7 & 0 & 166.67 \\ & & & 6.9 & 0 & 41.7 \\ & & & & 0 & 0 \\ & & \text{Symmetrical} & & & 333.33 \end{bmatrix}$;

$(S_m)_3^* = (S_m)_5^* = S_3 = S_5$

12.11 $S^* = EA \times 10^{-3} \begin{bmatrix} 56.84 & 6.76 & -40.02 & -16.01 & 0 & 0 \\ & 56.84 & -16.01 & -6.4 & 0 & 0 \\ & & 98.55 & 3.21 & -50.0 & 0 \\ & & & 25.6 & 0 & 0 \\ & & & & 62.8 & 9.6 \\ & & \text{Symmetrical} & & & 40.53 \end{bmatrix}$

12.13 $S^* = \begin{bmatrix} 50009.6 & -7.2 & 0 \\ & 33.53 & -112.5 \\ \text{Symmetrical} & & 600.0 \end{bmatrix}$

12.15 $S^* = EI \begin{bmatrix} 0.38 & 0.68 & -0.01 & -0.07 \\ & 2.87 & 0.07 & 0.43 \\ & & 0.03 & 0.01 \\ \text{Symmetrical} & & & 1.33 \end{bmatrix}$

12.17 $S^* = $

$$\begin{bmatrix} 480 & 0 & -36 & 240 & 0 & 0 & 0 & 0 \\ & 6000 & 0 & 0 & -3000 & 0 & 0 & 0 \\ & & 9.8 & 0 & 0 & 36 & 0 & -2.6 \\ & & & 960 & 0 & 240 & 0 & 0 \\ & \text{Symmetrical} & & & 3001 & 0 & -1.0 & -0.6 \\ & & & & & 480 & 0 & 0 \\ & & & & & & 2 & 0 \\ & & & & & & & 3.3 \end{bmatrix}$$

12.19 $\{D\} = \begin{Bmatrix} -0.00018 \\ -1.25702 \\ 0.15236 \end{Bmatrix} \begin{matrix} \text{m} \\ \text{m} \\ \text{rad} \end{matrix}$; $\{F_1\} = \begin{Bmatrix} 9.05 \\ 43.21 \\ -145.71 \\ -9.05 \\ 6.79 \\ 0 \end{Bmatrix}$; $\{F_2\} = \begin{Bmatrix} -11.31 \\ 0 \\ 11.31 \\ 0 \end{Bmatrix}$

12.21 See Problem 11.33.

12.23 See Problem 11.35.

12.25 $\{D\} = \dfrac{1}{EI}\{26.18, \quad -239.29, \quad 26.8, \quad -305.62, \quad -6.97, \quad -161.01,$
$-25.33, \quad -15.17, \quad 3.30, \quad 26.18, \quad 26.8, \quad -6.97, \quad -25.33, \quad -15.17\}.$

CHAPTER 13

13.1 $E_{pd} = [I]_{4,4}$; $E_{px} = \begin{bmatrix} 0 & 0 \\ 1 & 0 \\ 0 & 1 \\ 0 & 0 \end{bmatrix}$; $F_d = \dfrac{2}{3EI}\begin{bmatrix} 3 & 0 & 0 & 0 \\ 0 & 4 & 0 & 0 \\ 0 & 0 & 3 & -1.5 \\ 0 & 0 & -1.5 & 3 \end{bmatrix}$;

$F_x = \dfrac{2}{3EI}\begin{bmatrix} 3 & 0 \\ 0 & 4 \end{bmatrix}$; $F_{dx} = \dfrac{2}{3EI}\begin{bmatrix} -1.5 & 0 \\ 0 & -2 \\ 0 & 0 \\ 0 & 0 \end{bmatrix}$;   for P matrix see Problem 11.1.

13.3 $E_{pd} = \begin{bmatrix} 1 & 0 & 0 \\ 0 & 1 & 0 \\ -\frac{1}{16} & -\frac{1}{16} & -\frac{1}{16} \end{bmatrix}$; $E_{px} = \begin{bmatrix} 1 & 0 & 0 \\ 0 & 1 & 0 \\ 0 & 0 & -\frac{1}{16} \end{bmatrix}$;

$F_d = \dfrac{4}{3EI}\begin{bmatrix} 2 & 0 & 0 \\ 0 & 2 & -1 \\ 0 & -1 & 2 \end{bmatrix}$; $F_x = \dfrac{1}{3EI}\begin{bmatrix} 30 & -15 & 0 \\ -15 & 30 & 0 \\ 0 & 0 & 8 \end{bmatrix}$;

$F_{dx} = \dfrac{4}{3EI}\begin{bmatrix} 0 & 0 & -1 \\ 0 & 0 & 0 \\ 0 & 0 & 0 \end{bmatrix}$;   for P matrix see Problem 11.7.

13.5 $\{F'\} = \{0, \quad 43.83, \quad -43.83, \quad 30.33, \quad -30.33, \quad 0\}$ kN·m

13.7 $\begin{bmatrix} F'_{x1} \\ F'_{x2} \\ F'_{x3} \end{bmatrix} = \begin{bmatrix} -29.1 \\ 38.9 \\ -22.1 \end{bmatrix}$ k; $\begin{bmatrix} F_{d1} \\ F_{d2} \\ F_{d3} \end{bmatrix} = \begin{bmatrix} 29.1 \\ -38.9 \\ -32.1 \end{bmatrix}$ k

13.9 (a) See Problem 11.21.

(b) $\{F'\}_{5\times1} = EA\,\{-12.7, \quad -124.8, \quad 354.3, \quad -115.6, \quad 84.0\}$ k,

$\{D'\}_{4\times1} = \dfrac{1}{EA}\,\{-3249.8, \quad 191.3, \quad -1148.7, \quad -6997.1\}$ ft;

(c) $\{F'\}_{5\times1} = EA\,\{-0.026, \quad -0.013, \quad -0.0065, \quad 0.029, \quad 0.029\}$,

$\{D\}_{4\times1} = \dfrac{1}{EA}\,\{0, \quad 0.25, \quad 0, \quad -0.125\}$

13.13 Determinate structure; force method does not apply.

13.15 (b): $\{F'_x\}_{9\times1} = \{127.1, -74.7, -70.9, 134.8, -75.6, -59.2, -55.8, -84.3, 117.6\}$ k · ft;

$\{F_d\}_{7\times1} = \{74.7, \quad 90.2, \quad 56.3, \quad 59.2, \quad 55.8, \quad 28.7, \quad 55.6\}$ k · ft;

$\{D\}_{7\times1} = \dfrac{1}{EI}$

$\times \{-523.9, -446.7, -619.5, -11966.9, -417.8, -348.7, -24463.9\}$ rad and ft

13.17 (b): $\{F'_x\}_{3\times1} = \{-68.7, \quad -18.7, \quad -85.4\}$ k · ft; $\{F_d\}_{3\times1} = \{18.7, \quad 85.4, \quad 191.4\}$ k · ft; $\{D\}_{3\times1} = \dfrac{1}{EI}\,\{6661.5, \quad 393.4, \quad -727.3\}$ rad and ft

13.19 $M_B = 210$ k · ft

13.21 $F_{tie} = 231.1$ kN, $M_A = 12.3$ kN · m

13.23 $F_{OA} = -2.1$ kN, $F_{OB} = 2.9$ kN, $F_{OC} = 7.9$ kN

13.25 $F_1 = -9.3$ k, $F_3 = 0.8$ k, $F_4 = 1.9$ k, $F_6 = 5.7$ k

13.27 $H_A = 126.7$ k, $V_A = 50$ k

13.29 $H_A = 5.41$ kN, $M_B = 4.2$ kN · m

13.31 $F_{AB} = 7.93$ kN, $F_{CD} = 7.93$ kN, $F_{AC} = 7.93$ kN, $F_{BD} = -12.07$ kN, $F_{AD} = -11.21$ kN, $F_{BC} = 17.07$ kN; $\Delta_{BV} = 0.10$ mm, $\Delta_{BH} = 0.37$ mm

13.33 $F_{AB} = -3.16$ k, $F_{AC} = -17.06$ k, $F_{AD} = -5.79$ k, $F_{BC} = -0.46$ k, $F_{BD} = 5.79$ k

CHAPTER 14

For Problems 14.1–14.7, the following influence ordinates are given from the left end

14.1 1.5, 1, 0, −0.25 for left reaction;   −0.5, 0, 1, 1.25 for right reaction

14.3 1, 1, 0 for left reaction;   0, 0, 1 for right reaction

14.5 0, −4, 0, 4, 0 for horizontal left reaction;   1, −3, 0, 3, 0 for vertical left reaction;   0, 8, 1, −6, 0 for vertical interior reaction;   0, 4, 0, −4, 0 for horizontal right reaction; 0, −4, 0, 4, 1 for vertical right reaction

14.7 3, 2, 1, 0, −1, −2, −3 for vertical left reaction;   −2, −1, 0, 1, 2, 3, 4 for vertical right reaction

14.9 Influence ordinates for moment at G: 0, 5, 0;   influence ordinates for shear at G: 0, 0.5, 0

14.11 0, −0.33, 0.5, 0 for V_{AB};   0, 10, 15, 0 for M_B

14.13 0.5, 0, 1.0, 0, −0.5 for V_{AB};   −5, 0, 10, 0, −5 for M_B

14.15 0, −0.5, 0.5, 0, −0.5 for V_F;   0, 2.5, 0, −2.5 for M_F;   0, 0.5, 1, 1.5 for V_E;   0, −5, −10, −15 for M_E

14.17 0, −0.5, −1, 0, 0.5 for H_D;   0, 5, −10, 0, 5 for M_D.   Maximum M_D: **(a)** −300 k · ft;   **(b)** −1066.67 k · ft;   **(c)** −200 k · ft

14.19 0, 0, 0, −0.625, 0, 0, 0, 0, 0, 0, 0, 0, 0 for F_1;   0, −0.1, −0.21, −0.31, 0.83, 0.73, 0.62, 0.52, 0.42, 0.31, 0.21, 0.1, 0 for F_2;   0, 0.08, 0.16, 0.25, 0.33, −0.58, −0.5, −0.42, −0.33, −0.25, −0.17, −0.08, 0 for F_3;   0, −0.1, −0.21, −0.31, −0.42, 0.73, 0.63, 0.52, 0.42, 0.31, 0.21, 0.10, 0 for F_4;   0, −0.1875, −0.375, −0.5625, −0.75, −0.9375, −1.125, −0.9375, −0.75, −0.5625, −0.375, −0.1875, 0 for F_5;   0, 0.25, 0.5, 0.75, 1.0, 0.75, 0.625, 0.5, 0.375, 0.25, 0.125, 0 for F_6.   Maximum tensile forces: $F_1 = 0$, $F_2 = 257.36$ k, $F_3 = 0$, $F_4 = 181.75$ k, $F_5 = 0$, $F_6 = 450.0$ k;   maximum compression forces: $F_1 = 46.88$ k, $F_2 = 0$, $F_3 = 145.2$ k, $F_4 = 0$, $F_5 = 506.25$ k, $F_6 = 0$

14.21 0, −0.5, −1.0, −1.5, −1.0, −0.5, 0 for F_1;   0, −0.37, −0.75, 1.12, 0.75, 0.37, 0 for F_2;   0, 0, −1.12, 0, 0, 0, 0 for F_3;   0, 0, 1.0, 0, 0, 0, 0 for F_4

14.23 $H_A = 103.1$ kN, $M = -31$ kN · m, $T = 103.1$ kN

14.25 0, −0.276, −0.640, −1.144, −0.298, 0, 0.223, 0.405, 0.392, 0.239, 0.   Maximum $R_B = 29.52$ k;   maximum force in member 5–7 = −28.26 k

14.27 **(a)** $R_B = \dfrac{(18 - x)(x^2)}{432}$, x measured from A

  (b) 0, −0.52, 0.48, 0 influence ordinates for V_C;   **(c)** 0, 1.04, 0 influence ordinates for M_C;   **(d)** $R_B = 0.567$ kN

14.29 Influence line equation for H at $A = \dfrac{5 \times 10^{-10}}{8} \{(100 - a)(a^3)(400 - 3a) + (100 - a)^3 a(100 + 3a)\}$, a measured from A; $H = 3.906$ kN.

14.31 **(a)** 0, 0.5, 0.8, 1, 0.8, 0.5, 0 influence ordinates for R_B;   1, 0.58, 0.26, 0, −0.067, −0.083, 0 influence ordinates for R_A;   0, −0.466, −0.214, 0, 0.054, 0.066, 0 for F_a;   **(b)** $F_a = 11.46$ k

14.33 0, 4.09, 8.57, 3.8, 0.15, −2.48, −3.90, −4.17, −3.41, −1.92, 0 for M_F at 10 ft intervals;   0, −0.28 to left and 0.46 to right, 0.51, 0.30, 0.16, 0.031, −0.044, −0.078, −0.071, −0.041, 0 for V_F;   0, 0.57, 1.33, 1.72, 1.94, 2.00, 1.94, 1.62, 1.73, 0.61, 0 for T_F. $M_F = -99.5$ k · ft, $V_F = -1.0$ k, $T_F = -51.3$ k

14.35 0, 0.434, 0.759, 1, 0.759, 0.434, 0 influence ordinates for R_B;   1.0, 0.617, 0.289, 0, −0.045, −0.050, 0 influence ordinates for R_A;   0, −0.598, 0.272, 0, −0.041, −0.046, 0 influence ordinates for F_{DF}

14.37 Influence ordinates for horizontal thrust at 10 ft intervals: 0, 0.161, 0.286, 0.375, 0.429, 0.446, 0.429, 0.375, 0.286, 0.161, 0

CHAPTER 15

15.1 $M_{AB} = -22.92$ k · ft, $M_{BA} = -10.42$ k · ft

15.3 $M_{AB} = -2.78$ k · ft, $M_{BA} = -4.59$ k · ft, $M_{CB} = -7.65$ k · ft, $M_{DC} = 6.16$ k · ft

15.5 $M_{BA} = -13.65$ kN · m, $M_{CB} = 18.20$ kN · m

15.7 $M_{BA} = 121.3$ kN · m, $M_{CB} = -38.2$ kN · m, $M_{DC} = -78.5$ kN · m

15.9 $M_A = -18.33$ kN · m, $M_B = 0$, $K_A = 1.013$ EI, $K_B = 0.536$ EI, $C_{AB} = 0$, $C_{BA} = 0.905$

Index

ISBN 0-06-043634-4

90000

9 780060 436346

ISBN 0-06-043634-4

90000

9 780060 436346